Lecture Notes in
Computer Science

T0190002

Lecture Notes in Computer Science

Lecture Notes in Computer Science

Edited by G. Goos and J. Hartmanis

427

O. Faugeras (Ed.)

Computer Vision – ECCV 90

First European Conference on Computer Vision
Antibes, France, April 23–27, 1990
Proceedings

Springer-Verlag

Berlin Heidelberg New York London Paris Tokyo Hong Kong

CR Subject Classification (1987): I.2.4, I.2.6, I.2.9-10, I.3.5, I.3.7, I.4, I.5.4

ISBN 3-540-52522-X Springer-Verlag Berlin Heidelberg New York
ISBN 0-387-52522-X Springer-Verlag New York Berlin Heidelberg

Printing and binding: Druckhaus Beltz, Hemsbach/Bergstr.
2145/3140-543210 – Printed on acid-free paper

FOREWORD

Gathered in this volume are the 62 papers and 19 posters which were accepted for publication at the First European Conference on Computer Vision.

The selection was extremely difficult for the Program Committee because of the number of papers submitted and their generally excellent quality.

They show clearly that computer vision has emerged today as a mature discipline in Europe. This maturing is for a large part a consequence of the funding that has become available through the CEC and helped to establish strong research groups with critical masses. I hope that this will continue and expand in the future.

Many thanks to my colleagues on the Program Committee and to the reviewers for their hard work at short notice.

Many thanks also to the staff of the Public Relations Department of INRIA for their excellent work in making this conference really happen.

Antibes, France
April 1990   O. Faugeras

Conference

Chairperson

Olivier Faugeras   INRIA, Unité de Recherche Sophia-Antipolis

Program Committee

N. Ayache	INRIA, Rocquencourt
M. Brady	Oxford University
B. Buxton	G.E.C. Hirst Research Centre, Wembley
J. Crowley	L.I.F.I.A., Grenoble
J.O. Eklundh	Royal Institute of Technology, Stockholm
O. Faugeras	INRIA, Sophia-Antipolis
G. Granlund	Linköping University
D. Hogg	Sussex University, Brighton
I. Koenderink	Utrecht University
J. Mayhew	Sheffield University
R. Mohr	L.I.F.I.A., Grenoble
H.H. Nagel	Karlsruhe University, Fraunhofer Institute
B. Neumann	Hamburg University
G. Sandini	Genova University
V. Torre	Genova University

Workshop C.E.C.

Chairperson

Patrick Van Hove   Commission of the European Communities, DG XIII, ESPRIT, Brussels

Workshop Basic Research Action

Coordination

James L. Crowley   L.I.F.I.A. - I.M.A.G., Grenoble

Organisation

Institut National de Recherche en Informatique et en Automatique (INRIA)
Unité de Recherche Sophia-Antipolis

Th. Bricheteau	INRIA, France
S. Achard	INRIA, France
S. Gosset	INRIA, France
C. Juncker	INRIA, Sophia-Antipolis
M.L. Meirinho	INRIA, Sophia-Antipolis
J. Tchobanian	INRIA, Sophia-Antipolis

Referees

M. Andersson	Sweden	P. Marthon	France
N. Ayache	France	G. Masini	France
		S. Maybank	United Kingdom
F. Bergholm	Sweden	J. Mayhew	United Kingdom
M. Berthod	France	M. Mohnhaupt	Fed. Rep. of Germany
J. Bigün	Switzerland	O. Monga	France
M. Brady	United Kingdom	R. Mohr	France
B. Buxton	United Kingdom		
H. Buxton	United Kingdom	N. Navab	France
		H.H. Nagel	Fed. Rep. of Germany
A. Calway	Sweden	B. Neumann	Fed. Rep. of Germany
S. Carlsson	Sweden	K. Nordberg	Sweden
D. Castelow	United Kingdom		
L. Cohen	France	G. Olofsson	Sweden
J.L. Crowley	France/USA		
		J. Philip	Sweden
K. Danilidis	Fed. Rep. of Germany	P. Pousset	France
R. Deriche	France		
M. Dhome	France	L. Quan	France
L. Dreschler	Fed. Rep. of Germany		
C. Drewniak	Fed. Rep. of Germany	M. Richetin	France
		G. Rives	France
J.O. Eklundh	Sweden	L. Robert	France
O. Faugeras	France	G. Sagerer	Fed. Rep. of Germany
O. Fahlander	Sweden	P. Sander	France/Canada
		G. Sandini	Italy
J. Garding	Sweden	M. Schmitt	France
G. Granlund	Sweden	C. Schroeder	Fed. Rep. of Germany
P. Grossmann	United Kingdom	S. Stiehl	Fed. Rep. of Germany
E. Grosso	Italy	L. Svensson	Sweden
L. Haglund	Sweden	M. Thonnat	France
I. Herlin	France	H. Tistarelli	Italy
D. Hogg	United Kingdom	K. Tombre	France
R. Horaud	France	V. Torre	Italy
J.L. Jezouin	France	R. Vaillant	France
		F. Veillon	France
H. Knutsson	Sweden	T. Vieville	France
I. Koenderink	The Netherlands		
B. Kruse	Sweden	C.F. Westin	Sweden
		C.J. Westlins	Sweden
R. Lenz	Sweden	W. Winroth	Sweden
J. Levy-Vehel	France	J. Wiklund	Sweden
M. Li	Sweden		
C.E. Liedtke	Fed. Rep. of Germany	D.S. Young	United Kingdom
T. Lindeberg	Sweden		
Q.T. Luong	France	Z. Zhang	France
A. Lux	France		

TABLE OF CONTENTS

SHAPE DESCRIPTION

RECOGNITION - MATCHING

POSTERS

IMAGE FEATURES

On scale and resolution in the analysis of local image structure

Kjell Brunnström, Jan-Olof Eklundh, Tony Lindeberg

Computer Vision and Associative Pattern Processing Laboratory (CVAP)*
Royal Institute of Technology
S-100 44 Stockholm, Sweden
Email: kjellb@bion.kth.se, joe@bion.kth.se, tony@bion.kth.se

Abstract

Focus-of-attention is extremely important in human visual perception. If computer vision systems are to perform tasks in a complex, dynamic world they will have to be able to control processing in a way that is analogous to visual attention in humans.

In this paper we will investigate problems in connection with foveation, that is examining selected regions of the world at high resolution. We will especially consider the problem of finding and classifying junctions from this aspect. We will show that foveation as simulated by controlled, active zooming in conjunction with scale-space techniques allows robust detection and classification of junctions.

1 Introduction

A central theme of computational vision is the derivation of local features of image brightness. Existing computational models often make reference to and are inspired by biological vision, see e.g. Marr and Hildreth (1980)or Watt (1989). From this perspective local feature extraction is similar to foveation. Usually, one also tacitly assumes that the goal of computer vision research is to develop methods for analyzing visual information that perform as well as the human visual system. In this paper we argue that such performance can hardly be expected by most current techniques, simply due to limits on resolution. Standard camera systems usually have a visual angle of about 50° and give image matrices of say 500 × 500 up to 1000 × 1000 pixels. This should be compared to the 2° in foveal vision, which in view of visual acuity can be said to correspond to an image resolution of about 200 × 200 pixels. Obviously, this difference implies that multiresolution processing, like it is done in pyramids cannot be seen as analogous to foveation. The images coming out of a standard camera system are too much limited by resolution. In our experiments we overcome this limitation by doing controlled zooming and windowing. In this way we can simulate a system that performs foveation, that is *takes a closer look* at interesting regions in the field view.

An approach like this raises computational problems different from those appearing e.g. in edge detection or other general methods for searching for local structure in normal *overview*

---

*The support from the National Swedish Board for Technical Development, STU, is gratefully acknowledged.

pictures. Obviously, we need a method for deciding where to focus our attention. This will be discussed in Sections 2–3. Moreover, the increased resolution is likely to enhance the noise at least relative to the prominence of the structures we are looking for. This may call for a different type of algorithms for detecting local structure. In fact, we will show how very simple local statistical techniques can be used to obtain robust detection and classification of junctions and corners or high-curvature points. Finally, we have the problem of detecting structure when we dynamically vary resolution and window size. What we propose is a method of stability of responses. The rationale for this approach is the observation that local structure, in the highly resolved foveated images, generally will belong to one of a small set of generic types. Stability of responses is therefore simple to assess. Notwithstanding this, there are more sophisticated approaches to the problems, as we will get back to later.

2 Detecting local structure

We will in this paper apply our idea of simulating foveation by active focusing to the detection and analysis of junctions of two or more boundary segments in gray-level images. We are particularly trying to find T-junctions and images of 3-dimensional corners, often showing up as junctions with three or more edge directions. These give important cues to 3-dimensional structure, for instance the T-junctions indicate interposition and hence relative depth. Naturally, also L-junctions or image corners are of interest.

It is well-known that elaborate edge detection methods like those proposed by Hückel (1971), Marr and Hildreth (1980), Canny (1986) or Bergholm (1987), have problems at junctions. Zero-crossings, that is boundaries between positive and negative regions (of the second derivatives) will, of course, not correctly divide three regions in a junction. If first order derivatives are used other problems arise, e.g. at the computation of the gradient direction.

To overcome such problems direct methods for junction detection have been proposed, e.g. by Moravec (1977), Nagel (1986) and Kitchen and Rosenfeld (1982). The results have been applied to matching for stereo and motion analysis, but it is not clear that these approaches give any precise information about the nature of the junction. On the other hand, explicit use of the feature as a cue to scene structure would require some sort of classification or description of the junction.

Proposed edge and junction detectors like those cited above are based on precise assumptions about the local image structure. Usually, one assumes two or three regions, each with locally constant or linearly varying intensities, regions which in turn are clearly separated. It is obvious that these models are difficult to assess from noisy data with very few samples from each population. Marr and Hildreth (1980) pointed out that while the smallest receptive fields according to e.g. Wilson (1983) contain about 500 cones, most edge and corner detectors work with a few tens of samples. This indicates that one should indeed work with large operators, as some authors also do, see e.g. Canny (1986) and Bergholm (1987). However, the problem is not solved with this. First, we still have the relation to the size of the structure we are looking for. We can't assess, say, a *three regions* model by simply using a large operator, if the window contains many different instances of it, just because the window is large. Hence, the idea of using large operators should only be expected to give *good* (see above) performance if it could be applied so that, at least ideally, no more than one feature occurred in each field. Now this poses a second problem. Since we only can assume coarse knowledge or expectance about the scale of the features how do we find a reasonable operator size? Especially, how do we detect if two or a few nearby features appear in the same field? This can happen even if the resolution is increased. The proposed answer to this is to use the stability of the computed measures as the parameters, field size, resolution and location, are varied. More precisely, we have addressed the problem of finding the initial hypotheses about existing structure and its scale using scale-space representations. We will briefly review this theory in the next section. From these initial hypotheses zooming is performed and stability is verified or rejected.

3   Determining significant image structures and their scales

The prominent structures in the gray-level image can be determined using the scale-space primal sketch approach, developed in Lindeberg, Eklundh (1990) on the basis of the scale-space theory by Witkin (1983) and Koenderink (1984).

The main idea is to build a scale-space representation of blobs, i.e., peaks and valleys in the gray-level landscape and their *extent*. It is based on a conservative notion of a gray-level blob, which implies an inherent competition between parts, see Figure 1. These blobs are then linked between scales into higher order objects, called scale-space blobs, which have extent not only in space and gray-level but also in scale. The bifurcations occurring in scale-space are registered and a hierarchical data structure is created. There are several computational problems to solve in building this representation, we refer the reader to Lindeberg, Eklundh (1990) for the details.

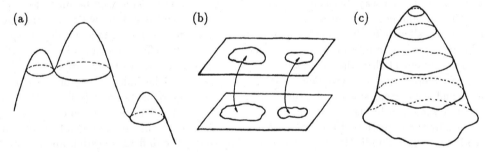

(a) (b) (c)

Figure 1: (a) Example illustrating the definition of a gray-level blob. (b) By linking gray-level blobs between scales one obtains (c) scale-space blobs which are objects having extent both in space, gray-level and scale. These objects are the fundamental primitives in the representation.

From measurements of significance and stability in scale-space the representation gives a qualitative description of the image structure, which enables determination of approximate location and extent of relevant regions in the images as well as appropriate levels of scale for treating those regions. The significance value of a scale-space blob is essentially the volume of the scale-space blob in the four-dimensional scale-space given by the spatial, gray-level and scale coordinates. See Figure 2 for an example.

Experimentally it has been demonstrated that the extracted information correspond to perceptually relevant regions in the image. Hence, we suggest the use of this output for generating hypothesis for the focusing procedure as well as for controlling the setting of initial values for window size and position.

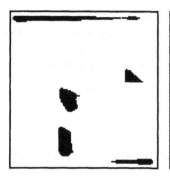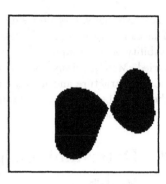

Figure 2: Illustration of the scale-space primal sketch output for the upper right block image in Figure 3. The dark regions correspond to significant blobs in the scale-space primal sketch. Note that they are not just plain regions in the image, but entities with well-defined scale information and an associated significance measure determined from the behaviour of the blob in scale-space. Observe that the individual blocks are extracted as single units and that the adjacent blocks become grouped into coarser scale objects.

It should be stressed that the intrinsic shape of the gray-level landscape is extracted in a completely bottom-up data-driven way without any a priori information about spatial location, scale or shape of the primitives.

An important problem in deriving this representation concerns measuring significance and scale in such a way that comparisons between significance values can be made for blobs existing at different levels of scale. This issue must be explicitly dealt with. For instance, a concept called "effective scale", which is the natural unit for measurements of scale-space lifetime, needs to be introduced. It is argued, that when the effective scale is increased with a small increment then the expected amount of image structure to be destroyed as well as the probability that an extremum point disappears should be independent of the current scale and the current amount of structure. From these requirements we have shown that this effective scale can be defined in, in principle, one way only. The resulting effective scale is essentially the logarithm of the expected number of local extrema at the given level of scale, see Lindeberg, Eklundh (1990) for the details as well as the difficult question about how to estimate the expected amount of structure in an image.

The significance values obtained from scale-space primal sketch also induce a well-defined ranking between the blobs, which empirically has been found to give an intuitively reasonable ordering of the events. Thus, we believe that such a module really can serve as a guide to the focus-of-attention.

4   Image structure at junctions

We will now turn to our problem of finding and classifying junctions. In the next section we will briefly review a technique for their detection. It turns out that most methods working on standard imagery will have thresholding problems. However, various additions to the basic algorithm can be used to limit the number of falsely detected corners, with a certain risk of loosing some of the true feature points. However, this trade-off seems inevitable and tends to occur in all proposed approaches. In particular, false responses tend to occur along edges. Also noise spikes, if they exist, will according to the model give responses that are strong local maxima.

There are, in fact, five situations in which strong responses mainly occur:

at noise spikes
along edges
at 2-junctions
at 3-junctions
at n-junctions, $n > 3$

Let us now consider an ideal case, in which the resolution is infinite and each region corresponds to a smoothly varying surface. Let us also assume that the illumination varies smoothly over each region. In that situation we could easily discriminate between the different cases by considering the distributions of gray-levels and edge directions in a sufficiently small neighbourhood of the possible (image) corner. In fact, this would be possible also if the surfaces contained some non-infinitesimal surface markings as well. The classification would be:

Case	Intensity	Edge direction
at noise spikes	uniform	*
along edges	bimodal	unimodal
at 2-junctions	bimodal	bimodal
at T-junctions	trimodal	bimodal
at 3-junctions	trimodal	trimodal
at n-junctions	*	*

where — * stands for inconclusive.

Experiments by Nagel and his co-workers, Nagel (1989), show that high quality intensity images indeed satisfy such conditions just as a straightforward model would predict.

The important question now is what happens in a realistic case, e.g. with direct and indirect illumination and noise and with finite resolution. One would still expect, as is indeed the basis for most of the low-level methods propose, that the given classification would be valid anyway. However, establishing this classification requires sufficiently many samples of the different distribution. Moreover, there is a need for a method of classification that is robust with respect to noise and variations due to the imaging process.

What this means in practice is that the resolution has to be high enough and that the window size should be appropriately chosen for correct classification. If we had a precise model for the imaging process and the noise characteristics, one could conceive deriving bounds on the resolution and the window size, at least in some simple cases. However, apart from the fact that this might be difficult in itself, such models are hardly available. What we propose instead is to use the process of focusing. Focusing means that we increase the resolution locally in a *continuous* manner (even though we still have to sample at discrete resolutions). We can simultaneously either keep the window size constant with respect to the original image, or vary it independently. Generally we want to decrease the window sizes. The contention we make about the focusing approach is that it avoids the problem of selecting an appropriate pair of resolution/window size, since the simple features we are looking for will show up in a stable manner during the variation of the parameters. An important reason behind this argument is that the classification can be based on simple features (the number of peaks of a histogram) of simple statistics. We can summarize the main principles of the approach as follows:

- *take a closer look* at candidate interest points by increasing the resolution, i.e. aquiring a *new* image with higher sampling density, and varying the window size

- detect stable classifications based on the simple features in the table above

So far, we have used no objective method to assess what we mean by *detecting the stable classifications*. However, we will show a number of experiments that demonstrate that our computational model indeed predicts what happens. Moreover, the approach described in Section 3 can be applied also to this problem.

5 Experimental technique

We will demonstrate our approach on a set of images at different resolutions of scenes containing simple man-made objects. Let us describe the processing steps.

We first perform a bottom-up pre-processing step, in which a set of candidate points are determined, as described in Sections 2 and 3. These points are tentative junctions, that is points where several regions meet. A crucial issue is to find a set of such points that is limited in size and is likely to contain true and significant junctions in the image.

The algorithm we use is based on the method of computing directional variances, suggested by Moravec (1977) and contains some additional constraints of *cornerity*. In this way significant junctions can be automatically obtained without any strong dependence on thresholding, with is difficult in Moravec's original algorithm. We will not discuss this further since we are not trying to analyze all junctions here. See Brunnström et al (1989) for details.

Having found a point of interest, we next simply complete the distribution of gray-levels and edge directions in a sequence of windows of varying resolution and size around the point. The edge direction statistics are obtained for pixels labelled as edge elements only. Hence, some edge detector has to be applied first. A simple detector like Roberts' cross might do, but the choice is of little importance. Something that may be important, though, is the effect of point of interest itself. As discussed above it is at the point that most low-level algorithms will have problems. Assuming that the resolution is high enough, so that we can appropriately sample the postulated simple underlying structure in the neighbourhood, we can treat the point as being a removable discontinuity. Hence, we compute the statistics in all but a subwindow around the point. Since we vary both the resolution and the window size and look for stability, we have chosen to let the subwindow cover a fixed part of the total window, say $\frac{1}{9}$ or $\frac{1}{16}$ of the area.

The decisions to make only concern how many peaks there are in the histograms, and if these numbers are stable and fit with any of the generic cases. We have used a method for sharpening the peaks in histograms proposed by Peleg (1978). This method is basically equivalent to smoothing and peak detection.

6 Results

In the toy block image in Figure 3, we focus on a corner of the bright block to the left. The possible existence of a corner has previously been established according to Section 5.

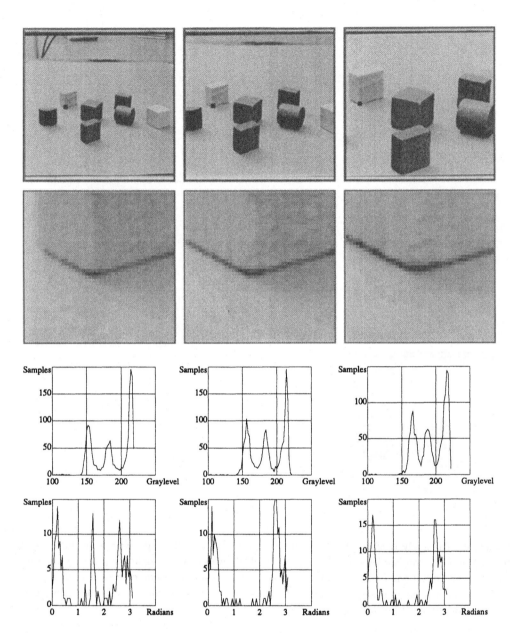

Figure 3: The figure illustrates the stability of the local statistics under resolution variations. In the left column, where the coarsest resolution is shown, the window is too large, which here shows up as an extra peak in the edge direction histogram.

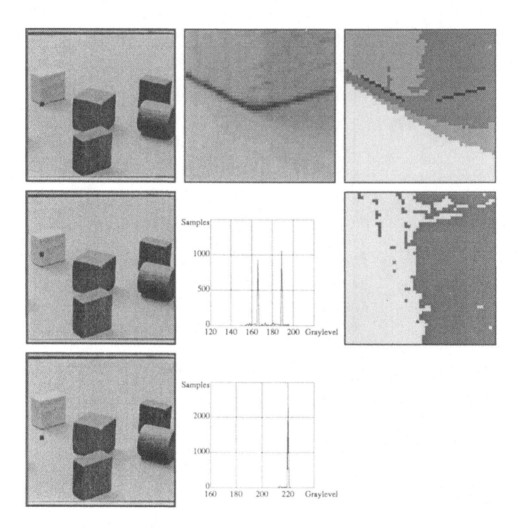

Figure 4: The analysis of the fixation point shown in the top left image, with a neighbourhood size shown in the top middle window, gives the result shown at the top right, i.e., three gray-levels and two edge directions (indicating an L-junction). This is not a generic case. For resolving the situation two new fixation points are analyzed, shown in the middle and in the bottom row. This results in the correct classification.

At coarse resolution three gray-levels and *three* directions are detected. This is because the outer boundary on the left is visible. As the focusing procedure continues, there is a stable response of three gray-levels and *two* directions. The vertical edge is very weak and cannot be detected, not even by e.g. Canny–Deriche's edge detector. However, backprojection into the image shows that this response is not generic. Therefore, it is assumed that some information is missing or incorrect. The point focused at is therefore shifted up and down the bisector of the detected L-junction, in a simulated eye-movement. The correct structure can then be assessed, as is illustrated in Figure 4. Note that in all cases both the window size and the resolution is varied, but only very few pictures can be shown here.

It should be stressed that the results shown in Figures 3–4 are obtained with simple tests in a case when e.g. sophisticated edge detectors fail.

Figure 5 shows some illustrative output in the case of a true T-junction and when one of the objects is curved. We are showing the final stable results which indicate a T-junction. Notably, a third but weakly supported direction is detected. This direction is, in fact, due to the lower part of the curved edge. However, backprojection of the predicted edges can again be used to characterize this type of response.

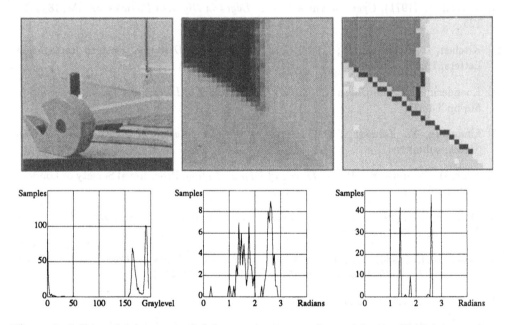

Figure 5: A T-junction between a slightly curved contour and a straight one. The left image shows the overview of the scene, the middle shows the local neighbourhood which has been processed and the right shows the result. The histograms are from left to right; gray-level, edge direction and peak sharpened edge direction.

7   Conclusions

We have argued that focus-of-attention mechanisms are necessary in computer vision systems, if they are to perform their tasks in a complex, dynamic world. We discussed how foveation, that is examining selected regions of the visual world at high resolution, can be incorporated in an active vision system. We studied the use of this capability in the task of finding and classifying junctions of edges. It was shown that the junctions in this context could be labelled into a small set of generic cases and that these cases could be established computationally by robust and simple algorithms.

References

[1] Bergholm, F., (1987), *Edge Focusing*, IEEE PAMI, 9:6, 726–741

[2] Brunnström, K., Eklundh, J.O., Kakimoto, A., (1989), *On Focus-of-Attention by Active Focusing*, Proc. NATO ASI on Robotics and Active Computer Vision, Springer Verlag, New York, in press.

[3] Canny, J.F., (1986), *A Computational Approach to Edge Detection*, IEEE PAMI, 8:6, 679–698

[4] Hückel, M., (1971), *Operator which Locates Edges in Digitized Pictures*, JACM, 18, 113–125

[5] Kitchen, L., Rosenfeld, R., (1982), *Gray-Level Corner Detection*, Pattern Recognition Letters, 1:2, 95–102

[6] Koenderink J.J., van Doorn A.J. (1984) *The Structure of Images*, Biological Cybernetics, **50**, pp363-370.

[7] Lindeberg, T., Eklundh, J.O., (1990), *On the Computation of a Scale-Space Primal Sketch*, submitted

[8] Marr, D., Hildreth, E., (1980), *Theory of Edge Detection*, Proc. Royal Society of London, B-207, 187–217

[9] Moravec, H.P., (1977), *Obstacle Avoidance and Navigation in the Real World by a Seeing Robot Rover*, Stanford AIM-340

[10] Nagel, H.H., (1986), *Image Sequences — The (Octal) Years — From Phenomenology towards a Theoretical Foundation*, Proc. 8th ICPR, 1174–1185

[11] Nagel, H.H., (1989), *Personal communication*

[12] Peleg, S., (1978), *Iterative Histogram Modification 2*, IEEE SMC, 8, 555–556

[13] Watt, R., (1988), *Visual Processing: Computational Psychophysical and Cognitive Research*, Laurence Erlbaum Associates, Publishers, London

[14] Wilson, H.R., (1983), *Psychophysical Evidence for Spatial Channels*, Braddick, O.J., Sleigh, A.C., eds., in Physical and Biological Proc. of Images, Springer Verlag, New York

[15] Witkin, A.P., (1983), *Scale-Space Filtering*, Proc. 8th IJCAI, 1019–1021

OPTIMAL FILTER FOR EDGE DETECTION METHODS AND RESULTS

Serge CASTAN, Jian ZHAO
Lab. IRIT-CERFIA, UPS
118, route de Narbonne
31062 Toulouse FRANCE

Jun SHEN
Southeast University
Nanjing, CHINA

Abstract

In this paper, we give a new demonstration in which it is proved that the symmetric exponential filter is the optimal edge detection filter in the criteria of the signal to noise ratio, localization precision and unique maximum. Then we deduce the first and the second directional derivative operators for symmetric Exponential Filter and realize them by first order recursive algorithm, and propose to detect the edges by maxima of Gradient (GEF), or by the zeros crossing of Second directional Derivative along the gradient direction (SDEF).

I. Introduction

Edge detection is one of the most important subjects in image processing, which finds wide applications in the pattern recognition, the scene analysis and the 3-D vision, because the edges correspond in general to the important changes of physical or geometrical properties of objects in the scene and they are widely used as primitives in the pattern recognition, the image matching etc.

The edges coincide, generally speaking, grey level transition, they can be detected by maxima of gradient or the zero-crossing of the second derivatives calculated by some differential operators. Because the differential operators are sensitive to noise, a preprocessing such as smoothing is often necessary to eliminate the noise. A well-known soomthing filter is Gaussian filter and the edges can therefore be detected by a Laplacien-Gaussian filter. But there is an essential difficulty of the Laplacien-Gaussian filter which is the contradiction between the smoothing effect and the precision of edge localization. To overcome this difficulty, We proposed the optimal linear filter based on one step model (a step edge and the white noise) and the multi-edge model [9][10][11]. This optimal smoothing filter is a symmetric exponential filter of an infinitely large window size and can be realized by very simple recursive algorithm. It is proved that the band limited Laplacien of an input image filtered by this filter can be calculated from the Difference between the input and the output of this Recursive Filter (DRF). The edges detected by DRF method are less noisy and with a much better precision of localization.

The maxima of gradient or zeros of the second directional derivative along the gradient are a natural definition of intensity edges. Zeros of the Laplacian are only extensively used for their computational convenience. However, we must stress here that the zeros crossing of the Laplacien are not always coincided with the maxima of gradient, for example, the zeros of the Laplacien are farther apart than the maxima of gradient for circularly symmetric patterns, this lack of localization by the Laplacien can also be seen in the fact that zeros of Laplacien "swing wide" of corners. Therefore, it had better to detect the edges by maxima of gradient or zeros of the second directional derivative along the gradient .

In this paper, we give a new demonstration in which it is proved that the symmetric exponential filter is the optimal edge detection filter in the criteria of the signal to noise ratio, localization precision and unique maximum. then we deduce the first and the second directional derivative operators for symmetric Exponential Filter, and realize them by the first order recursive algorithm. Using these operators, we propose two methods for edges detection, one uses maxima of Gradient (GEF), another uses the zeros crossing of Second directional Derivative along the gradient (SDEF). It is done for the theoretical analysis on comparing the performance of the filters. The performances of the GEF, SDEF methods and DRF method for edge detection are compared experimentally, the results show that the new methods are less sensitive to noise and have much better precision of edge localization.

II. Optimal Filter for Edges Detection

Let f(x) be the low-pass smoothing filter function that we want to find which gives the best results for step edge detection. The input signal is a step edge plus the white noise.[9,10]

We define a measure SRN of the signal to noise performance of the operator which is independent of the input signal:

$$SRN = f^2(0)/(\int_{-\infty}^{\infty} f^2(x)dx) \qquad (1)$$

Because the maxima of the first derivative correspond to the edge points, an optimal filter f'(x) should maximaze the ratio of the output in response to step input to the output in response to the noise, i.e. maximaze the criterion SRN (the CANNY's criterion Σ). Besides, the edge position is localized by the zero-crossing of the second derivative, therefore we should diminish noise energy in the second derivative of output signal, i.e. maximaze at the same time the other criterion as follows:

$$C = 1/\int_{-\infty}^{\infty} f''^2(x)dx \qquad (2)$$

To maximaze at the same time the criteria (1) and (2), we can maximaze the criterion Cr which is the product of the two criteria as follows:

$$Cr = f^2(0)/(\int_{-\infty}^{\infty} f^2(x)dx \cdot \int_{-\infty}^{\infty} f''^2(x)dx) \qquad (3)$$

In order to avoid the introduction of additional maxima at the positions other than the edge position, this maximum at x=0 should be unique, i.e. f(x) should satisfy[10]:

$$\text{Max} \quad f(x) = f(0) > 0 \qquad (4)$$
$$x \in (-\infty, \infty)$$

and : $f(x_1) > f(x_2)$ when $|x_1| < |x_2|$

and in order to maximaze Cr, f(x) should be an even function [10]. Eq.(3) can be written as :

$$Cr = f^2(0)/(2 \cdot \int_0^{\infty} f^2(x)dx \cdot \int_0^{\infty} f''^2(x)dx) \qquad (5)$$

Except signal to noise ratio and unique maximum, the other important performance of the filter for edges detection is the localization precision. For the localization criterion we proceed as follows.

Because the zero-crossing of the second derivative correspond to the edge position, we can find edge position x1 by the inequality as follows:

$$\begin{cases} S_0''(x_1^-) \geq 0 \\ S_0''(x_1^+) \leq 0 \end{cases} \quad \text{or :} \quad \begin{cases} S_0''(x_1^-) \leq 0 \\ S_0''(x_1^+) \geq 0 \end{cases} \qquad (6)$$

where, $S_0(x)$ is the output signal, $x_1^- = \lim_{\varepsilon \to 0}(x_1-\varepsilon)$ et $x_1^+ = \lim_{\varepsilon \to 0}(x_1+\varepsilon)$ $\quad \varepsilon > 0$

$$S_0''(x) = (S(x)*f(x))'' + (N(x)*f(x))'' = S_V(x) + S_N(x) \qquad (7)$$

If f'(x) and f''(x) are continuous at x=0, we have:

$$x_1 \approx -S_N(0)/ A \cdot f''(0)$$

We can notice that this ratio means CANNY's localization criterion [15], obtained from a more general approach.

From this formula, if the noise exist ($S_N(0) \neq 0$), the localization error ($x_1 \neq 0$) always exist. So, with this CANNY's criterion, it is theoretically impossible to find a filter without any localization error ($x_1 = 0$).

But if f'(x) is not continuous at x=0, and f(x) is an even function which satisfy the formula (4), we have :

$| f'(0^-)| = | f'(0^+)|$, $f'(0^-) > 0$ and $f'(0^+) < 0$, f'(x) is odd function.

Provided the absolute value of the noise is less than that of the signal, i.e. $S_N(0) < | f'(0^+)|$, we can satisfy the inequality (6) at $x_1=0$ (no error).

So, we should put in the condition as follows to the criterion (5) for obtaining the best localization precision ($x_1 = 0$).

$$f'(0^+) < 0 \qquad (8)$$

Besides, a stable infinite window size filter must satisfy the boundary conditions :

$$f(\infty) = 0, \ f'(\infty) = 0, \ f''(\infty) = 0 \qquad (9)$$

To sum up, the optimal edge detection filter has now been defined implicitly by maximazing the criterion (5) (signal to noise ratio) , and this filter must satisfy the conditions (4) (unique maximum), (8) (localization precision) and (9) (stability).

We consider at first a filter f(x) with a limited window size 2W, the equations (5) become:

$$Cr = f^2(0)/(2 \cdot \int_0^W f^2(x)dx \cdot \int_0^W f''^2(x)dx) \qquad (10)$$

In order to find the optimal function which maximazes the criterion (10), we can form a composite function $\psi(x,f,f')$, and finding a solution for this unconstrained problem is equivalent to finding the solution to the constrained problem.

$$C_N = \int_0^W \psi(x,f,f'')dx \qquad (11)$$

where: $\psi(x,f',f'') = f^2(x) + \lambda \cdot f''^2(x)$

This is an example of what is known as reciprocity in variational problems. The optimal solution should satisfy the Euler equation :

$$2 \cdot \lambda \cdot f''''(x) - 2 \cdot f'' = 0 \qquad (12)$$

The general solution f'(x) of this differential equation is :

$$f'(x) = C_1 e^{-\alpha x} + C_2 \cdot e^{\alpha x} + C_3 \qquad (13)$$

where $\alpha = 1/(\lambda)^{1/2}$ and C_1, C_2, C_3 are the constants.

For obtaining an infinite window size filter, we should consider the case where w→∞.

From Eq. (13), the constants C_1, C_2, C_3 can be determined as follows by the conditions (8) and

(9).

$C_2 = C_3 = 0$ and $C_1 < 0$

Because $f'(x)$ is an odd function, $f'(x)$ can be written in the form :

$$f'(x) = \begin{cases} C_1 e^{-\alpha x} & x>0 \\ \\ -C_1 e^{-\alpha x} & x<0 \end{cases} \qquad (14)$$

And we can equally get the low-pass filter $f(x)$ as follows, which satisfy the condition (4).

$$f(x) = C \cdot e^{-|x|} \qquad (15)$$

where the constant $C > 0$.

Thus far we find the optimal edge detection filter which is a symmetric exponential filter, in the criteria of the output signal to noise ratio, localization precision and unique maximum. The means of these criteria are similar to that of CANNY, but we search the filter with an infinite window size and no function continuity constraint is used. So, our result is more general than that of CANNY.

III. The First and the Second Directional Derivative Operators for Symmetric Exponential Filter

A normalized symmetric exponential filter on 1-D can be written :

$$f_L(x) = C \cdot a_0 \cdot (1-a_0)^{|x|} = f_1(x) * f_2(x) = C \cdot (f_1(x) + f_2(x) - a_0 \cdot \delta(x)) \qquad (16)$$

where: $C = 1/(2-a_0)$, $*$ means the convolution.

$$f_1(x) = \begin{cases} a_0 \cdot (1-a_0)^x & x \geq 0 \\ 0 & x<0 \end{cases} \qquad f_2(x) = \begin{cases} 0 & x>0 \\ a_0 \cdot (1-a_0)^{-x} & x \leq 0 \end{cases} \qquad (17)$$

we can write the first derivative operator of exponential filter:

$$f_L'(x) = f_2(x) - f_1(x) \qquad (18)$$

And we can obtain the normalized second derivative operator of exponential filter :

$$f_L''(x) \approx f_1(x) + f_2(x) - 2 \cdot \delta(x) \qquad (19)$$

Because the exponential fonction is separable, we can write out 2-D exponential filter:

$$f(x,y) = f_L(x) \cdot f_L(y) \qquad (20)$$

From the equations (18),(19) and (20), the first and the second directional derivative operators for symmetric exponential filter can be written like this:

$$f_x(x,y) = f_L(y) \cdot (f_2(x) - f_1(x)) \qquad (21)$$
$$f_y(x,y) = f_L(x) \cdot (f_2(y) - f_1(y)) \qquad (22)$$
$$f_{xx}(x,y) = f_L(y) \cdot (f_1(x) + f_2(x) - 2 \cdot \delta(x)) \qquad (23)$$
$$f_{yy}(x,y) = f_L(x) \cdot (f_1(y) + f_2(y) - 2 \cdot \delta(y)) \qquad (24)$$

IV. The Recursive Algorithm for realizing the These Directional Derivative Operators of Symmetric Exponential Filter

The exponential filter is an IIR filter corresponding to an infinite window size, so we should realize the fonctions $f_1(x)$ and $f_2(x)$ (see formula (17)) by a recursive algorithm.

Supposing $I(x,y)$ is the input image, $I_1(x,y) = I(x,y) * f_1(x)$ and $I_2(x,y) = I(x,y) * f_2(x)$, we have the recursive algorithm :

$$I_1(x,y) = I_1(x-1,y) + a_0 \cdot (I(x,y) - I_1(x-1,y)) \qquad (25)$$
$$I_2(x,y) = I_2(x+1,y) + a_0 \cdot (I(x,y) - I_2(x+1,y))$$

From the equations (16),(21),(22),(23) and (24), the band-limited first and second directional derivative of input image can be calculated by the recursive algorithm f_1 and f_2 as follows

$$I_x(x,y) = I(x,y) * f_1(y) * f_2(y) * (f_2(x) - f_1(x)) \qquad (26)$$
$$I_{xx}(x,y) = I(x,y) * f_1(y) * f_2(y) * (f_2(x) + f_1(x)) - 2 \cdot I(x,y) * f_1(y) * f_2(y) \qquad (27)$$
$$I_y(x,y) = I(x,y) * f_1(x) * f_2(x) * (f_2(y) - f_1(y)) \qquad (28)$$
$$I_{yy}(x,y) = I(x,y) * f_1(x) * f_2(x) * (f_2(y) + f_1(y)) - 2 \cdot I(x,y) * f_1(x) * f_2(x) \qquad (29)$$

With this algorithm, we can calculate at the same time the band-limited first and second directional derivative I_x and I_{xx} (or I_y and I_{yy}) of input image.

V. Edges Detection

The band-limited first and second directional derivative of input image can be obtained by the algorithms as stated above. Using them, we can then realize the edge detection for an image.

The maxima of gradient or zeros of the second directional derivative along the gradient are a natural way of characterizing and localizing intensity edges, so we present here to detect the edges from the maxima of gradient or zeros of the second directional derivative along the gradient by using the differential operators of exponential filter.

1. Edges from the maxima of gradient

Using the first directional derivative operator of the exponential filter, the two band-limited first directional derivatives I_x and I_y can be calculated, and the gradient vector can be therefore determined approximatively for every point in image. The gradient magnitude image is then non maxima suppressed in the gradient direction and thresholded with hysterisis, i.e. if the entire segment of the contour lies above a low threshold T1, and at least one of part of which is above a high threshold T2, that contour is output. The non maxima suppression scheme requires three points, one of which will be the current point, and the other two should be estimated of the gradient magnitude at points displaced from the current point by vector normal to the edge direction.

2. Edges from the zero-crossings of the second directional derivative along the gradient direction

Because edges detected from local gradient maxima can not be a pixel width (less good precision of localization), we propose an other method which detect the edges from the zeros crossing of the second directional derivative along the gradient direction.

We can calculate I_x, I_y, I_{xx} and I_{yy} by using the method shown in paragraphe IV, and therefore obtain approximatively the gradient vector and the second derivative in the gradient direction for every point in image. We extract at first the zero crossing of second derivative along the gradient direction on which the gradient magnitude must be above a low threshold, so an edges image is obtained. To this image, the entire segment of the contour will be kept, if the gradient magnitude on at least one part of this contour is above a high threshold.

VI. Comparision of Performance of the Filters

Filtering is a problem of estimation from noisy signal, and edge detection is a problem of estimating the position of maximal local signal change. Up to now, many works are done for edges detection in image, and different filters are proposed, for example, Gaussien filter, Canny filter, exponential filter, Deriche filter etc...

We appreciate the performance of the filters as follows :

(1) Precision of edge localization

According to our analysis [10], we can calculate the average localization error x_e for Gaussian filter, Canny filter [15], Deriche filter [16] and the exponential filter : $x_{eG} = 4 \cdot (2 \cdot e \cdot \pi)^{1/2}/\alpha$, $x_{ec} \approx 0.81/\alpha$, $x_{eD} = 4 \cdot e^{-1}/\alpha = 1.47/\alpha$, $x_{eE} = 0$, i.e. $x_{eG} > x_{eD} > x_{ec} > x_{eE} = 0$.

So, we can see that the exponential filter localizes edge points with the best precision.

(2) Signal/Noise ratio on the edge point detected

Because x_e is the average estimation for the position of the edge point detected, we propose to calculate Signal/Noise ratio (Eq.(7)) at the point x_e.

And the signal/noise ratio for the Gaussien filter, Canny filter, Deriche filter and the exponential filter is : $SNR_G = 2 \cdot \sigma \cdot e^{-32\pi\sigma}/\pi^{1/2}$, $SNR_C = 0.39/\alpha$, $SNR_D = 0.64/\alpha$, $SNR_E = 1/\alpha$, i.e. $SNR_E > SNR_D > SNR_C > SNR_G$.

Then, we see that the exponential filter has the best noise eliminating effect among the above four.

(3) Complexity of calculation

For the complexity of calculation, we only tell the difference from exponential filter and Deriche filter [16], because they are implemented by recursive algorithms which have a simpler calculation.

Because ISEF can be realized by first order recursive filter, the ISEF algorithms are much simpler than that of Deriche filter. Besides, the ISEF algorithms can be implemented independently to every line and every column, it can be easily realized by a parallel system.

According to the analysis results above, the ISEF filter is superior to the others at the 3 principal aspects of the performance of the filter.

VII. Experimental Results

Our new algorithms have been tested for different types of images and provided very good results. For comparision purpose, we take two examples : an indoor scenes image (Fig. 1) and a synthetic very noisy image (Fig. 2).

The experimental results show that GEF and SDEF methods are less sensitive to noise than DRF method, because the maxima of gradient or zero-crossing of directional second derivative rather than the Laplacian are used. The edges detected by GEF method are not always one pixel width, so its precision of localization is less good than that of SDEF method.

VIII. Conclusion

The symmetric exponential filter of an infinite large window size is an optimal linear filter deduced from one step edge model and the multi-edge model, now we further prove that the symmetric exponential filter is the optimal edge detection filter in the criteria of the signal to noise ratio, the localization precision and unique maximum. Obviously, the real images will be still more complicated than these models, however DRF method has already provided good results for different type of images. The results obtained through the two new methods further show the superior performance of this filter. The theoretical analysis for the performance of the filters shows also that the exponential filter is superior to the other current filters.

The first and second directional derivative operators can be realized by recursive algorithm and calculated at the same time. The new algorithms are therefore very simple as well as DRF algorithm, and they are also easy to implemente in a parallel way.

From the experimental results, the new methods show better effect than that of the other methods.

References
[1] W.K. PRATT, Digital Image Processing, New York, 1978.
[2] J. PREWITT, Object Enhancement and Extraction,Picture Processing and Psychopictories, Etd. by B. Lipkin and A. Rosenfeld, New York, pp. 75-149, 1970.
[3] M. HUECKEL, An Operator Which Locates Edges in Digitized Pictures. J.A.C.M., Vol. 18, pp 113-125, 1971.
[4] R.O. DUDA and P.E. HART, Pattern Classification and Scene Analysis. Wiley, New York, 1973.
[5] R. HARALICK,Edge and Region Analysis for Digital Image Data. C.G.I.P., Vol. 12, pp 60-73, 1980.
[6] R.HARALICK and L.WATSON, A Facet Model for Image Data. C.G.I.P., Vol. 15, pp 113-129, 1981.
[8] D. MARR and E.C. HILDRETH, Theory of Edge Detection. Proc. R. Soc. Lond. B, Vol. 207, pp 187-217, 1980.
[9] J. SHEN and S. CASTAN, Un nouvel algorithme de detection de contours, proceedings of 5th Conf. on P.R.&.A.I. (in French), Grenoble, 1985.
[10] J. SHEN and S. CASTAN, An Optimal Linear Operator for Edge Detection. Proc. CVPR'86, Miami.
[11] J. SHEN and S. CASTAN, Edge Detection Based on Multi-Edge Models.Proc. SPIE'87, Cannes, 1987.
[12] J. SHEN and S. CASTAN, Further Results on DRF Method for Edge Detection. 9th I.C.P.R., ROME, 1988.
[13] V.TORRE and T.A.POGGIO, On Edge Detection IEEE Transaction on Pattern Analysis and Machine Intelligence, Vol. Pami-8, N° 2, March 1986.
[14] J.S.CHEN and G.MEDIONI, Detection, Localization, and Estimation of Edges. IEEE Transaction on Pattern Analysis and Machine Intelligence, Vol. 11, N° 2, February 1989.
[15] J.F.CANNY, Finding Edges And Lines in Images. MIT Technical Raport N° 720, 1983.
[16] R. DERICHE, Optimal Edge Detection Using Recursive Filtering. In proc. First International Conference on Computer Vision, London, June 8-12 1987.

Fig. 1

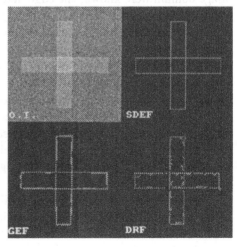

Fig. 2

Biased Anisotropic Diffusion —A Unified Regularization and Diffusion Approach to Edge Detection

Niklas Nordström

Department of Electrical Engineering and Computer Sciences
University of California, Berkeley, CA 94720

Abstract

We present a global edge detection algorithm based on variational regularization. The algorithm can also be viewed as an anisotropic diffusion method. We thereby unify these two from the original outlook quite different methods. The algorithm to be presented moreover has the following attractive properties: 1) It only requires the solution of a *single* boundary value problem over the *entire* image domain—almost always a very simple (rectangular) region. 2) It converges to a solution of interest.

1 Introduction

Edge detection can in short be described as the process of finding the discontinuities of the partial derivatives up to some order of an *image function* defined on an open bounded connected *image domain* $B \subseteq \mathbf{R}^2$. The image function can represent various kinds of data collected from the visible surfaces in the scene. We will be concerned with brightness data. In this case the image function is real-valued, and the discontinuities of interest appear in the zeroth order derivative, that is the image function itself. If the *true image function* one would obtain by pure projection of the brightness in the scene onto the image domain were known, the edge detection problem would be easy. However, because of imperfections in the image formation process, the *original image function* one is given, is distorted so that the discontinuities in the true image function are disguised into large gradients. Edge detection therefore essentially boils down to numerical differentiation—a problem well-known to be ill-posed (in the sense of Hadamard) due to its instability with respect to the initial data. Since measurement noise and other undesirable disturbances cannot be avoided, the edge detection problem thus has to be stabilized in order to have a meaningful solution. In more practical terms this means that the undesirable disturbances must be suppressed without the disappearance or dislocation of any of the edges. Over the last six years or so many attempts along these lines have appeared in the literature. One can distinguish between two seemingly quite different approaches, viz. regularization and anisotropic diffusion.

Regularization can be achieved in different ways. In probabilistic regularization the problem is reformulated as Bayesian estimation. In variational regularization it is posed as a cost (or energy) functional minimization problem leading to a variational principle. In spite of the different outlooks of these approaches they essentially end up with the same mathematical and computational problem; given the original image function $\zeta : B \to \mathbf{R}$, minimize a cost functional $C_\zeta(w, z)$ where w is some function representing the edges, and $z : B \to \mathbf{R}$ is the *estimated* (or reconstructed) *image function*. In each case the total cost can furthermore be divided into three components according to $C_\zeta(w, z) = \mathcal{E}(w) + \mathcal{D}(z, \zeta) + \mathcal{S}(w, z)$ where the *edge cost* $\mathcal{E}$ measures the extent of the edges, the *deviation cost* $\mathcal{D}$ measures the discrepancy between the estimated and the original image functions, and the *stabilizing cost* measures the unsmoothness or the a priori "unlikeliness" of the estimated image function. The "edge function" w can be defined in a variety of ways. Here we will only be concerned with edge functions of the form $w : B \to \mathbf{R}$.

Given a specific edge function w it is generally the case that there exists a unique optimal estimated image function $\check{z}_w$, which can be found by solving a linear partial differential equation. While most of the regularization approaches do take advantage of this condition, none of them is capable of solving for the optimal edges in a similar way. The optimality conditions for the edges do either not exist, or else they consist of unsolvable equations. For the minimization of $C_\zeta(w, z)$ with respect to w all of the regularization approaches referred to above therefore resort to some kind of stochastic or deterministic search method such as the "Metropolis algorithm" or "steepest descent". Because of the tremendous size of the solution space any such search method is by itself quite expensive. In addition the general nonconvexity of the cost function causes any converging search algorithm to get stuck at local minima. The common response to this unfortunate situation has been to solve whole sequences of minimization problems, as a mechanism for "gravitating" towards a good local possibly a global minimum. The GNC-algorithm introduced in [1, 2] and simulated annealing [3] are both examples thereof. In summary most global edge detection methods up to date involve some form of repeated iterative minimization process, and because of the high computational cost that this implies, the optimality of the solution is often compromised.

In contrast to the regularization-based methods, the anisotropic diffusion method presented in the literature does not seek an optimal solution of any kind. Instead it operates by repeatedly filtering the image function with a smoothing kernel of small support, thereby producing a sequence of *diffused image functions* of successively lower resolution. In order to retain the strength and correct location of the edges, the "smoothing power" of the filter kernel is made to depend (inversely) on the magnitude of the image function gradient in a heuristic fashion. At some stage in the iterated filtering process remarkably impressive edges can be extracted by postprocessing the diffused image function with a rudimentary local edge detector. In the limit, however, all edges disappear, and the diffused image function converges to a constant. Needless to say, any solution of interest therefore has to be selected from the sequence of diffused image functions way before convergence. This selection has so far been a matter of manual inspection. If automation is necessary, one can of course, in the absence of more sophisticated rules, simply prespecify a number of filter iterations. A more serious problem due to the necessity to select a solution prior to convergence, may arise in an analog circuit implementation where the diffusion process must be "latched" or halted in order to retrieve the diffused image function of interest.

In this paper we show how the variational regularization approach by Terzopoulos [4, 5] can be modified so that the calculus of variations yields useful optimality conditions, not only for the estimated image function, but for the edges as well. The result is a global edge detection algorithm, which as it turns out, also can be viewed as a (new) *biased anisotropic diffusion* method. This unification of the from the original outlook quite different regularization and diffusion approaches is in itself quite interesting. It also brings the anisotropic diffusion approach an appealing sense of optimality. Anisotropic diffusion is thus a method for solving a well-defined mathematical problem, not just an image processing technique, by which one image can be transformed into another more pleasing looking one. Even more exciting than the unification just mentioned, is the strong indications that the new algorithm shares the better properties of both the regularization-based methods and the anisotropic diffusion method. First of all, it does only require the solution of a *single* boundary value problem on the *entire* image domain—almost always a very simple region. In particular, no explicit search method is necessary. Neither is the solution of a sequence of minimization problems. Secondly, the algorithm converges to a solution of interest. The problem of manual selection of which one in the sequence of diffused image functions to be postprocessed with the local edge detector, is thereby removed from the anisotropic diffusion method.

Before we continue, some notation needs to be settled: $\mathbf{R}_-$ and $\mathbf{R}_+$ will denote the sets $]-\infty,0[$ and $]0,\infty[$ respectively. The norm $\|\cdot\|$ will always refer to the Euclidean norm in $\mathbf{R}^n$. Finally $\vee$, $\wedge$ and $\circ$ will denote the binary maximum, minimum and function composition operators respectively.

2 Controlled-Continuity Stabilization

The "classical" stabilizers that first appeared in early vision problems did not allow estimation or reconstruction of image functions with discontinuities. In order to improve on this for vision problems far too restrictive model Terzopoulos [4, 5] introduced a more general class of stabilizing functionals referred to as *controlled-continuity stabilizers*. These are of the form $S(w,z) \doteq \int_{\mathbf{R}^K} \sum_{i=1}^I w_i \sum_{k_1=1}^K \cdots \sum_{k_i=1}^K (\partial^i z / \partial x_{k_1} \cdots \partial x_{k_i})^2 \, dx$ where $w \doteq [w_1 \cdots w_I]^T$, and the weighting functions $w_1, \ldots, w_I : \mathbf{R}^K \rightarrow [0,1]$, referred to as *continuity control functions* are in general discontinuous. In particular they are able to make jumps to zero, and edges, where the partial derivatives of z of order $\geq j$ are allowed to be discontinuous, are represented by the sets $\bigcap_{i=j+1}^I w_i^{-1}(\{0\})$, $j = 0, \ldots, I-1$. For the edge cost Terzopoulos proposes the functional $\mathcal{E}(w) \doteq \int_{\mathbf{R}^K} \sum_{i=1}^I \lambda_i (1-w_i) \, dx$ where the constants $\lambda_1, \ldots, \lambda_I \in \overline{\mathbf{R}_+}$ satisfy $\sum_{i=1}^I \lambda_i > 0$. This paradigm apparently fails to support a genuinely variational technique for minimizing the total cost with respect to the continuity control function vector w. First of all, all the control functions that represent nonempty sets of edges, belong to the boundary of the continuity control function space. Secondly, the total cost functional is affine in w, whence it does not have any critical points. Terzopoulos resolves this problem by first discretizing the entire space of continuity control functions; w is defined on a finite subset D—a dual pixel grid—and only allowed to take the values 0 or 1. The edge cost is modified accordingly to $\mathcal{E}(w) \doteq \sum_{x \in D} \sum_{i=1}^I \lambda_i [1-w_i(x)]$ For a solution he then applies a descent method in the space of all possible continuity control function vectors.

3 Genuinely Variational Edge Detection

For our problem of detecting discontinuities of a bivariate image function, the appropriate deviation and stabilizing costs in the paradigm above are given by $\mathcal{D}(z, \zeta) \doteq \int_B (z - \zeta)^2 \, dx$ and $\mathcal{S}(w, z) \doteq \int_B w \|\nabla z^T\|^2 \, dx$. In order to remedy the difficulties with Terzopoulos' method, we propose the use of a smooth continuity control function $w : B \to \overline{\mathbf{R}_+}$. To avoid the problem with non-critical optimal continuity control functions, which are impossible to find by means of variational calculus, we will choose the edge cost so that for each estimated image function z the total cost $\mathcal{C}_\zeta(w, z)$ attains its minimum for exactly one optimal continuity control function $\check{w}_z$ whose range is confined to lie in $]0, 1]$. This idea is similar to the use of barrier functions in finite dimensional optimaization [6]. The uniqueness of $\check{w}_z$ for a given z also allows us to solve for $\check{w}_z$ in terms of z in a way similar to Blake and Zisserman's elimination of their "line process" [1, 2]. The edge costs we propose for this purpose are of the form $\mathcal{E}(w) \doteq \int_B \lambda f \circ w \, dx$ where the *edge cost coefficient* $\lambda > 0$ is constant, and the *edge cost density function* $f : \mathbf{R}_+ \to \mathbf{R}$ is twice differentiable. Our total cost functional is thus given by

$$\mathcal{C}_\zeta(w, z) \doteq \int_B [\lambda f \circ w + (z - \zeta)^2 + w \|\nabla z^T\|^2] \, dx \tag{1}$$

Setting the first variation of $\mathcal{C}_\zeta(w, z)$ to zero yields the Euler equations

$$z(x) - \zeta(x) - \nabla \cdot (w \nabla z)(x) = 0 \qquad \forall x \in B \tag{2a}$$

$$\lambda f'(w(x)) + \|\nabla z(x)^T\|^2 = 0 \qquad \forall x \in B \tag{2b}$$

$$w(x) \frac{\partial z}{\partial e_n}(x) = 0 \qquad \forall x \in \partial B \tag{2c}$$

where $\nabla \cdot$ denotes the divergence operator, and $\partial/\partial e_n$ denotes the directional derivative in the direction of the outward normal. The second variation of $\mathcal{C}_\zeta$ with respect to w is also easily found to be

$$\delta_{ww}^2 \mathcal{C}_\zeta(w, z) = \int_B \frac{\lambda}{2} (f'' \circ w)(\delta w)^2 \, dx \tag{3}$$

Together with the desired existence of a unique optimal continuity control function $\check{w}_z$ for each possible estimated image function z these equations put some restrictions on the edge cost density f. In fact from (2b) it follows that $f'|]0, 1] \to \overline{\mathbf{R}_-}$ must be bijective, and that $f'(]1, \infty[) \subseteq \mathbf{R}_+$. Likewise from (3) we see that f'' must be strictly positive on $]0, 1[$, and that $f''(1) \geq 0$. Two of the simplest functions satisfying these requirements are given by $f(\omega) \doteq \omega - \ln \omega$ and $f(\omega) \doteq \omega \ln \omega - \omega$, but there are of course many other possibilities [7].

Given that f satisfies the conditions above, $f'|]0, 1]$ is invertible. Since moreover w is strictly positive, we end up with the equations

$$z(x) = \zeta(x) + \nabla \cdot (w \nabla z)(x) \qquad \forall x \in B \tag{4a}$$

$$w(x) = g(\|\nabla z(x)^T\|) \qquad \forall x \in B \tag{4b}$$

$$\frac{\partial z}{\partial e_n}(x) = 0 \qquad \forall x \in \partial B \tag{4c}$$

where the function $g : \overline{\mathbf{R}_+} \to]0,1]$, referred to as the *diffusivity anomaly*, is defined by

$$g(\gamma) \doteq (f'|]0,1])^{-1} \left(-\frac{\gamma^2}{\lambda} \right) \qquad \gamma \geq 0 \tag{5}$$

The properties of the edge cost density f clearly imply that g is a strictly positive strictly decreasing differentiable bijection. For the two edge cost densities proposed above the diffusivity anomaly takes the form $g(\gamma) \doteq (1 + \gamma^2/\lambda)^{-1}$ and $g(\gamma) \doteq e^{-\gamma^2/\lambda}$ respectively.

Since our method necessarily yields continuity control functions for which $w^{-1}(\{0\}) = \emptyset$, Terzopoulos' edge representation is inadequate. The simplest and most reasonable modification is to consider the edges to consist of the set $w^{-1}(]0,\theta])$ where θ is a fixed threshold. Since the diffusivity anomaly g is strictly decreasing, we have $w^{-1}(]0,\theta]) = \|\nabla z^T\|^{-1}([g^{-1}(\theta),\infty[)$ whence the edges are obtained by thresholding the magnitude of the gradient of the estimated image function.

4 Biased Anisotropic Diffusion

Perona and Malik [8, 9] have introduced anisotropic diffusion as a method of suppressing finer details without weakening or dislocating the larger scale edges. The initial value problem governing their method is given by

$$\frac{\partial z}{\partial t}(x,t) = \nabla \cdot (w\nabla z)(x,t) \qquad \forall x \in B \qquad \forall t > 0 \tag{6a}$$

$$w(x,t) = g(\|\nabla z(x,t)^T\|) \qquad \forall x \in B \qquad \forall t > 0 \tag{6b}$$

$$\frac{\partial z}{\partial e_n}(x,t) = 0 \qquad \forall x \in \partial B \qquad \forall t > 0 \tag{6c}$$

$$z(x,0) = \zeta(x) \qquad \forall x \in B \tag{6d}$$

where the diffused image function z and the *diffusivity* w are functions of both position $x \in B$ and time $t \geq 0$, $\nabla \cdot$ and ∇ denote the divergence and the gradient respectively with respect to x, and the diffusivity anomaly $g : \overline{\mathbf{R}_+} \to \overline{\mathbf{R}_+}$ is a decreasing function.

The Euler equations we derived in the previous section are very similar to the initial value problem (6). In fact, a solution of (4) is given by the steady state of the initial value problem

$$\frac{\partial z}{\partial t}(x,t) = \zeta(x,t) - z(x,t) + \nabla \cdot (w\nabla z)(x,t) \qquad \forall x \in B \qquad \forall t > 0 \tag{7a}$$

$$w(x,t) = g(\|\nabla z(x,t)^T\|) \qquad \forall x \in B \qquad \forall t > 0 \tag{7b}$$

$$\frac{\partial z}{\partial e_n}(x,t) = 0 \qquad \forall x \in \partial B \qquad \forall t > 0 \tag{7c}$$

$$z(x,0) = \chi(x) \qquad \forall x \in B \tag{7d}$$

which is obtained from (6) by replacing the anisotropic diffusion equation (6a) by the closely related *"biased" anisotropic diffusion* equation (7a). Since our interest is in the steady state solution, the initial condition (6d) can also be replaced by an arbitrary initial condition (7d). The continuity control function w thus plays the role of the diffusivity, and will be referred to as such whenever the context so suggests. The bias term $\zeta - z$

intuitively has the effect of locally moderating the diffusion as the diffused image function z diffuses further away from the original image function ζ. It is therefore reasonable to believe that a steady state solution does exist. This belief is further substantiated by our experimental results.

The possibility of suppressing finer details while the more significant edges remain intact, or are even strengthened, is a consequence of the anisotropy, which in both the diffusions described above in turn is caused by the non-constancy of the diffusivity anomaly g. For our variational method governed by the boundary value problem (4) the choice of g was based on optimality considerations. Perona and Malik select their function g by demanding that the resulting *unbiased* anisotropic diffusion enhances the already pronounced edges while the less significant edges are weakened. Based on an analysis including only blurred linear step edges they vouch for diffusivity anomalies of the form $g(\gamma) \doteq c/[1 + (\gamma^2/\lambda)^a]$ where $c, \lambda > 0$ and $a > 1/2$ are constants. It is easy to check that, if these functions were substituted in the Euler equation (4b), the corresponding edge cost densities would satisfy the requirements of our variational method. Incidentally, for their experimental results Perona and Malik use exactly the diffusivity anomalies we proposed in section 3.

5 The Extremum Principle

The extremum principle is a common tool for proving uniqueness and stability with respect to boundary data for linear elliptic and linear parabolic problems [10]. For quasi-linear equations, such as the Euler equation (4a) and the biased anisotropic diffusion equation (7a), it is not quite as conclusive. Nevertheless it provides bounds on the solution and useful insight for convergence analysis of the numerical methods employed to find such a solution.

Theorem 5.1 *Let* $z : \overline{B} \times \overline{R_+} \to R$ *be a solution of the biased anisotropic diffusion problem (7) where it is assumed that* $g : \overline{R_+} \to \overline{R_+}$ *is continuously differentiable and* $\zeta : B \to R$ *is uniformly continuous. Assume further that* z *and its first and second order partial derivatives with respect to* x *are continuous (on* $\overline{B} \times \overline{R_+}$*). Then the following claims are true:*

(i) *If* $\pm y_\tau : \overline{B} \to R : x \mapsto \pm z(x, \tau)$ *has a local maximum at* $\xi \in \overline{B}$ *for some fixed* $\tau > 0$, *then* $\pm(\partial z/\partial t)(\xi, \tau) \leq \pm\zeta(\xi) \mp z(\xi, \tau)$.

(ii) *If* $\pm z$ *has a local maximum at* $(\xi, \tau) \in \overline{B} \times R_+$, *then* $\pm z(\xi, \tau) \leq \pm\zeta(\xi)$.

(iii) $\inf_{\xi \in B}[\zeta(\xi) \wedge \chi(\xi)] \leq z(x, t) \leq \sup_{\xi \in B}[\zeta(\xi) \vee \chi(\xi)] \qquad \forall x \in \overline{B} \qquad \forall t \geq 0$

Theorem 5.2 *Let* $z : \overline{B} \to R$ *be a solution of the boundary value problem (4) where it is assumed that* $g : \overline{R_+} \to \overline{R_+}$ *is continuously differentiable. Assume further that* z *and its first and second order partial derivatives are continuous (on* $\overline{B}$*). Then*

$$\inf_{\xi \in B} \zeta(\xi) \leq z(x) \leq \sup_{\xi \in B} \zeta(\xi) \qquad \forall x \in \overline{B}$$

According to the two theorems above the solutions of the biased anisotropic diffusion problem are well-behaved, in that they do not stray too far away from the original image function ζ unless forced to by the initial condition, and even if so, they eventually approach the range of ζ as $t \to \infty$. In particular, our variational edge detection method produces an estimated image function whose range is contained inside that of the original image function.

6 Edge Enhancement

It was mentioned earlier that the biased anisotropic diffusion (7) in similarity with its unbiased counterpart (6) has the important property of suppressing finer details while strengthening the more significant edges. In order to see this we define the edges to consist of the points in the image domain B at which the magnitude of the gradient of the diffused image function z has a strict local maximum along the direction perpendicular to the edge, that is the direction of ∇z. Letting $\partial/\partial e_\nu$ denote the directional derivative in this direction, Δ denote the Laplacian operator, and $\sigma \doteq \|\nabla z^T\|$ represent the strength of the edge, we furthermore observe that $\partial^2 \sigma/\partial e_\nu^2 \approx \Delta\sigma < 0$ at typical edge points of interest. At such edge points it can then be shown [7] that

$$\frac{\partial}{\partial t}(\sigma - \sigma_\zeta) = -(\sigma - \sigma_\zeta) + (g' \circ \sigma)\frac{\partial^2 \sigma}{\partial e_\nu^2}\sigma + (g \circ \sigma)\Delta\sigma \approx -(\sigma - \sigma_\zeta) + (\varphi' \circ \sigma)\Delta\sigma \quad (8)$$

where $\sigma_\zeta \doteq \partial\zeta/\partial e_\nu$, and $\varphi(\gamma) \doteq g(\gamma)\gamma$, $\gamma \geq 0$. From this equation it is clear that the bias term $-(\sigma - \sigma_\zeta)$ merely has a moderating effect on the enhancement/blurring of the edge while the decision between enhancement vs. blurring depends on the sign of the "driving" term $(\varphi' \circ \sigma)\Delta\sigma$ associated with the unbiased anisotropic diffusion. Since $\Delta\sigma < 0$, the desired performance of weakening the weak edges while strengthening the strong ones in a consistent manner, therefore requires that there exists an *edge enhancement threshold* $\gamma_0 \in \mathbf{R}_+$ such that $\varphi'^{-1}(\mathbf{R}_-) =]\gamma_0, \infty[$, $\varphi'^{-1}(\{0\}) = \{\gamma_0\}$ and $\varphi'^{-1}(\mathbf{R}_+) = [0, \gamma_0[$. If that is the case, the threshold γ_0 clearly determines the sensitivity of the edge detector, and one would hence expect it to be closely related to the intuitively similarly acting edge cost coefficient λ. Indeed, from (5) and the definition of φ it follows that γ_0 is proportional to $\sqrt{\lambda}$. It is also easy to verify that the two diffusivity anomalies proposed in section 3 satisfy the threshold condition above with $\gamma_0 = \sqrt{\lambda}$ and $\gamma_0 = \sqrt{\lambda/2}$ respectively.

If the diffused image function converges to a steady state solution, that is a solution of the boundary value problem (4), according to (8) we obtain the steady state edge enhancement $\sigma - \sigma_\zeta = (\varphi' \circ \sigma)\Delta\sigma$. Since the range of the steady state solution by the extremum principle is confined to lie within the range of the original image function, an amply enhanced edge strength σ can only be maintained along a very short distance across the edge. Such edges are therefore sharpened.

Besides being of vital importance for the edge enhancement mechanism, the existence of the edge enhancement threshold γ_0 also provides a natural choice for the threshold to be used in the postprocessing whereby the edges are finally extracted from the estimated image function. For our edge representation to be consistent with the edge enhancement mechanism, the edge representation threshold in section 3 should thus be given by $\theta \doteq g(\gamma_0)$.

7   Experimental Results

The experiments presented here were conducted with the edge cost density $f(\omega) \doteq \omega - \ln \omega$ corresponding to the diffusivity anomaly $g(\gamma) \doteq 1/(1 + \gamma^2/\lambda)$. The images involved were obtained by solving a discrete approximation of the boundary value problem (4) with a Gauss-Seidel-like iteration method.

As mentioned earlier, the iteration method converges to a solution of interest. This condition is illustrated by figure 1, which shows an original image and the estimated image obtained after convergence in the "sense of insignificant perceptible changes". With the initial image function $z^{(0)}$ set equal to ζ this convergence required about 100 iterations. For the purpose of edge detection, however, reasonably good results were obtained already before 50 iterations.

The variational edge detection method itself as well as the iteration method we employed to solve it appear to be remarkably robust with respect to changes in the initial image function. To demonstrate this behavior we tried the algorithm on the same original image function as in figure 1, but with the for convergence particularly unfavorable initial image function $z^{(0)} = 0$. Once again the algorithm converged if yet at a slower rate. The two solutions obtained with the two different initial image functions were not identical, but very similar. The significant differences were indeed limited to affect only small blobs of high contrast relative to the local background.

In order to extract a set of edges from the estimated image function z we followed the strategy outlined in section 3, and simply thresholded the gradient magnitude. Figure 2 shows the edges extracted from the estimated image function in figure 1 (b) with a gradient magnitude threshold ϑ 50% higher than the edge enhancement threshold γ_0. The edges obtained with ϑ 50% lower than γ_0 were almost identical. Since the edge enhancement/blurring mechanism discussed earlier depletes the set of points at which the gradient magnitude takes values close to γ_0, this ought to be expected.

8   Acknowledgements

The author is grateful to M. Singer for writing most of the user interface for the software implementation of the algorithm and to K. Pister for generous help with generating the images.

References

[1] A. Blake and A. Zisserman, "Some properties of weak continuity constraints and the GNC algorithm," in *Conference on Computer Vision and Pattern Recognition*, pp. 656–660, IEEE, 1986.

[2] A. Blake and A. Zisserman, *Visual Reconstruction*. The MIT Press, 1987.

[3] S. Geman and D. Geman, "Stochastic relaxation, Gibbs distributions, and the Bayesian restoration of images," *IEEE Transactions on Pattern Analysis and Machine Intelligence*, vol. 6, pp. 721–741, Nov. 1984.

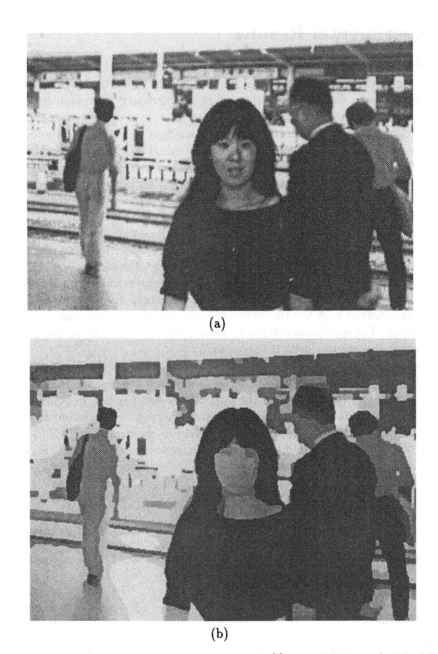

(a)

(b)

Figure 1: Estimated image after i iterations, when $z^{(0)} = \zeta$. (a) $i = 0$ (original image). (b) $i = 100$.

Figure 2: Extracted edges from estimated image in figure 1 (b).

[4] D. Terzopoulos, "Computing visible-surface representations," A.I. Memo 800, Massachusetts Institute of Technology Artificial Intelligence Laboratory, Cambridge, MA, Mar. 1985.

[5] D. Terzopoulos, "Regularization of inverse visual problems involving discontinuities," *IEEE Transactions on Pattern Analysis and Machine Intelligence*, vol. 8, pp. 413–424, July 1986.

[6] D. G. Luenberger, *Linear and Nonlinear Programming*. Reading, MA: Addison-Wesley, second ed., 1984.

[7] K. N. Nordström, "Biased anisotropic diffusion—a unified regularization and diffusion approach to edge detection," Report UCB/CSD 89/514, Computer Science Division University of California, Berkeley, CA, June 1989.

[8] P. Perona and J. Malik, "Scale space and edge detection using anisotropic diffusion," in *Workshop on Computer Vision—Miami*, pp. 16–22, IEEE Computer Society, June 1987.

[9] P. Perona and J. Malik, "Scale-space and edge detection using anisotropic diffusion," Report UCB/CSD 88/483, Computer Science Division University of California, Berkeley, CA, Dec. 1988.

[10] R. Courant and D. Hilbert, *Methods of Mathematical Physics*, vol. II. John Wiley & Sons, 1962.

HIERARCHICAL IMAGE ANALYSIS USING IRREGULAR TESSELLATIONS

Annick Montanvert
Equipe RFMQ - TIM3 (UA CNRS 397)
CERMO BP 53X . 38041 GRENOBLE cedex . FRANCE

Peter Meer ; Azriel Rosenfeld
Center for Automation Research. University of Maryland
College Park. MD 20742. USA

I. INTRODUCTION

An image pyramid is a multi-layer data structure in which the input is represented at successively reduced resolutions. The input picture, a N^2 array, is taken as level 0, the bottom of the pyramid. The upper levels are built recursively by a parallel process, usually using a four-fold reduction in resolution. The height of the image pyramid is thus *log N*. Many tasks can be accomplished in *O[log N]* on an image pyramid with a small increase of memory space [Rosenfeld, 1984 ; Uhr, 1987].

The dependence of the low resolution representations on the interaction between the sampling grid and the input image is of importance for image pyramid applications. The rigidity of the sampling structure must be taken into consideration, for example, many segmentation algorithms employ a delineation process in which the weights (parent-child links) are iteratively changed *after* the initial pyramid was built [Hong 84].

We present irregular tessellations to generate an adaptive multiresolution description of the input image ; the hierarchy is built bottom-up adapted to the content of the input image and most of the properties of the "classical" image pyramids are preserved. We employ a local stochastic process which can be associated with different feature fusion criteria to build lower resolution representations. A graph formulation is defined to achieve this target. Applied on labeled pictures, every connected component is reduced to a separate root, and the adjacency graph is simultaneously built. In gray level pictures we perform a segmentation of the initial image. This work is presented in more details in [Montanvert 1989].

II. IRREGULAR TESSELLATIONS AND STOCHASTIC PYRAMIDS

In the image pyramids based on regular tessellations the artifacts caused by the rigidity of the sampling structure are always present. Only a hierarchy of irregular tessellations can be molded to the structure of the input image ; however, the topological relations among cells on the different levels are no longer carried implicitly by the sampling structure. Thus it is convenient to use the formalism of graphs.

The ensemble of cells defining the level *l* of the pyramid are taken as the vertices V[*l*] of an undirected graph G[*l*] = (V[*l*], E[*l*]). The edges E[*l*] of the graph describe the adjacency relations between cells at level *l*. The graph G[0] describes the 8-connected sampling grid of the input, G[*l*] is called the *adjacency graph*. As the image pyramid is to be built recursively bottom-up we must define the procedure of deriving G[*l+1*] from G[*l*], we are dealing with a *graph contraction* problem. We must design rules for :

- choosing the new set of vertices V[*l+1*] from V[*l*] ,i.e., the survivors of the decimation process;
- allocating each non-surviving vertex of level *l* to the survivors, i.e., generating the parent-children links;
- creating the edges E[*l+1*], i.e., recognizing the adjacency relations among the surviving cells.

In order to have any cell in the hierarchy correspond to a compact region of the input, a node of $c[l+1] \in$ $V[l+1]$ must represent a connected subset of $V[l]$ which defines the children of $c[l+1]$. We want to assure that this process can employ only local operations in parallel to build $G[l+1]$. A solution is provided by searching a maximum independant set (MIS) of vertices [Luby, 1985] among $V[l]$. Because of the properties of the MIS, two important constraints will be respected :

(1) any non-surviving cell at level l has at least one survivor in its neighborhood,

(2) two adjacent vertices on level l cannot both survive and thus on level $l+1$ the number of vertices must decrease significantly yielding a pyramidal hierarchy.

The last step in defining the graph of level $l+1$ is the definition of the edges $E[l+1]$: two vertices are connected in $G[l+1]$ iff there exists at least one path between them in $G[l]$ of two or three edges long. The graph $G[l+1]$ of the next level is now completely defined by a parallel and local process.

The survivors will be spread on $V[l]$ in a more flexible way than on a regular pyramid structure. To obtain a multiresolution hierarchy adapted to the input image, this selection of parents and then the allocation of children will depend on the content of the image.

A probabilistic algorithm achieves the graph contraction satisfying the two constraints ; it is analyzed in more details in [Meer 1989, Meer and Connelly 1989]. It is different from other solutions proposed in the literature. The basic principle is to iteratively extract from $V[l]$ some vertices satisfying constraints (1) and (2), which then belong to $V[l+1]$, and so on until no more vertices can be added : each vertice owns a random value (the outcome of a random variable) and is kept iff this value is a local maxima on the subgraph induced by the current iteration. It converges after a few steps. If the process is applied to build a whole structure (until it remains just one vertex), a complete irregular tessellation hierarchy is defined. The power of this algorithm (compared to other algorithms to compute the MIS) comes from the local maxima principle which allows the process to be adapted to an information such as images.

III. LABELED IMAGES

In a labeled image every maximal set of connected pixels sharing the same label is a connected component of the image. Looking at its neighborhood a cell can see which cells share its label (they are said to be in the same class) which define the *similarity graph* $G'[l]=(V[l], E'[l])$ induced on $G[l]$. In the similarity graph the connected components become maximal connected subgraphs. The techniques describe precedently are adapted in the following way : to become a survivor the outcome of the random variable drawn by the cell must be the local maximum *among* the outcomes drawn by the neighbors in the same class. Thus the subgraphs of the similarity graph are processed independently and in parallel, each subgraph being recursively contracted into one vertex, the *root* of the connected component ; at each iteration a maximum independent set of the similarity graph is extracted to fix the new subgraph.(see Figure 1).

All the artifacts of rigid multiresolution hierarchies are eliminated. From each connected component a pyramidal hierarchy based on irregular tessellations is built in log(component_size) steps (the component_size of a connected component is its intrinsic diameter). Since random processes are involved in the construction of the irregular tessellations the location of the roots depends on the outcomes of local processes. Nevertheless, always the same root level adjacency graph is obtained at the top of the hierarchy.The famous connectedness puzzle of Minsky and Papert (Figure 1) can be solved in parallel with the help of the root level adjacency graph.

IV. GRAY SCALE IMAGES

In gray level images we have to analyse the difference between the values of two adjacent pixels. We have seen that in our technique to build the hierarchies the pixels in a neighborhood must be arranged into classes. The class membership induces the similarity graph on which the stochastic decimation takes place. Similar to the labeled images, class membership can be defined based on the gray level difference between the center cell c_0 and one of its neighbors c_i, $i = 1,...,r$, let g_i be their gray levels. Employing an absolute threshold T, c_0 decides that its neighbor c_i is in the same class iff $\delta_i = |g_0 - g_i| \leq T$.

Like in the labeled case this criterion is symmetrical. See Figure 2 for an illustration of a similarity graph on a gray scale picture. The stochastic decimation algorithm selects the survivor cells and the non-survivors are allocated to their most similar surviving neighbor. The survivors (parents) compute a new gray level value g based on their children. Hence the hierarchy is not built on some well-defined subgraphs of $G'[l]$ as it was the case on labeled pictures. Now the similarity subgraphs and then their meaning on the input picture evolved when we build the pyramid, since g_i values are recomputed. We conclude that a symmetric class membership criterion strongly influences the structure of the hierarchy and therefore the final representation of the image ; some artefacts can be created (see Figure 2).

To achieve satisfactory results in our irregular tessellations based multiresolution gray level image analysis a *non-symmetric* class membership criterion must be used. Let the cell c_0 have r neighbors. In this neighborhood we define the local threshold $S[c_0]$ such that $0 \leq S[c_0] \leq T$. Thus the cell c_0 declares as similar to itselve only its neighbors c_i for which $|g_0 - g_i| \leq S[c_0]$. The threshold $S[c_0]$ is specific to the neighborhood of cell c_0 and therefore the criterion is not symmetric. Indeed, in general $|g_0 - g_i| \leq S[c_0]$ does not imply $|g_0 - g_i| \leq S[c_i]$ since the two thresholds are computed based·on only partially overlapping neighborhoods. As a consequence of the non-symmetrical membership criterion the similarity graphs become *directed*. The neighborhood dependent local threshold $S[c_0]$ assures that every cell connects first to its neighbors with the most similar gray level values. Thus the individual rows in the image shown in Figure 2.b are reduced to single cells *before* two cells belonging to adjacent rows can become neighbors on the similarity graph.

The value of the local threshold $S[c_0]$ is computed based on a subset of cells in the neighborhood of the cell c_0 . The extreme case $S[c_0] = 0$ corresponds to connected component recognition. The other extreme, $S[c_0] = T$ yields the symmetric class membership criterion. Several approaches are available to determine the $S[c_0]$ value best dichotomizing the neighborhood into two classes. We will employ only gray level information in computing $S[c_0]$.

Let $\delta_{[i]}$, $i = 0,1,...,r$ be the ordered sequence of absolute gray level differences $\delta_i = |g_0 - g_i|$.

Thus $\delta_{[0]} = 0 \leq \delta_{[1]} \leq ... \delta_{[s]} \leq S[c_0] < \delta_{[s+1]} \quad ... \leq \delta_{[t]} \leq T < \delta_{[t+1]} \quad ... \leq \delta_{[r]}$.

We compute k as the *maximum averaged contrast* method in which the threshold $S[c_0]$ is set to the most significant step in the sequence of $\delta_{[i]}$ For all the t neighbors we compute

$$A_j = \frac{\sum_{i=1}^{j} \delta_{[i]}}{j} \qquad B_j = \frac{\sum_{i=j+1}^{t} \delta_{[i]}}{t-j} \qquad 1 \leq j \leq t-1$$

Let u = min arg (max$_j$ (B_j - A_j)) and s = max arg ($\delta_{[i]}$ = $\delta_{[u]}$). The threshold $S[\ c_0\ ]$ = $\delta_{[s]}$ is set by the first occurrence of the maximum averaged contrast between the two classes; it is automatically adjusted to the local gray level configuration. The gray level value of a parent is computed as the weighted average of the children's gray levels. See Figure 3 for an illustration of the result on an aerial image. Different outcomes for the employed random variables cause changes in the hierarchy structure : by changing the set of survivor cells the values attributed to these cells may also change slightly yielding changes in the similarity graphs of subsequent levels. As expected, regions with sharp boundaries in the input image achieve very similar representations (see Figure 3). Lastly we can notice that if in a labeled picture a root (an isolated vertex in the similarity graph G'[l]) remains a root at higher levels, this is not true in gray scale images while the similarity graph continues to evolve. That is, the root may disappear at subsequent levels, its receptive field being fused into a larger region.

The decimation process can be modified to be biased toward cells with high informational value. Jolion and Montanvert [1989] proposed an adaptive pyramid in which cells belonging to the most homogeneous regions have priority to become survivors. Such an approach, however, is not successful for labeled pictures in which many cells carry identical descriptions.

V. CONCLUSION

In this paper we have presented an image analysis technique in which a separate hierarchy is built over every compact object of the input. The approach is made possible by a stochastic decimation algorithm which adapts the structure of the hierarchy to the analyzed image. For labeled images the final description is unique. For gray level images the classes are defined by converging local processes and slight differences may appear. At the apex every root can recover information about the represented object in logarithmic number of processing steps, and thus the adjacency graph can become the foundation for a relational model of the scene.

REFERENCES

T.H. Hong and A. Rosenfeld (1984): Compact region extraction using weighted pixel linking in a pyramid. *IEEE Trans. on Pattern Analysis and Machine Intelligence.* PAMI-6, 222-229.

J.M. Jolion and A. Montanvert (1989): La pyramide adaptive: construction et utilisation pour l'analyse de scenes 2D. *Proceedings of the Seventh RFIA Conference* , November 1989, Paris, 197-206.

M. Luby (1985): A simple parallel algorithm for the maximal independent set problem. In *Proceedings of the Seventeenth Annual ACM Symposium on Theory of Computing*, 1-10.

P. Meer (1989): Stochastic image pyramids. *Computer Vision Graphics and Image Processing.* 45, 269-294.

P. Meer and S. Connelly (1989): A fast parallel method for synthesis of random patterns. *Pattern Recognition*, 22, 189-204.

A. Montanvert ; P. Meer and A. Rosenfeld (1989): Hierarchical image analysis using irregular tessellations. *Technical Report, University of Maryland, CS-TR-2322, September 1989.*

A. Rosenfeld, ed. (1984): *Multiresolution Image Processing and Analysis..* Springer Verlag, Berlin.

L. Uhr, ed. (1987): *Parallel Computer Vision* Academic Press, Boston.

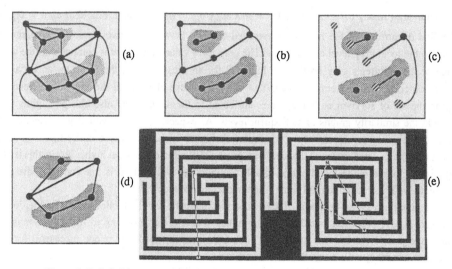

Figure 1. Labeled images. (a) G[l] : adjacency graph (b) G'[l] : similarity graph
(c) Allocation of non survivors (hashed circles) at level l (d) G[l+1]
(e) The root level projected on the bottom for the puzzle of Minsky and Papert

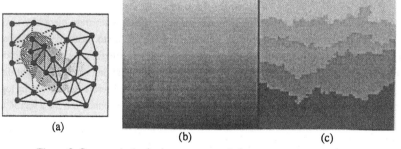

Figure 2. Symmetrical criterion on gray scale images
(a) Dot lines show the differences between G[l] and G'[l]
(b) The original image : a uniform gray level slope
(c) The result provides at the root level shows the artefacts provide by the
use of a symmetrical criterion

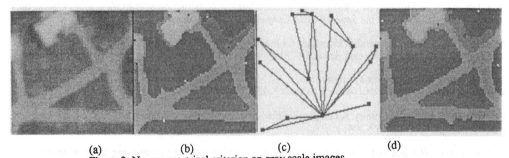

(a)   (b)   (c)   (d)

Figure 3. Non symmetrical criterion on gray scale images
(a) The original picture   (b) The result with the root locations superposed
(c) The adjacency graph of the root level (d) The result for another outcome
of the random variable

A BIT PLANE ARCHITECTURE FOR AN IMAGE ANALYSIS PROCESSOR IMPLEMENTED WITH P.L.C.A. GATE ARRAY

J.C. Klein, F. Collange & M. Bilodeau

Ecole des Mines de Paris
Centre de Morphologie Mathématique
35, rue Saint Honoré 77305 Fontainebleau Cedex
Tél. (1) 64 22 48 21, Téléfax (1) 64 22 39 03, Télex MINEFON 694 736.

Abstract

As Image Analysis resorts to increasingly powerful algorithms, the processing time is correspondingly extended. Consequently, system designers are constantly looking for new technologies and new architectures capable of improving processing speed without increasing the complexity and the cost of the machines.

To achieve this objective, the Centre de Morphologie Mathématique (Ecole des Mines de Paris) has designed and developed, a new image processor for Mathematical Morphology based on Programmable Logic Cell Array (PLCA) technology. This processor, incorporated into the Cambridge Instruments Quantimet 570, is capable of performing complex morphological transformations on 512 x 512 images of 8 bits per pixel at the rate of 27 msec per image. This speed, associated with extensive algorithmic software support, makes it an extremely powerful tool in the field of image analysis.

1. INTRODUCTION

Mathematical Morphology [1,2,3] is a technique for image analysis, which was initially based on set transformations. The results obtained within this framework have been extended to the study of functions [4]. The usual approach consists of transforming the image so as to simplify it, segment it and measure some of its components [5,6]. In order to achieve this, the pixels of the image are computed and modified according to their neighbourhood and, in some cases, according to information taken from another image. The transformation is obtained by the action of a structuring element composed of N x N points on each pixel. A value depending on the type of operation to be performed and on the relationships existing between the structuring element and the neighbourhood of the pixel is then assigned to the centre pixel. The architecture of the processor is based on the structuring element concept [7,17,18].

2. GENERAL STRUCTURE

The morphological processing unit [8,9,10] described in this paper has been implemented (Fig.1) into the Cambridge Instruments Quantimet 570 which is briefly presented below.

2.1 The Quantimet 570

The Quantimet 570 image processor consists of :
- 68000-based processor monitoring unit
- Graphic overlay and binary image memory
- Measurement processor
- Acquisition and display system
- Peripheral control interfaces
- 386-based PC compatible host computer serving as the user's interface.

2.2 The morphological processing unit

The morphological processing unit consists of :
- An image memory containing up to 8 banks of 8 512*512 images of 8

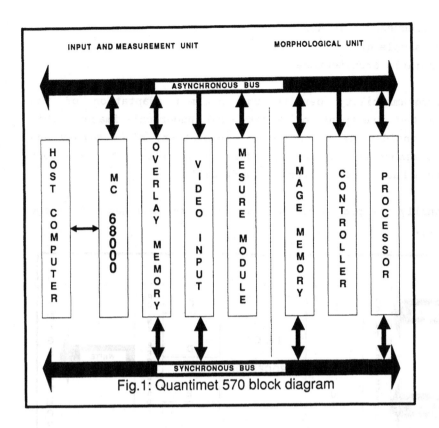

Fig.1: Quantimet 570 block diagram

bits. This represents a maximum capacity of 16 megabytes. Fast access time memory is used allowing 3 read-cycles and 1 write-cycle of 8 pixels each of 8 bits to be performed in 800ns. Two read-accesses and the write-access are reserved for the morphological processor, the third read-access is used to provide the synchronous display of a numerical image.

- The control and synchronization of the memory and the processor. The size of the image is programmable, up to 512 lines x 512 pixels.
- A morphological processor based on a programmable architecture, which gives it optimum adaptive properties for processing the different types of algorithms used in Mathematical Morphology.

3. THE MORPHOLOGICAL PROCESSOR

The use of Programmable Logic Cell Arrays allows considerable computing capacity to be obtained within a small volume. The main specifications of the processor are the following :

- Pixel frequency : 10 Mhz
- Square sample grid
- Programmable architecture

This programmability permits the optimum adaptation of the processor to the treatment of binary and numerical images. This represents a considerable technological innovation, since binary and numerical processors use the same hardware facilities. They are both designed from a basic cell called a binary cell.

Additionally the processor supports euclidean and geodesic image treatment [11].

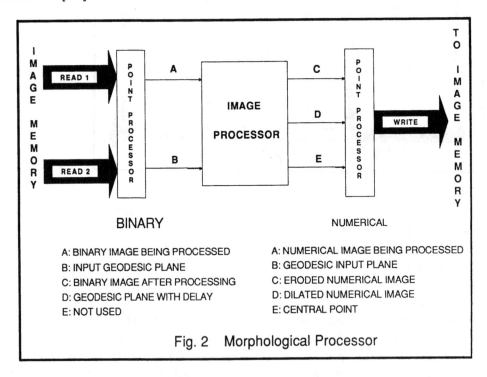

BINARY

A: BINARY IMAGE BEING PROCESSED
B: INPUT GEODESIC PLANE
C: BINARY IMAGE AFTER PROCESSING
D: GEODESIC PLANE WITH DELAY
E: NOT USED

NUMERICAL

A: NUMERICAL IMAGE BEING PROCESSED
B: GEODESIC INPUT PLANE
C: ERODED NUMERICAL IMAGE
D: DILATED NUMERICAL IMAGE
E: CENTRAL POINT

Fig. 2 Morphological Processor

In numerical or binary mode, the neighbourhood morphological processor is followed and preceeded by a point processor (Fig. 2). These point processors, which are not fully described here, are look-up tables and logic and arithmetic operators.

3.1 Programmable Logic Cell Arrays

Programmable Logic Cell Arrays are a family of integrated circuit digital products, which combine high performance with a high level of integration. These circuits are also versatile, since their programming is not irreversible. This allows the configuration to be modified and loaded in its new form, at any time.

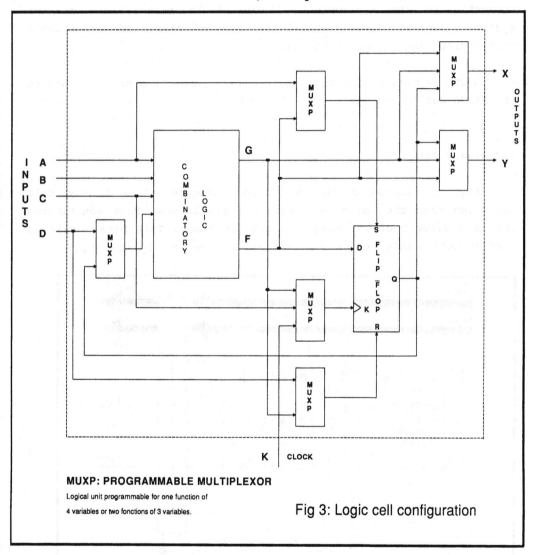

MUXP: PROGRAMMABLE MULTIPLEXOR
Logical unit programmable for one function of
4 variables or two fonctions of 3 variables.

Fig 3: Logic cell configuration

There are 3 main classes of programmable components contained in one such circuit. The number of components and their functions are dependent on the type.

- Logic blocks (numbering from 64 to 320) each contain one or two registers and a combinatory logic section. These blocks constitute the main functional element of the product. A logic block of the XILINX 2064 or 2018 series is presented in Fig. 3.

- Logic blocks take chargeof the I/O of the circuits (numbering from 58 to 144) and contain all the facilities required to interface the product with its environment. These blocks possess one or two registers to insure, if needed, the synchronization of I/O signals with a clock integrated in the gate array.

- Connections provide for the required interconnection network between the different logic blocks.

3.2 Binary cell

Fig. 4 represents the simplified diagram of a binary cell developed from the Logic Cell Arrays logic blocks. This constitutes the main element of the neighbourhood processor. This binary cell is used in both binary mode and numerical mode by the bin/num signal.

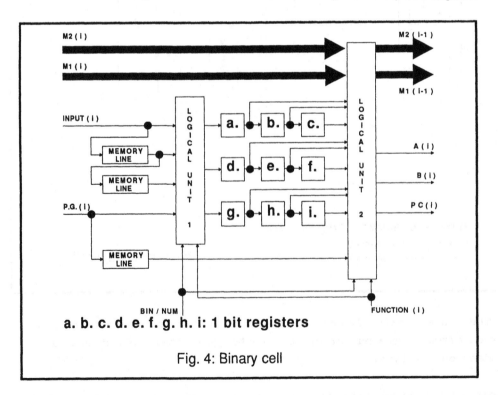

a. b. c. d. e. f. g. h. i: 1 bit registers

Fig. 4: Binary cell

The binary cell contains an input logical unit and an output logical unit. The part played by these logical units depends on the operating mode.

The input logical unit deals with the problems related to the field border in euclidean and geodesic mode [11].

The output logical unit enables the programming of the morphological operator and of the structuring element.

Binary mode:

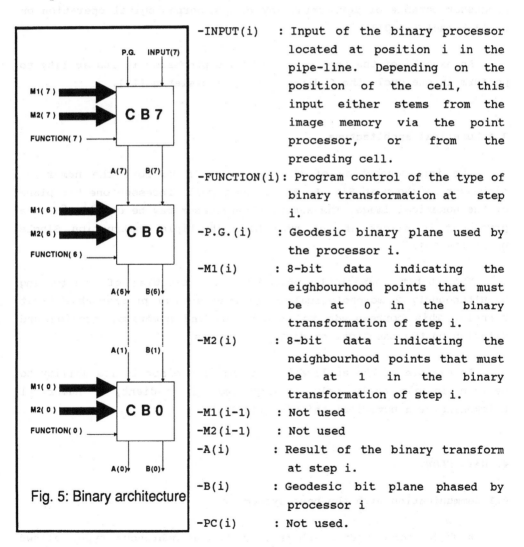

Fig. 5: Binary architecture

-INPUT(i)   : Input of the binary processor located at position i in the pipe-line. Depending on the position of the cell, this input either stems from the image memory via the point processor, or from the preceding cell.

-FUNCTION(i) : Program control of the type of binary transformation at step i.

-P.G.(i)   : Geodesic binary plane used by the processor i.

-M1(i)   : 8-bit data indicating the eighbourhood points that must be at 0 in the binary transformation of step i.

-M2(i)   : 8-bit data indicating the neighbourhood points that must be at 1 in the binary transformation of step i.

-M1(i-1)   : Not used

-M2(i-1)   : Not used

-A(i)   : Result of the binary transform at step i.

-B(i)   : Geodesic bit plane phased by processor i

-PC(i)   : Not used.

The input and output logical units serve to manage and compute the masks that indicate the neighbouring pixels that are still considered in the computation of the erosion and of the dilation. (Note-dilation corresponds to the examination of the neighbourhood points to determine the maximum value and erosion to the minimum).

3.3 Binary architecture

The interconnection of the binary cells is shown in Fig. 5. There are 8 elementary binary cells in a pipe-line. Each is a complete processor capable of performing any binary morphological operation on a 3*3 kernel [16,21].

An example of the application of this processor is its ability to perform in one cycle the 8 directions of a skeleton [12].

3.4 Numerical architecture

The interconnection of the binary cells to form the numerical processor is shown in Fig. 6. Each binary cell processes one bit plane of the numerical image. The numerical processor may be considered as a processor in slices of the type MISD (Multiple Instruction Single Data) [19,20].

The 8-bit numerical processor is capable of performing simultaneously 2 morphological operations on two programmable 9-bit kernels. This corresponds to computating the numerical erosion and dilation on two identical or distinct kernels.

An example of the application of this processor is its ability to perform in one cycle only a morphological gradient, a numerical thickening or a numerical thinning [12,1].

4. DATA FLOWS

4.1 Communication with the host system

A first communication channel of the asynchronous type, allows the Control Processor of the Quantimet to control the morphological processor via its 68000 microprocessor.

Numerical mode :

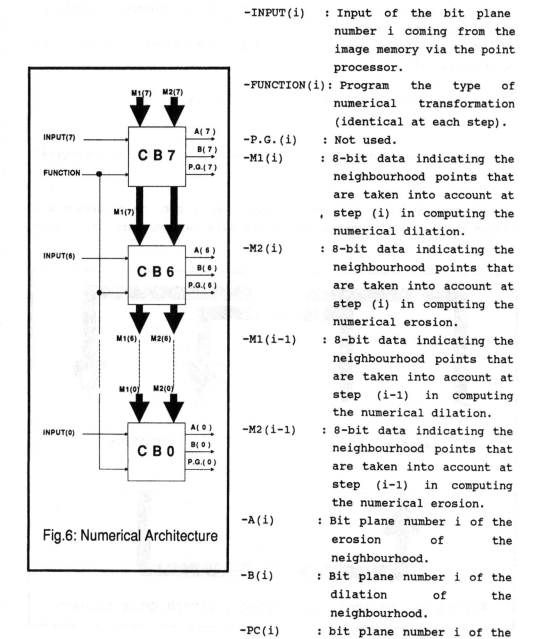

Fig.6: Numerical Architecture

-INPUT(i)   : Input of the bit plane number i coming from the image memory via the point processor.

-FUNCTION(i) : Program the type of numerical transformation (identical at each step).

-P.G.(i)   : Not used.

-M1(i)   : 8-bit data indicating the neighbourhood points that are taken into account at , step (i) in computing the numerical dilation.

-M2(i)   : 8-bit data indicating the neighbourhood points that are taken into account at step (i) in computing the numerical erosion.

-M1(i-1)   : 8-bit data indicating the neighbourhood points that are taken into account at step (i-1) in computing the numerical dilation.

-M2(i-1)   : 8-bit data indicating the neighbourhood points that are taken into account at step (i-1) in computing the numerical erosion.

-A(i)   : Bit plane number i of the erosion of the neighbourhood.

-B(i)   : Bit plane number i of the dilation of the neighbourhood.

-PC(i)   : bit plane number i of the central pixel.

A second communication channel, synchronous with the video
control signals, allows the transfer a 512x512 image in
40-milliseconds. This channel is used :
- to transmit numerical images, acquired by the Quantimet 570, to the
memory of the morphological processor.
- to transmit the images from the morphological processor memory to
the display of the Quantimet 570
- to exchange binary images between the morphological processor and
the Quantimet 570.

The communication channels are shown in Fig. 1.

**4.2 Communication between the image memory and the morphological
processor**

The communication between the image memory and the processor are
managed by a specific controller. Pixel data is propagated at a rate

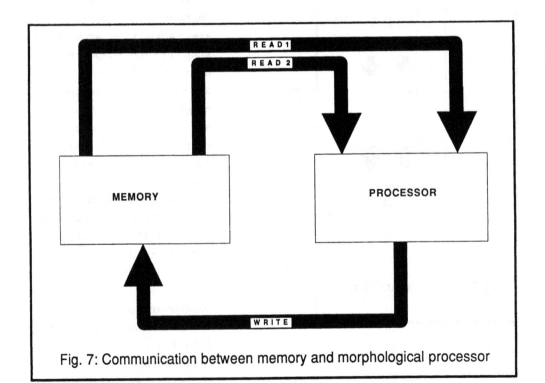

Fig. 7: Communication between memory and morphological processor

of 100ns per pixel on three buses (Fig. 7) which are asynchronous with respect to the video control signals. This architecture was chosen to permit the simultaneous reading (READ1 and READ2) of two 8-bit numerical images, as morphological transformations give a resulting image (WRITE) by using one or two input images.

5. THE SOFTWARE

The morphological processing unit is controlled from a library of procedures installed in the 68000 microprocessor of the Q570. This unit communicates with the PC-type Control Processor via a parallel port and executes the instructions sent by the host computer. Messages made of instructions may be sent by any user program, for example, the QUIC menu program used by the Q570 or the QBasic interpreter. During the development, we interfaced to the library with a Pascal compiler. It includes approximately 85 procedures classified into the following categories.

5.1 Image memory access

The images can be accessed by transferring pixels, or blocks of variable size, or complete images. For example, the procedure LOADIMAGE img. "fred" transfers the image "fred", stored on disk, into the image memory number "img". The acquisition of an image from a camera can be performed with or without summation. Display instructions allow the display binary and numerical images, with an option of superimposing the binary image.

Instructions for the transfer of images between the Q570 and the morphological processor, allow the user to directly use the facilities provided by each system.

5.2 Neighbourhood transformations

The library contains all the basic transformations in binary and numerical modes. Table 1 summarises the main procedures. The use of elementary transformations allows the creation of new algorithms. The watershed is a technique of image segmentation which is similar to the grey level skeleton. The Top Hat transformation is used to produce spatially adaptive thresholding. The example given in Fig. 8 is a sequential alternating algorithm, which alternates openings and closings of increasing size. The five parameters are : source and target images (img.1, img. 2), the order according to which transformations must be performed (opening-closing or closing-opening) determined by "ind", "elst", which defines the shape of the structuring element, and an integer (n) specifying the number of iterations (3).

```
procedure Asf(img1,img2,elst,ind,n:integer;
var i:integer;
begin
   Greymove(img1,img2);
   if ind=0 then for i:=1 to n do
         begin
               Greyopen(img2,img2,elst,i);
               Greyclose(img2,img2,elst,i);
         end;
   else for i:=1 to n do
         begin
               Greyopen(img2,img2,elst,i);
               Greyclose(img2,img2,elst,i);
         end;
end.
```

Fig. 8: Example of morphological filtering program

Laplacian, gradient and user-defined image convolution functions are integrated in the library.

Watershed	Watershed divide line
Maxima	Detection of local maxima
Tophat	Top hat transformation
Greybuild	Reconstruction of numerical image
Gradient	Morphological gradient
Greythresh	Numerical image thresholding
Greyskel	Numerical image skeleton
Greyprune	Numerical pruning
Binskel	Binary skeleton
Binprune	Binary pruning
Skiz	Skeleton by zone of influence

Table 1 - Examples of algorithms available in the library

5.3 Arithmetic transformations

All the functions : addition, subtraction, multiplication and division between two images or between an image and a constant are available for 8 and 16 bit images. Utility programs allow an image to be converted from the signed value to the absolute value.

6. PERFORMANCE

We have used the first prototype as a development tool for our research in image analysis. Mathematical morphology applies to numerous fields, let us mention, among others, biology [13], medicine [14] and robotics [15].

Table 2 gives some examples of the time needed to compute 512x512 images. The execution time for each transformation is proportional to the size of the image. Some algorithms cannot be characterized by a fixed execution time as they depend on the image content. For example, skeletonization algorithms are of this type.

Operation	Size	msec
Grey dilate/erode	3 x 3	27
Grey open/close	3 x 3	54
Tophat transform	3 x 3	81
Gradient transform	3 x 3	27
Binary dilate/erode	17 x 17	27
Binary open/close	9 x 9	27
Geodesic dilate/erode	17 x 17	27
Point transform	-	27
Image add/subtract	-	27

Table 2 - Timings of typical operations

In order to estimate the processing potential of the machine, we have used the system for a particular application: the segmentation of proteins in an electrophoresis gel. The algorithm was studied at the Centre de Morphologie Mathematique, by S. Beucher [6]. Fig. 9 illustrates the results obtained after each step and thus allows the performance of the processor to be evaluated.
- Initial step: On the filtered image, regional minima are detected. Time: 1.5 s (Fig. 9b).

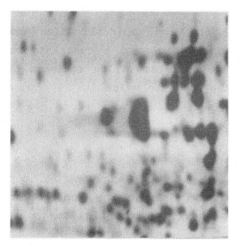

Fig. 9a: Initial image

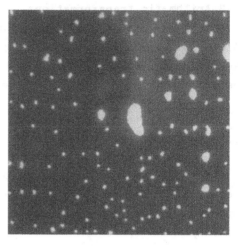

Fig. 9b: Minima of filtered image

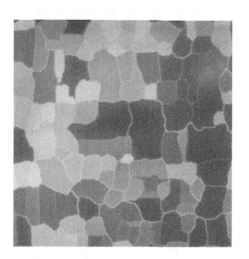

Fig. 9c: Catchment basins

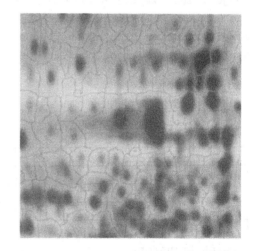

Fig.9d: Watersheds of filtered image

- Second step: The catchment basins of the filtered image are computed. This yields image 9c. Time: 35 s.
- Third step: The watersheds of the filtered image are obtained by an adaptive thresholding of image 9c (Fig. 9d). Time: 81 ms.
- Fourth step: The marker image is obtained by taking the union of the image of the minima and of the watersheds image (Fig.9e). Time: 27 ms.
- Final step: The marker image is used to modify the homotopy of the

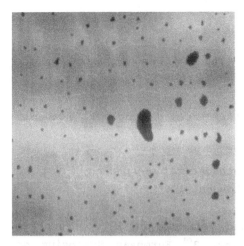
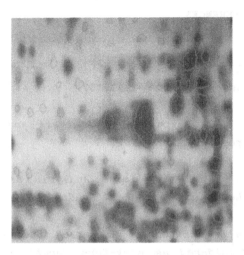

Fig. 9e: marker image   Fig. 9f: Contour of spots

gradient of the original image. Then, the watersheds of this modified gradient provide an accurate detection of the spot contours. This final result is displayed on Fig. 9f.

CONCLUSION

The binary processing cell, wich may be configured in numerical or binary mode, constitutes the core of the system. To develop it, we have used widely available. Programmable Logic Cell Arrays. This allowed an optimum ratio to be acheived between the degree of integration and the cost of the cell implementation.

The organization, programmed according to the processing mode being used, of eight cells in pipeline forms the morphological processor. This structure supports all morphological transformations. The time required for the execution of the basic transformations is much shorter than can be obtained from a classical computer.

The programming software is easy to use and provides the user with the whole range of the morphological tools for a moderate cost compared to its performance.

REFERENCES

[1] G. Matheron, *Eléments pour une Théorie des Milieux Poreux*, Masson, Paris, 1967.967.

[2] J. Serra, *Image Analysis and Mathematical Morphology*, Academic Press, London, 1982.

[3] J. Serra, *Image Analysis and Mathematical Morphology*, Vol 2: *Theoretical Advances*, Academic Press, London, 1988.

[4] S. Beucher, J. Serra, *Shapes and Patterns of Microstructures Considered as Grey-tone Functions*, Proc. 3rd European Symposium on Stereology, Ljubljana (Yougoslavie), june 22-26, 1981.

[5] M. Coster, J-L. Chermant, *Précis d'Analyse d'Images*, Ed. du CNRS, 1985.

[6] S. Beucher, L. Vincent, *Introduction aux Outils Morphologiques de Segmentation*, Journal de Microscopie et de Spectroscopie Electroniques, December 1989, to be published.

[7] M. Golay, *Hexagonal Pattern Transforms*, IEEE trans. on Comp., C18, 8, August 1969.

[8] J-C. Klein, *Conception et Réalisation d'une Unité Logique pour l'Analyse d'Images*, PhD Thesis, Université de Nancy, 1976.

[9] F.Collange, *Système de Vision "temps réel vidéo" Conception et Réalisation d'un Module d'Acquisition et de Mémorisation d'Images*, Phd Thesis, Université de Clermont, 1986.

[10] R. M. Lockeed, D. L. McCubbrey, S. R. Sternberg, *Cytocomputers: Architecture for Parallel Image Processing*, IEEE Worshop on Picture Data Description and Management, 1980.980.

[11] C. Lantuéjoul, S. Beucher, *On the Use of Geodesic Metric in Image Analysis*, J. of Microscopy, 121, 39-49, 1981.

[12] F. Meyer, *Skeletons and Perceptual Graphs*, Signal Processing, Special Issue in Advances on Mathematical Morphology, Vol.16, N°4, 335-364, April 1989.

[13] F. Meyer, *Quantitative Analysis of the Chromatin of Lymphocytes. An Essay on Comparative Structuralism*, Blood Cells 6, 159-172, 1980.

[14] B. Laÿ, *Analyse Automatique des Images Angiofluorographiques au Cours de la Rétinopathie Diabétique,* PhD Thesis, Ecole des Mines, Paris, 1983.

[15] S. Beucher, J. M. Blosseville and F. Lenoir, *Traffic Spacial Measurements Using Video Image Processing,* SPIE Cambridge Symp. on Advances in Intelligent Robotics Systems, November 1987.

[16] International Patent, Automatic scanning device for analyzing textures. Inventor: J. Serra. Assignee: Institut de Recherches de la Sidérurgie Française (I.R.S.I.D.). N°1.449.059, July 2, 1965.

[17] G. S. Gray, *Local Properties of Binary Images in Two Dimensions,* IEEE Trans. Comput. C-20, 551-561, 1971.

[18] International Patent, *Device for the Logical Analysis of textures.* Inventor: J. Serra. Assignees: Association pour la Recherche et le Developpement de Processus Industriels (A.R.M.I.N.E.S.) and J. SERRA. N°70 21 322, June 10, 1970.

[19] M.J.B. Duff, *CLIP4 A Large Scale Integrated Circuit Array Parrallel Processor,* 3 rd International Joint Conference on Pattern Recognition, 728-732, 1976.

[20] B. Kruse, *A Parallel Picture Processing Machine,* IEEE Trans. on Computers, Vol. C-22, N°. 12, 1075-1087.

[21] F.A. Gerritzen, L.G. Aardema, *Design and Use of a Fast Flexible and Dynamically Microprogrammable Pipe-lined Image Processor,* First Scandinavian Conference on Image Analysis, Linköping, Sweden, 1980.

Scale-Space Singularities

Allan D. Jepson and David J. Fleet

Department of Computer Science, University of Toronto, Toronto, Canada M5S 1A4

Introduction

In order to compute image velocity or binocular disparity it is necessary (in some sense) to localize structure in an image sequence and track it across frames, or to match it between left and right stereo views. For example, differential-based velocity techniques measure the translation of level contours of either constant intensity [6], or constant filter response [5], while zero-crossing approaches focus on the motion of zero-crossings in the output of band-pass filters [11]. Recently, the use of contours of constant phase has been suggested for the measurement of binocular disparity [3, 7, 8] and image velocity [2]. In choosing what type of structure to track it is important to consider its stability under common image deformations such as contrast variations, dilations, shears, and rotations in addition to simple translation (cf. [10]). One main advantage of phase information is that, except near certain points referred to here as *singularities*, phase is generally stable with respect to affine image deformations.

In this paper we illustrate the stability of phase information as compared to the amplitude of filter output. In addition, we discuss the existence of phase singularities, the neighbourhoods about them where phase is unreliable, and we present a simple method for their detection. Given this detection scheme, highly accurate and robust approaches to the measurement of optic flow and binocular disparity are possible. For example, based on the spatiotemporal gradient of phase information, Fleet and Jepson [2] reported a technique for the measurement of component (normal) image velocity for which approximately 90% of the accepted estimates had less than 5% relative error in cases of significant dilation and shear.

Finally, the results presented here are of general interest for several reasons. First, they also apply to zero-crossings of the filter output in that zero-crossings can be viewed as lines of constant phase. Second, similar results apply to 2-d signals. Third, the problems caused by deviations from image translation do not exist solely for phase-based techniques (cf. [9, 10]). The fact that these issues can be addressed within a phase-based framework is a major advantage for such approaches.

Gabor Scale-Space

To demonstrate the robustness of phase and its singularities we first consider a *Gabor scale-space* expansion of a 1-d signal that expresses the filter output as a function of spatial position and the principal wavelength to which the filter is tuned. It is defined by

$$S(x, \lambda) = Gabor(x; \sigma(\lambda), k(\lambda)) * I(x), \tag{1}$$

where $*$ denotes convolution, $I(x)$ is the input, $Gabor(x; \sigma, k) \equiv e^{ixk}G(x; \sigma)$ denotes the Gabor kernel [4] where $G(x; \sigma)$ is a Gaussian, and λ is the scale parameter (i.e. wavelength). The peak tuning frequency is given by $k(\lambda) = 2\pi/\lambda$, and the radius of support by $\sigma(\lambda) = (2^\beta + 1)/(2^\beta - 1)k(\lambda)$, where β denotes the octave bandwidth (usually taken to be near one). Because $Gabor(x; \sigma, k)$ is complex-valued, $S(x, \lambda)$ may be written as $\rho(x, \lambda)e^{i\phi(x, \lambda)}$ where the amplitude and phase components are given by

$$\rho(x, \lambda) = |S(x, \lambda)|, \qquad \phi(x, \lambda) = \arg(S(x, \lambda)) \equiv \mathrm{Im}\left[\log S(x, \lambda)\right]. \tag{2}$$

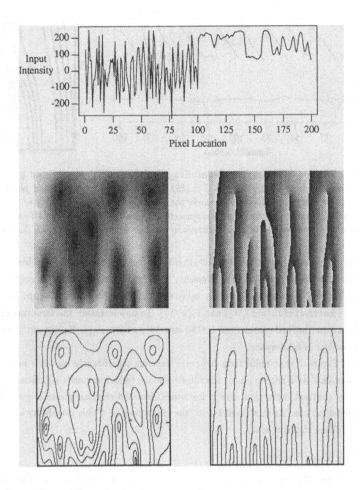

Figure 1. **Scale-Space Expansion:** *(top) 1-d input signal – a sample of white Gaussian noise concatenated with a scanline from a real image. (middle) $\rho(x, \lambda)$ and $\phi(x, \lambda)$ are shown with scale on the vertical axis spanning two octaves with $12 \leq \lambda \leq 48$ pixels. (bottom) Level contours of $\rho(x, \lambda)$ and $\phi(x, \lambda)$.*

The local frequency of $S(x, \lambda)$ can be defined as the spatial derivative of phase [12]: $\phi_x(x, \lambda)$. Although, $S(x, \lambda)$ is not expected to have constant frequency (linear phase), a first-order approximation to the spatial phase variation is generally accurate because of the band-pass filter tuning (cf. Fig. 2). This yields an amplitude-modulated, constant-frequency approximation to the local structure of $S(x, \lambda)$. Figure 1 shows a 1-d signal with the amplitude and phase components of its scale-space expansion.

As mentioned above, it is important that the signal property to be tracked is *stable* with respect to scale perturbations. This suggests that its level contours should be vertical in scale-space. To see this, consider two 1-d signals (e.g. left and right views) where one is a near-identity affine transformation of the other; i.e. let

$$I_r(a(x)) = I_l(x), \qquad \text{where} \quad a(x) = a_0 + a_1 x, \qquad (3)$$

with a_1 near 1. Because the filters have constant bandwidth the two output signals will satisfy

$$S_r(a(x), \lambda_1) = S_l(x, \lambda_2), \qquad \text{where} \quad \lambda_1 = a_1 \lambda_2. \qquad (4)$$

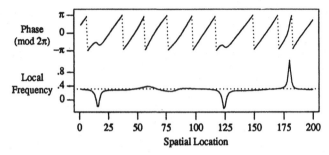 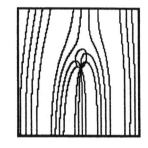

Figure 2. **Phase and Local Frequency Near Singularities:** *(left)* $\phi(x, \lambda)$ *and* $\phi_x(x, \lambda)$ *are shown for a slice of the scale-space in Figure 1 (with* $\lambda = 20$). *Vertical dotted lines denote phase wrapping (not discontinuities), and the horizontal dotted line marks the filter's peak tuning frequency* $k(\lambda) = 0.314$. *(right) Typical behaviour of level phase contours near a singularity. The singularity is the point in the centre through which several phase contours pass. The small ellipsoidal contour marks the retrograde boundary.*

That is, the two outputs would have similar structure if filters tuned to λ_1 and λ_2 had been applied to $I_r(x)$ and $I_l(x)$ respectively. However, in measuring disparity (or velocity) it is common to apply the same filters to $I_r(x)$ and $I_l(x)$ because the scale factor a_1 is unknown. In other words, we attempt to recover $a(x)$ by matching structure (features) of $S_r(a(x), \lambda_1)$ and $S_l(x, \lambda_1)$. To be successful, the structure of $S_l(x, \lambda_2)$ that is used for matching must be well represented by the structure of $S_l(x, \lambda_1)$. Equivalently, its level contours should be nearly vertical in scale-space.

It is clear from Figure 1 that amplitude structure depends significantly on scale in that its the level contours are *not* vertical. As a consequence, the filter response (which depends significantly on amplitude) is also unstable. By contrast, note that except for several isolated regions, phase is generally stable with respect to scale perturbations. As explained below, the major source of this instability is the occurrence of singularities in the phase signal $\phi(x, \lambda)$.

Singularity Neighbourhoods

For a general image $I(x)$, the scale-space defined by (1) is analytic, and contains a number of isolated zeros, where $S(x, \lambda) = 0$. In $\rho(x, \lambda)$ shown in Figure 1, zeros appear as black spots. The phase signal in (2) is also analytic, except at the zeros of $S(x, \lambda)$. The expected density of these *phase singularities* is proportional to the peak tuning frequency. Here, we describe the characteristic behaviour of $S(x, \lambda)$ in neighbourhoods about singular points. In what follows, let (x_0, λ_0) denote the location of a typical singularity.

The neighbourhoods just above and below singular points can be characterized in terms of the behaviour of phase $\phi(x, \lambda)$, and local frequency $\phi_x(x, \lambda)$. Above singular points (for $\lambda > \lambda_0$) they are characterized by local frequencies that are significantly below the corresponding peak tuning frequencies $k(\lambda)$. Within these neighbourhoods there exist *retrograde regions* where local frequencies are negative, i.e. $\phi_x(x, \lambda) < 0$. Along the boundaries of retrograde regions (which begin and terminate at singular points) the local frequency is zero; i.e. $\phi_x(x, \lambda) = 0$. The significance of this is that, where $\phi_x(x, \lambda) = 0$, the level phase contours are horizontal, and not vertical as desired. Nearby this boundary, both inside and outside the retrograde regions, the level contours are generally far from vertical, which, as discussed above, implies considerable phase instability. Below singular points (for $\lambda < \lambda_0$) the neighbourhoods are characterized by local frequencies of response that are significantly higher than the peak tuning frequencies. In addition, the local frequency changes rapidly as a function of spatial location.

To illustrate this behaviour Figure 2 *(left)* shows a 1-d slice of $\phi(x, \lambda)$ and $\phi_x(x, \lambda)$ from the scale-space in Figure 1 at a single scale ($\lambda = 20$). This slice (marked on the amplitude contour plot in Fig. 1) passes through three singularity neighbourhoods, two just above singularities (near locations 17 and 124), and one just below a singularity (near location 180). Notice the low (sometimes negative) and high local frequencies near the singularities. Figure 2 (*right*) shows the typical behaviour of level phase contours near a singularity. The phase singularity is the point in the middle through which several of the phase contours pass. The small elliptical contour marks the retrograde boundary where $\phi_x(x, \lambda) = 0$. The instability above the singular point is clear from the nearly horizontal level phase contours. Directly below the singular point, the high local frequencies are evident from the high density of phase contours.

Finally, the neighbourhoods spatially adjacent to singular points can be characterized in terms of amplitude variation. As we approach a singular point, $\rho(x, \lambda_0)$ goes to zero. Based on a simple linear model of $\rho(x, \lambda_0)$ at x_1 near x_0, the distance to the singularity $|x_0 - x_1|$ is approximately $\rho(x_1, \lambda_0)/|\rho_x(x_1, \lambda_0)|$. Therefore, as we approach the singularity $|\rho_x(x_1, \lambda_0)|/\rho(x_1, \lambda_0)$ increases.

Detection of Singularity Neighbourhoods

In order to use phase information reliably toward the measurement of image velocity or binocular disparity, singularity neighbourhoods must be detected so that measurements in them may be discarded. Here we introduce constraints on local frequency and amplitude that can be used to identity locations within singularity neighbourhoods, while avoiding the explicit localization of the singular points.

To detect the neighbourhoods above and below the singular points we constrain the distance between the local frequency of response and the peak tuning frequency. This can be expressed as a function of the extent of the amplitude spectrum (measured at one standard deviation $\sigma_k(\lambda)$) as follows:

$$\frac{|\phi_x(x, \lambda) - k(\lambda)|}{\sigma_k(\lambda)} < \tau_k , \qquad \sigma_k(\lambda) = k(\lambda) \frac{(2^\beta - 1)}{(2^\beta + 1)} . \tag{5}$$

The neighbourhoods adjacent to singular points can be detected with a local amplitude constraint:

$$\sigma(\lambda) \frac{|\rho_x(x, \lambda)|}{\rho(x, \lambda)} < \tau_\rho , \tag{6}$$

where $\sigma(\lambda)$ defines the radius of filter support. Level contours of (5) for different values of τ_k form 8-shaped regions with the singular points are their centres, while level contours of (6) form ∞-shaped regions. (see Figure 3 (*top row*)). As τ_k and τ_ρ decrease, the constraints become tighter and larger neighbourhoods are detected. Figure 3 (*top-right*) shows the combined behaviour of (5) and (6) as applied to the scale-space in Figure 1, with $\tau_k = 1.2$ (i.e. local frequencies are accepted up to 20% outside the nominal tuning range of the filters) and $\tau_\rho = 1.0$ (i.e. points within $\sigma(\lambda)$ of a singularity are discarded). These constraints typically remove about 15% the scale-space area. Finally, Figure 3 (*bottom row*) also shows the original level phase contours of Figure 1, the contours that survive the constraints, and the contours in those regions removed. Notice the stability of the contours outside the singularity neighbourhoods.

Measurement of Binocular Disparity

To illustrate the problems caused by phase instability and the rapid variation of local frequency that occur in singularity neighbourhoods, we compare the results of a technique for disparity measurement with and without their detection. Following Jenkin and Jepson [7] and Sanger [8], estimates of binocular disparity are computed as

$$d(x) = \frac{[\phi_l(x) - \phi_r(x)]_{2\pi}}{k_0} , \tag{7}$$

where $\phi_l(x)$ and $\phi_r(x)$ denote the phase responses of the left and right views, k_0 denotes the peak tuning

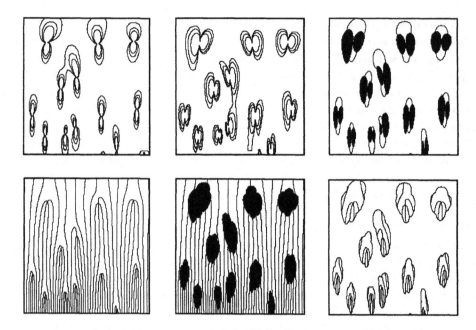

Figure 3. **Detection of Singularity Neighbourhoods:** *(top-right) Level contours of (5) for $\tau_k = 1$, 1.5 and 2 for the scale-space in Fig. 1. (top-middle) Level contours of (6) for $\tau_\rho = 0.75$, 1, and 1.5. (top-right) Neighbourhoods removed by (6) with $\tau_\rho = 1$ are shown in black, while the contours show the remaining regions marked by (5) with $\tau_k = 1.2$. (bottom) Level phase contours for scale-space (cf. Fig. 1), contours that survive the constraints, and the phase contours in those regions removed with $\tau_\rho = 1$ and $\tau_k = 1.2$.*

frequency of the filter, and $[\theta]_{2\pi} \in (-\pi, \pi]$ denotes the principal part of θ. This computation presumes a model of local phase given by $\phi(x) = k_0 x + \phi_0$; when the left and right signals are shifted versions of one another, and the filter outputs have constant frequency k_0, then (7) yields the exact result. Toward a more general model we can replace k_0 in (7) by the average local frequency in the left and right outputs $(\phi_l'(x) + \phi_r'(x))/2$. This allows frequencies other than k_0 with accurate results [3]. Making the local model explicit is important because the measurement accuracy and reliability depend on the appropriateness of the local model. For example, in neighbourhoods above and below singular points, which are characterized by a high variation in local frequency, the linear phase model is inappropriate and the numerical approximation of $\phi'(x)$ will be poor. The removal of these regions is therefore important.

To illustrate this, assume a simple situation in which the left and right views are shifted versions of the 1-d signal shown in Figure 1 (*top*). Let the disparity be 5 pixels, and let the Gabor filters be tuned to a wavelength of 20 pixels. Thus the left and right phase signals are shifted versions of the scale-space slice shown in Figure 2, which crosses three singularity neighbourhoods. Figure 4(*top*) shows the results of (7) with the crude linear model [7, 8], and without the removal of singularity neighbourhoods. Figure 4 (*middle*) shows the consequence of removing any disparity measurement for which the left or the right filter responses did not satisfy (5) or (6) with $\tau_\rho = 1.0$ and $\tau_k = 1.2$ (as in Fig. 3). In [8] a heuristic constraint on amplitude differences between left and right signals and subsequent smoothing were used to lessen the effects of such errors. Unfortunately, this smoothing will sacrifice the resolution and accuracy of nearby estimates. Finally, Figure 4 (*bottom*) shows the improvements obtained with the more general linear model.

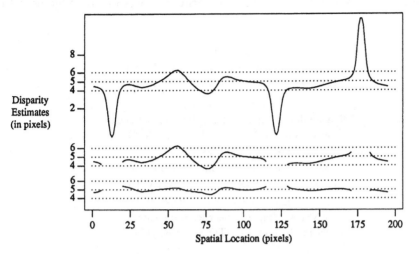

Figure 4. **Disparity Measurement:** *The top two plots show the disparity estimates based on (7) without, and then with, the removal of singularity neighbourhoods. Notice the substantial errors in the first case versus the second. The bottom plot shows the improved technique in which the local frequency is used instead of the peak frequency in (7). The same neighbourhoods have been removed.*

Summary

Phase-based techniques for the measurement of binocular disparity and image velocity are encouraging, especially because of the stability of band-pass phase information with respect to deviations from image translation that are typical in projections of 3-d scenes. Despite this stability, phase is unreliable in the neighbourhoods of phase singularities. This instability was described, and it was shown that singularity neighbourhoods may be detected using simple constraints on the local frequency and the amplitude of the filter output. Finally, these results were discussed briefly in the context of binocular disparity measurement.

References

1. Burt, P.J., et.al. (1989) Object tracking with a moving camera. *IEEE Motion Workshop*, Irvine, p 2-12
2. Fleet, D. and Jepson, A. (1989) Computation of normal velocity from local phase information. *Proc. IEEE CVPR*, San Diego, pp 379-386
3. Fleet, D. Jepson, A. and Jenkin M. (1990) Phase-based disparity measurement. *submitted*
4. Gabor, D. (1946) Theory of communication. *J. IEE* 93, pp. 429-457
5. Glazer, F. (1987) Hierarchical gradient-based motion detection. *Proc. DARPA IUW*, LA., pp 733-748
6. Horn, B.K.P. and Schunck, B.G. (1981) Determining optic flow. *Artif. Intel.* 17, pp. 185-204
7. Jenkin, M. and Jepson, A.D. (1988) The measurement of binocular disparity. *in* **Computational Processes in Human Vision**, (ed.) Z. Pylyshyn, Ablex Press, New Jersey
8. Sanger, T. (1988) Stereo disparity computation using Gabor filters. *Biol. Cybern.* 59, pp. 405-418
9. Schunck, B.G. (1985) Image flow: fundamentals and future research. *Proc. IEEE CVPR*, San Francisco, pp 560-571
10. Verri, A. and Poggio, T. (1987) Against quantitative optic flow. *Proc. IEEE ICCV*, London, p 171-179
11. Waxman, A.M., Wu, J., and Bergholm, F. (1988) Convected activation profiles: Receptive fields for real-time measurement of short-range visual motion. *Proc. IEEE CVPR*, Ann Arbor, pp 717-723
12. Whitham, G.B. (1974) **Linear and Nonlinear Waves**. John Wiley and Sons, New York

Recursive Filtering and Edge Closing : two primary tools for 3D edge detection

Olivier MONGA
INRIA - Domaine de Voluceau - B.P. 105
78153 LE CHESNAY CEDEX - FRANCE

Rachid DERICHE
INRIA Sophia-Antipolis - 2004, Route des Lucioles
06565 VALBONNE CEDEX - FRANCE

Grégoire MALANDAIN
INRIA - Domaine de Voluceau - B.P. 105
78153 LE CHESNAY CEDEX - FRANCE

Jean Pierre COCQUEREZ
ENSEA - Impasse des chênes pourpres
95014 CERGY PONTOISE - FRANCE

Abstract

This paper deals with edge detection in 3D images such as scanner, magnetic resonance (NMR), or spatio-temporal data. We propose an unified formalism for 3D edge detection using optimal, recursive and separable filters recently introduced for 2D edge detection. Then we obtain some efficient 3D edge detection algorithms having a low computational cost. We also show that 3D edge tracking/closing enables to extract many edges not provided by the filtering stage without introducing noisy edges. Experimental results obtained on NMR images are shown.

1   Introduction

Many industrial or medical applications provide three dimensional images representing volumic informations. Modern scanning techniques such as Computed Tomography (CT) produce 3D images where the grey level function is proportional to the local density [5]. In biomedicine for example, 3D images are produced by magnetic resonance imaging (NMR), computed tomography (CT), or positron emission tomography (PET). Thus, in such data the volumes presenting a homogeneous grey level distribution correspond to the various entities of the 3D structure. To extract these volumes we can either look directly for a partition of the 3D image into homogeneous area : region based approach, either search the surfaces forming their boundaries : edge based approach. The segmentation into regions of 2D or 3D images set the acute problem of finding some homogeneity properties suitable for the regions [7]. This is the main reason why we have chosen an edge based approach that can be split into two main stages : 3D edge detection by filtering, and 3D edge tracking/closing using morphological properties.

Then a first stage to identify these surfaces is to search their points using only signal information. This sets the classical edge detection problem widely studied for 2D images [1,3,2], but which has not yet received much attention in 3D. Basically 2D and 3D edge detection set the same problem ie how to detect the discontinuities of a discrete and noisy (2D or 3D) function.

Thus existing 3D edge detectors are issued from a generalization in 3D of 2D edge detectors [11,6,13,10,9]. Their aim is to approximate the gradient or the laplacian of the image thanks to convolution masks. Therefore, there is a trade-off to meet between the size of the convolution masks and their detection and localization performances. Particularly in the 3D case the huge size of data makes the algorithmic complexity and the storage requirement be key points.

Recently in order to get rid of that drawback recursive filtering has been introduced for 3D edge detection [8]. It allows to implement filters having an infinite impulse response with a low computational cost (about the same than a 3.3.3 convolution mask in the 3D case). The 3D edge detection operator proposed in [8] is issued from a 2D operator that is extended to 3D thanks to separable filtering [3].

In this paper we propose an unified formalism for 3D edge detection using recursive filters either for the extraction of gradient extrema, either for the determination of the zero crossings of the laplacian. This is achieved thanks to a direct extension of the 2D algorithms [2] for the first derivative approach. In the case of the second derivative approach the extension is less direct.

Due to the noise this filtering stage is generally not sufficient to obtain results to be used for surface and volume reconstruction. To improve the 3D edge map it is therefore useful to introduce morphological informations. For this design we propose a 3D edge tracking/closing algorithm derived from a 2D edge closing method [4].

The article is organized as follow :

In section 2 we show that separable filtering enables to reduce the nD (and especially 3D) edge detection problem to smoothing and derivation of a 1D signal. This is true both for first and second derivative approaches.

In section 3 we list some 1D optimal [2] and recursive filters aiming to smooth or to derive a 1D signal. Section 4 describes the algorithm in the 3D case. Section 5 deals with the 3D edge tracking/closing method. In sections 5 and 6 we present experimental results obtained on NMR data and conclude.

2 Statement of the nd edge detection problem

The goal of this section is to show that under reasonable assumptions, the nD edge detection (and particularly 3D) task can be reduced to smoothing and derivation of a 1D signal.

Let $I(x_1, x_2, \ldots, x_n)$ be a noisy signal of dimension n

Classically edge detection is tackled thanks to the following two schemes :

1. Gradient computation and extraction of the local extrema of its magnitude in the gradient direction : gradient approach.

2. Laplacian computation and determination of its zero crossings : Laplacian approach

Most of the techniques use linear filters to approximate the gradient or the laplacian. Given that the derivative of a convolution product is equal to the convolution of the signal by the derivative of the filter response, the schemes 1 and 2 can be reduced to the determination of a smoothing filter $L(x_1, x_2, \ldots, x_n)$. First and second derivatives with respect to x_i are respectively computed by convolving the image with $\frac{\partial L(x_1, x_2, \ldots, x_n)}{\partial x_i}$ and $\frac{\partial^2 L(x_1, x_2, \ldots, x_n)}{\partial x_i^2}$. We notice that the laplacian can be directly computed by convolution with the filter whose response is : $\frac{\partial^2 L(x_1, \ldots, x_n)}{\partial x_1^2} + \cdots + \frac{\partial^2 L(x_1, \ldots, x_n)}{\partial x_n^2}$. We will see in the next section that in some cases this expression can be simplified. Another solution for the computation of the laplacian consists in approximating it by difference between the image smoothed by two filters of different parameters [12]. However the results obtained by this solution are less satisfactory than these provided by a direct laplacian computing.

The definition of convolution masks to approximate gradient or laplacian sets an acute computing time problem. For example in dimension 2 the convolution of an image $d_x \times d_y$ by a convolution mask of size $p \times p$ costs $p^2 d_x d_y$. This leads to the use of filters having separable response with respect to directions $x_1, x_2, \ldots, x_n$:

$$L(x_1, x_2, \ldots, x_n) = L_{x_1}(x_1) L_{x_2}(x_2) \cdots L_{x_n}(x_n)$$

The main drawback of separable filtering is that we can obtain anisotropic filters along directions not parallel to $x_1, x_2, \ldots, x_n$. Gaussian filters are among the very few separable filters which are isotropic.

Therefore L can be implemented by the cascade of the filters : $L_{x_1}, \ldots, L_{x_n}$

If we suppose the noise homogeneous along any direction we can set :

$$L_{x_1} = L_{x_2} = \cdots = L_{x_n} = S$$

In the sequel we will suppose that the noise is isotropic but this method can be used for an anisotropic noise but homogeneous along directions $x_1, x_2, \ldots, x_n$. We will see that recursive filtering can be used only if the noise is homogeneous along each manifold of dimension 1 : $x_i = cste$, $i \neq j$.

3 How to smooth or to derive a 1D signal

In this section we present some optimal [1] and recursive filters to smooth or to derive one or two times a 1D signal. We strongly insist on recursive implementation because of the low computational cost it allows [2]. For more details concerning these filters one can refer to [2].

3.1 Shen filter

$$s_1(x) = ce^{-\alpha.|x|}$$

s_1 and s_1' are first order recursive filters. s_1 has the following realization [12] :

$$
\begin{aligned}
y^+(m) &= ax(m) + by^+(m-1) && \text{for } m = 1, \ldots, N \\
y^-(m) &= abx(m+1) + by^-(m+1) && \text{for } m = N, \ldots, 1 \\
y(m) &= y^-(m) + y^+(m) && \text{for } m = 1, \ldots, N
\end{aligned}
$$

This filter has been introduced by Shen and Castan to approximate the laplacian by difference between the original image and the smoothed image [12]. The derivative filter is an optimal solution for the first part of Canny criteria [12]. It corresponds to an optimal value for the product $\Sigma\lambda$ ie to the best trade-off detection-localization. The discontinuity of order 1 at point 0 avoids to delocalize the edges in the smoothed image when α is small. However this discontinuity can induce multiple edges.

3.2 Deriche filter

$$s_2(x) = (c|x| + 1)e^{-\alpha|x|}$$

s_2, s_2', and s_2'' are second order recursive filters [3].

This filter has been recently proposed by R. Deriche [3] and its derivative is an exact solution to Canny equation extended for infinite filters.

4 Algorithm in the 3D case

4.1 Filtering stage

This stage can be easily formalized for any dimension, thus we describe it for any dimension.

4.1.1 Choice of a 1D smoothing operator : $s(x)$

We strongly recommand to choice a filter that can be implemented recursively, mainly because of the computing time [2]. We can for example choose one of the two filters precedently described : s_1 or s_2.

Theoretically the derivation filter $s_2'(x)$ is better with respect to Canny's multiple response criterion, but $s_1'(x)$ meets the best trade-off detection-localization. For small values of α the shape of $s_2'(x)$ induce some delocalization problems. This drawback is shared by any filter whose impulse response is continuous at point 0. Neverless, the results do not significantly differ on many kinds of images.

4.1.2 Choice of the kind of approach : Gradient or Laplacian

Generally the computation of the second derivative is more sensible to noise. On the other hand its computational cost is lower, because of the simplifications that occur when computing the impulse response of the filter by multiplication and addition of smoothing and second derivative operators. However in the case of noisy images it is generally useful to threshold the zero crossing provided by the filtering stage , and this requires to compute the gradient magnitude at each zero crossing. The localization of the edges provided by the two kinds of methods is experimentally the same. It may be point out that Laplacian approach tends to smooth the right angles.

Ones we have chosen the filter and the kind of approach, we have to compute Gradient or Laplacian.

Let $I(x_1, \ldots, x_n)$ be an image of dimension n.

Let $G(x_1, \ldots, x_n)$ be the gradient of I.

$$G(I) = (\frac{\partial I}{\partial x_1}, \ldots, \frac{\partial I}{\partial x_n})^t$$

The computation of the Gradient components $\frac{dI}{dx_i}$ is done by computing images (D_i) corresponding to the partial derivatives with respect to x_i as follow :

$$
\begin{array}{l}
\text{for } i = 1, \ldots, n \text{ do} \\
\left[
\begin{array}{l}
D_i = I \\
\text{for } j \in [1, \ldots, n] \backslash \{i\} \text{ do} \\
\left[D_i = D_i \star s(x_j) \right. \\
D_i = D_i \star s'(x_i)
\end{array}
\right.
\end{array}
$$

For an image of size p (ie $d_1 \times \cdots \times d_p$) the computation of each gradient component needs to compute p convolutions per point. Then we obtain for the entire computation $p^2 . \prod_{i=1}^{p} d_i$ convolutions.

If we use a direct implementation of a convolution mask 1D of size k, we obtain the following complexity : $kp^2 \prod_{i=1}^{p} d_i$. A recursive filtering of order r allows to obtain a complexity of order : $rp^2 \prod_{i=1}^{p} d_i$.

For example in the 2D case we obtain for Deriche filter 13 multiplications and 12 additions per point and this for any value of α.

The laplacian can be computed by adding the second derivatives . The computation of the second derivative may be done with the algorithmic structure precedently described where the first derivative operator $s'(x)$ is replaced by the second derivative operator $s''(x)$. However for some cases calculus simplifications occur allowing to reduce the computational cost [2].

4.2   From gradient or laplacian to 3D edges

Although the computation of gradient or laplacian is the essential part of edge detection it does not provide directly the edge points. For first or second derivative approach two complementary stages have to be done. In this section we describe these stages in the 3D case.

4.3   Gradient approach

4.3.1   Extraction of the local gradient extrema

This section deals with the extraction of local gradient magnitude extrema in the gradient direction. The principle of this method is to move along the normal to the contour (approximated by gradient) and to select points having the highest gradient magnitude. We notice that this stage could also be generalized to any dimension.

Let $I(x, y, z)$ be a 3D image

Let $G(x, y, z)$ be the gradient of I at point (x, y, z)

Let M be a point of I

Let $G(M)$ be the gradient at point M

Let d be a given distance (eg $d = 1$)

Let M_1 and M_2 be the two points of the straight line including M and whose direction is $G(\vec{M})$, located at a distance d of M ; M_1 is taken in the gradient sense and M_2 in the opposite sense.

$$M_1 = M + d \frac{G(\vec{M})}{\|G(M)\|} \; ; \; M_2 = M - d \frac{G(\vec{M})}{\|G(M)\|}$$

We approximate the gradient at points M_1 and M_2 thanks to a linear approximation using the neighbours. An example is precisely described in [8]. The point M is selected if $N(M) > N(M_2)$ and $N(M) \geq N(M_1)$. To choose the strict maximum in the gradient direction and large maximum in the opposite sense is equivalent to select edge points on the brighter side of the border.

4.3.2   Thresholding of the gradient extrema

We have extended the hysteresis thresholding introduced by Canny for 2D edge detection [1] to 3D and slightly improved it by adding a constraint. The principle of hysteresis thresholding is to select among all extrema whose gradient magnitude is higher than a low threshold t_l these such that it exists a connected path of extrema whose gradient magnitude is higher than t_l between the involved point and an extremum whose gradient magnitude is higher than a high threshold t_h. Moreover we can deal only with connected components whose the length is more than a minimal length l_{min}.

This thresholding algorithm can be improved in the following way. The expansion in connected components could be performed in any direction (by using 8-connectivity in 2D or 26-connectivity in 3D). But we have an estimation of the direction orthogonal to the contour i.e. the gradient direction. The idea is to move along a direction orthogonal to the gradient i.e. within the hyperplane tangent to the contour. This may be done in the following way. Let M_0 be the edge point involved, $\vec{G}(M_0)$ be its gradient, V be the set of its neighbourhoods (in 26-connectivity for instance). The expansion is performed by examination of all points $M \in V$ such that the distance between M and the hyperplane tangent at point M_0 is less than a threshold s:

$$\frac{|\vec{MM_0} \times \vec{G}(M_0)|}{\|\vec{G}(M_0)\|} < s$$

The choice of s is related to the curvature of the contour we want to obtain. We notice that by choosing s enough high, the expansion in connected components is done for all neighbourhoods of M_0.

This improvement is essentially useful in the case where the images are very noisy. And unfortunately this is true for many medical 3D images. Particularly it allows to push down the low threshold without introducing too much false edge points.

This thresholding strategy is particularly efficient in the 3D case because it enables to get good connected edge points. This is of great interest to regroup these points in order to built surfaces.

4.4 Laplacian approach

In the case of a second derivative approach we have computed the laplacian value at each point. We suppose that the edge points are located at the laplacian zero crossing [11]. A last thresholding stage is also necessary to remove zero crossing whose gradient magnitude is too low.

5 Closing 3D edges

It is often very difficult to select adequate thresholds for the thresholding stage. High thresholds allows to remove noisy edge points but also true edge points ; low thresholds allow to obtain all true edge points but also noisy edge points. We can not avoid this compromise and the choice of thresholds defines a trade-off between true and noisy edges.

Generally it is easier to extend uncompleted contours than to validate true contours in a noisy edge image. Thus we choose high thresholds to remove false edge points and then we use a tracking/closing algorithm.

We have extended to 3D a 2D edge closing method proposed by Deriche and Cocquerez [4]. This 2D method supposes that it is possible to recognize endpoints of contours by the examination of a neighbourhood 3×3 of an edge point. Each kind of neighbourhood is attached to a labeling code. All possible configurations of contour endpoints are indexed. Moreover the topology of each endpoint allows to define an exploration direction to continue the contour. To each configuration is attached a list of neighbours to examine. The selected neighbours belong to the local gradient magnitude extrema points.

The implementation of this algorithm is easy. The image is scanned, if an edge point is identified as an extremity (ie the extremity configuration code is indexed) the algorithm is applied recursively to the involved extremity until a stop condition is verified.

Two choices have to be done :

- *Choice of the neighbourhood to continue the contour* : Deriche and Cocquerez select the point having the higher gradient magnitude. This works if the image is not too much noisy else we take interest in adding the condition that the candidate points have to belong to the gradient extrema image.

- *Stop conditions* : Many stop conditions may be used. Either there is no more candidates (if these ones are picked in the extrema image), either the path created recursively by the algorithm has reached an edge point or a given length.

The 3D extension of this algorithm is done by applying it on each plane XY, YZ, ZX and by adding the three edge images obtained. This can be justified by the assumption that the intersection of a 3D surface by at least one plane among XY,YZ and ZX is a curve on which the algorithm can be applied.

Despite of the result improvement due to the use of this tracking/closing algorithm, it remains some localized information lacks that cause holes of width 1 along X,Y or Z. These blanks are either due to the filter (a linear filter has not a good behavior near the multiple contours) either to the noise present in the image. To solve this problem we pick up the idea of the precedent algorithm, ie to attach a code to the neighbourhood of a point. Thus we select the codes corresponding to

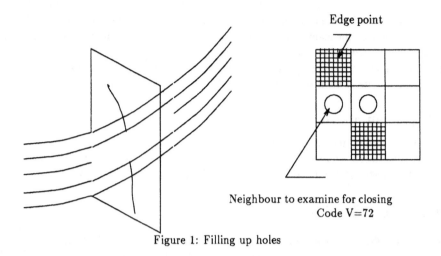

Figure 1: Filling up holes

holes of width 1 (see figure 1). The 2D implementation consists in scanning the image and to fill up each identified hole. The 3D extension is the same than precedently.

6 Experimental results

These algorithms have been implemented and tested on 3D magnetic resonance images of heart given by Hôpital Kremlin Bicêtre Paris. To obtain isotropic data (ie having the same resolution along X,Y,Z), we perform a linear interpolation along Z-axis. We suppose that the noise has the same parameters along X,Y,Z and thus we use the same operator width α to smooth or to derive in any direction.

Figures 2 to 5 present results provided by the chain of processes on a 3D NMR image of the chest ($256 \times 256 \times 46$, $256 \times 256 \times 16$ before interpolation, resolution 1.5 mm).

These algorithms have been implemented on a SUN 3 workstation. For a $256 \times 256 \times 46$ image, the filtering stage and the tracking/closing stage take respectively about 30' CPU and 5' CPU.

7 Conclusion

We have proposed a 3D edge detection scheme that can be split into two main stages :

- Filtering

- Edge tracking/closing

The key point of the filtering stage is to use optimal recursive and separable filters [2] to approximate gradient or laplacian. The recursive nature of the operators enables to implement infinite 3D impulse response with a computing time roughly similar to a $3 \times 3 \times 3$ convolution mask. This saving in computational effort is of great interest for 3D edge detection due to the huge size of 3D images. Then we obtain an algorithmic framework for the 3D edge detection filtering stage that combines better theoretical and experimental performances and a lower algorithmic complexity than the classical ones. Moreover, we stress that our approach could be used for any dimension.

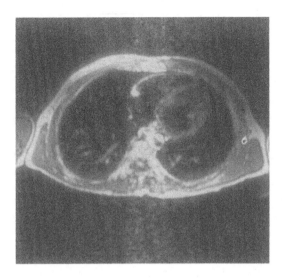

Figure 2: Original cross section of a RMN image of the chest corresponding to the diastolic cardiac phase

The principle of the edge tracking/closing is to select from the previous stage, only the more reliable edge points and then to apply an edge closing method derived from the idea developed in [4]. This enables to improve substancially the results provided by the filtering stage. We point out that our tracking/closing algorithm is limited mainly because it is not really 3D.

Currently we investigate true 3D closing edge methods using discrete topology and mathematical morphology.

8 Acknowledgement

The authors would like to thank Dr. J.M. Rocchisani from Hôpital Cochin Paris and Dr. J. Bittoun from Hôpital Kremlin Bicêtre Paris for having provided the RMN images.

References

[1] J.F. Canny. *Finding edges and lines in images*. Technical Report TR. 720, MIT, June 1983.

[2] R. Deriche. Fast algorithms for low level vision. *IEEE Transactions on Pattern Analysis and machine Intelligence*, 1989.

[3] R. Deriche. Using canny's criteria to derive a recursively implemented optimal edge detector. *International Journal of Computer Vision*, 1(2), May 1987.

[4] R. Deriche and J.P. Cocquerez. An efficient method to build early image description. In *International Conference on Pattern Recognition*, Rome, 1988.

[5] R. Gordon, G.T. Herman, and S.A. Johnson. Image reconstruction from projections. *Sci. Amer.*, 233:56–68, 1975.

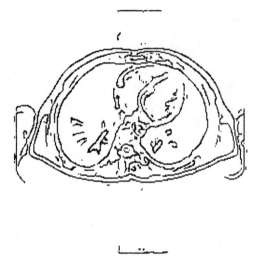

Figure 3: 3d edges after hysteresis thresholding, Shen filter, $\alpha = 0.6$

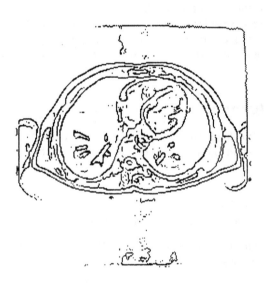

Figure 4: 3d edges after tracking/closing

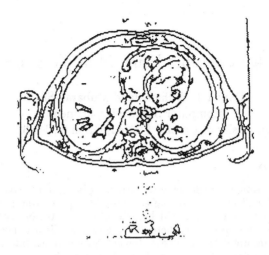

Figure 5: 3d edges after tracking/closing and filling up holes of width 1

[6] H.K. Lui. Two and three dimensional boundary detection. In *Comput. Graphics Image Process.*, pages 123–134, 1977. Vol. 6.

[7] O. Monga. An optimal region growing algorithm for image segmentation. In *International Journal of Pattern Recognition and Artificial Intelligence*, pages 351–376, December 1987.

[8] O. Monga and R. Deriche. 3d edge detection using recursive filtering. In *Computer Vision and Pattern Recognition*, IEEE, San Diego, Juin 1989.

[9] O. Monga and R. Deriche. A new three dimensional boundary detection. In *International Conference on Pattern Recognition*, Paris, 1986.

[10] M. Morgenthaler and A. Rosenfeld. Multidimensional edge detection by hypersurface fitting. *PAMI-3*, 4, July 1981.

[11] A. Rosenfeld and A. Kak. Digital image processing. *New York: Academic*, 1976.

[12] J. Shen and S. Castan. An optimal linear operator for edge detection. In *Conference on Vision and Pattern Recognition*, IEEE, USA, Juin 1986.

[13] S.W. Zucker and R.A. Hummel. A three dimensional edge operator. *IEEE Transactions on Pattern Analysis and Machine Intelligence*, 3(PAMI-3), May 1981.

Edge Contours Using Multiple Scales *

Donna J. Williams[†]and Mubarak Shah
Computer Science Department
University of Central Florida, Orlando, FL 32816

1 Introduction

The field of computer vision is concerned with extracting information contained in images about scenes they depict. The effectiveness of the early levels of processing is crucial in determining how successful higher level processing will be. Edge point detectors typically return an edge map identifying the location of points where the intensity gradient is high, together with some gradient and direction information. The next step is to group points into edge segments. Several algorithms have been developed for linking edge points, e.g., [5].

The choice of the scale to use in smoothing an image has been much studied. Smaller scales result in too much noise and fine texture while larger scales result in delocalization of edges and gaps. One approach to this problem is to use multiple scales, e.g. [4]. Witkin [9] introduced the concept of *scale space*, where the zero crossings of the second derivative are examined for a continuous spectrum of scales rather than a few discrete values. The properties of scale space have been examined by a number of authors, among them [1,2,6].

This paper presents a method of producing connected edge contours which are suitable for higher processing. The algorithm uses a gradient of Gaussian operator to determine gradient magnitude and direction, followed by non-maxima suppression to identify ridges in the gradient map. Canny[3] has shown this to be a near optimal edge detector. The resulting ridge is often more than 1 pixel wide, and may have small noisy spurs. In this paper the gradient maxima points are thinned to one pixel wide and linked into contours by an algorithm using weights to measure noise, curvature, gradient magnitude, and contour length. The set of points giving the largest average weight is chosen. This algorithm is then extended to one using multiple scales in the edge linking step and to a third algorithm which uses multiple scales during non-maxima suppression. Both multi-scale algorithms improve the detection of edge contours with little increase in the response to noise. The third also reduces the delocalization occurring at larger scales.

Further, in order to determine the size neighborhood where an edge point can appear at a different scale, a theoretical analysis of the movement of idealized edges is performed. Shah et al., developed equations for step pairs convolved with the Gaussian and its derivatives, and showed the general shape of the scale space curves. That work is extended to develop the equations of the scale space curves and analyze *quantitatively* the amount of the delocalization that occurs as images containing these steps are convolved with Gaussians having different values of σ.

2 Single Scale Edge Detection and Linking

In this section, we present an algorithm for finding a single good path through the set of gradient maximum points. In this method, the image is first convolved with a gradient of Gaussian operator. The set of gradient maximum points is placed in a priority queue with the point having largest magnitude on the top. Thus the strongest edge points will be extended into contours first.

The search for points to assign to a contour proceeds as follows. The first edge point is retrieved from the queue and gradient direction is used to determine the next edge point. The

*The research reported in this paper was supported by the Center for Research in Electro Optics and Lasers, University of Central Florida, grant number 20-52-043 and by NSF grant number IRI 87-13120.

[†]Now at Stetson University, DeLand, Florida 32720

point in the computed direction is examined first, then those in the adjacent directions on either side of it. Each branch is followed to the end and a weight assigned at each point based on four factors. The four factors are: (1) the difference between the gradient direction and the direction in which the point lies, (2) the difference between gradient direction of adjacent points, (3) gradient magnitude, (4) contour length. The weights are designed to favor the longest, strongest, straightest path. After points in the three primary directions have been examined, the path having the largest average weight is chosen and returned to the calling program. Contours are constructed in this manner until there are no more edges in the queue having magnitude greater than a given per cent of the maximum magnitude.

3 Maximum Movement of Edges

It is known that as edges undergo smoothing, they are delocalized. In this section the question of how far an edge undergoing Gaussian smoothing can move is analyzed. The edge model used is the ideal step edge, and the two cases considered are adjacent edges having the same and opposite parity. These are the *staircase* and the *pulse* edge types. If the unit step edge is $U(x)$ then the staircase and pulse with edges located at $-a$ and a have respective equations
$$S_a(x) = bU(x + a) + U(x - a) \text{ and } P_a(x) = bU(x + a) - U(x - a)$$
The relative heights of the two steps, b, satisfies $0 < b \leq 1$. Thus the weaker edge is at $x = -a$. After convolving with the derivative of the Gaussian, the equation for the staircase is
$$s_{\sigma,a}(x) = bg(x + a) + g(x - a) \text{ and that for the pulse is } p_{\sigma,a}(x) = bg(x + a) - g(x - a)$$
The equation of σ as a function of x at the maxima for the staircase is
$$\sigma = \left[\frac{2ax}{\ln(b(a+x)/(a-x))} \right]^{1/2} \text{ and for the pulse, } \sigma = \left[\frac{2ax}{\ln(b(x+a)/(x-a))} \right]^{1/2}$$
For derivations see [7]. The graphs for two values of b are given in Figure 1. The middle branch which appears for small values of σ in the graphs for the staircase represents a gradient minimum rather than a maximum so does not correspond to an edge. These will be referred to as the scale space images.

First we will consider the staircase. Notice that if $b = 1$ the edges move together until they meet when $\sigma = a$, then only one edge exists at $x = 0$ for $\sigma > a$. When $b < 1$ the stronger edge moves toward the middle and approaches the asymptote $a(1-b)/(1+b)$ as σ approaches ∞. The maximum corresponding to the weak edge on the left disappears when σ becomes sufficiently large. For example, when $b = .3$ the maximum movement of the weak edge occurs just before it disappears and is $(1 - .723)a = .277a$. The units on both axes are a.

In practice, when an image is being examined, σ is known, but a is not. Thus, σ can be fixed, a can be allowed to vary, and the movement (m) can be plotted versus a. This is shown in Figure 2(a) and (b). On both axes the units are σ. When $a > 2\sigma$ the amount of movement is negligible. This corresponds to the part of the scale space image where the curve is near vertical and the edges are far enough apart to have little interaction. The interaction begins slowly as the edges appear closer together, then increases rapidly to the maximum, then decreases until a reaches 0. When $b < 1$, the weaker edge disappears. The largest possible movement, σ, occurs for equal edges which are 2σ apart.

A similar analysis can be performed for a pulse. Figure 3(a) and (b) show the movement versus a for a pulse. The maximum movement, when $b = 1$, is σ as it was for the staircase, but this value is now the limiting value as the edges become closer together. When $b < 1$ the strong edge in the scale space image approaches the vertical asymptote $a(1 + b)/(1 - b)$ and displays a well-defined maximum movement as in the staircase, for a value of a between 0 and σ. But the weak edge can move indefinitely as a becomes smaller. In the scale space image the weak edge approaches the horizontal parabola $x = (1/2a)(\ln b)\sigma^2$. But $p_{\sigma,a}(\frac{\ln b}{2a}\sigma^2) = 0$ for all values of σ, thus the parabola gives the location where the gradient value is zero. The gradient value of the weak edge becomes small, and falls below any threshold being used as σ increases, and for $0 < a < 1/2$ the conditions of the sampling theorem are not met, thus in practice the movement of the weak edge is limited. Figure 3(c) gives a graph of maximum movement for the stronger edge of a pulse as b varies.

In summary, for a staircase greatest movement is σ and occurs when edges are 2σ apart and have equal contrast. Movement decreases rapidly for edges closer or farther away and those having unequal stepsize. For pulse, maximum movement is σ for equal edges and for the stronger of two unequal edges and occurs when edges are very close together. The weaker of two edges can exhibit unbounded movement, but gradient magnitude decreases, so an edge will usually not be detected farther than σ from its original location.

4 Multiple Scale Algorithms

This section extends the algorithm given in Section 2 to one using multiple scales, as follows. Initially the image is convolved with gradient masks at three or more scales. The search for a contour proceeds as for the single scale, using the largest scale, until a best partial contour at that scale has been found. Then the next finer scale is chosen and the neighborhood around the ends of the contour is searched to see if the edge can be extended. The neighborhood searched is only one pixel in each direction, based on the analysis in Section 3. The original algorithm is then followed for each point that gives a good continuation of the contour, and the best is chosen as an extension to the original edge. While extending the edge, if any point is discovered to be a possible edge point at a coarser scale, the search scale is increased to that value. When the contour cannot be extended further the scale is decreased to the next finer scale, and the process is repeated until the contour cannot be extended at the finest scale. This algorithm resulted in a considerable improvement in the detection of some of the incomplete edge contours, with almost no degradation due to inclusion of noisy edge points.

A second algorithm combines the gradient information computed at several scales during non-maxima suppression. Non-maxima suppression is performed in the usual manner for the coarsest scale and the possible edge points are marked. Then non-maxima suppression is performed at successively smaller scales. If a point is being marked as a gradient maximum and an adjacent point was a maximum at a coarser scale, but not at the present scale, then the label for the coarser scale is moved to the present point. This had the effect of shifting a delocalized edge point to its location at the finer scale. An additional weight counts the number of scales at which a point was detected, similar to the Marr-Hildreth spacial coincidence assumption.

5 Experimental Results

The algorithms were tested on several real images. The values of σ used were 1, $\sqrt{2}$, and 2 and a threshold of .08 was applied. For comparison, the images were also processed using the Canny operator. The results for two images, Part and Tiwanaku, are shown in Figure 5. The Canny operator is (a), the single scale algorithm is (b), the multiple scale algorithm is (c), and the multi-scale non-maxima suppression algorithm is (d).

The single scale edge linking algorithm cleans up the Canny edges and in addition produces a set of linked lists corresponding to the contours found. The multiple scale algorithm is able to improve detection of edges that are close together and interact at scales which are large enough to remove noise and fine texture. It also improves detection of weak, but well defined edges, such as those of the shadows in the Part image. Thus a number of fragmented contours have been completed. Best results in all cases occurred with the multiple scale non-maxima suppression algorithm. Edges which had been delocalized were moved back to their location at a smaller scale, separating edges which had become too close together to differentiate and some contours were extended farther than with the multiple scale linking algorithm.

The weights used in the linking were chosen heuristically. Experiments varying the weights indicated that the actual values were not critical as long as higher weights were given to the points in the primary direction having the same direction as the current point, high magnitude, and longer length contour. Experiments in which each one of the factors in turn was removed, however, indicated that no three gave as good results as using all four.

6 Conclusions

An edge linking agorithm is presented that first computes gradient magnitude then uses non-maxima suppression to reduce the search space. The gradient magnitude and direction information is used to assign weights to paths through the set of points, and the best path is chosen. The single scale algorithm uses a depth first search, but each point is allowed to occur in only one

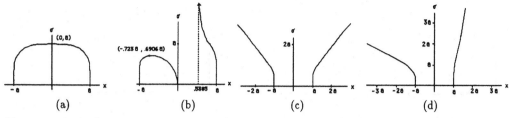

Figure 1: Scale Space Image: Location of Gradient Extrema for Staircase when (a) $b = 1$, (b) $b = .3$ and for Pulse when (c) $b = 1$, (d) $b = .3$

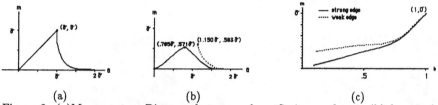

(a)   (b)   (c)

Figure 2: (a)Movement vs Distance between edges, Staircase, $b = 1$, (b) $b = .8$, (c) Maximum movement in terms of σ vs b for weaker and stronger edges.

subtree. This algorithm is extended to one which links edge points detected at multiple scales. A second variation is presented which uses the gradient information at multiple scales in the non-maxima suppression operation. A modified version of the single scale algorithm then links these edge points into contours. The first multi-scale method fills in gaps in single-scale contours, however the second method provides better localization and separation of nearby edges. Both improve detection of edges over the single-scale algorithm without introducing the noisy edges detected at the smaller scales.

References

[1] V. Berzins Accuracy of laplacian edge detectors. *CVGIP*, 27:195–210, 1984.

[2] F. Bergholm. Edge focusing. *IEEE Trans. on PAMI*, 9(6):726–741, 1987.

[3] J. F. Canny. *Finding Edges and Lines in Images*. Master's thesis, MIT, 1983.

[4] D. Marr and E. Hildreth. Theory of edge detection. In *Proc. R. Soc. Lond.*, pages 187–217, 1980.

[5] R. Nevatia and K. R. Babu. Linear feature extraction and description. *Computer Graphics and Image Processing*, 13:257–269, 1980.

[6] M. Shah, A. Sood, and R. Jain. Pulse and staircase edge models. *Computer Vision, Graphics, and Image Processing*, 34:321–341, June, 1986.

[7] D. Williams and M. Shah. *Edge Contours*. Technical Report CS-TR-88-18, University of Central Florida, Computer Science Department, Sept, 1988.

[8] D. Williams and M. Shah. Multiple Scale Edge Linking. In Trivedi, M. M., editor, *Applications of Artificial Intelligence VII*, pages 13–24, SPIE, 1989.

[9] A. Witkin Scale-space filtering. In *Internat. Joint Conf. on A.I.*, pages 1019–1021, 1983.

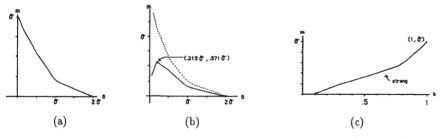

Figure 3: (a)Movement vs Distance between edges, Pulse, $b = 1$, (b) $b = .8$, (c) Maximum movement in terms of σ vs b for stronger edge.

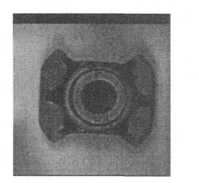

Figure 4: Original Images: Part, Tiwanaku.

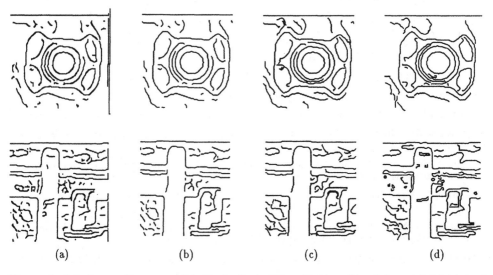

Figure 5: (a) Canny Operator, (b) Single Scale Edge Linking Algorithm, (c) Multiple Scale Algorithm, (d) Multi-Scale Non-maxima Suppression.

STEREO AND RECONSTRUCTION

Stereo Integration, Mean Field Theory and Psychophysics

A. L. Yuille (D.A.S., Harvard University), D. Geiger (A.I. Lab, M.I.T.)
H. Bülthoff (Dept. Cog. Sci, Brown)

Abstract

We describe a theoretical formulation for stereo in terms of the Markov Random Field and Bayesian approach to vision. This formulation enables us to integrate the depth information from different types of matching primitives, or from different vision modules. We treat the correspondence problem and surface interpolation as different aspects of the same problem and solve them simultaneously, unlike most previous theories. We use techniques from statistical physics to compute properties of our theory and show how it relates to previous work. These techniques also suggest novel algorithms for stereo which are argued to be preferable to standard algorithms on theoretical and experimental grounds. It can be shown (Yuille, Geiger and Bülthoff 1989) that the theory is consistent with some psychophysical experiments which investigate the relative importance of different matching primitives.

1 Introduction

In this paper we introduce a theoretical formulation for stereo in terms of the Bayesian approach to vision, in particular in terms of coupled Markov Random Fields. We show that this formalism is rich enough to contain most of the elements used in standard stereo theories.

The fundamental issues of stereo are: (i) what primitives are matched between the two images, (ii) what *a priori* assumptions are made about the scene to determine the matching and thereby compute the depth, and (iii) how is the geometry and calibration of the stereo system determined. For this paper we assume that (iii) is solved, and so the corresponding epipolar lines between the two images are known. Thus we use the epipolar line constraint for matching, some support for this is given by the work of Bülthoff and Fahle (1989).

Our framework combines cues from different matching primitives to obtain an overall perception of depth. These primitives can be weighted according to their robustness. For example, depth estimates obtained by matching intensity are sometimes unreliable since small fluctuations in intensity (due to illumination or detector noise) might lead to large fluctuations in depth, hence they are less reliable than estimates from matching edges. The formalism can also be extended to incorporate information from other depth modules (Bülthoff and Mallot, 1987, 1988) and provide a model for sensor fusion (Clark and Yuille, 1990). This framework was initially described in Yuille and Gennert (1988).

Unlike previous theories of stereo which first solved the correspondence problem and then constructed a surface by interpolation Grimson (1981), our theory proposes combining the two stages. The correspondence problem is solved to give the disparity field which best satisfies the *a priori* constraints.

Our model involves the interaction of several processes and is fairly complex. We will introduce it in three stages at different levels of complexity.

At the first level features (such as edges) are matched, using a binary matching field V_{ia} determining which features correspond. In addition smoothness is imposed on the disparity field $d(\vec{x})$ which represents the depth of the surface from the fixation plane. In this case the correspondence problem, determining the V_{ia}, is solved to give the smoothest possible disparity field. The theory is related to work by Yuille and Grzywacz (1988a, 1988b) on motion measurement and correspondence, and, in particular, to work on long–range motion.

At the second level we add line process fields $l(\vec{x})$ (which represents depth discontinuities) (Geman and Geman, 1984) to break the surfaces where the disparity gradient becomes too high.

The third level introduces additional terms corresponding to matching image intensities. Such terms are used in the theories of Gennert (1987) and Barnard (1986) which, however, do not have line process fields or matching fields. A psychophysical justification for intensity matching is given by the work of Bülthoff and Mallot (1987, 1988). Thus our full theory is expressed in terms of energy functions relating the disparity field $d(\vec{x})$, the matching field V_{ia}, and the line process field $l(\vec{x})$.

By the use of standard techniques from statistical physics, particularly the mean field approximation, we can eliminate certain fields and obtain effective energies for the remaining fields (see Geiger and Girosi, 1989; Geiger and Yuille, 1989). As discussed in Yuille (1989b) (following Wyatt - private communication) this can be interpreted as computing marginal probability distributions. We use this to show (Yuille, Geiger and Bülthoff 1989) that several existing stereo theories, such as the cooperative stereo algorithms (Dev, 1975; Marr and Poggio, 1976) and disparity gradient limit theories (Prazdny 1985; Pollard, Mayhew and Frisby, 1987), are closely related to versions of our model.

These techniques, however, also suggest novel algorithms for stereo computation. We argue that these algorithms incorporate constraints about the set of possible matches better than previous algorithms. They can also be directly related (Yuille 1989b) to analog methods for solving the travelling salesman problem. Moreover the greater empirical success of the elastic net algorithm (Durbin and Willshaw 1987) compared with the Hopfield and Tank method (1985) strongly suggests that our novel stereo algorithms will be more sucessful than some existing algorithms.

Our model can be related (Yuille, Geiger and Bülthoff 1989) to some psychophysical experiments (Bülthoff and Mallot, 1987, 1988; Bülthoff and Fahle, 1989) in which perceived depth for different matching primitives and disparity gradients are precisely measured. Their results suggest that several types of primitive are used for correspondence, but that some primitives are better than others. Our model is in good general agreement with the data from these experiments.

The plan of this paper is as follows: in Section 2 we review the Bayesian approach to vision and describe our theory. Section 3 introduces techniques from statistical physics and uses them to analyse the theory. The paper Yuille, Geiger and Bülthoff (1989) describes this work in more detail, in particular the comparison between theories and the relations to psychophysics.

2   The Bayesian Approach to Stereo

There has been a vast amount of work on stereo. Barnard and Fischler (1982) gives a good survey of the literature. We first review the problem of stereo and give an overview of our theory. Our theory is then described in terms of an energy function and finally put into a probabilistic framework.

2.1 The Matching Problem

There are several choices of matching primitives for stereo. Some theories use features such as edges or peaks in the image intensity (e.g., Marr and Poggio, 1976; Pollard, Mayhew and Frisby, 1987; Prazdny, 1985) while others match the image intensity directly (e.g., Barnard, 1986; Gennert, 1987). Yet another class of theory acts on the Fourier components of the images (e.g., Sanger 1988; Jepson and Jenkin, 1989) and hence is particularly sensitive to texture. It is unclear which primitives the human visual system uses. Current research (Bülthoff and Mallot, 1987, 1988; Bülthoff and Fahle, 1989) suggests that at least edges and image intensity are used as primitives.

It is desirable to build a stereo theory that is capable of using all these different types of primitives. This will allow to reduce the complexity of the correspondence problem and will enhance the robustness of the theory and its applicability to natural images. But not all primitives are equally reliable, however. A small fluctuation in the image intensity might lead to a large change in the measured disparity for a system which matches intensity. Thus image intensity tends to be less reliable than features such as edges.

Some assumptions about the scene being viewed are usually necessary to solve the correspondence problem. These can be thought of as natural constraints (Marr and Poggio 1976) and are needed because of the ill-posed nature of vision (Poggio and Torre, 1984). There are two types of assumption: (i) assumptions about the matching primitives, i.e., that similar features match (*compatibility constraint*), and (ii) assumptions about the surface being viewed (*continuity constraint*). For (ii) one typically assumes that either the surface is close to the fixation point (disparity is small) or that the surface's orientation is smoothly varying (disparity gradient is small) with possible discontinuities (we discuss possible smoothness measures in Section 2.2).

Our theory requires both assumptions but their relative importance depends on the scene. If the features in the scene are sufficiently different then assumption (i) is often sufficient to obtain a good match. If all features are very similar, assumption (ii) is necessary. We require that the matching is chosen to obtain the smoothest possible surface, so interpolation and matching are performed simultaneously (the next section formalizes these ideas).

2.2 The First Level: Matching Field and Disparity Field

The basic idea is that there are a number of possible primitives that could be used for matching and that these all contribute to a disparity field $d(x)$. This disparity field exists even where there is no source of data. The primitives we will consider here are features, such as edges in image brightness. Edges typically correspond to object boundaries, and other significant events in the image. Other primitives, such as peaks in the image brightness or texture features, can also be added. We will describe the theory for the one-dimensional case.

We assume that the edges and other features have already been extracted from the image in a preprocessing stage. The matching elements in the left eye consist of the features x_{i_L}, for $i_L = 1, ..., N_l$. The right eye contains features x_{a_R}, for $a_R = 1, ..., N_r$. We define a set of binary matching elements $V_{i_L a_R}$, the matching field, such that $V_{i_L a_R} = 1$ if point i_L in the left eye corresponds to point a_R in the right eye, and $V_{i_L a_R} = 0$ otherwise. A *compatibility field* $A_{i_L a_R}$ is defined over the range $[0, 1]$. For example, it is 1 if i_L and a_R are compatible (i.e. features of the same type), 0 if they are incompatible (an edge cannot match a peak),

We now define a cost function $E(d(x), V_{i_L a_R})$ of the disparity field and the matching elements. We will interpret this in terms of Bayesian probability theory in the next section. This will suggest several methods to estimate the fields $d(x), V_{i_L a_R}$ given the data. A standard estimator is to minimize $E(d(x), V_{i_L a_R})$ with respect to $d(x), V_{i_L a_R}$.

$$E(d(x), V_{i_L a_R}) = \sum_{i_L, a_R} A_{i_L a_R} V_{i_L a_R} (d(x_{i_L}) - (x_{a_R} - x_{i_L}))^2$$

$$+\lambda\{\sum_{i_L}(\sum_{a_R}V_{i_La_R}-1)^2+\sum_{a_R}(\sum_{i_L}V_{i_La_R}-1)^2\}+\gamma\int_M(Sd)^2dx. \qquad (1)$$

The first term gives a contribution to the disparity obtained from matching i_L to a_R. The third term imposes a smoothness constraint on the disparity field imposed by a smoothness operator S.

The second term encourages features to have a single match, it can be avoided by requiring that each column and row of the matrix $V_{i_La_R}$ contains only one 1. In Section 3 we will argue that it is better to impose constraints in this way, hence the second term will not be used in our final theory. However we will keep it in our energy function for the present since it will help us relate our approach to alternative theories.

Minimizing the energy function with respect to $d(\vec{x})$ and $V_{i_La_R}$ will cause the matching which results in the smoothest disparity field. We discuss ways of doing this minimization in Section 3.

The coefficient γ determines the amount of a priori knowledge required. If all the features in the left eye have only one compatible feature in the right eye then little a priori knowledge is needed and γ may be small. If all the features are compatible then there is matching ambiguity which the a priori knowledge is needed to resolve, requiring a larger value of γ and hence more smoothing. In Yuille, Geiger and Bülthoff (1989) we show that this gives a possible explanation for some psychophysical experiments.

The theory can be extended to two-dimensions in a straightforward way. The matching elements $V_{i_La_R}$ must be constrained to only allow for matches that use the epipolar line constraint. The disparity field will have a smoothness constraint perpendicular to the epipolar line which will enforce figural continuity.

Finally, and perhaps most importantly, we must choose a form for the smoothness operator S. Marr and Poggio (1976) proposed that, to make stereo correspondence unambiguous, the human visual system assumes that the world consists of smooth surfaces. This suggests that we should choose a smoothness operator which encourages the disparity to vary smoothly spatially. In practice the assumptions used in Marr's two theories of stereo are somewhat stronger. Marr and Poggio I (1976) encourages matches with constant disparity, thereby enforcing a bias to the fronto-parallel plane. Marr and Poggio II (1979) uses a coarse to fine strategy to match nearby points, hence encouring matches with minimal disparity and thereby giving a bias towards the fixation plane.

Considerations for the choice of smoothness operator (Yuille and Grzywacz 1988a, 1988b) are discussed in Yuille, Geiger and Bülthoff (1989). They emphasize the importance of choosing an operator so that the smoothness interaction decreases with distance. An alternative approach is to introduce discontinuity fields which break the smoothness constraint, see next section. For these theories the experiments described in Yuille, Geiger and Bülthoff (1989) are consistent with S being a first order derivative operator. This is also roughly consistent with Marr and Poggio I (1976). We will therefore use $S = \partial/\partial x$ as a default choice for our theory.

2.3 The Second and Third Level Theories: Adding Discontinuity and Intensity Fields

The first level theory is easy to analyse but makes the a priori assumption that the disparity field is smooth everywhere, which is false at object boundaries. The second level theory introduces discontinuity fields $l(x)$ to break the smoothness constraint (Blake 1983, Geman and Geman 1984, Mumford and Shah 1985). The third level theory adds intensity fields. Our energy function becomes

$$E(d(x), V_{i_La_R}, C) = \sum_{i_L,a_R} A_{i_La_R}V_{i_La_R}(d(x_{i_L}) - (x_{a_R} - x_{i_L}))^2$$

$$+\mu\int\{L(x) - R(x + d(x))\}^2dx$$

$$+\lambda\{\sum_{i_L}(\sum_{a_R}V_{i_L a_R}-1)^2+\sum_{a_R}(\sum_{i_L}V_{i_L a_R}-1)^2\}+\gamma\int_{M-C}(Sd)^2dx+M(C). \tag{2}$$

If certain terms are set to zero in (3) it reduces to previous theories of stereo. If the second and fourth terms are kept, without allowing discontinuities, it is similar to work by Gennert (1987) and Barnard (1986). If we add the fifth term, and allow discontinuities, we get connections to some work described in Yuille (1989a) (done in collaboration with T. Poggio). The third term will again be removed in the final version of the theory.

2.4 The Bayesian Formulation

Given an energy function model one can define a corresponding statistical theory. If the energy $E(d,V,C)$ depends on three fields: d (the disparity field), V the matching field and C (the discontinuities), then (using the Gibbs distribution – see Parisi 1988) the probability of a particular state of the system is defined by $P(d,V,C|g) = \frac{e^{-\beta E(d,V,C)}}{Z}$ where g is the data, β is the inverse of the temperature parameter and Z is the partition function (a normalization constant).

Using the Gibbs Distribution we can interpret the results in terms of Bayes' formula

$$P(d,V,C|g) = \frac{P(g|d,V,C)P(d,V,C)}{P(g)} \tag{3}$$

where $P(g|d,V,C)$ is the probability of the data g given a scene d,V,C, $P(d,V,C)$ is the *a priori* probability of the scene and $P(g)$ is the *a priori* probability of the data. Note that $P(g)$ appears in the above formula as a normalization constant, so its value can be determined if $P(g|d,V,C)$ and $P(d,V,C)$ are assumed known.

This implies that every state of the system has a finite probability of occurring. The more likely ones are those with low energy. This statistical approach is attractive because the β parameter gives us a measure of the uncertainty of the model temperature parameter $T = \frac{1}{\beta}$. At zero temperature ($\beta \to \infty$) there is no uncertainty. In this case the only state of the system that have nonzero probability, hence probability 1, is the state that globally minimizes $E(d,V,C)$. Although in some nongeneric situations there could be more then one global minimum of $E(d,V,C)$.

Minimizing the energy function will correspond to finding the most probable state, independent of the value of β. The mean field solution,

$$\bar{d} = \sum_{d,V,C} dP(d,V,C|g), \tag{4}$$

is more general and reduces to the most probable solution as $T \to 0$. It corresponds to defining the solution to be the mean fields, the averages of the f and l fields over the probability distribution. This enables us to obtain different solutions depending on the uncertainty.

In this paper we concentrate on the mean quantities of the field (these can be related to the minimum of the energy function in the zero temperature limit). A justification to use the mean field as a measure of the fields resides in the fact that it represents the minimum variance Bayes estimator (Gelb 1974).

3 Statistical mechanics and mean field theory

In this section we discuss methods for calculating the quantities we are interested in from the energy function and propose novel algorithms.

One can estimate the most probable states of the probability distribution (5) by, for example, using Monte Carlo techniques (Metropolis et al 1953) and the simulated annealing (Kirpatrick et al 1983) approach. The drawback of these methods are the amount of computer time needed for the implementation.

There are, however, a number of other techniques from statistical physics that can be applied. They have recently been used to show (Geiger and Girosi 1989, Geiger and Yuille 1989) that a number of seemingly different approaches to image segmentation are closely related.

There are two main uses of these techniques: (i) we can eliminate (or average out) different fields from the energy function to obtain effective energies depending on only some of the fields (hence relating our model to previous theories) and (ii) we can obtain methods for finding deterministic solutions.

There is an additional important advantage in eliminating fields - we can impose constraints on the possible fields by only averaging over fields that satisfy these constraints.

For the first level theory, see Section 3.1.1, it is possible to eliminate the disparity field to obtain an effective energy $E_{eff}(V_{ij})$ depending only on the binary matching field V_{ij}, which is related to cooperative stereo theories (Dev 1975, Marr and Poggio 1976). Alternatively, Section 3.1.2, we can eliminate the matching fields to obtain an effective energy $E_{eff}(d)$ depending only on the disparity. We believe that the second approach is better since it incorporates the constraints on the set of possible matches implicitly rather than imposing them explicitly in the energy function (as the first method does).

Moreover it can be shown Yuille (1989b) that there is a direct correspondence between these two theories (with $E_{eff}(V_{ij})$ and $E_{eff}(d)$) and analog models for solving the travelling salesman problem by Hopfield and Tank (1985) and Durbin and Willshaw (1987). The far greater empirical success of the Durbin and Willshaw algorithm suggests that the first level stereo theory based on $E_{eff}(d)$ will be more effective than the cooperative stereo algorithms.

We can also average out the line process fields or the matching fields or both for the second and third level theories. This leaves us again with a theory depending only on the disparity field, see Sections 3.1.2, and 3.1.3.

Alternatively we can use (Yuille, Geiger and Bülthoff 1989) mean field theory methods to obtain deterministic algorithms for minimizing the first level theory $E_{eff}(V_{ij})$. These differ from the standard cooperative stereo algorithms and should be more be more effective (though not as effective as using $E_{eff}(d)$) since they can be interpreted as performing the cooperative algorithm at finite temperature thereby smoothing local minima in the energy function.

Our proposed stereo algorithms, therefore, consist of eliminating the matching field and the line process field by these statistical techniques leaving an efective energy depending only on the disparity field. This formulation will depend on a parameter β (which can be interpreted as the inverse of the temperature of the system). We then intend to minimize the effective energy by steepest descent while lowering the temperature (increasing β). This can be thought of as a deterministic form of simulated annealing (Kirkpatrick et al 1983) and has been used by many algorithms, for example (Hopfield and Tank 1985), (Durbin and Willshaw 1987) (Geiger and Girosi 1989). It is also related to continuation methods (Wasserstrom 1973).

3.1 Averaging out Fields

In the next few sections we show that, for the first and second level theories, we can average out fields to obtain equivalent, though apparently different, formulations. As discussed in Yuille (1989b) (following Wyatt - private communication) this can be interpreted as computing marginal probability distributions.

3.1.1 Averaging out the Disparity Field for the first level theory

We now show that, if we consider the first level theory, we can eliminate the disparity field and obtain an energy function depending on the matching elements V only. This can be related (Yuille, Geiger and Bülthoff 1989) to cooperative stereo algorithms and it does impose the matching constraints optimally.

The disparity field is eliminated by minimizing and solving for it as a function of the V (Yuille and Grzywacz 1988a,b). Since the disparity field occurs quadratically this is equivalent to doing mean field over the disparity (Parisi 1988).

For the first level theory, assuming all features are compatible, our energy function becomes

$$E(d(x), V_{i_L a_R}) = \sum_{i_L, a_R} V_{i_L a_R} (d(x_{i_L}) - (x_{a_R} - x_{i_L}))^2 + \mu \sum_{i_L} (\sum_{a_R} V_{i_L a_R} - 1)^2$$

$$+ \mu \sum_{a_R} (\sum_{i_L} V_{i_L a_R} - 1)^2 + \lambda \int_M (Sd)^2 dx. \qquad (5)$$

Since the energy function is quadratic in the d's the Euler-Lagrange equations are linear in d. We can (Yuille, Geiger and Bülthoff 1989) solve these equations for the d's as functions of the matching fields and substitute back into the energy function obtaining:

$$E(V_{i_L a_R}) = \lambda \sum_{i_L, j_L} (\sum_{a_R} V_{i_L a_R}(x_{i_L} - x_{a_R}))(\lambda \delta_{i_L j_L} + G(x_{i_L}, x_{j_L}))^{-1}(\sum_b V_{j_L b_R}(x_{j_L} - x_{b_R}))$$

$$+ \lambda \sum_{i_L} (\sum_{a_R} V_{i_L a_R} - 1)^2 + \lambda \sum_{a_R} (\sum_{i_L} V_{i_L a_R} - 1)^2 \qquad (6)$$

where G is the Green function of the operator S^2. This calculation shows that the disparity field is strictly speaking unnecessary since the theory can be formulated as in (13), the connection to cooperative stereo algorithms is discussed in Yuille, Geiger and Bülthoff (1989). A similar calculation (Yuille and Grzywacz 1988a, 1988b) showed that minimal mapping theory (Ullman 1979) was a special case of the motion coherence theory.

A weakness of this formulation of the theory in (13), and the cooperative stereo algorithms based on it, is that the uniqueness constraints are imposed as penalties in the energy function, by the second and third terms on the right hand side of (13). As mentioned earlier we believe it is preferable to use mean field theory techniques which enforce the constraints strictly, see Section 3.1.2.

3.1.2 Averaging out the matching fields for the first level theory

We prefer an alternative way of writing the first level theory. This can be found by using techniques from statistical physics to average out the matching field, leaving a theory which depends only on the disparity field.

The partition function for the first level system, again assuming compatibility between all features, is defined to be $Z = \sum_{V_{i_L a_R}, d(x)} e^{-\beta E(V_{i_L a_R}, d(x))}$ where the sum is taken over all possible states of the system determined by the fields V and d.

It is possible to explicitly perform the sum over the matching field V yielding an effective energy for the system depending only on the disparity field d. Equivalently we could obtain the marginal probablity distribution $p(d|g)$ from $p(d, V|g)$ by integrating out the V field (Wyatt -personal communication).

To compute the partition function we must first decide what class of $V_{i_L a_R}$ we wish to sum over. We could sum over all possible $V_{i_L a_R}$ and rely on the $\lambda \sum_{i_L} (\sum_{a_R} V_{i_L a_R} - 1)^2 + \lambda \sum_{a_R} (\sum_{i_L} V_{i_L a_R} - 1)^2$ term to bias against multiple matches. Alternatively we could impose the constraint that each point has a unique match by only summing over $V_{i_L a_R}$ which contain a single 1 in each row and each column. We could further restrict the class of possible matches by requiring that they satisfied the ordering constraint.

For this section we will initially restrict that each feature in the left image has a unique match in the right image, but not vice versa. This simplifies the computation of the partition function, but it can be relaxed (Yuille, Geiger and Bülthoff 1989). The requirement of smoothness on the disparity field should ensure that unique matches occur, this is suggested by mathematical analysis of a similar algorithm used for an elastic network approach to the T.S.P. (Durbin, Szeliski and Yuille 1989).

Since we are attempting to impose the unique matching constraint by restricting the class of V's the $\lambda \sum_{i_L}(\sum_{a_R} V_{i_L a_R} - 1)^2 + \lambda \sum_{a_R}(\sum_{i_L} V_{i_L a_R} - 1)^2$ terms do not need to be included in the energy function. We can now write the partition function as

$$Z = \sum_{V,d} \prod_{i_L} e^{-\beta\{\sum_{a_R} V_{i_L a_R}\left(d(x_{i_L}) - (x_{a_R} - x_{i_L})\right)^2 + \int_M (Sd)^2 dx\}}. \tag{7}$$

For fixed i_L we sum over all possible $V_{i_L a_R}$, such that $V_{i_L a_R} = 1$ for only one a_R (this ensures that points in the left image have a unique match to points in the right image). This gives

$$Z = \sum_{d} \prod_{i_L} \{\sum_{a_R} e^{-\beta\left(d(x_{i_L}) - (x_{a_R} - x_{i_L})\right)^2}\} e^{-\beta \int_M (Sd)^2 dx}. \tag{8}$$

This can be written using an effective energy $E_{eff}(d)$ as

$$Z = \sum_{d} e^{-\beta E_{eff}(d)}, \text{ where } E_{eff}(d) = \frac{-1}{\beta} \sum_{i_L} log\{\sum_{a_R} e^{-\beta\left(d(x_{i_L}) - (x_{a_R} - x_{i_L})\right)^2}\} + \int_M (Sd)^2 dx. \tag{9}$$

Thus our first level theory of stereo can be formulated in this way without explicitly using a matching field. We are not aware, however, of any existing stereo theory of this form. Since it has formulated the matching constraints in computing the partition function we believe it is preferable to standard cooperative stereo algorithms.

3.1.3 Averaging out the matching and discontinuity fields for the third level theory

We can apply the same techniques to the second and third theories eliminationg both the matching field and the discontinuity fields (Yuille, Geiger and Bülthoff 1989). This gives an effective energy for the third level theory:

$$E_{eff}(d) = \frac{-1}{\beta} \sum_{i_L} log\{\sum_{a_R} e^{-\beta\left(d(x_{i_L}) - (x_{a_R} - x_{i_L})\right)^2}\} \frac{-1}{\beta} \sum_{a_R} log\{\sum_{i_L} e^{-\beta\left(d(x_{a_R}) - (x_{a_R} - x_{i_L})\right)^2}\}$$
$$-\frac{1}{\beta} ln(1 + e^{-\beta[\alpha(d_k - d_{k-1})^2 - \gamma]}) + \mu \int \{L(x) - R(x + d(x))\}^2 dx. \tag{10}$$

Again a deterministic annealing approach should yield good solutions to this problem.

Averaging out the discontinuity field from the second level theory will give (Yuille, Geiger and Bülthoff 1989) a theory, depending only on the disparity field and the matching field, which is reminiscent of disparity gradient limit theories (Pollard, Mayhew and Frisby, 1987; Prazdny, 1985).

The mean field theory approach can also yield deterministic algorithms for theories including the binary matching elements (although we believe these algorithms will be inferior to methods which eliminate the matching fields for the reasons discussed in Section 3.1.2) which should be superior to the cooperative stereo algorithms (the cooperative algorithms are the zero temperature limit of these equations, and hence are less able to escape local minima).

4 Conclusion

We have derived a theory of stereo on theoretical grounds using the Bayesian approach to vision. This theory is able to incorporate most of the desirable elements of stereo and it is closely related to a number of existing theories.

The theory can combine information from matching different primitives, which is desirable on computational and psychophysical grounds. The formulation can be extended to include monocular depth cues for stereo correspondence (Clark and Yuille. 1990).

A basic assumption of our work is that correspondence and interpolation should be performed simultaneously. This is related to the important experimental and theoretical work of Mitchison (1988) and Mitchison and McKee (1987).

The use of mean field theory enables us to average out fields, enabling us to make mathematical connections between different formulations of stereo. It also suggests novel algorithms for computing the estimators (due to enforcing the matching constraints while performing the averaging, see Section 3.1.2) and we argue that these algorithms are likely to be more effective than a number of existing algorithms.

Finally the theory agrees well with some psychophysical experiments (Bülthoff and Mallot, 1987, 1988; Bülthoff and Fahle, 1989). Though further experiments to investigate the importance of different stereo cues are needed.

Acknowledgements

A.L.Y. would like to acknowledge support from the Brown/Harvard/MIT Center for Intelligent Control Systems with U.S. Army Research Office grant number DAAL03-86-K-0171. Some of these ideas were initially developed with Mike Gennert. We would like to thank Mike Gennert, Manfred Fahle, Jim Clark, and Norberto Grzywacz for helpful conversations.

References

Barnard, S. *Proc. Image Understanding Workshop*, Los Angeles, 1986.

Barnard, S. and Fischler, M.A, "Computational Stereo". *Computing Surveys*, **14**, No. 4, 1982.

Blake, A. "The least disturbance principle and weak constraints," *Pattern Recognition Letters*, **1**, 393-399, 1983.

Blake, A. "Comparison of the efficiency of deterministic and stochastic algorithms for visual reconstruction," *PAMI*, Vol. 11, No. 1, 2-12, 1989.

Bülthoff, H. and Mallot, H-P. "Interactions of different modules in depth perception". In *Proceedings of the First International Conference on Computer Vision*, London, 1987.

Bülthoff, H. and Mallot, H-P. "Integration of depth modules: stereo and shading". *J. Opt. Soc. Am.*, **5**, 1749-1758, 1988.

Bülthoff, H. and Fahle, M. "Disparity Gradients and Depth Scaling". *Artificial Intelligence Memo 1175*, Cambridge, M.I.T., 1989.

Burt, P. and Julesz, B. "A disparity gradient limit for binocular fusion". *Science* **208**, 615-617, 1980.

Clark, J.J. and Yuille, A.L. **Data Fusion for Sensory Information Processing Systems.**, Kluwer Academic Press, 1990.

Dev, P. "Perception of depth surfaces in random-dot stereograms: A neural model". *Int. J. Man-Machine Stud.* **7**, 511-528, 1975.

Durbin, R., Szeliski, R. and Yuille, A.L. "The elastic net and the travelling salesman problem". Harvard Robotics Laboratory Technical Report. No. 89-3, 1989.

Durbin, R. and Willshaw, D. "An analog approach to the travelling salesman problem using an elastic net method". *Nature*, **326**, 689-691, 1987.

Duchon, J. *Lecture Notes in Mathematics*. 571. (Eds Schempp, W. and Zeller, K.), 85-100 (Berlin, Springer-Verlag, 1979).

Geiger, D. and Girosi, F., "Parallel and deterministic algorithms from MRFs :integration and surface reconstruction". *Artificial Intelligence Laboratory Memo 1114*. Cambridge, M.I.T., June 1989.

Geiger, D. and Yuille, A., "A common framework for image segmentation". Harvard Robotics Laboratory Technical Report. No. 89-7, 1989.

Geiger, D. and Yuille, A., "Stereopsis and eye movement", *Proceedings of the First International Conference on Computer Vision.* London, pp 306-314, 1987.

Gelb, A. **Applied Optimal Estimation.** M.I.T. Press. Cambridge, Ma., 1974.

Geman, S. and Geman, D. "Stochastic relaxation, Gibbs distributions and the Bayesian restoration of images". *IEEE Trans. PAMI,* **6,** 721-741, 1984.

Gennert, M. "A Computational Framework for Understanding Problems in Stereo Vision". M.I.T. AI Lab PhD. Thesis, 1987.

Grimson, W.E.L. **From Images to Surfaces: A computational study of the human early visual system.** M.I.T. Press. Cambridge, Ma., 1981.

Hopfield, J.J. and Tank, D.W. "Neural computation of decisions in optimization problems". *Biological Cybernetics,* **52,** 141-152, 1985.

Jepson, A.D. and Jenkin, M.R.M. "The fast computation of disparity from phase differences". *Proceedings Computer Vision and Pattern Recognition '89.* pp 398-403, San Diego, 1989.

Kirkpatrick, S., Gelatt, C.D. Jr. and Vecchi, M.P. "Optimization by simulated annealing". *Science,* **220,** 671-680, 1983.

Marr, D. and Poggio, T. "Cooperative computation of stereo disparity". *Science,* **194,** 283-287, 1976.

Marr, D. and Poggio, T. "A computational theory of human stereo vision". *Proc. R. Soc. Lond. B.* Vol 204, pp 301-328, 1979.

Marroquin, J. In *Proceedings of the First International Conference on Computer Vision.* London. 1987.

Metropolis, N. Rosenbluth, A., Rosenbluth, M., Teller, A., and Teller, E. "Equation of state calculations by fast computing machines". *J. Phys. Chem.* **21,** 1087-1091, 1953.

Mitchison, G.M. "Planarity and segmentation in stereoscopic matching. *Perception,* **17,** 753-782, 1988.

Mitchison, G.M. and McKee, S. "The resolution of ambiguous stereoscopic matches by interpolation". *Vision Research,* Vol 27, no 2. pp 285-294, 1987.

Mumford, D. and Shah, J. "Boundary detection by minimizing functionals, I", *Proc. IEEE Conf. on Computer Vision and Pattern Recognition,* San Francisco, 1985.

Parisi, G. **Statistical Field Theory.** Addison-Wesley, Reading, Mass. 1988.

Pollard, S.B., Mayhew, J.E.W. and Frisby, J.P. "Disparity Gradients and Stereo Correspondences". *Perception,* 1987.

Poggio, T. and Torre, V. "Ill-posed problems and regularization analysis in early vision". M.I.T. A.I. Memo No. 773, 1984.

Prazdny, K. "Detection of Binocular Disparities". *Biological Cybernetics,* **52,** 93-99, 1985.

Sanger, T. "Stereo disparity computation using Gabor filters," *Biological Cybernetics,* **59,** 405-418, 1988.

Ullman, S. **The Interpretation of Visual Motion.** Cambridge, Ma. M.I.T. Press, 1979.

Wasserstrom, E. "Numerical solutions by the continuation method". *SIAM Review,,***15,** 89-119, 1973.

Yuille, A.L. "Energy Functions for Early Vision and Analog Networks". *Biological Cybernetics,* **61,** 115-123, 1989a.

Yuille, A.L. Harvard Robotics Laboratory Technical Report 89-12. 1989b.

Yuille, A.L., Geiger, D. and Bülthoff, H. "Stereo Integration, Mean Field Theory and Psychophysics". Harvard Robotics Laboratory Technical Report 89-11.

Yuille, A.L. and Gennert, M. Preprint. 1988.

Yuille, A.L. and Grzywacz, N.M. "A Computational Theory for the Perception of Coherent Visual Motion". Nature, 1988a.

Yuille, A.L. and Grzywacz, N.M. "The Motion Coherence Theory". *Proceedings of the Second International Conference on Computer Vision.* pp 344-353. Tampa, Florida. 1988b.

A PYRAMIDAL STEREOVISION ALGORITHM BASED ON CONTOUR CHAIN POINTS

Aimé Meygret, Monique Thonnat, Marc Berthod
INRIA Sophia Antipolis, 2004 route des Lucioles
06565 Valbonne cedex, France

Abstract

We are interested in matching stereoscopic images involving both natural objects (vegetation, sky, reliefs,...) and man made objects (buildings, roads, vehicles,...). In this context we have developed a pyramidal stereovision algorithm based on "contour chain points." The matching process is performed at different steps corresponding to the different resolutions. The nature of the primitives allows the algorithm to deal with rich and complex scenes. Goods results are obtained for extremely fast computing time.

Introduction

The fundamental problem of stereovision is matching homologous visual characteristics extracted in several images of the same scene observed from different view points. These visual characteristics are also called images primitives. Marr and Poggio [Grim81] have noted that the difficulty of the correspondence problem and the subproblem of eliminating false targets is directly proportional to the range and resolution of disparities considered and to the density of matchable features in an image. It is also crucial for the subsequent triangulation process to localize these primitives very accurately. Thus, many stereo matching algorithms have been developed. Multi-resolution approaches provided an efficient way to limit the complexity of the matching process [Marr79], [Hann84]. In these algorithms ambiguous matches are solved by enforcing the continuity of disparity in the neighborhood of ambiguous points. However, the continuity constraint is no more available when the neighborhood used for resolving ambiguities, crosses an occluded contour. Mayhew and Frisby [Mayh81] partially solved this problem by enforcing continuity of disparity along edges in the image. This constraint which they called "figural continuity" is more realistic than the surface smoothness assumption since, while disparity varies discontinuously accross surface boundaries, it rarely varies discontinuously along such a boundary. Grimson [Grim85] implemented a new version of his earlier algorithm, incorporating this constraint to eliminate random matches.

Kim and Bovik [Kim86] have also used the continuity of disparity along contours for both disambiguation and matching control. In a first step they match extremal points (high curvature edge points, terminations and junctions) by enforcing continuity of disparity along contours. They check the consistency of the disparity at a matched extremal point by examining disparities at its neighbours which have been matched and propagate the disparities along the contours in order to match other edge points. Their algorithm gives good results but it has apparently only been tested on indoor stereo images with few objects.

We think that this constraint is an important continuity law stereo consequence and we have tried to include it in a pyramidal stereovision algorithm.

We can also facilitate the matching process by using richer primitives such as edge segments [Medi85], [Long86], [Ayac85], regions [Wrob88],...

Within Prometheus European Project we are interested in three-dimensional localization of obstacles in road scenes. The nature of images we deal with (diversity and complexity of shapes) made us choose contour chain points as primitives. For a pedestrian which is a typical obstacle, it is very difficult to extract linear features (as edge segments) because of the smoothness of the surface of human body. On the other hand it is difficult to match chains of contours for two reasons at least: the chain geometry description difficulty and problems involved by chains cutting out management. Moreover the chain as an entity is not localizable very accurately (same problem with regions). As against this, contours chains points are primitives with rich information, so easy to match, are very accurately localizable and are suitable to describe any type of scene, man made, natural or mixt. Moreover, contour chain points provide a studied three-dimensional world more complete description.

The hierarchical structure of Prometheus images (important objects, details), the existence of strong disparities and time processing constraint (for obstacles detection we must be able to take a decision very quickely) made us choose a coarse-to-fine approach.

Presentation of the method

Data structure: We use a pyramidal image data structure in which the search for objects starts at a low resolution and is refined at ever increasing resolutions until one reaches the highest resolution of interest. We consider a pyramid at the four highest resolution levels. The consolidation is made in a 2x2 neighborhood. We have chosen a pyramidal approach rather than a classical multi-resolution approach essentially for time computing reasons. In a pyramid of resolution the image size is reduced by the consolidation process.

Primitives: Contour chain points extraction is performed in three steps: gradient computation [Deri87], hysteresis thresholding to eliminate noise and contour points chaining [Gira87]. The contour chain is used on the one hand to eliminate false targets and on the other hand to propagate the disparity. The three-dimensional world description provided by contour chains is richer and facilitates the further recognition process.

Matching: Matching of the two stereo images is performed by optimizating a similarity function. For two contour points, (x_l, y_l) in the left image and (x_r, y_r) in the right image, we define the similarity function as:

$$f(x_l, y_l, x_r, y_r) = \frac{[G(x_l, y_l) - G(x_r, y_r)]^2}{S_G^2} + \frac{[\theta(x_l, y_l) - \theta(x_r, y_r)]^2}{S_\theta^2} \quad \text{where}$$

$G(\ x\ ,\ y\ )$ and $\theta(\ x\ ,\ y\ )$ respectively design the gradient norm and orientation at $(\ x\ ,\ y\ )$ point.

S_G and S_θ are respectively thresholds on the gradient norm and orientation difference. For all $(\ x_i\ ,\ y_i\ )$ in the left image $(\ x_j\ ,\ y_j\ )$ is a potential matching in the right image if:

- $|\ G(\ x_{li}\ ,\ y_{li}\ )\ -\ G(\ x_{rj}\ ,\ y_{rj}\ )|\ \leq\ S_G$
- $|\ \theta(\ x_{li}\ ,\ y_{li}\ )\ -\ \theta(\ x_{rj}\ ,\ y_{rj}\ )|\ \leq\ S_\theta$
- $f(\ x_{li}\ ,\ y_{li}\ ,\ x_{rj}\ ,\ y_{rj}\ )$ minimum with regard to j

$(\ x_{rjo}\ ,\ y_{rjo}\ )$ is a potential matching;

The pair $((\ x_{li}\ ,\ y_{li}\ )\ ,\ (\ x_{rjo}\ ,\ y_{rjo}\ ))$ is validated if $f(\ x_{li}\ ,\ y_{li}\ ,\ x_{rjo}\ ,\ y_{rjo}\ )$ is minimal with regard to i. The matching process is then symmetric, and uniqueness is guaranteed. Matched points should have similar properties, of course, because they are both projections of the same surface point , but in many cases there will be ambiguous. So we have to determine criterions allowing to decide which matches are correct. The three-dimensional spatial continuity of real world surfaces constrains the two-dimensional spatial distribution of disparity in the image plane. This is the second stereo law. The continuity of disparity over most of the image can be used to avoid false matches based on similarity alone, by suppressing matches in the absence of supporting local evidence. We have chosen contour chains as local support to check the matches consistency. Edge points along the chain belong to the same surface, except when the edge crosses an occluding contour: we have then to stop the chain; we avoid this problem by using Giraudon chaining algorithm which cuts chains when it finds triple points. It may happen that the occluded contour gradient is too weak to be detected or is eliminated as noise. The disparity continuity assumption along the chain does not hold, but we can limit this occurence with a coarse to fine algorithm. As the search of a potential matching is hierarchically governed, we can choose very low thresholds at the step of hysteresis thresholding and so increase the number of primitives without too much complicating the matching process. We can generally consider that the disparity varies smoothly along the chain. Then we avoid the problem of the region overlapping an occluding contour.

For each pair of matched points we determine the disparity vector (its norm L, and its orientation β) and we check the vectors continuity along contour chains. A couple i of matched points is validated if:

$$\sum_{j\ \in\ V_i} \frac{\frac{|\ L_i\ -\ L_j\ |}{(\ L_i\ +\ L_j\ )}\frac{1}{|\ i\ -\ j\ |^n}}{\sum_{j\ \in\ V_i}\frac{1}{|\ i\ -\ j\ |^n}}\ \leq\ S_L \qquad \text{and} \qquad \sum_{j\ \in\ V_i} \frac{\frac{|\ \beta_i\ -\ \beta_j\ |}{(\ \beta_i\ +\ \beta_j\ )}\frac{1}{|\ i\ -\ j\ |^n}}{\sum_{j\ \in\ V_i}\frac{1}{|\ i\ -\ j\ |^n}}\ \leq\ S_\beta$$

V_i : i neighborhood along the chain.

S_L et S_β are thresholds, n define the neighborhood size.

The hierarchy in the pyramid: After consolidation, contours (local maximas of the first derivate) are extracted on each stereo image and then, matched at finer and finer resolutions. The disparity information obtained at a given resolution is used to specify the search space for finding a matching point at a finer resolution. So we limit the matching algorithm complexity by controlling at each step, the number of primitives and the search window size. At the first step (coarsest resolution) the search window size is defined by the minimum and maximum depths estimated in the image. At the following steps it is defined by the size of the filter for the extraction of contours. The use of epipolar geometry permits us to reduce strongly the correspondence problem dimension: the search windows are reduced to strips along epipolar lines. At each step we interpolate the disparity along the chain. The disparity at a chain point is given by the average of the neighbours disparities weighted by the inverse of the distance between the current point and its neighbours. The disparity interpolation along the chain gives us a richer disparity map to specify the search space of a potential matching at finer resolution. At a given resolution, the search space of a potential matching is obtained by searching at the nearer coarse resolution the nearer neighbour which has been matched.

Results

We have treated two very different Prometheus scenes in order to test the algorithm robustness. The images were taken using two CCD cameras mounted parallel on top of a car with a height of about 1.60 m and a distance of 40 cm between the cameras. The first stereo pair (figure 1) is a countryside scene and is characterized by a lot of discontinuities. The figure 2 represents a part of the reconstructed scene in a three-dimensional space. The car profile, the dividing line and the post on the right appear clearly. There are two important sources of errors in the matching process:

- the lack of precision in cameras calibration which has constrained us to use epipolar bands and not epipolar lines
- matching errors which arise at horizontal lines due to the cameras relative geometry; we expect to limit this problem by swinging the cameras support in an appropriate direction.

For this scene, though we have chains, we have only represented points in three-dimensional space because few false matches (which often arise at horizontal lines) are sufficient to blur the reconstructed scene. The program was coded in C and implemented in SUN4 110. For 512x144 stereo images, the consolidation process, the edges detection and the chaining take less than 1 minute for the two images. At the last step 1445 points have been matched and after the disparity interpolation along chains we had 2522 points. The matching process (including interpolation) takes less than 30 secondes.

fig. 1: The stereo images (countryside scene).

fig. 2: The reconstructed scene, in a 3D space

The second scene takes place in town (many discontinuities in the scene) and shows a typical obstacle: a cyclist. We present in figure 3 the stereo images.The figures 4a, 4b and 4c represent the reconstructed chains in a three-dimensional space viewed by an observer turning around the scene. The software displays thready chains, so according to the view point some chains are seen by transparency; a surface modelisation could cope with this problem. In the figure 4a we clearly distinguish the file of cars on the left, the middle road line mark and the cyclist. As the observer moves towards the left, in the figure 4b, the cyclist appears very clearly. In the figure 4c we still distinguish the cyclist; the road right boundary and a parking car appear.

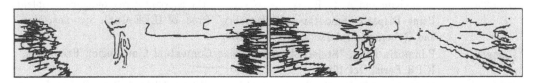

fig. 3: The stereo images (town scene).

fig. 4a: The reconstructed chains in a 3D space; the observer
has moved slightly towards top right.

fig. 4b: The reconstructed chains in a 3D
space, front view

fig. 4c: The reconstructed chains in a 3D
space; the observer has moved
slightly towards top left.

The presented multi-resolution approach gives very interesting results as well for the fullness of the reconstructed information as for the quality of this information. We have now to deal with three-dimensional data to provide a three-dimensional description of the environment seen from the car.

Conclusion

We have presented a matching stereovision algorithm reliable enough to provide a three-dimensional description of the environment seen from the car. The hierarchy permits to avoid aberrant matches by matching main structures before details.

Using a symetric similarity function garantees non ambiguity according to the stereo unicity law. The second stereo law, the continuity, is checked when validating the matches.

Using a chain contour as a local support for consistency makes the method indifferent to the types of handled scenes (man made, natural) since the contours density and nature (occluded or not) does not affect the method and so makes it more general than methods adapted to robotic scenes [Ayac85] or to natural scenes [Grim81] .

Finally, a pyramidal approach reduces greatly the computing time (by a factor between 3 and 4) with regard to classical multi-resolution approach. Although this algorithm permits to work with rich scenes (involving man made or natural objects), its computing time is nevertheless extremely fast.

Acknowledgements

We would like to thank Gerard Giraudon and John Fairfield for many valuable discussions and Professor Nagel who provided us the stereo images.

References

[Ayac85]   N. Ayache and B. Faverjon, "Fast Stereo Matching of Edges Segments Using Prediction and Verification of Hypotheses", CVPR, 1985.

[Deri87]   R. Deriche, Optimal Edge Detection Using Recursive Filtering, Proceedings of of the first International Conference on Computer Vision (ICCV), pp. 501-505, Londres, 1987.

[Gira87]   G. Giraudon, "Chainage rapide sur des images de contour," Rapport de recherche INRIA, 1987.

[Grim81]   W. E. L. Grimson, "From Images to surfaces," MIT Press, Cambridge, 1981.

[Grim85]   W. E. L. Grimson, "Computational Experiments with a Feature Based Stereo Algorithm," IEEE Trans. PAMI, vol. PAMI-7, pp. 17-34, January 1985.

[Hann84]   M. J. Hannah, "Decription of SRI's Baseline Stereo System," Technical Note No.342, Artificial Intelligence Center, SRI International, Oct 1984.

[Kim86]   N. H. Kim and A. C. Bovik, "A solution to the Stereo Correspondence Problem Using Disparity Smoothness Constraint," Proc. of IEEE Conf. Systems, Man, and Cybernetics, Atlanta, October 1986.

[Long86]   P.Limozin-Long, "Stereo Matching Using Contextual Line Region Primitives," ICPR Paris, Oct 1986.

[Marr79]   D. Marr and T. Poggio, "A theory of Human Stereo Vision", Proc. R. Soc. Lond.' vol. B 204, pp. 301-328, 1979.

[Mayh81]   J.E.W. Mayhew and J.P. Frisby, "Psychophysical and computationnal studies towards a theory of human stereopsis," Artif.Intell., 17, pp. 349-385, 1981.

[Medi85]   G. Medioni and R. Nevatia, "Segment-Based Stereo Matching", Computer Vision, Graphics, ang Image Processing, vol. 31, pp. 2-18, July 1985.

[Wrob88]   B. Wrobel-Dautcourt, "Coopération entre l'extraction et la mise en correspondance symbolique de régions," MARI Paris, pp 181-188, 1987

Parallel and deterministic algorithms from MRFs: surface reconstruction and integration

Davi Geiger[1,2] and Federico Girosi [1]

[1] Massachusetts Institute of Technology
Artificial Intelligence Laboratory, and

[2] Siemens Corporate Research, Inc,
755 College Road East, Princeton, NJ 08540

1 Introduction

In order to give a viewer information about a three dimensional scene many algorithms have been developed on several early vision processes, such as edge detection, stereopsis, motion, texture, and color . This information refers to properties of the scene as shape, distance, color, shade or motion. The input to these systems are usually noisy and some times sparse as well as its output and then more processing is necessary to extract the relevant information and fill in sparse data. In this way the problem of surface reconstruction starting from a set of noisy sparse data is prototypical for vision. In recent years many researchers[10][13][5] [3][4] have investigated the use of Markov random fields (MRFs) for early vision. They have also been used to integrate early vision processes to label physical discontinuities. Two fields are usually required in the MRFs formulation of a problem: one represents the function that has to be reconstructed, and the other is associated to its discontinuities.

The essence of the MRFs model resides on Bayes theory with local interaction between the fields, where the posterior probability distribution for the configuration of the fields, given a set of data, is given as a Gibbs distribution. The model is then specified by the a priori information about the system and a conditional probability of the data given the fields. In the standard approach an estimate of the field and its discontinuities is given by the configuration that maximizes the probability distribution. This becomes a combinatorial optimization problem, that can be solved by methods of the Monte Carlo type (simulated annealing[11], for example). The MRFs formulation has two main drawbacks: the amount of computer time needed for the implementation and the difficulty in estimating the parameters of the model.

In this paper we propose a deterministic approach to MRFs models. It consists in explicitly writing down a set of equations from which we can compute estimates of the mean values of the field f and the line process. We use the *mean field approximation* and the *saddle point approximation*, both well known statistical mechanics tools, to obtain an approximated solution, that is given in implicit form by a set of non linear equations. We call these equation *deterministic* to underline the deterministic character of the whole procedure.

An advantage of such an approach is that the solution of the deterministic equations is faster than the Monte Carlo techniques, fully parallelizable and feasible of implementation on analog networks. The possibility of writing a set of equations is also useful for a better understanding of the nature of the solution and of the parameters of the model.

We study the Weak Membrane model that has been already studied by several authors[2][12][14][13]. It is interesting to notice that the GNC algorithm, proposed by Blake and Zisserman [2], arises naturally in the framework of statistical mechanics. This estabilish a connection between MRFs and deterministic algorithm already used in vision. The model is applied to dense data and to sparse data as well. The problem of surface reconstruction (and image restoration) from sparse data is addressed and an algorithm to perform these tasks is obtained and implemented. We also outline an algorithm that solves the problem of aligning the discontinuities of different visual models with intensity edges that can be used for the integration of different modules.

The paper is organized in the following way: section 2 presents an overview of MRFs in vision. section 3 discusses the deterministic approximation of MRFs for the three energy functions mentioned above. Section 4 dicusses the issue of parameters estimation. In section 5 some results are described including sparse data. Section 6 shows applications for the integration of visual modules with intensity edges. Section 7 concludes the paper.

2 MRFs and the Weak Membrane model

Here we briefly summarize how MRFs are applied to the Weak Membrane model. A more extensive discussion is given in Geman and Geman [10], Marroquin[13], Chou[3], Gamble and Poggio[5], Gamble, Geiger, Poggio, Weinshall [4] and Geiger [9].

Consider the problem of approximating a surface given sparse and noisy depth data, on a regular 2D lattice of sites. We think the surface as a field (surface-field) defined in the regular lattice, such that the value of this field at each site of the lattice is given by the surface height at this site. The Markov property asserts that the probability of a certain value of the field at any given site in the lattice depends only upon neighboring sites. According to the Clifford-Hammersley theorem, the probability of a state of the field f has the Gibbs form:

$$P(f) = \frac{1}{Z_f} e^{-\beta U(f)} \tag{2.1}$$

where f is the field, e.g. the surface-field, Z_f is the partition function, $U(f) = \sum_i E_i(f)$ is an "energy function" that can be computed as the sum of local contributions from each lattice site i, and β is a parameter that is called the inverse of the natural temperature of the field. If a sparse observation g for any given surface-field f is given and a model of the noise is available then one knows the conditional probability $P(g|f)$. Bayes theorem then allows to write the posterior distribution:

$$P(f|g) = \frac{P(g|f)P(f)}{P(g)} \equiv \frac{1}{Z} e^{-\beta E(f|g)} . \tag{2.2}$$

Geman and Geman [10] introduced the idea of another field, the line process, located on the dual lattice, and representing explicitly the presence or absence of discontinuities that break the smoothness assumption (2.2).

As a simple example, when the surfaces (surface-fields) are expected to be smooth but not at the discontinuities and the noise is Gaussian, the energy, for the one dimensional case, is given by

$$E(f, l|g) = \sum_i \{\lambda_i (f_i - g_i)^2 + \alpha \sum_{j \epsilon N_i} [(f_i - f_j)^2 (1 - l_{ij})] + \sum_C V_C(l_{ij})\}, \tag{2.3}$$

where $\lambda_i = 1$ or 0 depending on whether data are available or not and N_i is a set of sites in an arbitrary neighborhood of the site i. l_{ij} is the element of the binary field l located between site i, j. The term $V_C(l_{ij})$, where C is a clique defined by the neighborhood system of the line process (binary field), reflects the fact that certain configurations of the line process are more likely to occur than others. Depth discontinuities are usually continuous and non-intersecting, and rarely consist of isolated points. These properties of physical discontinuities can be enforced locally by defining an appropriate set of energy values $V_C(l_{ij})$ for different configurations of the line process ([10], [14]). In our models the cliques will be simplified to the nearest neighbors.

2.1 The Line Process for two dimensions

In this case we define a horizontal line process h_{ij} and a vertical line process v_{ij}. We point out that another possible approach is considered by Geiger and Yuille [8] and Geiger[9], where the line process is a scalar quantity in two dimensions. Any how, the line process h_{ij} connects the site (i, j) to the site $(i, j - 1)$, while v_{ij} connects the site (i, j) to the site $(i - 1, j)$.

2.2 The Weak Membrane model

A special case of (2.3) is the Weak membrane model given in two dimensions, for sparse data, by

$$E_1(f, h, v) = E_{fg}(f) + E_{fl}(f, h, v) + E_l(h, v) \tag{2.4}$$

where

$$E_{fg}(f) = \sum_{i,j} \lambda_{ij}(f_{i,j} - g_{i,j})^2 \quad , \quad E_l(h, v) = \gamma \sum_{i,j}(h_{i,j} + v_{i,j}) \tag{2.4a}$$

$$E_{fl}(f, h, v) = \alpha \sum_{i,j} [(f_{i,j} - f_{i,j-1})^2(1 - v_{i,j}) + (f_{i,j} - f_{i-1,j})^2(1 - h_{i,j})] \tag{2.4b}$$

α and γ are positive valued parameters and λ_{ij} is 1 when there is data and zero otherwise.

The first term, as in the previous case, enforces closeness to the data and the second one contains the interaction between the field and the line processes: if the horizontal or vertical gradient is very high at site (i, j) the corresponding line process will be very likely to be active ($h_{i,j} = 1$ or $v_{i,j} = 1$), to make energy decrease and signal a discontinuity. The third term takes into account the price we pay each time we create a discontinuity and is necessary to prevent the creation of discontinuities everywhere.

The maximum of the posterior distribution (MAP) or other related estimates of the "true" data-field value can not be computed analytically, but sample distributions of the field with the probability distribution of (2.2) can be obtained using Monte Carlo techniques such as the Metropolis algorithm [15]. These algorithms sample the space of possible values of the surface-field according to the probability distribution $P(f|g)$.

3 A deterministic approximation of MRFs

3.1 Mean field theory and Weak Membrane

We assume that there is uncertainty in the model and that (2.4) should be understood within the context of (2.2). We then propose to estimate the mean field values from the statistical model. The mean field value of f, $\bar{f}$, is given by

$$\bar{f}_{lk} = \frac{1}{Z} \sum_{f,h,v} f_{lk} e^{-\beta(\sum_{i,j}[\lambda_{ij}(f_{ij}-g_{ij})^2 + E_{fl} + E_l])}$$

where $\sum_{f,h,v}$ represents the sum over all possible configurations of the fields f, h and v. A configuration of the field f is a possible set of values assumed by f in the lattice. Z is given by

$$Z = \sum_{f,h,v} e^{-\beta E_1(f,h,v)}$$

From these definitions the following equality is derived:

$$-2\beta\lambda_{lk}(\bar{f}_{lk} - g_{lk}) = \frac{1}{Z}\frac{\partial Z}{\partial g_{lk}} \tag{3.1}$$

We still have to compute the partition function Z. In the case of (2.4) Z becomes

$$Z = \sum_{\{f\}} e^{-\beta \sum_{i,j}[\lambda_{ij}(f_{i,j}-g_{i,j})^2 + \gamma]} \sum_{\{h,v\}} e^{-\beta \sum_{i,j}[(h_{i,j}-1)G^h_{i,j} + (v_{i,j}-1)G^v_{i,j}]} \tag{3.2}$$

where $G^h_{i,j} = \gamma - \alpha\Delta^h_{i,j}{}^2$, $G^v_{i,j} = \gamma - \alpha\Delta^v_{i,j}{}^2$, $\Delta^h_{i,j} = f_{i,j} - f_{i-1,j}$ and $\Delta^v_{i,j} = f_{i,j} - f_{i,j-1}$.

3.1.1 Averaging Out the Line Process

The contribution of the line process to the partition function can be exactly computed. Indeed the line process term in (3.2) is the partition function of two spin systems (h and v) in an external field (G^h and G^v) with no interaction between neighboring sites. Then each spin contributes to the partition function independently from the others and its contribution is $(1 + e^{\beta G^h_{i,j}})$ for the horizontal field and a similar factor for the vertical one. The partition function can then be rewritten as

$$Z = \sum_{\{f\}} e^{-\beta \sum_{i,j}[\lambda_{ij}(f_{i,j}-g_{i,j})^2 + \gamma]} \prod_{ij}(1 + e^{\beta G^h_{i,j}})(1 + e^{\beta G^v_{i,j}}) \tag{3.3} .$$

3.1.2 The Effective Potential

We discuss how the interaction of the field f with itself has changed after the line process has been eliminated from the partition function. From (3.3) we notice that the partition function can be rewritten as

$$Z = \sum_{\{f\}} e^{-\beta(E_{fg}(f) + E_{eff}(f))}$$

where

$$E_{eff}(f) = \sum_{i,j} \gamma - \frac{1}{\beta}ln[(1 + e^{\beta G^h_{i,j}})(1 + e^{\beta G^v_{i,j}})]$$

and $E_{fg}(f)$ is given by (2.4a). This is the partition function of a system composed of one continuous valued field, whose energy is $E_{fg} + E_{eff}$. We interpret this result as the **effect** of the interaction of the line processes with the field f. This effect can be simulated by modifying

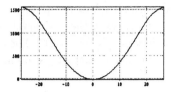 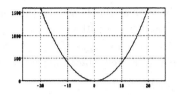

Figure 1: *A slice of the effective potential as a function of Δ_{ij}^h. a) For $\beta = 0.002$. b) Zero temperature limit ($\beta \to \infty$).*

appropriately the interaction of the field with itself, substituting the smoothing term in the energy function with a new temperature dependent potential.

In figure 1 the effective potential is depicted for different temperatures. It simulates the effect of the line processes on the field f. Notice that the energy function is still the sum of local interactions between first neighbors. For the zero temperature limit one can see in figure 1 that the smoothing term is active only when the gradient is smaller than a threshold, proportional to the ratio between γ and α.

3.1.3 The saddle point approximation

The saddle point approximation substitutes the partition function in (3.3) by its largest term, so

$$Z \approx max_f e^{-\beta \sum_{ij}[\lambda_{ij}(f_{ij}-g_{ij})^2 + E_{eff}(f)]} = e^{-\beta \sum_{ij}[\lambda_{ij}(<f_{ij}>-g_{ij})^2 + E_{ij}^{eff}(<f>)]} \tag{3.4}$$

where $< f >$ minimizes the free energy

$E_{fg} + E_{eff}(f) = \sum_{ij} \lambda_{ij}(f_{ij} - g_{ij})^2 + \gamma - \frac{1}{\beta}ln[(1 + e^{\beta G_{i,j}^h})(1 + e^{\beta G_{i,j}^v})]$. Under this approximation $< f >$ is the mean field of f.

3.1.4 MF equations for h, v, and f

Once we obtained the partition function in (3.4) we can now compute the mean field values of h, v, and f.

• *Solving for the line process, h and v*
In analogy to (3.1) we can derive the mean field equation for h to be

$$\bar{h}_{lk} = 1 - \frac{1}{\beta}\frac{\partial lnZ}{\partial G_{lk}^h}$$

and similarly we find the mean field equation for v. So after some algebra we derive

$$\bar{h}_{i,j} = \frac{1}{1 + e^{\beta(\gamma-\alpha(\bar{f}_{i,j}-\bar{f}_{i-1,j})^2)}} \quad \text{and} \quad \bar{v}_{i,j} = \frac{1}{1 + E^{\beta(\gamma-\alpha(\bar{f}_{i,j}-\bar{f}_{i,j-1})^2)}} \tag{3.5}.$$

where $\bar{f} =< f >$ is the mean field solution (under the saddle point approximation) that we calculate next. It is interesting to notice that the mean value of the line process can vary continuously from 0 to 1. Moreover the value of l that is defined everywhere in the lattice, stresses the strength of the edges. This can be used to decide the existence or not of an edge and to analyze its shape. In the zero temperature limit ($\beta \rightarrow \infty$) (3.5) becomes the Heaviside function (1 or 0) and the interpretation is simple: when the horizontal or vertical gradient ($\bar{f}_{i,j} - \bar{f}_{i,j-1}$ or $\bar{f}_{i,j} - \bar{f}_{i-1,j}$) are larger than a threshold ($\sqrt{\frac{\gamma}{\alpha}}$) a vertical or horizontal discontinuity is created, since the price to smooth the function at that site is too high.

- *Solving for the field f*

The mean field solutions for f are obtained by minimizing the free energy. The set of deterministic equations can be written as

$$\frac{\partial}{\partial f_{i,j}}(E_{fg} + E_{eff}(f)) = 0$$

and after some computation

$$\lambda_{ij}\bar{f}_{i,j} = \lambda_{ij}g_{i,j} - \alpha(\bar{f}_{i,j} - \bar{f}_{i,j-1})(1 - \bar{v}_{i,j}) + \alpha(\bar{f}_{i,j+1} - \bar{f}_{i,j})(1 - \bar{v}_{i,j+1})$$
$$-\alpha(\bar{f}_{i,j} - \bar{f}_{i-1,j})(1 - \bar{h}_{i,j}) + \alpha(\bar{f}_{i+1,j} - \bar{f}_{i,j})(1 - \bar{h}_{i+1,j}) \qquad (3.6)$$

where $\bar{h}_{i,j}$ and $\bar{v}_{i,j}$ are given by (3.5).

Equation (3.6) gives the field at site i,j as the sum of data at the same site, plus an average of the field at its neighbor sites. This average takes in account the difference between the neighbors. The larger is the difference, the smaller is the contribution to the average. This is captured by the term $(1 - l_{i,j})$, where $l_{i,j}$ is the line process. At the zero temperature limit ($\beta \rightarrow \infty$) the line process becomes 1 or 0 and then only terms smaller than a threshold must be taken in account for the average. This interpretation helps us in understanding the role of the α and γ parameters, as it will be discussed in section 4. Notice that the form of (3.6) is suitable for the application of a fast, parallel and iterative scheme of solution. In order to solve (3.6) we can introduce a damping like force ($\frac{\partial f}{\partial t}$) to the effective potential such that the fixed point of the dynamic equation

$$\frac{\partial f}{\partial t} = -\frac{\partial}{\partial f_{i,j}}(E_{fg} + E_{eff}(f)) \qquad (3.7)$$

is the mean field solution. This is a gradient descent algorithm.

3.1.5 The effective potential and the graduated non convexity algorithm

We have to point out that this energy function has been studied by Blake and Zisserman [2], in the context of edge detection and surface interpolation. They do not derive the results from the MRFs formulation but they simply minimize the Weak Membrane energy function. From a statistical mechanics point of view the mean-field solution does not minimize the energy function, but this becomes true in the zero temperature limit, so their approach must be recovered from the MRFs formulation in this limit. This is indeed the case, and it is easy to show that the effective potential becomes the Blake and Zisserman potential when β goes to infinity. In order to obtain the minimum of the energy function E_1 Blake and Zisserman introduce the GNC (graduated non convexity) algorithm that can be embedded in the MRFs framework in a natural way. Let us review briefly the GNC algorithm. The main problem with the Weak Membrane Energy is that is not a convex function and a gradient descent method can not be applied to obtain the minimum because one could be trapped in a local minimum. In order to solve this problem Blake and Zisserman introduce a family of energy functions $E^{(p)}$, depending continuously on a parameter p, $p\epsilon[0,1]$, such that $E^{(1)}$ is convex, $E^{(0)} \equiv E_1$ and

$E^{(p)}$ are non convex for $p\epsilon[0,1)$. Gradient descent, as in (3.7), is successively applied to the energy function $E^{(p)}$ for a prescribed decreasing sequence of values of p starting from $p = 1$, and this procedure is proved to converge for a class of given data. The construction of the family of energy functions $E^{(p)}$ is *ad hoc* and uses piecewise polynomials. In our framework, a family of energy functions with such properties is naturally given by $E_{eff}^{(T)}$ where T is the temperature of the system. The GNC algorithm can then be interpreted as the tracking of the minimum of the energy function as the temperature is lowered to zero (like a deterministic annealing). In this way the approach of Blake and Zisserman can be viewed as a deterministic solution of the MRFs problem.

4 Parameters

The parameters α, γ, and β must be estimated in order to develop an algorithm that smoothes and finds the discontinuities of the given data-field.

4.1 The parameter α

The parameter α controls the balance between the "trust" in the data and the smoothing term. The noisier are the data the less you want to "trust" it so α is larger, the less noisy are the data the more you "trust" it so α should be smaller. To estimate α various mathematical methods are available. The generalized cross validation method introduced by Wahba [17] and the standard regularization method described by Tikhonov [16, 1] give good pratical results. For a more detailed analysis see, for example, Geiger and Poggio [7].

4.2 The parameter γ

From (3.5) one can see that $\sqrt{\frac{\gamma}{\alpha}}$ is the threshold for creating a line in the Weak Membrane energy. For the stereo module, where the data field is a depth-field, $\sqrt{\frac{\gamma}{\alpha}}$ is the threshold for the changes in depth to be called a depth discontinuity. This value is determined according to the resolution of the stereo system available. For the intensity data, the parameter $\sqrt{\frac{\gamma}{\alpha}}$ represents the threshold for detecting edges. This value is somehow arbitrary, and probably context dependent. The exact value depends on the attention of the observer and/or the sensitivity of the system.

4.3 The parameter β

The parameter β controls the uncertainty of the model. The smaller is β the more inaccurate is the model. This suggests that for solving the mean field equations a rough solution can be obtained for a small value of β (high uncertainty) and thereafter we can increase β (small uncertainty) to obtain more accurate solutions. This can be called deterministic annealing.

5 Results

For the implementation the zero temperature limit equations have provided results as good as the deterministic annealing with a faster computational time. We do not have proof of

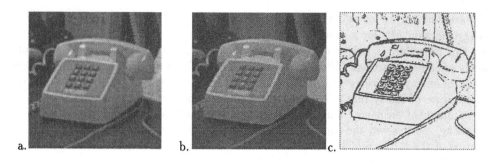

Figure 2: *a. 8-bit image of 256 X 256 pixels. b. The image smoothed with $\alpha = 1$, $\gamma = 19$, 10 iterations. c. The line process field for $\alpha = 0.1$, $\gamma = 19$, and 10 iterations.*

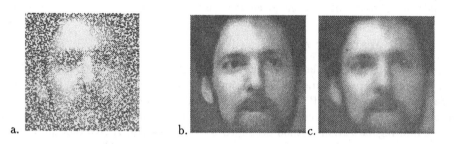

Figure 3: *a. Randomly chosen 50 % of an original image. For display the other 50% are filled with white dots. b. The algorithm described in (3.6) is applied to smooth and fill in at the same time with $\gamma = 208$ and $\alpha = 4$ for 10 iterations. c. For comparison and to stress the importance of the line process field we show the result of simply bluring the sparse data, when $h = v = 0$ everywhere*

convergence but only suggestive experimental results. In order to find the mean field solution we solved (3.6) using (3.7) together with (3.5) in a coupled and iterative way.

The algorithm is parallel depending upon the first neighbors. For a serial implementation we first update the even sites (like the white squares in a chess board) and then the odd sites (like the black squares). Typically the algorithm has converged in 10 iterations which takes about 1 minute for images of 64×64 pixels on a Symbolics 3600.

When we apply the algorithm to a real still life image the result is an enhancement of specular edges, shadow edges and some other contours while smoothing out the noise (see Figure 2).

From one face image we produced sparse data by randomly suppressing 50 % of the data (see Figure 3). We then applied the Weak Membrane model to sparse data. The parameters were kept the same as the other real image. We also compared the results with simply bluring the data (no line process).

The reconstruction from sparse data can be applied to depth data in which case it is usually called surface reconstruction.

6 Alignment of visual modules with intensity edges

The integration of different visual modules to improve the detection of the discontinuities can also be addressed in this scheme. As suggested by Gamble & Poggio [5] , we can add the term $\delta\,(v_{ij} + h_{ij})(1 - e_{ij})$ to the Weak Membrane model. Here e_{ij} is an external field, for example the edge map that is coupled with the stereo field. For implementation purposes the only consequence of adding this term is the change of the global parameter γ into the local parameter $\gamma'_{ij} = \gamma - \delta(1 - e_{ij})$.

7 Conclusion

- We have used statistical mechanics tools to derive deterministic approximations of Markov random fields models. In particular we have studied an energy model that is suitable for image reconstruction or any field reconstruction.

- We derived a deterministic solution for the mean values of the surface and discontinuity fields, consisting of a system of coupled nonlinear equations. The "key" step was the elimination (averaging out) of the binary fields (h and v) from the "energy" function. An algorithm has been implemented to obtain a solution for this system: it is fully parallelizable, iterative and recursive, allowing efficient computation.

- An understanding of the role of the parameters is possible,

- We have shown that the deterministic algorithm of GNC can be regarded as an approximation of the gradient descent method with a deterministic annealing schedule to solve the mean field equations.

- We extended the model to deal with sparse data and alignment of the discontinuities of different modules with the intensity edges.

This work suggests a unified framework to connect different methods used on image segmentation, restoration and surface reconstruction. We show in another paper[8] that several deterministic algorithms for image segmentation and reconstruction are approximations of two methods to solve the mean field equations: the gradient descent method discussed in this paper and the parameter space method discussed in [8]. In another paper [6] we analyze possible extensions to this model to include propagation of lines.

Acknowledgments We are grateful to Tomaso Poggio for his guidance and support.

References

[1] M. Bertero, T. Poggio, and V. Torre. Ill-posed problems in early vision. Technical report. Also Proc. IEEE, in press.

[2] A. Blake and A. Zisserman. *Visual Reconstruction*. MIT Press, Cambridge, Mass, 1987.

[3] P. B. Chou and C. M. Brown. Multimodal reconstruction and segmentation with Markov random fields and HCF optimization. In *Proceedings Image Understanding Workshop*, pages 214–221, Cambridge, MA, February 1988. Morgan Kaufmann, San Mateo, CA.

[4] E. Gamble, D. Geiger, T. Poggio, and D. Weinshall. Integration of vision modules and labeling of surface discontinuities. *Invited paper to IEEE Trans. Sustems, Man & Cybernetics*, December 1989.

[5] E. B. Gamble and T. Poggio. Visual integration and detection of discontinuities: The key role of intensity edges. A.I. Memo No. 970, Artificial Intelligence Laboratory, Massachusetts Institute of Technology, October 1987.

[6] D. Geiger and F. Girosi. Parallel and deterministic algorithms for mrfs: surface reconstruction and integration. A.I. Memo No. 1114, Artificial Intelligence Laboratory, Massachusetts Institute of Technology, May 1989.

[7] D. Geiger and T. Poggio. An optimal scale for edge detection. In *Proceedings IJCAI*, August 1987.

[8] D. Geiger and A. Yuille. A common framework for image segmentation and surface reconstruction. Harvard Robotics Laboratory Technical Report 89-7, Harvard, August 1989.

[9] Davi Geiger. *Visual models with statistical field theory*. PhD thesis, Massachusetts Institute of Technology, 1989.

[10] S. Geman and D. Geman. Stochastic relaxation, Gibbs distributions, and the Bayesian restoration of images. *IEEE Transactions on Pattern Analysis and Machine Intelligence*, PAMI-6:721–741, 1984.

[11] S. Kirkpatrick, C.D. Gelatt, and M.P. Vecchi. Optimization by simulated annealing. *Science*, 220:219–227, 1983.

[12] C. Koch, J. Marroquin, and A. Yuille. Analog 'neuronal' networks in early vision. *Proc. Natl. Acad. Sci.*, 83:4263–4267, 1985.

[13] J. L. Marroquin. Deterministic Bayesian estimation of Markovian random fields with applications to computational vision. In *Proceedings of the International Conference on Computer Vision*, London, England, June 1987. IEEE, Washington, DC.

[14] J. L. Marroquin, S. Mitter, and T. Poggio. Probabilistic solution of ill-posed problems in computational vision. In L. Baumann, editor, *Proceedings Image Understanding Workshop*, pages 293–309, McLean, VA, August 1985. Scientific Applications International Corporation.

[15] N. Metropolis, A. Rosenbluth, M. Rosenbluth, A. Teller, and E. Teller. Equation of state calculations by fast computing machines. *J. Phys. Chem*, 21:1087, 1953.

[16] A. N. Tikhonov and V. Y. Arsenin. *Solutions of Ill-posed Problems*. W.H.Winston, Washington, D.C., 1977.

[17] G. Wahba. Practical approximate solutions to linear operator equations when the data are noisy. *SIAM J. Numer. Anal.*, 14, 1977.

Parallel Multiscale Stereo Matching Using Adaptive Smoothing [1]

Jer-Sen Chen [2] *and Gérard Medioni*
Institute for Robotics and Intelligent Systems
University of Southern California
Los Angeles, California 90089-0273
Email: medioni@usc.edu

We present a multiscale stereo algorithm whose design makes it easily implemented on a fine grain parallel architecture. The matching at a given scale is performed in accordance with the algorithm proposed by Drumheller and Poggio, using edges as primitives. Using multiple scales is in general difficult and costly because of the correspondence problem between scales, but this problem is solved here using adaptive smoothing, a process which preserves edge location at different scales. The results are better than those obtained at a single scale, and the algorithm runs in a few seconds on a Connection Machine.

Introduction

Range or depth information has long been considered essential in the analysis of shape. High-resolution range information can be obtained directly so long as the range sensor is available. Binocular stereo has been widely used to extract the range information when such a high-resolution sensor is absent. Barnard and Fischler [Barnard82] define six steps necessary to stereo analysis: image acquisition, camera modeling, feature acquisition, image matching, depth reconstruction and interpolation. Multiple scale processing not only provides a description of the signal but only facilitate a coarse-to-fine hierarchical processing for various vision tasks. The correspondence problem in stereo matching is usually very tedious and can be alleviated through a multiple scale approach.

The correspondence problem between two images can be solved by matching specific features such as edges, or by matching small regions by the correlation of the image intensities. Edgel-based stereo matching techniques usually use the edges characterized by the derivatives of a smoothed version of the signal, for instance the zero-crossings of a Laplacian-of-Gaussian convolved image. The correlation-based stereo matching measures the correlation of the image intensity patches centered around the matched pixels. Our multiscale stereo matching is edgel-based and the matching primitives are edgels extracted with adaptive smoothing. Since adaptive smoothing provides accurate edge detection across different scales, it facilitates a straightforward multiscale stereo matching.

To identify corresponding locations between two stereo images, or among a sequence of motion images, is difficult because of the *false targets problem*. Certain constraints and assumptions have to be made in order to establish the correct pairings. The *Uniqueness constraint* [Marr82] states that there is at most one match from each line-of-sight since the depth value associated with each matching primitive, left or right, must be unique. The *Continuity constraint* [Marr82] states that the depth map of an image should be mostly continuous, except at those points where depth discontinuities occur. Therefore neighboring potential matches having similar disparity values should support each other. The *Opacity constraint* further limits the occurrence of the false targets. Extending the uniqueness constraint, which limits to only one match along each line-of-sight, the opacity constraint states that there is at most one match in the hourglass-shape forbidden zone bounded by two lines-of-sight. The *Compatibility constraint* [Marr82] limits the construction of potential matches from matching primitives, for example, the potential matches are allowed to occur only when two zero-crossings from the LoG convolved images have the same sign. Further restrictions can be made on the orientation and the gradient of the matching edgels.

The false targets problem can be alleviated either by reducing the range and resolution of the disparity or by reducing the density of the matching features in the image. One commonly used

---

[1] This research was supported by the Defense Advanced Research Projects Agency under contract and F33615-87-C-1436, and monitored by Wright-Patterson Air Force Base.

[2] current address: Wright State University, Department of Computer Science and Engineering, Dayton, Ohio 45435 (email: jschen@odin.wright.edu)

method to obtain both resolution and range of disparity information is to apply a multi-resolution algorithm. The information obtained from the matching at coarse resolution can be used to guide the matching at fine resolution.

Marr and Grimson [Marr82,Grimson81] use Gaussian kernels as the scaling filters for multiscale stereo matching. The matches obtained at coarser scale establish a rough correspondence for the finer scales, thereby reducing the number of false matches, but a vergence control is necessary because of the poor accuracy of the LoG edge detection.

This is overcome here by using the adaptive smoothing, which preserves edge locations across scales. The next section briefly reviews the major properties of adaptive smoothing, and section 3 presents the stereo algorithm and results on two sets of images.

Adaptive Smoothing

We have recently introduced a formalism called adaptive smoothing [Saintmarc89]. It keeps all the desirable properties of Gaussian smoothing [Perona87,Chen89] and preserves the location of edges across scales, making the correspondence problem trivial. Adaptive smoothing, in its basic formulation, assumes that the signal is piecewise constant inside a region and uses an ideal step edge model. It not only preserves edges but also enhances them. With a suitable choice of the scale parameter, an accurate edge detection scheme at different scales can be achieved by adaptive smoothing and therefore facilitate multiple scale signal processing. A multiscale representation of the signal can be easily derived by choosing the necessary number of scales without dealing with the tedious correspondence problem as encountered in the traditional Gaussian scale space [Witkin83, Asada86]. Edges corresponding to adaptive smoothing are presented in the results section.

On the Connection Machine with 16K physical processors, adaptive smoothing takes about 11 milliseconds per iteration in the case of one pixel per physical processor, compared to 10 seconds per iteration on a serial machine (Symbolics 3645) for a 128×128 image.

Multiscale Stereo

As mentioned in the introduction, multiscale processing is often used in a coarse to fine strategy to solve the matching problem, but one must solve the correspondence problem between scales. Adaptive smoothing can overcome this disadvantage with its accuracy of edges over scales. The matching results at coarser scale with adaptive smoothing are therefore much more reliable and the propagation of the disparity information between scales is straightforward.

We have used our adaptive smoothing and implemented a multiscale stereo matching algorithm to extract the matching features. It is based on Drumheller and Poggio's [Drumheller86] parallel stereo matching implementation on the Connection Machine [Hillis85]. The parallel stereo matching, as in most stereo matching algorithms, utilizes the *uniqueness* the *continuity* constraint on the surface and therefore the values on the disparity map. It also imposes the *opacity* constraint on the surfaces and the *compatibility* constraint on the matching of the edges.

We use three scales, namely coarse, intermediate and fine, in our multiscale stereo matching. We first extract edges at coarse scale for both images using adaptive smoothing. The stereo images are assumed to be epipolarly registered and the matching is performed scan-line by scan-line. A potential match is marked only when the corresponding edges from the two images have approximately the same orientation and gradient. Imposing the continuity constraint, the number of potential matches is counted over a flat uniformly-weighted square support (chosen for computational convenience) centered at each pixel. Enforcing the opacity and uniqueness constraints, there must be no more than one match in the forbidden zone, therefore a *winner-take-all* strategy is applied in the forbidden zone.

The matches at the edge locations from the coarse scale are then propagated to the intermediate scale. Each match at coarse scale generates a forbidden zone which forbids potential matches to be marked at intermediate scale. This greatly reduces the number of potential matches at intermediate scale and therefore facilitates producing more reliable matching results. After potential matches are formed at intermediate scale, the same continuity, opacity and uniqueness constraints are employed

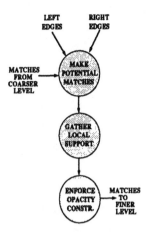

Figure 1: Flowchart of Multiscale Stereo Matching

to produce the matches which are then propagated to the fine scale. A flow chart of the process is shown in figure 1.

The multiscale stereo matching algorithm is implemented on the Connection Machine. The accurate edge detection by adaptive smoothing provides a very simple control mechanism for multiscale processing, and the simplicity is essential when considering parallel implementation.

We first show in figure 2 stereo matching of the aerial view of Pentagon ($256 \times 256 \times 8bits$, courtesy of Dr. W. Hoff). The pair is epipolarly registered, i.e. the two images correspond to each other scan-line by scan-line. Subfigures (a) and (b) show the left and right view of the original stereo pair. Part (c) shows the matching result using single (fine) scale. Subfigures (d), (e) and (f) show the left edgels, right edgels and the matching result at the coarse scale, while (g), (h) and (i) are for intermediate scale and (j), (k) and (l) are for fine scale. The scaling parameter k is set to 8, 4 and 0 (no smoothing) for the coarse, intermediate and fine scales respectively. The range of disparity for this example is -5 (farther) to 5 (nearer) pixels, and the brighter gray level values indicate a position closer to the viewer.

We show in figure 3 another example of a fruit scene ($256 \times 256 \times 8bits$, courtesy of Dr. T. Kanade, Carnegie-Mellon University). The scaling parameter k is set to 12, 6 and 0 for coarse, intermediate and fine scales respectively. The range of disparity is -15 to 22 pixels. Figures 3(a) and (b) show the stereo image pair. The results of single scale and multiscale matching are shown in figure 3(c) and (d) respectively.

Table 1 summarizes the statistics of the number of potential matches for the stereo matching. The column "multiple" stands for multiscale stereo matching and column "single" stands for stereo matching at each scale individually. Since the matches at coarser scale are used to form the forbidden zones when constructing potential matches at finer scale, the number of potential matches is greatly reduced at the finer scales as we can observe from the table.

Conclusion

We have shown a multiscale coarse-to-fine hierarchical matching of stereo pairs which uses adaptive smoothing to extract the matching primitives. The number of matching primitives at coarse scale is small, therefore reducing the number of potential matches, which in return increases the reliability of the matching results. A dense disparity can be obtained at a fine scale where the density of edgels is very high. The control strategy is very simple compared to other multiscale approaches such as the ones using Gaussian scale space, this results from the accuracy of edges detected by

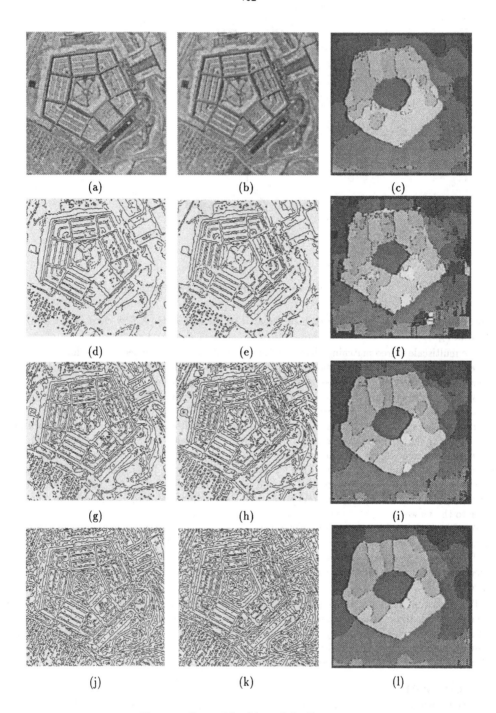

Figure 2: Stereo Matching of the Pentagon

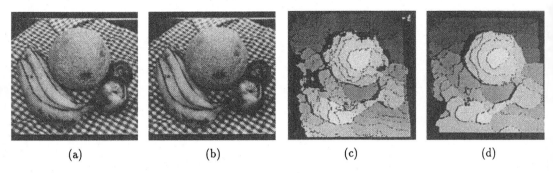

| | (a) | (b) | (c) | (d) |

Figure 3: Stereo Matching of the Fruit Scene

	Pentagon			Fruit		
	coarse	Interm.	fine	coarse	Interm.	fine
singlescale	22670	28892	33323	36885	80045	120026
multiscale	22670	11372	11279	36885	37588	35676

Table 1: Number of Potential Matches

adaptive smoothing at different scales. The simplicity of the control strategy is especially important for low-level processing, and makes parallel implementation quite simple.

References

[Asada86] H. Asada and M. Brady. "The Curvature Primal Sketch", *IEEE Tran. on Pattern Analysis and Machine Intelligence*, Vol. 8, No. 1, January 1986, pp 2-14.

[Barnard82] Barnard, S. and Fischler, M., "Computational Stereo", *ACM Computing Surveys*, Vol. 14, No. 4, December 1982, pp 553-572.

[Burt83] Burt, P.J.,"Fast Algorithm for Estimating Local Image Properties," *Journal of Computer Vision, Graphics, Image Processing*, Vol. 21, March 1983, pp. 368-382.

[Chen89] Chen, J.S., "Accurate Edge Detection for Multiple Scale Processing", Ph.D. thesis, University of Southern California, October 1989.

[Drumheller86] Drumheller, M., Poggio, T., "On Parallel Stereo", *Proceedings of the 1986 IEEE International Conference on Robotics and Automation*, April 1986, pp 1439-1488.

[Grimson81] Grimson, W.E.L., *From Images to Surfaces*, MIT Press, Cambridge, USA, 1981.

[Hillis85] D. Hillis. *The Connection Machine*. MIT Press, Cambridge, MA, 1985.

[Marr82] Marr, D., *Vision: A Computational Investigation into the Human Representation and Processing of Visual Information*. W. H. Freeman and Co., San Francisco, 1982.

[Perona87] Perona, P., Malik, J., "Scale Space and Edge Detection using Anisotropic Diffusion", *Proceedings of the IEEE Workshop on Computer Vision*, Miami Beach, Fl. 1987, pp 16-22.

[Saintmarc89] Saint-Marc, P., Chen J.S. and Medioni, G., "Adaptive Smoothing: A General Tool for Early Vision", *Proceedings of IEEE Conference on Computer Vision and Pattern Recognition*, San Diego, June 1989.

[Witkin83] Witkin, A.P., "Scale Space Filtering", *Proceedings of International Joint Conference on Artificial Intelligence*, pp 1019-1022, Karlsruhe, 1983.

OPTICAL FLOW

On the motion of 3D curves and its relationship to optical flow *

Olivier Faugeras

INRIA Sophia Antipolis

2004 Route des Lucioles

06565 Valbonne Cedex

FRANCE

Abstract

I establish fundamental equations that relate the three dimensional motion of a curve to its observed image motion. I introduce the notion of spatio-temporal surface and study its differential properties up to the second order. In order to do this, I only make the assumption that the 3D motion of the curve preserves arc-length, a more general assumption than that of rigid motion. I show that, contrarily to what is commonly believed, the full optical flow of the curve can never be recovered from this surface. I nonetheless then show that the hypothesis of a rigid 3D motion allows in general to recover the structure and the motion of the curve, in fact without explicitly computing the tangential optical flow.

1   Introduction

This article is a condensed version of a longer version which will appear elsewhere [7]. In particular, I have omitted all the proofs of the theorems. It presents a mathematical formalism for dealing with the motion of curved objects, specifically curves. In our previous work on stereo [1] and motion [9], we have limited ourselves to primitives such as points and lines. I attempt here to lay the ground for extending this work to general curvilinear features. More specifically, I study the image motion of 3D curves moving in a "non-elastic" way (to be defined later), such as ropes. I show that under this weak assumption the full apparent optical flow (to be defined later) can be recovered. I also show that recovering the full real optical flow (i.e the projection of the 3D velocity field) is impossible. If rigid motion is hypothesized, then I show that, in general, the full 3D structure and motion of the curve can be recovered without explicitly computing the full flow real. I assume that pixels along curves have been extracted by some standard edge detection techniques [3,4].

This is related and inspired by the work of Koenderink [14], the work of Horn and Schunk [13] as well as that of Longuet-Higgins and Prazdny [15] who pioneered the analysis

---

*This work was partially completed under Esprit P2502 and BRA Insight

of motion in computer vision, that of Nagel [16] who showed first that at grey level corners the full optical flow could be recovered, as well as by the work of Hildreth [12] who proposed a scheme for recovering the full flow along image intensity edges from the normal flow by using a smoothness constraint. This is also related to the work of D'Hayer [5] who studied a differential equation satisfied by the optical flow but who did not relate it to the actual 3D motion and to that of Gong and Brady [11] who recently extended Nagel's result and showed that it also held along intensity gradient edges. All assume, though, that the standard motion constrain equation:

$$\frac{dI}{d\tau} = \nabla I \cdot \mathbf{v} + I_\tau = 0 \tag{1}$$

is true, where I is the image intensity, $\mathbf{v}$ the optical flow (a mysterious quantity which is in fact defined by this equation), and τ the time. It is known that this equation is, in general, far from being true [19]. In my approach, I do not make this assumption and, instead, keep explicit the relation between image and 3D velocities.

In fact there is a big confusion in the Computer Vision literature about the exact meaning of the optical flow. I define it precisely in this paper and show that two flows, the "apparent" and the "real" one must be distinguished. I show that only the apparent one can be recovered from the image for a large class of 3D motions.

My work is also related to that of Baker and Bolles [2] in the sense that I also work with spatio-temporal surfaces for which I provide a beginning of quantitative description through Differential Geometry in the case where they are generated by curves.

It is also motivated by the work of Girosi, Torre and Verri [18] who have investigated various ways of replacing equation (1) by several equations to remove the inherent ambiguity in the determination of the optical flow $\mathbf{v}$.

2 Definitions and notations

I use some elementary notions from Differential Geometry of curves and surfaces. I summarize these notions in the next sections and introduce my notations.

2.1 Camera model

I assume the standard pinhole model for the camera. The retina plane $\mathcal{R}$ is perpendicular to the optical axis Oz, O is the optical center. The focal distance is assumed to be 1. Those hypothesis are quite reasonable and it is always possible, up to a good approximation, to transform a real camera into such an ideal model [17,10].

2.2 Two-dimensional curves

A planar curve (c) (usually in the retina plane) is defined as a C^2 mapping $u \to \mathbf{m}(u)$ from an interval of R into R^2. We will assume that the parameter u is the arclength s of (c). We then have the well known two-dimensional Frenet formulas:

$$\frac{d\mathbf{m}}{ds} = \mathbf{t} \quad \frac{d\mathbf{t}}{ds} = \kappa \mathbf{n} \quad \frac{d\mathbf{n}}{ds} = -\kappa \mathbf{t} \tag{2}$$

where $\mathbf{t}$ and $\mathbf{n}$ are the tangent and normal unit vectors to (c) at the point under consideration, and κ is the curvature of (c), the inverse of the radius of curvature r.

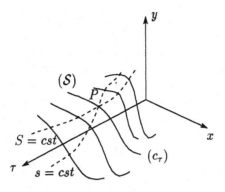

Figure 1: Definition of the spatio-temporal surface $(\mathcal{S})$

2.3 Surface patches

A surface patch $(\mathcal{S})$ is defined as a C^2 mapping $(u, v) \rightarrow \mathbf{P}(u, v)$ from an open set of R^2 into R^3. Such a patch is intrinsically characterized, up to a rigid motion, by two quadratic forms, called the two fundamental forms [6], which are defined at every point of the patch. The first quadratic form Φ_1 defines the length of a vector in the tangent plane T_P. More precisely, the two vectors $\mathbf{P}_u = \frac{\partial \mathbf{P}}{\partial u}$ and $\mathbf{P}_v = \frac{\partial \mathbf{P}}{\partial v}$ are parallel to this plane and define therein a system of coordinates. Each vector in the tangent plane can be defined as a linear combination $\lambda \mathbf{P}_u + \mu \mathbf{P}_v$. Its squared length is given by the value of the first fundamental form Φ_1. Moreover, the normal $\mathbf{N}_P$ to $(\mathcal{S})$ is parallel to the cross-product $\mathbf{P}_u \times \mathbf{P}_v$.

The second fundamental quadratic from Φ_2 is related to curvature. For a vector $\mathbf{x} = \lambda \mathbf{P}_u + \mu \mathbf{P}_v$ in the tangent plane, we can consider all curves drawn on $(\mathcal{S})$ tangent to $\mathbf{x}$ at P. These curves have all the same normal curvature, the ratio $\frac{\Phi_2(\mathbf{x})}{\Phi_1(\mathbf{x})}$.

It is important to study the invariants of Φ_2, ie. quantities which do not depend upon the parametrization (u, v) of $(\mathcal{S})$. Φ_2 defines a linear mapping $T_P \rightarrow T_P$ by $\Phi_2(\mathbf{x}) = \psi(\mathbf{x}) \cdot \mathbf{x}$. The invariants of Φ_2 are those of ψ. Those of interest to us are the principal directions, the principal curvatures from which the mean and gaussian curvatures can be computed.

3 Setting the stage: real and apparent optical flows

We now assume that we observe in a sequence of images a family (c_τ) of curves, where τ denotes the time, which we assume to be the perspective projection in the retina of a 3D curve (C) that moves in space. If we consider the three-dimensional space (x, y, τ), this family of curves sweeps in that space a surface $(\mathcal{S})$ defined as the set of points $((c_\tau), \tau)$ (see figure 1).

At a given time instant τ, let us consider the observed curve (c_τ). Its arclength s can be computed and (c_τ) can be parameterized by s and τ: it is the set of points $m_\tau(s)$. The corresponding points P on $(\mathcal{S})$ are represented by the vector $\mathbf{P} = (\mathbf{m}_\tau^T(s), \tau)^T$. The key observation is that the arclength s of (c_τ) is a function $s(S, \tau)$ of the arclength S

of the 3D curve (C) and the time τ, and that the two parameters (S, τ) can be used to parameterize $(\mathcal{S})$ in a neighborhood of P. Of course, the function $s(S, \tau)$ is unknown.

The assumption that s is a function of S and τ implies that S itself is not a function of time; in other words we do not consider here elastic motions but only motions for which S is preserved, i.e non-elastic motions such as the motion of a rope or the motion of a curve attached to a moving rigid object. We could call such motions *isometric* motions.

As shown in figure 1, we can consider on $(\mathcal{S})$ the curves defined by $s = cst$ or $S = cst$. These curves are in general different, and their projections, parallel to the τ-axis, in the (x, y)- or retina plane have an important physical interpretation, related to our upcoming definition of the optical flow.

Indeed, suppose we choose a point M_0 on (C) and fix its arclength S_0. When (C) moves, this point follows a trajectory (C_{M_0}) in 3-space and its image m_0 follows a trajectory $(c_{m_0}^r)$ in the retina plane. This last curve is the projection in the retina plane, parallel to the τ-axis, of the curve defined by $S = S_0$ on the surface $(\mathcal{S})$. We call it the "real" trajectory of m_0.

We can also consider the same projection of another curve defined on $(\mathcal{S})$ by $s = s_0$. The corresponding curve $(c_{m_0}^a)$ in the retina plane is the trajectory of the image point m_0 of arclength s_0 on (c_τ). We call this curve the "apparent" trajectory of m_0.

The mathematical reason why those two curves are different is that the first one is defined by $S = S_0$ while the second is defined by $s(S, \tau) = s_0$.

Let me now define precisely what I mean by optical flow. If we consider figure 2, point m on (c_τ) is the image of point M on (C). This point has a $3D$ velocity $\mathbf{V}_M$ whose projection in the retina is the *real optical flow* $\mathbf{v}_r$ (r for *real*); mathematically speaking:

- $\mathbf{v}_r$ is the partial derivative of $\mathbf{m}$ with respect to time when S is kept constant, or its total time derivative.

- The *apparent optical flow* $\mathbf{v}_a$ (a for *apparent*) of m is the partial derivative with respect to time when s is kept constant.

Those two quantities are in general distinct. To relate this to the previous discussion about the curves $S = S_0$ and $s = s_0$ of $(\mathcal{S})$, the vector $\mathbf{v}_a$ is tangent to the "apparent" trajectory of m, while $\mathbf{v}_r$ is tangent to the "real" one.

I now make the following fundamental remark. All the information about the motion of points of (c_τ) (and of the 3D points of (C) which project onto them) is entirely contained in the surface $(\mathcal{S})$. Since $(\mathcal{S})$ is intrinsically characterized, up to a rigid motion, by its first and second fundamental forms [6], they are all we need to characterize the optical flow of (c_τ) and the motion of (C).

4 Characterization of the spatio-temporal surface $(\mathcal{S})$

In this section, we compute the first and second fundamental forms of the spatio-temporal surface $(\mathcal{S})$. We will be using over and over again the following result.

Given a function f of the variables s and τ, it is also a function f' of S and τ. We will have to compute $\frac{\partial f'}{\partial S}$ and $\frac{\partial f'}{\partial \tau}$, also called the total time derivative of f with respect

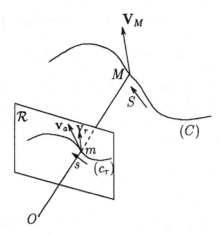

Figure 2: Definition of the two optical flows: the real and the apparent

to time, $\dot{f}$; introducing $u = \frac{\partial s}{\partial S}$ and $v = \frac{\partial s}{\partial \tau}$, we have the following equations:

$$\frac{\partial f'}{\partial S} = u\frac{\partial f}{\partial s} \quad \dot{f} = \frac{\partial f'}{\partial \tau} = v\frac{\partial f}{\partial s} + \frac{\partial f}{\partial \tau} \tag{3}$$

Following these notations, we denote by $\mathbf{P}(s, \tau) = (\mathbf{m}^T(s, \tau), \tau)^T$ the generic point of $(\mathcal{S})$ and by $\mathbf{P}'(S, \tau) = (\mathbf{m}'^T(S, \tau), \tau)^T$ the same point considered as a function of S and τ.

4.1 Computation of the first fundamental form

Using equations (3), we write immediatly:

$$\mathbf{P}'_\tau = v\mathbf{P}_s + \mathbf{P}_\tau = [v\mathbf{t}^T + \mathbf{v}_a^T, 1]^T \tag{4}$$

We now write the apparent optical flow $\mathbf{v}_a$ in the reference frame defined by $\mathbf{t}$ and $\mathbf{n}$:

$$\mathbf{v}_a = \alpha\mathbf{t} + \beta\mathbf{n} \tag{5}$$

We see from equations (4) and (5) that $\mathbf{P}'_\tau = [(v + \alpha)\mathbf{t}^T + \beta\mathbf{n}^T, 1]^T$; but by definition, $\mathbf{P}'_\tau = [\mathbf{v}_r^T, 1]^T$. Therefore $w = v + \alpha$ is the real tangential optical flow and β the normal real optical flow. Therefore, the real and apparent optical flows have the same component along $\mathbf{n}$, we call it the normal optical flow. The real optical flow is given by:

$$\mathbf{v}_r = w\mathbf{t} + \beta\mathbf{n} \tag{6}$$

Without entering the details and referring the reader to [7], we can write simple formulas for the time derivatives of a function f from $(\mathcal{S})$ into R, for example.

$$\frac{\partial f}{\partial \tau} = \alpha\frac{\partial f}{\partial s} + \partial_{\mathbf{n}_\beta} f \tag{7}$$

$$\dot{f} = w\frac{\partial f}{\partial s} + \partial_{\mathbf{n}_\beta} f \tag{8}$$

where $\partial_{\mathbf{n}_\beta} f$ means the partial derivative of f in the direction of $\mathbf{n}_\beta = [\beta\mathbf{n}, 1]^T$ of T_P, the tangent plane to $(\mathcal{S})$ at P.

These relations also hold for functions f from $(\mathcal{S})$ into R^p. We will be using heavily the case $p = 2$ in what follows.

From equations (4), and the one giving $\mathbf{P}'_S$, we can compute the coefficients of the first fundamental form. I skip the details of the computation and state the main result:

Given the normal $\mathbf{N}_P$ to the spatio-temporal surface $(\mathcal{S})$ whose coordinates in the coordinate system $(\mathbf{t}, \mathbf{n}, \tau)$ (τ is the unit vector defining the τ-axis) are denoted by N_1, N_2, N_3, we have:

$$\beta = -\frac{N_3}{N_2} \quad N_1 = 0$$

We have thus the following theorem:

Theorem 1 *The normal to the spatio-temporal surface $(\mathcal{S})$ yields an estimate of the normal optical flow β.*

4.2 Computation of the second fundamental form

Again, we skip the details and state only the results:

Theorem 2 *The tangential apparent optical flow α satisfies:*

$$\frac{\partial \alpha}{\partial s} = \kappa\beta \tag{9}$$

where κ is the curvature of (c_τ).

Equation (9) is instructive. Indeed, it shows that α, the tangential component of the apparent optical flow $\mathbf{v}_a$ is entirely determined up to the addition of a function of time by the normal component of the optical flow β and the space curvature κ of (c_τ):

$$\alpha = \int_{s_0}^{s} \kappa(t, \tau)\beta(t, \tau)dt \tag{10}$$

Changing the origin of arclengths from s_0 to s_1 on (c_τ) is equivalent to adding the function $\int_{s_0}^{s_1} \kappa(t, \tau)\beta(t, \tau)dt$ to α, function which is constant on (c_τ). This is the fundamental result of this section. We have proved the following theorem:

Theorem 3 *The tangential apparent optical flow can be recovered from the normal flow up to the addition of a function of time through equation (10).*

We now state an interesting relationship between κ and β which is proved in [7].

Theorem 4 *The curvature κ of (c_τ) and the normal optical flow β satisfy:*

$$\partial_{\mathbf{n}_\beta} \kappa = \frac{\partial^2 \beta}{\partial s^2} + \kappa^2\beta \tag{11}$$

4.3 What information can be extracted from the second fundamental form

The idea now is that after observing $(\mathcal{S})$, we compute an estimate of Φ_2 from which we attempt to recover the unknowns, for example u or v. We show that it is impossible without making stronger assumptions about the motion of (C).

We have seen that all invariants of Φ_2 are functions of the principal directions and curvatures. We omit the derivation and only state the result:

Theorem 5 *The invariants of the second fundamental form of the surface $(\mathcal{S})$ are not functions of u, v, w, the real tangential optical flow nor of α, the apparent tangential optical flow.*

4.4 Conclusions

There are three main consequences that we can draw from this analysis. Under the weak assumption of *isometric* motion:

1. The normal optical flow β can be recovered from the normal to the spatio-temporal surface,

2. The tangential apparent optical flow can be recovered from the normal optical flow through equation (10), up to the addition of a function of time,

3. The tangential real optical flow cannot be recovered from the spatio-temporal surface.

Therefore, the full real optical flow is not computable from the observation of the image of a moving curve under the isometric assumption. In order to compute it we <u>must</u> add more hypothesis, for example that the 3D motion is rigid. This makes me wonder what the published algorithms for computing the optical flow are actually computing since they are not making any assumptions about what kind of 3D motion is observed.

I show in the next section that if we assume a 3D rigid motion then the problem is, in general, solvable but that there is no need to compute the full real optical flow.

5 Assuming that (C) is moving rigidly

We are now assuming that (C) is moving rigidly; let $(\Omega, \mathbf{V})$ be its kinematic screw at the optical center O of the camera. We first derive a fundamental relation between the tangents $\mathbf{t}$ and $\mathbf{T}$ to (c_τ) and (C) and the angular velocity Ω. In this section, the third coordinate of vectors is a space coordinate (along the z-axis) whereas previously it was a time coordinate (along the τ-axis).

5.1 Stories of tangents

Let us denote by $\mathbf{U_t}$ the vector $\mathbf{Om} \times \begin{bmatrix} \mathbf{t} \\ 0 \end{bmatrix}$. This vector is normal to the plane defined by the optical center of the camera, the point m on (c_τ), and $\mathbf{t}$ (see figure (3)). Since this

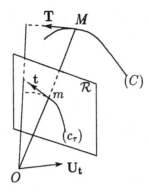

Figure 3: Relation between **t** and **T**

plane contains also the tangent **T** to (C) at M, the 3D point whose image is m, we have:

$$\mathbf{U_t} \cdot \mathbf{T} = 0 \tag{12}$$

But, because (C) moves rigidly, **T** must also satisfy the following differential equation (this equation is satisfied by any constant length vector attached to (C)):

$$\dot{\mathbf{T}} = \boldsymbol{\Omega} \times \mathbf{T} \tag{13}$$

Skipping once again the details, we obtain:

$$\mathbf{W} = \mathbf{U_t} \times (\mathbf{U_t} \times \boldsymbol{\Omega} + \dot{\mathbf{U}}_t) \tag{14}$$

$$\mathbf{T} = \epsilon \frac{\mathbf{W}}{\|\mathbf{W}\|} \tag{15}$$

Where $\epsilon = \pm 1$. We can assume that $\epsilon = 1$ by orienting correctly (c_τ) and (C).

Equations (14) and (15) are important because they relate in a very simple manner the tangent **T** to the unknown 3D curve (C) to the known vector $\mathbf{U_t}$, the angular velocity $\boldsymbol{\Omega}$ and and to $\dot{\mathbf{U}}_t$. Notice that this last vector contains the unknown tangential real optical flow w.

Furthermore, **W** itself satisfies a differential equation. Skipping the details, we state the result:

Theorem 6 *The direction* **W** *of the tangent to the 3D curve* (C) *satisfies the following differential equation:*

$$\mathbf{W} \times (\dot{\mathbf{W}} \times \mathbf{W} + (\mathbf{W} \cdot \mathbf{W})\boldsymbol{\Omega}) = \mathbf{0} \tag{16}$$

Equation (16) is fundamental: it expresses the relationship between the unknown geometry and motion of the 3D curve (C) and the geometry and motion of the 2D curve (c_τ).

In order to exploit equation (16), we have to compute **W**. This is done in [7] and we find that equation (16) involves w and $\dot{w}$, the real tangential optical flow and its total time derivative, as well as $\dot{\boldsymbol{\Omega}}$, the angular acceleration.

5.2 Obtaining more equations

We are now going to use the perspective equation:

$$ZOm = OM \tag{17}$$

to obtain a number of interesting relations by taking its total time derivative.

Taking the total derivative of equation (17) with respect to time, and projecting it on $\mathbf{t}$ and $\mathbf{n}$, we obtain two scalar equations:

$$Z(w + \mathbf{\Omega} \cdot \mathbf{b}) = V_t - (\mathbf{Om} \cdot \mathbf{t})V_z \tag{18}$$

$$Z(\beta - \mathbf{\Omega} \cdot \mathbf{a}) = V_n - (\mathbf{Om} \cdot \mathbf{n})V_z \tag{19}$$

where $\mathbf{a}$ and $\mathbf{b}$ depend only upon the image geometry.

These equations are the standard flow equations expressed in our formalism. They are fundamental in the sense that they express the relationship between the unknown 3D motion of a point and its observed 2D motion.

Notice that we can eliminate Z between (18) and (19) and obtain the value of the tangential real optical flow w as a function of $\mathbf{\Omega}$ and $\mathbf{V}$.

5.3 Closing the loop or finding the kinematic screw

The basic idea is to combine equation (16) which embeds the local structure of (C) at M (its tangent) and the fact that it moves rigidly, with the equation giving w which is a pure expression of the kinematics of the point M without any reference to the fact that it belongs to a curve.

We take the total time derivative $\dot{w}$ of w. In doing this, we introduce the accelerations $\dot{\mathbf{\Omega}}$ and $\dot{\mathbf{V}}$. If we now replace w and $\dot{w}$ by those values in equation (16), we obtain two polynomial equations in $\mathbf{\Omega}$, $\mathbf{V}$, $\dot{\mathbf{\Omega}}$, and $\dot{\mathbf{V}}$ with coefficients depending on the observed geometry and motion of the 2D curve (the two equations come from the fact that equation (16) is a cross-product). Two such equations are obtained at each point of (c_τ). Those polynomials are of degree 5 in $\mathbf{V}$, 1 in $\dot{\mathbf{V}}$, homogeneous of degree 5 in $(\mathbf{V}, \dot{\mathbf{V}})$, of degree 4 in $\mathbf{\Omega}$, 1 in $\dot{\mathbf{\Omega}}$, and of total degree 9 in all those unknowns.

This step is crucial. This is where we combine the structural information about the geometry of (C) embedded in equation (16) with purely kinematic information about the motion of its points embedded in equations (18) and (19). This eliminates the need for the estimation of the real tangential flow w and its time derivative $\dot{w}$. We thus have the following theorem:

Theorem 7 *At each point of (c_τ) we can write two polynomial equations in the coordinates of $\mathbf{\Omega}$, $\mathbf{V}$, $\dot{\mathbf{\Omega}}$ and $\dot{\mathbf{V}}$ with coefficients which are polynomials in quantities that can be measured from the spatio-temporal surface $(\mathcal{S})$:*

$$\beta \quad \frac{\partial \beta}{\partial s} \quad \frac{\partial^2 \beta}{\partial s^2} \quad \partial_{\mathbf{n}_\beta}\beta \quad \partial_{\mathbf{n}_\beta}\frac{\partial \beta}{\partial s}$$
$$\kappa \quad \frac{\partial \kappa}{\partial s} \quad \partial_{\mathbf{n}_\beta}\kappa$$

Those polynomials are obtained by eliminating w and $\dot{w}$ between equations (16), and the equations giving w and $\dot{w}$ [7]. They are of total degree 9, homogeneous of degree 5 in $(\mathbf{V}, \dot{\mathbf{V}})$, of degree 5 in $\mathbf{V}$, 1 in $\dot{\mathbf{V}}$, 4 in $\mathbf{\Omega}$, 1 in $\dot{\mathbf{\Omega}}$.

Thus, N points on (c_τ) provide $2N$ equations in the 12 unknowns $\mathbf{\Omega}$, $\mathbf{V}$, $\dot{\mathbf{\Omega}}$, and $\dot{\mathbf{V}}$. Therefore, we should expect to be able to find, in some cases, a finite number of solutions. Degenerate cases where such solutions do not exist can be easily found: straight lines, for example [8], are notorious for being degenerate from that standpoint. The problem of studying the cases of degeneracy is left for further research. Ignoring for the moment those difficulties (but not underestimating them), we can state one major conjecture/result:

Conjecture 1 *The kinematic screw $\mathbf{\Omega}$, $\mathbf{V}$, and its time derivative $\dot{\mathbf{\Omega}}$, $\dot{\mathbf{V}}$, of a rigidly moving 3D curve can, in general, be estimated from the observation of the spatio-temporal surface generated by its retinal image, by solving a system of polynomial equations. Depth can then be recovered at each point through equation (19). The tangent to the curve can be recovered at each point through equation (14).*

Notice that we never actually compute the tangential real optical flow w. It is just used as an intermediate unknown and eliminated as quickly as possible, as irrelevant. Of course, if needed, it can be recovered afterwards, from equation (18).

6 Conclusion

I have studied the relationship between the 3D motion of a curve (C) moving isometrically and the motion of its image (c_τ). I have introduced the notion of real and apparent optical flows and shown how they can be interpreted in terms of vector fields defined on the spatio-temporal surface (S) generated by (c_τ).

I have shown that the full apparent flow and the normal real flow can be recovered from the differential properties of that surface, but not the real tangential flow.

I have then shown that if the motion of (C) is rigid, then two polynomial equations in the components of its kinematic screw and its time derivative, with coefficients obtained from geometric properties of the surface (S), can be written for each point of (c_τ). In doing this, the role of the spatio-temporal surface (S) is essential since it is the natural place where all the operations of derivation of the geometric features of the curves (c_τ) take place. Conditions under which those equations yield a finite number of solutions have not been studied. Implementation of those ideas is under way. Some issues related to this implementation are discussed in [7].

I think that the major contribution of this paper is to state what can be computed from the sequence of images, under which assumptions about the observed 3D motions, and how. I also believe that similar ideas can be used to study more general types of motions than rigid ones.

References

[1] N. Ayache and F. Lustman. Fast and reliable passive trinocular stereovision. In *Proc. First International Conference on Computer Vision*, pages 422–427, IEEE, June 1987. London, U.K.

[2] H. Harlyn Baker and Robert C. Bolles. Generalizing Epipolar-Plane Image Analysis on the Spatiotemporal Surface. *The International Journal of Computer Vision*, 3(1):33–49, 1989.

[3] J. Canny. A computational approach to edge detection. *IEEE Transactions on Pattern Analysis and Machine Intelligence*, 8 No6:679–698, 1986.

[4] R. Deriche. Using Canny's Criteria to Derive an Optimal Edge Detector Recursively Implemented. *The International Journal of Computer Vision*, 2, April 1987.

[5] Johan D'Hayer. Determining Motion of Image Curves from Local Pattern Changes. *Computer Vision, Graphics, and Image Processing*, 34:166–188, 1986.

[6] M. P. DoCarmo. *Differential Geometry of Curves and Surfaces*. Prentice-Hall, 1976.

[7] Olivier D. Faugeras. *On the motion of 3-D curves and its relationship to optical flow*. Technical Report, INRIA, 1990. To appear.

[8] Olivier D. Faugeras, Nourr-Eddine Deriche, and Nassir Navab. From optical flow of lines to 3D motion and structure. In *Proceedings IEEE RSJ International Workshop on Intelligent Robots and Systems '89*, pages 646–649, 1989. Tsukuba, Japan.

[9] Olivier D. Faugeras, Francis Lustman, and Giorgio Toscani. Motion and Structure from point and line matches. In *Proceedings of the First International Conference on Computer Vision, London*, pages 25–34, June 1987.

[10] Olivier D. Faugeras and Giorgio Toscani. The calibration problem for stereo. In *Proceedings CVPR '86, Miami Beach, Florida*, pages 15–20, IEEE, 1986.

[11] S. Gong. Curve Motion Constraint Equation and its Applications. In *Proceedings Workshop on Visual Motion*, pages 73–80, 1989. Irvine, California, USA.

[12] Ellen C. Hildreth. *The Measurement of Visual Motion*. MIT Press, Cambridge, Mass., 1984.

[13] Berthold K. P. Horn and Brian G. Schunk. Determining Optical Flow. *Artificial Intelligence*, 17:185–203, 1981.

[14] Jan J. Koenderink. Optic Flow. *Vision Research*, 26(1):161–180, 1986.

[15] H. C. Longuet-Higgins and K. Prazdny. The interpretation of moving retinal images. *Proceedings of the Royal Society of London*, B 208:385–387, 1980.

[16] H.H Nagel. Displacement Vectors Derived from Second Order Intensity Variations in Image Sequences. *Computer Vision, Graphics, and Image Processing*, 21:85–117, 1983.

[17] Roger Tsai. An Efficient and Accurate Camera Calibration Technique for 3D Machine Vision. In *Proc. International Conference on Computer Vision and Pattern Recognition*, pages 364–374, IEEE, June 1986. Miami Beach, Florida.

[18] Alessandro Verri, F. Girosi, and Vincente Torre. Mathematical Properties of the 2D Motion Field: from Singular Points to Motion Parameters. In *Proceedings Workshop on Visual Motion*, pages 190–200, 1989.

[19] Alessandro Verri and Tomaso Poggio. Against quantitative optical flow. In *Proceedings First International Conference on Computer Vision*, pages 171–180, 1987.

Structure-from-Motion under Orthographic Projection

Chris Harris

Plessey Research Roke Manor, Roke Manor, Romsey, Hants, England

Abstract

Structure-from-motion algorithms based on matched point-like features under orthographic projection are explored, for use in analysing image motion from small rigid moving objects. For two-frame analysis, closed-form n-point algorithms are devised that minimise image-plane positional errors. The bas-relief ambiguity is shown to exist for arbitrary object rotations. The algorithm is applied to real images, and good estimates of the projection of the axis of rotation onto the image-plane are obtained.

1 Introduction

Structure-from-motion (SFM) algorithms are used in the analysis of image motion caused by relative three-dimensional (3D) movement between the camera and the (unknown) imaged objects. These algorithms attempt to recover both the 3D structure of the image objects (assumed rigid) and the 3D motion of each object with respect to the camera (or *vice versa*). The SFM algorithms explored in this paper use point image features, extracted independently from each image in the sequence by use of a 'corner' detection algorithm [1], and matched between images forming the sequence [2].

As the imaging mechanism of conventional cameras is perspective projection (ie. cameras behave as if they were 'pin-hole' cameras), most SFM algorithms have been based on perspective projection [3,4]. These algorithms have been found to provide acceptable solutions to the 'ego-motion' problem, where a camera (of relatively wide field-of-view) moves through an otherwise static environment. However, for the perspective SFM algorithms to be well-conditioned, the angle subtended by the viewed object (in the ego-motion problem, the viewed scene) must be large, and the viewed object must span a relatively large range of depths. Thus the perspective SFM algorithms are of little or no practical use for analysing everyday imagery of independently moving objects, such as driven cars and flying aircraft. It is algorithms for the analysis of such imagery that is the concern of this paper.

It is well-known that SFM algorithms are unable to produce an unambiguous solution from visual motion data alone, because of the speed-scale ambiguity. On the analysis of a pair of images, the speed-scale ambiguity dictates that the direction of translation of the camera (relative to the viewed object) may be determined, but not the magnitude of translation, ie. the speed. SFM algorithms are thus carefully constructed to avoid attempting to resolve this ambiguity. However, the current perspective SFM algorithms *do* attempt to solve for all the other motion parameters, no matter how ill-conditioned they may be. A prime example here is the bas-relief ambiguity, where, for example, an indented fronto-parallel surface (ie. viewed head-on) rotates about an axis lying in the surface. The problem of differentiating between a deeply indented surface rotating through a small angle, and a

The author gratefully acknowledges the European Commission who together with the Plessey Company supported this work under the Esprit Initiative project P2502.

shallowly indented surface rotating by a larger angle, is ill-conditioned. Current perspective SFM algorithms applied to such a scene produce diverse and incorrect solutions. This is not to say that such scenes are inherently intractable, but that the SFM algorithm is failing because it attempts to determine the value of ill-conditioned variables. What are needed are algorithms in which the ill-conditioned variables (or combinations of variables) are taken care of explicitly and analytically (just as the speed-scale ambiguity is), leaving only well-conditioned variables to be solved for.

The ill-conditioning that we wish to circumvent occurs for objects subtending a small range of depths, and generally subtending a small angle. In these circumstances, a good approximation to the imaging process is orthographic projection, in which the variation in object depths is assumed to be negligible with respect to the distance of the object from the camera. A further reason for using orthographic projection is that it is mathematically tractable for the analysis of the motion of point-like image features between two images of a sequence [5].

2   Structure-from-Motion Algorithm

Let there be n matches between the two frames, at image locations $\{x_i, y_i\}$ on the first frame, and at $\{x'_i, y'_i\}$ on the second frame. Define a coordinate system with the z-axis aligned along the optical axis, the x and y axes aligned with the image coordinate axes, and the origin at a distance L in front of the centre of projection (the camera pin-hole), so placing the centre of projection at $z = -L$. Let the i'th point on the moving object be located in 3D at $r_i = (X_i, Y_i, Z_i)$ at the time of the first frame, and at $r'_i = (X'_i, Y'_i, Z'_i)$ on the second frame. Below, the object will be assumed to be situated close to the coordinate origin, and be small compared to L. Without loss of generality, decompose the object motion between the two frames as a rotation about the origin, specified by the orthogonal rotation matrix R, followed by a translation t. Hence

$$r'_i = R \, r_i + t$$

Perspective projection onto a forward image plane a unit distance from the camera pin-hole gives

$$(x_i, y_i) = (X_i, Y_i) / (L + Z_i), \qquad\qquad (x'_i, y'_i) = (X'_i, Y'_i) / (L + Z'_i)$$

Substituting gives

$$x'_i = R_{11}x_i + R_{12}y_i + R_{13}z_i + t_x / L + O(L^{-2})$$

and similarly for y'_i. Dropping the $O(L^{-2})$ terms for large L (this is the orthographic limit), and without loss of generality setting L=1, gives

$$x'_i = R_{11}x_i + R_{12}y_i + R_{13}z_i + t_x, \qquad\qquad y'_i = R_{21}x_i + R_{22}y_i + R_{23}z_i + t_y$$

Now, for real data, the positions $\{x_i, y_i, x'_i, y'_i\}$ will be contaminated by measurement noise, so that the above equations will not hold true exactly. Assuming isotropic Gaussian noise on the observed image-plane locations, the maximum likelihood solution is found by minimising, E, the sum of the squares of the residuals of the above equations

$$E(R, t_x, t_y, \{z_i\}) = \sum_{i=1}^{n} [\, (R_{11}x_i + R_{12}y_i + R_{13}z_i + t_x - x'_i)^2 + (R_{21}x_i + R_{22}y_i + R_{23}z_i + t_y - y'_i)^2 \,]$$

Note that it is the actual residuals of the image-plane locations that are being minimised, and not some other mathematically convenient but less meaningful formulation. To minimise E, first define

$$u_i = R_{11}x_i + R_{12}y_i - x'_i$$

$$v_i = R_{21}x_i + R_{22}y_i - y'_i$$

Thus

$$E = \sum_i [(u_i + R_{13}z_i + t_x)^2 + (v_i + R_{23}z_i + t_y)^2]$$

Minimising with respect to z_i gives the optimal depth of each point. Substituting the optimal depths back gives E as a function of the motion variables alone

$$E(R,t_x,t_y) = \sum_i [R_{23}u_i - R_{13}v_i + (R_{23}t_x - R_{13}t_y)]^2 / [R_{13}^2 + R_{23}^2]$$

Note that the x and y translations enter the above equation in a fixed relationship, so that minimising with respect to t_x and to t_y would give identical results. To circumvent this problem, E is minimised with respect to the appropriate linear combination of the translations

$$\partial E/\partial(R_{23}t_x - R_{13}t_y) = 0 \implies R_{23}t_x - R_{13}t_y = - \sum_i (R_{23}u_i - R_{13}v_i)$$

Substituting back into E, and simplifying the notation by assuming that mean values have been removed from $x_i, y_i, x'_i, y'_i, u_i$ and v_i, enables E to be written as

$$E(R) = \sum_i [R_{23}u_i - R_{13}v_i]^2 / [R_{13}^2 + R_{23}^2]$$

Without loss of generality, write the rotation matrix, R, as

$$R = \begin{vmatrix} \cos\phi & -\sin\phi & 0 \\ \sin\phi & \cos\phi & 0 \\ 0 & 0 & 1 \end{vmatrix} \begin{vmatrix} 1 & 0 & 0 \\ 0 & \cos\eta & -\sin\eta \\ 0 & \sin\eta & \cos\eta \end{vmatrix} \begin{vmatrix} \cos\theta & \sin\theta & 0 \\ -\sin\theta & \cos\theta & 0 \\ 0 & 0 & 1 \end{vmatrix}$$

$$= \begin{vmatrix} \cos\phi \cos\theta + \sin\phi \sin\theta \cos\eta & \cos\phi \sin\theta - \sin\phi \cos\theta \cos\eta & \sin\phi \sin\eta \\ \sin\phi \cos\theta - \cos\phi \sin\theta \cos\eta & \sin\phi \sin\theta + \cos\phi \cos\theta \cos\eta & -\cos\phi \sin\eta \\ - \sin\theta \sin\eta & \cos\theta \sin\eta & \cos\eta \end{vmatrix}$$

Hence

$$E(\theta,\phi,\eta) = \sum_i [(x_i \cos\theta + y_i \sin\theta) - (x'_i \cos\phi + y'_i \sin\phi)]^2$$

Note that E is independent of η. This means that there is an irresolvable ambiguity in the rotational motion obtained from two frames in orthographic projection, as reported by Huang [5], and is present for arbitrarily large angles of rotation. This ambiguity is a generalisation of the bas-relief ambiguity found at small angles of rotation, that was discussed in the Introduction. We call the angle η the bas-relief angle. The analytic solution for θ and ϕ is subtle, and ends up by finding the zeros of an 8'th order polynomial. This is performed by use of a standard numerical algorithm, and the solution generated by each real root compared numerically, to see which provides the minimum value of E.

3 Axis of Rotation

The SFM algorithm results in an interpretation of the object motion in which the axis of rotation passes through some point along the optical axis. Were the axis of rotation chosen to pass through some other point (say, because the images were shifted slightly) then an equally satisfactory explanation of the data would result, with unchanged rotation, but with different values for the object translations. The above ambiguity in interpretation can be resolved by seeking a solution in which the object translations are zero; this is appropriate for a scenario where the camera is static, and the object is executing pure rotations.

Let $r = (x,y,z)$ be any point on the object in the first frame, and $r' = (x',y',z')$ be the equivalent point on the object in the second frame. These may be actual observed points, or for better conditioning, the centroids of all the matched points may be chosen. Let $c = (c_x, c_y, c_z)$ be a (3D) point on the axis of rotation. For a motion interpretation with no translation, the motion of the object point about the axis of rotation must be due to rotation alone, hence

$$r' - c = R (r - c)$$

Using the first two equations of this vector equation, and eliminating the term in $z - c_z$, results in the following equation linear in the unknowns c_x and c_y, and so specifies a straight line which is the projection of the axis of rotation onto the image

$$c_x \sin((\theta+\phi)/2) - c_y \cos((\theta+\phi)/2) = (x \cos\theta + y \sin\theta - x' \cos\phi - y' \sin\phi) / (2 \sin((\theta-\phi)/2))$$

That the projection of the rotation axis is independent of the bas-relief angle means that the one-dimensional continuum of possible object rotation axes lie in a plane passing through the camera pin-hole. These conclusions concerning the *orientation* of the rotation axis would still be valid if the object translations had not been chosen to be zero.

4 Affine Projection

The orthographic projection algorithm relies upon the object not significantly approaching or receding from the camera, as this would induce size changes to the image which are not catered for. The addition of such size changes to orthographic projection are called affine projection, and the orthographic algorithm may be modified to cater for them. The equation for the rigid object motion is augmented by a scale (or zoom) factor, s

$$r'_i = s (R r_i + t)$$

which leads to the following energy term to be minimised

$$E(s,R,t_x,t_y,\{z_i\}) = s^2 \sum_i [(R_{11}x_i+R_{12}y_i+R_{13}z_i+t_x-x'_i/s)^2 + (R_{21}x_i+R_{22}y_i+R_{23}z_i+t_y-y'_i/s)^2]$$

Proceeding as before, the minimisation variables are reduced to s, θ and ϕ

$$E(s,\theta,\phi) = \sum_i [s (x_i \cos\theta + y_i \sin\theta) - (x'_i \cos\phi + y'_i \sin\phi)]^2$$

Defining $p = s \cos\theta$ and $q = s \sin\theta$ enables E to be easily minimised using no more than an arctangent.

5 Results

The inability to resolve the bas-relief ambiguity makes the results of this algorithm difficult to appreciate, as both the rotation and the structure of the object depend upon the bas-relief angle. The SFM algorithms were applied to a sequence of 16 real images of a toy truck on a turntable. The images are 128 pixels square, and the truck subtends an angle of about 5° from the camera. Between each frame of the sequence, the truck was rotated by 10° about an axis passing through the centre of the turntable, and oriented some 3° clockwise of the vertical. Thus the true projection of the axis of rotation is a nearly vertical line in the image, about one third of the image width from the right-hand edge of the image.From each image feature-points were extracted using a corner detector [1], from 20 to 30 being extracted from each image; these are indicated by the black crosses in the Figures. The feature-points were matched by hand for expediency.

The Figures show the projection of the calculated axis of rotation for the analysis of various image pairs. In Figure 1, successive pairs of images are analysed, corresponding to a rotation of 10° between images. The flow-vector of each matched point is shown as a short white line, which terminates at the location of the feature-point in the later of the image pairs. The projection of the calculated axis of rotation for the orthographic algorithm is indicated by the white line spanning the image, and that for the affine algorithm by the black line (it is sometimes wholly or partly obscured by the white line). Figures 2 and 3 show the results for 20° and 40° rotation respectively, and in Figure 4 are shown four results each for 60°, 80°, 100° and 120° rotations in successive rows of the image.

For even the 10° rotations, the SFM algorithms perform well, and for larger angles of rotation the rotation axis is increasingly accurately positioned. Both algorithms produce quite similar results, mainly because there was no significant zooming of the object (the calculated zooms were all close to and consistent with a value of unity). The accuracy of the results increase less than proportionately with the angle of rotation. This is due to the number of matches decreasing as the angle increased, and due to the inconsistency in objective positioning of the feature-points with large object movement.

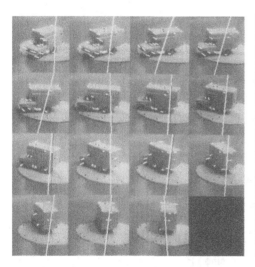

Figure 1. 10° rotation.

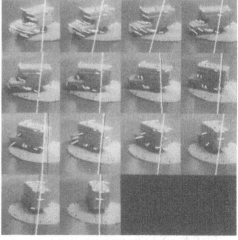

Figure 2. 20° rotation.

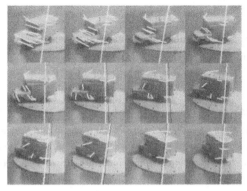

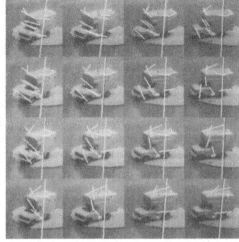

Figure 3. 40° rotation.   Figure 4. 60°, 80°, 100° and 120° rotation.

6 Conclusions

The two-frame orthographic and affine SFM algorithm have been demonstrated to be well-founded (since they minimise image-plane positional errors), closed-form in implementation, and produce good and well-conditioned results. The existence of the bas-relief ambiguity means that there is an ambiguity in interpreting both the rotational motion and the structure, resulting in only a limited number of readily interpretable outputs, such as the projection of the axis of rotation. To resolve the bas-relief ambiguity, matches from three frames in a sequence are needed, and algorithms for analysing such data are currently under development.

7 References

1 Harris, CG and MJ Stephens, *A Combined Corner and Edge Detector*, Proc. 4th Alvey Vision Conference (1988), pp.147-152.

2 Harris, CG and JM Pike, *3D Positional Integration from Image Sequences*, Proc. 3rd Alvey Vision Conference (1987), pp. 233-236.

3 Harris, CG, *Determination of Ego-Motion from Matched Points*, Proc. 3rd Alvey Vision Conference (1987), pp.189-192.

4 Faugeras, O, F Lustman and G Toscani, *Motion and Structure from Motion from Point and Line Matches*, Proceedings IEEE International Conference on Computer Vision, pp. 25-34, 1987.

5 Huang, TS and CH Lee, *Motion and Structure from Orthographic Projection*, IEEE Trans. PAMI, Vol 11, No 5, May 1989, pp. 536-540.

PARALLEL COMPUTATION OF OPTIC FLOW

Shaogang Gong

Computer Science Department

Queen Mary College, University of London

Mile End Road, London E1 4NS, UK

Michael Brady

Department of Engineering Science

Oxford University

Parks Road, Oxford OX1 3PJ, UK

Abstract

Both the tangential and normal components of the flow can be computed reliably where the image Hessian is well-conditioned. A fast algorithm to propagate flow along contours from such locations is proposed. Experimental results for an intrinsically parallel algorithm for computing the flow along zero-crossing contours are presented.

1 Introduction

An algorithm for the computation of optic flow should satisfy (at least) the following two conditions: (i) it should estimate flow vectors *accurately*, particularly if it is to be used to determine usable three-dimensional scene reconstructions; and (ii) it should provably and demonstrably compute flow vectors *fast*. As a result of research over the past decade or so, many methods have been proposed to compute optic flow. Most time and effort has been concentrated on achieving the first of the above properties: correctness and accuracy. Murray and Buxton (1989) forthcoming book, as well as Scott (1986) and Gong (1989b) survey previous work. Many algorithms operate on the basis of the motion constraint equation, which corresponds to a Taylor's series expansion of the image function $I(x, y, t)$ up to first order. In that case, the computation of optic flow is *under-constrained*, and additional smoothness constraints are imposed. These are implemented as local weighted averaging of flow estimates. Inevitably, the resulting algorithms are inherently *slow*, and this prevents them from being useful in practice.

In earlier work (Gong 1988; Gong 1989a) we proposed an additional constraint, which we call the *Curved Motion Constraint Equation*, to develop Hildreth's scheme (Hildreth 1984) for estimating the flow along zero-crossing contours. We restrict attention to zero crossings of a Laplacian operator. In that case, assume that: (i) the intensity function is spatio-temporally differentiable up to second order, and (ii) third order derivatives of the intensity function can be ignored. Then it can be shown that

$$(\mathbf{t}^\top \mathbf{H} \mathbf{n})(\mathbf{t}^\top \mathbf{H} \mu) + (\mathbf{t}^\top \mathbf{H} \mathbf{n})(\nabla I_t \cdot \mathbf{t}) + (\mathbf{t}^\top \mathbf{H} \mathbf{t})(\nabla I_t \cdot \mathbf{n}) + \frac{1}{3} det(\mathbf{H})(\mathbf{n} \cdot \mu) = 0 \qquad (1)$$

By expanding μ in the local $(\mathbf{t}, \mathbf{n})$ coordinate frame, we deduce that $(\mathbf{t} \cdot \mu)$ (and hence the full flow μ) can be estimated in a well-conditioned manner wherever $(\mathbf{t}^\top \mathbf{H} \mathbf{n})$ and $(\mathbf{t}^\top \mathbf{H} \mathbf{t})$ are significant. We extend Hildreth's scheme to take account of the initial well-conditioned estimates of the tangential flow. We minimise

$$\Theta = \int [(\frac{\partial \mu_x}{\partial s})^2 + (\frac{\partial \mu_y}{\partial s})^2] ds + \alpha \int [\mathbf{n} \cdot \mu - \mu^\perp]^2 ds + \beta \int [\mathbf{t} \cdot \mu - \mu^\top]^2 ds. \qquad (2)$$

Here $\beta(s)$ is a Lagrange multiplier that expresses confidence in the tangential flow estimates μ^T, and is a function of the local Hessian matrix, $\beta = \frac{det\mathbf{H}}{\epsilon}$, where ϵ is the condition number of the Hessian matrix.

In this paper, we show how to adapt a mixed wave-diffusion algorithm (Scott, Turner and Zisserman 1988) to propagate quickly the full flow from known edge loci to all other edge locations at which only the normal component is known initially. We demonstrate the robust performance of an implemented algorithm on a number of real image sequences and compare its performance with Hildreth's scheme. We have designed a parallel implementation of our algorithm for a network of Transputers; but for reasons of space it will be reported elsewhere.

2 Propagating Flow

The time required by Hildreth's algorithm (and by the improved version using the Curved Motion Constraint Equation that we described above), is mostly spent on iterations of the conjugate gradient subroutine, which is inherently sequential. To overcome this problem, we have developed a novel, intrinsically local, propagation mechanism, which is described in this Section.

Even using the Curve Motion Constraint Equation, full flow estimates are still restricted to a few locations along image curves, and so the minimisation computation suggested by Hildreth is ill-posed. An additional smoothness constraint is required to guarantee that a minimum is reached. Inspired by (Yuille 1988), we develop a modified one-dimensional Motion Coherence Theory which captures the essence of our problem. In order to have a smooth flow field, the functional to be minimised in terms of both flow components $\mu(s)$ along a curve C can be formulated as:

$$E_{1D} = \int [M(\mu(s)) - \mathbf{M}_\mu(s)]^2 ds + \lambda \int \sum_{m=0}^{\infty} c_m (\frac{\partial^m \mu(s)}{\partial s^m})^2 ds \qquad (3)$$

where $\lambda \geq 0$ is a Lagrange multiplier, the constants $c_m \geq 0$ weight the various derivative (smoothness) terms, and $\mathbf{M}_\mu(s)$ are the initial local measurements of the continuous flow field along a curve. (Note that in Yuille's original formulation, the data points $M(\mathbf{U}_i)$ are at sparse image locations.)

Equation (3) is reminiscent of Hildreth's scheme, which in fact, it generalises. In that case, $c_1 = 1$, and all other c_i are zero. Using the Curve Motion Constraint Equation, instead of $\mu(s) \cdot \mathbf{n}(s)$, as used by Hildreth, the local flow field measurement $M(\mu(s))$ is $\mu(s)$. For $c_1 = 1$, the Euler--Lagrange equation is:

$$\lambda(\frac{\partial^2 \mu(s)}{\partial s^2}) = [\mu(s) - \mathbf{M}_\mu(s)]$$

Now $\mathbf{M}_\mu(s)$ is the initial local measurement of the flow field, that is $\mu_0(s) = \mathbf{M}_\mu(s)$. Iterating,

$$\mu(s) - \mu_0(s) = d\mu(s) = \frac{\partial \mu(s,t)}{\partial t} \Delta t$$

As usual, we treat Δt as a unit time interval between successive iterations of the procedure. It follows that the Euler-Lagrange equation of E_{cf} can be re-written as the diffusion equation:

$$\lambda\left(\frac{\partial^2 \mu(s,t)}{\partial s^2}\right) = \frac{\partial \mu(s,t)}{\partial t} \tag{4}$$

In a diffusion equation, the physical interpretation of λ has dimensions of length^2 $\times$ time^{-1}. Therefore, a combination of variables s and t with λ gives a dimensionless parameter: $\eta = s^2/\lambda t$. Then

$$\mu(s,t) = K \int_0^{\frac{1}{2\lambda}(st^{-1/2})} G_\sigma(\nu)d\nu \tag{5}$$

where $K = \sqrt{\pi}B$ is a constant. It follows that to construct the flow field μ along a curve by minimising the above functional is to propagate the initial local measurements of flow by a Gaussian interaction. The problem with diffusion processes is that they are too slow. This is clear from the above equation, which shows that that the flow field is a function of $st^{-1/2}$. That is, points along s at time t, such that $st^{-1/2}$ has a particular value, should all have the same flow vector. In other words, at any specific time t, there is a particular flow vector which has moved along the positive (as $0 \leq s \leq N$ where $N \geq 0$) curve direction, with a distance proportional to *the square root of the time*. This means that the speed of propagation of a flow vector is $c = \frac{\varphi}{\sqrt{t}}$, where φ is a constant corresponding to the strength of the flow vector being propagated. This speed decreases according to the root of the time and, since bigger φ means faster c, larger flow vectors propagate faster. On the other hand, as the diffusion progresses, the φ of flow vectors being propagated gets smaller as time t elapses.

To address this problem, Scott, Turner and Zisserman (1988) have proposed the use of a mixed wave-diffusion process. They were primarily concerned with the computation of symmetry sets (such as the symmetric axis transform, and smoothed local symmetries) for two-dimensional visual shape representations. We show that their technique can be adapted to propagate quickly flow between locations at which the curved motion constraint equation is well-conditioned.

A wave propagates information at some constant speed C. The wave equation needs boundary conditions that specify $\mu(s,0)$ and its velocity $\partial \mu(s,0)/\partial t = \Psi(s)$ (say) at time $t = 0$. In our case, $\mu(s,0) = M_\mu(s) = \mu_0(s)$, and $\Psi(s) = 0$. The (d'Alembert) solution to the wave equation is given by:

$$\mu(s,t) = \frac{1}{2}[\mu_0(s-Ct) + \mu_0(s+Ct)] + \frac{1}{2C} \int_{x-Ct}^{x+Ct} \Psi(\xi)d\xi$$

which, in our case reduces to:

$$\mu(s,t) = \frac{1}{2}[\mu_0(s-Ct) + \mu_0(s+Ct)] \tag{6}$$

This means that the every flow vector that is measured initially is propagated in both directions along the curve, each at half the strength.

Theoretically, this is a very attractive property as the speed of the wave can be chosen to be much faster than the diffusion equation. However, a wave equation itself does not impose any constraint through the process and, furthermore, numerically, there is another problem associated with wave which can be seen from the following form of equation (6):

$$\mu(s, t+1)' = \mu(s, t)' + C^2 \frac{\partial^2 \mu(s, t)}{\partial s^2}$$
$$\mu(s, t+1) = \mu(s, t)' + \mu(s, t)$$
$$(t = 0, 1, 2, \ldots, \infty) \qquad (7)$$

Here $\mu(s, t)$ is the wave displacement, $\mu(s, t)'$ and $\mu(s, t+1)'$ represent the velocities for the wave equation at time t and $t+1$ at a curve location s. Because of quantisation errors and noise in the digital image, the second order partial derivatives $\frac{\partial^2 u_i}{\partial s^2}$ are unstable. This causes unpredictable pulses in the velocity, which are then propagated. Therefore, propagation according to a wave equation is unstable for digitised images unless the chosen propagation speed C is *small*. Smaller C compensates for errors in the computation of the partial derivatives.

The instability of wave processing comes from second order spatial differentiation and furthermore, this second order differentiation only affects non-linear changes in the displacement. Thus, in order to reduce the sensitivity of the velocity computation, we smooth any *large* non-linear changes in the displacement caused by noise. From the earlier discussion of the diffusion equation, it is clear that diffusion is well suited to that task. Following Scott, Turner and Zisserman (1988), we propose a combined procedure such that every iteration of a wave propagation is followed by an iteration of a diffusion process. This combined procedure has the following desirable features:

1. It quickly damps out the violent non-linear changes that are most likely to be caused by quantisation errors and noise in the displacement. This stablises the propagation process, and enables a reasonably fast wave speed to be used. On the other hand, a faster wave speed implies more smoothing between each successive iteration of the propagation.

2. It imposes a consistent smoothing on the flow vectors through the propagation. Although this side-effect of smoothing the flow vectors does not apply any constraint explicitly and it will not constrain the flow field to be interpolated under the smoothness constraint, the flow field is modified towards the desired distribution. It reduces the task of smoothing in the next step to be carried out by a diffusion.

3. It propagates a weaker influence from any source if the distance along the curve from this source is greater. This is certainly an improvement over a pure wave propagation which would propagate the same strength of data, half of the original, irrespective of the distance from the source. Actually, this is another way to see that the combined procedure carries out a weak smoothing interpolation.

The combined wave-diffusion propagation does not impose a smoothness constraint or a Gaussian interaction between the flow vectors. With fixed boundary values, it will carry on indefinitely. We need to stop the wave-diffusion propagation when the data at the two ends of the curve reach each other simultaneously. For a pure wave, the propagation speed is C. For a modified wave-diffusion, which smooths the flow while it is propagating values, the speed of propagation is faster than C. From experiments, we find that the propagation speed and the time at which to stop the processing can be approximated heuristically by:

$$C_{wd} = C + \frac{1}{log_{10}l} \text{ and } T_{wd} = \frac{l}{C + 1/log_{10}l} \tag{8}$$

where l is the length of the image curve.

3 Experimental Evaluation

In order to evaluate our algorithm, we have tested it on dozens of image sequences of real moving objects. We compare optic flows computed by our algorithm (curved motion constraint extension addition to Hildreth, plus mixed wave-diffusion propagation) with those computed by a reimplementation of Hildreth's method. First, we estimate the optic flow for two toy cars moving in the same direction, where the larger one moves with greater speed. This motion takes place parallel to the image plane (see (a) in figure 1).

It is clear from (b) and (d) in figure 1 that the derivative based local computations are noise sensitive, especially for the second order based local tangential flow. Median filtering is crucial for the interpolation of the tangential flow (see (c) and (e) in figure 1). The flow fields shown in (f) and (g) of figure 1 are computed respectively by our scheme and by Hildreth's. Both schemes detect the greater speed of the larger car, but Hildreth's scheme imposes stronger smoothness on the flow field. As a result, it computes more accurate magnitudes of the flow vectors in this case, but it is also more noise sensitive. It also makes errors at the points indicated by rectangles in the figure.

As a second example, in figure 2, the optic flow estimated for a hand moving in the image plane is given. The hand moves from the bottom line towards the top right corner of the image plane, while at the same time, the fingers spread out slightly.

Both schemes recover the flow vectors at the finger tips, where the motions are parallel to the edges (aperture problem). But notice that Hildreth's scheme propagates the flow field across the edge that joins the second finger and a texture line in the background (indicated by a rectangle). On the other hand, our scheme does not, and is more desirable for tasks such as flow based object segmentation. Again, Hildreth's scheme computes erroneous flow at points such as that indicated by a rectangle at the top right corner in (g).

In the third example, two static coffee mugs whose surfaces were marked with features were taken by a moving camera that moved from the left to the middle in the image plane coordinate (see figure 3).

Both schemes demonstrate an ability to recover smoothed flow fields, but Hildreth's scheme smooths the flow field more than ours. As a whole, the optic flows computed by both schemes are similar. Our scheme is, however, considerably faster than Hildreth's.

Finally, the optic flow of a hand moving in depth is estimated (see figure 4). In addition, the moving object is brightly lit, so that the image is much noisier. This is shown in the locally computed normal and tangential flow in (b) and (d) of figure 4. Without median filtering, some parts of the flow field computed by Hildreth's scheme are unstable and render the entire flow field almost useless.

For comparatively simple scenes giving rise to few edges, our scheme is about 3 times faster than Hildreth's (two cars moving in the image plane, a hand moving in the image plane, and a hand moving against the distinguished background). However, as the scene becomes more complex (more and longer edges in the image), the relative advantage of our scheme increases. This is strongly demonstrated by the results from coffee mug's ego-motion, an approaching hand, and the hand rotating in depth.

References

S.G. Gong, 1988 (September). Improved Local Flow. In *Alvey Vision Conference*, pages 129–134, University of Manchester, Manchester, England.

S.G. Gong, 1989a (March). Curve Motion Constraint Equation and its Applications in the Parallel Visual Motion Estimation. In *IEEE Workshop on Visual Motion*, University of California, Irvine, California.

S.G. Gong, 1989b (September). *Parallel Computation of Visual Motion*. PhD thesis, Departmant of Engineering Science, Oxford University.

E.C. Hildreth, 1984. *The Measurement of Visual Motion*. MIT Press, Cambridge, Massachusetts.

D.W. Murray and B.F. Buxton, 1989. *Experiments in the Machine Interpretation of Visual Motion*. MIT Press, Cambridge, Massachusetts.

G.L Scott, 1986. *Local and Global Interpretation of Moving Images*. PhD thesis, Cognitive Studies Program, University of Sussex, England.

G.L. Scott, S. Turner and A. Zisserman, 1988 (September). Using a Mixed Wave/Diffusion Process to Elicit the Symmetry Set. In *Alvey Vision Conference*, University of Manchester, Manchester, England.

A. Yuille, 1988 (December). A Motion Coherence Theory. In *IEEE International Conference on Computer Vision*, Tampa, Florida.

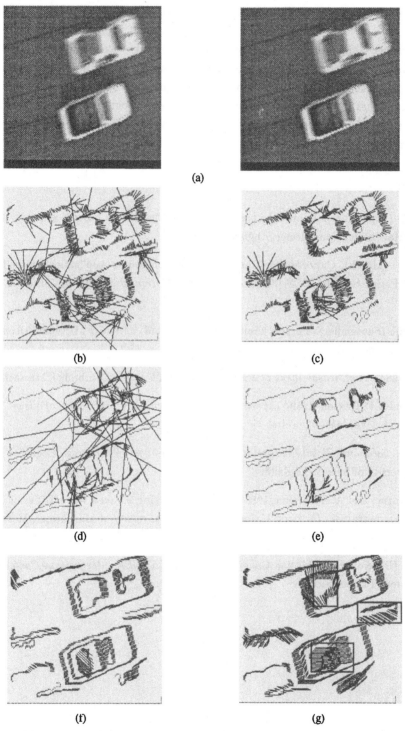

Figure 1: *(a) Two cars move in the image plane. (b) Normal flow before median filtering. (c) Median filtered normal flow. (d) Tangential flow before median filtering. (e) Median filtered tangential flow. (f) Diffusion smoothed full flow. (g) The associated full flow from Hildreth's scheme.*

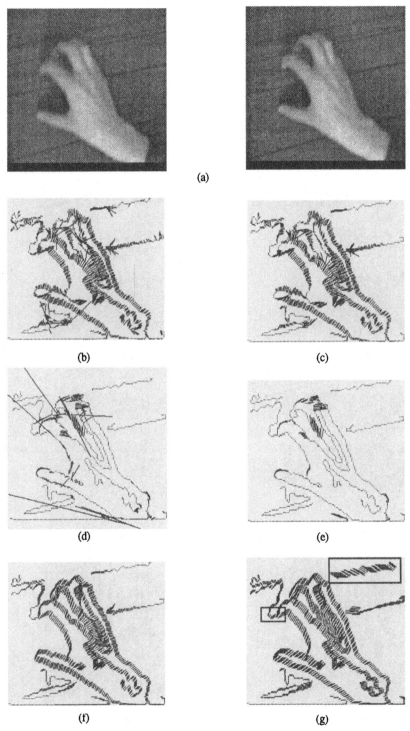

Figure 2: *(a) A hand moves in the image plane. (b) Normal flow before median filtering. (c) Median filtered normal flow. (d) Tangential flow before median filtering. (e) Median filtered tangential flow. (f) Diffusion smoothed full flow. (g) The associated full flow from Hildreth's scheme.*

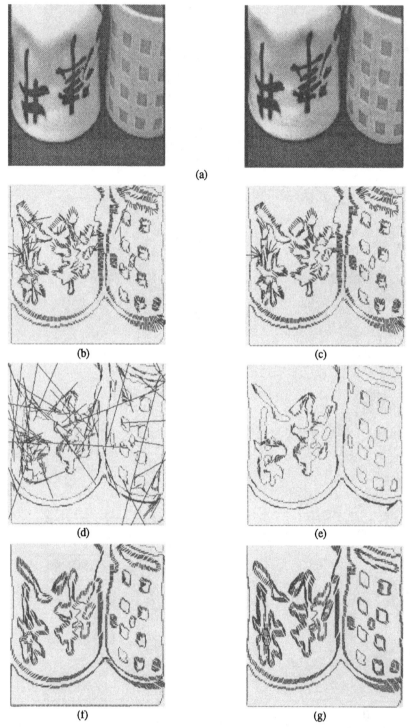

Figure 3: *(a) Ego-motion of cups. (b) Normal flow before median filtering. (c) Median filtered normal flow. (d) Tangential flow before median filtering. (e) Median filtered tangential flow. (f) Diffusion smoothed full flow. (g) The associated full flow from Hildreth's scheme.*

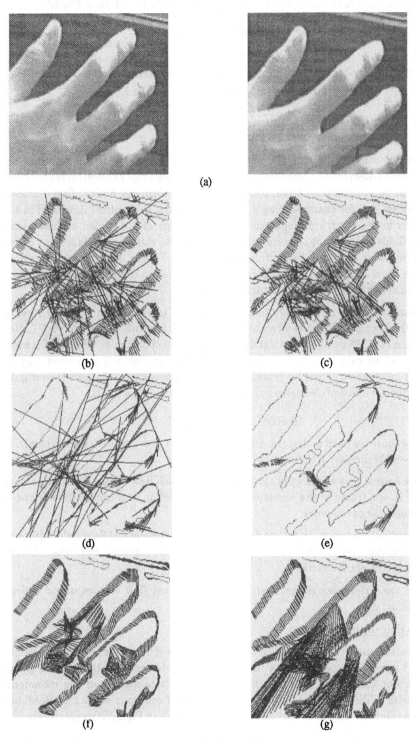

Figure 4: *(a) A hand approaches the viewer. (b) Normal flow before median filtering. (c) Median filtered normal flow. (d) Tangential flow before median filtering. (e) Median filtered tangential flow. (f) Diffusion smoothed full flow. (g) The associated full flow from Hildreth's scheme.*

Obstacle Detection by Evaluation of Optical Flow Fields
from Image Sequences

Wilfried Enkelmann

Fraunhofer-Institut für Informations- und Datenverarbeitung (IITB)

Fraunhoferstraße 1, D-7500 Karlsruhe 1

INTRODUCTION

Image sequences contain information about the dynamic aspects of the recorded scene. Optical flow fields describe the temporal shift of observable gray value structures in image sequences. Various approaches have been suggested to estimate optical flow fields from image sequences - see, for example, Nagel 87. In most cases, this optical flow field is a good approximation to the temporal displacement of the image of a depicted surface element. Optical flow fields contain not only information about the relative displacement of image points but also about the spatial structure of the recorded scene. Several investigations in the literature show how these 2D-fields can be interpreted to infer information about the 3D-environment (Aggarwal & Nandhakumar 88).

To detect obstacles, some authors try to segment single images (Olin et al. 87), evaluate range data (Daily et al. 87) or use divergence fields (Koenderink & van Doorn 77, Nelson 88). Storjohann et al. 88 evaluate spatial disparities to detect obstacles whereas Dickmanns & Christians 89 extract only a few edges and decide wether the extracted edges are projections of an obstacle or not by tracking them. In this contribution, we investigate an approach for the detection of obstacles by evaluation of optical flow fields.

OBSTACLE DETECTION

The basic procedure we used to detect stationary as well as non-stationary obstacles in front of a moving vehicle consists of three steps (Enkelmann 89):

a) Calculate optical flow vectors $u(x,t) = (u(x), v(x))^T$ which link the pixel at image location $x = (x,y)^T$ in one frame to the corresponding pixel position in a consecutive frame from the recorded image sequence.

b) Estimate model vectors $u_M(x,t)$ that describe the expected optical flow field without any obstacle in the scene. The current implementation assumes that the camera is translating on a planar surface.

c) Evaluate the differences u_D between calculated optical flow vectors u and estimated model vectors u_M.

Optical Flow Calculation

To test the obstacle algorithm we used the first five frames of an image sequence to calculate optical flow vectors. During this time interval the vehicle carrying the camera passed a distance of about 2.5 meters. The optical flow vectors calculated from the monotonicity operator (Kories & Zimmermann 86), correspondences of countour elements (Rehfeld 88), and with an analytical approach (Enkelmann 88) are shown in Figs. 1-3, respectively.

Estimation of a model vector field

The two-dimensional model vector field $u_M(x,t)$ assigns to each image location $x = (x,y)^T$ the maximal possible shift in the image plane that will be acceptable if the projected 3D-point does not belong to an obstacle. If we assume only a translational camera motion parallel to the road plane, the components of the model vector $u_M = (u,v)^T$ are given by:

$$u = x' - x = \left(\frac{X'_C f}{Z'_C} + p_x \right) - x \qquad\qquad v = y' - y = \left(\frac{Y'_C f}{Z'_C} + p_y \right) - y \qquad (1)$$

Variables in lower case denote image plane coordinates. Quoted variables correspond to the end of the time interval used to calculate the optical flow vectors. In general, the direction of sensor motion is not parallel to the optical axis, so that the focus of expansion (FOE) is different from the projection point (p_x, p_y) of the optical axis onto the image plane. Subscripts C correspond to the camera coordinate system.

Insertion of the coordinate transformation of 3D-scene points on the road plane and the camera translation into Eq. (1) results in the following equations (where $Z'_C = Z_C - v_{CZ}\Delta t$):

$$u = \frac{(x - p_x) - f v_{CX}/v_{CZ}}{Z_C/(v_{CZ}\Delta t) - 1} = \frac{(x - p_x) - (FOE_x - p_x)}{Z_C/(v_{CZ}\Delta t) - 1} \qquad v = \frac{(y - p_y) - f v_{CY}/v_{CZ}}{Z_C/(v_{CZ}\Delta t) - 1} = \frac{(y - p_y) - (FOE_y - p_y)}{Z_C/(v_{CZ}\Delta t) - 1} \qquad (2)$$

Equations (2) relate the components of the model vector $u_M(x) = (u,v)^T$ to the image location $x = (x,y)^T$, the FOE, the distance Z_C of the depicted 3D-scene point from the camera, and the component $v_{CZ}\Delta t$ of the sensor translation vector. The FOE was determined from the calculated optical flow field using the approach of Bruss & Horn 83.

The distance Z_C of a 3D-scene point on the road plane can be calculated from the intersection of a 3D-line of sight with the road plane if we know the transformation matrix between camera and vehicle coordinate system from a calibration procedure. At those image locations where the line of sight has crossed the horizon, the Z_C component of an intersection point with a virtual plane parallel to the road plane is used instead. This virtual plane is given a priori by the height of the moving vehicle.

The last unknown term in Eq. (2) is the component $v_{CZ}\Delta t$ of the camera translation vector. In the current implementation a vehicle state model is not yet available which can directly be used to estimate the model vector field. Therefore, we used in our experiments described in this contribution a 3D-translation vector which had been determined by backprojecting manually established point correspondences.

Evaluation of Differences

To detect obstacles, all differences between calculated optical flow and estimated model vectors have to be evaluated. Image locations where the length of model vector $|u_M|$ is larger than the length of the corresponding optical flow vector $|u|$ are not considered as an image of an obstacle, since the distance of the projected scene point to the moving vehicle is larger than those points of the volume where obstacles have to be detected. If the length of the model vector is less than the length of the optical flow vector, the ratio of the absolute values of difference vector $|u_D|$ and model vector $|u_M|$ will be compared to a threshold $\theta = 0.3$ for all experiments. If ratio $|u_D| / |u_M|$ is larger than threshold θ, the image location will be considered a projection of a 3D-obstacle point. In areas around the FOE this ratio would become larger than θ even if the absolute differences were small. Therefore, another test is performed to make sure that the denominator of the ratio is significantly different from zero. The detection results for optical flow vectors in Figs. 1-3 are shown in Figs. 4-6.

It can be seen clearly that the monotonicity operator (Fig. 4) extracted only few features in the image of obstacles. It seems very difficult to reliable detect obstacles using these features alone. The contour element correspondence approach uses much more information. Contours belonging to the box in the center of the road (Fig. 5) are clearly detected to be an image of an obstacle as well as the points in the upper right image. However, some false alarms were raised due to problems with optical flow components along straight line contours varying in length. With the analytic approach (Fig. 6) image areas which cover the parking cars, trees, and the box in the center are correctly detected to be an image of an obstacle. However, image locations in the vicinity of the box in the center are incorrectly marked as obstacles because the hierarchical multigrid approach spreads some information across contours into a small neighborhood around the image of an obstacle. We expect to overcome this problem with a theoretically well-founded optical flow estimation approach (Schnörr 89) which responds much more strong to discontinuities in the optical flow field.

In Figs. 4-6 we also see that obstacles beside the road are marked due to the fact that the corresponding scene points are not in the road plane. To concentrate the detection of obstacles to the volume that will be swept by the moving vehicle, additional limitations of the so-called 'motion tunnel' (Zimmermann et al. 86) have to be inserted. Two additional virtual planes perpendicular to the road plane are introduced to reduce the volume (Fig 7). The 3D-coordinates of these additional virtual planes are determined by the width of the moving vehicle. This approach assumes that the velocity vector of the vehicle remains constant. Using this bounded motion tunnel, a model vector field can be estimated with the procedure described above. Due to shorter distances to the camera this results in larger model vectors to the side of the road and, therefore, scene points outside this volume are not considered to be a projection of an obstacle (Fig. 8).

CONCLUSION

The approach discussed in this contribution is an example for the interpretation of temporal variations in image sequences recorded by a translating camera. The encouraging results show how obstacles can be detected in image sequences taken from a translating camera by evaluation of optical flow vectors estimated with independently developed approaches. Further developments are neccessary to extend the approach to more general motion and more complex environments.

ACKNOWLEDGEMENTS

This work has been partially supported by the "Bundesministerium für Forschung und Technologie" of the Federal Republic of Germany and the Daimler-Benz AG during the "Verbundprojekt Autonom mobile Systeme" as well as by the Eureka project PROMETHEUS. I thank Daimler-Benz AG for supplying the facility to record image sequences from a moving bus. I thank N. Rehfeld for providing the contour element correspondences shown in Fig. 3, and G. Hager, H.-H. Nagel and G. Zimmermann for their discussions and comments on a draft version of this paper.

REFERENCES

Aggarwal, J.K, N. Nandhakumar (1988).
On the Computation of Motion from Sequences of Images - A Review. Proceedings of the IEEE, Vol. 76, No. 8 (1988) 917-935

Bruss, A.R. , B.K.P. Horn (1983).
Passive Navigation. Computer Vision, Graphics, and Image Processing 21 (1983) 3-20

Daily, M.J., J. G. Harris, K. Reiser (1987).
Detecting Obstacles in Range Imagery, Image Understanding Workshop, Los Angeles/California February 23-25, 1987, 87-97

Dickmanns, E.D., Th. Christians (1989).
Relative 3D-State Estimation for Autonomous Guidance of Road Vehicles, Intelligent Autonomous Systems 2, T. Kanade, F.C.A. Groen, L.O. Hertzberger (eds.), 11-14 December, 1989, Amsterdam, 683-693

Enkelmann, W. (1988).
Investigations of Multigrid Algorithms for the Estimation of Optical Flow Fields in Image Sequences. Computer Vision, Graphics, and Image Processing (1988) 150-177

Enkelmann, W. (1989).
Interpretation of Traffic Scenes by Evaluation of Optical Flow Fields from Image Sequences, Control, Computers, Communications in Transportation, CCCT 89, Sept. 19-21, 1989, Paris, 87-94

Koenderink, J.J., A.J. van Doorn (1977).
How an ambulant observer can construct a model of the environment from the geometrical structure of the visual inflow, Kybernetik 1977, G. Hauske, E. Butenandt (eds.), Oldenbourg Verlag München Wien 1978, 224-247

Kories, R., G. Zimmermann (1986).
A Versatile Method for the Estimation of Displacement Vector Fields from Image Sequences. Workshop on Motion: Representation and Analysis, Kiawah Island Resort, Charleston/SC, May 7-9, 1986, IEEE Computer Society Press, 1986, 101-106

Nagel, H.-H. (1987).
On the estimation of optical flow: relations between different approaches and some new results. Artificial Intelligence 33 (1987) 299-324

Nelson, R.C. (1988)
Visual Navigation, PhD Thesis, Center for Automation Research, University of Maryland, August 1988, CAR-TR-380

Olin, K.E., F.M. Vilnrotter, M.J. Daily, K. Reiser (1987)
Development in Knowledge-Based Vision for Obstacle Detection and Avoidance, Image Understanding Workshop, Los Angeles/California February 23-25, 1987, 78-86

Rehfeld, N. (1988).
Dichte Verschiebungsvektorfelder entlang von Kantenzügen für zeitliche und stereoskopische Bildpaare. 10. DAGM Symposium, Zürich, 27.-29. September 1988, H. Bunke, O. Kübler, P. Stucki (Hrsg.), Mustererkennung 1988 Informatik-Fachberichte 180, Springer-Verlag Berlin Heidelberg New York, 1988, 97-103

Schnörr, C. (1989).
Zur Schätzung von Geschwindigkeitsvektorfeldern in Bildfolgen mit einer richtungs-abhängigen Glattheitsforderung, 11. DAGM Symposium, Hamburg, 2.-4.10. 1989, H. Burkhardt, K.H. Höhne, B. Neumann (Hrsg.), Mustererkennung 1989, Informatik-Fachberichte 219, Springer-Verlag Berlin Heidelberg New York 1989, 294-301

Storjohann, K., E. Schulze, W. v. Seelen (1988).
Segmentierung dreidimensionaler Szenen mittels perspektivischer Kartierung. 10. DAGM Symposium, Zürich, 27.-29. September 1988, H. Bunke, O. Kübler, P. Stucki (Hrsg.), Mustererkennung 1988 Informatik-Fachberichte 180, Springer-Verlag Berlin Heidelberg New York, 1988, 248-254

Zimmermann, G., W. Enkelmann, G. Struck, R. Niepold, R. Kories (1986).
Image Sequence Processing for the Derivation of Parameters for the Guidance of Mobile Robots. Proc. Int. Conf. on Intelligent Autonomous Systems, Amsterdam, Dec. 8-10, 1986, Elsevier Science Publisher B.V., Amsterdam, 1986, 654-658

Fig. 1: Optical flow vectors calculated from features determined by the monotonicity operator (Kories & Zimmermann 86).

Fig. 2: Optical flow vectors at contour elements which could be matched by the approach of Rehfeld 88.

Fig. 3: Optical flow field calculated using a multigrid approach (Enkelmann 88). The FOE is marked by a small cross in the upper center of the image.

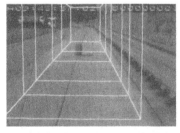

Fig. 7: The 'motion tunnel'. The lines perpendicular to the road axis have a distance of approximately 6m, 8m, 10m, 15m, 20m, and 30m, respectively.

Fig. 4: Tracked feature points extracted by the monotonicity operator (Fig.1) which are considered to be projections of 3D-obstacles are marked with a square. Otherwise, the image location is marked with a little dot.

Fig. 5: as Fig. 4, but optical flow vectors were calculated from contour correspondences in Fig. 2.

Fig. 6: Image locations which are considered to be projections of obstacles. The obstacles were detected by evaluating the optical flow field in Fig. 3.

Fig.8: as Fig. 6, but only those image locations are marked which are considered to be projections of 3D-objects located within the 'motion tunnel'.

Extending the 'Oriented Smoothness Constraint' into the Temporal Domain and the Estimation of Derivatives of Optical Flow

Hans-Hellmut Nagel

Fraunhofer-Institut für Informations- und Datenverarbeitung (IITB)
Fraunhoferstr. 1, D-7500 Karlsruhe 1 / Federal Republic of Germany
and
Fakultät für Informatik der Universität Karlsruhe (TH)

Abstract

Recent experimental results by *Schnörr 89* with an approach based on a simplified 'oriented smoothness constraint' show considerable improvement at expected discontinuities of the optical flow field. It thus appears justified to study whether the local gray value variation can be exploited in the temporal as well as in the spatial domain in order to achieve further improvements at discontinuities in the optical flow field associated with the image areas of moving objects in image sequences. An extension of the oriented smoothness constraint into the temporal domain is presented. In this context, a local estimation approach for the spatio-temporal partial derivatives of optical flow has been developed. This, in turn, is used to compare two approaches for the definition of optical flow.

1. Introduction

The notion of *optical flow* has been introduced in the course of studies of human perception. The 2-D optical flow can be an important clue both for the 3-D relative motion between camera and scene as well as for the relative depths of points in the scene. Since the concept of optical flow has originally been introduced only qualitatively, various possibilities exist for attempts to define it in such a manner that it becomes amenable to quantitative estimation.

Optical flow $\mathbf{u} = (\, u \,, v \,)^T$ is taken to describe the apparent shift of gray value structures $g(x,y,t) = g(\mathbf{x},t)$ in an image plane of a camera which moves relative to the depicted scene. Here, the 2-D vector $\mathbf{x} = (\, x \,, y \,)^T$ denotes a location in the image plane and t denotes time. One usually defines optical flow by the requirement that the gray value structure $g(\mathbf{x} + \mathbf{u}\delta t, t + \delta t)$ observed at time $t + \delta t$ at the location $\mathbf{x} + \mathbf{u}\delta t$ is the same as $g(\mathbf{x},t)$. This results in the so-called 'Brightness Change Constraint Equation (BCCE)' which expresses a single constraint between the components u and v of $\mathbf{u}$:

$$g_x u \; + \; g_y v \; + \; g_t = 0 \qquad (1.1)$$

The derivatives of $g(\mathbf{x},t)$ with respect to x, y, and t are denoted by the corresponding subscripts. Since eq. (1.1) does not allow the estimation of both components of $\mathbf{u}$, *Horn and*

Schunck 81 formulated a minimization approach in order to estimate u and v as a function of **x** :

$$\int\int d\mathbf{x}\left\{\left[g_x u + g_y v + g_t\right]^2 + \lambda^2\left[u_x^2 + u_y^2 + v_x^2 + v_y^2\right]\right\} => minimum \qquad (1.2)$$

In order to reduce the smoothing effect of the second term - multiplied by λ^2 - in eq. (1.2) across potential discontinuities in the optical flow field, Nagel suggested to let the contribution of this smoothness term be controlled by the local gray value variation. This resulted in the following modification of eq. (1.2) :

$$\int\int d\mathbf{x}\left\{\left[g_x u + g_y v + g_t\right]^2 + \lambda^2 trace\left[(\nabla\mathbf{u})^T W(\nabla\mathbf{u})\right]\right\} => minimum \qquad (1.3)$$

where the weight matrix W is given by (*Nagel 87*) :

$$W = \frac{1}{g_x^2 + g_y^2 + 2\gamma}\begin{pmatrix} g_y^2 + \gamma & -g_x g_y \\ -g_x g_y & g_x^2 + \gamma \end{pmatrix} \qquad (1.4)$$

Snyder 89 recently proved that - assuming general constraints for such an expression - the weight matrix given by eq. (1.4) is the only reasonable choice. In addition, *Schnörr 89* has just proved that the problem formulation according to eq. (1.3) with W according to (1.4) has a unique solution which depends continuously on the input data. *Schnörr 89* compared results obtained based on this oriented smoothness constraint with results based on eq. (1.2). This comparison supports the expectation that the oriented smoothness constraint contributes to a much better demarkation of the optical flow field arount moving object images than the isotropic smoothness constraint introduced by *Horn and Schunck 81*.

These encouraging results lead to the following consideration: just as the spatial gray value gradient may be used to constrain the strength and orientation of a spatial smoothness requirement for the optical flow estimates, strong temporal changes in the components of optical flow - i. e. potential discontinuities - could be estimated more reliably by the introduction of a temporal smoothness constraint which in turn would have to be controlled by the spatio-temporal gray value variation. This conjecture will be formulated quantitatively in the next section.

2. Extension of the oriented smoothness constraint into the temporal domain

The observed gray value function g(x,t) can be interpreted as a density in the three-dimensional space (**x**,t). The partial derivatives with respect to the variables (**x**,t) or (x,y,t) may be considered to form a three-component gradient vector $\nabla_{xt} = (\partial/\partial x, \partial/\partial y, \partial/\partial t)^T$. If we extend the definition of the optical flow vector **u** formally to $\acute{\mathbf{u}} = (u, v, 1)^T$ by a third component with the constant value 1, we may write the BCCE (1.1) in the form

$$(\nabla_{xt} g)^T \acute{\mathbf{u}} = 0 \qquad (2.1)$$

This equation can be interpreted as the requirement that the optical flow vector $\acute{\mathbf{u}}(\mathbf{x},t)$ is constrained to a plane defined by the normal vector $(\nabla_{xt} g)$ at (**x**,t). This leaves one degree of freedom to $\acute{\mathbf{u}}(\mathbf{x},t)$, namely its orientation within this plane. We may interpret the optical flow vector ú(x,t) as the tangent to a flow line through the point (x,t). If the gray value function g(x,t) does not vary with time, $\partial g / \partial t = 0$ and all tangent planes are parallel to the t-axis. In this case one would like to have flow lines which are parallel to the t-axis, too. If the gray value structure is translated with constant velocity in the image plane, the normal vector $(\nabla_{xt} g)$ will be slanted with respect to the t-axis, but one would like to retain flow lines which are still parallel to each other within tangent planes to g(x,t). This aim can be achieved by postulating that the direction of a flow line, i. e. the orientation of an optical flow vector, remains constant for infinitesimal displacements within the tangent plane to g(x,t). This is equivalent to the postulate that changes in ú occur at most in the direction of the gradient $(\nabla_{xt} g)$. This in turn implies that changes in the components of ú , i. e. the vectors $(\nabla_{xt} u)$ and $(\nabla_{xt} v)$, are collinear with $(\nabla_{xt} g)$. These latter requirements can be expressed by demanding that the vector products between $(\nabla_{xt} g)$ and both $(\nabla_{xt} u)$ as well as $(\nabla_{xt} v)$ vanish :

$$[(\nabla_{xt} g) \times (\nabla_{xt} u)] = \begin{pmatrix} 0 & -\dfrac{\partial g}{\partial t} & \dfrac{\partial g}{\partial y} \\[2mm] \dfrac{\partial g}{\partial t} & 0 & -\dfrac{\partial g}{\partial x} \\[2mm] -\dfrac{\partial g}{\partial y} & \dfrac{\partial g}{\partial x} & 0 \end{pmatrix} \begin{pmatrix} \dfrac{\partial u}{\partial x} \\[2mm] \dfrac{\partial u}{\partial y} \\[2mm] \dfrac{\partial u}{\partial t} \end{pmatrix} = 0 \qquad (2.2a)$$

and analogously for $(\nabla_{xt} v)$. Since such a requirement will in general be too strong for measurements of g(x,t) corrupted by noise, eq. (2.2a) will be replaced by the less stringent requirement that

$$\|[(\nabla_{xt} g) \times (\nabla_{xt} u)]\|^2 = \begin{pmatrix} \dfrac{\partial u}{\partial x} & \dfrac{\partial u}{\partial y} & \dfrac{\partial u}{\partial t} \end{pmatrix} \begin{pmatrix} g_y^2 + g_t^2 & -g_x g_y & -g_x g_t \\[2mm] -g_x g_y & g_x^2 + g_t^2 & -g_y g_t \\[2mm] -g_x g_t & -g_y g_t & g_x^2 + g_y^2 \end{pmatrix} \begin{pmatrix} \dfrac{\partial u}{\partial x} \\[2mm] \dfrac{\partial u}{\partial y} \\[2mm] \dfrac{\partial u}{\partial t} \end{pmatrix} => \text{minimum}$$

$$(2.2b)$$

It is seen that the upper left 2x2 submatrix on the right hand side of eq. (2.2b) corresponds to the oriented smoothness expression for ∇u, provided g_t^2 is set equal to zero, γI with I as the 2x2 identy matrix is added and the sum is normalized by the trace of the resulting weight matrix. The generalized oriented smoothness constraint for the first component of u can thus be written as

$$\lambda^2 \frac{1}{2(g_x^2 + g_y^2 + g_t^2) + 3\gamma} \begin{pmatrix} \dfrac{\partial u}{\partial x} & \dfrac{\partial u}{\partial y} & \dfrac{\partial u}{\partial t} \end{pmatrix} \begin{pmatrix} g_y^2 + g_t^2 + \gamma & -g_x g_y & -g_x g_t \\[2mm] -g_x g_y & g_x^2 + g_t^2 + \gamma & -g_y g_t \\[2mm] -g_x g_t & -g_y g_t & g_x^2 + g_y^2 + \gamma \end{pmatrix} \begin{pmatrix} \dfrac{\partial u}{\partial x} \\[2mm] \dfrac{\partial u}{\partial y} \\[2mm] \dfrac{\partial u}{\partial t} \end{pmatrix} \quad (2.3)$$

The expression (2.3) and a corresponding expression for ∇v can thus be considered to represent a generalization of the two-dimensional oriented smoothness constraint from eqs. (1.3) and (1.4) to the three-dimensional (x,y,t)-space. The same weight matrix as in expression (2.3) has been used by *Krämer 89* in an attempt to estimate structure and motion directly from monocular image sequences.

During the development of earlier forms of the oriented smoothness constraint (see *Nagel 87*), analytical investigations of u and v as a function of the local spatio-temporal gray value variation turned out to be very useful. It thus is expected that similar investigations will advance the understanding of the implications of the generalized oriented smoothness constraint proposed in this section. As a preparation for such studies, u and v as well as their partial derivatives with respect to x,y, and t should be obtained as explicit functions of the local spatio-temporal gray value variation. This is the topic of subsequent sections.

3. Explicating local spatio-temporal gray value variations

In order to explicate the dependency of partial derivatives of g on x and t, we choose the point of interest as the origin of a local coordinate system and consider the Taylor expansion of the gray value gradient components up to first order terms. Moreover, in order to restrict this approximation to an explicitly characterized local spatio-temporal environment around the origin, this Taylor expansion is weighted by a trivariate Gaussian

$$g_x(\mathbf{x},t) = \left(g_x(0,0) + g_{xx}(0,0)\,x + g_{xy}(0,0)\,y + g_{xt}(0,0)\,t \right) \frac{c}{\left(\sigma_x^2\, 2\pi\right)^{3/2}} e^{-\frac{x^2 + y^2 + c^2 t^2}{2\sigma_x^2}}$$

(3.1)

where the ratio $\sigma_x / \sigma_t = c$ has the dimension of a velocity. This approximation implies a choice for σ_t which allows to disregard contributions from the infinite past or future as being negligibly small. The spatial derivatives with respect to y and t can be written analogously. Similarly, we introduce the following representation for the first component u of the optical flow :

$$u(\mathbf{x},t) = \left(u(0,0) + u_x(0,0)\,x + u_y(0,0)\,y + u_t(0,0)\,t \right) \frac{c}{\left(\sigma_x^2\, 2\pi\right)^{3/2}} e^{-\frac{x^2 + y^2 + c^2 t^2}{2\sigma_x^2}}$$

(3.2)

and an analogous one for the second component v. It should be noted that eq. (3.2) comprises partial derivatives of u and v with respect to time in addition to those with respect to the spatial coordinates x and y.

In order to simplify subsequent derivations, we introduce a more compact notation where the arguments (0,0) of the partial derivatives have been dropped : $G_x = (g_x, g_{xx}, g_{xy}, g_{xt})^T$;

$G_y = (g_y \, , \, g_{yx} \, , \, g_{yy} \, , \, g_{yt} \,)^T; \quad G_t = (g_t \, , \, g_{tx} \, , \, g_{ty} \, , \, g_{tt} \,)^T.$ In analogy we introduce $U = (u \, , u_x \, , u_y \, , u_t \,)^T$ and $V = (v \, , v_x \, , v_y \, , v_t \,)^T.$ Similarly, we define

$$X = \frac{c}{\left(\sigma_x^2 \, 2\pi\right)^{3/2}} \begin{pmatrix} x \\ y \\ t \end{pmatrix} e^{-\dfrac{x^2 + y^2 + c^2 t^2}{2\sigma_x^2}} \tag{3.3}$$

Using these conventions, we may write the modified BCCE with explicated dependency on x, on t, and on the extent of the environment:

$$\left(G_x^T X\right)\left(U^T X\right) + \left(G_y^T X\right)\left(V^T X\right) + \left(G_t^T X\right) = 0 \tag{3.4}$$

Rather than requiring that this form of the BCCE should be valid at each space-time location (x,t), we postulate that the integral of the square of eq. (3.4) should be minimized:

$$\int_{-\infty}^{+\infty}\int_{-\infty}^{+\infty} dxdt \left\{ \left[G_x^T X \bullet U^T X + G_y^T X \bullet V^T X + G_t^T X \right]^2 \right\} => minimum \tag{3.5}$$

The trivariate Gaussian in X will enforce convergence of the integral without the necessity to introduce sharp boundaries for the integration region. It appears reasonable to estimate the spatio-temporal derivatives G_x , G_y , and G_t by a convolution of g(x,t) with the corresponding derivatives of the trivariate Gaussian in X. The choice of σ_x and σ_t thus determines the extent of the spatio-temporal domain which contributes significantly to the integral in eq. (3.5). 'Local estimation' of U and V is understood to refer to the domain around the origin determined by σ_x and σ_t .

4. Equations for the unknown components of U and V

The partial derivatives contained in G_x , G_y , and G_t , taken at the origin of the coordinate system, are observed constant values. The explication of the dependency on x and t through the expression for X defined by eq. (3.3) implies that the components of U and V are considered to be constant unknown values associated with the origin (0,0) of the local coordinate system. Setting the partial derivatives of the integral in eq. (3.5) with respect to the unknown components of U and V to zero results in the following system of eight linear equations, written here as a set of two (4x1)-vector equations:

$$\int\int_{-\infty}^{+\infty} dxdt\left\{ G_x^T X \bullet G_x^T X \bullet XX^T\right\} U + \int\int_{-\infty}^{+\infty} dxdt\left\{ G_y^T X \bullet G_x^T X \bullet XX^T\right\} V = -\int\int_{-\infty}^{+\infty} dxdt\left\{ G_t^T X \bullet G_x^T X \bullet X\right\} \tag{4.1a}$$

and

$$\int\int_{-\infty}^{+\infty} dxdt\left\{ G_x^T X \bullet G_y^T X \bullet XX^T\right\} U + \int\int_{-\infty}^{+\infty} dxdt\left\{ G_y^T X \bullet G_y^T X \bullet XX^T\right\} V = -\int\int_{-\infty}^{+\infty} dxdt\left\{ G_t^T X \bullet G_y^T X \bullet X\right\} \tag{4.1b}$$

The coefficients of this system of linear equations in U and V are integrals with each integrand consisting of a trivariate Gaussian, multiplied by monomials in x, y and t of up to fourth order. The expansion of the coefficient integrals yields for the general case on the left hand side with $\eta, \zeta \in \{x, y\}$ and $\{G_\eta^T X \bullet X^T G_\zeta \bullet X X^T\} = \{(G_\eta^T X X^T G_\zeta) \bullet X X^T\}$:

$$
\left\{ \frac{c^2}{(\sigma_x^2 2\pi)^3} \bullet \begin{pmatrix} (G_\eta^T X X^T G_\zeta)\bullet 1 & (G_\eta^T X X^T G_\zeta)\bullet x & (G_\eta^T X X^T G_\zeta)\bullet y & (G_\eta^T X X^T G_\zeta)\bullet t \\ (G_\eta^T X X^T G_\zeta)\bullet x & (G_\eta^T X X^T G_\zeta)\bullet x^2 & (G_\eta^T X X^T G_\zeta)\bullet xy & (G_\eta^T X X^T G_\zeta)\bullet xt \\ (G_\eta^T X X^T G_\zeta)\bullet y & (G_\eta^T X X^T G_\zeta)\bullet yx & (G_\eta^T X X^T G_\zeta)\bullet y^2 & (G_\eta^T X X^T G_\zeta)\bullet yt \\ (G_\eta^T X X^T G_\zeta)\bullet t & (G_\eta^T X X^T G_\zeta)\bullet tx & (G_\eta^T X X^T G_\zeta)\bullet ty & (G_\eta^T X X^T G_\zeta)\bullet t^2 \end{pmatrix} \bullet e^{-\frac{2(x^2 + y^2 + c^2 t^2)}{2\sigma_x^2}} \right\}
$$

(4.2)

The evaluation of the integral over each component of this 4x4 matrix presents no problems. Space limits prevent an illustration of intermediate results for this evaluation. The integrands on the right hand side of eqs. (4.1) can be treated analogously.

Using these coefficients, it is straightforward to write the system of linear equations (4.1) for the unknown entities u, v, and their partial derivatives. It turns out, however, that attempts at the symbolic solution of this system of equations result in rather involved expressions. In order to provide some insight into the structure of this system of linear equations, a smaller system will be discussed here. It is obtained by omitting the derivative with respect to time t in the Taylor expansion of eqs. (3.1) as well as (3.2) and by omitting the Gaussian weight function depending on t. The system of linear equations resulting from these steps are formally equal to those in eqs. (4.1) although the evaluation of the integrals yields slightly different values for the coefficients.

5. Equations for u, v, and their partial derivatives with respect to x and y

Since there will be no Gaussian weighting function with respect to time and thus no σ_t, the distinction between σ_t and σ_x will not be necessary. Henceforth, σ without a subscript will be used. Only u, v, u_x, u_y, v_x, and v_y are retained as unknowns. The integrals are evaluated in analogy to the procedure discussed previously.

In order to simplify the subsequent presentation, it is assumed that the local coordinate system has been aligned with the directions of principal curvatures of $g(x,t)$, i. e. $g_{xy} = 0$. Moreover, in all expressions it turns out that whenever a second partial derivative of $g(x,t)$ appears, it does so always in combination with a factor $\sigma/2$. It thus is advantageous to introduce the convention - again with $\zeta \in \{x, y\}$ - that $g_{\zeta\zeta}^* = (\sigma/2) g_{\zeta\zeta}$ and $g_{t\zeta}^* = (\sigma/2) g_{t\zeta}$. By this definition, the second partial derivatives with an asterisk have the same dimension - namely [gray value] $\bullet$ [length]$^{-1}$ or [gray value] $\bullet$ [time]$^{-1}$, respectively, - as the first partial derivatives. This can be useful while checking some of the formulas. For simplicity, the asterisk is henceforth dropped and all second partial derivatives are taken to carry implicitly a factor $\sigma/2$. Similarly, all first partial derivatives of u and v are taken to carry implicitly a factor $\sigma/2$. Using these conventions, the linear system of equations for the components of U and V can be written as follows, after having exchanged some rows and columns in order to group u and v as well as their first derivatives together:

$$
\begin{pmatrix}
g_x^2+g_{xx}^2 & g_x g_y & 2\dfrac{\sigma}{2}g_x g_{xx} & 0 & g_y g_{xx} & g_x g_{yy} \\[2mm]
g_x g_y & g_y^2+g_{yy}^2 & g_y g_{xx} & g_x g_{yy} & 0 & 2g_y g_{yy} \\[2mm]
2g_x g_{xx} & g_y g_{xx} & g_x^2+3g_{xx}^2 & 0 & g_x g_y & g_{xx}g_{yy} \\[2mm]
0 & g_x g_{yy} & 0 & g_x^2+g_{xx}^2 & g_{xx}g_{yy} & g_x g_y \\[2mm]
g_y g_{xx} & 0 & g_x g_y & g_{xx}g_{yy} & g_y^2+g_{yy}^2 & 0 \\[2mm]
g_x g_{yy} & 2g_y g_{yy} & g_{xx}g_{yy} & g_x g_y & 0 & g_y^2+3g_{yy}^2
\end{pmatrix}
\begin{pmatrix}
u \\ v \\ u_x \\ u_y \\ v_x \\ v_y
\end{pmatrix}
= -\frac{32}{9}\pi\sigma^2
\begin{pmatrix}
\frac{3}{4}g_t g_x + g_{tx}g_{xx} \\[2mm]
\frac{3}{4}g_t g_y + g_{ty}g_{yy} \\[2mm]
g_t g_{xx}+g_x g_{tx} \\[2mm]
g_x g_{ty} \\[2mm]
g_y g_{tx} \\[2mm]
g_x g_{yy}+g_y g_{ty}
\end{pmatrix}
$$

$$(5.1)$$

It turns out that the matrix in eq. (5.1) has only rank five rather than the required full rank of six. The eigenvector of this matrix corresponding to its eigenvalue 0 has the form

$$e_0 = (g_y, -g_x, 0, g_{yy}, -g_{xx}, 0)^T \qquad (5.2)$$

A test using the components of e_0 as factors reveals that not only the row vectors of the coefficient matrix, but both left and right hand sides of eqs. (5.1) are linearly dependent, i. e. either the first, second, fourth, or fifth equation can be deleted as redundant. Deletion of, say, the fifth equation leaves a system of five equations with six unknowns, i. e. five unknowns can be determined up to a linear function of the sixth one, say v_x.

Assume that a special solution is known for the unknowns. Since the eigenvector e_0 is orthogonal to all row vectors of the coefficient matrix of eq. (5.1), we may define the general solution of this system of equations by adding a multiple w of e_0 to this special solution, i. e.

$$(\hat{u}, \hat{v}, \hat{u}_x, \hat{u}_y, \hat{v}_x, \hat{v}_y)^T = (u, v, u_x, u_y, v_x, v_y)^T + w\,e_0. \qquad (5.3)$$

If we now treat the newly defined entities on the left hand side of eq. (5.3) as unknowns, we can set $\hat{v}_x = v_x - wg_{xx} = 0$ by choosing w appropriately. This removes the fifth column from the left hand side of eq. (5.1) and enables us to solve for the five remaining variables $\hat{u}$, $\hat{v}$, $\hat{u}_x$, $\hat{u}_y$, and $\hat{v}_y$. Since the resulting expressions in the derivatives of g are rather lengthy, space limitations force us to restrict the discussion of the results to a particularly interesting case.

Eq. (5.3) in combination with eq. (5.2) shows that the solutions for u_x and v_y do not depend on the free parameter w. It is possible, therefore, to determine the divergence of u directly in terms of the spatio-temporal derivatives of the grayvalue function $g(x,t)$:

$$div\,\mathbf{u} = u_x + v_y = \frac{-8\pi\sigma}{9\left[(g_x^2 g_{yy}^2 + g_y^2 g_{xx}^2)^2 + 4 g_{xx}^4 g_{yy}^4\right]}\left\{8\left(g_{tx} g_x g_{yy} + g_{ty} g_y g_{xx}\right)\left[g_{xx}^3(g_y^2 - g_{yy}^2) + g_{yy}^3(g_x^2 - g_{xx}^2)\right]\right.$$

$$+ g_t g_{xx}\left[(g_y^2 + g_{yy}^2)(g_x^2 g_{yy}^2 + g_y^2 g_{xx}^2) + 2 g_{yy}(g_x^2 g_{yy}^3 + g_y^2 g_{xx}^3)\right]$$

$$\left. + g_t g_{yy}\left[(g_x^2 + g_{xx}^2)(g_x^2 g_{yy}^2 + g_y^2 g_{xx}^2) + 2 g_{xx}(g_x^2 g_{yy}^3 + g_y^2 g_{xx}^3)\right]\right\} \tag{5.4}$$

In this expression, the implicit factor of $\sigma/2$ in the definition of u_x and v_y has already been made explicit again and has been compensated against the same factor on the right hand side of eq. (5.4).

Another possibility to cope with the degeneracy of the linear system of eq. (5.1) consists in supplementing the minimization problem by a regularization term. Then, the inverse of the correspondingly supplemented coefficient matrix according to eq. (5.1) exists. Using a symbolic algebra programming system like MAPLE, it can be computed without problem. The resulting expressions, however, are very lengthy and are not immediately amenable to significant simplifications. Therefore, they will not be presented here.

Alternatively, one could fix either one or a linear combination of the unknowns by an additional assumption. Since it is desirable to introduce any additional assumption in a manner invariant to rotations of the coordinate system, it is suggested to demand that the shear tensor should vanish.

$$shear(\mathbf{u}) = \frac{1}{2}\begin{pmatrix} u_x - v_y & u_y + v_x \\ u_y + v_x & -(u_x - v_y) \end{pmatrix} = 0 \tag{5.5}$$

This implies two additional constraint equations, leaving only four unknowns. The postulate expressed by eq. (5.5) can be incorporated into the minimization problem by adding the following terms to the integrand:

$$\mu^2(u_x - v_y)^2 + \nu^2(u_y + v_x)^2 \tag{5.6}$$

where μ^2 and ν^2 represent Lagrange multipliers. The contribution of these terms drop out of the equations obtained by forming the partial derivatives of the expanded minimization problem with respect to U and V since they contain a factor of either $(u_x - v_y)$ or $(u_y + v_x)$ which vanishes. One obtains a system of five equations of rank 5 for four unknowns. A pseudo-inverse formalism can be used to solve for the unknowns. The resulting expressions are involved.

During a recent discussion, I learned that Koenderink and coworkers also investigate the direct estimation of optical flow and its partial derivatives, but taking into account higher than second order spatio-temporal derivatives of the gray value distribution. In both cases, a Gaussian is used to localize the estimation procedure (*Koenderink 89*).

6. Comparison with the definition of optical flow by Girosi et al. 89

The following discussion concentrates on the direct estimation of u and v. The BCCE in itself does not provide sufficient constraints in order to estimate both u and v. If, however, the gray values vary sufficiently as a function of x and y, it could be shown that - by taking into account higher order spatial derivatives of g(x,t) - both components of u can be estimated directly (*Nagel 83*).

Recently, *Girosi et al. 89* discussed another possibility to directly estimate u and v which may be considered to be a special case of the approach investigated, for example, in *Nagel 83 + 87*. The difference consists in the fact that *Girosi et al. 89* define optical flow not by something like eq. (1.1) but as the solution of the vector equation

$$
\frac{d}{dt} \nabla g \;=\; \begin{pmatrix} \dfrac{d}{dt} g_x \\[2mm] \dfrac{d}{dt} g_y \end{pmatrix} \;=\; \begin{pmatrix} g_{xx}\dfrac{dx}{dt} + g_{xy}\dfrac{dy}{dt} + g_{xt} \\[2mm] g_{yx}\dfrac{dx}{dt} + g_{yy}\dfrac{dy}{dt} + g_{yt} \end{pmatrix} \;=\; 0
$$

(6.1)

which results in the following system of equations :

$$
g_{xx}u^* + g_{xy}v^* = -g_{xt} \qquad and \qquad g_{xy}u^* + g_{yy}v^* = -g_{yt}
$$

(6.2)

Here, $\mathbf{u}^* = (u^*, v^*)^T$ has been used in order to emphasize the difference between the definition of optical flow $\mathbf{u}^*$ according to eq. (6.1) and the one of u according to eq. (1.1). The solution to this system of equations is

$$
\begin{pmatrix} u^* \\[2mm] v^* \end{pmatrix} = -\frac{1}{g_{xx}^2 g_{yy}^2} \begin{pmatrix} g_{yy}^2 & 0 \\[2mm] 0 & g_{xx}^2 \end{pmatrix} \begin{pmatrix} g_{xt} \\[2mm] g_{yt} \end{pmatrix}
$$

(6.3)

where the convention of setting $g_{xy} = 0$ has been used. This approach captures only part of the situations which allow to estimate locally both components of the optical flow u, namely only those situations with a non-singular Hessian. For curved lines of maximum gray value slope, in particular for 'gray value corners' characterized as points of maximum curvature in such locus lines of maximum slope, the gradient is maximum which implies that the corresponding second partial derivative vanishes. In such cases, the Hessian becomes singular and the approach expressed by eq. (6.1) breaks down, whereas the one of *Nagel 83* provides a useful estimate in such cases.

7. Conclusion

The extension of the oriented smoothness constraint into the temporal domain is expected to facilitate a better detection and localization of discontinuities in the optical flow field.

It has been shown that, by appropriate modeling of the local gray value variation, it becomes possible - at least in theory - to estimate not only both components of optical flow, but in addition some linear combination of its partial derivatives with respect to x and y. In particular, it becomes possible to estimate div(u) directly from spatio-temporal gray value variations. If the isotropic smoothness term introduced by *Horn and Schunck 81* is included into the model developed here, one does not need to make the assumption of vanishing shear(u) in order to determine all spatial partial derivatives of u and v. Obviously, experimental investigations are needed in order to test the reliability of these approaches.

Acknowledgements

I thank J. Rieger for an introduction to the MAPLE program and help during its application to the problems of this investigation as well as Ch. Götze and Ch. Schnörr for a careful reading of the draft. Stimulating discussions with O. Faugeras, J.J. Koenderink, and V. Torre about this topic are gratefully acknowledged.

This investigation has been partially supported by the EC Basic Research Action project INSIGHT.

References

F. Girosi, A. Verri, and V. Torre, "Constraints for the Estimation of Optical Flow", Proc. Workshop on Visual Motion, March 20-22, 1989, Irvine / CA, pp. 116-124

B.K.P. Horn and B.G. Schunck, "Determining Optical Flow", Artificial Intelligence 17 (1981) 185-203

J.J. Koenderink, private communication, September 29-30, 1989

T. Krämer, "Direkte Berechnung von Bewegungsparametern und einer Tiefenkarte durch Auswertung monokularer Grauwertbildfolgen", Dissertation, Fakultät für Informatik der Universität Karlsruhe (TH), Karlsruhe / Bundesrepublik Deutschland, Mai 1989

H.-H. Nagel, "Displacement Vectors Derived from Second Order Intensity Variations in Image Sequences". Computer Vision, Graphics, and Image Processing 21 (1983) 85-117

H.-H. Nagel, "On the Estimation of Optical Flow: Relations between Different Approaches and Some New Results", Artificial Intelligence 33 (1987) 299-324

C. Schnörr, "Zur Schätzung von Geschwindigkeitsvektorfeldern in Bildfolgen mit einer richtungsabhängigen Glattheitsforderung", 11. DAGM-Symposium Mustererkennung 1989, 2.-4. Oktober 1989, Informatik-Fachberichte 219, H. Burkhardt, K.H. Höhne, B. Neumann (Hrsg.), Springer-Verlag Berlin Heidelberg New York 1989, pp. 294-301

M.A. Snyder, "On the Mathematical Foundations of Smoothness Constraints for the Determination of Optical Flow and for Surface Reconstruction", Proc. Workshop on Visual Motion, March 20-22, 1989, Irvine / CA, pp. 107-115

A COMPARISON OF STOCHASTIC AND DETERMINISTIC SOLUTION METHODS IN BAYESIAN ESTIMATION OF 2-D MOTION [†]

Janusz Konrad [‡] *and Eric Dubois*

INRS-Télécommunications, 3 Place du Commerce
Verdun, Québec, Canada, H3E 1H6

Abstract

A new stochastic motion estimation method based on the Maximum *A Posteriori* Probability (MAP) criterion is developed. Deterministic algorithms approximating the MAP estimation over discrete and continuous state spaces are proposed. These approximations result in known motion estimation algorithms. The theoretical superiority of the stochastic algorithms over deterministic approximations in locating the global optimum is confirmed experimentally.

1. INTRODUCTION

Substantial work has been carried out recently in the application of stochastic models to the estimation of two-dimensional motion. Based on Markov random field (MRF) models, the problem has been formulated using the MAP criterion and solved by stochastic [1],[2],[3],[4] and deterministic [5],[6] methods. Although both approaches have proved successful, no experimental comparison has been carried out. In this paper we extend our previous work by developing stochastic MAP estimation over a continuous state space of solutions. Also, by instantaneously "freezing" a Markov chain produced by a stochastic relaxation algorithm, we propose two deterministic estimation methods. These approximations result in known motion estimation algorithms.

2. FORMULATION

2.1 Terminology

Let u and g denote the true underlying and the observed time-varying images, respectively. Let g be a sample from a random field (RF) G, and be quantized in amplitude and sampled on a lattice Λ_g in R^3. Let $(\mathbf{x}, t)$ be a site in Λ_g, where $\mathbf{x}$ and t denote spatial and temporal positions, respectively. Let also d be the true (unknown) displacement field associated with u. Since it is not feasible to estimate d on a continuum of spatial positions, it will be estimated on a lattice $\Lambda_\mathbf{d}$ in R^3, which may be different than Λ_g as in the case of temporal interpolation.

It is assumed that Λ_g, $\Lambda_\mathbf{d}$ are rectangular lattices with horizontal, vertical and temporal sampling periods (T_g^h, T_g^v, T_g) and $(T_\mathbf{d}^h, T_\mathbf{d}^v, T_\mathbf{d})$, respectively. Each field of the image sequence contains M_g picture elements, and each motion field consists of $M_\mathbf{d}$ vectors.

The true displacement field d is assumed to be a sample (realization) from random field $\mathbf{D}$. Let $\mathbf{d}$ denote any sample field from $\mathbf{D}$ and allow $\hat{\mathbf{d}}$ be an estimate of the true displacement field d. Assuming a linear motion trajectory between two images we define a displacement field as follows:

[†] Work supported by the Natural Sciences and Engineering Research Council of Canada under Strategic Grant G-1357

[‡] This research was conducted when the author was also with the Department of Electrical Engineering, McGill University, Montreal, Canada, on leave from the Technical University of Szczecin, Poland

The displacement field d defined over $\Lambda_{\mathbf{d}}$ is a set of 2-D vectors such that for all $(\mathbf{x}_i, t) \in \Lambda_{\mathbf{d}}$ the *preceding* image point $(\mathbf{x}_i - \Delta t \cdot \mathbf{d}(\mathbf{x}_i, t), t_-)$ has moved to the *following* point $(\mathbf{x}_i + (1.0 - \Delta t) \cdot \mathbf{d}(\mathbf{x}_i, t), t_+)$, where $\Delta t = \frac{t}{T_g} - \left\lfloor \frac{t}{T_g} \right\rfloor$, $t_- = t - \Delta t \cdot T_g$ and $t_+ = t + (1.0 - \Delta t) \cdot T_g$.

To model abrupt changes in displacement vector length and/or orientation we use the concept of motion discontinuity. The true motion discontinuities l are defined over continuous coordinates $(\mathbf{x}, t)$, and are unobservable like the true motion fields. They can be understood as indicator functions for each $(\mathbf{x}, t)$. We assume that l is a sample from a RF L. Let $\hat{l}$ be an estimate of l. The RF L will be called a *line process*, its sample l will be called a *line field* while individual discontinuities from l will be named *line elements*. We will estimate l on a union of shifted lattices $\Psi_l = \psi_h \cup \psi_v$, where $\psi_h = \Lambda_{\mathbf{d}} + [0, T_{\mathbf{d}}^v/2, 0]^T$ and $\psi_v = \Lambda_{\mathbf{d}} + [T_{\mathbf{d}}^h/2, 0, 0]^T$ are orthogonal cosets of horizontal and vertical line elements, respectively.

We assume that the random field $\mathbf{D}_t$ is defined over the state space $\mathcal{S}_{\mathbf{d}} = (\mathcal{S}_{\mathbf{d}}')^{M_{\mathbf{d}}}$, where $\mathcal{S}_{\mathbf{d}}'$ is the single vector state space. Two cases of $\mathcal{S}_{\mathbf{d}}'$ are considered: a discrete state space (square 2-D grid) and a continuous state space R^2. It is also assumed that the random field L is defined over the discrete state space $\mathcal{S}_l = (\mathcal{S}_l')^{M_l}$, where $\mathcal{S}_l'$ is the single line element state space. Finally, let the subscript t denote the restriction of a random field (RF) or of its realization to time t.

2.2 MAP estimation criterion

To estimate the pair (d_t, l_t) of true displacement and line fields corresponding to image u on the basis of the observations g, a pair $(\hat{\mathbf{d}}_t, \hat{l}_t) \in \mathcal{S}_{\mathbf{d}} \times \mathcal{S}_l$ which maximizes the *a posteriori* probability $P(\mathbf{D}_t = \hat{\mathbf{d}}_t, L_t = \hat{l}_t | g_{t_-}, g_{t_+})$ must be found. Applying Bayes rule this probability can be factored as follows [2]

$$P(\mathbf{D}_t = d_t, L_t = l_t | g_{t_-}, g_{t_+}) = \frac{P(G_{t_+} = g_{t_+} | d_t, l_t, g_{t_-}) \cdot P(\mathbf{D}_t = d_t | l_t, g_{t_-}) \cdot P(L_t = l_t | g_{t_-})}{P(G_{t_+} = g_{t_+} | g_{t_-})}. \quad (1)$$

Note that since the probability in the denominator of (1) is not a function of d_t, it can be ignored when maximizing (1) with respect to $(\hat{\mathbf{d}}_t, \hat{l}_t)$. If displacement vectors are defined over a continuous state space $\mathcal{S}_{\mathbf{d}}' = R^2$, then Bayes rule for mixed random variables results in a similar probability distribution where *a priori* probability $P(\mathbf{D}_t = \hat{\mathbf{d}}_t | \hat{l}_t, g_{t_-})$ is replaced by the probability density $p(\hat{\mathbf{d}}_t | \hat{l}_t, g_{t_-})$.

2.2.1 Displaced pel difference model

To estimate motion from images a *structural* model relating motion vectors and image intensity values is needed. Disregarding illumination and occlusion effects we assume that over the time interval $[t_-, t_+]$ the intensity of image u along d is constant i.e., $u(\mathbf{x} + (1.0 - \Delta t) \cdot \mathbf{d}(\mathbf{x}, t), t_+) - u(\mathbf{x} - \Delta t \cdot \mathbf{d}(\mathbf{x}, t), t_-) = 0$. Extrapolating this relationship to the observed image g, which is a transformed and noise-corrupted version of u, we model the displaced pel differences (DPDs)

$$\tilde{r}(\mathbf{d}(\mathbf{x}_i, t), \mathbf{x}_i, t, \Delta t) = \tilde{g}(\mathbf{x}_i + (1.0 - \Delta t) \cdot \mathbf{d}(\mathbf{x}_i, t), t_+) - \tilde{g}(\mathbf{x}_i - \Delta t \cdot \mathbf{d}(\mathbf{x}_i, t), t_-)$$

by independent Gaussian random variables ($\tilde{g}(\mathbf{x}, t)$ denotes an intensity value at $(\mathbf{x}, t) \notin \Lambda_g$ obtained by interpolation). Consequently, the likelihood $P(G_{t_+} = g_{t_+} | d_t, l_t, g_{t_-})$ from (1) can be expressed as the following Gaussian distribution[†]

$$P(G_{t_+} = g_{t_+} | \hat{\mathbf{d}}_t, g_{t_-}) = (2\pi\sigma^2)^{-M_{\mathbf{d}}/2} \cdot e^{-U_g(g_{t_+} | \hat{\mathbf{d}}_t, g_{t_-})/2\sigma^2}, \quad (2)$$

[†] Note that d_t constitutes a complete description of motion and a line field l_t is only an aid in estimation of d_t. Hence, the conditioning on L_t in $P(G_{t_+} = g_{t_+} | \hat{\mathbf{d}}_t, \hat{l}_t, g_{t_-})$ can be dropped.

with energy U_g defined as follows

$$U_g(gt_+|\hat{\mathbf{d}}_t, gt_-) = \sum_{i=1}^{M_d} [\tilde{r}(\hat{\mathbf{d}}(\mathbf{x}_i, t), \mathbf{x}_i, t, \Delta t)]^2. \tag{3}$$

2.2.2 Displacement field model

Since motion fields are smooth functions of spatial position $\mathbf{x}$ (fixed t) except for occasional abrupt changes in vector length and/or orientation, we will model displacement fields $\mathbf{d}_t$ and displacement discontinuities l_t by vector and binary MRFs $(\mathbf{D}_t, L_t)$ [2],[4],[6].

Recall that in (1) the *a priori* displacement model is expressed by the probability (density) $P(\mathbf{D}_t = d_t|l_t, gt_-)$. Since the discontinuity model expressed by $P(L_t = l_t|gt_-)$ depends on the data gt_-, we assume that $\mathbf{D}_t$ can be described by the Gibbs distribution:

$$P(\mathbf{D}_t = d_t|l_t, gt_-) = P(\mathbf{D}_t = d_t|l_t) = \frac{1}{Z_d} e^{-U_d(d_t|l_t)/\beta_d}, \tag{4}$$

where Z_d, β_d are constants and $U_d(d_t|l_t)$ is an energy function defined as:

$$U_d(d_t|l_t) = \sum_{c_d = \{\mathbf{x}_i, \mathbf{x}_j\} \in \mathcal{C}_d} V_d(d_t, c_d) \cdot [1 - l(<\mathbf{x}_i, \mathbf{x}_j>, t)]. \tag{5}$$

c_d is a clique of vectors, while $\mathcal{C}_d$ is a set of all such cliques defined over lattice Λ_d. $(<\mathbf{x}_i, \mathbf{x}_j>, t) \in \Psi_l$ denotes a site of line element located between vector sites $\mathbf{x}_i$ and $\mathbf{x}_j$ which belong to Λ_d. V_d is a *potential function* crucial to characterization of the properties of displacement field d_t.

We specify the *a priori* displacement model by using $\|\mathbf{d}(\mathbf{x}_i, t) - \mathbf{d}(\mathbf{x}_j, t)\|^2$ as the potential function V_d for each clique $c_d = \{\mathbf{x}_i, \mathbf{x}_j\}$, as well as the first-order neighbourhood system $\mathcal{N}_d^1$ with 2-element horizontal and vertical vector cliques [2].

2.2.3 Line field model

Let the line field model be based on a binary MRF L_t with the Gibbs probability distribution

$$P(L_t = l_t|gt_-) = \frac{1}{Z_l} e^{-U_l(l_t|gt_-)/\beta_l}, \quad U_l(l_t|gt_-) = \sum_{c_l \in \mathcal{C}_l} V_l(l_t, gt_-, c_l), \tag{6}$$

where Z_l, β_l are the usual constants, c_l is a line clique and $\mathcal{C}_l$ is a set of all line cliques defined over Ψ_l. The line potential function V_l provides a penalty associated with introduction of a line element. Separate neighbourhood systems are associated with cosets ψ_h and ψ_v [4]. To model the smoothness and continuity of motion boundaries as well as to disallow formation of isolated displacement vectors inconsistent with their neighbours we chose the potential V_{l_4} defined over four-element cliques [4]. We also used potential V_{l_2} for two-element cliques to prevent formation of double contours.

Since the *a priori* probability of the line process (6) is conditioned on the observations, the image information gt_- should be considered when computing the line samples l_t. Similarly to Hutchison *et al.* [7] we assume that an introduction of a motion boundary coincides with an intensity edge. We use the following potential function for one-element cliques:

$$V_{l_1}(l_t, gt_-, c_l) = \begin{cases} \frac{\alpha}{(\nabla_v gt_-)^2} \cdot l_h(<\mathbf{x}_i, \mathbf{x}_j>, t) & \text{for horizontal } c_l = \{\mathbf{x}_i, \mathbf{x}_j\} \\ \frac{\alpha}{(\nabla_h gt_-)^2} \cdot l_v(<\mathbf{x}_i, \mathbf{x}_j>, t) & \text{for vertical } c_l = \{\mathbf{x}_i, \mathbf{x}_j\}, \end{cases} \tag{7}$$

where l_h, l_v are horizontal and vertical line elements, ∇_h, ∇_v are horizontal and vertical components of the spatial gradient at $(<\mathbf{x}_i, \mathbf{x}_j>, t)$, and α is a non-negative constant. The total line potential function $V_l(l_t, gt_-, c_l)$ is simply a sum of V_{l_4}, V_{l_2} and V_{l_1}.

2.3 A posteriori probability

Combining (2), (4) and (6) it follows that probability (1) is Gibbsian with energy function:

$$U(\widehat{\mathbf{d}}_t, \widehat{l}_t, g_{t_-}, g_{t_+}) = \lambda_g \cdot U_g(g_{t_+}|\widehat{\mathbf{d}}_t, g_{t_-}) + \lambda_{\mathbf{d}} \cdot U_{\mathbf{d}}(\widehat{\mathbf{d}}_t|\widehat{l}_t) + \lambda_l \cdot U_l(\widehat{l}_t|g_{t_-}). \tag{8}$$

The conditional energies are defined in (3), (5) and (6) respectively, and $\lambda_g = 1/(2\sigma^2)$, $\lambda_{\mathbf{d}} = 1/\beta_{\mathbf{d}}$, $\lambda_l = 1/\beta_l$. The MAP estimation can be achieved by minimization of energy (8) with respect to $(\widehat{\mathbf{d}}_t, \widehat{l}_t)$. Note that the minimized energy consists of three terms and can be viewed as regularization: U_g describes the ill-posed matching problem of the data g_{t_-}, g_{t_+} by the motion field $\widehat{\mathbf{d}}$, while $U_{\mathbf{d}}$ and U_l are responsible for conforming to the properties of the *a priori* displacement and line models.

3. SOLUTION TO MAP ESTIMATION

The minimization of energy (8) is very complex because of the number of unknowns involved and because of multimodality of the objective function (dependence on $\widehat{\mathbf{d}}_t$ via g). We will carry out the MAP estimation using *simulated annealing* and appropriate deterministic approximations.

3.1 Stochastic optimization via simulated annealing

To implement the MAP estimation using simulated annealing [8], samples from MRFs $\mathbf{D}_t$ and L_t are needed as well as an annealing schedule to control temperature $\mathbf{T}$. We will generate such samples using the *Gibbs sampler* [9] which, like any stochastic relaxation algorithm, produces states according to probabilities of their occurrence i.e., the unlikely states are also generated (however less frequently). This property, incorporated into simulated annealing, allows the algorithm to escape local minima unlike the case of standard methods.

We will use the Gibbs sampler based on the *a posteriori* probability (1) with energy (8). The displacement Gibbs sampler at location $(\mathbf{x}_i, t)$ is driven by a (Gibbs) marginal conditional probability characterized by the following energy function [4]

$$U_{\mathbf{d}}^i(\widehat{\mathbf{d}}(\mathbf{x}_i, t)|\widehat{\mathbf{d}}_t^c, \widehat{l}_t, g_{t_-}, g_{t_+}) = \lambda_g \cdot [\widetilde{r}(\widehat{\mathbf{d}}(\mathbf{x}_i, t), \mathbf{x}_i, t, \Delta t)]^2 +$$
$$\lambda_{\mathbf{d}} \cdot \sum_{j: \mathbf{x}_j \in \eta_{\mathbf{d}}(\mathbf{x}_i)} \|\widehat{\mathbf{d}}(\mathbf{x}_i, t) - \widehat{\mathbf{d}}(\mathbf{x}_j, t)\|^2 \cdot [1 - \widehat{l}(<\mathbf{x}_i, \mathbf{x}_j>, t)], \tag{9}$$

where $\widehat{\mathbf{d}}_t^c = \{\widehat{\mathbf{d}}(\mathbf{x}_j, t) : j \neq i\}$ and $\eta_{\mathbf{d}}(\mathbf{x}_i)$ is a spatial neighbourhood of displacement vector at $\mathbf{x}_i$. Corresponding local energy function U_l^i driving the Gibbs sampler for l_t can be found in [4].

3.1.1 Discrete state space Gibbs sampler

For each candidate vector $\widehat{\mathbf{d}}(\mathbf{x}_i, t) \in \mathcal{S}_d'$, the marginal probability distribution is computed from the local energy (9). Then, two vector coordinates are sampled from this distribution. The necessity to obtain the complete probability distribution of a displacement vector at each $\mathbf{x}_i$ is decisive in the computational complexity of the discrete state space Gibbs sampler. A similar procedure applies to line elements, except that the state space $\mathcal{S}_l'$ is binary.

3.1.2 Continuous state space Gibbs sampler for $\mathbf{D}_t$

We avoid a very fine quantization of $\mathcal{S}_d'$ (to obtain the continuous state space) by approximating the local energy (9) by a quadratic form in $\widehat{\mathbf{d}}_t$ so that the Gibbs sampler is driven by a Gaussian probability distribution.

Assume that an approximate estimate $\dot{\mathbf{d}}_t$ of the true displacement field is known, and that the image intensity is locally approximately linear. Then, using the first-order terms of the Taylor expansion the DPD $\tilde{r}$ can be expressed as follows:

$$\tilde{r}(\hat{\mathbf{d}}(\mathbf{x}_i,t),\mathbf{x}_i,t,\Delta t) \approx \tilde{r}(\dot{\mathbf{d}}(\mathbf{x}_i,t),\mathbf{x}_i,t,\Delta t) + (\hat{\mathbf{d}}(\mathbf{x}_i,t) - \dot{\mathbf{d}}(\mathbf{x}_i,t)) \cdot \nabla_{\mathbf{d}}\tilde{r}(\dot{\mathbf{d}}(\mathbf{x}_i,t),\mathbf{x}_i,t,\Delta t),$$

where the spatial gradient of $\tilde{r}$ is defined as

$$\nabla_{\mathbf{d}}\tilde{r}(\dot{\mathbf{d}}(\mathbf{x}_i,t),\mathbf{x}_i,t,\Delta t) = \begin{bmatrix} \tilde{r}^x(\dot{\mathbf{d}}(\mathbf{x}_i,t),\mathbf{x}_i,t,\Delta t) \\ \tilde{r}^y(\dot{\mathbf{d}}(\mathbf{x}_i,t),\mathbf{x}_i,t,\Delta t) \end{bmatrix}. \tag{10}$$

$\tilde{r}^x$ and $\tilde{r}^y$ are computed as an average of appropriate derivatives at the end points of vector $\dot{\mathbf{d}}(\mathbf{x}_i,t)$ [10]. Including the temperature $\mathbf{T}$ the local energy $U_{\mathbf{d}}^i$ can be written as follows

$$U_{\mathbf{d}}^i(\hat{\mathbf{d}}(\mathbf{x}_i,t)|\hat{\mathbf{d}}_t^c,\hat{l}_t,g_{t_-},g_{t_+}) \approx \frac{\lambda_g}{\mathbf{T}} \cdot [\tilde{r}(\dot{\mathbf{d}}(\mathbf{x}_i,t),\mathbf{x}_i,t,\Delta t) + (\hat{\mathbf{d}}(\mathbf{x}_i,t) - \dot{\mathbf{d}}(\mathbf{x}_i,t)) \cdot \nabla_{\mathbf{d}}\tilde{r}(\dot{\mathbf{d}}(\mathbf{x}_i,t),\mathbf{x}_i,t,\Delta t)]^2$$

$$+ \frac{\lambda_{\mathbf{d}}}{\mathbf{T}} \cdot \sum_{j:\,\mathbf{x}_j \in \eta_{\mathbf{d}}(\mathbf{x}_i)} \|\hat{\mathbf{d}}(\mathbf{x}_j,t) - \hat{\mathbf{d}}(\mathbf{x}_i,t)\|^2 \cdot [1 - \hat{l}(<\mathbf{x}_i,\mathbf{x}_j>,t)],$$

where $\dot{\mathbf{d}}$ is fixed. It can be shown that the conditional probability density with the above energy is a 2-D Gaussian with the following mean vector $\mathbf{m}$ at location $(\mathbf{x}_i,t)$ [10]:

$$\mathbf{m} = \overline{\mathbf{d}}(\mathbf{x}_i,t) - \frac{\varepsilon_i}{\mu_i}\nabla_{\mathbf{d}}^T\tilde{r}(\dot{\mathbf{d}}(\mathbf{x}_i,t),\mathbf{x}_i,t,\Delta t),$$

where the scalars ε_i and μ_i are defined as follows

$$\varepsilon_i = \tilde{r}(\dot{\mathbf{d}}(\mathbf{x}_i,t),\mathbf{x}_i,t,\Delta t) + (\overline{\mathbf{d}}(\mathbf{x}_i,t) - \dot{\mathbf{d}}(\mathbf{x}_i,t)) \cdot \nabla_{\mathbf{d}}\tilde{r}(\dot{\mathbf{d}}(\mathbf{x}_i,t),\mathbf{x}_i,t,\Delta t)$$

$$\mu_i = \xi_i \frac{\lambda_{\mathbf{d}}}{\lambda_g} + \|\nabla_{\mathbf{d}}\tilde{r}(\dot{\mathbf{d}}(\mathbf{x}_i,t),\mathbf{x}_i,t,\Delta t)\|^2, \tag{11}$$

and $\overline{\mathbf{d}}(\mathbf{x}_i,t)$ is an average vector

$$\overline{\mathbf{d}}(\mathbf{x}_i,t) = \frac{1}{\xi_i} \sum_{j:\,\mathbf{x}_j \in \eta_{\mathbf{d}}(\mathbf{x}_i)} \hat{\mathbf{d}}(\mathbf{x}_j,t) \cdot [1 - \hat{l}(<\mathbf{x}_i,\mathbf{x}_j>,t)], \tag{12}$$

with $\xi_i = \sum_{j:\,\mathbf{x}_j \in \eta_{\mathbf{d}}(\mathbf{x}_i)}[1 - \hat{l}(<\mathbf{x}_i,\mathbf{x}_j>,t)]$. Note that averaging is disallowed across a motion boundary, which is a desirable property. The horizontal and vertical component variances σ_x^2, σ_y^2, as well as the correlation coefficient ρ, which comprise the covariance matrix, have the following form

$$\begin{bmatrix} \sigma_x^2 \\ \sigma_y^2 \end{bmatrix} = \frac{\mathbf{T}}{2\xi_i\lambda_{\mathbf{d}}\mu_i} \begin{bmatrix} \xi_i\frac{\lambda_{\mathbf{d}}}{\lambda_g} + [\tilde{r}^y(\dot{\mathbf{d}}(\mathbf{x}_i,t),\mathbf{x}_i,t,\Delta t)]^2 \\ \xi_i\frac{\lambda_{\mathbf{d}}}{\lambda_g} + [\tilde{r}^x(\dot{\mathbf{d}}(\mathbf{x}_i,t),\mathbf{x}_i,t,\Delta t)]^2 \end{bmatrix}$$

$$\rho\sigma_x\sigma_y = \mathbf{T}\frac{-\tilde{r}^x(\dot{\mathbf{d}}(\mathbf{x}_i,t),\mathbf{x}_i,t,\Delta t)\tilde{r}^y(\dot{\mathbf{d}}(\mathbf{x}_i,t),\mathbf{x}_i,t,\Delta t)}{2\xi_i\lambda_{\mathbf{d}}\mu_i}.$$

The initial vector $\mathbf{d}$ can be assumed zero throughout the estimation process, but then with increasing displacement vector estimates the error due to intensity non-linearity would significantly increase. Hence, it is better to "track" an intensity pattern by modifying $\dot{\mathbf{d}}$ accordingly. An interesting result can be obtained when it is assumed that at every iteration of the Gibbs sampler $\dot{\mathbf{d}} = \overline{\mathbf{d}}$ i.e., the initial (approximate) displacement field is equal to the average from the previous iteration. Then, the estimation process can be described by the following iterative equation:

$$\hat{\mathbf{d}}^{n+1}(\mathbf{x}_i,t) = \overline{\mathbf{d}}^n(\mathbf{x}_i,t) - \frac{\varepsilon_i}{\mu_i}\nabla_{\mathbf{d}}^T\tilde{r}(\overline{\mathbf{d}}^n(\mathbf{x}_i,t),\mathbf{x}_i,t,\Delta t) + n_i, \tag{13}$$

where n is the iteration number, and ε_i, μ_i and the covariance matrix are defined as before except for $\dot{\mathbf{d}} = \overline{\mathbf{d}}$. At the beginning, when the temperature is high, the random term n_i has a large variance and the estimates assume quite random values. As the temperature $\mathbf{T} \to 0$, $\sigma_x^2, \sigma_y^2, \rho\sigma_x\sigma_y$ get smaller, thus reducing n_i. In the limit the algorithm performs a deterministic update. Note that the variance σ_x^2 of the horizontal vector component for fixed values of $\lambda_g, \lambda_\mathbf{d}, \lambda_l$ and $\tilde{r}^y$ decreases with growing $\tilde{r}^x$. It means that when there is a significant horizontal gradient (detail) in the image structure the uncertainty of the estimate in horizontal direction is small. The same applies to σ_y^2. Hence, the algorithm takes into account the image structure in determining the amount of randomness allowed at a given temperature.

Note the similarity of the iterative update equation (13) to the update equation of the Horn-Schunck algorithm [11]. Except for ε_i equal to DPD instead of the motion constraint equation and except for different image model used, they are identical for $\mathbf{T}=0$. It is interesting that similar update equations result from two different approaches: Horn and Schunck establish necessary conditions for optimality and solve them by deterministic relaxation, while here a 2-D Gaussian distribution is fitted into the conditional probability driving the Gibbs sampler.

3.2 Deterministic optimization using steepest descent method

3.2.1 Discrete state space

Note that for $\mathbf{T}=0$ (deterministic update) the discrete state space Gibbs sampler generates only states with minimum local energy (9). Hence, the final result is only an approximation to the MAP estimate.

Besag [12] proposed a similar approach called *iterated conditional modes* (ICM). He argued that since it is difficult to maximize the joint *a posteriori* probability over the complete field, it should be divided into a minimal number of disjoint sets (or colours) such that any two random variables from a given set are conditionally independent given the states of the other sets. Using this approach displacement vectors or line elements can be computed individually (e.g., exhaustive search) for each location $(\mathbf{x}_i, t)$ one colour at a time. Note that also this technique does not result in maximization of probability (1), but provides separate MAP estimates for joint probabilities defined over corresponding colours. The difference between the ICM method and the Gibbs sampler with $\mathbf{T}=0$ is only the update order of variables [10]. Both techniques can be classified as a (pel) matching algorithm with smoothness constraint.

3.2.2 Continuous state space

Let the displacement vector state space be continuous ($\mathcal{S}_\mathbf{d}' = R^2$). The energy function under minimization (8) is non-quadratic in $\hat{\mathbf{d}}_t$ as well as in $\hat{l}_t$. We will perform interleaved optimization with respect to $\mathbf{d}_t$ and l_t. If $\hat{l}_t$ is known, then U_l in (8) is constant and only minimization of $\lambda_g U_g + \lambda_\mathbf{d} U_\mathbf{d}$ must be performed. Using the linearization of the DPD $\tilde{r}$ and establishing necessary conditions for optimality at each location $\mathbf{x}_i$ as well as assuming that $\dot{\mathbf{d}} = \overline{\mathbf{d}}$, it follows that the iterative update for this deterministic method is [10]:

$$\hat{\mathbf{d}}^{n+1}(\mathbf{x}_i, t) = \overline{\mathbf{d}}^n(\mathbf{x}_i, t) - \frac{\varepsilon_i}{\mu_i} \nabla_\mathbf{d}^T \tilde{r}(\overline{\mathbf{d}}^n(\mathbf{x}_i, t), \mathbf{x}_i, t, \Delta t), \tag{14}$$

where ε_i and μ_i are defined in (11) with $\dot{\mathbf{d}} = \overline{\mathbf{d}}$. To resemble the Gibbs sampler as close by as possible, the Gauss-Seidel relaxation will be used in (14) rather than the Jacobi relaxation. Once an estimate $\hat{\mathbf{d}}_t$ is known, an improved estimate $\hat{l}_t$ should be obtained. For a fixed $\hat{\mathbf{d}}_t$ the minimized energy $\lambda_\mathbf{d} U_\mathbf{d} + \lambda_l U_l$ is non-linear in $\hat{l}_t$. Since $l(\mathbf{x}_i, t)$ is binary for each i, the ICM method reported above can be used.

The above approximation to the continuous state space MAP estimation is a spatio-temporal gradient technique which can be viewed as a modified version of the Horn and Schunck algorithm, except that:

1. the modified algorithm (14) allows computation of displacement vectors for arbitrary Λ_d unlike the original Horn and Schunck algorithm in which $\Lambda_d = \Lambda_g + [0.5, 0.5, 0.5]^T$,

2. the scalar ε_i is a displaced pel difference in the modified version rather than a motion constraint equation: no temporal derivative is needed,

3. the spatial intensity derivatives are computed from a separable polynomial model in both images and appropriately weighted (10), instead of the finite difference approximation over a cube as proposed in [11].

The ability to estimate motion for arbitrary Λ_d is crucial for motion-compensated interpolation of sequences (original Horn-Schunck algorithm would require 3-D interpolation of motion fields).

The use of $\tilde{r}$ instead of the motion constraint equation in ε_i is important because it allows intensity pattern tracking thus permitting more accurate intensity derivative computation, and also does not require the computation of the purely temporal derivative (actually, $\tilde{r}$ is an approximation to the directional derivative). The purely temporal derivative used in the Horn-Schunck algorithm is a reliable measure of temporal intensity change due to motion as long as small displacements are applied to linearly varying intensity pattern. Otherwise, significant errors may result, for example an overestimation at moving edges of high contrast.

The deterministic algorithm (14) together with the ICM method for l_t is related to the algorithm proposed in [7]. The major differences are those reported above for the Horn-Schunck algorithm as well as the line potentials: the potential V_{l_1} for single-element cliques is binary (0 or ∞) in [7] while here it varies continuously according to the local intensity gradient.

4. EXPERIMENTAL RESULTS

The algorithms described above have been tested on a number of images with synthetic and natural motion. Results for two of images with natural data and inter-field distance $\tau_{60} = 1/60$ sec. are presented below.

To provide a quantitative test we generated test image 1 (Fig. 1(a)) with stationary background provided by the test image from Fig. 1(b) and a moving rectangle $\mathcal{R}$ obtained from another image through low-pass filtering, subsampling and pixel shifting. This test pattern permits non-integer displacements so that there is no perfect data matching. Fig. 1(b) shows the test image 2 containing natural motion, acquired by a video camera.

The stochastic relaxation used was based either on the discrete state space $\mathcal{S}_d'$ with maximum displacement ± 2.0 pixels and 17 quantization levels in each direction or on the continuous state space R^2. The first-order displacement neighbourhood system and, if applicable, the line neighbourhood with four-, two- and one-element line cliques, as proposed in [4], have been used. The ratio $\lambda_d/\lambda_g = 20.0$ has been chosen experimentally, however, as pointed out in [10], even a change of 2 orders magnitude did not have an excessively severe impact on the estimate quality. The motion estimates presented in the sequel have been obtained from pairs of images (fields) separated by $T_g = 2\tau_{60}$. All estimates have been obtained with Keys bicubic interpolator [10] except for the discrete state space estimation applied to test image 2, when bilinear interpolation was used.

 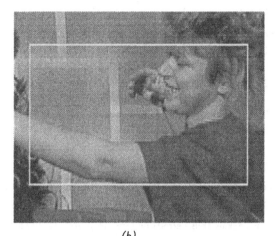

(a)   *(b)*

Fig. 1 Test images: (a) synthetic motion, $\mathbf{d}_s = [1.5, 0.5]$, $|\mathcal{R}| = 45 \times 20$, (b) natural motion (white frame outlines the area used in estimation: (a) 77×49, (b) 221×69).

$MSE = (0.1358, 0.0326)$, $b = (0.2008, 0.0831)$ $MSE = (0.9408, 0.1599)$, $b = (0.8100, 0.3467)$

(a) stochastic MAP   *(b)* deterministic ICM

Fig. 2 Discrete state space MAP and ICM estimates for the test image 1: $\Lambda_\mathbf{d} = \Lambda_g$.

Since the true motion field is known for the test image with synthetic motion (except for the occlusion and newly exposed areas), it is possible to assess the quality of motion field estimates. The Mean Squared Error (MSE) and the bias (b) measuring the departure of estimate $\hat{\mathbf{d}}$ from the known motion field $\mathbf{d}_s$, are computed within the rectangle and shown below appropriate estimates.

Fig. 2 shows the discrete state space MAP and ICM displacement estimates from the test image 1. The stochastic MAP estimate is superior to the ICM estimate both subjectively and objectively (MSE, b). In both cases the zero displacement field has been used as an initial state. In other experiments ML estimates ($\lambda_\mathbf{d}/\lambda_g = 0.0$) have been computed and used as a starting point (as suggested by Besag for ICM estimation). The ML estimates were characterized by substantial randomness in vector lengths and orientations, which can be explained by the lack of a prior model. As expected the initial state had no impact on the stochastic MAP estimate, but the final ICM estimate was inferior to the ICM estimate presented above both subjectively and in terms of MSE.

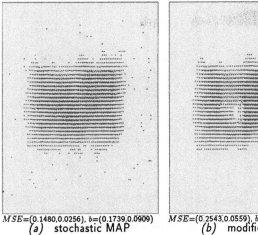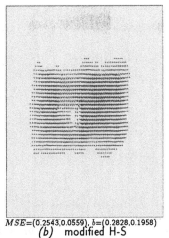

$MSE=(0.1480,0.0256)$, $b=(0.1739,0.0909)$ $MSE=(0.2543,0.0559)$, $b=(0.2828,0.1958)$ $MSE=(0.4499,0.0592)$, $b=(0.5205,0.1904)$
     (a) stochastic MAP   *(b)* modified H-S   *(c)* original H-S

Fig. 3 Continuous state space MAP, modified and original Horn-Schunck estimates for test image 1: $\Lambda_d = \Lambda_g + [0.5, 0.5, 0.5]^T$.

To compare the stochastic MAP estimate and its deterministic approximation (modified Horn-Schunck method) with the original Horn-Schunck algorithm (Fig. 3), the condition that $\Lambda_d = \Lambda_g + [0.5, 0.5, 0.5]^T$ was imposed. Note that the Horn-Schunck algorithm produces the worst result, both subjectively and in terms of MSE. The motion tends to be overestimated at strong edges (due to the purely temporal gradient), while it is underestimated in uniform areas. The deterministic approximation has produced a significantly lower MSE, and also subjectively the estimate is more uniform. Except for the visible triangle of underestimated displacements, the motion has been quite well computed. Superiority of the stochastic approach is clear from Fig. 3.a. Subjectively this estimate is closest to the true motion, MSE is the lowest of the three estimates and also the total energy is lower than for the deterministic approximation (original Horn-Schunck algorithm cannot be compared in terms of energy since is assumes different intensity model).

Fig. 4 shows the stochastic and deterministic estimates for the piecewise smooth motion model. The parameters used are the same as before. During experimentation we have observed that the ratio λ_l / λ_d had to be substantially lower for the deterministic algorithm in order to obtain results comparable with the stochastic MAP estimation. This may be explained by explicit averaging used in the deterministic algorithm. The continuous state space MAP estimation uses similar averaging, but it also involves a randomness factor thus allowing switching line elements off and on, even if motion discontinuity does not quite allow it. Note that both subjectively and in terms of MSE the deterministic estimate is clearly inferior.

Fig. 5 shows the discrete state space MAP and ICM displacement estimates for the test image 2. The ICM estimate is again subjectively poorer than the stochastic MAP estimate. The ICM algorithm failed to compute correctly the motion of the forearm and of the arm, except for the displacement vectors along the edge of the shirt sleeve. Also the vectors on the neck and parts of the face suggest that there is no motion, which is incorrect.

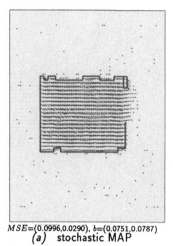

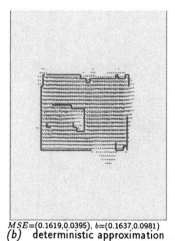

$MSE=(0.0996,0.0290), b=(0.0751,0.0787)$
(a) stochastic MAP

$MSE=(0.1619,0.0395), b=(0.1637,0.0981)$
(b) deterministic approximation

Fig. 4 Continuous state space stochastic and deterministic MAP estimates with piecewise smooth motion model for test image 1: $\Lambda_d = \Lambda_g$, $\lambda_l/\lambda_d = 0.8$ (a) and 0.15 (b), $\alpha = 10.0$.

Similarly, the three continuous state space methods have been applied to the test image 2 (Fig. 6). The original Horn-Schunck estimate shows some overestimated vectors (edge of shirt sleeve) and numerous underestimated ones (uniform area to the right). The deterministic approximation performs better: it is more uniform and has smaller edge effects. The stochastic estimate, however, is superior in terms of the total energy U as well as motion field smoothness and lack of edge effects.

5. CONCLUSION

In this paper two types of solution methods to the problem of 2-D motion estimation via the MAP criterion have been presented and compared: stochastic and deterministic. It has been demonstrated that, as an example of instantaneous freezing, the deterministic methods may be incapable of localizing the global minimum not only theoretically but also in practice. Higher value of the energy function for the deterministic solutions was confirmed by inferior subjective and objective (synthetic motion) quality. Such an improvement in estimate quality comes at a cost of increased computational effort, however. The computational overhead (per iteration) of the continuous state space stochastic estimation compared to its deterministic approximation is small (less than 25%) because it includes only the computation of the random update term. The number of iterations required to provide sufficiently slow annealing schedule, however, makes the stochastic method more involved computationally by about an order of magnitude.

REFERENCES

[1] D.W. Murray and B.F. Buxton, "Scene segmentation from visual motion using global optimization," *IEEE Trans. Pattern Anal. and Mach. Intell.*, vol. PAMI-9, pp. 220–228, March 1987.

[2] J. Konrad and E. Dubois, "Estimation of image motion fields: Bayesian formulation and stochastic solution," in *Proc. IEEE Int. Conf. on Acoust., Speech, and Signal Process. ICASSP'88*, 1988, pp. 1072–1074.

[3] J. Konrad and E. Dubois, "Multigrid Bayesian estimation of image motion fields using stochastic relaxation," in *Proc. IEEE Int. Conf. Computer Vision ICCV'88*, 1988, pp. 354–362.

[4] J. Konrad and E. Dubois, "Bayesian estimation of discontinuous motion in images using simulated annealing," in *Proc. Conf. Vision Interface VI'89*, 1989, pp. 51–60.

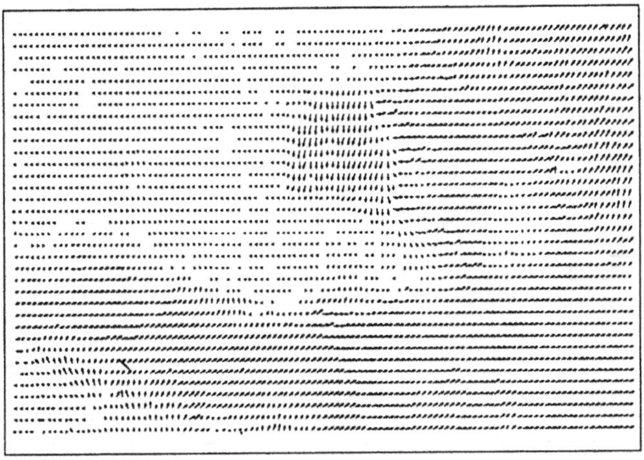

(a) stochastic MAP

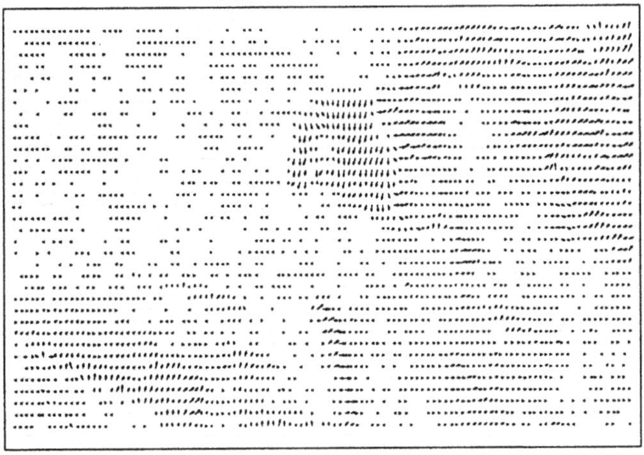

(b) deterministic ICM

Fig. 5 Discrete state space MAP and ICM estimates for the test image 2: $\Lambda_d = \Lambda_g$.

[5] P. Bouthemy and P. Lalande, "Motion detection in an image sequence using Gibbs distributions," in *Proc. IEEE Int. Conf. on Acoust., Speech, and Signal Process. ICASSP'89*, 1989, pp. 1651–1654.

[6] F. Heitz and P. Bouthemy, "Estimation et segmentation du mouvement: approche bayésienne et modelisation markovienne des occlusions," in *Proc. of 7-th Congress AFCET - RFIA*, 1989.

[7] J. Hutchinson, Ch. Koch, J. Luo and C. Mead, "Computing motion using analog and binary resistive networks," *Computer*, vol. 21, pp. 52–63, March 1988.

[8] S. Kirkpatrick, C.D. Gelatt, Jr. and M.P. Vecchi, "Optimization by simulated annealing," *Science*, vol. 220, pp. 671–680, May 1983.

[9] S. Geman and D. Geman, "Stochastic relaxation, Gibbs distributions, and the Bayesian restoration of images," *IEEE Trans. Pattern Anal. and Mach. Intell.*, vol. PAMI-6, pp. 721–741, November 1984.

[10] J. Konrad, "Bayesian estimation of motion fields from image sequences," Ph.D. Thesis, McGill University, Dept. of Electr. Eng., 1989.

[11] B.K.P. Horn and B.G. Schunck, "Determining optical flow," *Artificial Intelligence*, vol. 17, pp. 185–203, 1981.

[12] J. Besag, "On the statistical analysis of dirty pictures," *J. R. Statist. Soc.*, vol. 48, B, pp. 259–279, 1986.

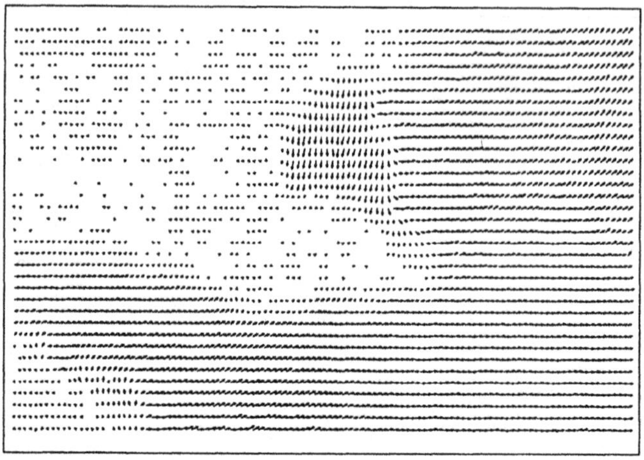

(a) stochastic MAP

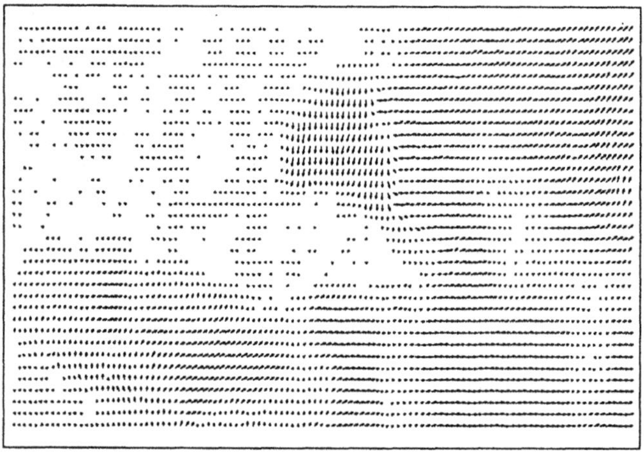

(b) modified H-S

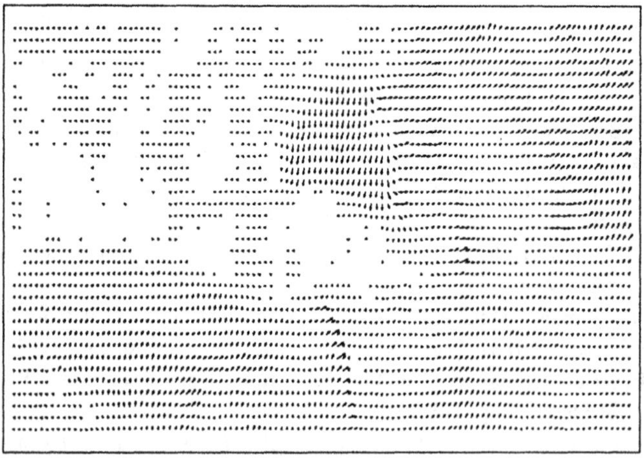

(c) original H-S

Fig. 6 Continuous state space MAP, modified and original Horn-Schunck estimates for test image 2: $\Lambda_d = \Lambda_g + [0.5, 0.5, 0.5]^T$.

Motion Determination in Space-Time Images

Bernd Jähne

Physical Oceanography Research Division A-030, Scripps Institution of Oceanography
La Jolla, CA 92093, USA
and
Institut für Umweltphysik, Universität Heidelberg, Im Neuenheimer Feld 366
D-6900 Heidelberg, Federal Republik of Germany

Abstract

A new approach to determine motion from multiple images of a sequence is presented. Motion is regarded as orientation in a three-dimensional space with one time and two space coordinates. The algorithm is analogous to an eigenvalue analysis of the inertia tensor. Besides the determination of the displacement vector field it allows the classification of four regions with regard to motion: a) constant regions, where no velocity determination is possible; b) edges, where the velocity component perpendicular to the edge is determined; c) corners, where both components of the velocity vector are calculated; d) motion discontinuities, which are used to mark the boundaries between objects moving with different velocities.

The accuracy of the new algorithm has been tested with artificially generated image sequences with known velocity vector fields. An iterative refinement technique yields more accurate results than the usage of higher order approximations to the first spatial and temporal derivatives. Temporal smoothing significantly improves the velocity estimates in noisy images. Displacements between consecutive images can be computed with an accuracy well below 0.1 pixel distances.

1 Introduction

Classical image sequence processing analyses motion from only two consecutive images of a sequence [16,17,18]. Since digital image processing hardware has become powerful enough to store, process, and display image sequences with many frames, considerable efforts have been made to extend these approaches to the simultaneous analysis of many images or the whole sequence. *Heeger* [6] used a set of similar but differently oriented space-time quadrature filter to determine the displacement vector field. A least square method is applied to compute the two-dimensional displacement vector field from the set of filter responses. In a similar approach, *Fleet and Jepson* [4] decompose an image sequence by a family of spatiotemporal velocity-tuned linear filters, but they calculate the velocity component normal to the filtered spatial orientation from local phase information. In a second step, the normal displacements are combined to gain the two-dimensional displacement vector.

Common to both approaches is the usage a large sets of velocity-tuned filters. It arises the question whether it is not possible to calculate the displacement more directly.

In this paper, a technique is discussed which originates from an analysis of the local orientation in the three-dimensional image sequence space with one time and two space coordinates. Such an approach to motion analysis has a long history in biological vision [1], but the developments presented here have rather been triggered by a complex physical application for image sequence processing, the analysis of the motions of small scale waves on a water surface [9,10,8,12].

In the first section of the paper the concept of motion determination by analysis of orientation is outlined and it is shown that determining the local orientation in the xt-space is equivalent to the eigenvalue analysis of the inertia tensor.

The rest of the paper is committed to the important issue of the accuracy of the velocity determination. Basically, three classes of errors can be distinguished:

- errors inherent to the algorithms used to compute the displacement vector field;
- errors caused by the imaging sensor such as signal noise, nonuniformity of the sensor elements, and geometrical distortions;
- errors due to the fact that the optical flow on the image plane and the two-dimensional motion field which is the perspective projection of the three-dimensional motion field in the observed scene are not identical. In a recent paper, *Verri and Poggio* [21] nicely demonstrate that both are in general different. *Nagel* [19] discusses additional terms to be included in the constraint equations for optical flow.

This paper deals only with the first class of errors. To separate algorithm related errors from sensor related errors, only computed image sequences have been used.

2 Motion and Orientation in xt-Space

2.1 The Concept

As an introduction to the concept, let us take an object $g(\boldsymbol{x})$ in the image sequence which moves with constant speed $\boldsymbol{u}$. In this case it can be described by

$$g(\boldsymbol{x}, t) = g(\boldsymbol{x} - \boldsymbol{u}t) \tag{1}$$

This equation is known as the general solution of the differential equation for waves in a non-dispersive medium in physics. The usage of the three-dimensional space with one time and two space coordinates offers the advantage that motion can also be analyzed in the corresponding Fourier space which will be denoted by $\boldsymbol{k}\omega$-space. Correspondingly, the abbreviation xt-space will be used.

An object moving with a constant velocity has a simple representation in the $\boldsymbol{k}\omega$-space. Fourier transformation of (1) gives

$$\hat{g}(\boldsymbol{k}, \omega) = \hat{g}(\boldsymbol{k})\,\delta(\boldsymbol{k}\boldsymbol{u} - \omega) \tag{2}$$

where δ is the Dirac distribution. With constant motion only one plane in the $\boldsymbol{k}\omega$-space is occupied by the wavenumber spectrum $\hat{g}(\boldsymbol{k})$ of the object which is given by

$$\omega = \boldsymbol{k}\boldsymbol{u} \tag{3}$$

This plane intersects the $k_1 k_2$-plane perpendicularly to the direction of the motion, because in this direction the scalar product $\boldsymbol{k}\boldsymbol{u}$ vanishes. The slope of the plane is proportional to the velocity.

The plane cannot be determined unambiguously if the wavenumber spectra lies on a line. In this case, the spatial structure of the object is oriented only in one direction (local spatial orientation respectively edge-like structure). Then only the velocity component perpendicularly to the edge can be determined. This problem is the well known *aperture problem* [16,7].

So far, the considerations have been limited to continuous xt- and $k\omega$-spaces. Yet all conclusions remain valid for discrete image sequences if the sampling theorem is satisfied. Then the image sequence can be exactly reconstructed from its samples. For image sequences there is a sampling condition both for the space and time coordinates [7] which requires that both the highest frequencies and wavenumbers are sampled at least twice per period and wavelength respectively.

2.2 The Analogy: Eigenvalue Analysis of the Inertia Tensor

In the previous section it has been discussed that motion and orientation analysis are equivalent. Therefore algorithms for orientation can also be used for motion analysis. One approach uses a set of directional quadrature filter pairs [14]. All filters show the same shape but only differ in the direction they select. Combining the filter outputs in a suitable way gives an estimate of the orientation. Such a procedure has been implemented by Heeger [6] for two-dimensional motion analysis using a set of xt-Gabor filters.

A more direct way has recently be presented by *Bigün and Granlund* [2]. They point out that the determination of local orientation in a multidimensional space is equivalent to a line fit through the origin in Fourier space. Local orientation in a neighborhood corresponds to the mathematical term of *linear symmetry* defined by

$$g(\boldsymbol{x}) = g(\boldsymbol{x}\boldsymbol{k}_0) \tag{4}$$

where $\boldsymbol{k}_0$ denotes the orientation of the greyvalue structure. For the sake of simplicity time is regarded as one component of the $\boldsymbol{x}$-vector and the frequency ω as one component of the $\boldsymbol{k}$-vector. The greyvalue is constant in a plane perpendicular to $\boldsymbol{k}_0$. Thus linear symmetry is equivalent to a line in the Fourier space.

The analysis of the distribution of the spectral energy in the $k\omega$-space can be performed by referring to a physical analogy. If we consider the spectral density as the density of a rotary body rotating about the axis $\boldsymbol{k}_0$, the inertia is given by

$$J = \int\limits_{-\infty}^{\infty} d^2(\boldsymbol{k}, \boldsymbol{k}_0)|\hat{g}(\boldsymbol{k})|^2 \, d\boldsymbol{k} \tag{5}$$

where d is the (Euclidian) distance function between the vector $\boldsymbol{k}$ and the line presented by $\boldsymbol{k}_0$.

Using this analogy, it will now be shown that an *eigenvalue analysis* of the inertia tensor will allow a motion analysis. The inertia tensor corresponding to the inertia defined by (5) has the following elements [5,2]

$$
\begin{aligned}
\text{diagonal elements} \qquad J_{ii} &= \sum_{j \neq i} \int\limits_{-\infty}^{\infty} k_j^2 |\hat{g}(\boldsymbol{k})|^2 d\boldsymbol{k} \\
\text{nondiagonal elements} \quad J_{ij} &= -\int\limits_{-\infty}^{\infty} k_i k_j |\hat{g}(\boldsymbol{k})|^2 d\boldsymbol{k}
\end{aligned}
\tag{6}
$$

Now let us consider different shapes of the spectral distribution:

- *Point at origin.* This corresponds to a region of constant greyvalues. The inertia is zero for rotation about all possible axes, consequently all eigenvalues of the inertia tensor are zero. No motion can be detected since no plane can be fitted through a point at the origin.
- *Line through origin.* In this case, a spatially oriented pattern is moving with constant speed. It is only possible to detect the velocity component perpendicularly to the spatial orientation. One eigenvalue of the inertia tensor is zero, since the rotation about the axis coinciding with the line has no inertia. The orientation of this line and thus the spatial orientation and the normal velocity are given by the eigenvector to the eigenvalue zero.
- *Plane trough origin.* This case corresponds to a region of constant motion with a spatially distributed pattern. Rotation about a axis normal to the plane has a maximum inertia, thus the eigenvector to the maximum eigenvalue gives the orientation of the plane and thus both components of the velocity.
- *Three-dimensional rotary body.* Now motion is no longer constant in the region.

2.3   Calculation of the inertia tensor in xt-space

The tensor elements (6) can readily be calculated in the $\boldsymbol{x}$-space since they contain scalar products of the form

$$k_j^2|\hat{g}(\boldsymbol{k})|^2 = |\mathrm{i}k_j\hat{g}(\boldsymbol{k})|^2 \tag{7}$$

and

$$k_ik_j|\hat{g}(\boldsymbol{k})|^2 = \mathrm{i}k_i\hat{g}(\boldsymbol{k})\,[\mathrm{i}k_j\hat{g}(\boldsymbol{k})]^* \tag{8}$$

where the superscript * denotes the conjugate complex. According to Parseval's theorem the integral (6) in $\boldsymbol{k}$-space can also be performed in the $\boldsymbol{x}$-space using the inverse Fourier transform of the corresponding expressions:

$$
\begin{aligned}
\text{diagonal elements} \qquad J_{ii} &= \sum_{j\neq i}\int_{-\infty}^{\infty}\left(\frac{\partial g}{\partial x_j}\right)^2\mathrm{d}\boldsymbol{x}\\[2mm]
\text{nondiagonal elements}\quad J_{ij} &= -\int_{-\infty}^{\infty}\frac{\partial g}{\partial x_i}\frac{\partial g}{\partial x_j}\,\mathrm{d}\boldsymbol{x}
\end{aligned}
\tag{9}
$$

Finally, a weighting function $w(\boldsymbol{x})$ is used to limit the determination of the inertia tensor to a certain local neighborhood in the image sequence:

$$
\begin{aligned}
\text{diagonal elements}\qquad J_{ii}(\boldsymbol{x}_0) &= \sum_{j\neq i}\int_{-\infty}^{\infty} w(\boldsymbol{x}-\boldsymbol{x}_0)\left(\frac{\partial g}{\partial x_j}\right)^2\mathrm{d}\boldsymbol{x}\\[2mm]
\text{nondiagonal elements}\quad J_{ij}(\boldsymbol{x}_0) &= -\int_{-\infty}^{\infty} w(\boldsymbol{x}-\boldsymbol{x}_0)\frac{\partial g}{\partial x_i}\frac{\partial g}{\partial x_j}\,\mathrm{d}\boldsymbol{x}
\end{aligned}
\tag{10}
$$

The width of the weighting function w determines the spatial resolution of the algorithm. In discrete images the operations contained in (10) can be performed as convolutions with appropriate operators and summarized in the operator expression

$$
\begin{aligned}
\text{diagonal elements ii}\qquad & \sum_{j\neq i}\boldsymbol{B}(\boldsymbol{D}_j\bullet\boldsymbol{D}_j)\\[2mm]
\text{nondiagonal elements ij}\quad & \boldsymbol{B}(\boldsymbol{D}_i\bullet\boldsymbol{D}_j)
\end{aligned}
\tag{11}
$$

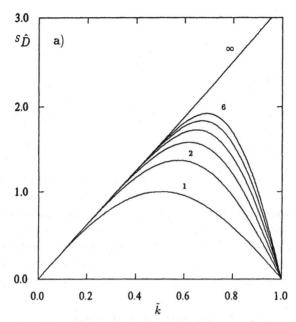

Fig. 1: Transfer function (imaginary part, the real part is zero) of the convolution masks (13) used to approximate the derivative operator. The straight line shows the transfer function of an ideal derivative operator.

The symbol D_i denotes a partial derivation with respect to the coordinate i, B a smoothing operator. Smoothing is performed over all coordinates. Finally, the symbol $\bullet$ denotes the pointwise multiplication of the filter results of the two derivations. This is a *nonlinear* operation which cannot be interchanged with the smoothing operation B. In the following, the results of the operations (11) applied on the image sequence G will be abbreviated with the notation

$$G_{ij} = [B(D_i \bullet D_j)]G \tag{12}$$

Efficient algorithms for the eigenvalue and eigenvector analysis in two and three dimensions, i. e. for one- and two dimensional velocity determination, are discussed in another paper [11].

3   Error considerations

Kearney et al. [13] give a detailed analysis of the gradient-based optical flow determination. They point out that highly textures surfaces, i. e. just the region with steep gradients where the flow can be detected most easily are most seriously effected by errors because of larger higher order spatial derivatives. Actually, this is not correct. It can be proven that any region with constant motion (1) yields an accurate velocity estimate in the one- and two dimensional case.

The problem is rather the discrete approximation of the spatial and temporal derivatives. The simple $1/2(1\ 0\ -1)$-operator is a poor approximation for higher wavenumbers (Figure 1). Therefore higher order approximations have been used with the following convolution masks [7]

$$
\begin{aligned}
{}^{(1)}\mathbf{D}_i &= & 1/2(1\ 0\ -1) \\
{}^{(2)}\mathbf{D}_i &= & 1/12(-1\ 8\ 0\ -8\ 1) \\
{}^{(3)}\mathbf{D}_i &= & 1/60(1\ -9\ 45\ 0\ -45\ 9\ -1)
\end{aligned}
\tag{13}
$$

The corresponding transfer functions in Figure 1 show that these masks approximate an ideal derivative operator better with increasing size of the mask, but even the largest mask shows considerable deviations for wavenumbers larger than 0.6 times the Nyquist wavenumber.

Because of this bad performance, also the idea of iterative refinement has been tested. The next determination is computed in a coordinate system which is moving with the previously estimated velocity. If the previous estimate has diminished the residual velocity, the temporal frequencies are smaller. Consequently, the temporal derivative will be calculated more accurately. It should be expected that the iteration finally converges to the correct solution provided a) the spatial interpolation necessary to calculate the temporal derivative is accurate enough and b) the image sequence includes no aliasing.

4 Experimental Results

4.1 Computed Image Sequences

The accuary of the algorithms has been tested with different computed image sequences. All experiments have been performed with 12-bit images. All convolutions and multiplications were computed in 16-bit integer arithmetic with appropriate scaling. All intermediate results were stored with 12-bit accuracy directly in a 12-bit frame buffer (FG-100 from Imaging Technology). Smoothing of the multiplications of the derivative operators in (11) was performed by a $17 \times 17 \times 5$ binomial kernel. Only the last steps of calculating the velocity components and the classification were performed in floating point arithmetic.

4.2 Accuracy: Dependence on spatial and temporal scales

The first set of experiments deals with the accuracy of the velocity estimate. Constantly moving sinusoidal patterns provided a means to test the influence of the wavenumber of the spatial structure on motion determination. To study subpixel accuracies non-integer values for the displacements and wavelengths have been used. Non-zero Gaussian noise with a standard deviation of 50 was added to the sinusoidal pattern with an amplitude of 500. A low displacement of only 0.137 pixels/frame was chosen to introduce no significant error in the temporal derivative. Figure 2a shows the computed displacement as a function of the wavelength using the three different approximations to the first derivative as discussed in section 3.

As expected, the deviations from the correct value increase with decreasing wavelength and are larger for lower order approximations. Towards larger wavelengths the estimated displacements converge to a value which deviates by less than 0.007 pixels from the correct value. This residual deviation is probably due to round-off errors caused by the 16-bit integer arithmetic and storage of intermediate results with only 12 bits accuracy. Taking this error limit, we can conclude that the first order approximation for the derivation

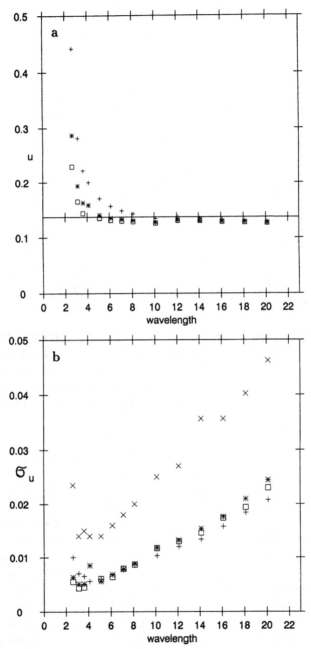

Fig. 2: a) Displacement estimate in an image sequence with a noisy sinusoidal pattern (amplitude 500, standard deviation of the Gaussian noise 50) moving with 0.137 pixels/frame as a function of the wavelength. Different discrete approximations for the first partial derivatives (13) have been used: $+$ $^{(1)}D_i$, $*$ $^{(2)}D_i$, $\square$ $^{(3)}D_i$. b) standard deviation of the displacement distribution in the image as a function of the wavelength, symbols as in a) but an additional case with $^{(1)}D_i$ where no temporal smoothing has been applied.

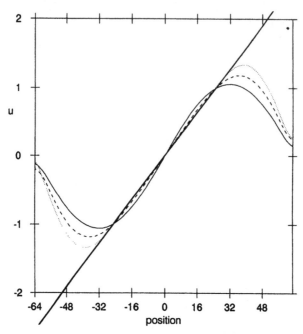

Fig. 3: A contracting sinusoidal pattern generates a linearly changing 1D-motion field. The contraction rate has been adjusted to result in displacements of half a wavelength at the left and right side of the image. The thick line marks the given velocity field, while the thin line show the computed results with different derivation operators (13): solid, 1/2(1 0 -1); dashed, 1/12 (-1 8 0 -8 1); dotted, 1/60 (1 -9 45 0 -45 9 -1). Wavelength of the pattern 5.10 pixels.

operator can be taken only if the wavelength is larger than 8 pixels, while the third order approximation yields about the same error at wavelengths as short as 3.5 pixels. This is still a factor of about two short to the smallest wavelength allowed by the sampling theorem with a wavelength of two pixels.

A linearly changing velocity nicely demonstrates the combined errors caused by both the spatial and temporal derivates. Since the displacement changes from $-\lambda/2$ to $\lambda/2$ over the image, just the frequencies are covered which are allowed by the sampling theorem. With low displacements, the error is dominated by the spatial derivative (Figure 3). The estimated displacements are too high and decrease with the order of approximation. For high displacements the error is dominated by the temporal derivative. Towards displacements of half a wavelength, the estimates even decrease and go towards zero. The computed behavior excellently agree with the expected one since it just resembles the transfer functions of the derivative operator shown in Figure 1. Interestingly, there is one point were the estimates are correct though both derivates are erroneous. This happens with a displacement of one pixel since then the temporal and spatial derivative are equal. Thus it has to be taken great care not to use integer displacements to test the accuracy of motion determination.

So far, there is the limitation that both the spatial and temporal frequencies must be well below the Nyquist limit. To overcome this problem the accuracy and convergence of the iterative refinement technique discussed in section 3 have been investigated. A 1/2 (1

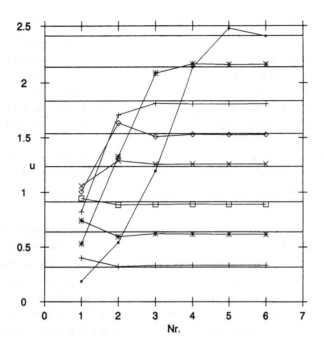

Fig. 4: Test of the method of iterative refinement with a sinusoidal pattern of a wavelength of 5.13 pixels. The computed displacements are shown as a function of the number of iterations for velocities between 0.317 and 2.413 pixels/frame. The thick lines mark the correct values.

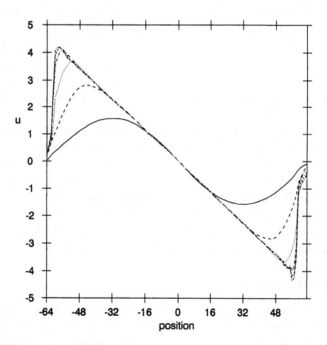

Fig. 5: Same as Figure 3 but displacements computed iteratively using the simple 1/2(1 0 -1) derivation operator. The iteration steps are indicated. Amplitude of the pattern 800; standard deviation of added zero mean Gaussian noise: 50.

σ_n	u	σ_u	u	σ_u	u	σ_u
	$\lambda = 4.13$		$\lambda = 6.13$		$\lambda = 8.13$	
10	0.1520	0.0017	0.1142	0.0021	0.1004	0.0027
20	0.1521	0.0027	0.1142	0.0032	0.1000	0.0041
50	0.1523	0.0058	0.1143	0.0067	0.1001	0.0087
100	0.1521	0.0116	0.1138	0.0135	0.1000	0.0174
200	0.1706	0.0321	0.1231	0.0354	0.1064	0.0430

TABLE 1: Dependence of the standard deviation of the displacement estimate σ_u on the standard deviation of the added Gaussian noise σ_n for a sinusoidal pattern of an amplitude of 500, moving with 0.1 pixels/frame.

0 -1) derivative operator and linear interpolation to estimate intergrid image points were applied. Despite the relative crude estimates in each step, the results are surprisingly good. Figures 4 and 5 show that the algorithm converges nearly in the whole possible displacement range of $\pm\lambda/2$. Of course the convergence is slower at higher displacements because the poor initial estimates allow only a slow decrease in the residual displacement but within 6 iterations the final values have been reached. The small deviations from the expected values are less than 0.03 pixels and obviously depend on the actual displacement.

4.3 Influence of Noise

The last section clearly showed that accurate estimates of the velocities can be computed. Now the standard deviation of the estimates is closer examined. First we take a look at the dependence of the standard deviation on the wavelength of the image structure at a constant noise level. Then standard deviation basically is proportional to the wavelength (Figure 2b). This fact results from the decrease of the spatial derivative with the wavelength. Only at small wavelengths were the estimate becomes erroneous (Figure 2a), it increases again.

In a second experiment the noise level was increased in a sequence with a sinusoidal pattern of 4.13 pixel wavelength, an amplitude of 500 and a displacement of 0.137 pixels/frame. Table 1 shows that standard deviation of the velocity estimate σ_u is roughly proportional to the noise level σ_n. Only at the highest noise level, with almost a signal-to-noise ratio of one, the increase is more than proportional and the estimate is biased, but well within the standard deviation. The fact that the bias is larger at smaller wavelength clearly indicates that it is caused by the local nonlinearity in the transfer function of the partial derivative which is larger at higher wavenumbers (Figure 1).

The robustness of the estimate using the iterative refinement technique in noisy images with linearly changing velocity field is demonstrated in Figure 5b.

4.4 Moving random pattern

So far rather unrealistic images with periodical patterns have been studied which helped to understand the dependence of the errors on various parameters. In contrast, a moving random pattern (Figure 6a) includes spatial scales of all sizes and should thus give a good estimate of the errors in real images.

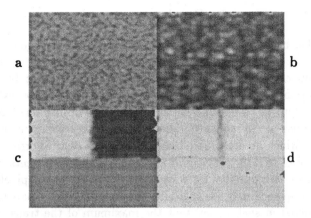

Fig. 6: Detection of motion discontinuities: a) One image of a sequence with random pattern which is moving 0.5 pixels/frame to the left and right in the upper left and right quadrants respectively. The lower half is not moving. b) Confidence level for motion detection. c) Computed 1D-velocity shown on a greyscale where zero velocity indicates a mean greyvalue. d) Measure indicating the degree of constant motion.

technique	u	σ_u
first order appr.	-0.555	0.015
second order appr.	-0.516	0.012
third order appr.	-0.501	0.019
iterative refinement	-0.494	0.020

TABLE 2: Displacement estimates for a random pattern ($\sigma_p = 362$) moving with -0.5 pixels/frame and $\sigma_n = 50$) using different techniques: different orders of approximation for the discrete partial derivatives or the iterative refinement technique.

Figure 6 shows several parameters. The sum of the squared derivatives is taken as a confidence level whether a velocity can be determined at all. The quantity γ measures the degree of constant motion and nicely shows the motion discontinuities (Figure 6c, d).

Table 2 shows that quite similar results as with the sinusoidal patterns are gained. The estimates get better for higher order approximations of the discrete derivatives. For the third order and the iterative refinement technique, the estimates agree within the standard deviation with the correct values. Despite the quite low signal-to-noise ratio of 7.2, the standard deviation is between 0.012 and 0.020 pixels. These values are three to four times higher than in comparable signal-to-noise ratios for sinusoidal pattern.

There is an easy explanation for this effect. In contrast to the sinusoidal pattern, the random pattern includes many regions with lower confidence levels for velocity estimates (Figure 6b). This measure averaged over the whole image is about four times lower for the random pattern than for the sinusoidal pattern. The noise sensitivity should be accordingly higher and this is exactly what has been observed. Thus the random pattern should give a realistic estimate of the influence of noise on the statistical error of the velocity estimate.

5 Conclusions

A new algorithm has been outlined for the analysis of motion in image sequences in the xt-space. Several tests performed with the 1D-algorithm indicate that a promising new approach to image sequence analysis opens up which is especially suitable for scientific and industrial applications. Accurate velocity determinations are possible even with noisy images. The algorithm proved to be superior to the standard optical flow approach using only two consecutive images of a sequence. The method of iterative refinements gives the most accurate results, which no detectable bias within the statistical error.

The algorithm is best used in a multigrid approach. The experiments demonstrated that the statistical errors are minimal for wavelengths between 3 to 6 pixel (Figure 2b). Therefore the image sequence may be spatially decomposed into a Laplacian pyramid [3,7] which is constructed in such a way that the maximum of the transfer function in each level coincides with the optimum wavenumber. In this way, also velocity information can also be gained from large scale structures, which otherwise would be lost. If no motion superimpostion is present, large displacements can be determined in a coarse-to-fine strategy.

Acknowledgements

The author gratefully acknowledges financial support by the German Science Foundation, the European Community (twinning contract with several Dutch organizations within the VIERS-1 project), the California Space Institute, and the Office of Naval Research.

References

1. E. H. Adelson, J. R. Bergen, Spatio-temporal energy models for the perception of motion, *J. Opt. Soc. America*, A2, 284–299 (1985).
2. J. Bigün, G. H. Granlund, Optimal orientation detection of linear symmetry, In *Proc. Int. Conference Computer Vision, London 1987*, , ed., pp. 433–438, IEEE Computer Society Press, Washington (1987).
3. P. J. Burt, E. H. Adelson, The Laplacian pyramid as a compact image code, *IEEE Trans. Comm.*, 31, 532–540 (1983).
4. D. J. Fleet, A. D. Jepson, Hierarchical construction of orientation and velocity selective filters, *IEEE Trans. Pattern Analysis and Machine Intelligence*, 11, 315–325 (1989).
5. H. Goldstein, *Klassische Mechanik*, Aula, Wiesbaden (1985).
6. D. J. Heeger, Optical flow using spatiotemporal filters, *Int. J. Comp. Vision*, 1, 279–302 (1988).
7. B. Jähne, *Digitale Bildverarbeitung*, Springer, Berlin, Heidelberg (1989).
8. B. Jähne, Energy balance in small-scale waves — an experimental approach using optical slope measuring technique and image processing, In *Radar scattering from modulated wind waves*, G. Komen, W. Oost, eds., pp. 105–120, Reidel, Dordrecht (1989).
9. B. Jähne, Image sequence analysis in environmental physics: water surface waves and air-sea gas exchange (in German), In *Mustererkennung 1986, Proc. 8. DAGM-Symposium, Paderborn*, G. Hartmann, ed., pp. 201–205, Springer, Berlin (1986).
10. B. Jähne, Image sequence analysis of complex physical objects: nonlinear small scale water surface waves, In *Proc. Int. Conference Computer Vision, London 1987*, pp. 191–200, IEEE Computer Society Press, Washington (1987).

11. B. Jähne, Motion determination in Space-Time Images, In *Image Processing III*, Conf. Proceedings 1135, SPIE, Washington (1989 in press).

12. B. Jähne, S. Waas, Optical measuring technique for small scale water surface waves, In *Advanced Optical Instrumentation for Remote Sensing of the Earth's Surface*, Conf. Proceedings 1129, SPIE, Washington (1989 in press).

13. J. K. Kearney, W. B. Thompson, D. L. Boley, Optical flow estimation: an error analysis of gradient based methods with local optimization, *IEEE Trans. Pattern Analysis and Machine Intelligence*, 9, 229–244 (1987).

14. H. Knutsson, *Filtering and reconstruction in image processing*, Dissertation, Linköping University (1982).

15. R. Lenz, Zur Genauigkeit der Videometrie mit CCD-Sensoren, In *Mustererkennung 1988, Proc. 10. DAGM-Symposium, Zürich*, pp. 179–189, Springer, Berlin (1988).

16. H. Nagel, Analyse und Interpretation von Bildfolgen I, *Informatik Spektrum*, 8, 178–200 (1985).

17. H. Nagel, Analyse und Interpretation von Bildfolgen II, *Informatik Spektrum*, 8, 312–327 (1985).

18. H. Nagel, Image sequences — ten (octal) years — from phenomenology towards a theoretical foundation, In *Proc. Int. Conf. Pattern Recognition, Paris 1986*, , ed., pp. 1174–1185, IEEE Computer Soc. Press, Washington (1986).

19. H. Nagel, On the constraint equation for the estimation of displacement rates in image sequences, *IEEE Trans. Pattern Analysis and Machine Intelligence*, 11, 13–30 (1989).

20. E. van Halsema, B. Jähne, W. A. Oost, C. Calkoen, P. Snoeij, First results of the VIERS-1 experiment, In *Radar scattering from modulated wind waves*, G. Komen, W. Oost, eds., p. , Reidel, Dordrecht (1989).

21. A. Verri, T. Poggio, Motion field and optical flow: qualitative properties, *IEEE Trans. Pattern Analysis and Machine Intelligence*, 11, 490–498 (1989).

MOTION

AMBIGUITY IN RECONSTRUCTION FROM IMAGE CORRESPONDENCES

Stephen J. Maybank

Long Range Laboratory, GEC Hirst Research Centre, East Lane,
Wembley, Middlesex HA9 7PP, UK

1 Introduction

The possibility of reconstructing the shape of the environment from the correspondences between two images first arose during the 19th century with the invention of photography. The methods developed for exploiting this possibility were based on projective geometry since projection provides a good model for image formation. At first the shape of the environment was reconstructed by linear methods [16], which did not make full use of the rigidity of the environment. Later methods incorporated the rigidity constraint [3,7], thus allowing reconstruction with fewer image correspondences, but at the cost of greatly increasing the complexity of the reconstruction algorithm.

More recently the reconstruction problem has been transformed by the advent of electronic cameras and computers. Large numbers of images are obtained by an electronic camera in a short space of time, the image correspondences are found automatically, and reconstruction is carried out by a computer algorithm. In this way a robot or an automatic vehicle can obtain useful information by passive means well suited to a wide variety of environments. The modern approach to reconstruction is based on Euclidean geometry and the vector calculus, rather than on projective geometry [3, 6, 8]. A very large number of algorithms for reconstruction have been published recently. An example of a linear algorithm may be found in [8], and examples of non-linear algorithms may be found in [6], together with further references to the literature.

In some cases reconstruction from image correspondences is ambiguous, in that two essentially different surfaces in space are obtained. Ambiguity has been studied previously using both the older projective geometric approach [3,4,17] and the newer Euclidean approach [3,5,6,9]. In this paper projective geometric and Euclidean techniques are used together to prove that ambiguous surfaces are invariant under a rotation through 180°. The rotation interchanges the two possible positions for the optical centre of the camera taking the second image. In consequence, a cubic polynomial constraint on ambiguous surfaces is obtained, which forms the basis of a new proof of Demazure's result [1] that there are in general exactly ten camera displacements compatible with five given image correspondences.

The invariance of an ambiguous surface under a rotation through 180° was first discovered in photogrammetry [4]. The cubic polynomial constraint on ambiguous surfaces is new. An extensive discussion of the projective geometric approach to ambiguity including alternative proofs of some of the results given here will appear in [14].

1.1 Notation

The necessary background in projective geometry is given in [15]. Euclidean three dimensional space $\mathbf{R}^3$ is regarded as a subset of projective space $\mathbf{P}^3$. Points of $\mathbf{R}^3$ are denoted by $[x_1, x_2, x_3]$, and points of $\mathbf{P}^3$ are denoted by (x_1, x_2, x_3, x_4). Two points of $\mathbf{P}^3$, (x_1, x_2, x_3, x_4) and (x_1', x_2', x_3', x_4') are the same if and only if there exists a non-zero scalar λ such that $x_i = \lambda x_i'$ for $1 \leq i \leq 4$. Thus, in $\mathbf{P}^3$,

$$(x_1, x_2, x_3, x_4) = \lambda(x_1, x_2, x_3, x_4) \qquad (1)$$

It follows from (1) that the zeros of a polynomial equation $f(\mathbf{x}) = 0$ are well defined in $\mathbf{P}^3$ if and only $f(\mathbf{x})$ is homogeneous in the coordinates x_i of $\mathbf{x}$.

The embedding $\mathbb{R}^3 \subset \mathbb{P}^3$ is the usual one $[x_1, x_2, x_3] \mapsto (x_1, x_2, x_3, 1)$. The same symbol x is used to denote both $[x_1, x_2, x_3]$ and $(x_1, x_2, x_3, 1)$, depending on context. The set $\mathbb{P}^3 \setminus \mathbb{R}^3$ is the plane at infinity, Π_∞. The equation of Π_∞ is $x_4 = 0$. The coordinates x_i of $\mathbb{R}^3$ (and $\mathbb{P}^3$) are chosen to be rectangular. In this case the x_i are referred to as Cartesian coordinates. The origin $(0, 0, 0, 1)$ of Cartesian coordinates is denoted by o.

Points of $\mathbb{P}^3$ are typically denoted by o, a, b, x, y, and lines of $\mathbb{P}^3$ are denoted by g, h, k. The line joining two distinct points a, b of P^3 is $< a, b >$, and similarly, the plane containing the line g and the point a not on g is $< g, a >$. This notation is extended to arbitrary numbers of lines and points. For example, $< g, h, a >$ is the smallest subspace of $\mathbb{P}^3$ containing the lines g, h and the point a. Planes in $\mathbb{P}^3$ are typically denoted by Π, Φ, Ξ. Quadric surfaces in $\mathbb{P}^3$ are typically denoted by ψ.

Each point x of $\mathbb{R}^3$ defines a vector, namely, the line segment from the origin o to x. This vector is denoted by the same symbol x as the point x. The dot product $x.y$ and the vector product $x \times y$ of vectors x, y are formed in the usual way. Each non-zero vector $x = (x_1, x_2, x_3, x_4)$ corresponds to a unique point $(x_1, x_2, x_3, 0)$ of Π_∞, which can be thought of as the direction of x. If x, y are points of Π_∞, then $x \times y$ is defined as the unique point of Π_∞ corresponding to the direction orthogonal to the directions of x and y.

The tensor product of two vectors x, y is denoted by $x \otimes y$. In the applications of the tensor product made in this paper x and y are points of $\mathbb{P}^3$ contained in Π_∞. Then $x \otimes y$ is the 3×3 matrix with i, jth entry equal to $x_i y_j$ for $1 \leq i, j \leq 3$. The coordinates of x and y are only defined up to a non-zero scalar multiple, thus $x \otimes y$ is also only defined up to a non-zero scalar multiple.

Invertible linear transformations, or collineations, are typically denoted by ω, τ. The value of ω at a point x is ωx (without brackets). If S is a set of points, for example a line, or a plane, then $\omega(S)$ (with brackets) is the set of ωx as x ranges over S. If $\omega(S) \subset S$ then S is said to be invariant under ω.

2 Reconstruction from image correspondences

The usual formulation of the reconstruction problem is employed [10]. Two images of the same set of scene points p_i are taken from distinct projection points, o and a. It is assumed that the p_i are fixed rigidly in space with respect to o and a. The point o is referred to as the optical centre of the first camera, and a is referred to as the optical centre of the second camera. The imaging surface of each camera is the unit sphere with centre at the optical centre of the camera. The image is formed by polar projection towards the optical centre. Each scene point p_i gives rise to points q_i, q'_i in the first and second images respectively. The correspondence between q_i and q'_i is denoted by $q_i \leftrightarrow q'_i$.

In practice, the camera projection is more complicated than polar projection onto the unit sphere. This discrepancy between theory and experiment is overcome by calibrating the camera. The acquired image is transformed in order to obtain the image that would have arisen from polar projection.

2.1 The Euclidean approach

In the Euclidean approach to reconstruction each camera has associated to it a coordinate frame in which the positions of the image points are measured. The displacement of the second camera with respect to the first is specified by giving both the translation from the first optical centre o to the second optical centre a, and the rotation R needed to bring the two camera coordinate frames into alignment after the translation. It is assumed that $\det(R) = 1$, thus excluding the possibility that the two coordinate frames differ by a reflection. When specifying the relative position of the cameras by a pair $\{R, a\}$ it is assumed that Cartesian coordinates have been chosen with the origin o at the optical centre of the first camera. The translation vector a is then identified with the point

a at the optical centre of the second camera.

If the pair $\{R, \mathbf{a}\}$ is known then the positions $\mathbf{p}_i$ of the scene points relative to the two cameras are easily calculated from the image correspondences. The reconstruction problem thus reduces to the problem of recovering $\{R, \mathbf{a}\}$ from the image correspondences. Each correspondence $\mathbf{q} \leftrightarrow \mathbf{q}'$ places a single linear constraint on $\{R, \mathbf{a}\}$ of the form $(R\mathbf{q} \times R\mathbf{a}).\mathbf{q}' = 0$.

In the ambiguous case there exist at least two camera displacements, $\{R, \mathbf{a}\}$ and $\{S, \mathbf{b}\}$ compatible with the same set of image correspondences such that **a** is not parallel to **b**. Let ψ be the surface giving rise to the image correspondences when the relative displacement of the two cameras is $\{R, \mathbf{a}\}$. It is shown in [9] that the equation of ψ is quadratic in **x** and of the form

$$(U\mathbf{x} \times \mathbf{x}).\mathbf{b} = (U\mathbf{a} \times \mathbf{x}).\mathbf{b} \tag{2}$$

where $U = S^\top R$. In (2) the points **x**, **a**, **b** are in $\mathbf{R}^3$, thus $\mathbf{x} = [x_1, x_2, x_3]$, etc. Equation (2) for the ambiguous surface ψ has the form

$$\mathbf{x}^\top M\mathbf{x} + \mathbf{l}.\mathbf{x} = 0 \tag{3}$$

where M is a symmetric 3×3 matrix and $\mathbf{l} = U\mathbf{a} \times \mathbf{b}$. It follows from (2) and (3) that there exist vectors **m**, **n** such that **n** is the axis of U, and such that M has the form

$$M = \frac{1}{2}(\mathbf{m} \otimes \mathbf{n} + \mathbf{n} \otimes \mathbf{m}) - \mathbf{m}.\mathbf{n}I \tag{4}$$

Let θ be the angle of rotation of U. Then

$$\mathbf{m} = \sin(\theta)\mathbf{b} + [1 - \cos(\theta)]\mathbf{b} \times \mathbf{n} \tag{5}$$

Equations (2) and (3) for ψ are appropriate for points **x** in $\mathbf{R}^3$. The equations are extended to Π_∞ by writing them in homogeneous form using the coordinate x_4. For example, (3) becomes

$$\mathbf{x}^\top M\mathbf{x} + x_4\mathbf{l}.\mathbf{x} = 0 \tag{6}$$

The point (x_1, x_2, x_3, x_4) is in ψ if and only if it satisfies (6). The points of $\mathbf{R}^3$ contained in ψ are unchanged because (6) reduces to (3) on setting $x_4 = 1$. The points of $\psi \cap \Pi_\infty$ are obtained by setting $x_4 = 0$ in (6).

The vectors **m**, **n** of (4) correspond to points $(m_1, m_2, m_3, 0)$ and $(n_1, n_2, n_3, 0)$ of Π_∞. Under this correspondence, **m** and **n** are points of ψ because $\mathbf{m}^\top M\mathbf{m} = 0$ and $\mathbf{n}^\top M\mathbf{n} = 0$. The points **m**, **n** are called the principal points of ψ.

A quadric surface in $\mathbf{P}^3$ for which the second order terms have the form $\mathbf{x}^\top M\mathbf{x}$ where M is given by (4) is known as a rectangular quadric. Ambiguous surfaces are examples of rectangular quadrics.

2.2 The projective geometric approach

The projective geometric approach to reconstruction is along the following lines. Let two images of the points $\mathbf{p}_i$ in space be obtained from cameras with optical centres at distinct points **o** and **a**. A point $\mathbf{q}_i$ in the first image defines a projection line $< \mathbf{o}, \mathbf{q}_i >$ such that all points on this line project to $\mathbf{q}_i$, and similarly, a point $\mathbf{q}_i'$ in the second image defines a projection line $< \mathbf{a}, \mathbf{q}_i' >$ such that all points on this line project to $\mathbf{q}_i'$. Thus, image formation is modelled as a linear transformation from the points of $\mathbf{P}^3$ to the two dimensional projective space of lines (sight rays) passing through the optical centre of the camera from which the image is obtained. The lines $< \mathbf{o}, \mathbf{q}_i >$ and $< \mathbf{a}, \mathbf{q}_i' >$ correspond if and only if they intersect at some point in space.

If the reconstruction is ambiguous then there exist points **a**, **b**, not both collinear with **o**, such that the camera taking the second image can have its optical centre either at **a** or at **b**. In this case, let q be the line of points projecting to a point **q** in the first image. Let r be the line of points projecting to the corresponding point $\mathbf{q}'$ in the second image when the optical centre of the second camera is at **a**, and let s be the line of points projecting to $\mathbf{q}'$ in the second image when the optical centre of the second camera is at **b**. The line q is the unique common transversal of r and s passing through **o**, as illustrated in Figure 1.

Figure 1:

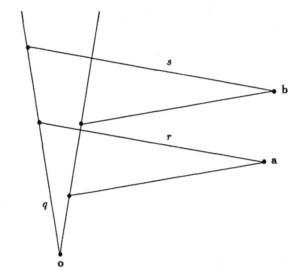

The image taken by the second camera is fixed independently of whether the camera is thought to be at **a** or **b**, thus the angles between pairs of lines r_i, r_j through **a** are equal to the angles between the corresponding pairs of lines s_i, s_j through **b**. It follows that there is a linear orthogonal (ie. angle preserving) transformation ω from the lines through **a** to the lines through **b** such that $\omega r = s$.

Each line through **a** intersects Π_∞ at a unique point, thus the space of lines through **a** is parameterised by Π_∞. Similarly, the space of lines through **b** is also parameterised by Π_∞. If these parameterisations are adopted, then ω becomes a linear transformation from Π_∞ to itself that preserves angles. Thus, ω is a rotation. In Cartesian coordinates the matrix of ω is orthogonal. In this case ω coincides with the rotation U appearing in (2).

2.3   Geometrical construction of an ambiguous surface

The projective geometric treatment of ambiguity leads to a geometrical construction of ambiguous surfaces, as given in [17]. The transformation ω from the lines through **a** to the lines through **b** induces a linear transformation, also denoted by ω, from the planes through **a** to the planes through **b**. If a line r through **a** is contained in a plane Φ, then the line ωr through **b** is contained in $\omega(\Phi)$. Let Π be any plane containing the line $h = \omega^{-1} < \mathbf{o}, \mathbf{b} >$. Then $\omega(\Pi)$ contains $< \mathbf{o}, \mathbf{b} >$. As Π varies through the one dimensional projective space of planes containing h, $\omega(\Pi)$ varies through the one dimensional projective space of planes containing $< \mathbf{o}, \mathbf{b} >$. The line $l = \Pi \cap \omega(\Pi)$ then sweeps out the ambiguous surface, ψ, associated with the camera displacement from **o** to **a**. This construction is illustrated in Figure 2.

The proof that l sweeps out ψ follows from the fact that, with the notation of Figure 1, q is the unique common transversal of r and s. In more detail, let $l = \Pi \cap \omega(\Pi)$, as shown in Figure 2. Let **p** be any point of l, let $r = < \mathbf{a}, \mathbf{p} >$, and let s be the line through **b** such that $\omega r = s$. Then s and $< \mathbf{o}, \mathbf{p} >$ are contained in Π, thus s intersects $< \mathbf{o}, \mathbf{p} >$. Hence $< \mathbf{o}, \mathbf{p} >$ is the unique common transversal of r, s passing through **o**. It follows that **p** is in ψ. The point **p** is an arbitrary point of l, thus l is contained in ψ as required.

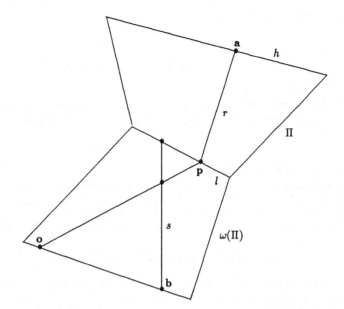

Figure 2:

3 Invariance of an ambiguous surface under rotation

It is shown in Theorem 3.1 below that an ambiguous surface is invariant under a rotation through 180°. Before proving this result the following fact about rotations is recalled. If ω is any rotation with axis $\mathbf{n}$, and τ is a rotation through 180° about an axis orthogonal to $\mathbf{n}$, then

$$\omega = \tau\omega^{-1}\tau \qquad (7)$$

To prove (7) note firstly that $\omega^{-1}\tau$ reverses the direction of $\mathbf{n}$. It follows that $\omega^{-1}\tau$ is a rotation through 180°.

Theorem 3.1. Let ψ be an ambiguous surface viewed by two cameras with optical centres at $\mathbf{o}$ and $\mathbf{a}$ respectively. Let $\mathbf{b}$ be the alternative optical centre for the camera at $\mathbf{a}$. Then ψ is invariant under a rotation through 180° that takes $\mathbf{a}$ onto the line $< \mathbf{o}, \mathbf{b} >$.

Proof. Let ψ be generated by the construction described in §2.3, and let $\mathbf{m}$, $\mathbf{n}$ be the principal directions of ψ, as defined in §2.1. The collineation ω of §2.3 is regarded as an orthogonal collineation of Π_∞, with axis $\mathbf{n}$ and angle of rotation θ. In this proof the point $\mathbf{b}$ serves only to fix the direction of the line $< \mathbf{o}, \mathbf{b} >$. It is convenient to regard $\mathbf{b}$ as the point of Π_∞ corresponding to the direction of $< \mathbf{o}, \mathbf{b} >$.

Let τ be the rotation of Π_∞ through 180° with axis $\mathbf{m} \times \mathbf{n}$. It follows from the definition of τ, on using (1), that

$$\tau\mathbf{b} = 2[\mathbf{b}.(\mathbf{m} \times \mathbf{n})]\mathbf{m} \times \mathbf{n} - \|\mathbf{m} \times \mathbf{n}\|^2\mathbf{b} \qquad (8)$$

It follows from (5) that

$$\mathbf{m} \times \mathbf{n} = \sin(\theta)\mathbf{b} \times \mathbf{n} + [1 - \cos(\theta)](\mathbf{b} \times \mathbf{n}) \times \mathbf{n}$$

thus

$$\mathbf{b}.(\mathbf{m} \times \mathbf{n}) = -(1 - \cos(\theta))\|\mathbf{b} \times \mathbf{n}\|^2 = -\frac{1}{2}\|\mathbf{m} \times \mathbf{n}\|^2 \qquad (9)$$

Equations (8) and (9) yield

$$\tau \mathbf{b} = \mathbf{m} \times \mathbf{n} + \mathbf{b} \tag{10}$$

Now

$$\omega^{-1}\mathbf{b} = -[1 - \cos(\theta)][\mathbf{b} - (\mathbf{b}.\mathbf{n})\mathbf{n}] + \sin(\theta)\mathbf{b} \times \mathbf{n} + \mathbf{b} \tag{11}$$

It follows from (10) and (11) that $\tau\mathbf{b} = \omega^{-1}\mathbf{b}$.

Let τ_ψ be the rotation of $\mathbf{R}^3$ through 180° with axis in the direction $\mathbf{m} \times \mathbf{n}$ such that $\tau_\psi(<\mathbf{o},\mathbf{b}>)$ contains $\mathbf{a}$. Let h be the line $\omega^{-1} <\mathbf{o},\mathbf{b}>$, as shown in Figure 2. The restriction of τ_ψ to Π_∞ is equal to τ. Thus

$$\tau_\psi\mathbf{b} = \tau\mathbf{b} = \omega^{-1}\mathbf{b} = h \cap \Pi_\infty$$

It follows that $\tau_\psi(<\mathbf{o},\mathbf{b}>) = h$.

Each plane Π containing h is uniquely defined by the intersection $\Pi \cap \Pi_\infty$. Thus, to find the planes $\omega(\Pi)$ and $\tau_\psi(\Pi)$, it is sufficient to consider only the actions of ω and τ_ψ on Π_∞. Let Π be any plane containing $<\mathbf{o},\mathbf{b}>$. Then $l = \Pi \cap \omega(\Pi)$ is a line in ψ. Let $\Xi = \tau_\psi\omega(\Pi)$. Then Ξ contains h. It follows from (7) that $\omega(\Xi) = \tau_\psi\omega^{-1}\tau_\psi(\Xi)$. The application of τ_ψ to l yields

$$\tau_\psi(l) = \tau_\psi(\Pi) \cap \tau_\psi\omega(\Pi) = \Xi \cap \tau_\psi\omega^{-1}\tau_\psi(\Xi) = \Xi \cap \omega(\Xi)$$

Thus $\tau_\psi(l)$ is contained in ψ. In consequence, ψ is invariant under τ_ψ. The result that $\tau_\psi\mathbf{a}$ is contained in $<\mathbf{o},\mathbf{b}>$ follows from the construction of τ_ψ. □

It is shown in [14] that τ_ψ is the unique rotation through 180° that leaves ψ invariant and that fixes the principal points of ψ. A converse to Theorem 3.1 is obtained as follows.

Theorem 3.2. Let ψ be a non-singular rectangular quadric and let τ_ψ be the rotation through 180° defined in the proof of Theorem 3.1. Let $\mathbf{o}$, $\mathbf{a}$, $\tau_\psi\mathbf{a}$ be distinct points of ψ such that $\tau_\psi\mathbf{a}$ is on a generator of ψ passing through $\mathbf{o}$. Then ψ is an ambiguous surface such that $\mathbf{o}$ is the optical centre for the camera taking the first image, and $\mathbf{a}$, $\tau_\psi\mathbf{a}$ are the two possible optical centres for the camera taking the second image.

Proof. Let $\mathbf{b}$ be the point at which $<\mathbf{o},\tau_\psi\mathbf{a}>$ meets Π_∞. Let $\mathbf{m}$, $\mathbf{n}$ be the principal points of ψ. Let τ be the restriction of τ_ψ to Π_∞, and let σ be a rotation of Π_∞ through 180° with axis $\mathbf{b} \times \mathbf{n}$. Define the rotation ω of Π_∞ by $\omega = \sigma\tau$. Then, on applying (1), $\omega^{-1}\mathbf{b} = \tau\sigma\mathbf{b} = \tau_\psi\mathbf{b}$.

Let $h = \tau_\psi(<\mathbf{o},\mathbf{b}>)$. Then ω defines a collineation from the planes containing h to the planes containing $<\mathbf{o},\mathbf{b}>$. Let ψ' be the ambiguous surface swept out by the lines $l = \Pi \cap \omega(\Pi)$ as Π varies through the planes containing h. To prove the theorem it suffices to show that $\psi = \psi'$.

Let $\mathbf{n}$, $\mathbf{i}_n$, $\mathbf{j}_n$ be the fixed points of ω in Π_∞. Let k_1, k_2 k_3 be the unique lines containing $\mathbf{n}$, $\mathbf{i}_n$, $\mathbf{j}_n$, respectively, such that each k_i meets both h and $<\mathbf{o},\mathbf{b}>$. It follows from (2) (with $U = \omega$) that $\mathbf{n}$, $\mathbf{i}_n$, $\mathbf{j}_n$ are contained in ψ. Each k_i meets ψ at three points, thus each k_i is contained in ψ. Let $\Pi = <h,\mathbf{n}>$. Then $\omega(\Pi) = <\mathbf{o},\mathbf{b},\mathbf{n}>$, thus

$$\Pi \cap \omega(\Pi) = <h,\mathbf{n}> \cap <\mathbf{o},\mathbf{b},\mathbf{n}> = k_1$$

thus k_1 is contained in ψ'. Similarly, k_2 and k_3 are contained in ψ'. It follows that $\psi \cap \psi'$ contains a (split) space curve of degree five comprising k_1, k_2, k_3, h, $<\mathbf{o},\mathbf{b}>$. Two distinct quadrics intersect in a space curve of degree four only [15], thus ψ and ψ' are not distinct. In other words, $\psi = \psi'$. □

An explicit expression is obtained for the rotation τ_ψ appearing in Theorem 3.1. The axis of τ_ψ has direction $\mathbf{m} \times \mathbf{n}$. Thus the action of τ_ψ on a general point $\mathbf{x}$ of $\mathbf{R}^3$ is given by

$$\tau_\psi\mathbf{x} = \mathbf{s} + \frac{2[\mathbf{x}.(\mathbf{m} \times \mathbf{n})]\mathbf{m} \times \mathbf{n}}{\|\mathbf{m} \times \mathbf{n}\|} - \mathbf{x}$$

where $\mathbf{s}$ is an unknown vector normal to $\mathbf{m} \times \mathbf{n}$. It follows from (3) that the vector normal to ψ at $\mathbf{o}$ has direction $\mathbf{l}$. Thus the vector normal to ψ at $\tau_\psi\mathbf{o}$ has direction $\tau_\psi\mathbf{l}$. It follows that $2M\mathbf{s} + \mathbf{l}$ is parallel to $\tau_\psi l$. Thus

$$s = -\frac{2[(n.l)m + (m.l)n]}{\|m \times n\|^2}$$

It follows that

$$\tau_\psi x = -\frac{2[(n.l)m + (m.l)n]}{\|m \times n\|^2} + \frac{2[x.(m \times n)]m \times n}{\|m \times n\|^2} - x \tag{12}$$

4 Two cubic constraints on ambiguous surfaces

Two cubic constraints on ambiguous surfaces are obtained. The first constraint is a new one arising from Theorem 3.1. The second cubic constraint is well known [5]. To obtain the first constraint the following three theorems are required.

Theorem 4.1. With the notation of Theorem 3.1, let ψ be a non-singular ambiguous surface with principal points m, n, and let Cartesian coordinates be chosen such that the origin o is the optical centre of the camera taking the first image. Let $r = m \times n$, let l be the normal to the tangent plane to ψ at o, and let a be a possible optical centre for the second camera not lying on a generator through o. Then

$$-4(m.l)(n.l) + 2(a.r)(l.r) - (a.l)(r.r) = 0 \tag{13}$$

Proof. Let τ_ψ be the unique non-trivial rigid involution of ψ that fixes both m and n. It follows from Theorem 3.1 that $\tau_\psi a$ lies in the tangent plane to ψ at o, thus $l.\tau_\psi a = 0$. The result follows on substituting the expression for $\tau_\psi a$ given by (12) into the equation $l.\tau_\psi a = 0$. $\square$

Theorem 4.2. Let m, n be vectors, and define the matrices N, L by

$$N = \frac{1}{2}(m \otimes n + n \otimes m) \tag{14}$$

$$L = (m \times n) \otimes (m \times n) \tag{15}$$

Let e_1^T, e_2^T, e_3^T be the rows of N. Then

$$L = 4 \begin{pmatrix} e_3^\mathsf{T} \times e_2^\mathsf{T} \\ e_1^\mathsf{T} \times e_3^\mathsf{T} \\ e_2^\mathsf{T} \times e_1^\mathsf{T} \end{pmatrix}$$

Proof. The result follows from (14) and (15) on expressing e_1, e_2, e_3 in terms of m and n. $\square$

Theorem 4.3. In the reconstruction problem, let two points in space be given as optical centres for the cameras taking the first and second images respectively. Then any ambiguous surface associated with these two optical centres satisfies a cubic polynomial constraint.

Proof. Cartesian coordinates are chosen, and the notation of Theorem 3.1 is employed. Let o, a be the two optical centres, and let an ambiguous surface ψ containing o, a have an equation of the form (3). Let m, n be the principal directions of ψ, and let N be the matrix defined by (14). It follows from (4) and (14) that

$$N = M - \frac{1}{2}\text{Trace}(M)I$$

The entries of N are thus linear functions of the entries of M.

The cubic polynomial constraint on ψ is obtained from (13). The term $-4(l.m)(l.n)$ on the left-hand side of (13) has the form

$$-4(l.m)(l.n) = -4l^\mathsf{T}Nl \tag{16}$$

The remaining two terms on the left-hand side of (13) have the form

$$2(a.r)(l.r) - (a.l)(r.r) = 2a^\mathsf{T}Ll - (a.l)\text{Trace}(L) \tag{17}$$

It follows from (16) and (17) that (13) is equivalent to

$$-4\mathbf{1}^\top N\mathbf{1} + 2\mathbf{a}^\top L\mathbf{1} - (\mathbf{a}.\mathbf{l})\mathrm{Trace}(L) = 0 \tag{18}$$

Let $\mathbf{e}_1^\top, \mathbf{e}_2^\top, \mathbf{e}_3^\top$ be the rows of N. Then, on applying Theorem 4.2, (18) takes the form

$$-\mathbf{1}^\top \begin{pmatrix} \mathbf{e}_1^\top \\ \mathbf{e}_2^\top \\ \mathbf{e}_3^\top \end{pmatrix}\mathbf{1} + 2\mathbf{a}^\top \begin{pmatrix} \mathbf{e}_3^\top \times \mathbf{e}_2^\top \\ \mathbf{e}_1^\top \times \mathbf{e}_3^\top \\ \mathbf{e}_2^\top \times \mathbf{e}_1^\top \end{pmatrix}\mathbf{1} - (\mathbf{a}.\mathbf{l})\mathrm{Trace}\begin{pmatrix} \mathbf{e}_3^\top \times \mathbf{e}_2^\top \\ \mathbf{e}_1^\top \times \mathbf{e}_3^\top \\ \mathbf{e}_2^\top \times \mathbf{e}_1^\top \end{pmatrix} = 0 \tag{19}$$

Equation (19) is the required cubic constraint on ψ. $\square$

The second, and well known, cubic polynomial constraint on the ambiguous surface ψ is

$$\det(N) = 0 \tag{20}$$

where N is defined by (14). Equation (20) follows from the fact that N has rank two. It is cubic in the coefficients of the equation (3) defining ψ because the entries of N depend linearly on the entries of the matrix M in (3).

5 The case of five image correspondences

An algebraic question arising in reconstruction is that of finding the number of essentially different camera displacements compatible with five given image correspondences. The number of unknown parameters in reconstruction is five, comprising three independent parameters for the rotation R, and two parameters for the direction of the translation $\mathbf{a}$. Each image correspondence imposes one constraint on $\{R, \mathbf{a}\}$, thus five image correspondences are sufficient to reduce the set of compatible camera displacements to a finite size. Determining the size of this set is of interest because the number thus obtained is a fundamental algebraic measure of the complexity of the reconstruction problem.

A comment on the method of counting solutions is required. If the camera displacement $\{S, \mathbf{b}\}$ is compatible with a given set of image correspondences, then so are $\{S, \lambda\mathbf{b}\}$ and $\{\sigma S, \mathbf{b}\}$, where λ is any non-zero scalar, and σ is a rotation of 180° with axis $S\mathbf{b}$. (See [6]). All these solutions are counted together as a single solution to the reconstruction problem. If $\{R, \mathbf{a}\}$ is an additional camera displacement compatible with the same image correspondences as $\{S, \mathbf{b}\}$ then the same ambiguous surface is obtained on pairing $\{R, \mathbf{a}\}$ with the $\{S, \lambda\mathbf{b}\}$ and $\{\sigma S, \mathbf{b}\}$ in turn.

Demazure [2] uses algebraic geometry to prove that the number of camera displacements compatible with five image correspondences is ten. In this context ten is high, indicating that the reconstruction problem is difficult. The following three theorems comprise a new proof of Demazure's result.

Theorem 5.1. Let five image correspondences be given compatible with a given camera displacement. Then a two dimensional space of quadrics can be constructed such that any ambiguous surface compatible with the five image correspondences and compatible with the given camera displacement is represented by a point in the two dimensional space of quadrics.

Proof. Let Cartesian coordinates be chosen with origin $\mathbf{o}$ at the optical centre of the camera from which the first image is obtained. Let $\mathbf{q}_i \leftrightarrow \mathbf{q}_i'$ be five image correspondences compatible with the given camera displacement $\{R, \mathbf{a}\}$, where R is an orthogonal matrix and $\mathbf{a}$ is the optical centre of the camera from which the second image is obtained. Let $\mathbf{p}_i$ be the points in $\mathbf{P}^3$ such that the image of $\mathbf{p}_i$ from $\mathbf{o}$ is $\mathbf{q}_i$, and the image of $\mathbf{p}_i$ from $\mathbf{a}$ is $\mathbf{q}_i'$.

An ambiguous surface, compatible with $\mathbf{q}_i \leftrightarrow \mathbf{q}_i'$ and compatible with $\{R, \mathbf{a}\}$ contains the five points $\mathbf{p}_i$, together with the points $\mathbf{o}$, $\mathbf{a}$. The space of all quadric surfaces contained in $\mathbf{P}^3$ is of dimension nine. The condition that a quadric contains a known point imposes a single linear constraint on the quadric, thus the quadric surfaces containing the $\mathbf{p}_i$, $\mathbf{o}$ and $\mathbf{a}$ form a $9 - 7 = 2$

dimensional space $\mathcal{S}^2$ contained in the space of all quadrics. A basis for $\mathcal{S}^2$ can be calculated from the $\mathbf{p}_i$, $\mathbf{o}$ and $\mathbf{a}$. $\square$

Not all the quadrics parameterised by points of $\mathcal{S}^2$ are ambiguous surfaces. The proof that there are ten essentially different camera displacements compatible with five image correspondences relies on selecting from $\mathcal{S}^2$ precisely those points corresponding to ambiguous surfaces.

Theorem 5.2. Let five image correspondences be given in general position, and compatible with a known camera displacement. Then there exist two cubic plane curves f_1, f_2 in the space $\mathcal{S}^2$ constructed in Theorem 5.1 such that each camera displacement compatible with the image correspondences, but different from the known camera displacement, arises from a common zero of f_1 and f_2. Conversely, each common zero of f_1 and f_2 gives rise to essentially only one camera displacement compatible with the given image correspondences, but essentially different from the known camera displacement.

Proof. Let $\{R, \mathbf{a}\}$ be the known camera displacement and let $\{S_i, \mathbf{b}_i\}$, $1 \leq i \leq n$ be the camera displacements essentially different from $\{R, \mathbf{a}\}$ but also compatible with the five image correspondences. Each pair $\{R, \mathbf{a}\}$, $\{S_i, \mathbf{b}_i\}$ yields an ambiguous surface ψ_i, with an equation of the form (2), where $\mathbf{b} = \mathbf{b}_i$ and $U = S_i^\top R$. Each ψ_i corresponds to a point $\mathbf{u}_i$ of $\mathcal{S}^2$. The cubic constraints (19) and (20) yield cubic plane curves f_1, f_2, respectively, in $\mathcal{S}^2$. Each $\mathbf{u}_i$ is an intersection point of f_1 and f_2. Conversely, by Theorem 3.2, each intersection point of f_1 and f_2 yields an ambiguous surface ψ. The surface ψ yields a unique camera displacement in the set $\{S_i, \mathbf{b}_i\}$, $1 \leq i \leq n$. $\square$

Theorem 5.3. Let five image correspondences be given in general position. Then there are exactly ten essentially different camera displacements compatible with the given image correspondences.

Proof. To prove the theorem it suffices to show that the cubic plane curves f_1, f_2 in $\mathcal{S}^2$ obtained in Theorem 5.2 have, in general, exactly nine distinct intersections. The property that f_1 and f_2 have nine distinct intersections is stable against small perturbations in the coefficients of f_1 and f_2, thus it is sufficient to consider the case in which the coefficients of f_1 (defined by (19)) involving $\mathbf{a}$ are negligibly small in comparison with the first term $-\mathbf{l}^\top N \mathbf{l}$. It is thus sufficient to consider the cubic curves in $\mathcal{S}^2$,

$$\mathbf{l}^\top N \mathbf{l} = 0 \qquad\qquad \det(N) = 0 \qquad\qquad (21)$$

In order to show that the two curves of (21) have, in general, nine distinct intersections, it is sufficient to produce a single example in which they have nine intersections. To this end, choose three of the reconstructed points in $\mathbf{P}^3$ to be $\mathbf{p}_1 = (1, 0, 0, 0)^\top$, $\mathbf{p}_2 = (0, 1, 0, 0)^\top$, $\mathbf{p}_3 = (0, 0, 1, 0)^\top$, and choose $\mathbf{a} = (1, -1, 1)^\top$. With these choices, the equation $\det(N) = 0$ splits into three linear factors. It can be shown by direct calculation that $\mathbf{p}_4$, $\mathbf{p}_5$ can be chosen in $\mathbf{R}^3$ such that each line comprising $\det(N) = 0$ meets the cubic plane curve $\mathbf{l}^\top N \mathbf{l} = 0$ at three distinct points. $\square$

In Theorem 5.3 the possibility is not ruled out that the intersections of the two cubic plane curves yield quadrics without real generators. Quadrics obtained in this way give rise to complex camera displacements compatible with the $\mathbf{q}_i \leftrightarrow \mathbf{q}'_i$, but which are not physically acceptable.

6 Conclusion

The reconstruction of the relative positions of points in space from the correspondences between two different images of the points is subject to ambiguity if the points lie on certain surfaces of degree two known as rectangular hyperboloids. The ambiguous case of reconstruction has been investigated using both the projective geometric methods developed by the photogrammetrists and the Euclidean methods developed in computer vision.

A new cubic polynomial constraint on ambiguous surfaces has been obtained. This leads to a new proof of the known result that there are, in general, exactly ten camera displacements compatible with five given image correspondences. It is conjectured that these results on ambiguity are relevent to the more general problem of describing the stability of algorithms for recovering camera displacement from image correspondences [2]. For example, instability may arise if the two cubic plane curves obtained in Theorem 5.2 are near coincident over a wide range of points.

Acknowledgements: This work is funded by ESPRIT P2502 (Voila). The author first learnt of the German work on the reconstruction problem during an extended visit to INRIA, Rocquencourt, at the invitation of Olivier Faugeras. The author thanks Tom Buchanan for discussions on projective geometry and on the work of the photogrammetrists dating from the last century and the first half of this century. Thanks are also due to Bernard Buxton for comments on an earlier draft of this paper.

References

1. M. Demazure (1988) *Sur deux problèmes de reconstruction.* INRIA Rapports de Recherche No. 882.

2. J.Q. Fang and T.S. Huang (1984) *Some experiments on estimating the 3-D motion parameters of a rigid body from two consecutive image frames.* IEEE Transactions on Pattern Analysis and Machine Intelligence, vol. 6, pp. 545-554.

3. O.D. Faugeras and S.J. Maybank (1990) *Motion from point matches: multiplicity of solutions.* Accepted by Int. J. Computer Vision.

4. W. Hofmann (1950) *Das Problem der "Gefärlichen Flächen" in Theorie und Praxis.* Dissertation, Fakultät für Bauwesen der Technischen Hochschule München, München, FR Germany.

5. B.K.P. Horn (1987) *Motion fields are hardly ever ambiguous.* Int. J. of Computer Vision, vol. 1, pp. 263-278.

6. B.KP. Horn (1989) Relative Orientation. Accepted by Int. J. Computer Vision.

7. E. Kruppa (1913) *Zur Ermittlung eines Objektes zwei Perspektiven mit innere Orientierung.* Sitz-Ber. Akad. Wiss., Wien, math. naturw. Kl., Abt. IIa., vol 122, pp. 1939-1948.

8. H.C. Longuet-Higgins (1981) *A computer algorithm for reconstructing a scene from two projections.* Nature, vol. 293, pp. 133-135.

9. H.C. Longuet-Higgins (1988) *Multiple interpretations of a pair of images.* Proc. Roy. Soc. Lond. A, vol. 418, pp. 1-15.

10. H.C. Longuet-Higgins and K. Prazdny (1980) *The interpretation of a moving retinal image.* Proc. Roy. Soc. Lond. B, vol. 208, pp. 385-397.

11. S.J. Maybank (1985) *The angular velocity associated with the optical flow field arising from motion through a rigid environment.* Proc. Roy. Soc. Lond. A, vol. 401, pp. 317-326.

12. S.J. Maybank (1987) *A theoretical study of optical flow.* PhD Thesis, Birkbeck College, University of London.

13. S.J. Maybank (1990) *Rigid velocities compatible with five image velocity vectors.* Accepted by Image and Vision Computing.

14. S.J. Maybank (1990) *The projective geometry of ambiguous surfaces.* In preparation.

15. J.G. Semple and G.T. Kneebone (1952) Algebraic Projective Geometry. Oxford: Clarendon Press (reprinted 1979).

16. R. Sturm (1869) *Das Problem der Projectivität und seine Anwendung auf die Flächen zweiten Grades.* Math. Annalen, vol. 1, pp. 533-573.

17. W. Wunderlich (1942) *Zur Eindeutigkeitsfrage der Hauptaufgabe der Photogrammetrie.* Monatsch. Math. Physik, vol. 50, pp. 151-164.

An Optimal Solution for Mobile Camera Calibration

P. Puget, T. Skordas

ITMI, Filiale de CAP SESA

ZIRST, 11 Chemin des Prés, BP 87, 38243 Meylan cédex - France

e-mail: puget@itmi.uucp, skordas@itmi.uucp

Abstract

This paper addresses the problem of determining the intrinsic and extrinsic parameters of a mobile camera. We present an optimal solution which consists of the following steps: first, the camera is calibrated in several working positions and for each position, the corresponding transformation matrix is computed using a method developed by Faugeras and Toscani [1]; next, optimal intrinsic parameters are searched for all positions; finally, for each separate position, optimal extrinsic parameters are computed by minimizing a mean square error through a closed form solution. Experimental results show that such a technique yields a spectacular reduction of calibration errors and a considerable gain relative to other existing on-site calibration techniques.

1 Introduction

This paper addresses the problem of determining the optical (internal) camera parameters (called *intrinsic parameters*) and the three-dimensional (3-D) position and orientation of the camera frame relative to some predefined world coordinate frame (*extrinsic parameters*).

This problem is a major part of the general problem of camera calibration and is the starting point for several important applications in Computer Vision and Robotics. 3-D object tracking and dynamic scene reconstruction, stereo vision, stereo calibration and stereo reconstruction, object recognition and localization from a single view, and sensory based navigation, are just a few situations where camera calibration is explicitly needed.

Calibrating a camera is determining the relationship between the 3-D coordinates of a point and the corresponding 2-D coordinates of its image. Such a relationship is usually expressed in terms of a 3×4 matrix M, which is called the *perspective transformation matrix* [2,7]. In other words, a camera is considered being calibrated if for each of its working positions, the corresponding matrix M can be derived. If the camera is fixed, the calibration problem is reduced to the problem of computing a single matrix M. If the camera moves arbitrarily, calibration involves computation of both intrinsic and extrinsic parameters, the latter being necessary for the hand/eye calibration [5,9,10].

Numerous techniques tackling the general calibration problem exist in the literature. An excellent survey of these techniques is given by Tsai in [8]. There are two main approaches for calibrating a moving camera. One approach is to compute the perspective transformation matrix from which the camera parameters are derived, [1,2,6]. In a second approach, that of Tsai [7], and Tsai and Lenz [9], no perspective transformation matrix is explicitly calculated. This method seeks the solution of a linear matrix equation in five unknowns merely consisting of extrinsic parameters, from which, the rest of the parameters can be easily recovered. This method is based on *a priori* knowledge of two intrinsic parameters. In [9], a method is proposed to compute all intrinsic camera parameters by performing a preliminary off-site calibration stage. This in turn, is likely to cause additional difficulties, since the camera parameters may be altered accidentally when fixing the camera on its working place. Their results in terms of accuracy are similar to those reported by Faugeras and Toscani.

We have performed camera calibration on several camera positions using the technique proposed by Faugeras and Toscani [1]. Among different camera positions, significant differences between homologous intrinsic camera parameters occurred. While the method turns out to be particularly accurate in the computation of the perspective matrix for a given camera position, nevertheless, when the intrinsic and extrinsic parameters are derived from this matrix, the results are disappointing. This is due to the noisy 2-D and 3-D data and to the instability of equations allowing to compute these parameters (see Section 3). Note also, the fact that the perspective matrix is accurate and the intrinsic parameters are sometimes inaccurate, implies that the extrinsic parameters are also affected by errors.

If one particular set of intrinsic parameters (corresponding to a given position) is assumed to be valid everywhere, and is associated with the extrinsic parameters corresponding to another camera position, the resulting errors are enormous (see section 5).

Our approach

Our technique can be summarized as follows: a camera is calibrated at several (N) positions and in each position, a perspective transformation matrix is computed by using the direct method (without applying extended Kalman filtering) of Faugeras and Toscani. Once the N matrices are available, the problem of computing correct intrinsic and extrinsic parameters is broken down into three successive steps: 1) a minimization criterion is formulated leading to the computation of optimal intrinsic parameters; 2) an optimal rotation is computed for each camera position; 3) an optimal translation is computed for each camera position.

The optimization of the rotational and translational parameters is achieved by minimizing a mean square error. All these optimizations are performed using closed-form algorithms. Such a technique yielded a spectacular reduction of errors and a gain in precision of a factor up to 50, relatively to the Faugeras-Toscani approach.

2   The camera model

The camera model that we use is the *pinhole* model. This is the most frequently used model existing in the bibliography [1,2,6,7]. The underlying mathematical model is the perspective transformation: a 3-D point $P = (X, Y, Z)^t$ is projected on a 2-D point $p = (u, v)^t$ in the image plane (see figure 1) where:

$$u = su/s \quad and \quad v = sv/s \tag{1}$$

with:

$$(su, \ sv, \ s)^t = M \ (X, \ Y, \ Z, \ 1)^t \tag{2}$$

Coordinates of P are expressed relatively to some scene coordinate frame, and coordinates of p are expressed relatively to the image frame in pixels. $M = (m_{ij})$ is a 3×4 matrix called the *perspective transformation matrix* and is defined up to a scale factor. Coefficients m_{ij} depend on 10 independent parameters which are the following:

- α_u and α_v, the scale factors on axes u and v of the image frame;

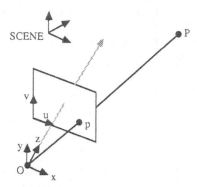

Figure 1: Mathematical model of the camera

- u_0 and v_0, the coordinates in the image frame of the intersection of the optical axis with the image plane;

- t_x, t_y and t_z, the coordinates of a translation vector $\vec{t}$;

- r_x, r_y and r_z, the coordinates of a rotation vector $\vec{r}$.

The first four parameters depend only on the physical characteristics of the camera and are called the *intrinsic camera parameters*. The last six parameters define the geometric transform T between the camera frame and a given scene coordinate frame. They depend on the position of the camera relative to the scene coordinate frame. They are called *extrinsic camera parameters*. Let us define a few notations:

$$M = \begin{pmatrix} \vec{m}_1 & m_{14} \\ \vec{m}_2 & m_{24} \\ \vec{m}_3 & m_{34} \end{pmatrix} \qquad T = \begin{pmatrix} \vec{r}_1 & t_x \\ \vec{r}_2 & t_y \\ \vec{r}_3 & t_z \end{pmatrix}$$

where $\vec{m}_i$ and $\vec{r}_i$ are (1×3) row vectors. Vectors $\vec{r}_i$ define a rotation matrix ρ. Let θ be the angle, $\vec{n} = (n_x, n_y, n_z)^t$ the unit vector defining the axis and $\vec{r} = (r_x, r_y, r_z)^t$ the vector of rotation. These variables are related by the expressions (see [3]):

$$\rho = \begin{pmatrix} \cos\theta + r_x^2 g(\theta) & r_x r_y g(\theta) - r_z f(\theta) & r_x r_z g(\theta) + r_y f(\theta) \\ r_x r_y g(\theta) + r_z f(\theta) & \cos\theta + r_y^2 g(\theta) & r_y r_z g(\theta) - r_z f(\theta) \\ r_x r_z g(\theta) - r_y f(\theta) & r_y r_z g(\theta) + r_z f(\theta) & \cos\theta + r_z^2 g(\theta) \end{pmatrix} \tag{3}$$

$$f(\theta) = \frac{\sin\theta}{\theta}, \qquad g(\theta) = \frac{1 - \cos\theta}{\theta^2}, \qquad \theta = \sqrt{r_x^2 + r_y^2 + r_z^2}, \qquad \vec{r} = \theta\vec{n}$$

Notice that $\vec{r}_i$ as rows of a rotation matrix define an orthonormal frame.

The expression of M as a function of the intrinsic and extrinsic parameters is:

$$M = \omega \begin{pmatrix} \alpha_u \vec{r}_1 + u_0 \vec{r}_3 & \alpha_u t_x + u_0 t_z \\ \alpha_v \vec{r}_2 + u_0 \vec{r}_3 & \alpha_v t_y + u_0 t_z \\ \vec{r}_3 & t_z \end{pmatrix} \tag{4}$$

ω is the scale factor of matrix M. As $\vec{r}_3$ is a unit vector, ω can be identified with $sqrt{\vec{m}_3.\vec{m}_3^t}$. In the sequel, M is considered with ω equal to 1.

The intrinsic and extrinsic parameters can be calculated by identifying coefficients m_{ij} with their expression given by (4), and by using properties of vectors $\vec{r}_i$. This method is presented and analyzed in the following section.

3 Instability in the computation of the parameters

The following equalities can be derived from the orthogonality of vectors $\vec{r_i}$:

$$\vec{r_3} = \vec{m_3} \tag{5}$$

$$\begin{cases} u_0 &=& \vec{m_1}.\vec{m_3} \\ v_0 &=& \vec{m_2}.\vec{m_3} \end{cases} \tag{6}$$

$$\begin{cases} \alpha_u &=& \|\vec{m_1} \times \vec{m_3}\| \\ \alpha_v &=& \|\vec{m_2} \times \vec{m_3}\| \end{cases} \tag{7}$$

$$\begin{cases} \vec{r_1} &=& 1/\alpha_u(\vec{m_1} - u_0\vec{m_3}) \\ \vec{r_2} &=& 1/\alpha_v(\vec{m_2} - v_0\vec{m_3}) \end{cases} \tag{8}$$

$$\begin{cases} t_x &=& 1/\alpha_u(m_{14} - u_0 m_{34}) \\ t_y &=& 1/\alpha_v(m_{24} - v_0 m_{34}) \\ t_z &=& m_{34} \end{cases} \tag{9}$$

The equations above are non linear and give rather unstable results, especially the first four ones which allow the computation of the intrinsic parameters and the rotational part of the extrinsic parameters.

A geometrical interpretation of the above calculation shows this instability. Figure 2 depicts the geometrical situation of the vectors $\vec{m_i}$ and $\vec{r_i}$. In this figure, we volunteerly magnified vectors $\vec{r_i}$. For the cameras commonly in use, the order of magnitude of α_u and α_v is 1000 and the order of magnitude of u_0 and v_0 is 250. The problem of finding the rotation and the intrinsic parameters is equivalent to the problem of properly "placing" the orthonormal vectors $\vec{r_1}$, $\vec{r_2}$ and $\vec{r_3}$ relatively to $\vec{m_1}$, $\vec{m_2}$ and $\vec{m_3}$ and of calculating the projections of $\vec{m_1}$ and $\vec{m_2}$ onto the vectors $\vec{r_i}$. This is done through the above equalities by:

1. identifying $\vec{r_3}$ with $\vec{m_3}$ (equality (5));

2. calculating $\vec{r_1}$ and $\vec{r_2}$ as the vectors orthogonal to $\vec{r_3}$ in the planes defined by $\vec{r_3}$ and $\vec{m_1}$ and $\vec{r_3}$ and $\vec{m_2}$, respectively (equality (8));

3. calculating the projections of $\vec{m_1}$ and $\vec{m_2}$ onto the vectors $\vec{r_i}$ (eq (6) and (7)).

The consequences of such a method are:

- even a small error on $\vec{m_3}$ badly propagates and can cause very significant errors in the calculation of vectors $\vec{r_i}$ and of the intrinsic parameters, (notice that especially u_0 and v_0 are extremely sensitive to $\vec{m_3}$),

- the resulting vectors $\vec{r_i}$ are not necessarily orthogonal; notice here that the "c" factor introduced by Toscani [1] not only reflects the non-orthogonality of the u and v axes of the image frame but also the errors produced in vectors $\vec{m_i}$ and $\vec{r_i}$ as well;

- even if the values of the parameters are wrong, they are mutually compatible: if the matrix M is recomputed from these values, results are quite good.

In order to give quantitative estimations of the numerical instability, the intuitive geometrical interpretation given above can be completed with calculations of error propagation through the equations (5) to (7). This is done in appendix A.

4 An optimal solution for computing the parameters

The basic principles of our method are the following:

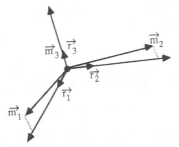

Figure 2: Geometrical interpretation of the calculation of the parameters

1. the intrinsic parameters are computed after having computed the perspective transformation matrix M on *several* positions. The underlying idea is to compute a unique set of intrinsic parameters that is correct for every position of the moving camera;

2. using the formerly computed intrinsic parameters for each calibration position, the extrinsic parameters are computed by two successive optimization algorithms: one for the rotational and one for the translational parameters.

4.1 Computing the intrinsic parameters

We suppose that we have computed N perspective transformation matrices $M^{(i)}$ corresponding to N different positions i of the camera:

$$M^{(i)} = \begin{pmatrix} \alpha_u \vec{r}_1^{(i)} + u_0 \vec{r}_3^{(i)} & \alpha_u t_x^{(i)} + u_0 t_z^{(i)} \\ \alpha_v \vec{r}_2^{(i)} + u_0 \vec{r}_3^{(i)} & \alpha_v t_y^{(i)} + u_0 t_z^{(i)} \\ \vec{r}_3^{(i)} & t_z^{(i)} \end{pmatrix} \tag{10}$$

The principle of the calculation is to look for invariants which are independent of the extrinsic parameters and which allow one to calculate the intrinsic parameters. The most obvious possibility is to choose for these non varying quantities the intrinsic parameters that we could calculate for each separate position. For each position i, we can write:

$$\begin{cases} u_0^{(i)} &= \vec{m}_1^{(i)} . \vec{m}_3^{(i)} \\ v_0^{(i)} &= \vec{m}_2^{(i)} . \vec{m}_3^{(i)} \end{cases} \qquad \begin{cases} \alpha_u^{(i)} &= \| \vec{m}_1^{(i)} \times \vec{m}_3^{(i)} \| \\ \alpha_v^{(i)} &= \| \vec{m}_2^{(i)} \times \vec{m}_3^{(i)} \| \end{cases}$$

The optimal value for u_0 is that which minimizes the criterion:

$$C = \sum_{i=1}^{N} (u_0 - u_0^{(i)})^2$$

This is the mean value of $u_0^{(i)}$:

$$u_0 = 1/N \sum_{i=1}^{N} u_0^{(i)}$$

The same reasoning holds for the other three intrinsic parameters.

4.2 Computing the extrinsic parameters

For each camera position, a perspective transformation matrix M is available, and a set of intrinsic parameters have been computed. Now we focus on the calculation of the extrinsic parameters.

Let X_j denote the 3-D points used for the calibration (expressed in the scene coordinate frame) and $(u_j, v_j)^t$ the 2-D coordinates of their projections onto the image plane. Eliminating s in equations (1) and (2) gives:

$$\begin{cases} (u_j \vec{m}_3 - \vec{m}_1).X_j + (u_j m_{34} - m_{14}) = 0 \\ (v_j \vec{m}_3 - \vec{m}_2).X_j + (v_j m_{34} - m_{24}) = 0 \end{cases} \tag{11}$$

Expressing $\vec{m}_j$ as functions of intrinsic and extrinsic parameters leads to:

$$\begin{cases} \alpha_u(\vec{r}_1.X_j + t_x) + (u_0 - u_j)(\vec{r}_3.X_j + t_z) = 0 \\ \alpha_v(\vec{r}_2.X_j + t_y) + (v_0 - v_j)(\vec{r}_3.X_j + t_z) = 0 \end{cases} \tag{12}$$

The problem is now to find $\vec{r}_1$, $\vec{r}_2$, $\vec{r}_3$, t_x, t_y, t_z so that equations (12) are verified for all points X_j. To separate the calculation of the translational part from the rotational part, we must first state the following lemma:

Lemma 1 *If X_j is considered as the unknown in equations (11), then the point $X_0 = -\rho^{-1}\vec{t}$ is a solution for all j.*

Proof: $(\vec{r}_1.X_0 + t_x)$, $(\vec{r}_2.X_0 + t_y)$, $(\vec{r}_3.X_0 + t_z)$ are the coordinates of the point $\rho X_0 + \vec{t}$. This vector is equal to $(-\rho\rho^{-1}\vec{t} + \vec{t})$, or the zero vector. The three coordinates are equal to zero and X_0 is then a solution of equations (12) for all j, and consequently a solution of the equations (11) which are equivalent. $\square$

Putting all equations (11) together gives a linear system. Its solution X_0 can be calculated by a least squares procedure.

The geometrical interpretation of lemma 1 is that X_0 represents the coordinates of the optical center of the camera, expressed in the scene coordinate frame. The equalities (11) considered as equations whose unknown is X_j are the equations defining the locus which is projected in $(u_j, v_j)^t$. This locus is the straight line passing through the optical center and the point $(u_j, v_j)^t$ of the image plane. When considering all equations for all j, the unique solution is the point which is the intersection of all these lines: this is the optical center.

4.2.1 Computing the rotation

Equalities (11) are now equivalent to:

$$\begin{cases} (u_j \vec{m}_3 - \vec{m}_1).X_j + (u_j m_{34} - m_{14}) = (u_j \vec{m}_3 - \vec{m}_1).X_0 + (u_j m_{34} - m_{14}) \\ (v_j \vec{m}_3 - \vec{m}_2).X_j + (v_j m_{34} - m_{24}) = (v_j \vec{m}_3 - \vec{m}_2).X_0 + (v_j m_{34} - m_{24}) \end{cases} \tag{13}$$

which leads to:

$$\begin{cases} \alpha_u \vec{r}_1.(X_j - X_0) + (u_0 - u_j)\vec{r}_3.(X_j - X_0) = 0 \\ \alpha_v \vec{r}_2.(X_j - X_0) + (v_0 - v_j)\vec{r}_3.(X_j - X_0) = 0 \end{cases} \tag{14}$$

As $\vec{r}_i.(X_j - X_0)$ is the ith coordinate of the vector $\rho(X_j - X_0)$, the first of equations (14) means that $\rho(X_j - X_0)$ is orthogonal to the vector $(\alpha_u, 0, (u_0 - u_j))$. For the same reason the second equation means that the vector $\rho(X_j - X_0)$ is orthogonal to the vector $(0, \alpha_v, (v_0 - v_j))$. This implies that the direction of the vector $\rho(X_j - X_0)$ is given by the cross-product of the two previous vectors:

$$N_j = \begin{pmatrix} \alpha_v(u_0 - u_j) \\ \alpha_u(v_0 - v_j) \\ -\alpha_u \alpha_v \end{pmatrix}$$

The geometrical interpretation of this result is that for every point X_j, the line joining the optical center and X_j must be colinear to the line joining the optical center and the image point (u_j, v_j).

Let us define:

$$S_j = \frac{X_j - X_0}{\|X_j - X_0\|} \quad \text{and} \quad Q_j = \frac{N_j}{\|N_j\|}$$

The problem of finding the rotation is finally equivalent to finding ρ which minimizes the criterion:

$$C_\rho = \sum_{j=1}^{M} \|\rho S_j - Q_j\|^2$$

To solve this problem, we consider the unit quaternion q associated with ρ. It is shown in appendix B that q is the unit eigenvector associated with the smallest eigenvalue of the positive symmetric matrix B where:

$$B = \sum_{j=1}^{M} (B_j^t B_j)$$

with B_j given by:

$$B_j = \begin{pmatrix} 0 & -S_{jx} + Q_{jx} & -S_{jy} + Q_{jy} & -S_{jz} + Q_{jz} \\ S_{jx} - Q_{jx} & 0 & S_{jz} + Q_{jz} & -S_{jy} - Q_{jy} \\ S_{jy} - Q_{jy} & -S_{jz} - Q_{jz} & 0 & S_{jx} + Q_{jx} \\ S_{jz} - Q_{jz} & S_{jy} + Q_{jy} & -S_{jx} - Q_{jx} & 0 \end{pmatrix} \quad (15)$$

4.2.2 Computing the translation

Equalities (12) can now be considered as equations whose unknowns are t_x, t_y, t_z:

$$\begin{cases} (\alpha_u \vec{r}_1 . X_j) t_x + (u_0 - u_j) t_z + \alpha_u \vec{r}_1 . X_j + (u_0 - u_j) \vec{r}_3 . X_j = 0 \\ (\alpha_v \vec{r}_2 . X_j) t_y + (v_0 - v_j) t_z + \alpha_v \vec{r}_2 . X_j + (v_0 - v_j) \vec{r}_3 . X_j = 0 \end{cases} \quad (16)$$

These equations are linear and we have two of them available for each calibration point X_j. The optimal value for $\vec{t} = (t_x, t_y, t_z)^t$ is computed with a classical least-squares algorithm.

5 Experimental results

Experimental setup

The experimental setup is shown on figure 3. A Pulnix TM 560 R camera is mounted on a robot wrist. This camera has a 2/3 inches, 500×582 elements, CCD sensor with a 16mm TV lens. We use a VME board for image acquisition. The calibration pattern is a grid of lines drawn with a laser printer. A 3-D calibration point is determined as the intersection of two lines. The calibration pattern is fixed on an horizontal metal plate moving along the z axis. The images are taken at an approximate distance from 40 to 50 cm. The number of points used for the calibration is about 100, lying on two to four distinct planes. The 2-D calibration points in the image are determined by extracting junctions. These junctions are found as intersections of lines segments detected in the image.

Error evaluation

To evaluate and compare our method with other methods, we used two criteria which are now presented. For each 3-D point X_j, let $p_j = (u_j, v_j)^t$ denote its observed projection on the image, and $p_j^{mod} = (u_j^{mod}, v_j^{mod})^t$ its projection given by the camera model corresponding to the parameters

Figure 3: The experimental setup

that have been found. For each point X_j, this gives an error vector $\vec{\delta_j} = (u_j^{mod} - u_j, v_j^{mod} - v_j)^t$. These vectors $\vec{\delta_j}$ constitute one criterion. Another criterion is the mean module of vectors $\vec{\delta_j}$.

Results
The following table reports the mean error module for eight different calibration positions, using five different methods:

	pos 1	pos 2	pos 3	pos 4	pos 5	pos 6	pos 7	pos 8
method 1	0.87	5.48	48.26	10.86	57.59	5.52	2.60	13.78
method 2	0.82	0.93	4.83	1.38	6.16	0.98	0.84	1.62
method 3	1.82	1.22	4.81	1.15	5.77	1.33	2.22	2.81
method 4	0.88	0.81	2.86	1.01	3.76	0.93	0.87	1.14
method 5	0.85	0.79	1.11	0.89	1.38	0.87	0.82	0.85

Method 1: This is the method of Faugeras and Toscani applied to each of the eight positions. Intrinsic parameters of position 1 are chosen as unique values for all positions and we associate them with the extrinsic parameters corresponding to each position. We see that the mean error module for position 1 is quite good: this shows that associating intrinsic and extrinsic parameters computed in the same position yields an excellent perspective transformation matrix. This is not the case when intrinsic and extrinsic parameters from different positions are associated.

Method 2: Intrinsic parameters computed in position 1 (also with Faugeras and Toscani's method) are again chosen, and optimal *extrinsic* parameters are computed with these selected values, using the method presented in section 4.2. The errors are dramatically reduced. This shows that, for a given set of intrinsic parameters, even if they are noisy, it is possible to find compatible extrinsic parameters which yield a positive result, i.e., give a relatively accurate perspective transformation matrix. The associated extrinsic parameters are also somewhat inaccurate.

Method 3: Optimal intrinsic parameters are computed taking into account all eight positions with the method presented in section 4.1. The extrinsic parameters are simply computed using equations (5), (8) and (9). With reference to method 2, the errors are reduced to a similar level.

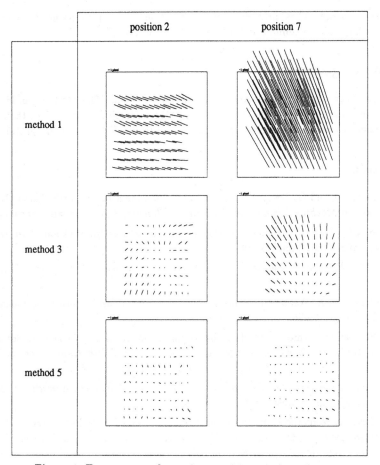

Figure 4: Error vectors for various positions and methods

Here, intrinsic and extrinsic parameters are less corrupted than for the first two methods. Notice also that the mean error in position 1 is worse here than the two methods above. The new values for intrinsic parameters are globally optimal for all the positions but not for this specific position itself.

Method 4: Intrinsic parameters are the same as in method 3 and extrinsic *translational* parameters are optimized. *Rotational* parameters are computed using equations (5) and (8).

Method 5: We use the complete optimal method that has been presented in this paper. This produces the best results. Compared with method 1, we have a gain factor varying between 1.02 (pos 1) and 43.47 (pos 3). The average gain factor is 16.4 .

A qualitative comparison of the various methods can be done by observing the error vectors. These are shown in figure 4 for positions 2 and 7, and methods 1, 3 and 5 respectively. The error vectors are magnified by a factor of 10 for position 2 and a factor of 5 for position 7.

The CPU time needed for a whole camera calibration process (computation of the perspective transformation matrices and optimization of parameters) is 13 seconds on a SUN 4/110 workstation for eight positions and 100 calibration points per position. This time does not include the low level image processing and the computation of 2-D calibration points.

Acknowledgements

This work has been partially supported by Esprit project P940. The authors would like to thank Philippe Bobet from LIFIA and the members of ITMI's research department for their contribution to the implementation of this work, and Radu Horaud from LIFIA for his insightful comments.

References

[1] O. D. Faugeras and G. Toscani. Camera Calibration for 3D Computer Vision. In *Proc of Int. Workshop on Machine Vision and Machine Intelligence*, Tokyo, Japan, February 1987.

[2] S. Ganapathy. Decomposition of transformation matrices for robot vision. *Pattern Recognition Letters*, 2:401–412, December 1984.

[3] R. P. Paul. *Robot Manipulators: Mathematics, Programming, and Control*. The MIT Press, 1981.

[4] E. Previn and J. A. Webb. *Quaternions in computer vision and robotics*. Technical Report CS-82-150, Carnegie-Mellon University, 1982.

[5] Y. C. Shiu and S. Ahmad. Finding the Mounting Position of a Sensor by Solving a Homogeneous Transform Equation of the Form AX=XB. In *IEEE Conference on Robotics and Automation*, pages 1666–1671, Raleygh, North Carolina, USA, April 1987.

[6] T. M. Strat. *Recovering the Camera Parameters from a Transformation Matrix*, pages 93–100. Morgan Kaufmann Publishers, Inc, 1987.

[7] R. Y. Tsai. A Versatile Camera Calibration Technique for High-Accuracy 3D Machine Vision Metrology Using Off-the-Shelf TV Cameras and Lenses. *IEEE Journal of Robotics and Automation*, RA-3(4):323–344, August 1987.

[8] R. Y. Tsai. Synopsis of Recent Progress on Camera Calibration for 3D Machine Vision. In Oussama Khatib, John J. Craig, and Tomás Lozano-Pérez, editors, *The Robotics Review*, pages 147–159, The MIT Press, 1989.

[9] R. Y. Tsai and R. K. Lenz. Techniques for Calibration of the Scale Factor and Image Center for High Accuracy 3-D Machine Vision Metrology. *IEEE Trans on Pattern Analysis and Machine Intelligence*, 10(5):713–720, september 1988.

[10] R.Y. Tsai and R.K. Lenz. Real Time Versatile Robotics Hand/Eye Calibration using 3D Machine Vision. In *IEEE International Conference on Robotics and Automation*, Philadelphia, Penn, USA, April 1988.

A Numerical estimation of errors in the computation of camera parameters

In this appendix, Δu_0, Δv_0, $\Delta \alpha_u$, $\Delta \alpha_v$ denote the maximal errors on u_0, v_0, α_u and α_v respectively. $\Delta \vec{m}_i$ denotes the maximal error vector on $\vec{m}_i$, *i.e.* the vector $(\Delta m_{i1}, \Delta m_{i2}, \Delta m_{i3})^t$. $|\vec{m}_i|$ denotes the vector $(|m_{i1}|, |m_{i2}|, |m_{i3}|)^t$. As for any two vectors $\vec{u}$ and $\vec{v}$, the following relationship holds:

$$\Delta(\vec{u}.\vec{v}) = \Delta\vec{u}.|\vec{v}| + \vec{u}.|\Delta\vec{v}|$$

we get the expression of Δu_0 and Δv_0 from equations (6):

$$\begin{cases} \Delta u_0 = \Delta \vec{m}_1 . |\vec{m}_3| + |\vec{m}_1| . \Delta \vec{m}_3 \\ \Delta v_0 = \Delta \vec{m}_2 . |\vec{m}_3| + |\vec{m}_2| . \Delta \vec{m}_3 \end{cases} \tag{17}$$

We have also:

$$\begin{cases} \Delta(\|\vec{u}\|) = \frac{1}{\|\vec{u}\|}(|\vec{u}| . \Delta \vec{u}) \\ \Delta(\vec{u} \times \vec{v}) = \Delta \vec{u} \otimes |\vec{v}| + |\vec{u}| \otimes \Delta \vec{v} \end{cases}$$

with $\otimes$ being the operation:

$$\vec{u} \otimes \vec{v} = \begin{pmatrix} u_y v_z + u_z v_y \\ u_z v_x + u_x v_z \\ u_x v_y + u_y v_x \end{pmatrix}$$

This leads to the expression of $\Delta \alpha_u$ and $\Delta \alpha_v$:

$$\begin{cases} \Delta \alpha_u = \frac{1}{\alpha_u}[|\vec{m}_1 \times \vec{m}_3| . (\Delta \vec{m}_1 \otimes |\vec{m}_3| + |\vec{m}_1| \otimes \Delta \vec{m}_3)] \\ \Delta \alpha_v = \frac{1}{\alpha_v}[|\vec{m}_2 \times \vec{m}_3| . (\Delta \vec{m}_2 \otimes |\vec{m}_3| + |\vec{m}_2| \otimes \Delta \vec{m}_3)] \end{cases}$$

These formulae allow one to calculate the maximal errors on the intrinsic parameters when the errors Δm_{ij} are known. However, due to the algorithm used to compute the perspective transformation matrix, it is difficult to calculate Δm_{ij} as a function of the errors on the 3-D points X_j and their projections p_j. Consequently, to estimate the errors Δm_{ij}, we did a Monte-Carlo simulation. Random errors on 2-D data whose standard deviation is 1.0 pixel leads to the following results:

calibration type	Δu_0	Δv_0	$\Delta \alpha_u$	$\Delta \alpha_v$
4 planes	7.42	12.54	6.42	11.99
2 upper planes	8.80	19.62	16.03	26.41
2 lower planes	17.81	25.02	24.94	40.19

These results show that the noise on 2-D data is amplified. With real data, we found differences of about 20.0 pixels on u_0 and v_0 for different camera positions.

B Finding the optimal rotation

We saw that the optimal rotation ρ is the rotation that minimizes the criterion:

$$C_\rho = \sum_{j=1}^{M} \|\rho S_j - Q_j\|^2 \tag{18}$$

Let q be the quaternion associated with the rotation ρ (notations were introduced in section 2):

$$q = \cos \frac{\theta}{2} + \sin \frac{\theta}{2}(n_x i + n_y j + n_z k)$$

It is possible also to consider a vector or a point as S_j as a quaternion (see [4] for more details):

$$S_j = S_{jx} i + S_{jy} j + S_{jz} k$$

Our demonstration is based on the three following properties:

Property 1 *The quaternion q associated with the rotation ρ is a unit quaternion. This means that:*

$$\|q\|^2 = q * \bar{q} = 1$$

where $*$ denotes the product of quaternions and $\bar{q}$ the conjugate of q:

$$\bar{q} = \cos\frac{\theta}{2} - \sin\frac{\theta}{2}(n_x i + n_y j + n_z k)$$

This implies that q^{-1} is equal to $\bar{q}$.

For the proof, calculate $\|q\|^2$ using the fact that $(n_x, n_y, n_z)^t$ is a unit vector and $\cos^2\frac{\theta}{2} + \sin^2\frac{\theta}{2}$ is equal to 1.

Property 2 *Representing both point S_j and rotation ρ with quaternions, the result of applying ρ to S_j is represented by the quaternion:*

$$\rho S_j = q * S_j * q^{-1}$$

For the proof, see [4].

Property 3 *For any quaternion p and any* unit *quaternion q, we have:*

$$\|p * q\|^2 = \|p * \bar{q}\|^2 = \|p\|^2$$

Proof:

$$\|p * q\|^2 = (p * q) * \overline{(p * q)} = p * q * \bar{q} * \bar{p} = p * \bar{p} = \|p\|^2 \,\square$$

Using successively properties 2, 1, and 3, (18) leads to:

$$C_\rho = \sum_{j=1}^M \|q * S_j * \bar{q} - Q_j\|^2 = \sum_{j=1}^M \|(q * S_j - Q_j * q) * \bar{q}\|^2 = \sum_{j=1}^M \|(q * S_j - Q_j * q)\|^2$$

$(q * S_j - Q_j * q)$ is a quaternion which is a linear function of q. More precisely, this quaternion is equal to $B_j q$ where B_j is the matrix introduced in equation (15). This leads to:

$$C_\rho = \sum_{j=1}^M (B_j q)^t (B_j q) = q^t (\sum_{j=1}^M B_j)^t (\sum_{j=1}^M B_j) q = q^t B q$$

$q^t B q$ is minimal for q being equal to the unit eigenvector associated with the smallest eigenvalue of matrix B. Notice that B, as a 4×4 symmetric positive matrix has four positive eigenvalues.

Analytical Results on Error Sensitivity of Motion Estimation from Two Views

Konstantinos Daniilidis, Hans-Hellmut Nagel
Fraunhofer Institut für Informations- und Datenverarbeitung
Fraunhoferstr. 1, D-7500 Karlsruhe 1, FRG

Abstract

Fundamental instabilities have been observed in the performance of the majority of the algorithms for three dimensional motion estimation from two views. Many geometric and intuitive interpretations have been offered to explain the error sensitivity of the estimated parameters. In this paper, we address the importance of the form of the error norm to be minimized with respect to the motion parameters. We describe the error norms used by the existing algorithms in a unifying notation and give a geometric interpretation of them. We then explicitly prove that the minimization of the objective function leading to an eigenvector solution suffers from a crucial instability. The analyticity of our results allows us to examine the error sensitivity in terms of the translation direction, the viewing angle and the distance of the moving object from the camera. We propose a norm possessing a reasonable geometric interpretation in the image plane and we show by analytical means that a simplification of this norm leading to a closed form solution has undesirable properties.

1 Introduction

The problem of estimating relative motions between the camera and objects in the scene space from monocular image sequences has been an intensive research area in machine vision during the last years (see the recent survey by [Aggarwal & Nandhakumar 88]). Three different groups of approaches have been developed for the recovery of three dimensional motion from monocular image sequences.

The *discrete* approach is based on the extraction and tracking of features in the image corresponding to three dimensional features in the scene space. The input to the motion algorithm is a set of interframe feature correspondences. The estimated parameters are the rotation and the direction of translation of a three dimensional feature configuration relative to the camera. In the case of points as features, the discrete case is equivalent to the problem of relative orientation in photogrammetry. When the time interval between successive frames is short, the point correspondences yield displacement vectors which may be considered as approximations of the displacement rates used in the continuous approach.

The *continuous* approach is based on an evaluation of the motion or displacement rate field representing the two dimensional velocity of the projections of the three dimensional points. Under certain conditions ([Nagel 89,Girosi et al. 89]) this motion field is equivalent to the optical flow field which describes the apparent instantaneous shift of gray value structures in the image plane. The output of the continuous approaches is the relative angular velocity and the direction of the relative translational velocity of a component of the scene with respect to the camera. A unifying description of approaches for optical flow estimation was presented by [Nagel 87] who also pointed out the relation between the continuous formulation and the discrete approach in case of simple features such as gray value extrema, corners, or edge elements. Underlying the majority of the approaches is the assumption of rigidity of the moving scene component or of the stationary environment in case of a moving camera.

Although the problem of motion estimation from point correspondences or displacement rates from two views was mathematically formulated long ago [Longuet-Higgins 81] there is still no algorithm which has a robust behaviour regardless of motion and scene configuration. In this paper, we study the relation between error sensitivity, the type of motion and the geometry of the problem. It is of major interest to recognize whether the instability depends on the formulation of the problem or if it is inherent in the problem. Our main contribution consists in the analyticity of our results which are consistent with what has already been stated intuitively or observed experimentally. The objective of our study is to point out an instability depending on the formulation of the problem. We begin therefore, with a review –based on unified notation– of the error norms proposed by the existing 3D-motion algorithms.

2 Unifying description of the existing error norms

We denote matrices by capital italics and vectors by boldface symbols. We choose a coordinate system $OXYZ$ with the origin at the center of projection (camera pinhole) and the Z axis coinciding with the optical axis of the camera. The unit vector in the direction of the optical axis is denoted by $\mathbf{z_0}$. We assume that the focal length is unity so that the equation of the image plane xy is $Z = 1$. All the geometric quantities are measured in focal length units. The position vectors of points in the scene space are denoted by uppercase letters $\mathbf{X} = (X, Y, Z)^T$ whereas position vectors of points in the image plane are denoted by lowercase letters $\mathbf{x} = (x, y, 1)^T$. They are related to each other by the equations of perspective projection under the assumption of unit focal length. The notation for the motion parameters is as follows: $\mathbf{t} = (t_x, t_y, t_z)^T$ for the translation, $\mathbf{v} = (v_x, v_y, v_z)^T$ for the translational velocity, R for the rotation and $\omega = (\omega_x, \omega_y, \omega_z)^T$ for the angular velocity. We denote by $\mathbf{x1}, \mathbf{x2}$ and $\mathbf{X1}, \mathbf{X2}$ the points in the image plane and in the scene space at the time instants t_1 and t_2, respectively. The 3D velocity of a point and the displacement rate of an image point are denoted by $\dot{\mathbf{X}}$ and $\dot{\mathbf{x}}$, respectively.

In the discrete case, the rigid motion of an object can be described as a rotation around an axis passing through the origin of the fixed camera coordinate system, followed by a translation. The motion equation of a point on this object reads $\mathbf{X2} = R\,\mathbf{X1} + \mathbf{t}$. The same equation describes the case of an object being stationary and a camera undergoing a rotation R^T about an axis through the origin followed by a translation $-\mathbf{t}$. We can derive directly a geometric relation between the measurements $\mathbf{x1}$ and $\mathbf{x2}$ and the unknown motion parameters by observing that the two line-of-sight vectors must lie in the same plane with the translation vector $\mathbf{t}$. This coplanarity condition can be written

$$\mathbf{x2}^T(\mathbf{t} \times R\mathbf{x1}) = 0 \tag{1}$$

[Longuet-Higgins 81] and [Tsai & Huang 84] first proposed linear solution methods based on (1) and using *essential parameters* and [Horn 88] proposed the direct minimization of the sum of the squares $\|\mathbf{x2}^T(\mathbf{t} \times R\mathbf{x1})\|$ with respect to the parameters of translation and rotation. All these approaches use the same error term in their minimization function, leading to highly sensitive estimates for the motion parameters as we shall see in the next section.

We continue with the problem formulation in the continuous case. Using the notation given at the beginning of this section, the three dimensional instantaneous velocity of a point on a moving object in scene space is given by

$$\dot{\mathbf{X}} = \mathbf{v} + \omega \times \mathbf{X} \tag{2}$$

After differentiating the equation of the perspective projection $\mathbf{x} = \mathbf{X}/(\mathbf{z_0}^T\mathbf{X})$ and making use of (2), we obtain the following expression for the displacement rate:

$$\dot{\mathbf{x}} = \frac{1}{\mathbf{z_0}^T\mathbf{X}}\mathbf{z_0} \times (\mathbf{v} \times \mathbf{x}) + \mathbf{z_0} \times (\mathbf{x} \times (\mathbf{x} \times \omega)) \tag{3}$$

We restrict our description and instability considerations to approaches that do not make use of the spatial or temporal derivatives of the image velocities. The displacement rate given in equ. (3) consists of a translational component dependent on the 3D scene structure and the translation, and a rotational component depending only on the angular velocity. If we rewrite the latter as $(\mathbf{z_0}^t(\mathbf{x} \times \omega))\,\mathbf{x} - \mathbf{x} \times \omega$ and take the scalar product of (3) with $(\mathbf{v} \times \mathbf{x})$ we obtain the following relation which is free from an explicit dependence on 3D scene structure:

$$(\mathbf{v} \times \mathbf{x})^T (\dot{\mathbf{x}} - \omega \times \mathbf{x}) = 0 \tag{4}$$

The same relation is valid when the model of spherical projection $\mathbf{x} = \mathbf{X}/\|\mathbf{X}\|$ is used. In this case [Maybank 86] proposed as an error norm the following residual linear in ω:

$$\epsilon^2(\mathbf{v}) = \min_{\omega} \sum_{i=1}^{m} \left\{ (\mathbf{x}_i \times (\mathbf{v} \times \mathbf{x}_i))^T \omega - (\mathbf{v} \times \mathbf{x}_i)^T \dot{\mathbf{x}}_i \right\}^2 \tag{5}$$

A search on the unit hemisphere for the direction of translation minimizing $\epsilon(\mathbf{v})$ was employed to complete the motion estimation. [Zhuang et al. 88] used (4) to formulate a minimization problem based on *essential parameters* (analogous to the discrete case). [Bruss & Horn 83] started by minimizing the discrepancies between the measured and the expected displacement rates with respect to the 3D velocities and the depths:

$$\iint_D \left\{ \dot{\mathbf{x}} - \frac{1}{\mathbf{z}_0^T \mathbf{X}} \mathbf{z}_0 \times (\mathbf{v} \times \mathbf{x}) - \mathbf{z}_0 \times (\mathbf{x} \times (\mathbf{x} \times \omega)) \right\}^2 \, d\mathbf{x} \Rightarrow \min \tag{6}$$

After eliminating the depths they obtained the objective function

$$\iint_D \left\{ \frac{(\mathbf{v} \times \mathbf{x})^T (\dot{\mathbf{x}} - \omega \times \mathbf{x})}{\|\mathbf{z}_0 \times (\mathbf{v} \times \mathbf{x})\|} \right\}^2 \, d\mathbf{x} \tag{7}$$

which is also proposed by [Scott 88] and is equivalent to the norm we propose. In order to derive a closed form solution, [Bruss & Horn 83] proposed another error norm by weighting the error terms of (7) with $\|\mathbf{z}_0 \times (\mathbf{v} \times \mathbf{x})\|$ yielding

$$\iint_D \left\{ (\mathbf{v} \times \mathbf{x})^T (\dot{\mathbf{x}} - \omega \times \mathbf{x}) \right\}^2 \, d\mathbf{x} \tag{8}$$

which can be derived directly from (4). This norm simplifies the minimization but causes a bias in the solution as we demonstrate in the next section.

Although the performance of most algorithms for motion recovery have been tested with synthetic or real data, the dependence of their instability on the form of the objective function, the kind of motion, and the geometry of the problem has been seldom investigated thoroughly. Likewise, very few error analyses lead to explicit formulations of the error sensitivity in terms of the input error, the motion and the structure of the scene. [Adiv 89] pointed out inherent ambiguities in determining the motion of a planar patch. In particular he discovered that, when the field of view is small, it is impossible to distinguish a pure translational motion parallel to the image plane ($\mathbf{v} = (p_1, p_2, 0)^T$) from a pure rotational motion about an axis parallel to the image plane ($\omega = (-p_2/d, p_1/d, 0)^T$). Furthermore, he analytically demonstrated the importance of depth variation and of the ratio of the translation magnitude to the distance of the object from the camera. The instability caused by these factors can be illustrated as a flattening of the error function in the neighborhood of the global minimum. Explicit error analysis has been carried out by [Maybank 86] concerning the error norm (5). He found out that the minima of the residual (5) should lie in the neighborhood of a particular line on the unit sphere and that the ratio v_x/v_y as well as the component of the angular velocity parallel to $(v_x, v_y, 0)^T$ could be reliably estimated in the presence of noise. The influence on the error sensitivity of the translation direction has been emphasized by [Weng et al. 89b], [Horn & Weldon 88] by geometric arguments. The importance of the translation direction has been shown experimentally by [Mitiche et al. 87], [Adiv 89] and [Weng et al. 89b]. [Weng et al. 89b] carried out an analysis of the errors in the estimated eigenvectors in order to compare it with the actual error. However, they did not show any explicit dependence of the error on the motion and structure parameters. [Weng et al. 89a] recognized the importance of a correct error norm and introduced a second step containing a nonlinear minimization of the discrepancies between observed and expected value in a maximum likelihood scheme. They have been able to show that this algorithm achieved the Cramer-Rao lower bounds on the error covariance. They also proposed as an alternative the minimization with respect to the parameters of rotation and translation of the following objective function ([Weng et al. 89a]):

$$\sum_{i=1}^{m} \frac{\left(\mathbf{x}2_i^T (\mathbf{t} \times R\mathbf{x}1_i) \right)^2}{\sigma^2 \left(\|\mathbf{z}_0 \times R^T (\mathbf{t} \times \mathbf{x}2_i)\|^2 + \|\mathbf{z}_0 \times (\mathbf{t} \times R\mathbf{x}1_i)\|^2 \right)} \tag{9}$$

where σ^2 is the noise variance of the image coordinates. The denominator is the variance of the numerator where the first square error norm in the denominator reflects the uncertainty of **x1** and the second one the uncertainty of **x2**.

3 Error sensitivity concerning the form of the error norm

In this section we show that the sensitivity to the direction of translation and the field of view is not an inherent instability in motion recovery from two views, but depends on the error norm used in the minimization. We explicitly prove that the form of the objective function

$$\iint_D \left\{ (\mathbf{v} \times \mathbf{x})^T (\dot{\mathbf{x}} - \omega \times \mathbf{x}) \right\}^2 d\mathbf{x} \tag{10}$$

extracted from condition (4) is the reason for the instability in the case of large deviation of the translation direction from the position vector to the object. In (10) we assume that the field of displacement rates $\dot{\mathbf{x}}$ is dense so that we can integrate over an area D with size proportional to the field of view in case of a moving camera or to the size of the projection of the moving object if the camera is stationary.

We are interested in the error of the translation direction, so we restrict ourselves to the case of pure translational motion. Our arguments are also valid in the case of general motion with known rotation or in the case of general motion when the used algorithm estimates translation and rotation sequentially as in [Spetsakis & Aloimonos 88]. When another general motion algorithm is used our arguments build only cues for the experimentally observed instability in the translation direction.

We begin with a geometric justification of our conjecture. Referring to Fig. 1 we can express the deviation of the displacement rate $\dot{\mathbf{x}}$ from the line joining the focus of expansion and the point **x** as

$$d' = \frac{|\dot{\mathbf{x}}^T (\mathbf{v} \times \mathbf{x})|}{\|\mathbf{z}_0 \times (\mathbf{v} \times \mathbf{x})\|}$$

We see that what is really minimized when (10) is used is a multiple of the real error distance d'. In fact, the real distance is weighted by the distance of the focus of expansion from the point **x**. In case of a small field of view, we expect the estimated translation to be biased to the direction of the line of sight to the object. The same geometric justification is valid in the discrete translational case if we assume that only the points $\mathbf{x2}_i$ in the second image are corrupted by noise. Then the distance of **x2** from the epipolar line in the second image plane is given by

$$d'' = \frac{|\mathbf{x2}^T (\mathbf{t} \times \mathbf{x1})|}{\|\mathbf{z}_0 \times (\mathbf{t} \times \mathbf{x1})\|}$$

which agrees with the norm used by [Weng et al. 89a] in (9) under the assumption of noise-free **x1**.

The minimization of (10) in the translational case can be formulated as an eigenvalue problem ([Bruss & Horn 83, Spetsakis & Aloimonos 88] and [Zacharias et al. 85]). The form to be minimized is

$$\iint_D \left\{ (\mathbf{v} \times \mathbf{x})^T \dot{\mathbf{x}} \right\}^2 d\mathbf{x} = \mathbf{v}^T \left\{ \iint_D (\mathbf{x} \times \dot{\mathbf{x}})^T (\mathbf{x} \times \dot{\mathbf{x}}) \, d\mathbf{x} \right\} \mathbf{v} \Rightarrow \min_{\|\mathbf{v}\|=1} \tag{11}$$

The solution for **v** is the eigenvector of the matrix

$$A = \iint_D (\mathbf{x} \times \dot{\mathbf{x}})^T (\mathbf{x} \times \dot{\mathbf{x}}) \, d\mathbf{x}$$

corresponding to the smallest eigenvalue. The instability of the problem can be expressed in terms of the relative positions between the smallest eigenvalues. In the noise-free case the smallest eigenvalue must be equal to zero. In the presence of noise, the eigenvectors of A can be interpreted as the directions of the axes of the ellipsoid $\mathbf{v}^T A \mathbf{v} = \lambda_{min}$ (see [Bruss & Horn 83]).

The lengths of the axes are 1, $\sqrt{\lambda_1/\lambda_2}$, and $\sqrt{\lambda_1/\lambda_3}$ if we assume without loss of generality that λ_1 is the smallest eigenvalue. Hence, the ellipsoid is circumscribed by a unit sphere with two tangential points on the axis corresponding to the solution eigenvector. The sensitivity to error in the direction of the eigenvector (and hence of the translation) grows when the two smallest eigenvalues come close to each other and the ellipsoid is near to become tangential to the unit sphere in two positions. In this extremely unstable case, a small deformation of the data matrix A can cause a change of 90 degrees in the eigenvector corresponding to the smallest eigenvalue. The instability due to purely isolated eigenvalues has already been investigated in numerical analysis [Wilkinson 65,Golub & van Loan 83]. The perturbation of eigenvectors with respect to the relative position of the eigenvalues is given by [Peters & Wilkinson 74]. We reformulate their result for the case of a symmetric matrix A perturbed by E:

$$\mathbf{x}'_k \approx \mathbf{x}_k + \sum_{j \neq k} \frac{\mathbf{x}_j^T E \mathbf{x}_k}{(\lambda_j - \lambda_k)} \mathbf{x}_j \tag{12}$$

where $\mathbf{x}'_k$ denotes the perturbed eigenvector. From (12) follows that the smaller the difference between the eigenvalues is, the greater is the perturbation in the eigenvector. This result is of general interest since it can be used for the estimation of the error covariance of the solution by all eigenvector problems.

We have been able to prove that this extreme situation of a 90 degrees error in the estimated translation direction can happen in our problem of motion recovery from two views as follows:

We model the projection of the moving object as a rectangle with sides of length α and β in the image plane with its center on the Z axis. Let $\mathbf{v} = (v_x, v_y, v_z)^T$ be the translation of a frontal plane lying at a distance $1/d$ from the origin. By introducing an additional noise term $(\xi, \eta, 0)^T$, the measured displacement rates can be written

$$\dot{\mathbf{x}} = \begin{pmatrix} d(v_x - xv_z) + \xi \\ d(v_y - yv_z) + \eta \\ 0 \end{pmatrix} \tag{13}$$

Using this model we integrate over D and we obtain the elements of A

$$a_{11} = \alpha\beta \left(d^2 v_y^2 + d^2 v_z^2 \frac{\beta^2}{12} + \eta^2 + 2\eta dv_y \right)$$

$$a_{12} = \alpha\beta \left(-d^2 v_x v_y - dv_x \eta - dv_y \xi - \eta\xi \right)$$

$$a_{22} = \alpha\beta \left(d^2 v_x^2 + d^2 v_z^2 \frac{\alpha^2}{12} + \xi^2 + 2\xi dv_x \right)$$

$$a_{13} = -\frac{\alpha\beta^3}{12} (d^2 v_x v_z + dv_z \xi)$$

$$a_{23} = -\frac{\alpha^3\beta}{12} (d^2 v_y v_z + dv_z \eta)$$

$$a_{33} = \alpha\beta \left(\frac{\alpha^2}{12} (dv_y + \eta)^2 + \frac{\beta^2}{12} (dv_x + \xi)^2 \right) \tag{14}$$

Suppose that the additional error term $(\xi, \eta, 0)^T$ is an unbiased random variable with second moments $E[\xi^2] = E[\eta^2] = \sigma^2$ and $E[\xi\eta] = 0$. Hence, the matrix A is a random variable with mean

$$A' = E[A] = \alpha\beta \begin{pmatrix} d^2 v_y^2 + d^2 v_z^2 \frac{\beta^2}{12} + \sigma^2 & -d^2 v_x v_y & -d^2 v_x v_z \frac{\beta^2}{12} \\ -d^2 v_x v_y & d^2 v_x^2 + d^2 v_z^2 \frac{\alpha^2}{12} + \sigma^2 & -d^2 \frac{\alpha^2}{12} v_y v_z \\ -d^2 v_x v_z \frac{\beta^2}{12} & -d^2 \frac{\alpha^2}{12} v_y v_z & \frac{\sigma^2}{12}(\alpha^2 + \beta^2) + \frac{d^2}{12}(\alpha^2 v_y^2 + \beta^2 v_x^2) \end{pmatrix} \tag{15}$$

If $\sigma^2 = 0$ then the matrix A' has rank equal to two except in the case when the area D degenerates to a line. In the presence of noise, the matrix has full rank and the smallest eigenvalue is different from zero. The question is whether the two smallest eigenvalues of this matrix are well isolated. Since all the quantities are measured in focal length units, the noise variance is much smaller than unity and the largest value for the sides of the rectangle α and β is two units which happens only when the rectangle covers the whole image and the field of view is 90 degrees.

In the case $v_x = v_y = 0$ the matrix A' takes the diagonal form

$$A' = \alpha\beta \begin{pmatrix} d^2 v_z^2 \frac{\beta^2}{12} + \sigma^2 & 0 & 0 \\ 0 & d^2 v_z^2 \frac{\alpha^2}{12} + \sigma^2 & 0 \\ 0 & 0 & \frac{\sigma^2}{12}(\alpha^2 + \beta^2) \end{pmatrix} \tag{16}$$

and its diagonal elements are identical with the eigenvalues. We take the differences between them

$$\begin{aligned} \lambda_1 - \lambda_3 &= \frac{\alpha\beta^3}{12} d^2 v_z^2 + \frac{\alpha\beta}{12}\sigma^2(12 - \alpha^2 - \beta^2) > 0 \\ \lambda_2 - \lambda_3 &= \frac{\alpha^3\beta}{12} d^2 v_z^2 + \frac{\alpha\beta}{12}\sigma^2(12 - \alpha^2 - \beta^2) > 0 \end{aligned} \tag{17}$$

and observe that the smallest eigenvalue is λ_3 and the corresponding eigenvector is $(0, 0, 1)^T$ as expected. The differences are significantly greater than zero, the eigenvalues are well isolated and thus the solution is robust.

In the case of a translation parallel to the image plane ($v_z = 0$) we obtain

$$A' = \alpha\beta \begin{pmatrix} d^2 v_y^2 + \sigma^2 & -d^2 v_x v_y & 0 \\ -d^2 v_x v_y & d^2 v_x^2 + \sigma^2 & 0 \\ 0 & 0 & \frac{\sigma^2}{12}(\alpha^2 + \beta^2) + \frac{d^2}{12}(\alpha^2 v_y^2 + \beta^2 v_x^2) \end{pmatrix} \tag{18}$$

The eigenvalues $\lambda_1 < \lambda_2$ are roots of the equation

$$\lambda^2 - \alpha\beta(d^2 v_y^2 + d^2 v_x^2 + 2\sigma^2)\lambda + \alpha^2\beta^2\sigma^2(d^2 v_y^2 + d^2 v_x^2 + \sigma^2) = 0 \tag{19}$$

and read as follows:

$$\lambda_1 = \sigma^2\alpha\beta \qquad \lambda_2 = \alpha\beta(d^2 v_x^2 + d^2 v_y^2 + \sigma^2) \tag{20}$$

We inspect the differences again:

$$\begin{aligned} \lambda_2 - \lambda_1 &= \alpha\beta d^2(v_y^2 + v_x^2) > 0 \\ \lambda_3 - \lambda_1 &= \frac{\alpha\beta}{12}\left(d^2(\alpha^2 v_y^2 + \beta^2 v_x^2) + \sigma^2(\alpha^2 + \beta^2 - 12)\right) \end{aligned} \tag{21}$$

and set the second difference equal to zero in order to obtain the condition for the extreme instability:

$$\sigma^2 = \frac{d^2(\alpha^2 v_y^2 + \beta^2 v_x^2)}{12 - (\alpha^2 + \beta^2)} \tag{22}$$

We show by example that this situation is a realistic one: Let the ratio of the translation magnitude to the distance of the object from the camera be 1/10 and the sides of the rectangle be one tenth of the focal length. It turns out that the noise variance should be about $\sigma^2 = 10^{-4}/12$ — for instance a uniform distribution in the interval $(-0.005, 0.005)$ measured in focal length units. In this case, λ_3 becomes the smallest eigenvalue and we obtain $(0, 0, 1)^T$ as solution for the translation direction which is wrong by 90 degrees. Thus we have proved that

an appropriate combination of noise level, viewing angle, and ratio of the translation magnitude to the distance of the object from the camera can cause an error of 90 degrees in the translation direction if the object is moving parallel to the image plane.

This error is not to be confused with the two-fold ambiguity in recovering the motion and the normal of the planar patch. This ambiguity allows, as a second interpretation, the interchanging of the translation direction with the normal to the frontal plane accompanied by a rotational motion. This ambiguity does not concern our error analysis since the considered objective function (11) possesses a unique minimum in the unit hemisphere due to the assumption of pure translation.

We now examine what happens between the stable and the extremely unstable situation. We restrict the translation direction to the XZ plane. We set v_y equal to zero in (15) and obtain

$$A' = E[A] = \alpha\beta \begin{pmatrix} d^2 v_z^2 \frac{\beta^2}{12} + \sigma^2 & 0 & -d^2 v_x v_z \frac{\beta^2}{12} \\ 0 & d^2 v_x^2 + d^2 v_z^2 \frac{\alpha^2}{12} + \sigma^2 & 0 \\ -d^2 v_x v_z \frac{\beta^2}{12} & 0 & \frac{\sigma^2}{12}(\alpha^2 + \beta^2) + \frac{d^2}{12}\beta^2 v_x^2 \end{pmatrix} \qquad (23)$$

Let λ_2 be the eigenvalue equal to element a'_{22} of the matrix A'. We are interested in the eigenvalues λ_1 and λ_3 which are roots of the characteristic equation

$$\lambda^2 - (a'_{11} + a'_{33})\lambda + a'_{11}a'_{33} - a'^2_{13} = 0 \qquad (24)$$

and give rise to eigenvectors lying in the XZ plane. In order to investigate whether they are well isolated, we form the difference between them and after some manipulation we obtain

$$|\lambda_1 - \lambda_3| = \frac{\alpha\beta}{12}\left((d\beta)^4(v_z^2 + v_x^2)^2 + \sigma^4(12 - \alpha^2 - \beta^2)^2 + 2d^2\beta^2\sigma^2(12 - \alpha^2 - \beta^2)(v_z^2 - v_x^2)\right)^{\frac{1}{2}}$$

Let ϕ be the angle between the Z axis and the direction of translation, $G = d^2\beta^2\|\mathbf{v}\|^2$ and $F = \sigma^2(12 - \alpha^2 - \beta^2) > 0$. Then we have

$$|\lambda_1 - \lambda_3| = \frac{\alpha\beta}{12}\left(G^2 + F^2 + 2FG\cos 2\phi\right)^{1/2} \qquad (25)$$

We find that the absolute difference of the eigenvalues decreases as ϕ increases from zero ($\mathbf{v}$ parallel to the optical axis) to 90 degrees ($\mathbf{v}$ parallel to the image plane). The smaller this difference, the greater the perturbations in the estimated eigenvector due to (12). Hence, we can state that the stability of the translation estimation is explicitly related to the translation direction as given by (25).

4 Appropriate Error Norms

To remove the instability proved in the last section, we propose the use of the correct error norm as illustrated in Fig. 1. This norm represents the projection of the translational component of the displacement rate in the direction of the normal to the line passing through the focus of expansion and the point:

$$\iint_D \left\{ \frac{(\mathbf{v} \times \mathbf{x})^T(\dot{\mathbf{x}} - \omega \times \mathbf{x})}{\|\mathbf{z}_0 \times (\mathbf{v} \times \mathbf{x})\|} \right\}^2 d\mathbf{x} \qquad (26)$$

If the spherical projection model is used, the error term should be the component orthogonal to the plane defined by the translation vector and the position vector of the point considered. Due to the high nonlinearity we have not been able to prove analytically that (26) can lead to an unbiased result with respect to the translation direction. The same error norm is suggested by [Spetsakis & Aloimonos 88], [Harris 87] and [Scott 88]. Other error norms having a geometric interpretation on the image plane have been discussed in [Toscani & Faugeras 87a], [Weng et al. 89a] and [Horn 88]. The drawback of (26) and all these error norms is that they lead to a highly nonlinear minimization problem that needs a good initial guess to converge.

Only [Spetsakis & Aloimonos 88] proposed a simplification of the problem in order to obtain an eigenvector solution. Exploiting the fact that the instability occurs when the viewing angle is small, they replaced the position vector of every point $\mathbf{x}$ in the denominator with the position vector $\mathbf{c}$ of the centroid of the projection area. In our notation, this yields

$$\iint_D \left\{ \frac{(\mathbf{v} \times \mathbf{x})^T(\dot{\mathbf{x}} - \omega \times \mathbf{x})}{\|\mathbf{v} \times \mathbf{x}\|} \right\}^2 d\mathbf{x} \approx \iint_D \left\{ \frac{(\mathbf{v} \times \mathbf{x})^T(\dot{\mathbf{x}} - \omega \times \mathbf{x})}{\|\mathbf{v} \times \mathbf{c}\|} \right\}^2 d\mathbf{x} = \frac{\mathbf{v}^T A \mathbf{v}}{\mathbf{v}^T C \mathbf{v}} \qquad (27)$$

where $C = \mathbf{c}^T \mathbf{c} I - \mathbf{c}\mathbf{c}^T$ and the change in the denominator compared to equ. (26) is due to the spherical projection used by [Spetsakis & Aloimonos 88]. The last fraction would be a generalized Rayleigh quotient with an eigenvector solution if the matrix C were not singular.

We will point out that the simplification (27) causes the stable case of motion parallel to the viewing direction to the object to become unstable. This is due to the observation that the denominator in (27) becomes zero when the translation is parallel to $\mathbf{c}$ whereas the numerator is different from zero due to the presence of noise. Hence, the objective function does not have a minimum in the expected position. We use the same model as in the previous section to prove our objection to this simplification explicitly. For $\mathbf{c} = (0, 0, 1)^T$ and the case of perspective projection onto an image plane, (27) becomes

$$\iint_D \left\{ \frac{(\mathbf{v} \times \mathbf{x})^T \dot{\mathbf{x}}}{\|\mathbf{z}_0 \times (\mathbf{v} \times \mathbf{c})\|} \right\}^2 d\mathbf{x} = \frac{\lambda_1 v_x^2 + \lambda_2 v_y^2 + \lambda_3 v_z^2}{v_x^2 + v_y^2} \tag{28}$$

where $\lambda_i (i = 1 \ldots 3)$ are the diagonal elements of A' in (16). It is evident that the correct solution $(0, 0, 1)^T$ causes the value of the objective function to become unbounded. Spetsakis and Aloimonos proposed a deformation of the matrix $C_\delta = \mathbf{c}^T \mathbf{c} I - (1 - \delta)\mathbf{c}\mathbf{c}^T$ in order to overcome its singularity. In case of motion towards the object, the matrix C_δ becomes

$$C_\delta = \begin{pmatrix} 1 & 0 & 0 \\ 0 & 1 & 0 \\ 0 & 0 & \delta \end{pmatrix} \tag{29}$$

and the minimization of $\mathbf{v}^T A \mathbf{v} / \mathbf{v}^T C \mathbf{v}$ is reduced to the estimation of the eigenvector corresponding to the smallest eigenvalue of

$$A'_\delta = \alpha\beta \begin{pmatrix} d^2 v_z^2 \frac{\beta^2}{12} + \sigma^2 & 0 & 0 \\ 0 & d^2 v_z^2 \frac{\alpha^2}{12} + \sigma^2 & 0 \\ 0 & 0 & \frac{\sigma^2}{12\delta}(\alpha^2 + \beta^2) \end{pmatrix} \tag{30}$$

By building the differences one more time

$$\begin{aligned} \lambda_1 - \lambda_3 &= \frac{\alpha\beta^3}{12} d^2 v_z^2 + \frac{\alpha\beta}{12}\sigma^2 (12 - \frac{\alpha^2 + \beta^2}{\delta}) \\ \lambda_2 - \lambda_3 &= \frac{\alpha^3\beta}{12} d^2 v_z^2 + \frac{\alpha\beta}{12}\sigma^2 (12 - \frac{\alpha^2 + \beta^2}{\delta}) \end{aligned} \tag{31}$$

we observe that the smallest eigenvalue is no longer λ_3 since the deformation parameter δ takes very small values. Consequently, the error norm (26) must be used in its original nonlinear form so that we avoid the appearance of new unstable configurations.

5   Concluding Remarks

We emphasized the influence of the form of the error norm in minimization approaches for the recovery of motion parameters from two views. We presented the error norms used by existing solution methods in a unifying notation and we referred to sensitivity results already extracted by other authors. We proved analytically that the objective function used in the majority of the approaches leads to an error of 90 degrees in the estimated translation direction when the viewing angle is small, the translation magnitude is small relative to the distance of the moving object from the camera and the motion is parallel to the image plane. We formulated the instability problem in terms representing the isolation of the computed smallest eigenvalue from the other eigenvalues. By using a model for the displacement rates arising from an arbitrary translating frontal plane, we expressed the error sensitivity as a function of the translation direction, the field of view and the ratio of the translation magnitude to the depth of the viewed object.

The situation we modeled in order to obtain analytical results is realistic. Suppose that a camera fixed on a vehicle is viewing an object at a distance with only one planar side visible (for example a stationary truck or bus). In case the camera is moving towards the object the motion can be robustly estimated but if the camera is moving in a lateral direction the estimated parameters become unreliable. Of course the navigation would not rely on the measurements from only two views. Two frames offer such unreliable information, particularly when the whole amount of motion is small or the situation is inherently ambiguous, that even the use of a correct error term would not improve considerably the results. However, before we proceed with the formulation of multiple-frame motion algorithms, it is important to understand the reasons for the instabilities in the two-view motion estimation. New results about the intrinsic ambiguities [Faugeras & Maybank 89] may even lead to the interpretation of specific unstable situations as motion and structure configurations lying in the neighborhood of multiply interpretable situations.

Acknowledgements

The financial support of the first author by the German Academic Exchange Service (DAAD) is gratefully acknowledged. We would like to thank Dr. Greg Hager for helpful comments on a draft of this paper.

References

[Adiv 89] G. Adiv, Inherent ambiguities in recovering 3-D motion and structure from a noisy flow field, *IEEE Trans. Pattern Analysis Machine Intelligence* PAMI-11 (1989) 477-489.

[Aggarwal & Nandhakumar 88] J.K. Aggarwal, N.Nandhakumar, On the computation of motion from sequences of images - a review, *Proceedings of the IEEE* 76 (1988) 917-935.

[Bruss & Horn 83] A. Bruss, B.K.P. Horn, Passive navigation, *Computer Vision, Graphics, and Image Processing* 21 (1983) 3-20.

[Faugeras & Maybank 89] O.D. Faugeras, S. Maybank, Motion from point matches: multiplicity of solutions, in *Proc. IEEE Workshop on Visual Motion*, March 20-22, 1989, Irvine, CA, pp. 248-255.

[Girosi et al. 89] F. Girosi, A. Verri, V. Torre, Constraints for the computation of optical flow, in *Proc. IEEE Workshop on Visual Motion*, March 20-22, 1989, Irvine, CA, pp. 116-124.

[Golub & van Loan 83] G.H. Golub, C.F. van Loan, *Matrix Computations*, The Johns Hopkins University Press, Baltimore, Maryland, 1983.

[Harris 87] C.G. Harris, Determination of ego-motion from matched points, in *Proc. Alvey Vision Conference*, Sept. 15-17, 1987, Cambridge, UK, pp. 189-192.

[Horn 88] B.K.P. Horn, Relative orientation, in *Proc. DARPA Image Understanding Workshop*, April 6-8, 1988, Cambridge, MA, pp. 826-837.

[Horn & Weldon 88] B.K.P. Horn, Direct methods for recovering motion, *Int. Journal of Computer Vision* 2 (1988) 51-76.

[Longuet-Higgins 81] H.C. Longuet-Higgins, A computer program for reconstructing a scene from two projections, *Nature* 293 (Sept. 1981) 133-135.

[Maybank 86] S.J. Maybank, Algorithm for analyzing optical flow based on the least squares method, *Image and Vision Computing* 4 (1986) 38-42.

[Mitiche et al. 87] A. Mitiche, X. Zhuang, R. Haralick, Interpretation of optical flow by rotation decoupling, in *Proc. IEEE Workshop on Computer Vision*, Nov. 3 - Dec. 2, 1987, Miami Beach, FL, pp. 195-200.

[Nagel 87] H.-H. Nagel, On the estimation of optical flow: Relations between different approaches and some new results, *Artificial Intelligence* 33 (1987) 299-324.

[Nagel 89] H.-H. Nagel, On a constraint equation for the estimation of displacement rates in image sequences, *IEEE Trans. Pattern Anal. Machine Intell.* PAMI-11 (1989) 13-30.

[Peters & Wilkinson 74] G. Peters, J.H. Wilkinson, Accuracy of computed eigensystems and invariant subspaces, in B.K.P. Scaife (Ed.), *Studies in Numerical Analysis*, Academic Press, London and New York 1974.

[Scott 88] G.L. Scott, *Local and global interpretation of moving images*, Pitman Publishing, London, 1988.

[Spetsakis & Aloimonos 88] M.E. Spetsakis, J. Aloimonos, Optimal computing of structure from motion using point correspondences, in *Proc. Int. Conf. on Computer Vision*, Dec. 5-8, 1988, Tampa, FL, pp. 449-453.

[Toscani & Faugeras 87a] G. Toscani, O.D. Faugeras, Structure and motion from two perspective views, in *Proc. IEEE Int. Conf. on Robotics and Automation*, Mar.31 - Apr. 2, 1987, Raleigh, North Carolina, pp. 221-227.

[Tsai & Huang 84] R. Y. Tsai, T. S. Huang, Uniqueness and estimation of 3-D motion parameters of rigid bodies with curved surfaces, *IEEE Trans. Pattern Anal. Machine Intell.* PAMI-6 (1984) 13-27.

[Weng et.al. 89a] J. Weng, T.S. Huang, N. Ahuja, Optimal motion and structure estimation, in *Proc. IEEE Conf. on Computer Vision and Pattern Recognition*, June 4-8, 1989, San Diego, CA, pp. 144-152.

[Weng et al. 89b] J. Weng, T.S. Huang, N. Ahuja, Motion and structure from two perspective views: algorithms, error analysis, and error estimation, *IEEE Trans. Pattern Anal. Machine Intelligence* PAMI-11 (1989) 451-476.

[Wilkinson 65] J.H. Wilkinson, *The Algebraic Eigenvalue Problem*, Oxford Science Publications, Oxford, 1965.

[Zacharias et al. 85] G.L. Zacharias, A.K. Caglayan, J.B. Sinacori, A model for visual flow-field cueing and self-motion estimation, *IEEE Trans. Systems, Man, and Cybernetics* SMC-15 (1985) 385-389.

[Zhuang et al. 88] X. Zhuang, T.S. Huang, N. Ahuja, R.M. Haralick, A simplified linear optical flow-motion algorithm, *Computer Vision, Graphics, and Image Processing* 42 (1988) 334-344.

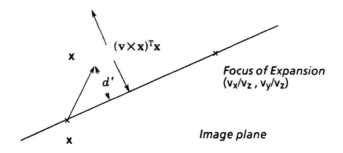

Fig. 1: Geometric interpretation of the error distance d'.

STRUCTURE FROM MOTION

ON THE ESTIMATION OF DEPTH FROM MOTION
USING AN ANTHROPOMORPHIC VISUAL SENSOR

Massimo Tistarelli and Giulio Sandini

University of Genoa – Department of Communication, Computer and Systems Science

Via Opera Pia 11A – I16145 Genoa, Italy

E–mail: tista@dist.dist.unige.it

ABSTRACT

In this paper the application of an anthropomorphic, retina–like visual sensor for optical flow and depth estimation, is presented. The main advantage, obtained with the non–uniform sampling, is the considerable data reduction, while a high spatial resolution is preserved in the part of the field of view corresponding to the focus of attention.

As for depth estimation a tracking egomotion strategy is adopted which greatly simplifies the motion equations, and naturally fits with the characteristics of the retinal sensor (the displacement is smaller wherever the image resolution is higher). A quantitative error analysis is carried out, determining the uncertainty of range measurements.

An experiment, performed on a real image sequence, is presented.

1. Introduction

Among the sensors for robots, visual sensors are those that require the gratest computational power to process the acquired data, but also provide a great deal of information [1,2,3]. A purposively planned acquisition strategy, performed using an appropriate visual sensor, can considerably reduce the complexity of the processing. Some researchers [4,5,6] proposed a motion strategy which greatly simplifies the problem of visual navigation of a mobile robot and also makes the depth–from–egomotion paradigm a well–posed problem, reducing the dimensionality of the equations. A retina–like sensor embeds many advantages for dynamic image processing and shape recognition. This potentiality, which is mainly related to the topology of the space–variant structure, can be considerably augmented defining a set of visual algorithms to be performed directly on the sensor, avoiding the delay to send the data to external devices. The computation of the optical flow from an image sequence and the estimation of depth from motion are among them. In this paper we define an algorithm to estimate depth from motion adopting a retina−like visual sensor, in the case of active, tracking egomotion. The errors deriving from the method are analyzed, defining the uncertainty associated to depth estimates.

1.1. Structure of the Retina−like Sensor

An interesting feature of the space−variant sampling, like that described in [7], is the topological transformation of the retinal image into its cortical projection. Such transformation is described as a conformal mapping of the points on the polar (retinal) plane (ρ, η), onto a Cartesian (cortical) plane $(\xi = \log\rho, \gamma = \eta)$ [8]. The cortical projection is invariant to linear scalings and rotations, relative to the center of the fovea, on the retinal image.

The prototype CCD sensor [9], has the structure sketched in Figure 1. It is divided into 3 concentric areas, each consisting of 10 circular rows, and a central fovea. Each circular row consists of 64 sensitive elements [10]. The central fovea is covered by a squared array of 102 sensitive elements. In the experiments, the information coming from the fovea is not used. As for the extra−foveal part of the sensor, the retino−cortical

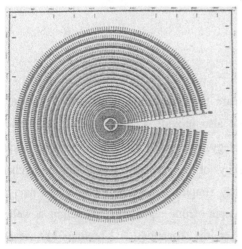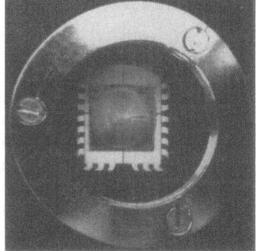

Figure 1   (a): Outline of the retinal CCD sensor. (b): Picture of the CCD sensor

transformation becomes:

$$\begin{cases} \xi = \log_a \rho - q \\ \gamma = q\,\eta \end{cases}$$

where q and a are constants determined by the physical layout of the CCD sensor.

2. Depth from Optical Flow

As for the acquisition of range data, active movements can induce much structural information in the flow of images, acquired by a moving observer. Performing a tracking egomotion only 5 parameters are involved determining camera velocity instead of 6. We consider the distances of the camera from the fixation point D_1 and D_2 in two successive time instants, and the rotations of the camera ϕ, θ and ψ, referred to its coordinate axes, shown in Figure 2. The translational velocity of the camera is then:

$$W_X = D_2 \cos\phi \sin\theta \tag{1}$$

$$W_Y = D_2 \sin\phi$$

$$W_Z = D_1 - D_2 \cos\phi \cos\theta$$

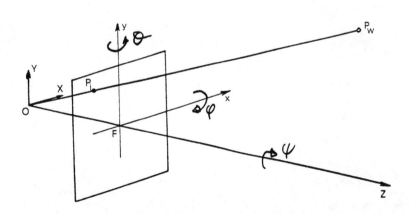

Figure 2   Diagram of the camera coordinate system

The optical flow can be expressed as function of the camera parameters, and decomposed into two terms depending on the rotational and translational components of camera velocity respectively [6]:

$$\vec{V}_r = \left[\frac{x y \phi - \left[x^2 + F^2 \right] \theta + F y \psi}{F} \ , \ \frac{\left[y^2 + F^2 \right] \phi - x y \theta - F x \psi}{F} \right] \tag{2}$$

$$\vec{V}_t = \left[\frac{x \left[D_1 - D_2 \cos\phi \cos\theta \right] - F D_2 \cos\phi \sin\theta}{Z} \ , \ \frac{y \left[D_1 - D_2 \cos\phi \cos\theta \right] - F D_2 \sin\phi}{Z} \right]$$

where Z is the distance of the world point from the image plane. The rotational part of the flow field $\vec{V}_r$ can be computed from the camera rotational angles and the focal length, while $\vec{V}_t$ requires also the knowledge of environmental depth. Once the global optic flow $\vec{V}$ is computed, $\vec{V}_t$ and consequently Z, is determined subtracting $\vec{V}_r$ from $\vec{V}$.

As explained in section 1.1, the images acquired with the retina–like sensor are the result of the retino–cortical topological transformation. The image velocity is then computed applying an algorithm for the estimation of the optical flow to a sequence of cortical images. In Figure 3 the first and last images of a sequence of 11 are shown. The images have been acquired with a conventional CCD camera and digitized with 256x256 pixels and 8 bits per pixel. The sequence represents 3 planar surfaces, actually 3 books; it has been acquired during the translation of the camera along a linear trajectory. The rail on which

Figure 3   First and last image of the sequence

the camera was mounted can be seen in the foreground. A point on the surface of the book in the middle, was tracked during the movement. The motion of the sensor was then a translation plus a rotation ϕ around its horizontal axis X. In Figure 4, the result of the retinal sampling applied to the images in Figure 3 is shown. These images have been obtained simulating the non−uniform sampling operated by the retinal sensor, according to the characteristics of the chip. In Figure 5 the simulated output of the retinal sensor is shown; the images are 30x64 pixels.

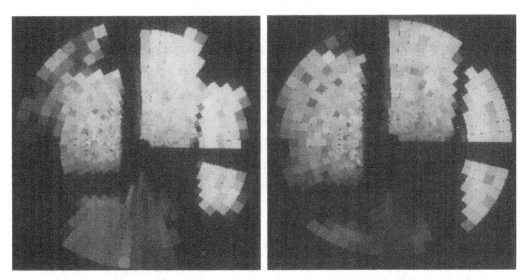

Figure 4   Output of the retinal sampling applied to the images in Figure 3

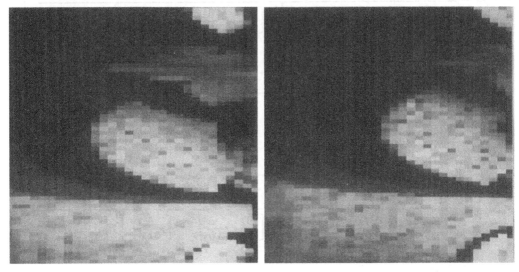

Figure 5   Simulated output of the retinal sensor for the images in Figure 3

The optical flow is computed solving an over–determined system of linear equations in the unknowns $(u,v) = \vec{V}$, imposing the constancy of the image brightness over time [11] and the stationarity of the image motion field [12]:

$$
\begin{cases}
u \dfrac{\partial I}{\partial x} + v \dfrac{\partial I}{\partial y} = -\dfrac{\partial I}{\partial \iota} \\[2mm]
u \dfrac{\partial^2 I}{\partial x^2} + v \dfrac{\partial^2 I}{\partial x \, \partial y} = -\dfrac{\partial^2 I}{\partial x \, \partial \iota} \\[2mm]
u \dfrac{\partial^2 I}{\partial x \, \partial y} + v \dfrac{\partial^2 I}{\partial y^2} = -\dfrac{\partial^2 I}{\partial y \, \partial \iota}
\end{cases}
\tag{3}
$$

where I represents the image intensity of the point (x,y) at time ι. The least squares solution of (3) can be found, for each point on the cortical plane, as:

$$
\vec{V}(x = \xi, y = \gamma, \iota) = (A\,'A)^{-1}\, A\,' \vec{b}
\tag{4}
$$

$$
A = \begin{bmatrix}
u \dfrac{\partial I}{\partial x} & v \dfrac{\partial I}{\partial y} \\[2mm]
u \dfrac{\partial^2 I}{\partial x^2} & v \dfrac{\partial^2 I}{\partial x \, \partial y} \\[2mm]
u \dfrac{\partial^2 I}{\partial x \, \partial y} & v \dfrac{\partial^2 I}{\partial y^2}
\end{bmatrix}
\qquad
\vec{b} = \begin{bmatrix}
-\dfrac{\partial I}{\partial \iota} \\[2mm]
-\dfrac{\partial^2 I}{\partial x \, \partial \iota} \\[2mm]
-\dfrac{\partial^2 I}{\partial y \, \partial \iota}
\end{bmatrix}
$$

$$
u = \frac{
\left[\dfrac{\partial I}{\partial x}\dfrac{\partial I}{\partial y} + \dfrac{\partial^2 I}{\partial x^2}\dfrac{\partial^2 I}{\partial x \, \partial y} + \dfrac{\partial^2 I}{\partial y^2}\dfrac{\partial^2 I}{\partial x \, \partial y} \right]
\left[\dfrac{\partial I}{\partial y}\dfrac{\partial I}{\partial \iota} + \dfrac{\partial^2 I}{\partial y^2}\dfrac{\partial^2 I}{\partial y \, \partial \iota} + \dfrac{\partial^2 I}{\partial x \, \partial \iota}\dfrac{\partial^2 I}{\partial x \, \partial y} \right] +
}{
\left[\dfrac{\partial I}{\partial x}^2 + \dfrac{\partial^2 I}{\partial x^2}^2 + \left[\dfrac{\partial^2 I}{\partial x \, \partial y}\right]^2 \right]
\left[\dfrac{\partial I}{\partial y}^2 + \dfrac{\partial^2 I}{\partial y^2}^2 + \left[\dfrac{\partial^2 I}{\partial x \, \partial y}\right]^2 \right] +
}
$$

$$
\frac{
- \left[\dfrac{\partial I}{\partial y}^2 + \dfrac{\partial^2 I}{\partial y^2}^2 + \left[\dfrac{\partial^2 I}{\partial x \, \partial y}\right]^2 \right]
\left[\dfrac{\partial I}{\partial x}\dfrac{\partial I}{\partial \iota} + \dfrac{\partial^2 I}{\partial x^2}\dfrac{\partial^2 I}{\partial x \, \partial \iota} + \dfrac{\partial^2 I}{\partial y \, \partial \iota}\dfrac{\partial^2 I}{\partial x \, \partial y} \right]
}{
- \left[\dfrac{\partial I}{\partial x}\dfrac{\partial I}{\partial y} + \dfrac{\partial^2 I}{\partial x^2}\dfrac{\partial^2 I}{\partial x \, \partial y} + \dfrac{\partial^2 I}{\partial y^2}\dfrac{\partial^2 I}{\partial x \, \partial y} \right]^2
}
$$

$$
v = \frac{
\left[\dfrac{\partial I}{\partial x}\dfrac{\partial I}{\partial y} + \dfrac{\partial^2 I}{\partial x^2}\dfrac{\partial^2 I}{\partial x \, \partial y} + \dfrac{\partial^2 I}{\partial y^2}\dfrac{\partial^2 I}{\partial x \, \partial y} \right]
\left[\dfrac{\partial I}{\partial x}\dfrac{\partial I}{\partial \iota} + \dfrac{\partial^2 I}{\partial x^2}\dfrac{\partial^2 I}{\partial x \, \partial \iota} + \dfrac{\partial^2 I}{\partial y \, \partial \iota}\dfrac{\partial^2 I}{\partial x \, \partial y} \right] +
}{
\left[\dfrac{\partial I}{\partial x}^2 + \dfrac{\partial^2 I}{\partial x^2}^2 + \left[\dfrac{\partial^2 I}{\partial x \, \partial y}\right]^2 \right]
\left[\dfrac{\partial I}{\partial y}^2 + \dfrac{\partial^2 I}{\partial y^2}^2 + \left[\dfrac{\partial^2 I}{\partial x \, \partial y}\right]^2 \right] +
}
$$

$$-\left[\frac{\partial I^2}{\partial x} + \frac{\partial^2 I^2}{\partial x^2} + \left[\frac{\partial^2 I}{\partial x\,\partial y}\right]^2\right]\left[\frac{\partial I}{\partial y}\frac{\partial I}{\partial \iota} + \frac{\partial^2 I}{\partial y^2}\frac{\partial^2 I}{\partial y\,\partial \iota} + \frac{\partial^2 I}{\partial x\,\partial \iota}\frac{\partial^2 I}{\partial x\,\partial y}\right]$$
$$-\left[\frac{\partial I}{\partial x}\frac{\partial I}{\partial y} + \frac{\partial^2 I}{\partial x^2}\frac{\partial^2 I}{\partial x\,\partial y} + \frac{\partial^2 I}{\partial y^2}\frac{\partial^2 I}{\partial x\,\partial y}\right]^2$$

where $(x = \xi, y = \gamma)$ represent the point coordinates on the cortical plane.

In Figure 6(a) the optical flow of the sequence of Figure 5 is shown, together with the error in 6(b), computed as stated in section 2.1. It is worth noting that in the area of the image near the fovea, which maps on the right side of the cortical plane, the velocity field is almost zero. This fact is due to the fact that the center of the image is stabilized during the tracking egomotion. As it can be noticed, velocity has been computed only at the image contours; the other image areas are almost uniform, lacking of the structural information needed to obtain significant values from the derivatives of image brightness.

In (2) we defined the relation between camera velocity and optical flow; we now develop the same equations for the velocity field induced onto the cortical plane.

Firstly we derive the motion equations on the cortical plane:

Figure 6   (a) Optical flow of the sequence in Figure 5. (b) Variance of the optical flow

$$\begin{cases} \dot{\xi} = \dfrac{\dot{\rho}}{\rho}\log_a e \\[2mm] \dot{\gamma} = q\dot{\eta} \end{cases} \tag{5}$$

where e is the natural logarithmic base, and a, q are constants, related to the eccentricity and the density of the receptive fields of the retinal sensor. The retinal velocity $(\dot{\rho}, \dot{\eta})$ can be expressed as function of the retinal coordinates relative to a Cartesian reference system (as $\rho = \sqrt{x^2 + y^2}$ and $\eta = atan(\frac{y}{x})$) :

$$\begin{cases} \dot{\rho} = \dfrac{x\dot{x} + y\dot{y}}{\sqrt{x^2 + y^2}} \\[4mm] \dot{\eta} = \dfrac{x\dot{y} - y\dot{x}}{x^2 + y^2} \end{cases} \tag{6}$$

$(\dot{x}, \dot{y})$ is the retinal velocity of the image point, referred to a Cartesian reference system centered on the fovea. Substituting (6) in (5) and expliciting the retinal velocity $\vec{V} = (u, v)$:

$$\vec{V} = \vec{V}_t + \vec{V}_r = \begin{bmatrix} x\,\dot{\xi}\log_e a - \dfrac{y\dot{\gamma}}{q} \\[3mm] y\,\dot{\xi}\log_e a + \dfrac{x\dot{\gamma}}{q} \end{bmatrix} \tag{7}$$

$(\dot{\xi}, \dot{\gamma})$ is the velocity field computed from the sequence of cortical images. Substituting the expression of $\vec{V}_r$ from (2) in (7), we obtain the translational flow $\vec{V}_t$ referred to the retinal plane:

$$\vec{V}_t = \begin{bmatrix} x\,\dot{\xi}\log_e a - \dfrac{y\dot{\gamma}}{q} - \dfrac{xy\phi - \left[x^2 + F^2\right]\theta + Fy\psi}{F} \\[4mm] y\,\dot{\xi}\log_e a + \dfrac{x\dot{\gamma}}{q} - \dfrac{\left[y^2 + F^2\right]\phi - xy\theta - Fx\psi}{F} \end{bmatrix} \tag{8}$$

It is worth noting that, expressing the Cartesian coordinates, centered on the fovea, of a retinal point in microns, also the focal length and the retinal velocity are expressed in the same units.

The depth of all the image points on the retinal plane, is computed as:

$$\frac{Z}{W_z} = \frac{D_f}{|\vec{V}_t|} \tag{9}$$

D_f is the displacement of the considered point from the focus of the translational field on the image plane, and W_z is the translational component of the sensor velocity along the optical (Z) axis. W_z can be easily determined from (1) if the distance of the fixation point from the camera is known at two successive time instants. The measurement unit of Z depends on the unit of W_z. If W_z is set equal to 1, the time to collision of the robot with respect to the world point, is computed. The location of the FOE is estimated computing the least squares fit of the pseudo intersection of the set of straight lines determined by the velocity vectors $\vec{V}_t$.

In Figure 7 the depth map of the scene in Figure 4, and in Figure 8 the associated uncertainty are shown. It has been computed, from the optical flow in Figure 6 and the known camera rotational velocity, applying equation (8) to estimate the translational velocity, then finding the location of the FOE with a least squares fit and finally evaluating equation (9) for each point on the retinal plane.

2.1. Error Analysis

The optical flow field is computed solving the overdetermined system of linear equations stated in (4). The optical flow error, on the retinal plane, is obtained differentiating (7) with respect to $\dot{\xi}$ and $\dot{\gamma}$:

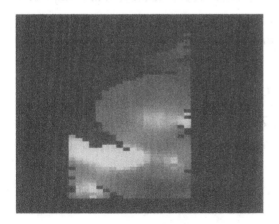 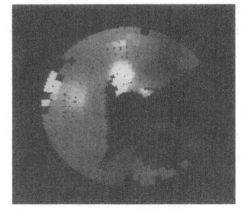

Figure 7   (a) Depth map on the cortical plane. (b) Depth map on the retinal plane

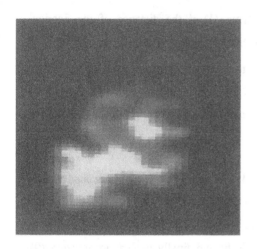 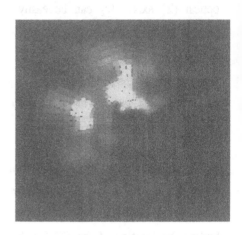

<u>Figure 8</u>   (a) Depth uncertainty on the cortical plane. (b) Depth uncertainty on the retinal plane

$$\begin{cases} \delta u = \delta\dot{\xi}\, x \log_e a \, - \, \delta\dot{\gamma}\,\dfrac{y}{q} \\ \delta v = \delta\dot{\xi}\, y \log_e a \, + \, \delta\dot{\gamma}\,\dfrac{x}{q} \end{cases} \tag{10}$$

$(\delta\dot{\xi},\delta\dot{\gamma})$ represent the error of the least squares fit.   We now derive a quantitative measure for the absolute and percentage error on the estimated depth.   The depth function can
be written as:

$$Z = f(x,y,u,v,FOE_x,FOE_y,\phi,\theta,\psi)$$

(x,y) is the position of the point on the retinal plane, referred to a Cartesian coordinate
system centered on the fovea; $\vec{V} = (u,v)$ is the estimated retinal velocity; (FOE_x,FOE_y) are
the coordinates of the focus of expansion on the retinal plane; (ϕ,θ,ψ) are the rotation
angles performed by the camera during the tracking motion. The error in depth can be
computed as:

$$\delta Z = \vec{J} \cdot \vec{\Delta}$$

$\vec{J}$ is the vector of the partial derivatives of the Z function and $\vec{\Delta}$ is the vector of the errors
in the parameters $(x,y,u,v,FOE_x,FOE_y,\phi,\theta,\psi)$. Computing the partial derivatives of (10) and
posing W_z equal to 1, obtain:

$$\delta Z = \left[\frac{(x - FOE_x) \left| \vec{V_t} \right| - D_f u_t}{D_f \left| \vec{V_t} \right|^2} \right] \delta x + \left[\frac{(y - FOE_y) \left| \vec{V_t} \right| - D_f v_t}{D_f \left| \vec{V_t} \right|^2} \right] \delta y$$

$$+ \left[\frac{(FOE_x - x)}{D_f \left| \vec{V_t} \right|^2} \right] \delta FOE_x + \left[\frac{(FOE_y - y)}{D_f \left| \vec{V_t} \right|^2} \right] \delta FOE_y \tag{11}$$

$$- \left[\frac{D_f u_t}{\left| \vec{V_t} \right|^3} \right] \delta u - \left[\frac{D_f v_t}{\left| \vec{V_t} \right|^3} \right] \delta v$$

$$+ \left[D_f \frac{u_t x y + v_t (y^2 + F^2)}{F \left| \vec{V_t} \right|^3} \right] \delta \phi - \left[D_f \frac{u_t (x^2 + F^2) + v_t x y}{F \left| \vec{V_t} \right|^3} \right] \delta \theta + \left[D_f \frac{u_t y - v_t x}{\left| \vec{V_t} \right|^3} \right] \delta \psi$$

Substituting the expression of Z, as from (9), in (11) and grouping similar terms, obtain:

$$\frac{\delta Z}{Z} = \frac{1}{D_f^2 \left| \vec{V_t} \right|^2} \left\{ \left| \vec{V_t} \right| \left[\left[x - FOE_x - D_f u_t \right] \delta x + \left[y - FOE_y - D_f v_t \right] \delta y \right] \right. \tag{12}$$

$$+ \left| \vec{V_t} \right|^2 \left[\left[x - FOE_x \right] \delta FOE_x + \left[y - FOE_y \right] \delta FOE_y \right] - D_f^2 \left[u_t \delta u + v_t \delta v \right]$$

$$\left. + \frac{D_f^2}{F} \left[\left[u_t x y + v_t (F^2 + y^2) \right] \delta \phi + F \left[u_t y - v_t x \right] \delta \psi - \left\{ v_t x y + u_t (F^2 + x^2) \right\} \delta \theta \right] \right\}$$

which gives an estimate of the percentage error. As it can be noticed, the greater contribution stems from the errors in the rotations of the camera. In fact, these errors are quadratic in the retinal coordinates (x,y) and are multiplied by the square distance from the FOE, D_f^2, which, in the analyzed egomotion, is generally very large. The errors $(\delta x, \delta y)$ are set equal to the radius of the sensitive cell on the CCD array, which depends on the spatial position of the point within the field of view. The errors in the rotational angles are set equal to a constant, which is determined by the positional error of the driving motors. Actually these values are set to $(0.1, 0.1, 0.1)$ degrees, which is a reasonable error for standard positioning devices. The errors in the optical flow components (u,v) and of the position of the FOE, are determined from a standard formula of the least squares error [13]:

$$H = \begin{bmatrix} \sigma_0 & r_{01} & \cdots\cdots\cdots & r_{0N} \\ r_{10} & \sigma_1 & \cdots\cdots\cdots & r_{1N} \\ & & \cdots\cdots\cdots & \\ r_{N0} & \cdots\cdots & r_{N(N-1)} & \sigma_N \end{bmatrix}$$

$\sigma_0, \sigma_1, \cdots, \sigma_N$ are the variances of the parameters estimated from the least squares fit. The matrix on the right is obtained first dividing each element of matrix A in (4), by the corresponding standard deviation. If B is the matrix obtained, then H is computed as:

$$H = \begin{bmatrix} B\,B^t \end{bmatrix}^{-1}$$

The standard deviation of the image derivatives used to compute the optical flow are estimated assuming a uniform distribution and unitary quantization step of the gray levels [14]. Using a 5 points (frames in time) derivative operator, a value of σ_I equal to 0.0753 for the first derivative, and σ_{II} equal to 0.8182 for the second derivative are obtained.

As for the computation of depth uncertainty, the algorithm is modeled as a stochastic process, where the parameters are uncorrelated probabilistic variables. Assuming the process to be Gaussian, with a set of variables whose mean values are equal to the measured ones, then:

$$\sigma_z^2 = J \cdot Q \cdot J^t$$

J is the Jacobian of the depth function and Q is the diagonal matrix of the variances of the independent variables. The variances in matrix Q, $(\sigma_x^2, \sigma_y^2, \sigma_\phi^2, \sigma_\theta^2, \sigma_\psi^2)$ are considered equal to the squared errors, while the variances in the optical flow and the position of the FOE, $(\sigma_u^2, \sigma_v^2, \sigma_{FOEx}^2, \sigma_{FOEy}^2)$ are determined as for the variances of the optical flow. Writing explicitly the variance in depth, obtain:

$$
\begin{aligned}
\sigma_z^2 = {} & \left[\frac{W_z}{D_f \left| \vec{V}_t \right|} \right]^2 \left\{ (FOE_x - x)^2 \, \sigma_{FOEx}^2 + (FOE_y - y)^2 \, \sigma_{FOEy}^2 \right\} \\[2mm]
& + \left[\frac{W_z}{D_f \left| \vec{V}_t \right|^2} \right]^2 \left\{ \left[(FOE_x - x) \left| \vec{V}_t \right| + D_f \, u_t \right]^2 \sigma_x^2 + \right. \\[2mm]
& \left. + \left[(FOE_y - y) \left| \vec{V}_t \right| + D_f \, v_t \right]^2 \sigma_y^2 \right\}
\end{aligned}
\tag{13}
$$

$$+ \left[\frac{W_z D_f}{\left| \vec{V}_t \right|^3} \right]^2 \left\{ u_t^2 \sigma_u^2 + v_t^2 \sigma_v^2 + \frac{\left[u_t x y + v_t (F^2 + y^2) \right]^2}{F^2} \sigma_\phi^2 \right.$$

$$\left. + \frac{\left[v_t x y + u_t (F^2 + x^2) \right]^2}{F^2} \sigma_\theta^2 + \left[u_t y - v_t x \right]^2 \sigma_\psi^2 \right\}$$

The last 2 rows represent the higher term, as it is multiplyed by D_f. Consequently, the higher errors are those due to the computation of image velocity and the estimation of the rotational angles of the cameras (or conversely, the positioning error of the driving motors).

The terms containing the errors in the rotational angles are also quadratic in the image coordinates, hence the periphery of the visual field will be more effected then the fovea. Nevertheless all these terms are divided by the cubic module of velocity, therefore if the amplitude of image displacement is sufficiently large, the error drops very quickly. As a matter of fact the amplitude of the optical flow is of crucial importance reducing the uncertainty in depth estimation. This can be achieved using a long motion baseline, for example sub–sampling the sequence over time, or, alternatively, cumulating the optical flows over multiple frames. As for the last option, it has been shown [15,16] the disadvantage of tracking image features, because errors are cumulated over time. Nevertheless these errors can be considerably reduced if, for each computed optical flow, also the previous measurements are taken into account to establish pointwise correspondence. A method to perform this temporal integration has been already presented [16]; another method, based on a local spatio–temporal interpolation of image velocities, is being developed.

3. Conclusions

The application of a retina–like, anthropomorphic visual sensor for dynamic image analysis has been investigated. In particular the case of a moving observer, undertaking active movements has been considered to estimate environmental depth. The main advantages obtained with the retina–like sensor, are related to the space–variant sampling structure, which features image scaling and rotation invariance and a variable resolution. Due to this topology, the amount of incoming data is considerably reduced, but a high resolution is preserved in the part of the image corresponding to the focus of attention.

Adopting a tracking egomotion strategy, the computation of the optical flow and depth—from—motion, is simplified. Moreover, as the amplitude of image displacements increases from the fovea to the periphery of the retinal image, almost the same computational accuracy is achieved throughout the visual field, minimizing the number of pixels to be processed.

In human beings, most "low—level" visual processes are directly performed on the retina or in the early stages of the visual system. Simple image processes like filtering, edge and motion detection must be performed quickly and with minimal delay from the acquisition stage, because of vital importance for survival (for example to detect static and moving obstacles). These processes could be implemented directly on the sensor, as local (even analogic, using the electrical charge output of the sensitive elements) parallel operations, avoiding the delay for decoding and transmitting data to external devices.

References

1. D. Marr, *Vision,* Freeman and Co., San Francisco (1982).

2. D. H. Ballard and C. M. Brown, *Computer Vision,* Prentice—Hall, New Jersey (1982).

3. E. C. Hildreth, *The Measurement of Visual Motion,* MIT Press, Cambridge, USA (1983).

4. A. Bandopadhay, B. Chandra, and D. H. Ballard, `Active Navigation: Tracking an Environmental Point Considered Beneficial'', *Proc. of "Workshop on Motion: Representation and Analysis",* pp. 23—29 , Kiawah Island ResortIEEE Computer Society, (May 7—9, 1986).

5. G. Sandini, V. Tagliasco, and M. Tistarelli, `Analysis of Object Motion and Camera Motion in Real Scenes'', *Proc. IEEE Intl. Conference on "Robotics & Automation",* pp. 627—633 , San FranciscoIEEE—CS, (April 7—10, 1986).

6. G. Sandini and M. Tistarelli, `Active Tracking Strategy for Monocular Depth Inference over Multiple Frames'', *IEEE Trans. on PAMI* PAMI—12, No. 1 pp. 13—27

Acknowledgements: This research was supported by the special project on robotics of the Italian National Council of Research

(January 1990).

7. G. Sandini and V. Tagliasco, "An Anthropomorphic Retina–like Structure for Scene Analysis", *Comp. Graphics and Image Proc.* **14 No.3** pp. 365–372 (1980).

8. E. L. Schwartz, "Spatial Mapping in the Primate Sensory Projection: Analytic Structure and Relevance to Perception", *Biol. Cybernetics* **25** pp. 181–194 (1977).

9. J. Van der Spiegel, G. Kreider, C. Claeys, I. Debusschere, G. Sandini, P. Dario, F. Fantini, P. Bellutti, and G. Soncini, "A Foveated Retina–Like Sensor Using CCD Technology", in *Analog VLSI and Neural Network Implementations*, ed. C. Mead and M. Ismail,De Kluwer (1989).

10. I. Debusschere, E. Bronckaers, C. Claeys, G. Kreider, J. Van der Spiegel, P. Bellutti, G. Soncini, P. Dario, F. Fantini, and G. Sandini, "A 2D Retinal CCD Sensor for Fast 2D Shape Recognition and Tracking", *Proc. 5th Int. Solid–State Sensor and Transducers*, , Montreux(June 25–30 1989).

11. B. K. P. Horn and B. G. Schunck, "Determining Optical Flow", *Artificial Intelligence* **17 No.1–3** pp. 185–204 (1981).

12. S. Uras, F. Girosi, A. Verri, and V. Torre, "Computational approach to Motion perception", *Biological Cybernetics*, (1988).

13. P.R. Bevington, *Data Reduction and Error Analysis,* McGraw–Hill, New York (1969).

14. B. Kamgar–Parsi and B. Kamgar–Parsi, "Evaluation of Quantization Error in Computer Vision", *Proc. of DARPA Workshop on "Image Understanding"*, (1988).

15. S. Bharwani, E. Riseman, and A. Hanson, "Refinement of Environmental Depth Maps Over Multiple Frames.", *Proc. of "Workshop on Motion: Representation and Analysis"*, pp. 73–80 , Kiawah Island ResortIEEE Computer Society, (May 7–9, 1986).

16. M. Tistarelli and G. Sandini, "Uncertainty Analysis in Visual Motion and Depth Estimation from Active Egomotion", *Proc. of IEEE/SPIE Intl. Conference on Applications of Artificial Intelligence VII* , pp. 333–343 , Orlando, Florida, (March 28–30, 1989).

THE DERIVATION OF QUALITATIVE INFORMATION
IN MOTION ANALYSIS

Edouard François and Patrick Bouthemy
IRISA / INRIA-Rennes
Campus de Beaulieu, 35042 Rennes Cedex, France

1. Introduction

Scene understanding using image sequence is a very challenging task [5]. Substantial works have been devoted to this topic, in particular to the recovery of quantitative 3D motion and structure parameters in the scene from optic flow, [1]. Because of noisy measurement of apparent motion, problems of numerical instability and estimation errors of the 3D measures appear. Then, as primarily suggested by Thompson et al, [11], it becomes attractive to follow a qualitative approach, in order to obtain stable and robust descriptions, which still keeps enough richness of information in many situations. Indeed theoretical studies have pointed out that the geometry of the apparent velocity field contains by itself significant useful information, [4,6]. This paper is concerned with a qualitative description of the kinematic behaviours of the objects in the scene (including the sensor itself), comprising both the kind of motion (e.g. translation, rotation), and the type of trajectory (e.g. parallel or perpendicular to the image plane). Nagel presents several approaches to provide conceptual descriptions of the scene, based on different typical situations, [7]. Burger and Bhanu in [3], try to reason using symbolic entities and multiple simultaneous qualitative interpretations of the scene. More recently, obstacle avoidance using a divergence cue has been treated according to a qualitative approach, by Nelson and Aloimonos, [8].

The problem is here addressed as follows : first, through a first order development, pertinent cues of the apparent velocity vector field in the 2D image are defined, each of them describing a particular aspect of this field (Section 1); then we establish the relation between these terms and the 3D motion parameters, and define possible sets of labels associated with different kinematic behaviours (Section 2); the label (model) validation step is solved using statistical approaches, the first one consisting in a likelihood test tree, the other one based on an information criterion (Section 3); Results are presented in Section 4.

2. 2D vector field description

The basic information we consider is the 2D apparent velocity vector field in the image, resulting from the relative motion between camera and objects in the scene. As emphasized later, this does not mean that an explicit estimation of this field is necessarily required. In this section, we will mathematically justify the introduction of particular cues on this field, which are very relevant for interpretation purposes, [8,11]. After some studies, it has appeared that first order description of a vector field can be sufficient to interpretation (second-order informations are likely to be very noisy while bringing few significant supplementary qualitative information).

Let us consider the first order development of the velocity vector $\underline{\omega}$ around a point $p(x_p, y_p)$ in the image :

$$\underline{\omega}(x,y) \;=\; (u,v)^T \;=\; (a_p,b_p)^T \;+\; M\cdot(x-x_p,\,y-y_p)^T \qquad (1)$$

the coefficients of the matrix M are given by : $m_{11}=\frac{\partial u}{\partial x}$, $m_{12}=\frac{\partial u}{\partial y}$, $m_{21}=\frac{\partial v}{\partial x}$, $m_{22}=\frac{\partial v}{\partial y}$. M can be decomposed into, [10] :

$$M = \tfrac{1}{2}\,(\text{trace } M)\,I + \tfrac{1}{2}\left(M - M^T\right) + \tfrac{1}{2}\left[M + M^T - (\text{trace } M)\,I\right] \qquad (2)$$

which leads to this formulation :

$$M = \frac{1}{2}\,div.\begin{bmatrix}1 & 0\\ 0 & 1\end{bmatrix} + \frac{1}{2}\,rot.\begin{bmatrix}0 & -1\\ 1 & 0\end{bmatrix} + \frac{1}{2}\,hyp1.\begin{bmatrix}1 & 0\\ 0 & -1\end{bmatrix} + \frac{1}{2}\,hyp2.\begin{bmatrix}0 & 1\\ 1 & 0\end{bmatrix} \qquad (3)$$

The terms $div = \frac{\partial u}{\partial x}+\frac{\partial v}{\partial y}$ (for divergence), $rot = \frac{\partial v}{\partial x}-\frac{\partial u}{\partial y}$ (for rotational), $hyp1 = \frac{\partial u}{\partial x}-\frac{\partial v}{\partial y}$ and $hyp2 = \frac{\partial u}{\partial y}+\frac{\partial v}{\partial x}$ (for hyperbolic terms) are much more convenient than $\frac{\partial u}{\partial x},\frac{\partial u}{\partial y},\frac{\partial v}{\partial x}$ and $\frac{\partial v}{\partial y}$ because they correspond to particular fields, which can be quite naturally interpretated. The interpretation of any vector field becomes much more easily based on these four first order terms. Moreover, they provide independant and complementary information, since the new decomposition is still obtained in an orthogonal basis.

3. Definition of the label set

When the analyzed field is a velocity vector one, the physical link between these four terms and the 3D structure and motion parameters can be explicitly proven. In a cartesian coordinate system $OXYZ$ where O is the projection centre and OZ is along the optical axis, we consider the 3D relative motion of an object consisting of translational velocity $\underline{T}=(U,V,W)^T$, and rotational velocity $\underline{\Omega}=(A,B,C)^T$, i.e. for any point P of this object: $\underline{V_P}=\underline{T}+\underline{\Omega}\text{x}\underline{OP}$. Using the classical perspective projection equations and deriving them with respect to time, it can be easily shown that :

$$\begin{cases} div = -2\frac{W}{Z_0} - \gamma_1\frac{U}{Z_0} - \gamma_2\frac{V}{Z_0} & rot = 2C - \gamma_1\frac{V}{Z_0} + \gamma_2\frac{U}{Z_0} \\[2mm] hyp1 = -\gamma_1\frac{U}{Z_0} + \gamma_2\frac{V}{Z_0} & hyp2 = -\gamma_1\frac{V}{Z_0} - \gamma_2\frac{U}{Z_0} \end{cases} \qquad (4)$$

where Z_0,γ_1 and γ_2 are respectively the depth and first order structure parameters of the object surface patch, whose projection in the image lies around point p.

For a qualitative interpretation of motion, what is important is the comparison of these terms to zero. Accordingly, we associate to each quantitative term $div, rot, hyp1, hyp2$ a qualitative (boolean) variable $V_{div}, V_{rot}, V_{hyp1}, V_{hyp2}$, equal to 0 if its quantitative value is non significant, and respectively $D, R, H1, H2$ otherwise. Now, we are no more reasoning on quantitative values, but symbols the associations of which correspond to particular motion configurations.

From the relations (4), it is clear that several typical dynamic situations can be expressed. For instance, in the $(V_{div}, V_{rot}, V_{hyp1}, V_{hyp2})$ basis, the $(D,0,0,0)$ association or label is the qualitative description of a motion along the optical axis (it is besides usual to base the obstacle detection on a divergence cue). $(0,R,0,0)$ describes a rotation around this axis, and so on. To show it is of great interest in practical situations, let us consider very concrete cases. In a car driving situation, recognition of relative axial and transversal motion to the camera is crucial. The first one would be labeled $(D,0,0,0)$; the second one $(0,0,0,0)$ if the camera is static, $(D,R,H1,H2)$ otherwise. In an industrial context, supervising a rotating device may be an useful task. By setting the optical axis of the camera along the rotation axis, the label would be $(0, R, 0, 0)$.

It is important to remark that the motion description is only based on the linear terms of the velocity field. These terms do not depend on the reference point in the image. Therefore we do not need to know the position of the optical center, or to estimate the FOE. Moreover, as we aim at obtaining a qualitative description, and not the exact estimation of 3D parameters, a complete camera calibration is not required.

4. Labeling process

Now, the key point is to properly achieve the numerical-to-symbolic step : that is deriving symbols from numerical data. Given an area in the image, we must determine which label (i.e. symbol association) is the most representative inside it, that is, decide which cue values are significant or not. Merely comparing the magnitudes of the quantitative terms $div, rot, hyp1, hyp2$ (or function of them) to a threshold, as initialy described in [11], remains very difficult and tricky. On one hand, the threshold choice would be very dependent on the context and objectives of the analysis. On the other hand, we cannot introduce noise model with this approach, though the considered observation (optical flow) is actually noisy. That is the reason why we resort to a statistical approach to deal with this problem. The first point is to define what we take as observation.

Let us denote $\underline{\omega}$ the true velocity field, $\underline{\omega}_{\Theta_m}$ the velocity field generated by the linear model m parametrized by Θ_m. For instance, for the $(D, 0, 0, 0)$ label, $\Theta_m = (div, 0, 0, 0)$ and $\underline{\omega}_{\Theta_m} = (a_p + \frac{1}{2}.div.x, b_p + \frac{1}{2}.div.y)^T$. I is the image intensity, $\underline{\nabla}I$ and I_t its spatial and temporal derivatives. $\underline{\omega}$ and I are linked by the well-known image flow constraint equation, [1] : $\quad \underline{\omega}.\underline{\nabla}I + \frac{\partial I}{\partial t} = 0$ (5). If the true field has been estimated, we consider as observations the vectorial random variables : $\quad \underline{e}_{\Theta_m}(x, y) = \underline{\omega}_{\Theta_m}(x, y) - \underline{\omega}(x, y)$ (6)

If the true velocity vector field is not available, we consider the following scalar random variables : $\quad e_{\Theta_m}(x, y) = (\underline{\omega}_{\Theta_m}(x, y) - \underline{\omega}(x, y)).\underline{\nabla}I(x, y)$

i.e. using relation (5), $\quad e_{\Theta_m}(x, y) = \underline{\omega}_{\Theta_m}(x, y).\underline{\nabla}I(x, y) + \frac{\partial I}{\partial t}(x, y)$ (7)

In both cases, the considered variables are supposed to be independant gaussian zero-mean variables. The second point now is to choose a proper decision criterion to realize the numerical-to-symbolic step.

A first method corresponds to the determination of each qualitative variable state by a generalized likelihood ratio (GLR) test. For instance, V_{div} will be determined by testing hypothesis $(D, R, H1, H2)$ against hypothesis $(0, R, H1, H2)$, which means comparing to a threshold the ratio of the likelihood of each label (i.e. the joint density probability of the corresponding parametrized observations, $f(e_{\Theta_m})$). If the ratio is greater than the threshold, V_{div} is declared to be equal to D, 0 otherwise. With this approach, we do not test directly the magnitude of the quantitative terms. Yet a threshold has still to be chosen; but in this case it is not a critical matter. The different tests for each qualitative variable must be considered according to a reasonable order depending on the application, which leads to a sequential decision tree.

A more attractive method consists in defining a significant sub-set of labels $\{L_m\}$ well suited to the application at hand, and in testing all *together* these labels. The optimal label $L_{\widehat{m}}$ is found using a statistical information criterion, of the kind:

$$\widehat{m} = \arg \min_{m} \left[-log[f(e_{\Theta_m})] + \Psi(n).dim(\Theta_m) \right] \qquad (8)$$

where Θ_m denotes the parameter vector of the model m; $dim(\Theta_m)$ is the vector dimension; f is again the likelihood of the model m; n, the number of points of the considered area; and Ψ a given function. There exists a family of criteria, corresponding to different choices of Ψ, [9]. After preliminary tests, it appeared that the most interesting criterion for this application is the Rissanen criterion (RIC). In this case, $\Psi(n) = \log(n)$. By the way, this version of Ψ can also be derived from a Bayesian approach. Before comparing the models, the optimal parameter vector $\hat{\Theta}_m$ has to be found for each model. The second term of the criterion (8) acts as a penalization term on the model complexity. Hence, if the likelihood of a given model is not really better than a simpler one, the more complete model will be eliminated because of the penalization term. This means that in fact, it does not bring any significant additional information.

5. Results and Conclusion

We present in Fig.1 some results of the two approaches on a real sequence. Labeling is undertaken after a spatio-temporal segmentation step which gives an image partition in areas in which an unique motion model is supposed to be present, [2]. In these experiments, the velocity field has not been estimated, and the observation variables are those of expression (7). A set of five labels has been considered (Fig.1d). The vehicles in the scene are moving toward the camera, which is itself fixed on a moving car, Fig.1a. All the movements in the scene are parallel to its optical axis. The corresponding regions in the image should be labeled as $(D, 0, 0, 0)$. As shown in Fig.1b-c, both methods give good and coherent results on the first car, which is the object of interest in this example. This has been confirmed by others experiments carried out with several real sequences acquired at the laboratory, corresponding to a static scene but different sensor motion. The type of the apparent resulting motion was correctly labeled. However, it sometimes happens that the second method fails, that is, selects a more complete model than the true one (for instance, instead of $(D, 0, 0, 0)$ label, the complete one $(D, R, H1, H2)$ is chosen). This may occur when the estimation of the model parameters is too rough. In this case, the likelihood term increases, which makes the penality term become negligible. Therefore, given advantages of each method (i.e. robustness for the GLR; convenience and generality for the RIC), we seek for a combination of them which in particular would avoid as often as possible thresholding steps. Besides, up to now, labeling is realized instantaneously (i.e. considering motion information between only two images). We are now integrating the temporal axis in the labeling process, which should improve the richness of the qualitative description and then the efficiency of the RIC method (the range of model dimensions should increase, which would reinforce the role of the penalty term).

This paper has dealt with the derivation of a qualitative description of the kinematic behaviours of the objects in the scene. It represents a real alternative to quantitative estimations of the 3D motion and structure parameters. The interests of this study are two-fold. First, we have explicitly and analytically linked the description cues to the 2D apparent motion and the 3D motion. Second, we have described model-based statistical decision methods to achieve the numerical-to-symbolic step. They enable to address any determination of qualitative motion information in a non ad-hoc and unified manner.

References:

[1] **J.K. Aggarwal and N. Nandhakumar**, On the Computation of Motion from Sequences of Images- A Review, *Proc. of the IEEE*, Vol. 76, No 8, August 1988, pp 917–935
[2] **P. Bouthemy and J. Santillana Rivero**, A hierarchical likelihood approach for region segmentation according to motion-based criteria, *Proc. 1st ICCV*, 1987, pp 463–467
[3] **W. Burger and B. Bhanu**, Dynamic scene understanding for autonomous mobile robot, *Proc. CVPR*, 1988, pp 736–741
[4] **S. Carlsson**, Information in the geometric structure of retinal flow field, *Proc. 2nd ICCV*, December 1988, pp 629–633
[5] **J.J. Koenderink**, Optic Flow, *Vision Research*, Vol. 26, No 1, 1986, pp 161–180
[6] **J.J. Koenderink and J.J. Van Doorn**, Invariant properties of the motion parallax field due to the movement of rigid bodies relative to an observer, *Optica Acta*, Vol. 22, No 9, 1975, pp 773–791
[7] **H.H. Nagel**, From image sequences towards conceptual descriptions, *Image and vision computing*, Vol. 6, No 2, May 1988, pp 59–74
[8] **R. C. Nelson and J. Aloimonos**, Using flow field divergence for obstacle avoidance in visual navigation, *Proc. 2nd ICCV*, 1988, pp 548–559
[9] **R. Shibata**, Criteria of statistical model selection, *Research Report*, KSTS/RR-86/009, Keio Univ., Dept. of Math., Yokohama, August 1986
[10] **P.Y. Simard and G.E. Mailloux**, A projection operator for the restoration of divergence-free vector-fields, *IEEE Trans. on PAMI*, Vol. PAMI-10, No 2, 1988, pp 248–256
[11] **W.B. Thompson, V.A. Berzins and K.M. Mutch**, Analyzing object motion based on optical flow, *Proc. ICPR*, 1984, pp 791–794

This work is supported by MRT (French Ministery of Research and Technology) in the context of the EUREKA european project PROMETHEUS, under PSA-contract VY/85241753/14/Z10. We thank Dr Enkelman for providing the image sequence.

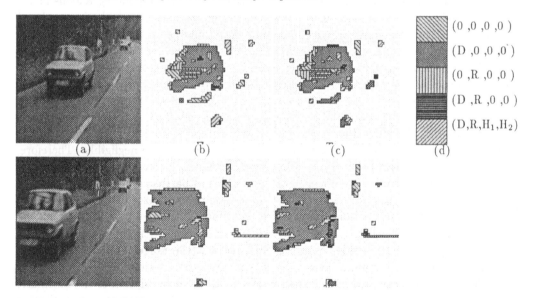

Figure 1: *Results of labeling at two different times in the sequence : (a) original image (first one of the considered pair), (b) labeling with the 1st method (GLR), (c) labeling with the 2nd method (RIC), (d) the set $\{L_m\}$ of the five considered labels*

ROAD FOLLOWING ALGORITHM USING A PANNED PLAN-VIEW TRANSFORMATION

R. A. Lotufo, B. T. Thomas and E. L. Dagless
Faculty of Engineering, University of Bristol
Bristol BS8 1TR, UK

1 Introduction

Driving a robot vehicle along a road using computer vision is a goal that is receiving attention worldwide from many research groups [1-5]. The purpose of the work is to interpret the image sensed by the vehicle in order to drive safely along a road. The major components of an autonomous vehicle consist of perception, reasoning, path planning and vehicle control.

Bristol University has been funded since 1986 to investigate the main components of these tasks and to work on a computer hardware architecture capable of performing the various tasks concerned. A Real-Time Image Processing System based on Inmos transputers has been developed and built. A small tracked vehicle has been used to test in a real environment the algorithms proposed. The first outdoor experiments were performed in December 1988 in which the vehicle followed part of a gravel path in the University gardens. This work is concerned with the vision perception and the reasoning task using temporal and spatial constraints to guide a robot vehicle along a road.

Two approaches for road segmentation have been described in the literature: pixel classification [1-3] and edge detection [4-6]. The technique presented here is a different approach to road segmentation. It is based on the observation that a road is a large homogeneous feature separated by parallel boundaries. Two parameters make the feature extraction difficult in typical road scenes: the perspective problem (making the edge detection fail) and the various possible inclinations of the road boundary line in the image (making shape recognition a more difficult task). The proposed method, first introduced in [7], consists of building a subsampled image using a *panned plan-view transformation* where the resultant image has the perspective view corrected, with all pixels representing the same spatial dimension in the world. The transformation has a pan angle which is adjusted to make the road edges vertical in the plan-view image. Assuming that the road has no sharp corners, a curved portion of the road can appear as two parallel circular arcs with predominantly vertical edges. The panned plan-view image is a powerful and compact representation of the road scene. It preserves all the main features to enable an accurate road boundary extraction. A typical compression factor is 32, reducing the raw image from 256 x 256 pixels to 64 x 32 in the panned plan-view image.

2 Vision System Architecture

Figure 1 shows a block diagram of the vision algorithm. The algorithm extracts the road boundaries and updates a road model to drive a vehicle along the road.

Once the monochrome image is digitised, the panned plan-view transformation is applied, resulting in a spatially reduced subsampled image representing a portion of the view in front of the vehicle. This image is used to extract the road boundaries using a large vertical edge detecting operator. The size of the operator has been chosen based on the fact that the road represents a large predominant feature in the image. A thinning and linking edge procedure filters the edges into a list of edge segments. These linked edge points are then transformed from the plan-view image coordinate system to the vehicle coordinate system. A straight line fitting algorithm converts the edge segments into a list of line segments. Then, the reasoning module makes use of geometric and temporal constraints to classify the segments lines into left, right or non-road edge segments. The classification criteria are based on the previous road reference and on the assumption that the road consists of parallel boundaries with only a small variation in width from cycle to cycle. Finally, the new road reference is updated and a predicted pan angle is determined for use in the plan-view transformation in the next cycle. The reasoning system is designed to cope with situations in which only one side of the

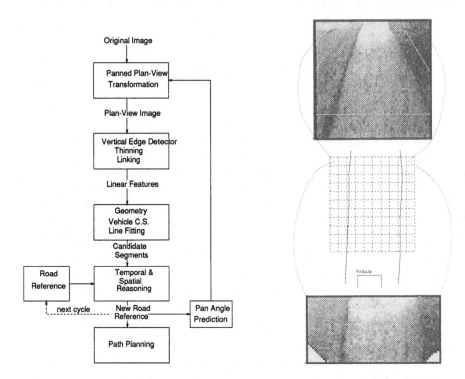

Figure 1: The vision algorithm   Figure 2: The panned plan-view image

road is visible. It keeps tracking the single road edge seeking for the other side to reappear using the information of the last road width extracted.

3   Panned Plan-View Transformation

The proposed method consists of selecting a rectangular area in front of the vehicle where a perspective transformation is applied. The position of the area is specified by the pan angle parameter. It orients the rectangular area to include the road edges in situations where the road is not straight. For each road image there is a pan angle which enables the road edges to appear as near-vertical parallel features. The transformation is applied to a number of equally spaced points in the selected area, reducing the amount of pixel data to be processed in the road feature extraction. The data reduction and the fact that the road always appears with approximately the same shape makes the method robust, relieving later algorithms from dealing with too much unstructured data. The pan angle is predicted by determining the angle between the central axis of the road and the vehicle trajectory. This method compresses the raw image while preserving all the main road features of interest. The panned plan-view image has each pixel representing a uniform area in the road plane and the road edges are represented as vertical or nearly vertical features.

Generating a panned plan-view image

The panned plan-view image of the road is generated by applying the transformation on a rectangular regular grid of points over the road plane. The plan-view transformation uses the assumption that the road is locally flat and all road features belong to the road plane. Although this requires a locally plane terrain, the method can cope with hills and valleys with small slopes.

Figure 2 illustrates the process of building the panned plan-view image. The top of Fig 2 shows the grid boundary used in the transformation superimposed on the original image. The centre of Figure 2 shows a schematic view of the road in the vehicle coordinate system and the bottom of

Fig 2 shows the resultant panned plan-view image. Its dimensions are 64 by 32 pixels, reducing the raw image data of 256 x 256 pixels by a factor of 32. The lateral dimension of the image is greater than the longitudinal direction giving a higher resolution laterally in order to make the extraction of the road edges more accurate.The pan angle θ performs an important part in ensuring that a useful panned plan-view image is generated with the road edges predominantly vertical. The choice of the pan angle is done automatically by the reasoning module after it has built the road model. The pan angle is determined by measuring the angle between the central axis of the road and the vehicle's direction of the motion.

Noise in subsampling

The reduced panned plan-image is generated by subsampling the individual pixels specified by the transformation without any previous smoothing so the resultant image is slightly noisy. To study the effect of this noise on the edge detector operator applied later, two methods of smoothing the original image are compared to the unfiltered method described.

The results of the edge detector operator applied to a non-smoothed, a gaussian smoothed and a pixel averaging smoothed images were compared visually.Although the gaussian smoothing has shown the best noise figure of the three methods, the unfiltered image is used to achieve real-time processing since the degradation effect of its edge detector output is comparatively small.

4 Road Edge Detection

The road edge extraction algorithm consists of two modules: feature extraction and geometric modules. The Feature Extraction module is responsible for detection of the candidate road features in the panned plan-view image. The output of the feature analysis is sent to the geometric module which transforms those features from the plan-view image pixel coordinates into the vehicle coordinate system, converting them into a sequence of straight line segments.

Feature Extraction

The process of extracting road edges from the plan-view road image involves several steps: applying a vertical convolution mask, thresholding and thinning the edges and finally linking. All the above algorithms assume that the road in the plan-view image is a large predominant feature with boundaries edges nearly vertical. This assumption makes it possible to improve the algorithms to extract the road edges.

A vertical gradient edge detector followed by a non-maxima suppression thinning technique is used to detect the predominantly vertical edges.

Linking edges can be a complex task when the image contains edges in all directions. In this case the task of linking segments is easier as the edges are near vertical and are well separated. The search process looks for its successor cell amongst the five neighbouring cells in the row below. If more than one is found, the nearest edge is chosen. The output of the linking algorithm is a list of edge segments in the panned plan-view image pixel coordinates.

Geometric module

The Geometric module receives as input a list of edge segments from the feature extraction module. The main steps performed by the geometric module are: translation from the panned plan-view image pixel coordinate system to the vehicle coordinate systems, linking possible gaps between segments, and application of a line fitting algorithm to those segments. The output of this module consists of a list of linked vector candidates on the right and the left road boundaries.

The line fitting uses the iterative endpoint fit algorithm. A list of curves approximated by straight lines is the output of the geometric module. The curves are candidates for right and left road boundaries.

5   Spatial and Temporal Reasoning

The reasoning module classifies each segment candidate for left and right road boundaries using spatial and temporal constraints. Normally, for an image with a well defined road with a discriminant background, the output of the geometric module consists of two road edges representing the road and the classification is trivial. In conditions where the feature extraction cannot cope with noise or imperfections on the road surface or when only one road edge is visible (a T junction or a sharp curve), or where the background is complex, the list of segments extracted may be fragmented or may not correspond directly to the left and right road edges.

The reasoning module uses both spatial and temporal constraints. The spatial constraints use the assumption that the road edges should be parallel segments separated by a predicted road width. The temporal constraints use a zeroth order prediction to match the best left and right segments against the previous road description. The zeroth order prediction is used since the vehicle used to test the algorithms has no sensors to feedback the actual movement displacements commanded. As the vehicle's maximum speed is about walking pace and the cycle processing time is just over two seconds, the road edges are within a reasonable distance to enable a correct matching correspondence. In the case where the vehicle moves at a faster speed, sensors would give important clues in predicting the road position. In this case it would be possible to take into account the vehicle movements and use a first order prediction.

There are two basic measures used in the reasoning module from which all other criteria are derived: The degree of overlap and the segment matching measures. The degree of overlap gives the amount of support the two segments exhibit in the horizontal direction. The segment matching measure gives the average distance and variation between two overlapping segments. A good road description consists of two segments with a large overlap, an average distance representing the road width and a small variation in the width reflecting the parallelism of road edges. A good correspondence matching requires a large overlap, a minimum width and a minimum width variation. The implementation details of these two measures are described later.

The algorithm uses a road reference data structure consisting of the right and the left road edge segments and the road width. The edge segments are linked lists of straight lines. The right or the left segment may be null if there is not enough evidence of its road segment pair.

6   Predicting the Pan Angle for The Panned Plan-View Transformation

The panned plan-view transformation requires the pan angle parameter to determine the correct rectangular grid. Properly selecting the grid ensures the road edges are strong vertical features in the plan-view image.

The pan angle used for the transformation is the angle that the axis of the movement of the vehicle makes with the central axis of the road at a fixed distance (D_n) ahead of the vehicle. The predicted pan angle is taken as weighted average of the inclination of both road segments from the new road reference extracted. The average is weighted by the length of each road segment. In the case of initialization or where no road edges are extracted, the current pan angle is set to zero, i.e. looking straight ahead of the vehicle.

7   Implementation and Tests

The algorithms have been implemented in Parallel C on a transputer based image processing system. The system has been developed and built at the University of Bristol as part of its Autonomously Guided Road Vehicle project. The basic system consists of a monochrome video A/D converter capturing 256 x 256 pixels of 8 bits in a non-interlaced standard TV 50 Hz format. The frame grabber is double-buffered and is located in the transputer address space. The frame can also be displayed, assisting the development of real-time image processing algorithms.

A small tracked battery operated vehicle is used for testing. The road used for testing is a gravel path in the University gardens. The road edges are sometimes ill defined. Grass and other vegetation defines the sides of the path. Simple algorithms based on pixel classification or simple edge detections were unable to properly segment the path. There are also circumstances where only one edge of the road is visible which constitutes a severe test for any road following algorithm.

8 Conclusions

We have presented a real-time vision system for vehicle road following using a panned plan-view transformation together with temporal and spatial reasoning. The panned plan-view transformation has proved to be a significant feature for improving both speed and robustness in the road feature extraction tasks. The road is seen as a large predominant feature in typical scenes. The panned plan-view transformation successfully compresses the raw image while preserving the primary road characteristics. The transformation also converts the road edges into predominant near-vertical components of the image. Therefore a more specialised and simple algorithm can be used for feature extraction. The temporal and reasoning system implemented has demonstrated another major improvement compared with previous road following algorithms. It can deal with missing road edges in situations where they can be masked by shadows, puddles, road marks or in situations where only one side of the road is visible.

Overall, the algorithm has shown some very robust behaviour due mainly to the two new components of the method. Although the algorithm has very powerful characteristics, its implementation does not require dedicated image processing hardware and has been implemented in a single transputer. This provides opportunity to explore parallel processing technology with the addition of more complex interpretation tasks such as obstacle avoidance, road junctions and path planning.

9 Acknowledgments

This work has been funded by RARDE (Chertsey) under MoD contract ER1/9/4/2034/085. R. A. Lotufo is supported in part by the Conselho Nacional de Desenvolvimento Cientifico e Tecnologico - CNPq, Brazil, under grant n.200172/86-EE, and in part by the Universidade Estadual de Campinas - Unicamp, Brazil.

10 References

1. SHARMA, U.K. and KUAN, D.: 'Real-Time Model Based Geometric Reasoning For Vision-Guided Navigation'. Machine Vision and Applications, n.2, pp.31-44, 1989.

2. TURK, M.A. et al.: 'VITS – A Vision System for Autonomous Land Vehicle Navigation', IEEE Trans. on Pattern Analysis and Machine Intelligence, vol. 10, n. 3, 1988.

3. WALLACE, R., et al.: 'Progress in Robot Road Following'. Proc. 1986 IEEE International Conf. Robotics & Automation, 1986.

4. WAXMAN, A.M. et al.: 'A Visual Navigation System for Autonomous Land Vehicles'. IEEE Journal of Robotics and Automation, vol RA-3, n.2,1987.

5. DICKMANNS, E. D. and GRAEFE, V.: 'Applications of Dynamic Monocular Machine Vision'. Machine Vision and Applications, n.1,pp.241-261,1988.

6. MORGAN, A.D., et al.: 'Road Edge Tracking for Robot Road Following'. Proc. of the Fourth Alvey Vision Conference, pp. 179-184, Manchester 1988.

7. LOTUFO, R.A., et al.: 'Road Edge Extraction Using a Plan-View Image Transformation'. Proc. of the Fourth Alvey Vision Conference, pp. 185-190, Manchester 1988.

3D-vision-based robot navigation: first steps

Luc Robert, Régis Vaillant and Michel Schmitt
INRIA
Sophia-Antipolis 2004 Route des Lucioles
06565 Valbonne Cedex France

Abstract

This article shows a way of using a stereo vision system as a logical sensor to perform mobile robot navigation tasks such as obstacle avoidance. We describe our system, from which the implementation of a task described by an automaton can be done very easily. Then we show an example of a navigation task.

1 Introduction

Many approaches have been proposed to the problem of mobile robot navigation. Some aim at following trajectories by servoing on feature that might be detected and tracked by sensors (the edges of a road [7], predefined beacons [1]). Our main goal is to obtain a passive-vision-based system that can *today* navigate in an unknown environment, avoid obstacles and collect some information without being constantly looked after. To achieve this, we use a trinocular stereovision system, which is a powerful and reliable vision sensor device. This system avoids developing sophisticated algorithms to detect obstacles in 2D images.

2 Hardware and software tools

2.1 Description

Our mobile robot is a *robuter* (fig 1), with two driving wheels and a front turret that supports the set of cameras and allows rotation around a vertical axis.

The primitives we use to describe a robot move are directly correlated to the amount of rotation of each driving wheel to achieve this move. Actually, the low-level implementation invokes a P.I.D. supplied by an odometric sensor on each driving wheel.

Computation is processed on the *Capitan* parallel machine and its host machine, a *Sun* workstation. The communication between these computers and the robot is established through a video and radio links.

2.2 The vision sensor

The processing of a triplet of stereoscopic images goes through the following steps: computation of the intensity gradient using recursive filtering [4], non-maxima suppression, thresholding by hysteresis, edge linking [5], polygonal approximation [2], trinocular stereovision [6].

This sensor device provides us with 3D segments in a frame associated to the system of cameras. Its implementation on *Capitan* has been performed by Régis Vaillant [8], and today it takes about 15s to process a triplet of 256×256 images.

3 Navigation

Before moving the robot, we check whether an object lies on the planned trajectory by sensing in the direction of this trajectory and analysing the set of reconstructed 3D edges.

3.1 Controlling moves

Knowledge of the robuter's geometry and of the orders for executing a robot move provides an estimate of the displacement involved within this operation. However, due to hardware reasons, this estimate is not very accurate in comparison to the degree of accuracy of the odometry on each wheel. Indeed, the final position of the robot highly relies on the synchronization of the two driving wheels.

Since successive executions of the same order lead to a relatively invariant displacement, we decided to allow a finite set of moves, and for each move to rely on a value of the displacement which has been previously determined during a calibration phase.

To each possible move we attach one orientation of the turret, so that the camera system faces the motion space, i.e. the portion of space to be occupied by the robot while performing this move (cf fig 1). To check whether the motion space is free, we apply the vision process and build a local map of the 3D space in the view direction. In practice, the view angle is not wide enough to encompass the entire motion space associated to one move, and the way we circumvent this problem is described in paragraph 3.3. Three cases may occur:

- There are 3D edges in the motion space or close to it.

- There is no such edge, and there are 3D edges further in the view direction.

- Very few edges have been reconstructed.

Only in the second case is the move authorized. The first case corresponds to the presence of an object in the target position. The third case usually occurs when the robot is so close to an object that there is no edge in the view angle or that disparity is too high for the edges of this object to be matched.

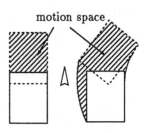

motion space

Figure 1: The mobile robot and its motion space

3.2 Behaviour automation

The behaviour of the robot is described by an automaton, each state of which corresponds to one move, and involves a complete cycle:

- determine if the move can be performed
  - move the turret in the adequate direction
  - activate the vision sensor
  - find out whether the motion space is free
- move the robot if possible
- go into the following automaton state, according to the fact that the move was or was not authorized.

The advantages of modelizing behaviours with finite state automata are described by Brooks [3]. Using this structure, we implemented the following task:

move straight as long as possible, otherwise turn right

Because of the geometry of the robuter, it is sometimes impossible to make the robot turn right while avoiding obstacles, for instance when the left side of the robuter is very close to a wall. Then, moving away from the wall requires maneuvers. The automaton which deals with the whole task is described in fig 2. For each possible move, the orientation of the turret to perform sensing is represented on the right side.

3.3 Building a 3D map

Since we know the displacement generated by each robot or turret move, we can project several 3D local maps in a global frame to get a larger 3D map of the environment.

We used those 3D maps to solve the view angle problem raised in paragraph 3.1. A solution to that problem would have been to turn the turret and sense several times, then merge the 3D results *before each robot move*. To enlarge the view angle while avoiding spending too much time in the checking step, we decided to allow only one sense per move (in a direction that allows to see an important portion of the motion space) and to merge the results provided by the last previous sensing processes.

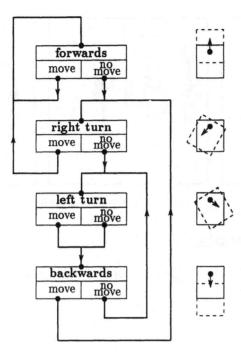

Figure 2: The behaviour automaton

4   Results

Moving every 20s, the robot is able to travel in a room for a few hours while avoiding obstacles. In figure 3, we show one local 3D map, and two different views of a merge of several local 3D maps.

Our system still has two main drawbacks:

- The turret can only move around a vertical axis, so there may always be objects out of the view area, for instance objects of small height located close to the robot.

- Our system is not fast enough to deal with rapid moving objects. Anyway, the use of a real-time hardware architecture for vision will soon help solve that problem.

5   Conclusion

The system we developed allows to implement behaviour automata very easily, and to run them directly on a mobile robotics system. An example of automaton was given that modelizes a simple navigation task. Its execution in different environments led to long and exciting excursions, which tended to prove that much more complex navigation tasks can be implemented within the same structure.

Future work will involve the insertion of a fusion algorithm to build more precise 3D maps and a better model of the environment, then deal with recognition of primitives such as corners, walls, or doors.

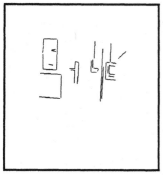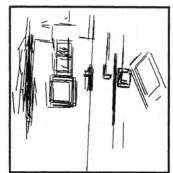

Figure 3: Several views of a corner

Acknowledgements

We would like to thank Pr. Olivier Faugeras for his helpful comments, and Hervé Mathieu for taking care of our robot.

References

[1] J. Amat and A. Casal. Environment Recognition for Automatic Guidance of Agricultural Machines. In *First Workshop on Multi-sensor Fusion and Environment Modelling*, October 1989.

[2] M. Berthod. Approximation polygonale de chaînes de contours. Programmes C, 1986. INRIA.

[3] R.A. Brooks and A.N. Flynn. Robot Beings. In *IEEE/RSJ International Workshop on Intelligent Robots and Systems*, pages 2–10, 1989.

[4] R. Deriche. Using Canny's Criteria to Derive an Optimal Edge Detector Recursively Implemented. In *The International Journal of Computer Vision*, pages 15–20, April 1987.

[5] G. Giraudon. *Chaînage efficace de contour*. Rapport de Recherche 605, INRIA, February 1987.

[6] F. Lustman. *Vision Stéréoscopique Et Perception Du Mouvement En Vision Artificielle*. PhD thesis, Université de Paris-Sud Centre d'Orsay, December 1988.

[7] C.N. Thorpe, S. Shafer, and T. Kanade. Vision and Navigation for the Carnegie Mellon Navlab. *IEEE Transactions on PAMI*, 362–373, 1988.

[8] Régis Vaillant, Rachid Deriche, and Olivier Faugeras. 3D Vision on the Parallel Machine CAPITAN. In *International Workshop on Industrial Application of Machine Intelligence and Vision*, April 1989.

Dynamic World Modeling Using Vertical Line Stereo

James L. Crowley, Philippe Bobet

LIFIA (IMAG), Grenoble, France

Karen Sarachik

MIT - AI Lab, Cambridge, USA

Abstract

This paper describes a real time 3-D vision system which uses stereo matching of vertical edge segments. The system is designed to permit a mobile robot to avoid obstacles and to position itself within an indoor environment. The system uses real time edge tracking to lock onto stereo matches. Stereo matching is performed using a global version of dynamic programming for matching stereo segments.

1. Introduction

Indoor man-made environments contain many vertical contours. Such contours correspond to environmental structures which a mobile robot may perceive to position itself and to navigate. This system was inspired by the system of Kriegman, Treindl and Binford [Kriegman et. al. 1989]. The stereo system is organized as a pipeline of relatively simple modules, as illustrated in figure 1.1.

2. Detecting and Linking Vertical Edges

The first module in the system is concerned with detecting vertical edges. A cascade of simple filters is used to first smooth the image and then approximate a first vertical derivative. Filters in this cascade are composed of binomial kernel for smoothing, and a first difference kernel for calculating the derivative.

2.1 The Filter for First Vertical Derivatives

Our first derivative filter is composed from a cascade of k convolutions of a circularly symmetric binomial filter, m convolutions of a vertical low pass filter followed by a convolution with a first difference filter.

$$m_{km}(i, j) = \begin{bmatrix} 1 & 2 & 1 \\ 2 & 4 & 2 \\ 1 & 2 & 1 \end{bmatrix}^{*k} * \begin{bmatrix} 1 \\ 2 \\ 1 \end{bmatrix}^{*m} * \begin{bmatrix} 1 & 0 & -1 \end{bmatrix}$$

The results presented below are based on the empirically obtained values of filter of $k = 4$ and $m = 2$. This filter will give a negative response for transitions from dark to light and a positive response for transitions from light to dark. In order to detect points which belong to vertical edges, each row of the filtered image is scanned for extrema. An extrema, or edge point is any pixel $e(i, j)$ that is a local maximum and has more than twice the absolute value of neighbors 3 pixels away.

2.2 Raster Scan Edge Chaining

The second module in the system is responsible for edge chaining and straight line approximations. Raster scan based chaining algorithms are well suited to real time implementation. Raster scan edge chaining is greatly complicated by the presence of edges near the scan direction [Discours 89]. By restricted edges to directions perpendicular to the scan direction the process becomes quite simple. Edge chains are converted to edge segments by the well known "recursive line splitting" process [Duda-Hart 73]. This algorithm is known to exhibit instabilities when representing curved edges. In experiments with curved objects, these instabilities have not proved a problem for subsequent stages in our system.

This work was sponsored by Société ITMI and Project EUREKA EU 110 : MITHRA.

242

2.4 The MDL Edge Segment Representation

As edge segments are detected they are transformed to a parametric representation composed of the mid-point, direction and length. We refer to this representation as MDL. This representation is designed to facilitate the matching step in the segment tracking phase, and to permit Kalman filter tracking of the center point as two independent parameters. The MDL parametric representation for segments is described in [Crowley-Stelmaszyk 90] in this conference. A segment is represented by a vector $S = \{c, d, \theta, h\}$ of parameters:

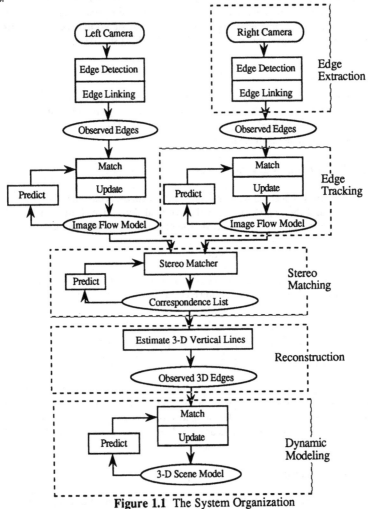

Figure 1.1 The System Organization

3. Measuring Image Flow by Tracking Edge Segments

Tracking allows us to preserve the correspondence between an observed edge and information in the 3-D scene model. This tracking process is well debugged and has been used in a number of projects. Real time hardware has recently been constructed using this algorithm [Chehikian 88]. This process has been described at the second I.C.C.V. [Crowley et. al. 88]. Performing stereo correspondence on the flow model provides cleaner data for stereo matching. In particular, this technique also permits the system to function in the presence of simple occlusions.

Correspondence of edges is maintained by a very simple tracking process based on a Kalman filter. The tracking process maintains a list of "active" edge segments composed of the parameter vector, $S = \{c, d, \theta, h\}$. The flow model also contains a confidence factor, CF, represented by a state from the set $\{1, 2, 3, 4, 5\}$ and a unique identity, ID. The identity of a segment permits the process to preserve the association between a segment, its corresponding segment in the other image, and the resulting 3-D segment.

4. Correspondence Matching using Dynamic Programming

Stereo matching is performed by a single pass of a dynamic programming algorithm over the entire image. The process "locks on" to correspondences, by feeding the previously discovered disparity for each segment pair into the matching process.

Line segments from the flow models from the left and right images are combined with predictions from the previous match to produce a new stereo correspondence list. The contents of the stereo correspondence list gives a list of vectors each containing 4 values: (Left ID, Right ID, Disparity, CF). The Left ID and Right ID are the ID's of the matching segments. The Disparity is the most recently observed horizontal disparity in pixels. The CF is the number of times which this correspondence has been found in the last 5 images. IF CF goes to zero, then the correspondence is removed from the list.

4.1 Matching by Dynamic Programming

Stereo matching is performed by a dynamic programming algorithm [Kanade-Ohta 85]. The dynamic programming algorithm calculates the best global match provided that the order of the segments is the same in the two images. The algorithm works by propagating matching costs in a grid. After propagating the cost from the bottom to the top of the grid, the least cost past is traced back to the far corner of the grid. The cells in this path provide the most likely globally consistent set of correspondences of segments from the left and right flow models.

The edge lines in each flow model are ordered from left to right based on column of the midpoint. When more than one segment has the same midpoint, the segments are ordered from top to bottom using the row of the midpoint. Thus, the columns of the DP array grid correspond to the segments from the left flow model, and the rows correspond to the segment from the right flow model. Paths through the DP grid correspond to possible matches of segments. A "lawn-mower" style algorithm is then used to propagate the cumulative cost for every possible path, based on a cost function for each pair of matches.

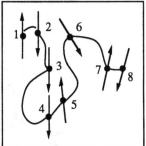

Figure 4.1 The segments in the flow models are sorted by row and column.

4.2 Cost Functions

The individual cost of matching a segment from the left image to a segment in the right image is based on the Mahalanobis distance between the attributes of the segments. Let us refer to S_n as the nth segment from the left flow model and S_m as the mth segment from the right flow model. The individual cost for matching S_n to S_m is given by:

$$C(n, m) = C_o(n, m) + C_\theta(n, m) + C_d(n, m)$$

where:

$$C_o(n, m) = ABS(d_n - d_m) / (h_n + h_m) \qquad C_d(n, m) = (c_n - c_m - D_o) / \sigma_o$$
$$C_\theta(n, m) = (\theta_n - \theta_m) / 10°$$

The term $C_O(n, m)$ is based on the overlap of the segments, as computed from the parameters d and h. If the segments do not overlap, then the normalized difference is greater than 1 and the match is rejected by setting the cost to infinity.

The term $C_\theta(n,m)$ is based on the similarity of orientation of the segments, as computed from the parameter θ. An experiment was performed in which it was observed that the largest difference in angles occurred when a 3-D line segment is tilted away from the cameras by 45°. In this case the observed difference in angle is 10°. Thus the cost for the difference in image plane orientation between the left and right images is given by the difference normalized by 10°.

The term C_d (n, m) is the difference between an expected disparity and the observed disparity. A nominal disparity is initially determined from a fixation distance. This distance is a parameter which can be dynamically controlled during matching. Whenever a stereo match exist from the previous frame has been determined with a CF > 1, the nominal disparity is reset to this previous disparity. In this way the process is biased to prefer existing matches. The cost is determined by dividing the difference from the fixation disparity by an uncertainty, σ_o , which is also a parameter which can be controlled in the process. The cost of skipping a match is equal to the cost from a 1 standard deviation difference on all three measures, that is C_{skip} = 3.

The system has been regularly operated using live images during debugging over the last few months. In a typical experiment, a sequence of 5 to 10 pairs of stereo images is a made as the mobile robot moves in a straight line. Matching statistics have been improved from early results of around 80% correct to nearly 100% correct by improving the stability of edge segments that are detected. This additional stability was achieved by increasing the degree of smoothing from m=2 to m=4, as well to enlarging the tolerance for recursive line fitting to 2 pixels.

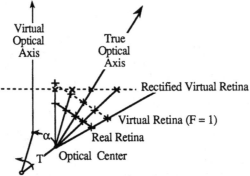

Figure 5.1 Cameras can be mounted with a vergence angle α. Rectification corrects for this angle by projecting back to virtual co-planar retinas.

5 Recovery of 3-D Position from Stereo Information

The 3-D inference process begins by rectification of the position of segments which are found to correspond. Segments are then limited to their overlapping parts, followed by a the calculation of the depth and uncertainty for the end points in a camera centered coordinate system. Segment parameters are then computed from the end-points, and the representation is transformed to scene centered coordinates.

5.1 Projection to a Virtual Retina at F=1

When the retina's of stereo cameras are co-planer, the depth equations reduce to a trivial form. However, it is impossible to mechanically mount cameras such that their retinas are sufficiently close to parallel for this simplification to apply. This problem is avoided by calibrating a transformation which projects points in stereo images to a pair of virtual co-planar retina's as illustrated in figure 5.1. We call these the "virtual rectified retinas. We use a technique inspired by Ayache [Ayache 88] to perform this transformation.

5.2 Depth from Co-planar Stereo Cameras

For coplanar retinas, the depth equation falls directly from a difference in similar triangles. For cameras separated by a base line distance of B, the depth, D, is given by the disparity by $D = F B / \Delta x$. If we observe that our edge lines are rectified to a virtual retina where $F = 1$, then the depth, D, is given by

$$D = B / \Delta x.$$

The linear term B is determined as a final step in calibration by calculating the distance between the optical centers of the left and right cameras as provided by calibration.

Projection to 3-D is performed using the segment end-points. Because segment length is not always reliable, we must first determine the overlapping part of the corresponding segments. For each of the end points in the corresponding segments, we compute the depth, $D = B / \Delta x$. We then compute the corresponding point in the scene, in the coordinate system of the stereo pair of camera as:

$$x_c = D \, x_{rr} \, (d_i/F) + B/2 \qquad y_c = D \, y_{rr} \, (d_j/F) \qquad z_c = D$$

The uncertainty of each recovered point is modeled as having two independent components: an uncertainty in x, σ_x, and an uncertainty in D, σ_D.

6 Sample Results

This system is the subject of ongoing tests and refinements at our laboratory. A sample of the results form the system is presented in the following figures. This is the second pair from a set of 5 pairs taken at 10 cm displacements. Matches were 100% correct in 4 of the 5 pairs in this sequence, with one incorrect match in the first stereo pair. Figure 6.1 shows the raw images and the vertical edge lines which are detected. Positive edges are shown in white, negative in black. Figure 6.2 shows the correspondance between the segments in the left and right flow models. Figure 6.3 shows an overhead view of the 3D segments which were projected into scene coordinates.

Acknowledgements

The architecture for this system was developed during discussion with Patrick Stelmaszyk and Haon Hien Pham of ITMI, and Alain Chehikian of LTRIF, INPG. The edge detection and chaining procedures are based on code written by Per Kloor of Univ of Linkoping during a post-doctoral visit at LIFIA. The first version of the stereo matching was constructed by Stephane Mely and Michel Kurek. This system is the result of a continuous team effort, and the authors thank all involved for their mutual support and encouragement.

Bibliography

[Ayache 88] Ayache, N. "Construction et Fusion de Représentations Visuelles 3D", Thèse de Doctorat d'Etat, Universtié Paris-Sud, centre d'Orsay, 1988.

[Chehikian et. al. 88] Stelmaszyk, P., C. Discours and , A. Chehikian, "A Fast and Reliable Token Tracker", In IAPR Workshop on Computer Vision, Tokyo, Japan, October, 1988.

[Crowley et. al. 88] Crowley, J. L., P. Stelmaszyk and C. Discours, "Measuring Image Flow by Tracking Edge-Lines", ICCV 88: 2nd International Conference on Computer Vision, Tarpon Springs, Fla., Dec. 1988.

[Crowley-Stelmaszyk 90] Crowley, J. L. and P. Stelmaszyk, "Measurement and Integration of 3-D Structures By Tracking Edge Lines", First E. C. C. V. (these proceedings), Antibes, April, 1990.

[Discours 89] Discours, C., "Analyse du Mouvement par Mise en Correspondance d'Indices Visuels", thèse de doctorat de nouveau régime, INPG, Novembre, 1989.

[Duda-Hart 73] Duda, R. O. and P. E. Hart, "Pattern Classification and Scene Analysis", Wiley, 1973.

[Ohta-Kanade 85] Ohta, Y. and T. Kanade, "Stereo by Intra and Inter Scanline Search using Dynamic Programming, IEEE Trans. on PAMI, 7:139-154, 1985.

[Kriegman et. al. 89] Kriegman, D. J., E. Triendl, and T. O. Binford, "Stereo Vision and Navigation in Buildings for Mobile Robots", IEEE Transactions on Robotics and Automation, Vol 5(6), Dec. 1989.

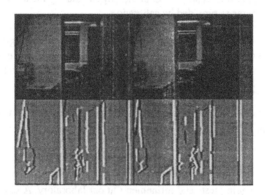

Figure 6.1 An example of the vertical edge segments detected in a laboratory scene.

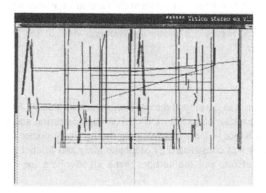

Figure 6.2 Stereo Correspondence for segments extracted from the stereo images.

Figure 6.3 Overhead view of the vertical segments reconstructed from the correspondences in figure 6.2.

TRACKING

TRACKING IN A COMPLEX VISUAL ENVIRONMENT

John (Yiannis) Aloimonos[1]   Dimitris P. Tsakiris[2]

(1) Computer Vision Laboratory, Center for Automation Research and
Institute for Advanced Computer Studies and Computer Science Department
(2) Electrical Engineering Department and Systems Research Center
University of Maryland, College Park, MD 20742-3411

Abstract

We examine the tracking of 3-dimensional targets moving in a complex (e.g. highly textured) visual environment, which makes the application of methods relying on static segmentation and feature correspondence very problematic, even under specific assumptions about the trajectory of the target. From the system kinematics and optical flow formalism we derive a general criterion, the *Tracking Constraint*, which specifies the optimal camera reorientation as the one that minimizes the optical flow in the center of the camera in the sense of an appropriate metric. A correspondence-free scheme is devised which employs dynamic segmentation and linear features of the image in order to capture global information about the scene and bypass numerical differentiation of the image intensity.

1. INTRODUCTION

Visual tracking is an important problem in Computer Vision, not only because of its applications in areas such as aircraft and missile tracking, robot manipulation of objects, navigation, traffic monitoring, cloud tracking in meteorology, cell motion and tracking of moving parts of the body (e.g. the heart) in biomedicine, but also because it increases the robustness of Early Vision process solutions [1], allows more efficient use of our sensory and computational resources by focusing them on an area of interest and avoids blurring from the motion of the target by stabilizing it in the center of the camera.

The Vision Module of a tracking system consists of the hardware and software necessary for the processing of the visual information. It gets the analog image, digitizes it, creates an array of intensities for all points of the field of view and from this information locates the target and computes the camera positioning that will achieve tracking. Locating the target is the part of the tracking process referred to as the *Target Acquisition* phase and corresponds in the human oculomotor system to the fixation of a target using saccadic eye movements. The computation and execution of a smooth and continuous camera motion that will keep the target foveated is called the *Tracking* phase and corresponds in the human visual system to the smooth pursuit eye movements. During this phase, the target tends to drift slowly away from the center of the image and, after some time, it may be lost. Then a new acquisition phase is required.

In the following we will assume that during the acquisition phase the target was brought to the center of the image plane and we will consider the problem of keeping the target foveated. We are mainly interested in the most general instance of the problem, involving a 3-D target moving in 3-dimensional space. Previous approaches [5], [6], [7], [8], [11], [13], [14], [16], [19], [20] to the problem are composed of three main steps: First, they perform segmentation of the image or of appropriately selected parts of it, in order to locate the target and extract the new position of some characteristic points of its image (centroid, feature points, etc.). Second, they match these new characteristic points with their corresponding ones in previous or current (in the binocular case) frames and thus extract information about the displacement of the target. Third, from this displacement they compute the necessary camera rotation that will achieve tracking.

In our approach we consider a general criterion (one not restricted to a specific target or visual environment) for 3-D tracking, the *Tracking Constraint*, which, if satisfied, guarantees tracking in the sense of keeping an initially foveated target stationary in the center of the camera. This formulates tracking as an Optimization problem and allows the use of powerful tools from Optimization Theory for the solution of the problem. We follow the continuous motion approach and use the concept of optical flow to formulate this constraint. The Tracking Constraint uses global information from entire areas of the image and does not require feature extraction or matching. Therefore it is expected to be more robust in the presence of noise than previous methods. To solve the problem in a correspondence-free manner (without computing the optical flow field), we assume that the shape of the target is known, so that we can estimate the kinematic parameters of its motion. Obviously, a requirement for knowledge of the shape in a general tracking scheme is by far less restrictive than an ad-hoc tracking scheme for a specific target or class of targets, since the latter may be used only for a specific target, while the former may be used for *any* tracking problem where the shape is known exactly or approximately.

Finally, and most importantly, the theory described here considers tracking in 3-D and suggests several research avenues on how tracking interacts with modules such as shape from "x", structure from motion, etc.; such interactions are necessary in our effort to integrate various vision modules [2], [12].

2. COMPUTATIONAL THEORY OF TRACKING

In this section we devise a model of the image formation process and the motion of the target, based on the system kinematics and the optical flow formalism. (For details see [3], [18].)

Optical Flow is the velocity field generated on the image plane from the projection of objects moving in 3-D, from the motion of the observer with respect to the scene or from apparent motion, when a series of images gives the illusion of motion. The image intensity $s(x, y, t)$ at a point (x, y) of the image plane at time t is related to the instantaneous velocity $(u(x, y), v(x, y))$ at that point by the Optical Flow Constraint equation ([9],[17]):

$$s_x(x, y, t)u(x, y) + s_y(x, y, t)v(x, y) + s_t(x, y, t) = 0, \tag{1}$$

Let $\vec{T} = (T_1, T_2, T_3)$ and $\vec{\omega} = (\omega_1, \omega_2, \omega_3)$ be respectively the translational and angular velocities of the target at time t, relative to the camera coordinate system (Figure 1). The pinhole camera model and the system kinematics can be used to show that, under perspective projection, the optical flow field (u, v) induced at the point (x, y) of the image plane by the motion of the target, is:

$$u(x, y) = \frac{fT_1 - xT_3}{Z} - \omega_1 \frac{xy}{f} + \omega_2 \frac{(x^2 + f^2)}{f} - \omega_3 y \tag{2}$$

$$v(x, y) = \frac{fT_2 - yT_3}{Z} - \omega_1 \frac{(y^2 + f^2)}{f} + \omega_2 \frac{xy}{f} + \omega_3 x. \tag{3}$$

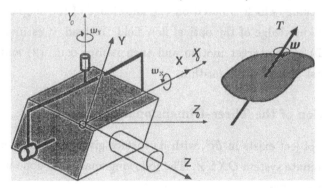

Figure 1.

Since the optical flow expresses the relative motion of the target and the camera, the problem of stabilizing the target at the origin of the image plane by a rotation (ω_x, ω_y) of the camera, can be expressed equivalently as making the optical flow field zero near the origin. When both the camera and the target are moving, the corresponding optical flow field will be the superposition of two components: One induced by the target (u^t, v^t) and one induced by the camera motion (u^{cam}, v^{cam}). By observing that a rotation of the camera by (ω_x, ω_y) has the same effect on the image formation process as a rotation of the target by $(-\omega_x, -\omega_y)$, both components of the optical flow can be related to the corresponding kinematic parameters by the equ. (2) and (3).

The *tracking problem* can then be set as the specification of the camera rotation (ω_x, ω_y) that will minimize the optical flow field, in the sense of an appropriate metric $(\mathcal{L}_2)$, in a neighborhood $\mathcal{B}$ of the origin of the image plane :

$$\min_{\omega_x, \omega_y} \iint_{\mathcal{B}} (u^2(x, y) + v^2(x, y)) dx\, dy, \tag{4}$$

where $u = u^t + u^{cam}$ and $v = v^t + v^{cam}$. This is called the **Tracking Constraint**. The neighborhood $\mathcal{B}$ is specified by the dynamic segmentation process.

An **algorithm** for tracking the motion of an object moving in 3-D then suggests itself: *Step 1 :* Estimate (u^t, v^t). *Step 2 :* Solve the Tracking Constraint (eq. (4)) and derive the desired camera angular velocities that will achieve tracking.

Typically Step 1 involves the Optical Flow Constraint. If we assume that the optical flow (u^t, v^t) can be computed from the image, then tracking is achieved through minimization of (4) or, if (4) is combined with (1), tracking is achieved through solution of a penalized least-squares problem. The reader interested in this approach is referred to [18]. However, due to the fact that the computation of the optical flow field still remains a very hard problem especially for the case of complex visual environments involving discontinuities, we chose to seek an alternative approach. The next section describes our algorithm for tracking, which is based on the knowledge of the shape of the target, but does not assume knowledge of the optical flow field. Instead, it estimates the kinematic parameters $(\vec{T}, \vec{\omega})$ of the target motion and then it uses equ. (2) and (3) to compute (u^t, v^t) for the first step of the algorithm.

2.1. Estimation of the Three-Dimensional Motion

Suppose that an object exists in $I\!R^3$, with its surface given as a function $Z = g(X, Y)$ in the camera coordinate system $OXYZ$. The following analysis is done with respect to the camera coordinate frame.

Theorem:

The optical flow (u, v), generated from the motion by $(\vec{T}, \vec{\omega})$ of a rigid target relative to our camera, is a linear function of the target motion parameters, of the form:

$$\begin{pmatrix} u \\ v \end{pmatrix} = \mathcal{U} \begin{pmatrix} \vec{T} \\ \vec{\omega} \end{pmatrix} = \frac{\partial P}{\partial \vec{X}} (P_g^{-1}(x, y)) \vec{V}(P_g^{-1}(x, y)) \begin{pmatrix} \vec{T} \\ \vec{\omega} \end{pmatrix}. \tag{5}$$

where the matrix $\mathcal{U}$ contains known functions of the shape, the projection function and the retinal coordinates and where $\frac{\partial P}{\partial \vec{X}}(X, Y, Z)$ is the Jacobian matrix of the projection function $P(X, Y, Z)$, $P_g^{-1}(x, y)$ is the inverse projection mapping from the image to the target surface, which has shape $Z = g(X, Y)$, and $\vec{V}(X, Y, Z) = \begin{pmatrix} 1 & 0 & 0 & 0 & Z & -Y \\ 0 & 1 & 0 & -Z & 0 & X \\ 0 & 0 & 1 & Y & -X & 0 \end{pmatrix}$. For a proof of the theorem, a discussion of its consequences, and examples see [3].

In the sequel, the motion parameters $(\vec{T}, \vec{\omega})$ will be denoted by $m_i, i = 1, \ldots, 6$.

The problem now is to compute the motion parameters $m_i, i = 1, \ldots, 6$ of (5), without actually computing the optical flow or the spatial derivatives of the image intensity function. For this purpose, we use the concept of *linear features*, which are sets of functionals over an area S of the image plane $F(t) = \{f_i(t) = \iint_S s(x, y, t))\mu_i(x, y) dx\, dy\, , i = 1, \ldots, n\}$, where $\mu_i(x, y)$ is called the measuring function of the ith linear feature. Consider the temporal derivative of the ith linear feature:

$$\dot{f}_i = \iint_S \frac{\partial s}{\partial t} \mu_i dx\, dy \stackrel{(1)and(5)}{=} -\sum_{k=1}^{6} m_k \iint_S \mu_i (u_{k_1} s_x + u_{k_2} s_y) dx\, dy\, ,$$

where u_{ki} are the elements of the matrix $\mathcal{U}$ in (eq. 5). Defining $h_{ik} = -\iint_S \mu_i(u_{k_1} s_x + u_{k_2} s_y) dx\, dy$, we get $\dot{f}_i = \sum_{k=1}^{6} m_k h_{ik}$ and then

$$\mathbf{H}\vec{m} = \dot{\vec{f}}, \tag{6}$$

where $\mathbf{H}$ is the $n \times 6$ matrix of the coefficients h_{ik} and $\vec{m} = (m_1, \ldots, m_6)$. The vector of derivatives of linear features $\dot{\vec{f}}$ and the matrix $\mathbf{H}$ depend on measurable quantities. We can therefore compute them, solve the linear system (6) and obtain the motion parameter vector $\vec{m}$. Since the number n of linear features will normally be greater than 6, a least-squares solution will be used. It is well known that the system (6) has the unique minimal norm solution $\vec{m} = \mathbf{H}^\dagger \dot{\vec{f}}$, where $\mathbf{H}^\dagger$ is the Moore-Penrose inverse of $\mathbf{H}$.

In our target motion parameter estimation scheme no explicit calculation of the optical flow field is needed, as we have seen up to this point. In the next section we will see that neither is it needed for the tracking scheme developed there. Moreover, the only calculation involving the derivatives of the image intensity function s is done in order to obtain the parameters h_{ik}. Spatial differentiation of s is needed there, but, since numerical differentiation is an ill-posed problem, we tried to bypass it by using Green's Theorem on the boundary ∂S of the area S:

$$\iint_S [\frac{\partial}{\partial x} Q(x, y) - \frac{\partial}{\partial y} P(x, y)] dx\, dy = \int_{\partial S} P(x, y) dx + Q(x, y) dy.$$

Then, from the definition of h_{ik}:

$$h_{ik} = -\int_{\partial S} \mu_i(x,y) u_{k_1}(x,y) s(x,y)\, dy + \int_{\partial S} \mu_i(x,y) u_{k_2}(x,y) s(x,y)\, dx$$
$$+ \iint_S s(x,y) \frac{\partial}{\partial x}(\mu_i(x,y) u_{k_1}(x,y)) dx\, dy + \iint_S s(x,y) \frac{\partial}{\partial y}(\mu_i(x,y) u_{k_2}(x,y)) dx\, dy$$

Therefore, spatial differentiation of s was replaced by differentiation of μ_i and u_{ki}, which can be done analytically. Now, if we consider as S the projection of the target on the image plane, we can use the dynamic segmentation procedure to specify ∂S and then the above procedure will specify the kinematic parameters of the target.

2.2. Tracking

Using eq. (2), (3), (5), and the target motion parameters m_i that we just computed, the Tracking Constraint can be brought to the form:

$$J(\omega_x, \omega_y) = \alpha \omega_x^2 + \beta \omega_y^2 + \gamma \omega_x + \delta \omega_y + \epsilon \omega_x \omega_y + \zeta\,,$$

where $\alpha, \beta, \ldots, \zeta$ are known functions of the data.

Obviously $\alpha, \beta \geq 0$. Experimental results show that they are not simultaneously zero and that $4\alpha\beta - \epsilon^2 \neq 0$. It is easy to prove that under those conditions the Tracking Constraint has a unique minimizer: $\omega_x^* = \frac{\delta\epsilon - 2\beta\gamma}{4\alpha\beta - \epsilon^2}$ and $\omega_y^* = \frac{\gamma\epsilon - 2\alpha\delta}{4\alpha\beta - \epsilon^2}$, which is the desired camera rotation that achieves tracking.

3. EXPERIMENTAL RESULTS

3.1. Synthetic experiments

The previous tracking algorithm was implemented on SGI IRIS graphics workstations for a textured planar target moving in front of a randomly textured background (Figure 2).

In this paper only the diagram of the distance oP_W versus time is presented as a measure of the quality of tracking. (P_w is the projection on the image plane of the "center" W of the target; o is the origin of the image plane coordinate system.) The simulation of the tracking process consists of a sequence of a target motion followed by a camera tracking action at each time instant. Therefore, motion at even time instants on the diagrams corresponds to target motion, while motion at odd time instants corresponds to camera tracking motion.

Figure 2.

Very good results were obtained for this tracking scheme fed with "perfect" estimates of the target motion (accurate values for the direction of translation and rotation) for various general motions of the target (Figure 3). The results degenerate "gracefully" as the quality of estimates decreases due to very fast motion of the target or noisy images. A sequence of "noisy" target motion estimates was generated in order to test the performance of the algorithm under inaccurate motion estimates. The algorithm performs well with 30% to 60% noise in target motion estimation (Figure 4).

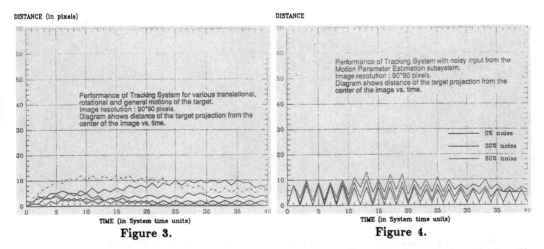

Figure 3.   **Figure 4.**

Errors are introduced in this tracking scheme from discretization effects, which will affect mainly the temporal derivatives of the linear features, numerical instabilities in the least-squares solution of the linear system $\mathbf{H}\vec{m} = \vec{f}$ and the coarse resolution of the image used in simulations.

Several sets of linear features were considered before establishing experimentally that the best 3-D motion parameter estimation is obtained with moments of the image, which have the advantage that they are easily implementable in hardware, therefore the above scheme can be used for real-time applications. (See [4] for a theoretical justification of this choice.) Moreover, the Connection Machine can be used for the real-time computation of the matrix H and the vector of linear features $\vec{f}$, which are the computational bottlenecks of the above algorithm.

3.2. A robotic implementation

In order to demonstrate the feasibility of this approach, we implemented the tracking algorithm in a robotic environment built in our laboratory. The task was the following: a programmable Mitsubishi robot arm held a soft drink can which it could move towards any direction against a textured background. On the hand, a 6-dof American Merlin robot arm equipped with a Sony DC-37 CCD camera was used to emulate the tracking system. The task then was to have the American Merlin robot arm track the can held by the first robot, which could move towards any direction (see Figure 5, which shows part of the "seeing" arm on the left "watching" the arm that holds the object to be tracked). Figure 6 shows the image of the object to be tracked as seen by the tracker.

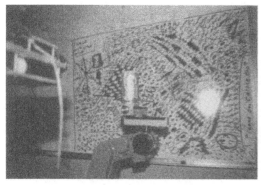 

| Figure 5. | Figure 6. |

Because the analysis of the images taken by the tracker had to be done on a VAX 785 and the primitive image processing operations on a VICOM image processor, all linked through a network, a real time implementation was impossible due to the heavy communication bottleneck. For this reason, we chose to present results from the experiments in the same format as in the case of the synthetic simulations. Again, the vertical axis of the plots represents the distance oP_w as before, and the horizontal axis represents time and motion at even time instants on the diagrams corresponds to target motion, while motion at odd time instants corresponds to camera tracking motion. Figure 7 shows the results from three experiments. In the first (solid line), the actual motion was translation along the x-axis and under the assumption that the object in view was planar (instead of cylindrical) in order to examine the deviation of the results as a function of deviation from the local shape assumption. Clearly the results are not sensitive. The next two experiments (dotted and dashed lines) show results under horizontal motion (dotted) and motion in the $x - z$ plane (dashed). It is quite clear that the results are very satisfactory.

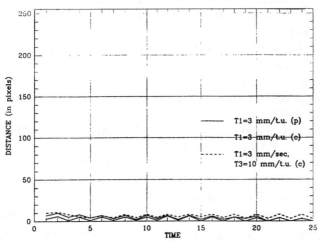

Figure 7.

Acknowledgements

The support of the Defense Advanced Research Projects Agency and the U.S. Army Engineer Topographic Laboratory under Contract DACA76-89-C-0019 is gratefully acknowledged, as is the help of Barbara Burnett in preparing this paper.

REFERENCES

1. J. Aloimonos, I. Weiss and A. Bandopadhay, "Active vision", *Int'l. J. of Comp. Vis.*, 333–356, 1988.
2. J. Aloimonos and D. Shulman, *Integration of Visual Modules: An Extension of the Marr Paradigm*, Academic Press, Boston, 1989.
3. J. Aloimonos and D. Tsakiris, "On the Mathematics of Visual Tracking", Center for Automation Research TR-2102, September 1988.
4. S. Amari, "Feature Spaces which Admit and Detect Invariant Signal Transformations", *Proc. 4th Intl. Joint Conf. on Pattern Recognition*, pp. 452–456, Tokyo, 1978.

5. P. Bouthemy and A. Benveniste, "Modelling of Atmospheric Disturbances in Meteorological Pictures", *IEEE Trans. PAMI*, PAMI-6, 5, 1984.

6. J.J.Clark and N.J.Ferrier "Control of Visual Attention in Mobile Robots", Proc. 1989 IEEE Conf. on Robotics and Automation.

7. K. Cornog, "Smooth Pursuit and Fixation for Robot Vision", M.S. Thesis, MIT AI Laboratory, 1985.

8. A.L. Gilbert, M.K. Giles, G.M. Flachs, R.B. Rogers and Y.H. U, "A Real-Time Video Tracking System", *IEEE Trans. PAMI*, PAMI-2, 1, 1980.

9. B.K.P. Horn and B.G. Schunk, "Determining Optical Flow", *Artificial Intelligence*, 17, pp. 185–203, 1981.

10. J. Loncaric, F. Decommarmond, J. Bartusek, Y.Pati, D. Tsakiris, R. Yang "Modular Dextrous Hand", Systems Research Center, TR 89-31, 1989.

11. S. Nagalia, "Real-Time Acquisition of Objects in Motion", Center for Automation Research, CS-TR-1398, 1984.

12. T. Poggio, E.B. Gamble, and J.J. Little, "Parallel integration of visual modules", *Science* **242**, 436–440, 1989.

13. S.A. Rajala, A.N. Riddle and W.L. Snyder, "Application of the One-Dimensional Fourier Transform for Tracking Moving Objects in Noisy Environments", *CVGIP*, 21, pp. 280–293, 1983.

14. J.W. Roach and J.K. Aggarwal, "Computer Tracking of Objects Moving in Space", *IEEE Trans. PAMI*, PAMI-1, 2, 1979.

15. A. Rosenfeld and A. Kak, *Digital Picture Processing*, Academic Press, 1982.

16. R.J. Schalkoff and E.S. McVey, "A Model and Tracking Algorithm for a Class of Video Targets", *IEEE Trans. PAMI*, PAMI-4, 1, 1982.

17. B.G. Schunk, "The Image Flow Constraint Equation", *CVGIP*, 35, pp. 20–46, 1986.

18. D.P. Tsakiris, "Visual Tracking Strategies", Systems Research Center, TR 88-66, 1988.

19. T.Y.Young and S.Gunasekaran "A regional approach to tracking 3-D motion in an image sequence", in *Advances in Computer Vision and Image Processing*, Vol. 3, pp. 63–99, JAI Press Inc., 1988.

20. B. Wilcox, D.B. Gennery, B. Bon and T. Litwin, "Real-Time Model-Based Vision System for Object Acquisition and Tracking", *SPIE*, Vol. 754, 1987.

Tracking Line Segments *

Rachid DERICHE - Olivier FAUGERAS

INRIA Sophia-Antipolis

2004 Route des Lucioles

06565 Valbonne Cedex France

deriche@rhodes.inria.fr

Abstract

This paper describes the development and the implementation of a line segments based token tracker. Given a sequence of time-varying images, the goal is to track line segments corresponding to the edges extracted from the image being analyzed. We will present a tracking approach that combines a prediction and a matching steps. The prediction step is a Kalman filtering based approach that is used in order to provide reasonable estimates of the region where the matching process has to seek for a possible match between tokens. Correspondence in the search area is done through the use of a similarity function based on *Mahalanobis* distance between attributes carefully chosen of the line segments. The efficiency of the proposed approach will be illustrated in several experiments that have been carried out considering noisy synthetic data and real scenes obtained from the INRIA mobile robot.

1 Introduction

The problem we want to adress is to detect and track features in a sequence of time varying images acquired by a camera on a robot moving in an indoor environment. At this end, edges are first detected through the use of an optimal operator developed in one of our previous work (see [1]). An edge linking step and a polygonal approximation yield the line segments that best approximate the extracted edges. A tracking approach based on a close cooperation between a prediction and a matching step is then developed. The prediction step utilizes a Kalman filter to estimate the position in the next image of each primitive in the current image. The matching step then need to be applied only within the predicted region. Matching employs a normalized distance based on the most stable parameters of each primitive.

The organization of the paper is as follows : A first section presents the problem of the representation for the line segments, and shows in particular why the mid point representation has been chosen. Section 2 gives a general overview of the developed approach and presents how the algorithm works. Section 3 explains in particular why

---

*This work was partially completed under Esprit P940

adding an error to the model is strongly recommended in our approach and how this remark can be taken into account in the Kalman filtering equations. Section 4 is devoted to the presentation of the matching process. A last section, before the conclusion, is then devoted to the experiments carried out on synthetic and real data.

2   What representation for the line segments ?

An important sub problem is to choose an appropriate representation for the line segments since the tracking will be based on this representation . It is clear for example that tracking both endpoints of each segment will be very difficult, since they are not at all reliable due to the fact that segments can be broken from one frame to another. For this reason, two types of representations have been considered.

2.1   Two Representations for Line Segments

In the first representation, a line segment having endpoints at points $P1(x_1, y_1)$ and $P2(x_2, y_2)$ is characterized by the vector $\mathbf{v_1} = [c, d, \theta, l]^T$ where θ represent the orientation of the line segment, l its length, the parameter c denotes the distance of the origin to the line segment and the parameter d, the distance along the line from the perpendicular intersection to the midpoint of the segment. These parameters are derived from the endpoints as follows :

$$
\begin{cases}
\theta = \arctan\left(\frac{y_2 - y_1}{x_2 - x_1}\right) \\[2mm]
l = \sqrt{((x_2 - x_1)^2 + (y_2 - y_1)^2)} \\[2mm]
c = \frac{(x_2 * y_1 - x_1 * y_2)}{l} \\[2mm]
d = \frac{(x_2 - x_1) * (x_1 + x_2) + (y_2 - y_1) * (y_2 + y_1)}{2 * l}
\end{cases}
\tag{1}
$$

The second representation that we considere is the midpoint representation and characterizes a line segment by the vector $\mathbf{v_2} = [x_m, y_m, \theta, l]^T$ where the point (x_m, y_m) defines the coordinates of the mid point Pm of the segment. Figure 1 illustrates both representations.

2.2   c,d,θ,l or midpoint representation ?

In order to find what representation is more appropriate for our tracking algorithm, we have to calculate the covariance matrices associated to the vectors $\mathbf{v_1}$ and $\mathbf{v_2}$ that define both representations from those of the endpoints. At this end we use the nonlinear relations that give theses vectors function of the vector $\mathbf{p} = [x_1, y_1, x_2, y_2]^T$ and the following relation :

$$
\Lambda_\mathbf{v} = \frac{\partial \mathbf{v}}{\partial \mathbf{p}} \Sigma \frac{\partial \mathbf{v}}{\partial \mathbf{p}}^T
\tag{2}
$$

Where $\frac{\partial v}{\partial p}$ is a the 4*4 Jacobian matrix and Σ is the 4*4 covariance matrix of the vector $p=[x_1, y_1, x_2, y_2]^T$. Assuming no correlation between the endpoints and the same covariance matrix Λ for both endpoints leads to the following covariance matrix Σ :

$$\Sigma = \begin{pmatrix} \Lambda & 0 \\ 0 & \Lambda \end{pmatrix} \tag{3}$$

where Λ is the 2*2 covariance matrix associated to each endpoint.

$$\Lambda = \begin{pmatrix} \sigma_x^2 & \sigma_{xy}^2 \\ \sigma_{xy}^2 & \sigma_y^2 \end{pmatrix} \tag{4}$$

Due to the fact that line segments may be broken differently from one image to another, an endpoint is not reliable. We model this segmentation noise, introduced by the polygonal approximation, as follows : We assume that Λ is diagonal in the coordinate system defined by $u_\parallel$ and $u_\perp$, two units vectors parallel and perpendicular to the line segment, respectively. In this coordinate system, Λ is written :

$$\Lambda_{\perp\parallel} = \begin{pmatrix} \sigma_\parallel^2 & 0 \\ 0 & \sigma_\perp^2 \end{pmatrix} \tag{5}$$

Noting that the coordinate system defined by $u_\parallel$ and $u_\perp$ is obtained using a rotation of an angle θ around the origin, leads to the following relations for the covariance matrix Λ :

$$\begin{cases} \sigma_x^2 = \sigma_\parallel^2 * cos(\theta)^2 + \sigma_\perp^2 * sin(\theta)^2 \\ \sigma_y^2 = \sigma_\perp^2 * cos(\theta)^2 + \sigma_\parallel^2 * sin(\theta)^2 \\ \sigma_{xy}^2 = (\sigma_\parallel^2 - \sigma_\perp^2) * sin(\theta) * cos(\theta) \end{cases} \tag{6}$$

Applying these results to the vectors $v_1 = [c, d, \theta, l]^T$ and $v_2 = [x_m, y_m, \theta, l]^T$, we find after some algebra that their covariance matrices Λ_{v_1} and Λ_{v_2} respectively are given as follows :

$$\Lambda_{v_1} = \begin{pmatrix} \frac{\sigma_\perp^2}{2} + \frac{2*d^2*\sigma_\perp^2}{l^2} & \frac{-2*c*d*\sigma_\perp^2}{l^2} & \frac{-2*d*\sigma_\perp^2}{l^2} & 0 \\ \frac{-2*c*d*\sigma_\perp^2}{l^2} & \frac{\sigma_\parallel^2}{2} + \frac{2*c^2*\sigma_\perp^2}{l^2} & \frac{2*c*\sigma_\perp^2}{l^2} & 0 \\ \frac{-2*d*\sigma_\perp^2}{l^2} & \frac{2*c*\sigma_\perp^2}{l^2} & \frac{2*\sigma_\perp^2}{l^2} & 0 \\ 0 & 0 & 0 & 2*\sigma_\parallel^2 \end{pmatrix} \tag{7}$$

$$\Lambda_{v_2} = \begin{pmatrix} \frac{\sigma_\parallel^2*cos(\theta)^2+\sigma_\perp^2*sin(\theta)^2}{2} & \frac{(\sigma_\parallel^2-\sigma_\perp^2)*sin(\theta)*cos(\theta)}{2} & 0 & 0 \\ \frac{(\sigma_\parallel^2-\sigma_\perp^2)*sin(\theta)*cos(\theta)}{2} & \frac{\sigma_\perp^2*cos(\theta)^2+\sigma_\parallel^2*sin(\theta)^2}{2} & 0 & 0 \\ 0 & 0 & \frac{2*\sigma_\perp^2}{l^2} & 0 \\ 0 & 0 & 0 & 2*\sigma_\parallel^2 \end{pmatrix} \tag{8}$$

From these results, one can say that the c, d, θ, l representation leads to a covariance matrix that depends strongly on the position of the associated line segment into the image through the parameters c and d that appear on the covariance matrix Λ_{v_1} Therefore two given segments with the same length and orientation will have their uncertainty on the (c, d) parameters completely differents depending on their position within the image. This is not the case for the mid point representation since the uncertainty associated to the mid point (x_m, y_m) depends only on the uncertainty of the endpoints.

A second important point to note for the mid-point representation is the decorrelation that exists between the parameters (x_m, y_m) and the parameters θ and l. This decorellation between the parameters does not exist for the representation c, d, θ, l. Adding to that, it is easy to check that for the particular case where $\sigma^2 = \sigma_\perp^2 = \sigma_\parallel^2$, then the four parameters x_m, y_m, θ, l are completely decorellated, while it is not the case for the c, d, θ, l representation, where we have to assume that $\sigma_\perp$ is equal to zero in order to decorellate the paramaters. This question of decorellation is important if we want to use for efficiency consideration, differents Kalman filters on each parameters.

From these remarks, it is clear that the mid point representation (x_m, y_m, θ, l) is more appropriate to our tracking algorithm where each segment is represented by four points in a 1 dimensional space : the x and y-position information and the length and orientation information. When a given segment moves in the image, these four points follow a trajectory in the 1D space. The kinematics of the motion of the given line segment is the kinematics of the fourth points i.e trajectory, velocity and acceleration. Therefore we will run four Kalman filters independently on each parameter.

3 Tracking

The tracking approach we have developed is based on the following steps :

1. Assign a kinematics model to each parameter of the representation of the line segment.

2. Use the model to predict the position of the given parameter in the next frame and the associated uncertainty.

3. Use the uncertainty to determine the search area around the predicted position.

4. Inside the search area, use a normalized distance to determine the best match.

5. If a match is found, use it as a new measure to update the kinematics model.

Initially a $t = 0$, no information is available about the kinematics of the line segments in the image. This is called the *bootstrapping stage*. We assume zero velocities and accelerations with large uncertainties indicating that we do not trust these guesses. We say that a line segment of the current frame is inside a searching area if the distance of each parameter to its predicted position is less than a given threshold. A similarity function that combines the fourth distances is then used in order to compute a score for each correspondance between the predicted position and the position of the line segments lying inside the searching area. Once a line segment from the current frame has been matched to a line segment of the model, its parameters are then used as a new measure in order to update the kinematics of the model.

A Kalman filter is used to perform tracking by providing reasonable estimates of the region where the matching process has to seek for a possible match between tokens. Kalman filtering is a statistical approach to estimate a time-varying state vector X_t from noisy measurements Z_t. Consider the estimation of X_{t+k} from the measurements up to the instant k, Kalman filtering is a recursive estimation scheme designed to match the dynamic system model, the statistics of the error between the model and reality, and the uncertainty associated with the measurements.

In our application, four Kalman filters are applied independently on each parameter of the representation (x_m, y_m, θ, l). Each state vector is just the position of the given parameter x (i.e x_m, y_m, θ, l), its velocity $\dot{x}$ and its acceleration $\ddot{x}$. The following discrete time steps notation is used for the state vector at the tth time step $X_t = [x_t, \dot{x}_t, \ddot{x}_t]^T$. The model of the system dynamics and the measurements model are given as follows in our application :

$$
\begin{cases}
X_{t+1} = \Phi_{t+1,t} X_t + \xi_t \\
\\
V_t = C_t X_t + \eta_t
\end{cases}
\tag{9}
$$

where ξ_k is assumed to be a zero mean Gaussian noise sequence of covariance Q_t representing the error of the model, $\Phi_{t+1,t}$ is the matrix which evolves the position x, the velocity $\dot{x}$ and the acceleration $\ddot{x}$ from one time sample to another. V_t is a vector of measurements with an uncertainty η_t, assumed to be a zero-mean Gaussian noise sequence of covariance R_t. In our application, $\Phi_{t+1,t}$ assumes a motion with constant acceleration and the measurement model assumes that the position x is measurable from the matching process while the velocity $\dot{x}$ and the acceleration $\ddot{x}$ are not.

The classical Kalman filtering equations (see [3] for example) allows to compute the optimal estimates of the state vector $\hat{X}_t = \hat{X}_{t/t}$ of X_t recursively from the data $V_0, V_1, ..., V_t$ and the initial estimation $E(X_0)$ and $Var(X_0)$.

4   Error Modelling

Implementing the classical theoretical Kalman filter equations leads to results strongly linked to the accuracy of the model. When there is a discrepancy between the true system model and the model assumed by the Kalman filter, the resulting estimation simply does not correspond to that predicted by theory. The Kalman filtering approach developed for our tracking algorithm is based on an assumed model of the trajectory (i.e constant velocity or acceleration). It is clear that this model can be considered as correct only *locally*. This means that we assume that the trajectory of each parameter of the line segment can always be approximated *locally* by a first or second degree polynomial. This is a more reasonable assumption than the one that considers that all the trajectory can be fitted by a first or second degree polynomial. Among all the solutions that have been proposed in the literature (see reference [3]) in order to deal with this discrepancy problem, we have chosen and implemented the following two techniques :

- **Add process noise.**

  This solution, consisting in the addition of process noise to the system model as done in equation 9 with the term ξ_t, is attractive for preventing divergence. In our application, the covariance Q_t of ξ_k in 9 is taken as :

$$
\mathbf{E}(\xi_t \xi_t^T) =
\begin{pmatrix}
\sigma_p^2 & 0 & 0 \\
0 & \sigma_v^2 & 0 \\
0 & 0 & \sigma_a^2
\end{pmatrix}
\tag{10}
$$

It can be shown that in such case, the Ricatti equation associated to the system model 9 can be resolved to get the following *steady state* Kalman filtering equations :

$$\hat{X}_{t|t} = \hat{X}_{t|t-1} + G(V_t - C_t\hat{X}_{t|t-1}) \tag{11}$$

Thus the filter will always track the data. This has to be compared to the case where no errors on the model were assumed. In such case the Kalman gain G_t will tend to zero as t tends to the infinity and thus the filter will no longer continue to track. The elements of Q (the values of σ_p^2, σ_v^2 and σ_a^2) depends to a great extent upon what is known about the unmodeled states. In our application all the used values have been derived from the simulation experiments. However the choice of σ_m^2, the variance on the position, is done in a manner reflecting our a priori estimate of the amount of noise to be expected from the previous step (Digitizing effects, edge detection and polygonal approximation).

It should be pointed out that the Ricatti equation may be solved before the filtering process is being performed. Therefore if we assign to the Kalman Gain G the following constant values $G = [\alpha, \beta]^T$ in our application that deals with a constant velocity model, then it can be shown that we obtain the following decoupled equations :

$$\begin{cases} x_{t/t} = -(\alpha + \beta - 2) * x_{t-1/t-1} - (1 - \alpha) * x_{t-2/t-2} + \alpha * v_t + (-\alpha + \beta) * v_{t-1} \\ \dot{x}_t = -(\alpha + \beta - 2) * \dot{x}_{t-1/t-1} - (1 - \alpha)\dot{x}_{t-2/t-2} + \beta * (v_t - v_{t-1}) \end{cases} \tag{12}$$

and the following covariance matrix **P** :

$$\mathbf{P} = \frac{\sigma_m^2}{1 - \alpha} \begin{pmatrix} \alpha & \beta \\ \beta & \beta * (\alpha + \beta) \end{pmatrix} \tag{13}$$

α is a real positive scalar less than 1 and β a real positive scalar less than $\frac{\alpha^2}{1-\alpha}$.

This limiting Kalman filtering is the well known near optimal α,β tracker [3]. It has been implemented in our application and found that it is extremely efficient since all the coefficients of the recursive equations given above can be calculated before the tracking starts.

- **Elimination of old data**

The basic idea in this approach is to consider the old data as no more meaningful and therefore to discard them. A simple way to accomplish this elimination is then to weight the old data according to when they occured. This means that the covariance of the measurement noise must somehow be increased for past measurement. It can be shown that one simple manner of accomplishing this is to multiply at each new measure the covariance prediction for the state vector with the age weighting scalar factor α greater than or equal to one [3].

5 Matching

The search area is determined through a simple set of attribute tests using the result of the Kalman filtering. For each token of the image flow model, represented by a feature vector of 4 components, we wish to know which token might correspond to it. In selecting a cost function for correspondences, we wanted to take into account the distance between the expected parameter value with its uncertainty and the current value of the measure with its uncertainty. This leads to calculate the so called *Mahalanobis* distance, for each components and to declare a token of the new frame inside a search area if all theses distances are less than a fixed threshold. Inside the search area, the correspondence is then controlled by the value of the sum of these distances. It is calculated for each possible match and the best score is used to validate the most consistent.

Let each new token, issued from the matching process, be represented by a feature vector of N components denoted T_m with a covariance matrix Γ. Let the estimated token represented by T_e with a covariance matrix Λ. It is then easy to find that in the case where no correlation exists between both vectors T_m and T_p, then the covariance matrix S of the vector difference $V = T_m - T_e$ is just the sum of Γ and Λ. The *Mahalanobis* distance is then defined to be:

$$d_{\chi^2} = (T_m, T_e) = (T_m - T_e)^T (S)^{-1} (T_m - T_e) \qquad (14)$$

This distance has a χ^2 distribution with N degrees of freedom, where N=1 in our case. This distance deliminates the upper bound on the variation of V from its mean. The probability that d_{χ^2} is less than a given threshold ϵ_{χ^2} may be obtained from a χ^2 distribution table. In order to deal with a search area where we have a probability of 95% to find the measure, we set the value of the threshold to 3.84.

6 Experimental Results

Many experiments have been carried out considering several noisy synthetic data and differents real scenes, however due to the limited number of pages, this part have benn considerably reduced from its original version. Figure 2 illustrates a noisy trajectory that have been synthetized in order to simulate an inaccurate constant+ramp model and the results given by the prediction without error modelling. Note that the results have not been superimposed but displaced in position in order to better compare the two trajectories. Figure 3 shows the predicted trajectories taking into account an error modelling through the value assigned to σ_v^2 (0.01 and 0.1). It appears that the lower the variance σ_v^2 is, the more serious the divergence problem is. On the other hand, the higher the variance is, more noisy the prediction is, because the filter takes into account few data.

The tracking on real data is illustrated in Figures 4 and 5 through the numbers assigned to each line segment. A close look at the results reveals how some line segments can appear or disappear. A new label is affected as soon as a new segment appears and the process continues without affecting the tracking algorithm. Through the number of real experiments, it has been demonstrated that the approach developped works well and gives satisfatory results. This approach is now an integral part of the standard project demonstration and our industrial partner ITMI is incorporating it into hardware for the DMA machine of the Esprit project P940.

7   Conclusion

A line segments based token tracker has been developed and implemented. Given a sequence of time-varying images, it allows to track line segments corresponding to the edges extracted from the scene being analyzed. The results obtained through many experiments on real data seems very promising. We are currently working on the exploitation of these results for computing 3D motion and structure [2].

References

[1] R.Deriche. Using canny's criteria to derive a recursively implemented optimal edge detector. *International Journal of Computer Vision*, 1(2):167–187, May 1987.

[2] O.D.Faugeras. R.Deriche. N.Navab From Optical Flow of Lines to 3D Motion and structure. *Proceedings IEEE Int. Work. on Intell. Syst.* pp 646-649, Sept 1989. Tsukuba, Japan.

[3] A.Gelb and al Applied Optimal Estimation *The Analytic Sciences Corporation* ed. Arthur Gelb. M.I.T Press

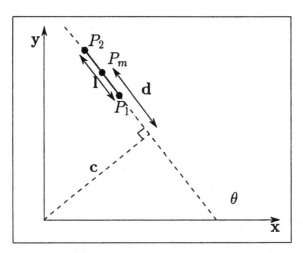

Figure 1: *Representations c,d,θ,l and xm,ym,θ,l*

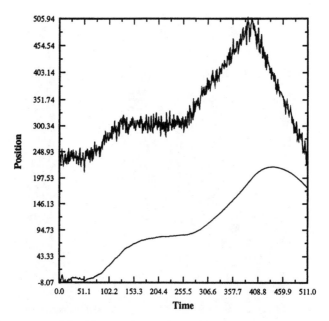

Figure 2: Noisy ramp model and its prediction by a Kalman without errors on the model

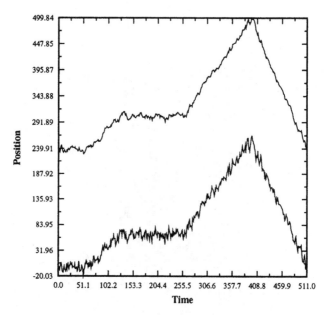

Figure 3: Prediction by a Kalman filtering with error modelling. $\sigma_v^2=0.001$ (Up) and $\sigma_v^2=0.1$ (Down)

Figure 4: Hall scene 2

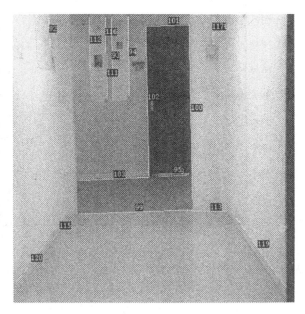

Figure 5: Hall scene 1

Measurement and Integration of 3-D Structures By Tracking Edge Lines

James L. Crowley   Patrick Stelmaszyk

LIFIA (CNRS) - IMAG   I.T.M.I.

Abstract

This paper describes a technique for building a geometric description of a scene from the motion of a camera mounted on a robot arm. The movements of edge-lines in a sequence of image are tracked to maintain an image plane "flow model". Tracking perserves the correspondance of segments, even when the camera displaces, makeing possible a inexpensive form of motion stereo. Three dimensional structure is computed using the matches provided by the segment tracking process and the displacement parameters provided by the robot controller. By fusion of 3D data from different view points, we obtain an accurate and complete representation of the scene.

Results from a sequence of 80 images taken from a camera mounted on a robot arm are presented to illustrate the technique. These results are used for an experimental evaluation ito illustrate the accuracy and the robustness of the technique.

1. Introduction

This paper describes a technique for reconstructing and modeling the 3-D geometry of a scene by tracking edge lines taken from a moving camera. This technique avoids both the cost of stereo correspondence matching, and the cost of matching recovered 3-D segments to update a 3-D model, by tracking. The computational and conceptual simplicity of this approach has made possible the development of an inexpensive real time hardware implementation.

A basic idea behind this work is that structures from a dense set of images may be matched with a simple linear complexity algorithm. Our work was partly inspired by Generey, who has shown that measurement of the motion of points in an image sequence could be based on a Kalman filter [Generey 82]. Matthies et. al. have recently demonstrated recovery of depth from lateral displacement of points in a dense image sequence using a Kalman filter [Matthies et. al 87].

2. Measuring Image Flow by Tracking Edge Segments

Correspondence of edges in a dense temporal sequence of images can be maintained by a very simply tracking process based on a Kalman filter. Such a process has been described in [Crowley et. al. 88]. This tracking process is well debugged and has been used in a number of our projects. Real time hardware has recently been constructed using this algorithm [Chehikian 88].

2.1 A Parametric Representation for Edge Segments

The first step in the tracking process is to express edge segments in a parametric representation. Our tracking algorithm is based on the use of an MDL parametric representation [Crowley-Ramparany 87], illustrated in figure 2.1. This representation is designed to facilitate the matching step in the segment tracking phase. As we shall see below, this representation is the 2-D analog of the 3-D representation which we apply to 3-D contours. For each segment, we also save the midpoint, P_m, expressed in image coordinates (i, j) as well as the end points, P_1, and P_2.

This work was performed as part of Project ESPRIT P940

2.2 Representation for the Image Flow

The tracking process maintains a list of "active" edge segments composed of the parameter vector, $S = \{c, d, \theta, h\}$ as well as the temporal derivative, a' and the covariance between the attribute and its temporal derivative for each attribute A of S. That is, for each $A \in S$:

$$A = \begin{bmatrix} a \\ a' \end{bmatrix} \qquad C_A = \begin{bmatrix} \sigma_a^2 & \sigma_{aa'} \\ \sigma_{aa'} & \sigma_{a'}^2 \end{bmatrix} \qquad \text{where} \qquad a' = \frac{\partial a}{\partial t}$$

The flow model also contains a confidence factor, CF, represented by a state from the set $\{1, 2, 3, 4, 5\}$ and a unique identity, ID. The identity of a segment permits the process to preserve the association between a segment, its corresponding segment in the other image, and the resulting 3-D segment.

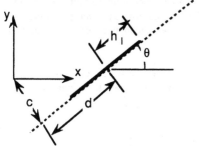

Line Segment Parameters: $S = \{c, d, \theta, h\}$.

c   The perpendicular distance of the segment from the origin.

d   The distance from the perpendicular intercept of the origin to the midpoint of the segment.

θ   The orientation of the segment.

h   the half-length of the segment.

Figure 2.1 The MDL Parametric Representation for Line Segments

2.3 Maintenance of a Dynamic Flow Model

The segments in the flow model are tracked by a three phase process illustrated in figure 3.1. The phases of this cyclic process are:
1) Prediction
2) Correspondence
3) Model Update

Performing stereo correspondence on the flow model provides cleaner data for stereo matching. In particular, this technique also permits the system to function in the presence of temporary occlusions.

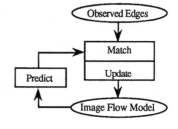

Figure 2.2 The Flow Measurement Process

3. Structure from Motion

Tracking edge lines from a moving camera provides a number of useful capabilities. Two important properties are that the flow model provides an image description which is less sensitive to image noise than any individual image. A second property is that the ID of the edge lines in the flow model provides a correspondence of image features for different views of the same scene.

The system described below is illustrated in figure 3.1. "Snapshots" of the flow model are saved after the camera has moved approximately 5 cm. In our data this corresponds to every 5th image. The ID attribute of the tokens in the model provide a correspondence between the tokens in each snapshot. Knowledge of the camera location at the time of each update permits us to compute the three dimensional locations for corresponding edge lines. These observed 3-D edge lines are then used to update the 3-D composite model.

The 3-D composite model uses a 3-D edge line primitive described in [Crowley 86]. These 3-D edge-line primitives contain an explicit estimate of the uncertainty of the 3-D positions. As with the flow model, a Kalman filter update equation is used to refine the estimated position and uncertainty. The resulting 3-D edge line segments are more reliable and more precise than individual observations of 3-D segments obtained from pairs of "snapshots" of the flow model.

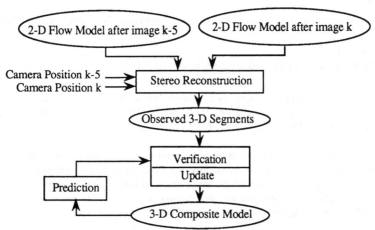

Figure 3.1 The components of our system for maintaining a composite model of the geometry of a scene.

3.1 Coordinate Systems and Notation

The derivation of the equations for recovery of 3-D structure from the movement of a camera on a robot arm requires that we define a set of coordinate systems, as well as homogeneous coordinate transformations between these coordinate systems. We represent a homogeneous coordinate transformation matrix as a bold letter. The coordinate system from which the transformation is applied is represented by a lower case subscript which appears after the matrix symbol. The resulting coordinate system is denoted by lower case superscript which precedes the symbol. Vectors will be illustrated with a capital letter followed by a lower case subscript which represents the coordinate system in which they are expressed.

The transformations with which we are concerned are:

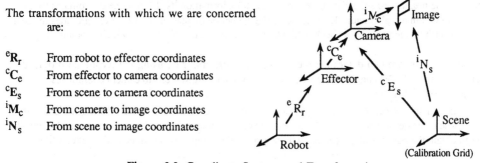

$^e\mathbf{R}_r$ From robot to effector coordinates

$^c\mathbf{C}_e$ From effector to camera coordinates

$^c\mathbf{E}_s$ From scene to camera coordinates

$^i\mathbf{M}_c$ From camera to image coordinates

$^i\mathbf{N}_s$ From scene to image coordinates

Figure 3.2 Coordinate Systems and Transformations

Notice that the transformations $^i\mathbf{M}_c$ and $^i\mathbf{N}_s$ are not square matrices and thus have no inverse. These coordinate systems and transformations are illustrated in figure 3.2.

Constructing an estimate of the position and uncertainty of 3-D contours in scene coordinates requires a model of the image formation process. Such a model is expressed as a composition of the intrinsic and extrinsic parameters of the camera. These parameters are obtained using homogeneous coordinate transformation as described in this section. The intrinsic parameters form a transformation, $^i\mathbf{M}_c$, which describes the projection of the point P_c in camera parameters to a point in the image, P_i.

$$P_i = {}^i\mathbf{M}_c\, P_c.$$

The extrinsic parameters describe a projection, ${}^{c}E_s$, of a point from scene coordinates, P_s to camera center coordinates, P_c.

$$P_c = {}^{c}E_s P_s$$

Together the intrinsic and extrinsic parameters describe a projection, ${}^{i}N_s$, from scene coordinates to image coordinates,

$$^{i}N_s = {}^{i}M_c \, {}^{c}E_s,$$

so that a point in the scene P_s is projected to a point in the image P_i by

$$P_i = {}^{i}N_s \, P_s.$$

For the intrinsic camera parameters we employ a standard "central projection" model of image formation. This model expresses the projection using the ration of the pixel sizes to focal length, D_x/F and D_y/F (expressed in mm/pixels) and the optical center of the image, C_x and C_y (expressed in pixels).

The extrinsic camera parameters may be estimated as a composition of the position and orientation of the robot arm "tool" coordinates, A, and the transformation from the tool coordinates to the camera B. An estimate of the position and orientation of the robot arm tool coordinates is provided to us by the arm controller. The estimate of the position and orientation with respect to tools coordinates is a rigid transformation which can be calibrated when the system is intialized.

3.2 Estimation of the Intrinsic Camera Parameters

We adopted a camera calibration technique due to Faugeras and Toscani [Faugeras-Toscani 86]. Errors in 3-D recovery due to errors in the calibration are included in our estimate of the uncertainty of the recovered 3-D contours. The explicit estimation of uncertainty permits our system to function despite the presence of such uncertainties.

The calibration of the intrinsic camera parameters requires observation of a set of 25 or more points arrayed on a grid. We use points extracted from the intersection of line segments in a grid of lines. A 15 by 10 grid with a 2 cm separation was prepared. Four images are taken of this pattern placed precise 3cm displacements heights above the work space. The first image defines a coordinate system at height Z = 0. Any three of the images are sufficient to determine the transformation ${}^{i}N_s$ from scene coordinates to image coordinates to within a scale factor. Knowledge of the the height displacement $\Delta z = 3$ cm provides the scale factor. The fourth image provides a check with which to verify the transformation which is obtained.

Having obtained ${}^{i}N_s$ we can determine the intrinsic parameters, ${}^{i}M_c$, using the technique provided in [Faugeras-Toscani 86]. We can then calculate the transformation from scene to camera coordinates as:

$$^{c}E_s = {}^{i}M_c^{-1} \, {}^{i}M_c \, {}^{c}E_s$$

Such estimate of the extrinsic parameters is obtained for three viewing positions during the calibration process. These estimates are combine with the position of the effector obtained from the robot arm controller in order to obtain the position of the camera with respect to the gripper as described in the next section.

3.3 Estimation of the Camera Position Relative to the Robot End Effector

For an arbitrary viewing position, k, estimation of the extrinsic camera coordinates, ${}^{c}E_{sk}$, requires knowledge of the position and orientation of the end effector as well as knowledge of the position of the camera with respect to the effector. The position and orientation of the robot end effector is obtained directly from the robot controller as a homogeneous coordinate transform ${}^{r}R_{ek}$.

The transformation from camera coordinates to effector coordinates is a rigid transformation, ${}^{e}C_c$. A technique for estimating the transformation ${}^{e}C_c$ has been developed by Shui and Ahmad [Shui-Ahmad 87]. This technique yields a homogeneous transform equation of the form $AX = XB$. In our notation, this relation has the form

$$A \, {}^{e}C_c = {}^{e}C_c \, B$$

The matrix $\mathbf{A}$ is the transformation of the effector position obtained by calibration at the two positions. For positions number 0 and 1, this is:

$$\mathbf{A}_{01} = {}^r\mathbf{R}_{e0}{}^{-1}\,{}^r\mathbf{R}_{e1}$$

The matrix $\mathbf{B}$ is the difference in the calibrated extrinsic camera parameters for the two camera positions. For positions 0 and 1 this is:

$$\mathbf{B}_{01} = {}^c\mathbf{E}_{s0}{}^{-1}\,{}^c\mathbf{E}_{s1}$$

By calibration at three camera positions (0, 1 and 2), such that the translations $\mathbf{A}_{01}$ and $\mathbf{A}_{12}$ are neither parallel nor anti-parallel, we obtain a set of equations of the form:

$$\mathbf{A}_{01}\,{}^e\mathbf{C}_c = {}^e\mathbf{C}_c\,\mathbf{B}_{01}$$
$$\mathbf{A}_{12}\,{}^e\mathbf{C}_c = {}^e\mathbf{C}_c\,\mathbf{B}_{12}$$

Solving these equations yields the transformation from camera to effector coordinates, ${}^e\mathbf{C}_c$. This transformation permits us to relate the extrinsic camera parameters for an arbitrary viewing position, to the calibration grid.

3.4 Estimation of the Camera Extrinsic Parameters for Arbitrary Robot Positions

Modeling the scene from multiple view points requires that the recovered 3-D structure be expressed in a common coordinate system. This common coordinate system may be any of the individual camera coordinate systems or the external scene coordinates. For convenience, we have chosen to use the calibrated scene coordinate system, defined by the calibration grid.

Let ${}^r\mathbf{R}_{e0}$ represent the robot position at the time of the calibration of the first image, and let ${}^r\mathbf{R}_{ek}$ represent the robot position at the time at which the flow model was updated from the kth image. The extrinsic camera transformation for the kth image may be computed from the robot effector position by:

$$ {}^c\mathbf{E}_{sk} = {}^c\mathbf{C}_c{}^{-1}\,{}^r\mathbf{R}_{ek}{}^{-1}\,{}^r\mathbf{R}_{e0}\,{}^e\mathbf{C}_c\,{}^c\mathbf{E}_{s0}. $$

This computation requires only one matrix inversion and two matrix multiplications. The terms ${}^e\mathbf{C}_c{}^{-1}$ and ${}^r\mathbf{R}_{e0}\,{}^e\mathbf{C}_c\,{}^c\mathbf{E}_{s0}$ can be computed at the time of calibration.

Tracking edge lines as the camera moves produces a flow model composed of 2-D line segments. The robot arm controller provides us with the tool position at the time at which the flow model is updated with from each image. Knowing the offset of the camera from the tool coordinates permits us to calculate the extrinsic camera parameters. Knowing the intrinsic camera parameters permits us to treat snapshots from the flow model as stereo images to recover three dimensional position in the scene. This recovery process is described in the next section.

3.5 Recovery of 3-D Scene Position By Motion Stereo

The extrinsic camera parameters for arbitrary viewing positions permit us to use standard stereo reconstruction equations. This process is illustrated in figure 3.3. The composition of intrinsic and extrinsic camera parameters describes the projection of a scene point to an image point.

$$ P_i = {}^i\mathbf{N}_s\,P_s. $$

Using homogeneous coordinates, this relation has the form:

$$ \begin{bmatrix} w\,x_i \\ w\,y_i \\ w \end{bmatrix} = \begin{bmatrix} N_{11} & N_{12} & N_{13} & N_{14} \\ N_{21} & N_{22} & N_{23} & N_{24} \\ N_{31} & N_{32} & N_{33} & N_{34} \end{bmatrix} \begin{bmatrix} x_s \\ y_s \\ z_s \\ 1 \end{bmatrix} $$

where x_i and y_i are the image coordinates and w is the homogeneous variable. Because we can not invert this matrix, we are obliged to deal algebraically with the individual relations.

For two observations (1 and 2) of the a scene point we obtain the image points $P_{i1} = (x_{i1}, y_{i1})$ and $P_{i2} = (x_{i2}, y_{i2})$. By algebra we can deduce the relations [Toscani-Faugeras 86] for (x_{i1}, y_{i1}):

$$(N_{11} - x_{i1} N_{31}) X_s + (N_{12} - x_{i1} N_{32}) Y_s + (N_{13} - x_{i1} N_{33}) Z_s = x_{i1} N_{34} - N_{14}$$
$$(N_{21} - y_{i1} N_{31}) X_s + (N_{22} - y_{i1} N_{32}) Y_s + (N_{23} - y_{i1} N_{33}) Z_s = y_{i1} N_{34} - N_{24}$$

and for (x_{i2}, y_{i2}):

$$(N_{11} - x_{i2} N_{31}) X_s + (N_{12} - x_{i2} N_{32}) Y_s + (N_{13} - x_{i2} N_{33}) Z_s = x_{i2} N_{34} - N_{14}$$
$$(N_{21} - y_{i2} N_{31}) X_s + (N_{22} - y_{i2} N_{32}) Y_s + (N_{23} - y_{i2} N_{33}) Z_s = y_{i2} N_{34} - N_{24}.$$

This provides us with a set of four equations with three unknowns. We select four sets of three equations to solve for (X_s, Y_s, Z_s) four times. We then calculate the average of these four values to cancel small errors in the measurement of the image points.

A common problem with edge line segments is the phenomena of "breaking". Although the token tracking process reduces this phenomena, we must still assure that the end-points of the segments correspond to the same physical point. To do this, we project the epipolar line from each end point into the other image, to determine the part of the two segments which is common to the two segments. The stereo reconstruction equations are applied to the end points of the common part of the segment to recover an observed 3-D edge segment which is represented as a pair of 3-D points and their uncertainty.

3.6 Representation of 3-D Structure

A line segment in the 2-D flow model corresponds to a line segment in the 3-D scene. In order to apply Kalman filter estimation techniques to 3-D reconstruction we require a representation which expresses a segment in terms of a minimum of parameters A 3-D form of the MDL represenation is used: 3-D segments are represented as a midpoint, a direction, a half length, the 2-D uncertainty of the midpoint (perpendicular to the segment) and a 2-D uncertainty of the direction.

Both observed and composite model 3-D segments are represented by a minimal representation and a set of redundant parameters. The minimal representation consists of a pair of end-points, P_1 and P_2, as well as an ID. Each end point is expressed as an estimated position expressed in scene coordinates (x, y, z) and its covariance, C, as illustrated in figure 3.4. This representation has the advantage of being both minimal and simple.

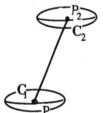

Thus the representation is expressed as

P_1 : The position of the first end point
C_1: The 3-D covariance in the position of the end-point.
P_2 : The position of the first end point
C_2: The 3-D covariance in the position of the end-point.
ID: The Identity of the Segment (From the Flow Model).
CF: Confidence Factor from the set {1, 2, 3, 4, 5}.

Figure 3.4 The End-point representation for a 3D LIne Segment

An MDL expression of the line segment is kept as a set of redundant parameters. These parameters are used to verify the match of observed segments to the corresponding composite model segment. This redundant representation is composed of the parameters shown in figure 3.5.

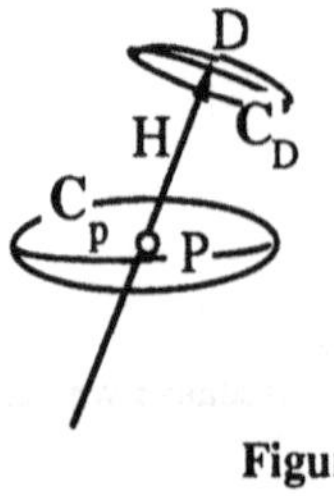

The 3-D MDL Parameters are:

P The center point of the segment (x, y, z)
C_p The 3x3 covariance of the center point
D The direction (expressed as Δx, Δy, and Δz)
C_D The 3x3 covariance of the direction
H The half length of the segment.

Figure 3.5 An MDL Reprentation for a 3D line segment.

These parameters are calculated by

$$P = \frac{1}{4}(P_1 + P_2) \qquad D = \frac{(P_2 - P_1)}{\| P_2 - P_1 \|} \qquad H = \frac{\| P_2 - P_1 \|}{2}$$

$$C_p = \frac{1}{4}(C_1 + C_2) \qquad C_D = \frac{(C_1 + C_2)}{\| P_2 - P_1 \|^2}$$

4. Integration of Geometric Structure

The integration of geometric information from independent sources is a fundamental problem in perception. This problem is often made difficult by the fact that different observations tend to have varying noise statistics. An explicit model of uncertainty, coupled with a model of the sensing process provides a powerful tool for this problem. In this section we illustrate how 3-D Segment observations from the motion stereo system are dynamically integrated into a composite model.

The integration process is illustrated in figure 4.1. Observed 3-D segments from the motion stereo process are compared to the corresponding segments from the 3-D composite model. The correspondence is provided by the ID attribute of the segments which is inherited from the segment tracking process.

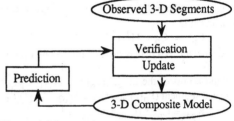

Figure 4.1 Integration process for 3-D segments.

4.1 Verification of Correspondences

The segment ID from the tracking process permits us to avoid searching for a 3-D correspondence for observed segments. However this technique assumes that the token tracking process provides rigorously true matches. If a false match occurs during tracking, the stereo reconstruction process may produce a 3-D observation with a non-realistic value.

While the effect of such false value will be limited due to the use of a Kalman filter in the composite model update process, such errors will decrease the precision of the composite model. It is possible to detect and eliminate tracking errors during the update phase of composite modeling by testing that the observed 3-D segment has a spatial position which is similar to the 3-D composite model segment.

Each observed 3-D segment is compared to the corresponding model segment by a 3-D analog for the 2-D test used in token tracking. That is, the 3-D segments are compared for similar orientation, for co-linearity and for overlap. These test employ the redundant parameters of the segments. We express these tests between observed 3-D segment parameters $S_o = \{P_o, C_{po}, D_o, C_{D0}, H_o\}$ and model segment parameters $S_m = \{P_m, C_{pm}, D_m, C_{Dm}, H_m\}$. These three tests are a form of Mahalanobis distance with a threshold of 1 standard deviation.

The test for orientation is performed by comparing the difference in the direction vectors to the sum of the covariances in direction is:

$$(\Delta D^T (C_{D0} + C_{Dm})^{-1} \Delta D) \leq 2 \qquad \text{Where} \qquad \Delta D = D_m - D_o$$

If the directions are found to be similar, the segments are tested for alignment. Alignment and overlap are based on a vector ΔP.

$$\Delta P = P_m - P_o$$

The test for alignment is based on the component of ΔP which is orthogonal to the model segment. This component is determined by subtracting the inner product of ΔP with the direction vector D_m. This component is compared to the covariance of the central point. by the test:

$$(\Delta P - \Delta P\, D_m)^T\, (C_{pm} + C_{po})^{-1}\, (\Delta P - \Delta P\, D_m) \leq 1$$

If the segments are found to have to be aligned, they are tested for overlap. The test for overlap compares the length of the vector ΔP to the sum of the half lengths of the segments

$$\| \Delta P \| \leq H_o + H_m$$

If any of these tests are not passed then the observed segment is rejected and the confidence of the composite model segment is reduced by 1.

4.2 Fusing an Observed 3-D Segment to the Composite Model

If an observed segment has been found to have a similar orientation, alignment, and overlap with the corresponding model segment, it can be used to update the parameters and the uncertainty of the model segment. Fusion is based on the end-points of the segments. However, it may be the case that the end-points do not correspond to the same physical point in the scene. To model this possibility, we treat half length of the segment as an uncertainty in position of the end-point, in the direction of the segment.

Using the attribute H defined in the previous section as, the uncertainty of each of the points is enlarged by the calculating

$$C_p^* = C_p + H^2\, D\, D^T$$

The term DD^T is a matrix defined by the outer product of the direction vector D.

The end-points of the model segment (P_{m1}, C_{pm1}^*) and (P_{m2}, C_{pm2}^*) are fused to the end-points of the observed segment (P_{o1}, C_{po1}^*) and (P_{o2}, C_{po2}^*) by calculating a kalman gain matrix :

$$K_{pm} = C_{pm}^*\, (C_{pm}^* + C_{po}^*)^{-1}$$

The estimated variance is then updated by

$$C_{pm}^+ = C_{pm}^* - K_{pm}\, C_{pm}^*$$

and the estimated point position is computed by

$$P_m^+ = P_m + K_{pm}\, (P_m - P_o).$$

After the parameters have been updated, new values are computed for the redundant parameters. The uncertainty in the direction of the segment is then removed from the uncertainty of the end points.

$$C_{pm}^+ = C_{pm} - H_m^2\, D_m\, D_m^T$$

The fusion process is completed by incrementing the confidence factor for the model segment.

4.3 Managing the Confidence of Composite Model Segments

The confidence factor in the 3-D composite model is maintained in a manner which is similar to that in the flow model, with one important difference. This difference concerns the elimination of segments which are occluded. Segments which reach a confidence value of CF_{max} (a value of 5 in our current implementation) do not have their CF reduced when they are no longer observed. In this way, the system can construct a model of a 3-D object which contains faces which are not simultaneously visible.

At the end of the update phase for a set of observed segments, any segment for which no correspondence was observed, and for which the CF has a value of less than CF_{max} has its confidence reduced by 1. If the CF of a segment drops below 1, the segment is removed from the composite model. Thus a segment must be present in at least 5 observations in order to be considered reliable enough to be preserved in the model.

5. Experimental Evaluation

We have performed a number of experiments in 3-D scene modeling with our system, using a camera mounted in a robot arm. The results of some of these experiments is described in this section.

5.1 Experimental Set-up

Our experiments used a CCD camera equipped with a 16 mm lens and mounted in the gripper of the arm. Video signals from the camera are digitized using a frame buffer/digitizer board mounted in the bus of a work-station. A six axis robot arm is linked with a robot controller capable of providing the location of the robot gripper simultaneously with the acquisition of each image. This information allows us to reconstruct the 3-D geometry of the scene using motion stereo.

The object used for each experiment is located at a distance of about 30 cm from the camera. The camera follows a roughly circular pre-programmed trajectory which passes over the object. Camera displacements are less than 1 cm per image, with rotations under 5 degrees per image.

Edge points were detected by a version of the Canny operator designed and programmed by R. Deriche of INRIA Rocquencourt [Canny 86], [Deriche 87]. Edge points were chained and segmented by a chaining program realized by G. Giraudon of INRIA Sophia-Antipolis. Near video rate hardware for these processes is currently under construction as part of the same project.

5.2 An Example of the Complete Process

One of the image sequences with which we have debugged the system is composed of 80 images of an electronic switch box. Figure 5.1 illustrates the description at different phases in the processing.

Each token in the flow model is assigned an identification number when it is created. This number identify tokens in snapshots of the flow model. Matches in each couple of images are represented by similar identification number and results are showed on the first line of figure 5.1 for images (30-35) and (65-70). For each couple, we proceed to a 3D reconstruction displayed on the second line of the figure 5.1. Each reconstruction corresponds to the same physical scene viewed by a different point of view but represented in the same frame coordinates. One can check that the same physical segment is represented by a similar label in the 3D reconstructed images.

The superposition of the 14 files obtained by reconstructing each 5 images along the sequence of 70 ones, is represented on the last line of figure figure 5.1 (left figure). This view points out a slight dispersion of the results but demonstrates the coherence of the obtained data. The last image (right) is the final result obtained by merging all the previous 3D files.

5.3 Validation

One possible verification of the acquired 3D model consists in reprojecting this model in one of the view of the sequence. By superimposing this projection on the corresponding image, we have a qualitative estimation of the error. But a qualitative evaluation can be only done by making some direct measurements on this object and comparing them with information provided by our process. Instead of dealing with the result of the process after combination of the whole sequence, we illustrate the evolution of both measurement and merging values at several snapshots of the sequence. Such an approach allows a better comprehension on the evolution of the process.

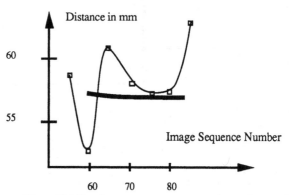

Figure 5.2 Depth measurement at several snapshots and fusion.

As an illustration of the precision of the process, consider the width of the switch box, as measured by the perpendicular distance between segment 87 and 156. The object's real width is 58 mm. The perpendicular distance between these segments, obtained from dumps of the model every 5 images between images 55 and

278

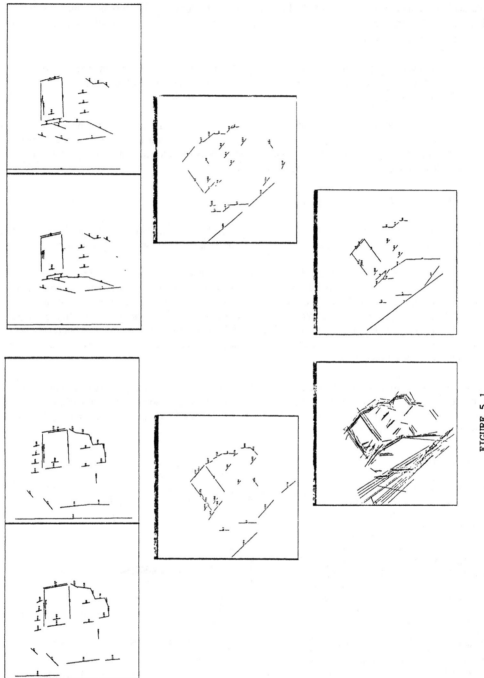

FIGURE 5.1

80 is {58.8, 53.9, 61.0, 58.0, 57.3 and 62.8}. When these values are integrated into the composite model using a Kalman filter, the distance is observed to be{57.7, 57.2, 57.4, 57.4 , 57.0 and 57.0} (see figure 5.2). The error in the integrated model is less that 2%.

Figure 5.2 shows a plot of the individual width values obtained by measurement after every 5th image, from images 55 to 80. We observe that for some camera orientations, the individual reconstructions may have errors which are on the order of 5%. We note that the precision is best when segments are located near the image center and when the 3D segments are closest to the cameras. Outside this set of views, the accuracy progressively degrades.

Nevertheless, Kalman filter the integration process copes with inaccurate data and is able to provide a good estimated value. Furthermore, this merging value is not drastically affected when an aberration in one measurement occurs.

6. Discussion and Improvements

The system described in this paper provides results which are sufficient for use in processes such as recognition by matching to a data-base. Nevertheless, the accuracy of the process can be improved by using a more sophisticated estimate of the uncertainty attached to each source of error and by solving some mechanical aspects of the demonstration. In particular, the system could be improved by the computation of the covariance matrix due to the calibration. An even more important improvement could be obtained if the robot arm controller were to accompany its estimateed position with an error estimate.

Our largest source of uncertainty is from the position and orientation of the robot end effector furnished by the robot arm controlled. The present computations consider the displacement matrix provided by the robot controller as accurate. However, after having obtained the results presented above, a verification of the robot calibration has shown that the the robot arm was very poorly calibrated. This fact points out the robustness of the technique and demonstrates that by over-estimated the covariance attached to each measurement, one can provide good results.

Acknowledgments

The work described in this paper was performed as part of project ESPRIT P940. The Token Tracker was implemented by Christophe Discours. The 3-D reconstruction algorithm was developed by Fano Ramparany and was integrated and tested by Jean-Noel Soulier and Laurent Lefort. Parts of this paper were written using the facilities of the laboratory of Prof. Tsuji of Univ. of Osaka, in Japan.

References

[Ayache 87]   Ayache, A. Faugeras, O. Maintaining Representation of the Environment of a Mobile Robot. In proc. International Symposium on Robotics Research, Santa Cruz, California, USA, August 1987.

[Ayache 89]   Ayache, N. "Construction et Fusion de Représentations Visuelles 3D", Thèse de Doctorat d'Etat, Universtié Paris-Sud, centre d'Orsay, 1988

[Brammer 89]   Brammer K. and G. Siffling, Kalman Bucy Filters, Artech House Inc., Norwood MA, USA, 1989.

[Canny 86] Canny, J. "A Computational Approach to Edge Detection", IEEE Trans. on P.A.M.I., Vol PAMI-8, No. 6, Nov. 1986.

[Chehikian 88] Stelmaszyk, P., C. Discours, and , A. Chehikian, "A Fast and Reliable Token Tracker", In IAPR Workshop on Computer Vision, Tokyo, Japan, October, 1988.

[Chehikian 89] Chehikian, A., S. Stelmaszyk and P. Depaoli, "Hardware Evaluation Process for tracking edges lines.", Workshop on Industrial Applications of Machine Intel. and Vision.Tokyo, 1989.

[Crowley 85] Crowley, J. L.,"Navigation for an Intelligent Mobile Robot", IEEE Journal on Robotics and Automation, 1 (1), March 1985.

[Crowley 86] Crowley, J. L., "Representation and Maintenance of a Composite Surface Model", IEEE International Conference on Robotics and Automation, San Francisco, Cal., April, 1986.

[Crowley-Ramparany 87] Crowley, J. L. and F. Ramparany, Mathematical Tools for Manipulating Uncertainty in Perception", AAAI Workshop on Spatial Reasoning and Multi-Sensor Fusion", Kaufmann Press, October, 1987.

Crowley et. al. 88] Crowley, J. L., P. Stelmaszyk and C. Discours, "Measuring Image Flow by Tracking Edge-Lines", ICCV 88: 2nd International Conference on Computer Vision, Tarpon Springs, Fla., Dec. 1988.

[Deriche 87] Deriche, R., "Using Canny's Criteria to Derive a Recursively Implemented Optimal Edge Detector", International Journal of Computer Vision, Vol 1(2), 1987.

[Durrant-Whyte 87] Durrant-Whyte, H. F., "Consistent Integration and Propagation of Disparate Sensor Observations", Int. Journal of Robotics Research, Spring, 1987.

[Faugeras, et. al. 86] Faugeras, O. D. , N. Ayache, and B. Faverjon, "Building Visual Maps by Combining Noisey Stereo Measurements", IEEE International Conference on Robotics and Automation, San Francisco, Cal., April, 1986.

[Faugeras and Toscani] Faugeras, O. D. ,G. Toscani, "The Calibration Problem for Stereo. Computer Vision and Pattern Recognition, pp 15-20, Miami Beach, Florida, USA, June 1986.

[Gennery, 82] Gennery, D. B., "Tracking Known Three Dimensional Objects", Proc. of the National Conference on Artificial Intelligence (AAAI-82), Pittsburgh, 1982.

[Harris 1988] Harris C.G, "Using a Sequence of More than 2 Images", IEEE Workshop on Motion and Stereopsis in Machine Vision", Digest 7, pp 44-49, 1989.

[Huang 83] Huang T. H., Image Sequence Processing and Dynamic Scene Analysis, Springer Verlag, Berlin, 1983.

[Hildreth 82] Hildreth, E. C., The Measurement of Visual Motion, 1983 ACM Distinguished Dissertation, MIT Press, Cambridge Mass. 1982.

[Matthies et. al. 87] Matthies, L., R. Szeliski, and T. Kanade, "Kalman Filter-based Algorithms for Estimating Depth from Image Sequences", CMU Tech. Report, CMU-CS-87-185, December 1987.

[Ramaparany 89] Ramparany, F., "Perception Multi-sensorielle de la Structure Geometrique d'une Scene", Thèse de Doctrat, INPG, Feb 1989.

[Roach and Aggarwal 80] Roach, J. W. and J. K. Aggarwal, "Determining the Movement of Objects in a Sequence of Images", IEEE Transactions on P.A.M.I., PAMI-2, No. 2, 1980.

[Shui-Ahmad 87] Shui, Y. C. and S. Ahmad, "Finding the mounting position of a sensor by solving a homogeneous transformation equation of the form AX = XB", in Proc. of the IEEE International Conference of Robotics and Automation, San Francisco, June, 1986.

[Suzuki and Yachida 89] Suzuki k. and Yachida M, "Establishing Correspondence and Getting 3-D Information in Dynamic Images", On Electronic Information and Communication Journal, Vol J72-D-II No5, pp 686-695 May 1989 (In Japenese).

[Tsai 85] Tsai, R. Y., "An Efficient and Accurate Camera Calibration Technique for 3-D Machine Vision", Technical Report, IBM T. J. Watson Reearch Center, 1986.

Estimation of 3D-motion and structure from tracking 2D-lines in a sequence of images

T. VIEVILLE

INRIA Sophia, BP109 06561 Valbonne, France

We establish the motion equations for rigidly moving 3D lines, and the structure equations that relate a temporal match of 2D lines in three consecutive pictures in an image sequence. We also analyse in details the numerical stability of such estimations.

1 Introduction

The aim of this study is to develop a method of recovery of structure from motion and of camera displacement estimation in a situation where the following features are available :
(1) A sequence of view with a stationary background and one or more rigid objects in motion,
(2) Only one camera,
(3) A real-time acquisition every 20 *msec*,
(4) An odometric estimation of ego-motion.

This study extends previous works [3,2], on the problem of recovery of structure and motion from two or three instantaneous views, and is directly related with these studies.

Basis of the study

In a temporal sequence of views, early vision provides the estimation of 2D line segments in the picture, with their statistical covariance. The line support of each line segment (unbounded lines), will be considered, as elementary tokens. The *token-tracker Algorithm* designed by R. Deriche [5] provides *temporal matches* between the same 2D line segment in two or more consecutive views, a kind of "temporal stereo" algorithm. It is then well known [3] that a match between at least 3 views is required to have a constraint on the motion on lines, as it is the case here. Given such matches the structure **and** the motion of the line can be computed.

Then, we can formulate the problem studied in this paper as follows : *Given 3 consecutive views in a temporal sequence of images closely related in time, and matches between 2D lines, in these views, how do we compute the related 3D line parameters, and the instantaneous rigid 3D-motion of this line (angular velocity ω and linear velocity $\mathbf{v}$).*

What is this paper about.

This paper introduce the data representation and the equations used for the estimation of 3D motion and structure of 2D lines. We discuss the precision, stability and sensitivity of these equations when using real data.

2 Equations of 3D lines and 3D motion

2.1 Representation of lines and rigid motion

Camera geometry and camera motion

The camera is calibrated, and the geometric quality of actual CCD sensors, legitimates the use of pinhole model for the camera, with a unit focal distance. Every quantity is referred to the camera intrinsic coordinates. The origin of this frame of reference is the optical center of the camera (its image nodal point, in fact) and the $(\mathbf{X}, \mathbf{Y}, \mathbf{Z})$ vectors are oriented as shown on Fig. 1, the $\mathbf{Z}$ axis corresponding to the optical axis of the camera.

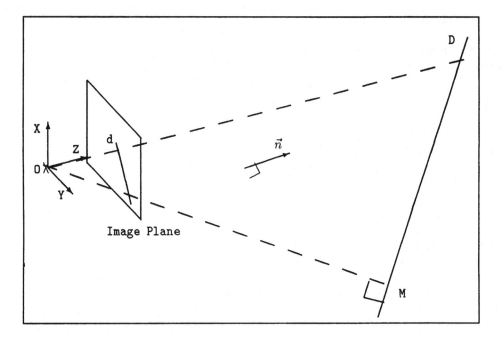

Figure 1: 2D and 3D lines representations

The camera motion is given by the kinematic screw $\{\mathbf{v}, \omega\}$. This 3D-motion will be assumed to be locally constant in time. The temporal sequence of views being taken at a high rate, the translation between two pictures will be approximated by :

$$\mathbf{t} = \Delta T \cdot \mathbf{v}$$

where ΔT is the sampling period between two pictures. The rotation between two pictures will be approximated by [1] :

$$R = e^{\Delta T \cdot \tilde{\omega}} \simeq I + \Delta T \cdot \tilde{\omega}$$

2D and 3D lines representations

A straight line D, in 3D space, is represented by its unitary direction vector δ and a point M on the line D. The point M is chosen to be the closest point to the optical center O of the camera (Fig.1) . It is equivalent to say that M is chosen such that $\vec{OM}$ is orthogonal to δ.

A 2D-line d in the retina plane is represented by a unitary vector $\mathbf{n} = (u, v, w)^T$ giving its equation :

$$u\,x + v\,y + w = 0 \tag{1}$$

The interpretation of $\mathbf{n}$ is that it is the normal to the plane defined by the 2D line and the optical center of the camera, as represented on Fig.1. The vector $\mathbf{n}$ is then orthogonal to δ and $\vec{OM}$.

The unitary vector $\mathbf{n}$ is the output of our version of the token-tracker. Each $\mathbf{n}$ estimation is - in fact - the mean value of a statistical estimation. A covariance matrix noted Wn is also provided by the token-tracker.

Summarizing, the unitary vector δ and the point M are constrained by the following equations :

$$\begin{aligned}
\mathbf{n}^T \cdot \mathbf{n} &= 1 \\
\mathbf{n}^T \cdot \delta &= 0 \\
\delta^T \cdot \delta &= 1 \\
\delta^T \cdot \vec{OM} &= 0 \\
\mathbf{n}^T \cdot \vec{OM} &= 0
\end{aligned}$$

Matching between 3 views

Let us consider three views: the *present view* (subscript 0), the *previous view* (subscript $-$), the *next view* (subscript $+$), as represented on Fig.2. The $\mathbf{n}$ vectors components, as computed by the token-tracker, are given in a frame of reference attached to each view, while we want all quantities to be expressed in the same frame of reference, let us say, in the present view.

We are going to use the subscript $/0$ (only when necessary), to state that a quantity is expressed in the present view frame of reference, as shown on Fig.1.

In order to simplify the notations let us take $\Delta T = 1$. This will not modify the nature our derivations.

---

[1] I will denote the 3×3 identity matrix, and $\tilde{\omega}$ the antisymmetric matrix such that :

$$\forall \mathbf{x}, \tilde{\omega} \cdot \mathbf{x} = \omega \wedge \mathbf{x}$$

where $\wedge$ denotes the cross product of two vectors.

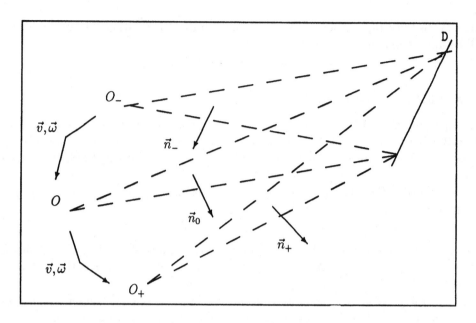

Figure 2: Schematic representation of a segment matching over 3 views

The transformations are then given by :

$$
\begin{aligned}
\mathbf{n}_{-/0} &= R \cdot \mathbf{n}_- \simeq (I + \tilde{\omega}) \cdot \mathbf{n}_- = \mathbf{n}_- + \omega \wedge \mathbf{n}_- \\
\mathbf{n}_{+/0} &= R^{-1} \cdot \mathbf{n}_+ \simeq (I - \tilde{\omega}) \cdot \mathbf{n}_+ = \mathbf{n}_+ - \omega \wedge \mathbf{n}_+ \\
\mathbf{n}_{0/0} &= \mathbf{n}_0 \\
\vec{OO}_{-/0} &= -\mathbf{v} \\
\vec{OO}_{+/0} &= +\mathbf{v} \\
O_{/0} &= O
\end{aligned}
$$

2.2 Constraints on the angular velocity and 3D-line direction

Derivating equations

All our developments will be based on the following geometrical property : *since in the three views the n vectors are orthogonal to the 3D-line direction δ, these three vectors are in the same plane.* This condition can be written as :

$$
\mid \mathbf{n}_0, \, \mathbf{n}_{-/0}, \, \mathbf{n}_{+/0} \mid = 0 \tag{2}
$$

Expanding equation 2 as a function of ω, we have [2] :

$$\mid n_0, n_-, n_+ \mid +(\omega, \beta) - (\omega, n_0) \cdot (\omega, n_- \wedge n_+) = 0 \tag{3}$$

where the vector $\beta = A \cdot n_0$. The A matrix is defined as :

$$A = n_- n_+^T + n_+ n_-^T - 2(n_+, n_-)I$$

Looking at the previous transformation equation 2 can be equivalently formulated as follows : the orientation of the 3D line (δ), computed from the previous and next views $(\delta = n_{-/0} \wedge n_{+/0})$, should be orthogonal to n_0. This statement is, of course, not dependent upon permutations of the n vectors. It is, in fact, the only one equation one can obtain on ω from the n vectors, in three views. The vector β defines the direction along which ω can be computed using equation 2.

The matrix A is a symmetric matrix. Its eigen values are easy to compute, since we have :

$$A \cdot x = \lambda \cdot x \Leftrightarrow (n_-, x)n_+ + (n_+, x)n_- = (\lambda + 2(n_-, n_+))x$$

and the three solutions are :

$$
\begin{array}{rclrcl}
x &=& n_- \wedge n_+, & \lambda_1 &=& -2(n_-, n_+) \\
x &=& n_- - n_+, & \lambda_2 &=& -1 - (n_-, n_+) \\
x &=& n_- + n_+, & \lambda_3 &=& 1 - (n_-, n_+)
\end{array}
$$

This result will be useful to study the stability of equation 2, and the conditioning of A. In addition, one can see that A is the sum of two projections. Since the matrix :

$$P_{u^\perp \| v} = uv^T - (u, v)I$$

defines the projection on to the vector plane orthogonal to u along the direction of v (Note that we have : $P_{u^\perp \| v}^T = P_{v^\perp \| u}$), we have : $A = P_{n_-^\perp \| n_+} + P_{n_+^\perp \| n_-}$

Equation 2 is a quadratic equation on ω and the related quadric Q. Taking the non-orthogonal frame of reference (x, y, z) defined by the equations :

$$
\begin{array}{lll}
(x, n_0) = 1 & (y, n_0) = 0 & (z, n_0) = 0 \\
(x, n_- \wedge n_+) = 0 & (y, n_- \wedge n_+) = 1 & (z, n_- \wedge n_+) = 0 \\
(x, \beta) = 0 & (y, \beta) = 0 & (z, \beta) = 1
\end{array}
$$

which have a unique solution if only if :

$$\mid n_0, n_- \wedge n_+, \beta \mid = ((n_0, n_+) + (n_0, n_-)) \cdot ((n_0, n_+) - (n_0, n_-)) \neq 0$$

---

[2] We used the following notations and relations in our computations :

$$
\begin{array}{c}
(x, y) = x^T y = y^T x \quad (x, x) = \|x\|^2 = x^2 \\
(x \wedge y, z) = \mid x, y, z \mid = \mid y, z, x \mid = - \mid y, x, z \mid = \ldots \\
x \wedge (y \wedge z) = (x, z)y - (x, y)z \\
(x \wedge y) \wedge (x \wedge z) = \mid x, y, z \mid \cdot x \\
\tilde{x} \cdot y = x \wedge y \\
(x \wedge y, x \wedge z) = (y, z)(x, x) - (x, y)(x, z)
\end{array}
$$

and the multilinearity of $\mid x, y, z \mid$.

and if we note $\omega = (a, b, c)^T$ in this frame of reference we simply get :

$$\mathbf{x} = (a, b, c)^T \in Q \Leftrightarrow a \cdot b + c + q_0 = 0$$

This is obviously the equation of an hyperbolic cone (see for example [4], Chap XI), a quadric of rank 3. In the case where there is not a unique solution, the degenerate case, the quadric is of rank 2, and it is known to be a pair of two planes.

Finally, one should notice that *only the component of ω orthogonal to the 3D-line direction δ can be computed*. The component of ω aligned with δ, induces a rotation for which δ is invariant, and the components of the n vectors in their local frame of references are not modified by it. (One should also notice that, since we have no information a priori on the relative location of two views, we have no information on the relative location of the n vectors in space, but only on their location with respect to each view frame of reference. This is a fundamental difference with stereo).

One can then assume ω to be orthogonal to δ, that is in the plane of the n vectors. These additional constraints are :

$$
\begin{aligned}
| \omega, \mathbf{n}_{-/0}, \mathbf{n}_{+/0} | &= (1 - \omega^2)(\omega, \mathbf{n}_- \wedge \mathbf{n}_+) + 2(\omega \wedge \mathbf{n}_-, \omega \wedge \mathbf{n}_+) &= 0 \\
| \omega, \mathbf{n}_0, \mathbf{n}_{-/0} | &= | \omega, \mathbf{n}_0, \mathbf{n}_- | + (\omega \wedge \mathbf{n}_0, \omega \wedge \mathbf{n}_-) &= 0 \quad\quad (4) \\
| \omega, \mathbf{n}_0, \mathbf{n}_{+/0} | &= | \omega, \mathbf{n}_0, \mathbf{n}_+ | - (\omega \wedge \mathbf{n}_0, \omega \wedge \mathbf{n}_+) &= 0
\end{aligned}
$$

Numerical stability of the previous equations

It is useful to consider the order of magnitude of the terms in the expansions of equation 2. Since the instantaneous rotation between the three views has a small angle, and is assumed to be locally constant, the relative angles between the n vectors, and the norm of ω are small quantities. We are now going to used Taylor expansions to study the numerical stability of the previous equations.

Our discussion will be based on the following statements :

1. Since $\mathbf{n}_0$, $\mathbf{n}_{-/0}$, and $\mathbf{n}_{+/0}$ are in the same plane we have :

$$(\widehat{\mathbf{n}_{-/0}, \mathbf{n}_{+/0}}) = (\widehat{\mathbf{n}_{-/0}, \mathbf{n}_0}) + (\widehat{\mathbf{n}_0, \mathbf{n}_{+/0}})$$

2. Since the translation is small, the angles $(\widehat{\mathbf{n}_{-/0}, \mathbf{n}_0})$ and $(\widehat{\mathbf{n}_0, \mathbf{n}_{+/0}})$ have the same order of magnitude. This is illustrated on Fig.3. The angles $(\widehat{O_-, M, O_0})$ and $(\widehat{O_0, M, O_+})$ are roughly equal, since v is small with respect to $\|\vec{O_-M}\|$, $\|\vec{O_0M}\|$, and $\|\vec{O_+M}\|$. Since $\mathbf{n}_{-/0}$, $\mathbf{n}_0$, and $\mathbf{n}_{+/0}$ are orthogonal respectively to $\vec{O_-M}$, $\vec{O_0M}$ and $\vec{O_+M}$, there relative angles are equal. We can then conclude that $(\widehat{\mathbf{n}_{-/0}, \mathbf{n}_0})$ and $(\widehat{\mathbf{n}_0, \mathbf{n}_{+/0}})$ are roughly equal. Precisely this means that they differ only at the second order. We will use this result though the following notations :

$$
\begin{aligned}
(\widehat{\mathbf{n}_{-/0}, \mathbf{n}_0}) &= \epsilon_- = \epsilon + \epsilon^2 \\
(\widehat{\mathbf{n}_0, \mathbf{n}_{+/0}}) &= \epsilon_+ = \epsilon - \epsilon^2
\end{aligned}
$$

3. Since ω is in the same plane P of the vectors n, its orientation is entirely defined one angle, tel us say $(\widehat{\omega, \mathbf{n}_0})$. Since the norm of ω is the angle of the rotation, and since $\mathbf{n}_{+/0}$ is related to $\mathbf{n}_0$ by this rotation, and $\mathbf{n}_0$ is also related to $\mathbf{n}_{0/-}$ by this rotation,

the norm of ω has the same order of magnitude as $(\widehat{n_{-/0}, n_0})$ and $(\widehat{n_0, n_{+/0}})$. We will summarize these two points by the following notations :

$$(\widehat{\omega, n_0}) = \alpha$$
$$\|\omega\| = w\varepsilon$$

where $w = \dfrac{2\|\omega\|}{(n_{-/0}, n_{+/0})}$ is the ratio between $\|\omega\|$ and ε. We have :

$$w = o(1)$$

since $\|\omega\|$ and ε have a similar order of magnitude.

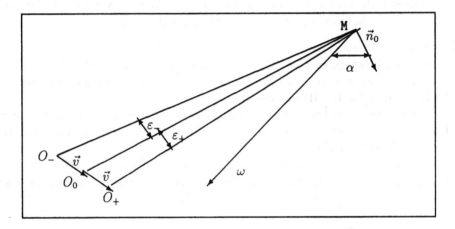

Figure 3: Schematic representation of the angles in the plane P defined by the vectors **n**

With these notations, and using equations 4 one can easily derive the orders of magnitude of all quantities related to equation 2. The following results are then obtained :

- The quadratic term in equation 2 is not negligible since :

$$\|\beta\| = \sqrt{3 + (2 - 10\cos(\alpha)^2)w^2}\varepsilon + o(\varepsilon^2)$$
$$| n_0, n_-, n_+ | = o(\varepsilon^2)$$
$$(\omega, n_- \wedge n_+)(\omega, n_0) = o(\varepsilon^2)$$

and we will have to deal with a quadratic equation, in term of ω, even for small angles of rotation.

- The vectors **n** do not form a stable frame of reference, since their relative angles are very small.

$$(\widehat{n_-, n_0}) = (\widehat{n_0, n_+}) = \sqrt{1 + 2\sin(\alpha)^2 w^2}\varepsilon + o(\varepsilon^2)$$
$$(\widehat{n_-, n_+}) = \sqrt{(6 + 2*\cos(\alpha)^2)\|\omega\|} + o(\varepsilon^2)$$

- The matrix A is not well conditioned since one eigen-value is very small with respect to the others, while the two other eigen-values are very similar. We precisely have :

$$\begin{aligned}
\lambda_3 &= (3 + \cos(\alpha)^2)\|\omega^2\| + o(\varepsilon^3) \\
\lambda_1 &= -2 + 2\lambda_3 + o(\varepsilon^3) \\
\lambda_2 &= -2 + \lambda_3 + o(\varepsilon^3)
\end{aligned}$$

- The direction of the vector β is not trivial, but we have :

$$\begin{aligned}
(\beta, \widehat{\mathbf{n}_- \wedge \mathbf{n}_+}) &= o(\varepsilon) \\
(\beta, \widehat{\mathbf{n}_{-/0} \wedge \mathbf{n}_{+/0}}) &= \tfrac{\Pi}{2} + o(\varepsilon) \\
(\widehat{\beta, \mathbf{n}_0}) &= \Phi\, o(\varepsilon) \ \ with \ \ 0 << \Phi << \tfrac{\Pi}{2}
\end{aligned}$$

This defines the direction along which ω can be estimated which is roughly aligned with $\mathbf{n}_- \wedge \mathbf{n}_+$ but not with δ or $\mathbf{n}_{-/0} \wedge \mathbf{n}_{+/0}$, as expected.

This has two consequences :
(1) On one hand, we are going to use the found quadratic constraint to recursively estimate ω. This equation can be used as a measurement equation in an extended Kalman Filter, used for the recovery of the 3D line motion.
(2) On the other hand, we are not going to try to reconstruct ω using five, or more than five, views, since even if the reconstruction is, in principle, possible, the computation will not be numerically stable.

In addition, the use of a first estimate of ω from odometric cues will be very useful in this approach, and the previous constraint will be used only to correct the a priori estimate.

Estimation of the 3D-line direction

Since $\mathbf{n}_0$ is constrained to be in the same plane as $\mathbf{n}_{-/0}$ and $\mathbf{n}_{+/0}$, as discussed previously, and since δ is orthogonal to this plane, the 3D-line direction is simply given by :

$$\delta \parallel \mathbf{n}_{-/0} \wedge \mathbf{n}_{+/0} = \mathbf{n}_- \wedge \mathbf{n}_+ + A\omega + o(\omega^2).$$

In the previous section, we studied in details the properties of the matrix A, and the direction β along which ω is estimated. Let us remind that β is very close to the direction of $\mathbf{n}_- \wedge \mathbf{n}_+$ which is a eigen direction of A. Then, ω estimation for this line is mainly performed in the direction on $\mathbf{n}_- \wedge \mathbf{n}_+$, only. We then are going to consider $\omega \parallel \mathbf{n}_- \wedge \mathbf{n}_+$, and we have $A\omega \parallel \mathbf{n}_- \wedge \mathbf{n}_+$ also, and finally :

$$\delta \wedge (\mathbf{n}_- \wedge \mathbf{n}_+) \ = \ 0 \tag{5}$$

or $\delta = \frac{\mathbf{n}_- \wedge \mathbf{n}_+}{\|\mathbf{n}_- \wedge \mathbf{n}_+\|}$.
However, since we have

$$\|\mathbf{n}_- \wedge \mathbf{n}_+\| \ = \ \sqrt{2 * \lambda_3} + o(\varepsilon^2)$$

where λ_3 has been defined previously, this direct estimate is not numerically stable, and equation 5 is to be used instead.

2.3 Constraints on the translation and the 3D-line distance

The translation velocity $\mathbf{v}$ and the 3d-line distance characterized by the location of the point M can be computed from the following relations :

$$\left\{ \begin{array}{ccc} O_{-}\vec{M} & \perp & \mathbf{n}_{-/0} \\ O_{+}\vec{M} & \perp & \mathbf{n}_{+/0} \\ \vec{OM} & \perp & \mathbf{n}_0 \\ \vec{OM} & \perp & \delta \\ & where : & \\ O_{-}\vec{M} & = & \mathbf{v} + \vec{OM} \\ O_{+}\vec{M} & = & \mathbf{v} - \vec{OM} \end{array} \right.$$

It is immediate to derive :

$$\begin{array}{ccc} (\mathbf{v}, \mathbf{n}_{-/0}) & = & -(\vec{OM}, \mathbf{n}_{-/0}) \\ (\mathbf{v}, \mathbf{n}_{+/0}) & = & +(\vec{OM}, \mathbf{n}_{+/0}) \end{array}$$

Since $\vec{OM}$ is orthogonal to δ, $\vec{OM}$ is in the plane P of the $\mathbf{n}$ vectors. Since $\vec{OM}$ is also orthogonal to $\mathbf{n}_0$, M is on the unique line of the plane P, orthogonal to $\mathbf{n}_0$, and going through the origin. Using this remark, and after some algebra, the previous set of relations is equivalent to :

$$(\mathbf{v}, \gamma) = 0 \tag{6}$$

and :

$$\begin{array}{c} \vec{OM} \parallel \zeta \\ (\vec{OM}, \mathbf{n}_{-/0} - \mathbf{n}_{+/0}) = -(\mathbf{v}, \mathbf{n}_{-/0} + \mathbf{n}_{+/0}) \end{array} \tag{7}$$

or :

$$\vec{OM} = \frac{(\mathbf{v}, \mathbf{n}_{-/0} + \mathbf{n}_{+/0})}{(\zeta, \mathbf{n}_{-/0} - \mathbf{n}_{+/0})} \zeta$$

The vectors γ and ζ are two vectors of the plane P containing the three $\mathbf{n}$ vectors, precisely :

$$\begin{array}{ccc} \gamma & = & a\mathbf{n}_{-/0} - b\mathbf{n}_{+/0} \\ \zeta & = & (\mathbf{n}_{+/0}, \mathbf{n}_0)\mathbf{n}_{-/0} - (\mathbf{n}_{-/0}, \mathbf{n}_0)\mathbf{n}_{+/0} \end{array}$$

where :

$$\begin{array}{ccc} a & = & (\mathbf{n}_{-/0}, \mathbf{n}_0) - (\mathbf{n}_{-/0}, \mathbf{n}_{+/0})(\mathbf{n}_{+/0}, \mathbf{n}_0) \\ b & = & (\mathbf{n}_{+/0}, \mathbf{n}_0) - (\mathbf{n}_{-/0}, \mathbf{n}_{+/0})(\mathbf{n}_{-/0}, \mathbf{n}_0) \end{array}$$

In equations 6 and 7 the estimation of $\mathbf{v}$ and M are decoupled. We have one equation for $\mathbf{v}$, which constrains its direction but not its amplitude, as expected in a monocular system. One additional constraint could be : $\mathbf{v}^2 = 1$. In our case, since we have an odometric estimation of $\mathbf{v}$, we are going to use equation 6 only to correct this estimate.

An estimation of $\mathbf{v}$ being provided the point M is uniquely defined using the pair of linear equations 7. The numerical stability of these equations can be studied as previously, and we have the following results :

- The vectors $\mathbf{n}_{-/0} + \mathbf{n}_{+/0}$ and $\mathbf{n}_{-/0} - \mathbf{n}_{+/0}$ are orthogonal, and can be used as an orthogonal frame of reference in the plane P.

- The vector ζ is roughly aligned with $n_{-/0} - n_{+/0}$ and its norm has the order of magnitude of the n vectors angles :

$$\|\zeta\| = 2\epsilon + o(\epsilon^4)$$
$$(\zeta, \widehat{n_{-/0} - n_{+/0}}) = o(\varepsilon)$$

- As for δ, the direct estimation of M is not numerically stable. However using equation 7, since ζ is almost aligned with $n_{-/0} + n_{+/0}$, the second line of equation 7 provides a stable constrains on the norm of $\vec{OM}$.

- The vector γ is roughly aligned with $n_{-/0} + n_{+/0}$ and its norm is very small which is acceptable, since it is used in a homogeneous equation (eq 6) :

$$\|\gamma\| = 4\epsilon^2\varepsilon + o(\epsilon^5)$$
$$(\gamma, \widehat{n_{-/0} + n_{+/0}}) = o(\tfrac{\epsilon^2}{\epsilon^{(1/2)}})$$

3   Conclusion

Given a match between three 2D-lines in three consecutive views, we can compute one quadratic equation on the angular velocity (eq 3), and one linear equation on direction of the linear velocity (eq 6). These equations are based on quantities having a small order of magnitude ($o(\varepsilon)$) but which are numerically stable since these order of magnitude are compatible.

The computation of the velocity torque should be done either in cooperation with other matches and/or in cooperation with odometric cues, since the velocity torque is evaluated only along a given direction (β or γ).

Given a match between three 2D-lines in three consecutive views, with an estimation of the velocity torque, the direction of the 3D-line and its location with respect to the optical center of the camera can be directly evaluated from equations 5 and 7.

In this study we do not use estimations of parameters velocity, as it was done in [2]. We then avoided the computations of time derivative, which have the drawbacks to be noise-sensitive, while the choice of a good derivative estimator is a complex problem. One can consider our study as a "discrete version" of the approach in [2], where we implicit estimate velocity and acceleration related quantities, since we use 3 consecutive views. However, the derivated equations are slightly different and seem to be much stable in our case.

In comparison to the study of [5], where a similar problem has been investigated, we made profit of the fact that pictures are very closed, while the rotation between two pictures can be approximated by a simple cross-product. We then come to linear equations, and could study in details the precision of our method. In addition, we are here dealing with the line support of the segments, instead of their extremities, or other points of interest. Line-tokens are less noisy geometrical primitives in a picture, since they are estimated from several points, and correspond to recognizable features in the visual environment.

Other methods proposed in the literature reach their limit very quickly as the noise in the data increases [1], or are based on rather heavy computations [6], or implies important restrictions on the type of visual environment [7].

The use of odometric cues in cooperation with one camera provides a solution to the scale factor problem, for a stationary background. In the future the cooperation between vision and odometry will be developed from this initial study.

Acknowledgments This work has been initiated by **Olivier Faugeras**. It has been realized thanks to its powerful advices and kind corrections. It is a direct continuation of some of his previous work in the field.

Rachid Deriche and Nassir Navab are gratefully acknowledge for the fruitful discussions we had during the progress of this work.

Formal computations have been derived using the Maple software, and its mpls package.

This work was partially completed under **Esprit Project P2502/VOILA**.

References

[1] J. Fang and T. Huang.
Some experiments on estimating the 3-d motion parameters of a rigid body from two consecutive image frames.
IEEE Transactions on Pattern Analysis and Machine Intelligence, 6:547–554, 1984.

[2] O. D. Faugeras, R. Deriche, and N. Navab.
From optical flow of lines to 3D motion and structure.
In *Proceedings of the IROS*, 1989.
to appear.

[3] O. D. Faugeras, F. Lustman, and G. Toscani.
Motion and structure from point and line matches.
In *Proceedings of the First International Conference on Computer Vision, London*, pages 25–34, June 1987.

[4] G. T. K. J. G. Semple.
Algebrical Projective Geometry.
Oxford, The Clarendon Press, 1979.

[5] G. Toscani, R. Deriche, and O. Faugeras.
3D motion estimation using a token tracker.
In *Proceedings of the IAPR Workshop on Computer Vision (Special Hardware and Industrial Applications), Tokyo, Japan*, pages 257–261, October 1988.

[6] H. Trivedi.
Estimation of stereo and motion parameters using a variational principle.
Image and Vision Computing, 5, May 1987.

[7] R. Tsai, T. Huang, and W. Zhu.
Estimating three-dimensional motion parameters of a rigid planar patch. ii: singular value decomposition.
IEEE Transactions on Acoustic, Speech and Signal Processing, 30:525–534, August 1982.

The Incremental Rigidity Scheme for Structure from Motion : The Line-Based Formulation[†]

Daniel Dubé[‡]   **Amar Mitiche**

INRS Télécommunications

3, Place du Commerce, Ile des Soeurs, Qc, Canada H3E 1H6

Abstract: This paper presents an extension to lines of Ullman's incremental rigidity scheme, originally formulated for a set of points. The formulation is based on the angular and distance invariance of rigid configurations of lines. It is shown that the line structure can be recovered incrementally from its motion.

1. Line-based incremental rigidity scheme

The incremental rigidity scheme for the recovery of structure from motion, as proposed by Ullman [1] for a structure of points, constructs an internal estimated model of the structure which is continuously updated, as rigidly as possible, at each time a new image is available. The current estimated model is modified by the minimal "physical change" that is sufficient to account for the observed transformations in the new image. According to Ullman's results, the proposed incremental rigidity scheme converges to the correct structure. The goal of this study is to examine a line-based formulation of Ullman's scheme. Albeit with an additional weak constraint, we show that results similar to those of Ullman with point structures can be obtained with configurations of lines.

The viewing system is modelled as in Figure 1 (cartesian reference system and parallel projections, the Z-axis pointing to the observer). The line-based formulation we propose consists of two successive steps, the first step being the recovery of the orientation of the lines and the second being the complete recovery of the structure. Using the principle of angular invariance, the scheme estimates the orientations of the lines; recovery of structure is completed using distance invariance.

We will take the model $M(t)$ of the line structure at time t to be the set $\{U_i, Z_i\}$, $i = 1, ..., N$, where U_i is the unit orientation vector of line L_i, Z_i is the depth of *any*

[†] This work was supported in part by the Natural Sciences and Engineering Research Council of Canada under grant NSERC-A4234

[‡] D.Dubé is now T/Associate Director - Network Systems Research, Bell Canada

point on line L_i and N is the number of lines in the structure. At $t = 0$, the model is taken to be "flat" (all the lines lying in a plane parallel to the image plane).

1.1 Angular invariance: recovery of orientations

For a rigid motion of a set of lines, the principle of angular invariance states that the angles between the lines do not change as a result of this motion [2]. Let L_i and L_j be two lines in space, having projections l_i and l_j, respectively (see Figure 1). Let unit vectors on L_i and L_j be U_i, U_j at time t and U_i', U_j' at time t'. Then, the principle of angular invariance states that (we assume that correspondence between lines, and their direction, has been established):

$$U_i \cdot U_j = U_i' \cdot U_j' \tag{1}$$

In expanded form, with unnormalized vectors V:

$$\frac{v_{1,i}v_{1,j} + v_{2,i}v_{2,j} + v_{3,i}v_{3,j}}{\|V_i\|\|V_j\|} = \frac{v_{1,i}'v_{1,j}' + v_{2,i}'v_{2,j}' + v_{3,i}'v_{3,j}'}{\|V_i'\|\|V_j'\|} \tag{2}$$

Equation 2 is written for each pair of lines (some equations may be redundant). Given the current estimate of the orientations and a new 2-D image, the problem is to determine the unknown parameters $v_{3,i}'$, $i = 1, ..., N$ so as to minimize the overall deviation from a rigid transformation as prescribed by the incremental rigidity scheme. The following function Ψ is considered for minimization:

$$\Psi = \sum_{i=1}^{N-1} \sum_{j=i+1}^{N} \psi_{ij} \tag{3}$$

$$\psi_{ij} = \left[\frac{v_{1,i}v_{1,j} + v_{2,i}v_{2,j} + v_{3,i}v_{3,j}}{\|V_i\|\|V_j\|} - \frac{v_{1,i}'v_{1,j}' + v_{2,i}'v_{2,j}' + v_{3,i}'v_{3,j}'}{\|V_i'\|\|V_j'\|} \right]^2 \tag{4}$$

After function Ψ is minimized, resulting in a new set of $v_{3,i}'$, $i = 1, ..., N$, the corresponding orientations become part of the current version of model M.

1.2 Distance invariance: recovery of structure

The distances between each pair of lines remain constant during rigid motion. If we observe two lines L_i and L_j at t and t', then $d_{ij} = d_{ij}'$. The objective here is to incorporate this rigidity constraint into the incremental rigidity scheme, using the line orientations obtained at the previous step.

At time t, let $U_i = (u_{1,i}, u_{2,i}, u_{3,i})$, $U_j = (u_{1,j}, u_{2,j}, u_{3,j})$ be unit vectors on L_i and L_j respectively, and $P_i = (x_i, y_i, z_i)$ and $P_j = (x_j, y_j, z_j)$ be points on L_i and L_j. At time t', let the unit vectors be $U_i' = (u_{1,i}', u_{2,i}', u_{3,i}')$, $U_j' = (u_{1,j}', u_{2,j}', u_{3,j}')$ and points be $P_i' = (x_i', y_i', z_i')$, $P_j' = (x_j', y_j', z_j')$ (P_i, P_j and P_i', P_j' are arbitrary and unrelated; no point correspondences are assumed).

According to the principle of distance invariance, the relation for lines L_i and L_j is:

$$\left| \frac{(x_j - x_i)\rho - (y_j - y_i)\sigma + (z_j - z_i)\tau}{\sqrt{\rho^2 + \sigma^2 + \tau^2}} \right| = \left| \frac{(x_j' - x_i')\rho' - (y_j' - y_i')\sigma' + (z_j' - z_i')\tau'}{\sqrt{\rho'^2 + \sigma'^2 + \tau'^2}} \right| \tag{5}$$

$$\rho = (u_{2,i}u_{3,j} - u_{3,i}u_{2,j}) \qquad \sigma = (u_{1,i}u_{3,j} - u_{3,i}u_{1,j}) \qquad \tau = (u_{1,i}u_{2,j} - u_{2,i}u_{1,j}) \tag{6}$$

$$\rho' = (u_{2,i}'u_{3,j}' - u_{3,i}'u_{2,j}') \qquad \sigma' = (u_{1,i}'u_{3,j}' - u_{3,i}'u_{1,j}') \qquad \tau' = (u_{1,i}'u_{2,j}' - u_{2,i}'u_{1,j}') \tag{7}$$

Equation 5 is written for each pair of lines (some of these may be redundant). Given the current estimate of distances and a new image, the problem is to find the unknown depths z_i', $i = 1, ..., N$ in accordance with the incremental rigidity scheme. The following function Λ is minimized:

$$\Lambda = \sum_{i=1}^{N-1} \sum_{j=i+1}^{N} \lambda_{ij} \tag{8}$$

$$\lambda_{ij} = \left[\left| \frac{(x_j - x_i)\rho - (y_j - y_i)\sigma + (z_j - z_i)\tau}{\sqrt{\rho^2 + \sigma^2 + \tau^2}} \right| - \left| \frac{(x_j' - x_i')\rho' - (y_j' - y_i')\sigma' + (z_j' - z_i')\tau'}{\sqrt{\rho'^2 + \sigma'^2 + \tau'^2}} \right| \right]^2$$

When Λ is optimized, a new set of z_i', $i = 1, ..., N$ is obtained and included in the current version of the model. A new image is acquired and the process is repeated: estimation of the new orientations using angular invariance, estimation of the positions using distance invariance, and update of the model.

2. Experimental results

For minimization, we used a quasi-Newton method. The initial values of $v_{3,i}'$, $i = 1, ..., N$, were set all to $+1$ or all to -1 [3]. Subsequent optimization steps used the current values of $v_{3,i}$. We have experimented with several arbitrary line structures in motion [3]. The motion reported here is a rotation about an axis parallel to the Y-axis, and located on the Z-axis. For the orientations, performance is measured with:

$$\Delta = \sum_{i} (1 - \cos \theta_i)/N \tag{9}$$

where $\cos \theta_i$ is the scalar product of the exact and estimated unit vectors of line i. For the positions of the lines in space, error function Γ is used:

$$\Gamma = \sum_{i,j} |d_{ij} - d'_{ij}|/N_p \tag{10}$$

where N_p is the number of pairs of lines and d_{ij}, d'_{ij} are respectively the exact and estimated distances for lines i and j.

With a rigid structure of 6 lines, typical results for the error functions Δ and Γ are shown in Figures 2 and 3 (rotations of 20°).

The scheme was experimented with *different rotation angles* (Figures 4 and 5). In general, results indicate that the convergence is accelerated with larger angular differences (also an observation in [1]). However when a sufficient level of difference is reached, greater angular variations do not necessarily increase the convergence rate.

We also examined the scheme with a *varying number of lines*. For structures containing 2 or 3 lines, the scheme exhibits an oscillatory error behavior. Results indicate that the performance gradually increases with the number of lines (Figures 6 and 7).

Experimental results also reveal that the estimation model can infer and maintain the exact structure in the presence of *small deviations from rigidity*. Performance deteriorates for larger perturbations.

Finally, better convergence results are obtained if *constraints* can be imposed on the depth parameters $v'_{3,i}$ (in practice, this amounts to knowing roughly the maximum size of the observed objects).

Summary: The incremental recovery scheme developed by Ullman [1] for point structures has been extended to line structures. The process relies on the minimization of two functions which are based on the maximum rigidity hypothesis and which involve the principles of angular and distance invariance. Albeit an additional weak constraint is met, the model eventually converges towards the exact structure which is then maintained.

References

[1] S.Ullman, "Maximizing Rigidity: The Incremental Recovery of 3-D Structure From Rigid And Nonrigid Motion," *Perception*, vol. 13, pp. 255-274, 1984.

[2] A.Mitiche, O.Faugeras, J.K.Aggarwal, "Counting Straight Lines," *Computer Vision, Graphics, and Image Processing*, vol. 47, pp. 353-360, 1989.

[3] D.Dubé, "Récupération de la structure 3-D de droites en mouvement selon une méthode d'inférence incrémentielle," Master's Thesis, INRS-Télécommunications, August 1989.

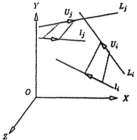

Fig. 1 *Parallel projections of lines on the X − Y image plane*

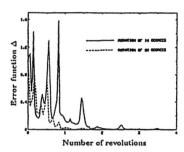

Fig. 2 *Error function Δ in terms of the number of rotations of 20°*

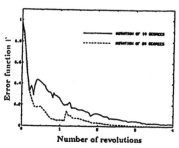

Fig. 3 *Error function Γ in terms of the number of rotations of 20°*

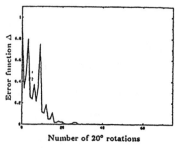

Fig. 4 *Error function Δ in terms of the number or revolutions for rotation angles of 10° and 20°*

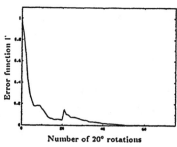

Fig. 5 *Error function Γ in terms of the number or revolutions for rotation angles of 10° and 20°*

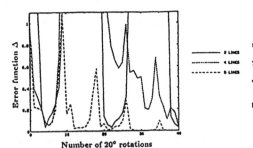

Fig. 6 *Impact of the number of lines (3,4,5) on the error function Δ in terms of the number of rotations of 20°*

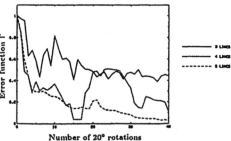

Fig. 7 *Impact of the number of lines (3,4,5) on the error function Γ in terms of the number of rotations of 20°*

Object detection using model based prediction and motion parallax

Stefan Carlsson
Telecommunication Theory

and

Jan-Olof Eklundh
Dep. of Numerical analysis and Computing Science
Royal Institute of Technology , S-100 44 Stockholm, Sweden

1. Motion parallax and object background separation

When a visual observer moves forward the projections of the objects in the scene will move over the visual image. If an object extends vertically from the ground its image will move differently from the immediate background. This difference is called motion parallax [1,2]. Much work in automatic visual navigation and obstacle detection has been concerned with computing motion fields, or more or less complete 3-D information about the scene [3-5]. These approaches in general assume very unconstrained environments and motion. If the environment is constrained, e.g. motion occurs on a planar road, then this information can be exploited to give more direct solutions to e.g. obstacle detection.[6]

Fig. 1.1 shows superposed the images from two successive times for an observer translating relative a planar road. The arrows show the displacement field, i.e. the transformation of the image points between the successive time points.

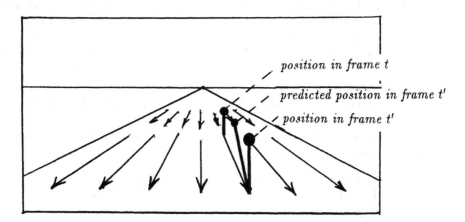

Fig. 1.1 Displacement field from road with predicted and actual position of vertically extended object

Fig. 1.1 also shows a vertically extended object at time t and t'. Note that the top of the object is displaced quite differently from the immediate road background. This effect is illustrated by using the displacement field of the road to displace the object. A clear difference between the actual image and the predicted image is observable for the object. This fact forms the basis of our approach to object detection. (Fig. 1.2) For a camera moving relative a planar surface the image transformation of the surface is computed and used to predict the whole image. All points in the image that are not on the planar surface will then be erroneously predicted. If there is intensity contrast at those parts we will get an error in the predicted image intensity. This error then indicates locations of vertically extended objects.

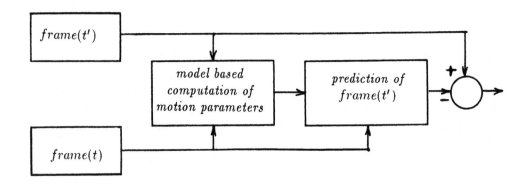

Fig. 1.2 Block diagram of processing for vertical object detection

2. Image transformation for motion relative a planar surface

With a moving camera each point in the scene will map to a different point in the image at different times. The transformation of the mapped image point over time is determined by the motion of the camera and the position of the point in the 3-dimensional scene. If the point is on a planar surface the transformation can be computed using the camera motion and position of the surface in space. Fig. 2-1 shows the coordinate system of the camera and the image plane. A rigid displacement of the camera can be decomposed into a translation with components D_X, D_Y, D_Z along the coordinates and a rotation around an axis passing through the point of projection, which can be decomposed into rotations around the axis of the coordinate system ϕ_X, ϕ_Y, ϕ_Z. Assuming small rotations, a point in the scene with coordinates X, Y, Z is then transformed to the point X', Y', Z' where:

$$\begin{pmatrix} X' \\ Y' \\ Z' \end{pmatrix} = \begin{pmatrix} 1 & -\phi_Z & \phi_Y \\ \phi_Z & 1 & -\phi_X \\ -\phi_Y & \phi_X & 1 \end{pmatrix} \begin{pmatrix} X \\ Y \\ Z \end{pmatrix} + \begin{pmatrix} D_X \\ D_Y \\ D_Z \end{pmatrix} \qquad [2.1]$$

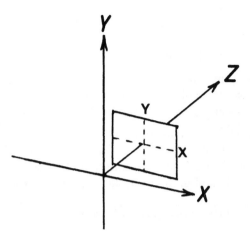

Fig. 2.1 Coordinate system of camera and image plane

If the image plane is located at unit distance from the point of projection the image coordinates (x, y) of a point X, Y, Z under perspective projection are

$$x = \frac{X}{Z} \qquad y = \frac{Y}{Z} \qquad\qquad [2.2]$$

The transformation of the projected image point of a point in the scene with depth Z is therefore [7]:

$$x' = \frac{x - \phi_Z y + \phi_Y + D_X/Z}{1 - \phi_Y x + \phi_X y + D_Z/Z} \qquad y' = \frac{y + \phi_Z x - \phi_X + D_Y/Z}{1 - \phi_Y x + \phi_X y + D_Z/Z} \qquad [2.3]$$

If the point X, Y, Z is located on a planar surface with equation $K_X X + K_Y Y + K_Z Z = 1$, the transformation in the image plane the becomes:

$$x' = \frac{(1 + K_X D_X)x + (K_Y D_X - \phi_Z)y + \phi_Y + D_X K_Z}{(1 + K_X D_Z - \phi_Y)x + (K_Y D_Z + \phi_X)y + K_Z D_Z} \qquad [2.4-a]$$

$$y' = \frac{(1 + K_Y D_Y)y + (K_X D_Y - \phi_Z)x - \phi_X + D_Y K_Z}{(1 + K_X D_Z - \phi_Y)x + (K_Y D_Z + \phi_X)y + K_Z D_Z} \qquad [2.4-b]$$

This is a nonlinear transformation of the image coordinates, determined by 9 parameters. The actual number of degrees of freedom of the transformation is however just 8, since parameters $\bar{K}$ and $\bar{D}$ always occur as products. which means that their absolute values are irrelevant.

3. Estimation of parameters by minimisation of prediction error

The transformation of the projected image points due to the motion of the camera will manifest itself as a transformation of the image intensity $I(x, y)$. If t and t' are the time instants before and after the transformation, we shall assume that:

$$I(x', y', t') = I(x, y, t) \qquad [3.1]$$

where x, y and x', y' are related according to eq. 2.4.

I.e. we assume that the transformation of the image intensity is completely determined by the geometric transformation of the image points. This is not strictly true in general since we neglect factors as changing illumination etc.

For points on a planar surface our assumption implies that the transformation of the intensity is determined by the 3 vectors $\bar{\phi}, \bar{D}$ and $\bar{K}$ charcterising the camera motion and surface orientation respectively. The determination of these parameters can therefore be formulated as the problem of minimising the prediction error:

$$P(\bar{\phi}, \bar{D}, \bar{K}) = \sum_{x,y} [\, I(x'(x, y, \bar{\phi}, \bar{D}, \bar{K}), y'(x, y, \bar{\phi}, \bar{D}, \bar{K}), t') - I(x, y, t)\,]^2 \qquad [3.2]$$

where the summation is over image coordinates containing the planar surface.

For the minimisation we use gradient descent, i.e. the values of the parameters are adjusted iteratively according to:

$$\bar{\phi}^{(i+1)} = \bar{\phi}^{(i)} - \mu_1 \frac{\partial P^{(i+1)}}{\partial \bar{\phi}}$$

$$\bar{D}^{(i+1)} = \bar{D}^{(i)} - \mu_2 \frac{\partial P^{(i+1)}}{\partial \bar{D}} \qquad [3.3]$$

$$\bar{K}^{(i+1)} = \bar{K}^{(i)} - \mu_3 \frac{\partial P^{(i+1)}}{\partial \bar{K}}$$

where i denotes iteration index. The derivatives with respect to the parameter vectors are taken componentwise.

The convergence properties of the gradient descent minimisation depend heavily on the structure of the image intensity function $I(x, y)$. Assume that we have computed approximate values $\hat{\phi}, \hat{D}, \hat{K}$ for the parameters. These values transform the point (x, y) to the point $(\hat{x}', \hat{y}') = (x' + \delta x', y' + \delta y')$ If the approximate parameter values are close enough to the true values, $\delta x'$ and $\delta y'$ will be small. For the prediction error we then have:

$$e \;=\; I(\hat{x}', \hat{y}', t') - I(x, y, t) \;=\; I(x' + \delta x', y' + \delta y', t') - I(x, y, t) \;\approx$$

$$\approx \frac{\partial I}{\partial x'} \delta x' + \frac{\partial I}{\partial y'} \delta y' \qquad [3.5]$$

In order to compute well defined values of motion and surface orientation parameters the prediction error e should be sensitive to small changes in these. From eq.3.5 we see that the size of the image intensity gradient is very important in this respect since it directly amplifies any variations in $\delta x', \delta y'$. Preferably image points x, y with high intensity gradient should be used in the computation of the prediction error in eq. . The selective use of points with high gradient also reduces the volume of the computations involved. For our application we therefore first applied a version of the Canny-Deriche edge detector to the image [8,9], and used only the edge points. The choice of edge detector for this problem is probably not critical. What is needed is just a selection of points with high intensity gradient.

The derivatives of the prediction error P with respect to the parameters were computed by systematically varying the parameters. If the difference between the transformed coordinates x', y' and the original x, y is small, e.g. by choosing the time interval $t' - t$ to be small, these derivatives could be computed more efficiently using the relation 3.5. In principle direct methods of determination of the parameters can be used. [10].

4. Sequential estimation using recorded sequence

Since the algorithm assumes motion relative a planar surface we first have to select points in the image, projected from the road , to be used for parameter estimation. Under normal driving conditions, the part of the image immediately in front of the car can be assumed to project from the planar road surface. The position of the car realtive to the road boundaries can also be considered as relatively stable. The points to be used in the algorithm are therefore selected from a rectangular window in the image choosen so that points from the road immediately in front of the car are contained in the window as shown in fig. 4-1. This window is fixed in time relative to the image. As obstacles are detected the window could be made adaptive so that these objects are not included in the pixels used for parameter estimation.

Fig. 4.1 Window for selection of points on planar road surface

The gradient descent algorithm for estimation of parameters can now be applied to succesive image frames in the sequence. If the time between the succesive frames is short enough, we can expect a high correlation in time between computed parameter values. This can be exploited in the gradient descent algorithm by using parameter values computed in the previous frame pair as start values for the next pair. For an ideal planar road surface there will in fact be a coupling between motion parameters, $\bar{\phi}, \bar{D}$ and surface orientation $\bar{K}$, since the change in surface orientation over time is determined by the motion. This can be introduced as an extra constraint in the algorithm in order to build a true spatio-temporal model of the position and orientation of the camera relative to the road. At this stage however we did not consider this coupling between parameters.

5. Object detection using prediction error

If the estimated parameters $\hat{\phi}$, $\hat{D}$ and $\hat{K}$ are correct and the road conforms to the planar surface model, the error in the prediction of the image intensity acc. to eq. 3.5 will be 0. Any errors in the estimated parameter values or errors in the model will however give rise to a non zero prediction error. Any vertically extended object in the scene will obviously violate the planar surface model and therebye cause a prediction error. The prediction error is thus an important variable to be used for deciding whether any objects are present in front of the car.

The effect of a vertically extended object on the prediction error is however highly dependent on its position in the scene. If we consider the ideal case of a camera aligned with optical axis parallell to the planar surface, translating in the direction of the optical axis only with no rotation, we have for a point in the scene at depth Z the following transformation in the image:

$$x' = \frac{x}{1 + D_Z/Z} \qquad y' = \frac{y}{1 + D_Z/Z} \qquad [5.1]$$

If we choose units so that the height of the camera over the ground, $-1/K_Y = 1$ we have for points on the planar road surface:

$$x^m = \frac{x}{1 - D_Z y} \qquad y^m = \frac{y}{1 - D_Z y} \qquad [5.2]$$

A point in the image with coordinates x, y at height h above the road will be at depth $Z = -1 + h/y$. For this point the difference between the actual coordinates and thoose predicted by the model is:

$$x' - x^m = \frac{h D_Z x y}{(1 - D_Z y - h)(1 - D_Z y)} \qquad y' - y^m = \frac{h D_Z y^2}{(1 - D_Z y - h)(1 - D_Z y)} \qquad [5.3]$$

Note that this applies only to points x, y in the image that project to the road surface. This means that they are below the horison, i.e. $y < 0$ and $h < 1 - D_Z y$.

We see that the error in the prediction of the coordinates grows monotonically with height h above the ground. However it also grows with distance from focus of expansion $x = 0, y = 0$. Objects close to the F.O.E. will therefore give rise to very small errors in the coordinates. This is natural since motion at the F.O.E. is zero, independent of the objects coordinates in space. If we use scene coordinates X and Z instead of image coordinates we get for the coordinate error:

$$x' - x^m = \frac{hD_ZX}{(Z+D_Z)(Z+D_Z-D_Zh)} \qquad y' - y^m = \frac{h(h-1)D_Z}{(Z+D_Z)(Z+D_Z-D_Zh)} \qquad [5.4]$$

From this we see that the error scales inversely with depth, i.e. distance in front of the camera. It also grows with the distance X to the side of the optical axis, and with D_Z the translatory displacement along the optical axis.

For small errors in the coordinates we can estimate the error in the predicted image intensity by projecting the coordinate errors on the intensity gradient acc. to eq. 3.5 . This means that errors in the predicted image intensity will only show up at the edges of the vertically extended objects, unlesss the errror in the predicted coordinates are large enough. Important to note is also the fact that the orientation of the edges of the vertically extended objects influence the size of the error in the predicted image intensity. From eq.5.3 we see that $\delta x'$, $\delta y'$ will be oriented radially out from the focus of expansion. For maximum prediction error acc to eq. 3.5 the gradient shouls be parallell to this orientation, i.e. the edges should be orhogonal to the lines radiating out from the focus of expansion. However we must emphasise that this only applies when coordinate errors $\delta x'$, $\delta y'$ are small

Another cause of prediction error is errors in the estimated parameters. For these errors we also have the dependence on the distance from the F.O.E. This is important to consider in the evaluation of false alarm detections e.g.

6. Experimental results and conclusions

The algorithm for sequential model based motion estimation and object detection was simulated using a digitised video tape recorded from a camera placed on top of a moving car. 100 frames with a frame rate of 25/sec was selected from the video sequence. In this sequence the car approaches a roadwork where the right lane of the road is blocked by warning signs and fences. In the first frame the car is about 150 m from the roadwork.

In order to get sufficiently large prediction errors for objects close to the F.O.E. the prediction was made 3 frames ahead. Parameters were computed for every frame however, i.e. the sequence of frame pairs used were 1-4, 2-5, 3-6, etc. A special problem was the initiation of the algorithm. Since only the pixels at the edges from the Canny-Deriche edge detector were used in the prediction error computation the convergence of the algorithm depended on the initial parameters not beeing too far away from the correct values. For the first frame pair therefore, all the pixels in the window were used for computation of the prediction error.

For each frame pair the gradient descent algorithm was iterated 30 times. No complexity considerations were considered in choosing this number. For more iterations the improvement of the prediction was found to be negligible.

Fig.6.1.a-b shows the unpredicted difference and the prediction error respectively for frames 97-100. We see that a clear reduction of the difference image is obtained in the prediction error image for image points projecting from the ground. For vertically extended objects the reduction is significantly less, depending on verical extent and distance from F.O.E. The diagrams in fig 6.1.c-f illustrates this in more detail. The curves show the image intensity from the difference and prediction error image respectively. The intensity along two different horisontal lines, indicated in fig.6.1.a-b are plotted. The first line containing white marking from the road is clearly reduced in the prediction error image, while the second containing the vertically extended objects in the roadwork shows comparatively less reduction from difference to prediction error image.

In fig.6.2 is shown with white markings thresholded prediction error images from several different times. The images were thresholded at the edge points only and the threshold was increased systematically with distance from F.O.E. The threshold was choosen so that at most one marking was obtained from the part of the road containing the parameter estimation window. This means that the threshold was adapted to any errors in the estimated parameters.

From fig.6.2 we see that as expected the sensitivity of the algorithm increases with distance from the F.O.E and height over ground. A very important factor is also the intensity contrast of the objects relative background. Some vertically extended objects are not over the threshold in every frame. By combining detections from several frames the performance should be increased however. In order to do this the detections from the different objects have to be grouped and anlysed separately. The main purpose of this prediction error based detection can therefore be seen as a preprocessing mechanism which directs resources of the system for further processing.

Acknowledgements

The video recording of the road sequence was made by Lars J. Olsson, Mercel AB and digitised data was provided by the Computer Vision Laboratory, Linkoping University. The work was performed under contract within the Prometheus-Sweden project. It was also supported by STU, the Swedish national board for technical developement under contract 88-01749.

References

[1] H. von Helmholtz, Physiological optics, vol.3, 1866

[2] J.J. Gibson, The perception of the visual world, *Houghton, Muffin, Boston 1950*

[3] H.C. Lounguet-Higgins and K.Prazdny, "The interpretation of a moving retinal image ", *Proc. Roy. Soc. London B-208, pp. 385 - 397. 1980*

[4] R.C. Nelson and J. Aloimonos, "Using flow field divergence for obstacle avoidance: Towards qualitative vision ", *Proc. of 2:nd Int. Conf. on ComputerVision, Tampa Florida, pp.188 - 196 1988*

[5] W.B.Thompsom, K.M.Much and V.A.Berzins, "Dynamic occlusion analysis in optical flow fields", IEEE Trans. Pattern Anal. Machine Intelligence, Vol. PAMI-7, No.4, July 1985

[6] H.A. Mallot, E Schulze and K. Storjohann, "Neural network strategies for robot navigation " *Proc.of nEuro'88, Paris, June 6-9, pp. 560 - 569, 1988*

[7] R.Y. Tsai and T.S. Huang, "Estimating three-dimensional motion parameters of a rigid planar patch ", IEEE-ASSP, Vol ASSP-29, No. 6, December 1981.

[8] J. Canny, A computational approach to edge detection, IEEE Trans. Pattern Anal. Machine Intelligence, Vol. PAMI-8, 679-698, 1986,

[9] R. Deriche, Using Canny's criteria to derive a recursively implemented optimal edge detector,Int. J. of Comp. Vision , 1, 167-187, 1987.

[10] B.K.P. Horn and E.J. Weldon, " Direct methods for recovering motion ", Int. J. of Comp. Vision 2, 51-76, 1988.

Fig. 6.1.a Difference frames 97-100

Fig. 6.1.b Prediction error frames 97-100

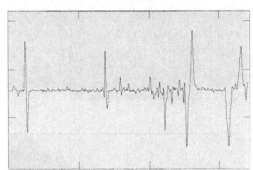

Fig. 6.1.c Intensity of line 1 in fig. 6.1.a

Fig. 6.1.d Intensity of line 1 in fig. 6.1.b

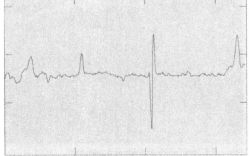

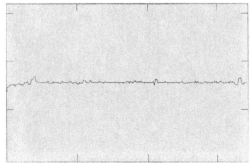

Fig. 6.1.e Intensity of line 2 in fig. 6.1.a

Fig. 6.1.f Intensity of line 2 in fig. 6.1.b

Fig. 6.2.a Thresholded residual frame 10

Fig. 6.2.b Thresholded residual frame 30

Fig. 6.2.c Thresholded residual frame 50

Fig. 6.2.d Thresholded residual frame 70

Fig. 6.2.e Thresholded residual frame 80

Fig. 6.2.f Thresholded residual frame 90

DETECTION AND TRACKING OF MOVING OBJECTS BASED ON A STATISTICAL REGULARIZATION METHOD IN SPACE AND TIME

Patrick Bouthemy and Patrick Lalande
IRISA / INRIA-Rennes
Campus de Beaulieu, 35042 Rennes Cedex, France

1. Introduction

The function often devoted to a vision system is to detect and track moving objects appearing in its field of view, [1-3]. A variety of applications are concerned with this problem, such as traffic control, remote surveillance of industrial areas, biomedical studies or target tracking. When the camera is static, moving regions in the image plane necessarily correspond to moving objects in the scene. This paper addresses this basic issue of motion detection in an image sequence, and describes an original framework based on statistical models, namely *spatio-temporal Markov fields*. This method generalizes a first attempt of this kind we have recently presented in [4]. Substantial modifications have been introduced, which contribute to solve several shortcomings of the previous algorithm. This new version is able to handle textured moving objects and overlapping cases. By this last term, we mean both, situations where successive projections in time of a moving object overlap each other in the image plane, and occlusion situations between different moving objects. The ability to cope with overlapping cases also avoid to pay attention to the time sampling of the processed image sequence with respect to the size and speed of moving objects of interest. Besides the new way of modeling and using temporal contextual information enables to easily track moving regions through the image sequence.

2. Problem statement

If moving objects are present in the scene, obviously changes in time will occur in the image intensity array. In turn, when the camera is static, temporal intensity changes can be related mainly to motion. Nevertheless, motion detection cannot be reduced to temporal change detection. In particular a moving object gives raise to three kinds of change regions; first one corresponding to uncovered background, second to covered background, third to the overlap of object projections (by the way, this last sub-class is often very partially perceptible). As a matter of fact what is only but completely sought for in every image are projections of moving objects, (also called moving object masks). Usual methods first extract successive temporal change maps, then try to recover projections of moving objects by applying some heuristics, [2,3]. We have adopted a quite different approach. The motion detection issue is considered as a whole and is stated as a *statistical bayesian labeling* problem based on Markov field models. Such an approach has already been proved relevant to other issues related to motion analysis, as reported in [5] for scene segmentation according to motion information, and in [6,7] for optic flow estimation.

In a labeling problem, two sets of elements must be defined: *observations*, (i.e. data to be considered); *labels* (i.e. primitives to be extracted). The remarks of the paragraph above naturally lead to take as observation the temporal derivative of the intensity function f. Let o_t denote the observation array at time t, and $p = (x, y)$ a pixel, we have: $o_t(p) = \partial f(p)/\partial t$. In fact temporal derivatives will be approximated by finite differences between time t and time $t - dt$, denoted by $\tilde{f}_t$. Moreover another set of information is taken into account: the binary logical map of temporal changes between time t and time $t - dt$: $\bar{o}_t$. It is obtained by an operator able to detect even weak temporal changes; the intensity is locally modeled by a linear function with an additive gaussian noise of constant variance and changes in the model parameter values are validated by a likelihood test as in [8].

On the other hand we consider primitives directly tied to the type of image content we want to delineate. Therefore the label set consists of two symbols, $\Omega = \{a, b\}$, a for moving object masks, b for static background. A priori spatio-temporal models are associated with these primitives; these models must express what properties the solution is supposed to have (that forms the *regularization effect*). Let us first give some intuitive insights to the modeling step. All masks of moving objects obviously share some intrinsic basic spatio-temporal properties. They must show sufficient spatial coherence and their successive positions in time obey a certain law. This can be expressed in terms of required spatio-temporal contextual configurations. Then Markov field models represent very efficient and well-posed means to mathematically formulate this problem, [9]. Let us denote the label field at time t by e_t. This field is modeled as a Markov field in space *and* time; this will be explained in details in the next section.

3. The modeling step and the decision criterion

The markovian property in time is assumed in both directions along the time axis. Indeed we need contextual information from the close past and from the near future to identify the label field at time t. That is the reason why we consider label fields in pairs in the identification process. The solution to the labeling problem is formulated according to the maximum a posteriori (MAP) criterion:

$$max_{e_{t-dt}, e_t} \quad P(e_{t-dt}, e_t/o_t, \bar{o}_t) \tag{1}$$

The best interpretation in terms of moving object projections must have the greatest a posteriori probability given the observations at hand. This statistical approach also permits to properly deal with noise-corrupted observations. Maximizing expression (1) with respect to e_{t-dt} and e_t is equivalent to maximizing the joint probability $\xi = P(e_{t-dt}, e_t, o_t, \bar{o}_t)$. One attractive aspect of this approach is that we can build an explicit version of ξ using the equivalence between Gibbs distributions and Markov fields, as primarily emphasized in the context of image processing in [9].

Local contextual models can then be related to potential functions defined on so-called cliques; cliques are subsets of sites (here sites are pixels) which are mutual neighbors. We consider the following spatio-temporal neighborhood system: a 3x3 spatial neighborhood centered in (p, t) and one-to-one connections from (p, t) towards $(p, t-dt)$ and $(p, t+dt)$. As far as spatial cliques c_s are concerned, we only take into account the four ones comprising two sites (horizontal, vertical and two diagonal cliques). Spatial potentials V_{c_s} have been defined in such a way as to favouring homogeneity of the label field, that is to have a spatial regularization effect, (e.g. to eliminate isolated points). They are of logistic kind; that is equal to a predefined level, (resp. $-\beta_s$ and β_s), according to the label configuration of the clique at hand, (resp. same labels and different labels), knowing that a negative value encourages the corresponding configuration. It is not necessary here to introduce a complementary edge-site system to take into account discontinuities as in [9], because of the existence of the temporal cliques. These potentials summed over spatial cliques form a first energy function W_s. Of more specificity to the problem at hand, are potential functions tied to the temporal clique also containing two sites. Again they are of logistic kind. They contribute to determine which temporal configurations between (a,a), (a,b), (b,a), (b,b), at pixel p, at two successive times, are encouraged and which are discouraged, according to $\bar{o}_t(p)$ considered as a deterministic external information. They are described in Table 1. In particular overlapping situations are thus correctly handled. These temporal potentials V_{c_r} lead to a second energy function W_r. The third energy function W_ϵ will express the adequacy between observations and current estimates of the primitives. We assume that the relation between these two sets is defined by:

$$o_t(p) = \psi(e_{t-dt}(p), e_t(p)) + n(p) \qquad (2)$$

where ψ can take a value among three possible ones which can be either predefined or locally estimated on-line; and n is a white (in space and time) zero-mean Gaussian noise of constant variance σ^2. The corresponding energy function W_ϵ is then given by:

$$W_\epsilon = \frac{1}{2\sigma^2} \sum_p [\tilde{f}_t(p) - \psi(e_{t-dt}(p), e_t(p))]^2 \qquad (3)$$

The noise variance is estimated once at the beginning of the processed image sequence. Finally we get the total energy function

$$W = W_s + W_r + W_\epsilon \qquad (4)$$

We can assume that ξ is proportional to $\exp(-W)$. The optimal map of moving object masks will correspond to the lowest value of energy W.

4. The optimization procedure, the tracking scheme, and results

To minimize W, we use an iterative deterministic method as in the early version described in [4]. It yields a very good trade-off in our case between computation speed and result quality. Moreover it is completed by an efficient procedure for iteratively selecting sites to be visited, as suggested in [10]. Let us outline that all computations are very local. The optimization of every label field e_t is done in a two-pass manner. First we derive a first estimate $\check{e}_t$ when considering pair (e_{t-dt}, e_t); second, considering pair (e_t, e_{t+dt}), we update $\check{e}_t$ and we get the final estimate $\hat{e}_t$; of course the first estimate $\check{e}_{t+dt}$ is simultaneously obtained. Afterwards the same holds for e_{t+dt} and so on. This prediction scheme, associated with the markovian property of the label field in time, explains that we can now easily implement a tracking procedure of the detected moving regions through the image sequence.

This is essentially a matter of recursive number allocation. First, an initializing step is needed; that is, given the first estimated label field $\hat{e}_{t_0}$ a different number is assigned to each connected subset of a-labeled points. Then let us assume that the tracking process has been achieved until time $t - dt$, it is pursued at time t as follows. Let q_k be a reference point among points with number k in the image at time $t - dt$. The same number k is assigned to the point p_q in the image t belonging to the temporal clique of q_k. This assumes that the intersection of two successive projections in time of a given moving object is not empty. Then starting from point p_q, the assignment of number k in the image t is propagated step-by-step to the a-labeled points connected through the spatial cliques. When this is achieved, number $k + 1$ is taken into account, and so on. Two specific cases may happen: the merging and the splitting of moving masks. They can correspond for instance to the crossing of two different moving objects. The first case will be detected when two different numbers k and k' are present in a spatial clique. This conflict situation is easily solved by "equaling" k' to k (i.e. by creating a link). The second case will provide the same situation as the one encountered when a new moving object is appearing. A subset of connected a-labeled points in the image at time t will remain without any number assigned at the end of the process. It will then receive a new number. This tracking stage has been separately presented to make the explanation easier. Actually numbers k can be considered as supplementary primitives which can be straightforwardly included in the modeling step. Therefore detection and tracking can be nearly simultaneously carried out.

Numerous experiments with several image sequences depicting outdoor scenes have been processed. Results are quite fine. We present here one example. Fig.1 shows three (not successive) images from the input sequence taken at times t_1, t_2, t_3. The camera is static and some cars are moving down and another one is moving up. The binary map of the change regions supplied by a simple temporal intensity change detector depicts very incomplete moving object masks and a lot of spurious isolated points all over the image. In the results obtained with the markovian approach, Fig.2, the stationary background is free of spuriously detected points and the extracted regions rather well correspond to the real masks of the moving objects. This is confirmed throughout the entire processed sequence. Let us outline that the images are not of high quality and that the part missing in the moving object mask at the bottom of the image is due to the complete lack of contrast between background intensity and car windscreen intensity. It has also been found that the parametrization of the model is not a critical problem for this motion detection issue; the same set of parameter values has been used for different image sequences. Moreover the number of parameters is rather small.

We have described a general, model-based and robust method for motion detection in an image sequence, which besides leads to a simple tracking procedure. This method does not suppose any a priori knowledge on the respective intensity values of moving object projections and background; it does not require anymore any identification of the background intensity distribution. As all the computations are local, an efficient fast implementation is indeed reachable, which allows an effective use in practical situations.

References:

[1] **H.-H. Nagel**, From image sequences towards conceptual descriptions, *Image and Vision Computing*, Vol. 6, No 2, May 1988, pp.59–74

[2] **R. Jain**, Dynamic scene analysis, in *Progress in Pattern Recognition 2*, L. Kanal and A.

Rosenfeld (eds.), Mach. Intell. and Patt. Rec. Series, Vol. 1, North–Holland, 1985, pp.125–167

[3] **J. Wiklund** and **G.H. Granlund**, Image sequence analysis for object tracking, *Proc. 5th Scandinavian Conf. on Image Analysis*, Stockholm, June 1987, pp.641–648

[4] **P. Bouthemy** and **P. Lalande**, Determination of apparent mobile areas in an image sequence for underwater robot navigation, *Proc. IAPR Workshop on Computer Vision: Spec. Hardw. and Ind. Applic.*, Tokyo, Oct. 1988, pp.409–412

[5] **D.W. Murray** and **B.F. Buxton**, Scene segmentation from visual motion using global optimization, *IEEE Trans. on PAMI*, Vol. 9, No 2, March 1987, pp.220–228

[6] **J. Konrad** and **E. Dubois**, Multigrid Bayesian estimation of image motion fields using stochastic relaxation, *Proc. 2nd Int. Conf. on Computer Vision*, Dec. 1988, pp.354–362

[7] **F. Heitz** and **P. Bouthemy**, Motion estimation and segmentation using a global Bayesian approach, *Int. Conf. on Acoustics, Speech and Signal Processing*, Albuquerque, April 1990

[8] **Y.Z. Hsu, H.–H. Nagel** and **G. Rekers**, New likelihood test methods for change detection in image sequences, *Computer Vision, Graphics and Image Processing*, Vol. 26, 1984, pp.73–106

[9] **S. Geman** and **D. Geman**, Stochastic relaxation, Gibbs distributions and the Bayesian restoration of images, *IEEE Trans. on PAMI*, Vol. 6, No 6, Nov. 1984, pp.721–741

[10] **P.B. Chou** and **R. Raman**, On relaxation algorithms based on Markov random fields, *Technical Report* No 212, Computer Science Dpt, Univ. of Rochester, July 1987, 28p.

This work was partly supported by the French CNRS Program, "PRC Communications Homme–Machine, pôle Vision". The image sequence was provided by the Laboratoire d'Electronique, University of Clermont-Ferrand

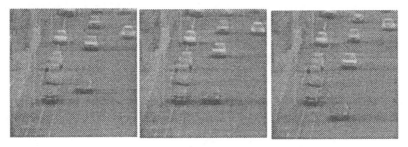

Figure 1: *Three original images (not successive) out of the sequence at times t_1, t_2, t_3.*

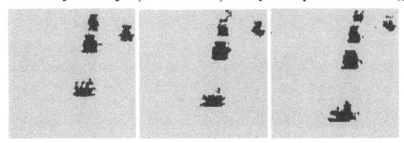

Figure 2: *The moving object masks at times t_1, t_2, t_3; ($\beta_\tau = 100$, $\beta'_\tau = 1000$, $\beta_s = 10$) (The time interval dt indeed corresponds to the video rate).*

$(e_{t-dt}, e_t, \bar{o}_t)$	(b,b,0)	(b,b,1)	(a,b,0)	(a,b,1)	(b,a,0)	(b,a,1)	(a,a,0)	(a,a,1)
$V_\tau(e_{t-dt}, e_t, \bar{o}_t)$	$-\beta_\tau$	$+\beta_\tau$	$+\beta'_\tau$	$+\beta'_\tau$	$+\beta_\tau$	$-\beta_\tau$	$+\beta_\tau$	$-\beta_\tau$

Table 1: *The temporal potentials ($\bar{o}_t = 1$, means temporal change, $\bar{o}_t = 0$ no change)*

STEREO AND MOTION

Vertical and Horizontal Disparities from Phase

K.Langley *

T.J.Atherton R.G.Wilson M.H.E.Larcombe †

Dept. Experimental Psychology
University of Oxford
Oxford

We apply the notion that phase differences can be used to interpret disparity between a pair of stereoscopic images. Indeed, phase relationships can also be used to obtain probabilistic measures both edges and corners, as well as the directional instantaneous frequency of an image field. The method of phase differences is shown to be equivalent to a Newton-Raphson root finding iteration through the resolutions of band-pass filtering. The method does, however, suffer from stability problems, and in particular stationary phase and aliasing. The stability problems associated with this technique are implicitly derived from the mechanism used to interpet disparity, which in general requires an assumption of linear phase and the local instantaneous frequency. We present two techniques. Firstly, we use the centre frequency of the applied band-pass filter to interpret disparity. This interpretation, however, suffers heavily from phase error and requires considerable damping prior to convergence. Secondly, we use the derivative of phase to obtain the instantaneous frequency from an image, which is then used to improve the disparity estimate. These ideas are extended into 2-D where it is possible to extract both vertical and horizontal disparities.

1 Introduction

Horizontal disparities provide important depth cues for passive range finding systems. Unfortunately, the computation of disparities has proven particularly difficult because of the *correspondence* problem. Indeed, there have been several attempts to formulate solutions to the problem which may categorised into feature and area based correspondence solutions [2]. Both methods, however, suffer from similar forms of the uncertainty principle [6]. With edge based stereoscopic algorithms, the presence of edge information alone yields considerable ambiguity, since edges themselves are difficult to distinguish, and can also occur spuriously within an image owing to the presence of noise, and the complexity of natural image data. In terms of the uncertainty principle we know *where* but not *what*. In contrast, area based correspondence algorithms usually apply normalised cross-correlation techniques to obtain measures of similarity between image functions. These techniques suffer from poor definition of window sizes that are used during the cross-correlation proceedure. This may also be considered as providing information equivalent to *where* but not *what*. We form the premise that at the lowest level of vision processing, *what* refers to both the local spectral properties (instantaneous frequency) and orientation of our image data and *where* is obtained through the resolutions of band-pass filtering. Marr [13], was also aware of the uncertainty principle. He proposed a coarse to fine edge based matching algorithm using the Laplacian of the Gaussian kernel whose zero-crossings can be interpreted

---

*This research was supported by a joint SERC Fellowship to the Dept. of Engineering Science and Exp. Psychology at the the University of Oxford.

†Dept. Computer Science, University of Warwick, U.K.

as edges. The approach presented here, is an extension of Marr's work but instead of using one operator, we apply quadrature filter pairs and track phase differences. Fundamental to Marr's work, lies the notion of eye vergence. His work was criticised because of the nature of his vergence mechanism [2]. We propose an explicit method to drive optical vergence. Indeed, we find that many of the criticisms directed towards Marr's work can be explained by our theory, which we also extend to include mechanisms for computing both vertical and horizontal disparities.

2 The Gabor representation

The problem that Gabor addressed, was the simultaneous representation of a signal in both space and frequency, which has received extensive application within the field of image processing (e.g [18]). The measure of duration that Gabor used to formulate his Uncertainty hypothesis was based upon a minimisation of the second moment of an arbitrary signal in both space and frequency by an expansion of *elementary* signals:

$$\Psi(x, \omega_g) = \frac{1}{(2\pi\sigma^2)^{\frac{1}{4}}} \exp[-\frac{(x - x_o)^2}{4\sigma^2}] \, cis(\omega_g x + \phi) \tag{1}$$

2.0.1 Phase from Gabor functions

Consider a pair of images, one of which has experienced a phase shift owing to horizontal disparities, we obtain the phase difference between image pairs $I_r(x)$ and $I_l(x)$ at x_o[9] by solving:

$$I_s = \mathcal{R}e[\Psi(x_o, \omega_g) * I_l(x_o)] \quad I_a = \mathcal{I}m[\Psi(x_o, \omega_g) * I_l(x_o)] \tag{2}$$

$$I_{s\phi} = \mathcal{R}e[\Psi(x_o, \omega_g) * I_r(x_o)] \quad I_{a\phi} = \mathcal{I}m[\Psi(x_o, \omega_g) * I_r(x_o)] \tag{3}$$

Let $I_l(X) = \cos\omega_o x$ and $I_r(x) = \cos\omega_o x + \phi$ represent our 1-D image function, then expanding equations (2) and (3), phase may be represented in terms of the rotation matrix: $R_{\theta\zeta}$ where $\zeta = \tanh(8\pi^2\sigma^2 u_o u_g)$ i.e:

$$\begin{bmatrix} I_{s\phi} \\ I_{a\phi} \end{bmatrix} = \begin{bmatrix} \cos\phi & -\frac{\sin\phi}{\zeta} \\ \sin\phi \, \zeta & \cos\phi \end{bmatrix} \begin{bmatrix} I_s \\ I_a \end{bmatrix} \tag{4}$$

Notice that the non-linearity in phase incorporates a shearing of the transformation matrix, with a phase difference found from:

$$\phi = \tan^{-1}(\frac{\frac{I_{a\phi}}{\zeta}}{I_{s\phi}}) - \tan^{-1}(\frac{\frac{I_a}{\zeta}}{I_s}) \tag{5}$$

and the disparity estimate given by:

$$D_{est} \approx \frac{\phi}{2\pi u_g} = \frac{u_o}{u_g} d \tag{6}$$

Naturally, this interpretation of disparity contains an implicit error based on the differences between the centre frequency of the Gabor function and the image signal [11]. We can ensure that the Gabor function approximates a linear phase filter by assuming $u_o = u_g$ and $8\pi^2 u_o u_g \sigma^2 \geq 8.0$ Consider the argument from $\Psi(x) * I(x)$ which may be defined as the instantaneous phase of a signal expressed as:

$$F(x) = \tan^{-1}(\frac{I_a}{I_s}) = \tan^{-1}(\frac{\tan(2\pi u_o x + \phi)}{\zeta}) \tag{7}$$

Since the right hand side of the above equation is independent of the centre frequency from the Gabor filter. It can easily be shown that:

$$\frac{d[F(x)]}{dx} = \frac{2\pi u_o}{\frac{\sin^2(2\pi u_o x)}{\zeta} + \zeta \cos^2(2\pi u_o x)} \tag{8}$$

Providing $\zeta = 1.0$ then the derivative of the phase response may be considered as an estimate of the instantaneous frequency of a signal. This may be used to improve the estimation of phase differences with little extra cost in computation, and without regard to the energy response.

3 The method of phase differences in 1-D

The method of phase differences has been applied to stereoscopic computation by several authors (Wilson and Knuttson,1987;Larcombe,1984; and Langley and Atherton,1988). Miller [15] , also implemented the same principle in hardware by the application of a Phase-Locked Loop. The approach is based upon the transformation of a real intensity signal into its analytic form by the addition of its quadrature counterpart. We therefore form the complex 1-D image function ($z(x) = \Psi(x, \omega_o) * I(x)$):

$$z(x) = E(x) \exp[j\phi(x)]$$

which at any one instance may be represented more succinctly in terms of the rotation matrix (R_ϕ) as:

$$z_l = R_\phi z_r + n \tag{9}$$

and n represents additive quadrature noise which receives contributions from sensor noise, and discontinuities in the disparity field. General closed-form solutions to functions of the type presented in equation (9) requires the probability density function (PDF) of the phase error. Unfortunately such a PDF is not yet known, but may be approximated by the Tichonov density [5], which is derived from the first order phase-locked loop whose input is the sum of Gaussian noise and a sinusoid. While it is possible to make quantitative statements regarding sensor noise, it unlikely that discontinuities in the disparity field may be equally represented. We must therefore, resort to smoothing to reduce the effects of noise under these conditions. We define the phase difference between two image functions at a spatial position (x) as the roots of:

$$\phi_l(x) - \phi_r(x_o) = 0 \tag{10}$$

Suppose that x is an exact root of the problem we require to solve, then expanding the above equation as a Taylor series with $x = x_o + d$ as an exact solution we have:

$$\phi_l(x_o + d) - \phi_r(x_o) = 0$$

$$= \phi_l(x_o) - \phi_r(x_o) + d\phi_l'(x_o) + \ldots\ldots$$

let us then make d_1 an approximation to d in which case we have:

$$d_1 = -\frac{[\phi_l(x_o) - \phi_r(x_o)]}{\phi_l'(x_o)}$$

from which we iterate to find the root of our initial equation by:

$$x_n = x_{n-1} - \frac{\phi_l(x_{n-1}) - \phi_r(x_o)}{\phi_l'(x_{n-1})} \tag{11}$$

which is in fact Newton-Raphson convergence problem with d_n as our current disparity estimate. We can immediately state that convergence from this method can only occur providing the new estimation of the root to x_n lies between the previous estimate x_{n-1} and the exact solution x. In addition, we can also state that the **sgn** of the derivative of phase must also correspond in each image pair, otherwise the method does not converge into a stable solution. We propose to reduce the phase error by iteration in the same manner as a Phase-Locked Loop and additionally resolve the aliasing problem. By the aliasing problem we refer to the difficulty in obtaining large disparity estimates from filter pairs tuned to high spatial frequencies.

4   A mechanism for eye vergence

For our purposes, eye vergence holds some interesting properties. By verging the eyes, we bring features of small spatial extent into nearer correspondence, with the reduced possibility of aliasing and disparity error. Initiating a vergence also increases the effective disparity range that each filter can detect, and thereby reducing the need for filters tuned to the very lowest spatial frequency elements. Vergence is simply obtained from coarse to fine resolutions by:

- Obtain the mean disparity estimates from the responses by convolving an Image function with quadrature Gabor filters and weighting the disparities by their associated energy responses summed from both image pairs.

- Induce a vergence mechanism based upon the mean disparity and progress to the next resolution of filtering.

Our measured mean disparirity is therefore taken from:

$$D_{mean} = \frac{\sum_{x=-N}^{x=N} P_x d_x}{\sum_{x=-N}^{x=N} P_x} \tag{12}$$

Where $P_x = \sqrt{E_l(x) E_r(x)}$ represents the product of energy from the left and right image image pairs respectively, and we convolve from -N to N pixels at a given resolution from the optical centre. A useful indication for the choice of N is the spatial standard deviation of the applied filter, at the central point of the image. For compuational purposes, we may rather our image pairs were fused at some mean disparity. To this end, we suggest increasing the size of N to incorporate the whole image array for all resolutions of filters. Thus we would anticipate arriving at the mean least squares estimate of disparity between the two image functions.

5   Reducing the effect from quadrature noise

The mechanism of subtracting phase, is unfortunately highly unstable, and particularly sensitive to aliasing problems. Wilson and Knuttson [19], suggest that phase can be damped by a modification of the Willsky [17] error measure on a circle:

$$\eta = \frac{1}{2}[1 + \cos(\phi_d)]$$

This term, however, is not sufficient with a 1-D filter, particularly when iterating phase diffences as a phase-locked loop. Fundamental to stability lies the notion of subsequently smoothing the disparity estimates at a given resolution. We know from Papouillis [16] that if we have a signal:

$$g(x) = g_f(x) + g_n(x)$$

where $g_n(x)$ is the output from a linear system, whoes input n(x) is white noise, then for a constant $g_f(x)$ signal, the minimum mean-square estimation error of g_f smoothed with a window $w(x)$, is obtained if $w(x)$ is the truncated parabola given by:

$$w(x) = \frac{3}{4X}[1 - (\frac{x - x_o}{X})^2]p_X(x - x_o)$$

where $p_X(x - x_o)$ is a pulse of width 2X. We argue that this is indeed a suitable filter to apply to the disparity estimates because shear and compression/expansion differences between stereo images significantly alter the local spectral properties, which can be difficult to interpret by this method alone. We are therefore only relying on this method to obtain approximate correspondence. Higher orders of disparity differences requires the examination of both orientation and curvature differences between image pairs [8] with the former reducing to differences in the first two moments taken from a band of filters applied in a circle. We must also consider the presence of negative frequency which is an unusual form of aliasing [10] (e.g fig. 1). Its occurence may be reduced by either applying a first order quadrature pair (e.g the derivative of the Gabor function) or by removing the mean d.c level from the image data. In addition, subtracting the mean intensity also reduces the large energy bias often observed at the extrema of image data, and reduces the instability of phase owing to edge effects. We therefore propose to add a recursive weighting to our phase locking iteration: We form a measure Q based upon:

$$Q(x) = \frac{1}{4}[1 + \cos(\phi_d)]^2 \tag{13}$$

As the square of the Willsky error measure. We choose the higher order to reduce the effects from noise. At a given scale of the iteration, we define our disparity as:

$$\bar{D}_{k+1}(x) = \frac{Q_k(x)\bar{D}_k(x) + Q_{k+1}(x)(D_{k+1}(x) + \bar{D}_k(x))}{Q_{k+1}(x) + Q_k(x)} \tag{14}$$

Where the measure k refers to the k_{th} resolution of bandpass filtering, and $\bar{D}_{k+1}(x)$ refers to the measurement of disparity from the recent update measured by $D_k(x)$. Thus we are updating our measurements based on the *goodness of fit*, at successive resolutions. To preserve a recursive nature, we re-arrange and modify the above expression to obtain:

$$\bar{D}_{k+1}(x) = \bar{D}_k + \frac{Q_{k+1}(x)}{Q_{k+1}(x) + Q_k(x)}D_{k+1}(x) \tag{15}$$

Where we redefine Q_{k+1} to function recursively as:

$$Q_{k+1} = \frac{Q_k}{2} + \frac{\bar{Q}_{k+1}}{2}$$

Which we propose holds the properties that we require. The combination of both damping at a given resolution of filtering, and between scales can overdamp the phase locking proceedure. To compensate for this, we propose incorporating the vergence mechanism.

5.0.2 Disparity interpretation from local spectral analysis

An examination of equation (11) indicates that the local derivative of phase (instantaneous frequency) from either image function can be used to interpet disparity. To resolve this ambiguity, we suggest taking a weighted mean (f_{av}) from both image pairs to interpret disparity:

$$f_{av}(x) = \frac{E_l^2(x)f_l(x) + E_r^2(x)f_l(x)}{E_l^2(x) + E_r^2(x)} \tag{16}$$

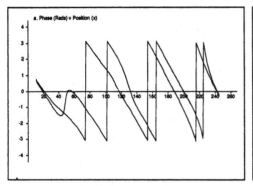
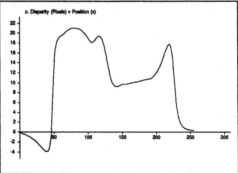

Figure 1: *Phase response from convolution with filters tuned to 1/64 cpp of raster line 110 from figure 2. Actual disparity varied from 25 pixels (lamp) to 10 pixels (background). (a) Left and Right image response superimposed. (b) Disparity interpreted from centre frequency of Gabor function.*

Where f_l and f_r correspond to the instantaneous frequency of both left and right images. This interpretation does suffer from the problems associated with stationary phase since we place a pole into our equation for disparity interpretation. At present, we threshold the data based upon the upper and lower cut-off frequency characteristics of our filter pairs. In addition, we also impose a further threshold restricting the difference between image spectral properties of 1.25 octaves, which we obtain from the disparity gradient hypothesis. In this case, smoothing also acts as a crude interpolator within regions that are not-analytic by this method.

6   The Stereoscopic Aperture Problem

Extending the work presented here into the 2-D case is interesting. In particular, we apply orientationally selective filters using the Compact Pyramid [4] code to reduce our computational load. Unfortunately, extending the method of phase differences into 2-D poses a dilemma. While in principle, it is possible to obtain unrotated phase differences which are resolved both horizontally and vertically [11], this implicitly also introduces a similar form of the motion aperture problem, which in this case is a direct consequence of orientationally selective filtering. Similarly, we can only resolve this ambiguity in image regions which are not restricted to singly directional signals. As is well established [1], corners provide the best image regions for resolving the aperture problem. Fortunately[12], there are methods for determining corner and edge confidence measures using directional filtering. We merely require the linear Fourier transform taken from the circle of energy responses from our orientationally selective filters. We have by Parseval's theorem a probabilistic measure for both "cornerness" ($P_c - P_e$) and "edgeness" (P_e) from:

$$P_e = \frac{[\sum_{i=0}^{N} E_i^2 \cos 2\theta_i]^2 + [\sum_{i=0}^{N} E_i^2 \sin 2\theta_i]^2}{[\sum_{i=0}^{N} E_i^2]^2} \quad ; \quad 0 \le P_e \le 1 \tag{17}$$

$$P_c = \frac{[\sum_{i=0}^{N} E_i^2 \cos 4\theta_i]^2 + [\sum_{i=0}^{N} E_i^2 \sin 4\theta_i]^2}{[\sum_{i=0}^{N} E_i^2]^2} \quad ; \quad 0 \le P_c \le 1 \tag{18}$$

which may also be modified to obtain the orientation of edges and corners. Edge orientation from directional filtering is attributed to Knuttson et al [7]. Equations (17) and (18) normalise the energy responses from the band of filters for both corner and edge detection. However, to distinguish between both corners and edges, we must also apply:

$$C_{orner} = P_c - P_e \tag{19}$$

Because the linear Fourier transform from the circle of edge energies also gives significant energy contributions to the corner measure. Fortunately, the converse is not the case. The calculation of disparity may be obtained by solving a weighted least-squares fit from a band of filter's energy responses, and phase differences applied in a circle at the same pixel location, i.e;

$$P \, \Phi D = P \, d \tag{20}$$

Where P is the MxM diagonal matrix of the energy responses from the ith filter in a circle whose leading elements are formed from $P_{ii} = \sqrt{E_{il}E_{ir}}$. $\Phi = [\cos\phi_i, \sin\phi_i]$ is the Mx2 matrix of directional orientation, and d represents the vector of measured disparity at all M orientations. E_{il} and E_{ir} represent the power response of the ith filter in both image domains, and: $D = [D_x, D_y]^T$ represents the unknown disparity estimate with both horizontal and vertical components. The above equation may be solved by numerical methods related to over determined sets of equations. We apply equation (20) to obtain the horizontal component of disparity. In the presence of dominant edge signals, there will be an interpretive error in horizontal disparity estimation that varies as a cosine of edge orientation. In this case, tracking horizontal disparities on the assumption of the epipolar constraint does not retain useful vertical disparities. This is the consequence of the aperture problem. It is possible to extract both vertical and horizontal disparities with great confidence at image regions exhibiting measures of cornerness. We turn to the Neurophysiologists for some assistance to this problem. Maske et al [14], studied the response patterns from orientationally selective cells in the striate cortex of the cat. They were interested in investigating the claims by Bishop and Pettigrew [3], that only cells with preferred stimulus orientation near to the vertical can make significant horizontal disparity interpretations. They showed that cells that were sufficiently end-stopped can make precise horizontal disparity discriminations, independent of the optimal stimulus orientation. They also showed, that end-free cells were only *effective* for the measurement of horizontal disparity providing their preferred orientations were near to the vertical. Alternatively, the end-stopped cells that were sensitive to disparity measurement showed no such preferences for orientation in their ability to interpret disparity information. Interestingly, a receptive field profile similar to the end-stopped cell can be produced from the linear sum of two orthogonal but orientationally selective Gabor functions [12], which is therefore ideally suited to respond in image regions from which the stereoscopic aperture problem can be solved. Thus it might appear, that the human visual system has also evolved to deal with this issue. There are many schemes that can be proposed to model these observations. We present the following:

$$D_h = P_e D_{hx} + (1 - P_e)D_x \tag{21}$$

Where D_h is the measure for horizontal disparity at a given scale and position in the image function. D_{hx}, D_x represent the horizontal estimates for disparity in the presence and absence of edge information respectively. We obtain D_{hx} from a separate calculation forming:

$$D_{hi} = \frac{d_i}{\cos(\theta_i)} \; ; \; \theta_i \neq \frac{n\pi}{2} \, , \, n = 1, 3... $$

Where $\cos\theta_i$ represents the orientation (frequency domain) of the ith filter from which disparity is interpreted. We would then anticipate estimating the final disparity estimate in a least squares sense from:

$$D_{hx} = \frac{\sum_{i=-\frac{\pi}{4}}^{\frac{\pi}{4}} P_{ii} Dhi}{\sum_{i=-\frac{\pi}{4}}^{\frac{\pi}{4}} P_{ii}} \tag{22}$$

We choose this form, because we have observed that edge information gives rise to a particularly large energy response. Since the aperture problem is a consequence of directional image energy, we suggest it is sufficient to detect the presence and absence of edge information only. This has the advantage of not requiring a description to any arbitrary response pattern (textures) that may occur from the circle of filters. The vertical component of disparity is then similarly obtained from:

$$D_v = (1 - P_e)D_y \tag{23}$$

where the subscript 'v' denotes a vertical component. Thus vertical components of disparity are only evaluated in the absence of edges.

7   Results

We present results from our algorithms to the "room" image for both 1-D techniques, and the extension into 2-D. It is apparent, that the process is capable of obtaining very large disparity estimates with little difficulty. However, the algorithm is severely dependent on the spectral properties contained within the image pairs. In particular with a 1-D filter, it is difficult to obtain phase locking from features of high spatial frequency elements but large disparity differences. We observe that coarse features can easily be identified in disparity space with very large disparity differences. We also present a depth image (intensity proportional to disparity) based on the interpretation of disparity from the local derivative of phase and the threshold constraints that we have also proposed (fig. 4). Our final depth image was obtained from orientationally selective 2-D filters (fig. 5), using the least squares measure to interpret horizontal disparity only.

Figure 2: *Images taken from a room with a 512x512 CCD camera. Data was compressed to 256x256 using the compact Pyramid. (a) Left image (b) Right image. Disparity differences ranged from 25 (central lamp) to 4 pixels (outside window).*

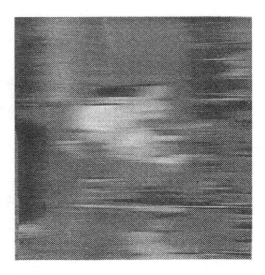

Figure 3: *Intensity depth image produced using the centre frequency of the Gabor function to interpret disparity.*

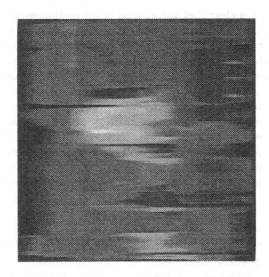

Figure 4: *Intensity depth image produced using the derivative of phase as a measure of instantaneous frequency to interpret disparity.*

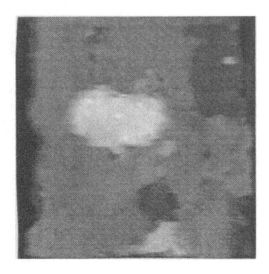

Figure 5: *Intensity depth image produced using centre frequency of the Gabor function to interpret disparity with a least squares estimate of disparity taken from 8 orientationally selective filters.*

8 Discussion and conclusion

From our results, it is clear that the method of phase differences is particularly suited for obtaining stereoscopic correspondence from large image features, or constant disparities accross an image field. Under these conditions, accurate results can be obtained for very large disparities, limited only by aliasing. To reduce this problem, we have proposed a vergence mechanism. In the presence of noise, the results cannot be exact since noise affects the response of a phase sensitive process. Indeed, even with noiseless image data, the method of phase differences has difficulty should the local spectral properties between image pairs differ significantly as a result of the transformation between image pairs. Features with small spatial extent, but large disparity differences also provide difficulty with this method. These problems may be reduced by the vergence mechanism, which brings features into approximate correspondence, and also by applying 2-D filters with orientation preferences. By the nature of applying orientationally selective filters, we are able to obtain estimates for vertical disparity. However, this implicitly also introduces a stereoscopic aperture problem. One way in which this problem may be disambiguated is by the detection of corner features. Fortunately, this is also possible with the same filters that we use to interpret disparity. In the case of single edges, however, the aperture problem is particularly undesirable for edges that approach a horizontal orientation. We have therefore proposed a scheme which obtains a horizontal disparity in the presence of edge information alone, while vertical disparity is restricted to image regions absent of single edges.

References

[1] Adelson.E.H. and Movshon.J.A. Phenominal coherence of moving visual patterns. *Nature*, 300:523–525, 1982.

[2] Baker.H.H. *Depth from Edge and Intensity based stereo*. Standforth University, 1982.

[3] Bishop.P.O. and Pettigrew.J.D. Neural mechanisms of binocular vision. *Vis.Res.*, 1986.

[4] Burt.P.J. and Adelson.E.H. The Laplacian pyramid as a compact image code. *IEEE Trans.Comm.*, COM-31.4:532–540, 1983.

[5] Foschini.G.J., Gitlin.R.D., and Weinstein.S.B. On the selection of a two-dimensional signal constellation in the presence of phase jitter and Gaussian noise. *Bell Sys. Tech. J.*, 927 – 965, 1973.

[6] Gabor.D. Theory of communication. *J.Inst.Elec.Eng*, 93:429–459, 1946.

[7] Knuttson.H., Wilson.R.G., and Granlund.G. Estimating the local orientation of local anistropic 2-d signals. *IEEE.A.S.S.P.Spectral Estimation Workshop.Florida*, 234–239, 1983.

[8] Koenderink.J.J. and Van Doorn.A.J. Geometry of binocular vision and stereopsis. *Biological Cybernetics*, 21:29–35, 1976.

[9] Langley.K. *Phase from Gabor filters*. Univ.Warwick. Tech.Rep.R119., 1987.

[10] Langley.K. Ph.D thesis. *University of Warwick*, 1990.

[11] Langley.K. and Atherton.T.J. A computational theory of stereopsis in the mammalian striate cortex. *submitted to IVC*, 1988.

[12] Langley.K. and Atherton.T.J. A confidence measure for the aperture problem. *Submitted to IVC*, 1988.

[13] Marr.D. and Hildreth.E. Theory of edge detection. *Proc.R.Soc.Lond*, 207:187–217, 1980.

[14] Maske.R., Yamane.S., and Bishop.P.O. End-stopped cells and binocular depth discrimination in the strite cortex of cats. *Proc.Soc.Lond.B.*, 229:257–276, 1986.

[15] Miller.W.M. Video image stereo matching using phase-locked loop techniques. *IEEE Proc. Int. Conf. on Robotics and Automation.*, 1:112–117, 1986.

[16] Papouillis.A. *Signal Analysis*. McGraw-Hill Int. Editions, 1987.

[17] Willsky.A.S. Fourier series and estimation on the circle with applications to synchronous communication. part 1:analysis. *IEEE Trans.Info.Theory*, IT-20,5, 1974.

[18] Wilson.R.G. and Granlund.G. The uncertainty principle in image processing. *I.E.E.E trans. on Pattern Analysis and Machine Intelligence*, 758–767, 1984.

[19] Wilson.R.G. and Knuttson.H. A multiresolution stereopsis algorithm based on the Gabor representation. *Proc. 3rd IEE Conf. on Image Processing.*, 19–22, 1989.

STEREO CORRESPONDENCE FROM OPTIC FLOW

Valérie Cornilleau-Pérès and Jacques Droulez

Laboratoire de Physiologie Neurosensorielle, CNRS

15 rue de l'Ecole de Médecine, 75270 Paris cedex 06, France

The cooperation between motion and stereopsis in the perception of 3-D structure has received several contributions [4,6,7,8]. These studies are motivated by the similarity between the algorithms of motion processing and stereopsis (the 2-D correspondence between image points feeds a 3-D interpretation stage), and by the need for reducing their sensitivities to noise. Most of them use both the optical flow and the intensity-based correspondence between each pair of stereo images as the inputs for the search of 3-D parameters. In [10] it is shown that motion and stereopsis can cooperate earlier, namely during the processing of the correspondence between stereo images. However this approach concerns only rigid surfaces of small curvature (locally approximated by their tangent plane), and its robustness under noise was not tested. In this paper, we show that the retinal velocity can be used directly for stereo matching, for any type of 3-D environment. The case of rigidity is used for predicting the performance of the method. We then present some computer simulations, and finally discuss the relevance of **stereo correspondence from optic flow (SCOF)** for biological vision.

1. The theoretical approach to SCOF.

Let us consider two optical systems of centers O_1 and O_2, with coplanar optical axes. The projection surfaces S_1 and S_2 are hemispheric, of centers O_1 and O_2. O is the middle of the baseline $[O_1O_2]$. In the orthogonal system (OIJK), I is a vector of the line (O_1O_2), and K is in the plane of the 2 optical axes. The coordinate system (OIjk) is obtained from (OIJK) by a rotation of angle $\emptyset$ of J and K around (OI) (Fig.1a), j is chosen unitary.

The points M_2 of S_2 that are possible correspondents of a point M_1 of S_1 lie on the epipolar line $E(M_1)$, which is the intersection of S_2 with the plane $(O_1O_2M_1)$. Given M_1 on S_1, there is a unique angle $\emptyset$ such that the plane (OIk) contains M_1, O_1 and O_2, and the search for the correspondent M_2 to M_1 is restrained to this plane which is the plane of Fig.1b.

Θ_1 is the angle between (O_1O_2) and (M_1O_1). In the orthogonal coordinate system $(O_1i_1jk_1)$, k_1 is a vector of the line (M_1O_1). Θ_2, i_2 and k_2 are defined accordingly for M_2 (Fig.2).

Let M be an object point moving with a 3-D velocity U, P_1 and P_2 the reciprocal of the distances O_1M and O_2M respectively, v_1 the velocity component along j of the image M_1 of M on the left image. The same notations hold for M_2, image of M on S_2. If $<\ ,\ >$ refers to the inner product, we have: $\qquad v_1 = P_1.<U, j> \qquad$ and $\qquad v_2 = P_2.<U, j>$·

Considering that $P_1.\sin\Theta_2 = P_2.\sin\Theta_1$, it implies that:

$$v_1/\sin\Theta_1 = v_2/\sin\Theta_2 \qquad (1)$$

Given M_1 on S_1, any point M_2 of $E(M_1)$ verifying eq. (1) is a possible correspondent of M_1. More generally, for any pair of viewing systems of optic centers O_1 and O_2, the *static epipolar constraint* expresses that correspondent points are coplanar with O_1 and O_2, while the *dynamic epipolar constraint*, in the form of eq. (1) here, *states that this coplanarity must be preserved during motion.* Therefore stereo-correspondence from optical flow (SCOF) consists in the pairing of points that satisfy both the static and dynamic epipolar constraints. SCOF is then equivalent to the classical process of stereo matching, where the luminous intensity is replaced by the function f defined along an epipolar line as: $f(\Theta_i) = v_i/\sin\Theta_i$ (i=1 or 2). If the noise which perturbates the measure of velocity is proportional to velocity-magnitude, the performance of SCOF must increase with the ratio $f'(\Theta)/f(\Theta)$ (we drop index i, and f' is the Θ-derivative of f along an epipolar line).

The case of rigidity. For a rigid surface, the motion can be decomposed in a rotation around O and a translation, of coordinates (w_X, w_Y, w_Z) and (t_X, t_Y, t_Z) in (IJK), and we have

$$f(\Theta) = [t_Y.\cos\emptyset - t_Z.\sin\emptyset + a.(w_Y.\sin\emptyset + w_Z.\cos\emptyset)]/z - (w_Y.\sin\emptyset + w_Z.\cos\emptyset)/tg\Theta - w_X \qquad (2)$$

where z is the k-coordinate of M and $\emptyset$ the vertical excentricity of M. Eq. (2) indicates that SCOF is not always possible (for instance if w_Y and w_Z are null and the surface is such that z does not vary with Θ), and that the performance of SCOF should be optimal when the depth variations are large in the direction of (O_1O_2). Finally t_X and w_X are useless for the SCOF, but w_X should play a role for noisy data, when the noise is proportional to $|f|$.

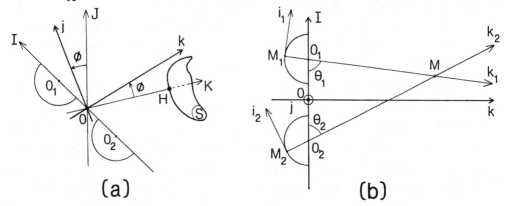

Figure 1. The coordinate systems.
(a) O_1 and O_2 are the optical centers of the left and right viewing systems. O is the middle of $[O_1O_2]$, (ÔK) is orthogonal to the line (O_1O_2) represented by the axis (OI). The axis J forms an orthogonal system with I and K. The system (OIjk) is obtained from (OIJK) by a rotation of angle $\emptyset$ around (OI). The 2 arcs of circles represent the hemispherical retinae. S is a surface intersecting (OK) in point H.
(b) M is an object point of images M_1 and M_2 on the left and right retinae respectively. The local referential $(M_1i_1jk_1)$ and $(M_2i_2jk_2)$ and the angles Θ_1 and Θ_2 are defined in text.

2. Computer simulations.

The first computations concern foveal vision, when the influence of each motion component is uniform over the image ($|\emptyset|$ small), while the second group handles large planar images where the influence of motion components varies over the image.

Initially the surfaces intersect the axis (OK) in a point H, 72 cm from O (Fig.1a). The *time unit tu* is arbitrary. The 3-D velocities are chosen so as to limit the image velocity to 1 image diameter per tu. Different types of motions are used; among them, motion R is a rotation around the axis (HJ) (i.e., a translation along (OI) associated to a rotation around (OJ)), and motion O is a rotation around (OK).

The velocity coordinate v along j is perturbated by 2 Gaussian noises, one with standard deviation proportional to $|v|$ ($0.03|v|$), the other of constant standard deviation (1 pix/tu). The noisy velocity $\underline{v}$ is filtered with a 11X11 pix Gaussian of radius 3 pix. A linear interpolation is then applied between epipolar lines. Given a pixel p_l of the left image, the algorithm searches for consecutive pixels p_r and p_r' in the right epipolar such that $f(p_l)$ ranges between $f(p_r)$ and $f(p_r')$ (with straightforward notations). p_l is termed 'high-confidence point' if it satisfies two conditions: (i) $|f(p_l)|$ is higher than 4 times the background noise; (ii) the variation of f between the two nearest neighbours of p_l exceeds 1%. When (ii) is not verified, or when no correspondent to p_l is found, the depth is set to that found in the previous epipolar pixel. The absolute depth is calculated by triangulation and the error thus performed, relative to the theoretical depth, is averaged over the high-confidence points. We also plot the depth profile in the plane (OIK) (horizontal profile).

Small field simulations. The field of view is 8° diameter (40 pixels). As $|\emptyset|$ remains small, eq. (2) indicates that t_Z and w_Y play a minor role relative to t_Y and w_Z, and motions R and O represent the two basic motions to be tested in this condition.

For motion R (Fig. 2a,b) the performance of the algorithm improves as the surface is more tilted in the direction of (O_1O_2) (75% of high-confidence points when the tilt-angle reaches 10°). For different surfaces the mean depth error is less than 1% whenever more than 30% points are matched confidently. *For motion O* (Fig. 2c,d) more than 73% of the points could be matched confidently, with a mean depth error smaller than 0.7%. These two motions yield a minimal retinal slip (null velocity in the center of the image). This condition was found to be critical here; for instance a 12% reduction of w_X in motion R impaired greatly the performance.

Large field simulations. The 120X120 pix images cover 60° viewing angle. For *motions R and O* (Fig. 3) the results are the same as for small field simulations, except that the depth-error is 8 to 12 times higher (this being partly due to the loss of resolution). For a wide variety of surfaces and motions, when the high-confidence area is larger than 70%, the depth error ranges generally between 1.5 and 15%. The algorithm was also much less sensitive to an increase of the retinal slip (without change of the maximal image velocity) than in the case of small viewing angle.

Overall, our results indicate that the robustness of SCOF to the type of noise that we used is optimal when w_X minimizes the overall image velocity. For any surface, motion O (rotation around the sagittal axis (OK)) provides reliable velocity information for the SCOF which then presents robustness under noise. In addition, motion R, as well as pure translations along (OK) for large viewing angles, allow a good reconstruction of the parts of the environment with steep variations in depth.

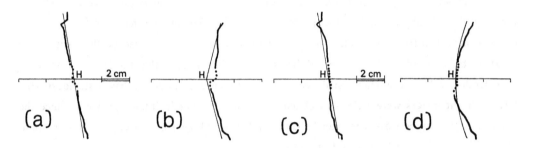

Figure 2. Results of small field simulations. The plane of the figure is (OIK) and the graduated axis is (OK). The thin line indicates the exact section of the surface, while the reconstructed section is in thick line. The dashed line shows the low-confidence points (see text). Point O is located 72 cm from H in the left direction of the figure. The scale is identical for all pannels in the horizontal and vertical directions. (a) Motion R, plane of tilt angle 10° in the direction (O_1O_2). (b) Motion R, 160° dihedron. (c) Motion O, same plane as in (a). (d) Motion O, sphere of radius 10 cm.

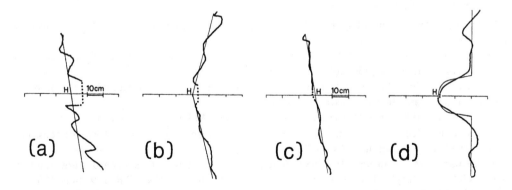

Figure 3. Results of large field simulations. Same legend as in Fig.2, except (d) where the sphere of radius 10 cm is located in front of a planar background.

3. Discussion.

Our simulations show that stereo-correspondence can be established by using the velocity field. However, except for particular 3-D motions, the SCOF is generally not sufficient for reconstructing the depth-map of a whole scene. Rather, it is likely to cooperate with intensity-based stereopsis and motion parallax, essentially because it can be applied to non-rigid environments and to 3-D motions that are useless for most structure-from-motion algorithms (translation in depth, rotation around the sagittal axis).

Several neurophysiological studies [3,9,11] have demonstrated the existence of neurons coding specifically the binocular disparity of retinal motion in the visual pathway of the cat and monkey. In addition, two psychophysical studies [1,5], suggest that motion disparity can be used by the visual system for the perception of *3-D motion*. We have performed some experiments [2] to find out whether *3-D structure* could also be perceived on the basis of the disparity between left and right optical flows. Our results suggest that the SCOF is not used as a depth-cue by the human visual system in the absence of intensity-based stereopsis (the left and right images were stereo-correlated in motion, but not in intensity). We advance the hypothesis that motion disparity can be used by the visual system as a cue for 3-D motion, but not for the fine analysis of 3-D structure.

REFERENCES

(1) Beverley K.I., Regan D., 1973: Evidence for the existence of neural mechanisms selectively sensitive to the direction of movement in space. J. Physiol., 235, 17-29.
(2) Cornilleau-Pérès V., Droulez J., 1990: Motion disparity in the absence of position disparity. A binocular visual cue for depth perception? Perception, 18, 535.
(3) Cynader M., Regan D., 1978: Neurons in cat parastriate cortex sensitive to the direction of motion in three-dimensional space. J. Physiol., 274, 549-569.
(4) Jenkin M.R.M., 1984: The stereopsis of time-varying imagery. Technical Report RBCV-TR-84-3. University of Toronto.
(5) Lee D.N., 1970: Binocular stereopsis without spatial disparity. Percept. Psychophys., 9 (2B), 216-218.
(6) Mitiche A., 1984: On combining stereopsis and kineopsis for space perception. Proc. First Conf. on AI Applications, 156-160.
(7) Mitiche A., 1988: Three-dimensional space from optical flow correspondence. Computer Vision, Graphics and Image Processing, 42, 306-317.
(8) Richards W., 1983: Structure from stereo and motion. M.I.T., A.I. Memo n^o731.
(9) Toyama K., Kozasa T., 1982: Responses of Clare-Bishop neurons to three-dimensional movement of a light stimulus. Vision Res., 22, 571-574
(10) Waxman A.M., Duncan J.H., 1985: Binocular image flows: steps toward stereo-motion fusion. Univ. of Maryland, Computer Vision Laboratory, Report CAR-TR-119. May 1985.
(11) Zeki S.M., 1974: Cells responding to changing image size and disparity in the cortex of the rhesus monkey. J. Physiol., 242, 827-841.

We model the responses of directionally selective cells in primary cortex to an image, $I(\vec{x},t)$, as the nonlinear filter

$$N\left(\vec{x},t: \Omega,\vec{n},\Omega_t,\sigma,\sigma_t\right) = \left|F\left(\vec{x},t: \Omega,\vec{n},\Omega_t,\sigma,\sigma_t\right) * I\left(\vec{x},t\right)\right|^2, \qquad (2.1)$$

where $*$ stands for convolution. This definition is similar to the one proposed by Adelson and Bergen (1985) who call it motion energy.

The Fourier tranform of s a Gabor is a Gaussian centered on $(\Omega\vec{n},\Omega_t)$. This shows that the nonfilter N is not directly tuned to velocity. However, although one filter cannot estimate the velocity, $\vec{v}$, as we will show in the next section, the set of filters responding most vigorously can.

A physiological interpretation for the model is developed in Grzywacz and Yuille (1990). A consequence of this interpretation is that the bandwidth of the temporal frequency tuning curves is relatively wide compared to the spatial bandwidth. Precisely, the assumption states that for all velocities, $\vec{v}$, to which the cells respond the following relationship holds: $(|\vec{v}|\sigma_t)^2 \ll \sigma^2$. Informally, it was verified by literature inspection that typically $3 \leq (\sigma/(|\vec{v}|\sigma_t))^2 \leq 60$.

3 Velocity Estimation

The spatiotemporal power spectrum of a translating image lies on the plane $\vec{\omega}\cdot\vec{v} + \omega_t = 0$ in the frequency domain (Watson & Ahumada 1985; Heeger 1987; Daugman 1988). This suggests using the combination of the outputs of cells tuned to specific spatiotemporal frequencies to detect this plane. Our results show how to combine these cells' responses in a computationally sensible way.

The following theorem shows that, if the filters are Gabor with constant standard deviation, then the maximal response lies on a plane in the space of cell parameters. Knowledge of this plane determines the velocity. This result does *not* follow trivialy from the knowledge that the spatiotemporal power spectrum of a translating image lies on a plane. Indeed, one can show that filters other than Gabor filters do not have the same property (this is related to the scale–space theorems; Yuille & Poggio 1986). The theorem is strictly correct only when the receptive field sizes and temporal windows are constant for all cells. However, in Theorem 3 and its corollary, we show that this constancy requirement can be relaxed under physiological conditions.

Theorem 1. *If σ and σ_t are constants, then the local maxima of $N(\vec{x},t: \Omega,\vec{n},\Omega_t,\sigma,\sigma_t)$ as a function of $(\Omega,\vec{n},\Omega_t)$ lie on the plane $\Omega\vec{n}\cdot\vec{v} + \Omega_t = 0$ for all images that move with a constant velocity $\vec{v}$.*

This results follows from a corollary of a stronger result: Theorem 2.

Theorem 2 will provide the response distribution in the three–dimensional space defined by the cells' optimal spatial and temporal frequencies.

Theorem 2. *The response $N(\vec{x},t: \vec{\Omega},\Omega_t,\sigma,\sigma_t)$ is weakly separable as follows: A function p exists such that $N(\vec{x},t: \vec{\Omega},\Omega_t,\sigma,\sigma_t) = p(\vec{x},t: \sigma^2\vec{\Omega} - \sigma_t^2\Omega_t\vec{v},\sigma,\sigma_t)\exp(-(\sigma_t^2\sigma^2)\left(\Omega_t + (\vec{\Omega}\cdot\vec{v})\right)^2/2(\sigma^2 + \sigma_t^2 v^2))$. Hence the only dependence of N on the spatial characteristics of the stimuli occurs within the function p.*

Proof: See Grzywacz and Yuille (1990).

The following corollary shows that if the receptive field sizes and temporal windows are constant, then the responses follow a known Gaussian distribution centered on the plane of Theorem 1. Thus, the claim in Theorem 1 follows from this corollary.

Corollary 1. *If σ and σ_t are constants, then the variation of $N(\vec{x},t: \vec{\Omega},\Omega_t,\sigma,\sigma_t)$ in the $(\vec{\Omega},\Omega_t)$ space in the direction $((\vec{v}\sigma_t)/\sigma^2, 1/\sigma_t)$ is a Gaussian function centered on the plane $\Omega\vec{n}\cdot\vec{v} + \Omega_t = 0$, and dependent only on $\vec{v}$, σ, and σ_t.*

Proof: The arguments $(\sigma^2\vec{\Omega} - \sigma_t^2\Omega_t\vec{v})$ of the function p do not vary in this direction. The only variation is therefore due to the Gaussian function $\exp(-(\sigma_t^2\sigma^2)\left(\Omega_t + (\vec{\Omega}\cdot\vec{v})\right)^2/2(\sigma^2 + \sigma_t^2 v^2))$. This Gaussian is centered on the plane $\Omega\vec{n}\cdot\vec{v} + \Omega_t = 0$.

A Model for the Estimate of Local Velocity

N. M. Grzywacz (C.B.I.P., M.I.T.) and A. L. Yuille (D.A.S. Harvard)

Abstract

Motion sensitive cells in the primary visual cortex are not selective to velocity, but rather are directionally selective and tuned to spatiotemporal frequencies. This paper describes physiologically plausible theories for computing velocity from the outputs of spatiotemporally oriented filters and proves several theorems showing how to combine the outputs of a class of frequency tuned filters to detect local image velocity. Furthermore, it can be shown (Grzywacz and Yuille 1990) that the filters' combination may simulate "Pattern" cells in the middle temporal area (MT), while each filter simulates primary visual cortex cells. This suggests that MT's role is not to solve the aperture problem, but to estimate velocities from primary cortex information. The spatial integration that accounts for motion coherence may be postponed to a later cortical stage.

1 Introduction

This paper gives a brief summary of a theory for motion estimation. We concentrate here on the mathematical aspects of the theory. The reader is referred to Grzywacz and Yuille (1990) for proofs of the theorems, comparisons of the theory to neurobiology and detailed references to the literature (these references alone take up over five pages).

Motivated by neuroscientific experiments the theory assumes that the motion is first filtered by spatiotemporally tuned filters and only later is velocity explicitly computed. It is, therefore, related to several existing models (Hassenstein and Reichardt 1956; Poggio & Reichardt 1976; van Santen and Sperling 1984; Watson & Ahumada 1985; Jasinschi 1988; Bulthoff, Little & Poggio 1989, Fleet & Jepson 1989) and, in particular, to theories involvinge spatiotemporally oriented motion energy filters (Adelson & Bergen 1985). Our model is closely related to the elegant model of Heeger (1987) that computes velocities through the spatiotemporal integration of the outputs of Gabor motion energy filters (Gabor 1946; Daugman 1985). Unfortunately his method of computing the velocities assumes that the image's power spectrum is flat, which is often incorrect. Our theoretical results show that this assumption is unnecessary.

The model has two stages: The first measures motion energies (the output of motion energy filters) and the second estimates velocity from these energies. Section 2 describes the first stage and Section 3 gives mathematical results used as a basis for the second stage in Section 4.

2 Model Description

Following previous work (Adelson & Bergen 1985, Heeger 1987) we use a (complex) spacetime Gabor filter (Gabor 1946; Daugman 1985) $F(\vec{x}, t : \Omega, \vec{n}, \Omega_t, \sigma, \sigma_t)$ where the σ's are the standard deviations and the Ω's are the frequencies of the filter.

Theorem 3 shows that the response distribution in the optimal–frequency space is of a simple form. This result is important, because the receptive field sizes and temporal windows may depend on the cells' optimal frequencies (Section 2). We denote these dependencies by $\sigma(\Omega) = K(|\vec{\Omega}|)/|\vec{\Omega}|$ and $\sigma_t(\Omega_t) = K_t(\Omega_t)/|\Omega_t|$, where K and $K_t(\sigma_t)$ are functions that are mildly dependent , or perhaps independent, of σ and σ_t respectively. More precisely, Theorem 3 assumes that for all velocities, $\vec{v}$, to which the cells respond, $(|\vec{v}|\sigma_t)^2 \ll \sigma^2$. An informal literature study seems to justify this assumption.

Theorem 3. *Given the approximation* $|\vec{v}|^2 \ll (\sigma/\sigma_t)^2$, *and remembering that* $\sigma_t = \sigma_t(\Omega_t)$ *and* $\sigma = \sigma(\Omega)$, *we find that the response* $N(\vec{x}, t : \vec{\Omega}, \Omega_t, \sigma, \sigma_t)$ *is weakly separable in the sense that there exists a function* r, *independent of* Ω_t *and* σ_t, *such that* $N(\vec{x}, t : \vec{\Omega}, \Omega_t, \sigma, \sigma_t) \approx r(\vec{x}, t : \vec{\Omega}, \sigma)$ $\times \exp\{-\sigma_t^2(\Omega_t + (\vec{\Omega} \cdot \vec{v}))^2/2\}$. *Proof:* See Grzywacz and Yuille (1990).

Corollary 2 shows that in the three–dimensional space of optimal frequencies, the response distributions as function of temporal frequency have maxima on the plane defined in Theorem 1. This means that the overall distribution has a maximal ridge on the plane. Under the approximation $(|\vec{v}|\sigma_t)^2 \ll \sigma^2$, the distribution of motion energies is oriented parallel to temporal frequency axis.

Corollary 2. *With the same assumptions and approximations as Theorem 3, along a one-dimensional line parallel to* Ω_t *axis, the maximum of* $N(\vec{x}, t : \vec{\Omega}, \Omega_t, \sigma, \sigma_t)$ *lies on the plane* $\vec{\Omega} \cdot \vec{v} + \Omega_t = 0$.

Proof: Consider the set of lines parallel to the Ω_t axis. The only variation of $N(\vec{x}, t : \vec{\Omega}, \Omega_t, \sigma, \sigma_t)$ is due to the $\exp\{-\sigma_t^2(\Omega_t + (\vec{\Omega} \cdot \vec{v}))^2/2\}$ term, which is unimodal with maximum at $\vec{\Omega} \cdot \vec{v} + \Omega_t = 0$.

Strictly speaking these theorems assume the velocity is constant over the whole image. Since, however, the filters have limited spatiotemporal range (determined by σ and σ_t) the velocity need only be approximately constant over this range.

4 Strategies and Neural Implementations for Velocity Estimation

We now describe three related methods for finding the velocity of the stimulus using the mathematical results of the previous section, and discuss possible neural implementations. The computational, psychophysical, and implementational aspects of this problem are described in Grzywacz and Yuille (1990).

The Ridge Strategy: This strategy uses Corollary 2 as a starting point and proposes to make excitatory connections from each motion–energy cell to the velocity selective cells most consistent with it. These connections should weakly prefer velocities with small components perpendicular to the preferred direction, so as to give a unique answer for the aperture problem in the large. Suppose we have a set of M motion–energy cells $(\vec{\Omega}^\mu, \Omega_t^\mu, \sigma^\mu, \sigma_t^\mu)$ with $\mu = 1, ..., M$. A possible implementation is to define the response, $R(\vec{x}, t : \vec{v})$, at time t of a velocity selective cell tuned to velocity $\vec{v}$, and whose receptive field is centered at position $\vec{x}$, by

$$R(\vec{x}, t : \vec{v}) = A \sum_\mu N(\vec{x}, t : \vec{\Omega}^\mu, \Omega_t^\mu, \sigma^\mu, \sigma_t^\mu) e^{-(\sigma_t^\mu)^2(\Omega_t^\mu + (\vec{\Omega}^\mu \cdot \vec{v}))^2/2} e^{-(\vec{v} \cdot \vec{\Omega}^{\mu*}/k)^2}, \qquad (4.1)$$

where $\vec{\Omega}^*$ is orthogonal to $\vec{\Omega}$, and A and k are constant parameters.

This equation suggests that the strength of the connection between cell $(\vec{\Omega}^\mu, \Omega_t^\mu, \sigma^\mu, \sigma_t^\mu)$ and the velocity selective cell tuned to the velocity $\vec{v}$ should be $\exp\{-(\sigma_t^\mu)^2(\Omega_t^\mu + (\vec{\Omega}^\mu \cdot \vec{v}))^2/2\} \exp\{-(\vec{v} \cdot \vec{\Omega}^{\mu*}/k)^2\}$.

This method is similar to correlation and template matching methods in computer vision. If we fix $\vec{\Omega}$ and let Ω_t vary, then from Theorem 3, we know that the form of the variation of the filtered response is $\exp\{-\sigma_t^2(\Omega_t + (\vec{\Omega} \cdot \vec{v}))^2/2\}$; this defines our template. The largest value of the correlation of this template with $N(\vec{x}, t : \vec{\Omega}^\mu, \Omega_t^\mu)$, as we vary the value of $\vec{v}$ while fixing $\vec{\Omega}$, gives an estimate for the velocity. To combine the results as $\vec{\Omega}$ varies, we simply add the magnitude of the responses for each $\vec{\Omega}$. The factor $\exp{-(\frac{\vec{v}\cdot\vec{\Omega}^\mu}{k})^2}$ is designed to prevent the aperture problem in the large (if the image motion is consistent with an infinite set of possible velocities, then the smallest velocity is perceived). The parameter k should be sufficiently large to maintain the validity of the results of Section 3. A number of velocity selective cells will be excited and the one with the largest response corresponds to the velocity estimate. A winner–take–all mechanism may then select the maximally responding cell.

The Estimation Strategy: This strategy attempts to estimate the image's spatial characteristics and compute the velocity simultaneously by minimizing a goodness–of–fit criterion. It is based on Theorem 3 which shows that the response N is the product of two functions, the first of which, r, is independent of Ω_t while the second depends only on the velocity of the image and the filter parameters. Thus several filters with the same Ω_t willput strong constraints on the possible velocity (since r will be constant for these filters).

A robust way of exploiting this idea is to minimize a goodness–of–fit criterion $E\left(\vec{v}, r(\vec{\Omega})\right)$, both with respect to $\vec{v}$ and $r(\vec{\Omega})$, given a set of measurements $N(\vec{x}, t : \vec{\Omega}^\mu, \Omega_t^\mu, \sigma^\mu, \sigma_t^\mu)$ for $\mu = 1, ..., M$. We choose the standard least–squares fit criterion

$$E\left(\vec{v}, r(\vec{\Omega}), \sigma, \sigma_t\right) = \sum_\mu \left(N(\vec{x}, t : \vec{\Omega}^\mu, \Omega_t^\mu, \sigma, \sigma_t) - r(\vec{x}, t : \vec{\Omega}^\mu)e^{-\sigma_t^2(\Omega_t^\mu + (\vec{\Omega}^\mu \cdot \vec{v}))^2/2}\right)^2, \qquad (4.2)$$

Suppose we have several lines of filters with constant Ω_t. Denote the values of $r(\vec{\Omega})$ on the lines as r^ν for $\nu = 1, ..., L$. This gives $E\left(\vec{v}, r^\nu\right) = \sum_\mu \left(N(\vec{x}, t : \vec{\Omega}^\mu, \Omega_t^\mu, \sigma, \sigma_t) - r^\nu(\mu)e^{-\sigma_t^2(\Omega_t^\mu + (\vec{\Omega}^\mu \cdot \vec{v}))^2/2}\right)^2$ One of the ways to find the velocity $\vec{v}$ that minimizes this equation is as follows. Since the goodness–of–fit criterion, $E\left(\vec{v}, r^\nu\right)$, is quadratic in r^ν, we can obtain by differentiation a system of L linear equations and L variables, whose solution gives the best r^ν as a function of $\vec{v}$. By substituting back for r^ν one obtains a cost function $\overline{E}(\vec{v})$. This function may be fed to velocity selective cells, that is, a cell selective to velocity $\vec{v}$ would receive input $\overline{E}(\vec{v})$. Among these cells, the one with the smallest response corresponds to the velocity estimate.

The Extra Information Strategy: This strategy uses the outputs of purely spatial frequency tuned cells to calculate the spatial characteristics of the image. This information can then be used to modify the Estimation Strategy by giving estimates for the form of $r(\vec{\Omega})$. We do not discuss this method in detail here.

5 Summary

The Gabor function is, strictly speaking, the only filter for which we can guarantee that the extrema of responses in the cells' optimal–frequency space lie on a ridge (unpublished calculations). This can be traced to the fact that the Gaussian is the only separable rotationally invariant function. If, however, the filters are similar, but not exactly, like Gabors, then we expect the results of Section 3 to be true most of the time. This expectation is confirmed by the velocity computation in real images with filters that were built by a self–organizing developmental model, and which resemble Gabor functions only roughly (Yuille & Cohen 1989).

In Grzywacz and Yuille (1990) we show that our model is consistent with four experimental phenomena in the primary visual cortex and the middle temporal area. There are, however, three psysiological problems with Gabor models (discussed in Grzywacz and Yuille 1990), despite their nice mathematical properties. We hope that the substance of our mathematical analysis will remain when we replace Gabors with more realistic filters and make our theory satisfy psysiological constraints (Grzywacz & Poggio 1989).

Our theory provides local estimations of velocity and hence gives a partial solution to the aperture problem. It does not, however, globally integrate these estimates to give a coherent motion flow. We therefore suggest that these estimates should be input to a motion coherence theory (such as Yuille & Grzywacz 1988) which might be implemented in later cortical areas that perform spatial integration over large receptive fields.

Acknowledgements

Some of the motivations for this work were provided by discussions with Dave Heeger. N.M.G. was supported by grants BNS-8809528 IRI-8719394 from the NSF. A.L.Y. was supported by the Brown–Harvard–MIT Center for Intelligent Control Systems with the U.S. Army Research Office grant DAAL03-86-K-0171.

References

Adelson, E.H. and Bergen, J. 1985 Spatiotemporal energy models for the perception of motion. *J. Opt. Soc. Am.* A **2**, 284–299.

Bulthoff, H., Little, J. and Poggio, T. 1989 A parallel algorithm for real–time computation of optical flow *Nature* **337**, 549–553.

Daugman, J.G. 1985 Uncertainty relation for resolution in space, spatial frequency, and orientation optimized by two dimensional visual cortical filters *J. Opt. Soc. Am.* A **2**, 1160–1169.

Daugman, J.G. 1988 Pattern and motion vision without Laplacian zero–crossings *J. Opt. Soc. Am.* A **5**, 1142–1148.

Fleet, D.J. and Jepson, A.D. 1989 Computation of normal velocity from local phase information *Technical Reports on Research on Biological and Computational Vision* **RBCV–TR–89–27**, Department of Computer Science, University of Toronto, Toronto, Canada.

Gabor, D. 1946 Theory of communication *J. Inst. Electr. Eng.* **93**, 429–457.

Gaddum, J.H. 1945 Lognormal distributions *Nature* **156**, 463–466.

Grzywacz, N.M. and Poggio, T. 1989 Computation of motion by real neurons. In *An Introduction to Neural and Electronic Networks* (eds. S.F. Zornetzer, J.L. Davis and C. Lau) Orlando, FL, USA: Academic Press. In Press.

Grzywacz, N.M. and Yuille, A.L. 1990 A model for the estimate of local image velocity by cells in the visual cortex. *Proceedings of the Royal Society of London.* In press.

Hassenstein, B. and Reichardt, W.E. 1956 Systemtheoretische analyse der zeit-, reihenfolgen- und vorzeichenauswertung bei der bewegungsperzeption des russelkafers *chlorophanus, Z. Naturforsch.* **11b**, 513–524.

Heeger, D. 1987 A model for the extraction of image flow, *J. Opt. Soc. Am.* A **4**, 1455–1471.

Hildreth, E.C. 1984 *The Measurement of Visual Motion.* Cambridge, MA: MIT Press.

Jasinschi, R.S. 1988 Space–time sampling with motion uncertainty: Constraints on space–time filtering. In *Proc. Second Int. Conf. Computer Vision* Tampa, FL, USA, pp. 428–434. Washington, DC: IEEE Computer Society Press.

Poggio, T. and Reichardt, W.E. 1976 Visual control of orientation behaviour in the fly: Part II: Towards the underlying neural interactions *Q. Rev. Biophys.* **9**, 377–438.

van Santen, J.P.H. and Sperling, G., 1984 A temporal covariance model of motion perception *J. Opt. Soc. Am.* A **1**, 451–473.

Watson, A.B. and Ahumada, A.J. 1985 Model of human visual-motion sensing *J. Opt. Soc. Am.* A **2**, 322–341.

Yuille, A.L. and Cohen, D.S. 1989 The Development and training of motion and velocity sensitive cells. Harvard Robotics Laboratory Technical Report 89–9.

Yuille, A.L. and Grzywacz, N.M. 1988. The motion coherence theory. In *Proc. Second Int. Conf. Computer Vision*, Tampa, Florida, USA. pp. 344–353. Washington, DC: IEEE Computer Society Press.

Yuille, A.L. and Poggio, T. 1986 Scaling theorems for zero–crossings *PAMI-8* **1**, 15–25.

Direct Evidence for Occlusion in Stereo and Motion

James J. Little
University of British Columbia
Vancouver, BC, Canada V6T 1W5

Walter E. Gillett
MIT
Cambridge, MA, USA 02139

Abstract: Discontinuities of surface properties are the most important locations in a scene; they are crucial for segmentation because they often coincide with object boundaries. Standard approaches to discontinuity detection decouple detection of disparity discontinuities from disparity computation. We have developed techniques for locating disparity discontinuities using information internal to the stereo algorithm of [2], rather than by post-processing the stereo data. The algorithm determines displacements by maximizing the sum, at overlapping small regions, of local comparisons. The detection methods are motivated by analysis of the geometry of matching and occlusion and the fact that detection is not just a pointwise decision. Our methods can be used in combination to produce robust performance. This research is part of a project to build a "Vision Machine" [7] at MIT that integrates outputs from early vision modules. Our techniques have been extensively tested on real images. [1]

1 Introduction

This investigation describes a component of the MIT Vision Machine [7], which integrates outputs from early vision modules for tasks such as recognition and navigation. The integration stage computes maps of scene properties augmented by an explicit representation of scene discontinuities, identifying their physical origin. Our major achievement in this paper is the development of techniques for locating disparity discontinuities using information internal to the stereo and motion modules, rather than by post-processing the output. Later processing to detect discontinuities [6] can then operate with substantially more information about their location. We have devised techniques for discontinuity location, based on an analysis of patchwise matching scores internal to the algorithm, and based on the effects of occlusion.

Stereo and motion both compute similar quantities – image displacements of image elements. We use a dense set of overlapping matching operators to compute displacements between the two images. Both stereo and motion apply uniqueness and continuity constraints. Scene geometries differ, however, and so do interpretations of ordering constraints.

1.1 The parallel stereo algorithm

The Drumheller-Poggio algorithm [2] served as an experimental testbed for the research described here. Stereo matching is an ill-posed problem [1] that cannot be solved without taking advantage of natural constraints. The continuity constraint asserts that the world consists primarily of piecewise smooth surfaces. If the scene contains no transparent objects, then there can be only one match along the left or right lines of sight (uniqueness). The ordering constraint [11] states that any two points must be imaged in the same relative order in the left and right eyes.

The specific assumption used is that the disparity of the surface is locally constant in a small region surrounding a pixel. It is restrictive, but may often be a satisfactory local approximation (it can be extended to more general surface assumptions in a straightforward way but at high computational cost). Let $E_L(x,y)$ and $E_R(x,y)$ represent the left and right image of a stereo pair or some transformation of the images. We look for a discrete disparity $d(x,y)$ at each location (x,y) in the image that minimizes

$$\|E_L(x,y) - E_R(x + d(x,y),y)\|_{N_{(x,y)}} \tag{1}$$

where the norm is a summation over a local neighborhood $N(x,y)$ centered at each location (x,y); $d(x,y)$ is assumed constant in the neighborhood. The algorithm actually implemented is somewhat more complicated, since it involves geometric constraints (ordering and uniqueness) that affect the way the maximum operation is performed (see [2]). The algorithm is composed of the following steps:

---

[1]This report describes research done within the Artificial Intelligence Lab at MIT as well as at UBC, supported, at MIT, by DARPA under Army contract DACA76–85–C–0010 and in part under ONR contract N00014-85-K-0124. Primary support for Gillett came from NIGMS Training Grant T32-GM07484, under by the MIT Dept. of Brain and Cognitive Sciences. This research was also supported by NSF Contract No. MIP-8814612, and by a grant from the Natural Sciences and Engineering Research Council of Canada.

1. Compute features for matching (edge detection or band-pass filtering).
2. Compute matches scores between features.
3. Determine the degree of continuity around each potential match.
4. Identify disparities based on the constraints of continuity, uniqueness, and ordering.

Potential matches between features are computed as follows. The images are registered so that the epipolar lines are horizontal. We compute match score planes, one for each horizontal disparity. Let $p(x, y, d)$ denote the value of the (x, y) entry of the match score plane at disparity d. For edge-based tokens, the results of comparison are binary. We set $p(x, y, d) = 1$ if there is a token at (x, y) in the left image and a compatible token at $(x - d, y)$ in the right image; otherwise set $p(x, y, d) = 0$. For brightness-based matching, the matching score continuously varies (E_L and E_R vary over some finite range and the norm of their difference assumes a range of values, not just 0 and 1 – see Equation 1).

The value computed by Equation 1 measures the degree of continuity around each potential match at (x, y, d). For edge-based matching, the summation counts the "votes" for the disparity d in the dth match plane. If the continuity constraint is satisfied near (x, y, d) then $N(x, y)$ will contain many votes and the score $s(x, y, d)$ will be high (see Equation 1). We mostly will discuss the edge-based methods in stereo and therefore will maximize the normalized correlation and will speak of peaks in the measured values. Finally, we select the correct matches by applying the uniqueness and ordering constraints. Under the uniqueness constraint, a match suppresses all other matches along the left and right lines of sight with weaker scores. To enforce the ordering constraint, if two matches are not imaged in the same relative order in left and right views, we discard the match with the smaller support score. Each match suppresses matches with lower scores in its forbidden zone [11][8] (see Section 2.2).

The matching scores of the stereo algorithm are valuable information. They provide a confidence level for each match that can discriminate between competing matches, as in forbidden zone suppression (using the ordering constraint). Matching scores are computed everywhere with no additional computation (because of homogeneous computation in SIMD machines), both for edge-base and brightness-based matching, producing dense information. The scores also help to suppress bad matches within occluded areas of the scene (Section 2.2).

2 Disparity discontinuities

We describe two discontinuity detection techniques, arising from analysis of the behavior of matching methods near occluding boundaries. One method is based on an analysis of matching scores for different disparities and the other arises from the effects of geometric constraints near occlusions.

2.1 Close winners

The close winners technique analyses matching scores. For each point $p = (x, y)$ in the left image and $q = (x+d, y)$ in the right image, the matcher computes a score $s(x, y, d)$ indicating the likelihood that p matches q, i.e., that p and q are images of the same physical point in the scene. The score at a point, $s(x, y, d)$, is the sum of pointwise match scores in a region $N(x, y)$ (see Equation 1). The matcher examines only disparities in the fixed interval $[id, fd]$, where id and fd are the initial and final disparities. Define the score vector $v(p) = \{s(id), s(id + 1), ..., s(fd)\}$, the sequence of matching scores for point p.

We begin with a simple example, a random-dot stereogram (RDS) which fuses to yield the impression of a 192×192 square floating in front of the background (256×256). Figure 1 shows a schematic representation of the scene; the dark strip on the left-hand side is an occluded part of the background seen in the left view but not the right. Point B is located on the boundary of the square. The local support neighborhood of point B, N_B, is divided between the square and the background. Approximately half of the edges in N_B will vote for the wrong disparity, namely the background disparity. The graph of $v(B)$ is bimodal, with one peak at the foreground disparity and another peak at the background disparity. Spoerri and Ullman[9] use a similar observation to derive a different scheme for motion segmentation. Matching scores vary over possible disparities (displacements) and will be maximal at the two displacements of the foreground and background. In contrast, $v(A)$ and $v(C)$ are unimodal, since their support regions cover constant disparity regions. Figure 2 shows score vectors computed for the RDS – high scores represent best matches. It is critical that the diameter of the support region be larger than the largest disparity gap in the image

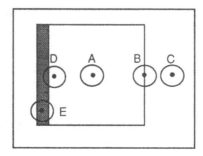

Figure 1: Line drawing of scene: floating square.

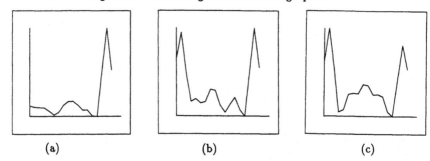

| (a) | (b) | (c) |

Figure 2: Score vectors for RDS. (a) A: (128,128). (b) B: (192,128). (c) D: (85,128).

– else the two peaks will not be detected using close winners. Also, the maximum value for the match score at B will at most be *half* that of the score at points such as A and C; this leads to a method for discontinuity identification using local spatial extrema of the match score (see [4]).

We call point B a close winner because the "winning" disparity has a close competitor; such points are likely to be located at disparity discontinuities. For all points p in the left image, use the following procedure to determine whether p is a close winner:

1. Identify peaks in $v(p) = \{s_{id}, s_{id+1}, \ldots s_{fd}\}$.
2. If $v(p)$ has two or more peaks, pick the two largest, α and β, $\alpha \geq \beta$. Let the margin $m = (\alpha - \beta)/\alpha$. If $m \leq M$ (0.2 for the results here), then p is a close winner.

Figure 3 shows close winners for several stereo scenes. Note that close winners can be correctly located for point B, but for point E, they identify locations in the center of the occluded area. These can be corrected by using a symmetric matching scheme, combined with mapping close winners into a common coordinate system.

2.2 Suppression Using Ordering Constraint

When one surface lies in front of another, the foreground surface occludes a portion of the background surface. The location of the occluded region depends on the viewpoint. Since the boundary on the near side of an occluded region is the discontinuity contour, identifying an occluded region leads us directly to the associated disparity discontinuity. This technique can be used to locate any disparity discontinuity except extended horizontal boundaries, which are not associated with occlusion.

Let us consider right-occluded areas, i.e., areas visible from the left but not the right view (see Figure 4b). Such an area does not have a match in the right image. Every potential match is surrounded by an hourglass-shaped region extending through the d and x dimensions, the forbidden zone (see [11]), as pictured in Figure 4a.

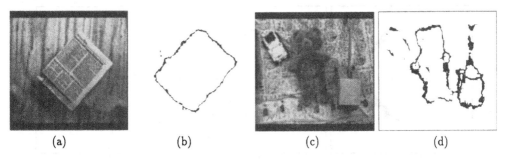

Figure 3: Close winners. (a) Newspaper on wood: left view. (b) Close winners. (c) Left view of truck, teddy bear, and crane. (d) Close winners for teddy.

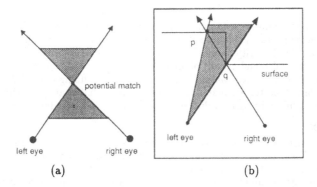

Figure 4: (a) The forbidden zone (shaded) for a particular potential match. (b) The shaded region is contained within the union of the forbidden zones for points p and q, showing that no match will be permitted there.

Consider a simple step discontinuity (Figure 4b) where the portion of the surface between points p and q is right-occluded. The shaded region contains all points that are imaged between p and q in the left view. Note that the shaded region is contained entirely within the union of the forbidden zones for p and q: the area above the line joining p and q is in the forbidden zone for q, and the area below the line is in the forbidden zone for p. Therefore all possible matches in the left view between the images of p and q will be suppressed. Match suppression is the key to locating occluded areas.

2.2.1 The Mechanics of Match Suppression

While computing matches and applying the ordering constraint, we can keep track of suppressed matches. Since matching scores are computed at all points, stereo produces dense suppression of competing matches at occlusions.

A point (x, y) in the left image is suppressed if, for all disparities d, the potential match at (x, y, d) has been suppressed. Suppressed points collectively determine regions of suppression that correspond to right-occluded areas. Disparity discontinuities are points on the right-hand side of suppressed regions, the near side in the case of right-occlusion. Others [10] have noted the connection between matching and identification of occlusions, but do not tie it in to the full ordering constraint. Figure 5 shows the suppressed regions for the newspaper scene. Some suppressed points are part of significant occluded regions and others result from incorrect matches or disparity quantization effects. As a simple measure to select significant regions, we threshold the width of contiguous strips of suppressed points. Figure 5 shows the suppressed regions and filtered suppressed regions for the left-occluded regions of the newspaper-on-wood scene. Left-occluded areas (visible in right image

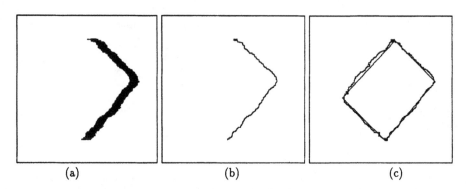

(a)	(b)	(c)

Figure 5: Left-occluded regions for newspaper. (a) Filtered suppressed points (left-occlusion). (b) Merged discontinuities (left/right). (c) Merged discontinuities over newspaper silhouette.

not from left image) are found by a symmetric analysis of the right image. The associated disparity discontinuities then lie on the left-hand side of the occlusion, and are mapped from the right into the left image, using the disparity value. The results of the analysis are shown in Figure 5.

Finally, there is an additional benefit of identifying occluded areas. Knowledge of occlusion can improve naive interpolation. Interpolation blurs discontinuities, filling in occluded areas with depth data from both sides. A better approach assumes that an occluded area has the same disparity as the background, e.g., filling in right-occluded regions with disparity values from left to right [3].

3   Conclusion

We have addressed the detection of discontinuities in stereo and motion, within the context of efficient, parallel implementation. The techniques we have examined all use information internal to the correspondence process to identify discontinuities. Any later processing to determine the figure/ground relation and to improve surface description begins with an almost complete description of the location of discontinuities. Further, these techniques all can easily be implemented on a SIMD parallel computer and simple circuits. A detailed discussion may be found in [5].

References

[1] M. Bertero, T. Poggio, and V. Torre. Ill-posed problems in early vision. *Proc. of the IEEE*, 76(8):869–889, Aug. 1988.

[2] M. Drumheller and T. Poggio. On parallel stereo. In *Proc. of IEEE Conf. on Rob. and Auto.*, pages 1439–1448, Washington, DC, 1986. IEEE.

[3] W. Gillett. Issues in parallel stereo matching. Master's thesis, Massachusetts Institute of Technology, 1988.

[4] J. J. Little, H. H. Bülthoff, and T. Poggio. Parallel optical flow using local voting. In *Proc. Int. Conf. on Comp. Vision*, pages 454–459, Tarpon Springs, Florida, Dec. 1988. IEEE, Washington, DC.

[5] J. J. Little and W. E. Gillett. Direct evidence of occlusion in stereo and motion. Tr-90-5, UBC Dept. of Computer Science, Vancouver, BC, 1990.

[6] T. Poggio, E. B. Gamble, and J. J. Little. Parallel integration of vision modules. *Science*, 242(4877):436–440, October 21 1988.

[7] T. Poggio and the staff. The MIT Vision Machine. In *Proc. Image Under. Work.*, Cambridge, MA, April 1988. Morgan Kaufmann, San Mateo, CA.

[8] S. B. Pollard, J. E. W. Mayhew, and J. P. Frisby. Disparity gradients and stereo correspondences. *Perception*, 1987.

[9] A. Spoerri and S. Ullman. The early detection of motion boundaries. In *Proc. Int. Conf. on Comp. Vision*, pages 209–218, London, England, June 1987. IEEE, Washington, DC.

[10] J. Weng, N. Ahuja, and T. S. Huang. Two-view matching. In *Proc. Int. Conf. on Comp. Vision*, pages 64–73, Tarpon Springs, Florida, Dec. 1988. IEEE, Washington, DC.

[11] A. L. Yuille and T. Poggio. A generalized ordering constraint for stereo correspondence. A.I. Memo No. 777, Art. Intell. Lab, MIT, 1984.

On the use of trajectory information to assist stereopsis in a dynamic environment

Michael R. M. Jenkin

Department of Computer Science, York University

Toronto, Ontario, Canada

Abstract

If stereopsis is to be used in a dynamic environment, it makes little sense to re-compute the entire representation of disparity space from scratch at each time step. One simple approach would be to use the results from the current solution to "prime" the algorithm for the next solution. If three dimensional trajectory information was available, this information could be used to first update the previous solution, and then this updated solution could be used to "prime" the algorithm for the following stereo pair. Recent work[4, 5] has demonstrated that it is possible to measure such trajectory information very quickly without complex token or feature extraction. This paper demonstrates how raw disparity measurement made by this earlier technique can be integrated into a single trajectory measurement at each image point. A mechanism is then proposed that updates a stereopsis algorithm operating in a dynamic environment using this trajectory information.

1   Introduction

For the most part, the application of computers to the task of stereopsis has been restricted to the static analysis of images, although some dynamic stereopsis algorithms have been developed. A static pair of images is presented to the cameras, and then some process for determining scene heights is applied to the two images. Even when applied to what are inherently dynamic tasks such as pick and place, or mobile robotics tasks, a static algorithm is applied over and over again. In light of this, some efforts have been made to embed stereopsis within temporal processing. For example, optical flow can be used to limit possible correspondences between the left and right images[12, 13], and point based correspondences can be considered in space-time, rather just in space or time[3, 7]. Another possibility is to use active vision techniques to reformulate stereopsis as an active task. Olson and Potter[9] have used Cepstral filtering and deconvolution to drive camera vergence movements. A similar philosophy drives this paper, but at a more local level. Rather than obtaining a single (or small number) of values that indicate the relative vergence and version that will be required to gaze onto structure in a scene, the technique presented in this paper obtains local measurements of three dimensional motion (stereomotion), and uses these measurements coupled with an earlier solution to the stereo problem to *prime* the stereo algorithm for the following time step. More explicitly, given a solution to the stereo problem at time t and a measurement of the local trajectory from time t to $t + \delta t$, a local approximation to the solution at time $t + \delta t$ can be computed. This first approximation can then be refined by the static stereopsis algorithm when it is applied at $t + \delta t$.

A complete implementation of a time-varying stereopsis algorithm is beyond the scope of this paper.

This paper addresses how trajectory information can be recovered and suggests a mechanism by which disparity estimates can be updated. A full algorithm which recovers surface and object descriptions in a dynamic environment is the subject of ongoing research.

2 Measuring a Unique Trajectory

Previous results have established that;

- The disparity between two bandpass signals can be determined by computing the relative phase difference between the two signals[6, 11]. This measurement can be performed directly from Sine and Cosine Gabor filtered versions of the input images.

- Space-time Sine and Cosine Gabor filters are available that are also selective for particular monocular image velocities[1].

- By combining the above two results it is possible to build detectors that measure particular left-right combinations of image velocity[4, 5]. These detectors are a very simple model of the stereomotion detectors found in biological binocular systems[2, 10].

The three dimensional trajectory detectors respond to a particular left-right velocity pair. In order to obtain a unique trajectory at a particular image point the responses of these detectors must be combined. Suppose that a number of detectors with the same speed and spatial frequency specificity are operating at a given point in an image but with different preferred trajectories. Then if structure passes through that point, each detector will respond with a projection of the true trajectory onto the particular trajectory the detector is tuned to. The response of the detectors can be modeled as $R(z) = A_0 + A_1 \cos(z + A_2)$ where $R(z)$ is the response of the z'th detector (tuned for three dimensional trajectory z). Least squares can be used to fit the detector responses to $R(z)$ and obtain estimates for A_1 (the strength of the response), and A_2 (the trajectory). Note that the responses cannot be fit to a least squares solution directly, and that $R(z)$ must be reformulated as $R(z) = \alpha + \beta \cos(z) + \gamma \sin(z)$ where $A_0 = \alpha$, $A_1^2 = \beta^2 + \gamma^2$ and $A_2 = \tan^{-1}(-\gamma/\beta)$. By recovering A_1 and A_2 we have a measure of the three dimensional trajectory and the strength of the response by pooling a number of detectors. One interesting special case of this formulation occurs when only four detectors are used tuned to motion in the frontoparallel plane to the left (R_1), motion directly away from the observer (R_2), motion in the frontoparallel plane to the right (R_3), and motion directly towards the observer (R_4). α, β and γ can then be recovered as $\alpha = (R_1 + R_2 + R_3 + R_4)/4$, $\beta = (R_1 - R_3)/2$, and $\gamma = (R_2 - R_4)/2$. β and γ then correspond to the opponent mechanism suggested in earlier papers[4, 5]. They also correspond to using two orthogonal energy detectors[1] and then using the projection of the two detectors to obtain the three dimensional trajectory. Note, however, that fitting the response to a sinusoid to obtain the trajectory is a more general technique and can easily be extended to obtain a trajectory measurement with more than four (or even four different) detectors.

Once a unique trajectory has been obtained by a given velocity channel, the outputs of the different velocity channels must be combined in order to obtain a final three dimensional trajectory at each point. As each velocity channel is returning both a direction and a strength for the preferred velocity, a winner take all approach based on the strength of the response is used to choose at each point the appropriate trajectory and velocity. A threshold is used to ignore detectors which have very low strength responses.

In order to show the promise of the technique, the trajectory measurement technique presented here is embedded within a very simple stereopsis algorithm. This implementation is designed only for exposition of the use of trajectory information and is not designed to be a complete stereo algorithm. Suppose that D_i are a collection of disparity demons, and that these demons tile the 'x' dimension. The demons are all tuned to the same spatial frequency, and each demon is tuned to zero disparity. At each time step each demon computes a new lock and certainty value, and updates its disparity lock position to the detected disparity. Associated with each demon is a trajectory measurement, and for the demons which have a valid lock on structure, their home disparity is updated by the disparity component of the recovered trajectory. Note that this is a very simple technique. Much more sophisticated methods are available for the disparity demons when used in a static environment[8].

3   A simple simulation

Figure 1 shows the left and right view of a random dot target oscillating in depth about zero disparity. Detectors were applied to the image selective for three dimensional trajectories with speeds of 1, 2 and 4 pixels/frame, and with directions of "towards", "away", "to the left", and "to the right". The response of the four detectors are then fitted to a Sine curve and a strength R_1 and phase (trajectory) R_2 is computed at each point. Four such families were constructed with speeds of 1, 2, and 4 pixels/frame. For each demon at each time step, the detector with the highest response was chosen, and a single three dimensional trajectory was obtained. For points with no acceptable trajectory measurement, a speed of zero was assumed, and the disparity and positional component of the recovered trajectory is shown in Figure 2. Integration of the three trajectory channels resulted in a very strong and easily recognizable trajectory measurement in depth. It is important to remember that each trajectory measurement has been made independently, and no mechanism has been applied to smooth the responses. Such a mechanism could be used to identify the incoherent responses obtained in the position component. Finally the trajectory information is used to update the home position of the disparity detectors. The recovered disparity of the detectors is shown in Figure 3. Once again, raw responses are plotted and no process has been used to smooth the responses or to identify responses of low certainty.

4   Discussion

This paper proposes that in order to avoid the computational expense involved in recomputing the stereo solution from scratch at each time step, it would be more useful to use the previous solution plus known trajectory information as a first approximation to the new solution. Recent results have shown that with very little computational expense (low level image filtering which can be performed in parallel, and often by specialized hardware), raw trajectory measurements can be made. These trajectory demons are modeled after computational process suggested by Beverley and Regan[2], and they have many similar properties. By fitting the responses of the detectors to a Sine wave, it is possible to pool detectors located at the same spatial position with similar speed and frequency tuning into a single response. A number of such pooled responses can be aggregated together in order to make a final measurement of the trajectory that structure with a particular spatial frequency has when it passes near a particular point in three space. This final trajectory measurement can be used in a Taylor series expansion of the known disparity at a given time, to predict the disparity at a slightly later time. This prediction can then be used to prime the more expensive disparity task.

References

[1] E. H. Adelson and J. R. Bergen. Spatiotemporal energy models for the perception of motion. *J. Optical Society of America A*, 2(2):284–299, 1985.

[2] K.I. Beverley and D. Regan. Evidence for the existance of neural mechanisms selectively sensitive to direction of movement. *J. Physiology*, 193:17–29, 1973.

[3] M. R. M. Jenkin. Tracking three-dimensional moving light displays. In *1983 ACM SIGART-SIGGRPAH Workshop on Motion*, pages 66–70, Toronto, 1983.

[4] M. R. M. Jenkin and A. Jepson. Measuring trajectory. In *IEEE Workshop on Visual Motion*, pages 31–37, Irvine, California, 1989.

[5] M. R. M. Jenkin and A. Jepson. Response profiles of trajectory detectors. *IEEE Transactions on Systems, Man and Cybernetics*, 19(6):1617–1622, 1989.

[6] M. R. M. Jenkin, A. D. Jepson, and J. K. Tsotsos. Techniques of disparity measurement. Technical Report RBCV-TR-87-16, Researches in Biological and Computational Vision, Department of Computer Science, University of Toronto, 1987.

[7] M. R. M. Jenkin and J. K. Tsotsos. Applying temporal constraints to the dynamic stereo problem. *Computer Vision, Graphics, and Image Processing*, 33:16–32, 1986.

[8] A. Jepson and M. R. M. Jenkin. The fast computation of disparity from phase differences. In *CVPR 89*, pages 398–403, San Diego, California, 1989.

[9] T. J. Olson and R. D. Potter. Real-time vergence control. In *CVPR 89*, pages 404–409, San Diego, California, 1989.

[10] T. Poggio and W. H. Talbot. Mechanisms of static and dynamic stereopsis in foveal cortex of rhesus monkey. *J. Physiology*, 315:469–492, 1981.

[11] T. D. Sanger. Stereo disparity computation using gabor filters. *Biol. Cybern.*, 59:405–418, 1988.

[12] A. M. Waxman and J. H. Duncan. Binocular image flows: steps towards stereo-motion fusion. Technical Report CAR-TR-119, Centre for Automation Research, University of Maryland, May 1985.

[13] A. M. Waxman and S. S. Sinha. Dynamic stereo: Passive ranging to moving objects from relative image flows. *IEEE Transactions on Pattern Analysis and Machine Intelligence*, 8(4):406–412, 1986.

These are views of a 1D (spatial) time varying stereogram. Intensity is encoded as height. A central region oscillated in disparity.

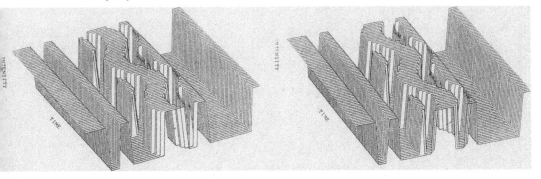

Figure 1: Left and right views of random dot stereogram

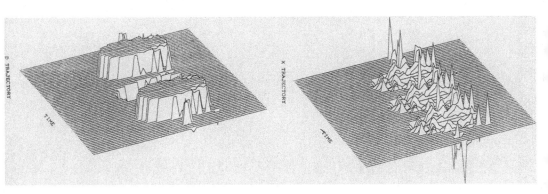

(a) Disparity component   (b) Position component
Figure 2: Recovered three dimensional trajectories. Speed is encoded as height.

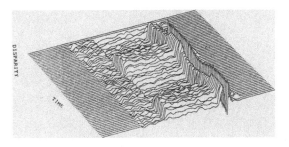

Figure 3: Disparity Demons response: Disparity is encoded as height

FEATURES / SHAPE

Distributed Learning of Texture Classification

John R. Sullins
Center for Automation Research
University of Maryland
College Park, MD 20742

Abstract

A large number of statistical measures have been postulated for the description and discrimination of textures. While most are useful in some situations, none are totally effective in all of them. An alternative approach is to *learn* which measures are best for particular circumstances. In this paper the distributed learning system of *constraint motion* is used to learn relevant texture descriptors from a set of well-known first and second order grey-level statistics. Using this system, a network of distributed units partitions itself into sets of units that detect one and only one of the given classes of textures. Each of these sets is further partitioned into individual units that detect natural subtypes of these texture classes, ones which do not necessarily produce the same types of statistics at the local level. Together, these units form a network capable of determining the texture classification of an image.

1. Introduction

"Texture" is an important cue for segmentation and for the description of objects that do not always have fixed shapes (such as "trees" or "grass"). However, defining our notions of texture is a difficult problem. Consider the textures in Figure 1. To our eyes they are obviously different from one another, but it is not easy to say exactly why. We can give vague descriptions like "rough", "uniform", "chaotic", etc., but unless we can define these terms mathematically they are of no use for machine vision.

Many statistical measures that attempt to simulate such descriptions have been proposed for texture classification; these include uniform density of image features (Aloimonos 1988), texture energy templates (Laws 1980), and second order grey-level dependencies (Kruger et al. 1974; Hall et al. 1971; Haralick et al. 1973). The general idea is to create a measurement or small set of measurements that describe textures. Theoretically, textures that "look alike" to our eyes should receive similar scores when such a measure is applied to them. Unfortunately, there is no such single measure that will accomplish this for all of the many different textures that exist in the world, so most classification systems use a combination of several of these features. Each of the n measures in the feature vector is applied to the texture sample in question, giving a point in the n-dimensional *feature space*. In the best case, points from similar textures will form distinguishable *clusters* in the feature

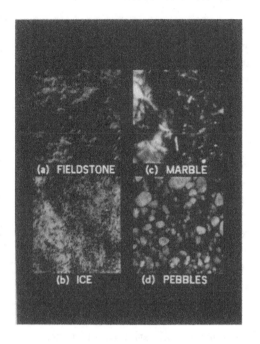 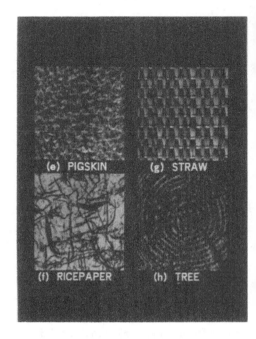

Figure 1

space, groups of points easily separable from other groups. If so, methods (Fukunaga 1972; Tou and Gonzolas 1974; Fu 1974) exist to *learn* a suitable partitioning of the feature space from a set of example texture images.

Such simple clustering may not be possible, however, especially when the image size is small and we can see only a small portion of the texture. Consider the textures shown in Figure 2. The smaller squares give several dissimilar samplings of the textures "marble", "pebbles", and "tree". It may well be that no single range of features is sufficient to discriminate them from other textures. We may need different detectors for each subtype of a given texture. Learning systems that use simple clustering may fail to partition the problem in this manner.

In this paper we take a different approach, applying the general learning system we call *motion in constraint space* (Sullins 1988) to the problem of texture classification. Textures are described in terms of a large number of simple first and second order statistics rather than a few complicated ones. The learning process is more complex, however. Each processor in the network focuses on a *portion* of a particular target texture class. That processor generalizes from a set of example textures by determining which of the statistics are relevant to the task of separating that subtexture from the other non-target texture classes, and the ranges of those statistics within which the subtexture lies. When combined into a network, those

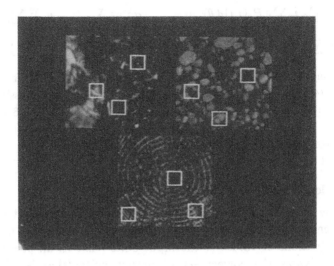

Figure 2

subtexture processors detect those and only those textures that are members of the target class.

2. Texture classification statistics

Since this learning algorithm is designed to run on a distributed system, first and second order grey level statistics were deemed especially appropriate for the problem. The co-occurrence matrices needed to compute the second order statistics may be accumulated in parallel.

The first order statistical properties of the individual grey levels used were the *mean* grey level $\mu \equiv \frac{1}{N}\sum i \, P_1(i)$, and the *variance* $\sigma^2 \equiv \frac{1}{N}\sum (i - \mu)^2 \, P_1(i)$, where $P_1(i)$ are the number of pixels in the image with grey level i and N is the total number of pixels in the image. Second order grey level statistics are relationships between grey levels at nearby pixels. These are generally computed in term of *co-occurrence matrices* $P_2(i,j|d,\theta)$, the number of points (x,y) and $(x+d\cos\theta,\ y+d\sin\theta)$ that have grey levels i and j respectively. The second order statistics (Haralick et al. 1973) used were

Energy:	$E(d,\theta) \equiv \sum_i \sum_j [P_2(i,j|d,\theta)]^2$
Inertia:	$I(d,\theta) \equiv \sum_i \sum_j (i - j)^2 \, P_2(i,j|d,\theta)$
Correlation:	$C(d,\theta) \equiv \dfrac{\sum_i \sum_j i \, j \, P_2(i,j|d,\theta) - \mu_x \mu_y}{\sigma_x \sigma_y}$ where

$$\mu_x \equiv \sum_j i \sum_i P_2(i,j \,|\, d,\theta) \qquad \sigma_x^2 \equiv \sum_j (i - \mu_x)^2 \sum_i P_2(i,j \,|\, d,\theta)$$

$$\mu_y \equiv \sum_i j \sum_j P_2(i,j \,|\, d,\theta) \qquad \sigma_y^2 \equiv \sum_i (j - \mu_y)^2 \sum_j P_2(i,j \,|\, d,\theta)$$

The second order statistics were computed with values of 1 and 2 for d and values of 0, $\pi/4$, $\pi/2$, and $3\pi/4$ for θ (statistics for $\theta + \pi$ were classified with those of θ). Counting the two first order statistics, this gave a total of 26 statistics from an example image. The second order statistics were scaled logarithmically in order to give an approximately linear distribution, and all values were normalized to lie between 0 and 8.

3. Defining learning

We define *learning* as the duplication of a given input-output behavior by some system. Given a set of binary-valued inputs and outputs, this means that for all possible combinations of values of the input units (that is, all possible *input vectors*) the system activates the correct set of output units (that is, the correct *output vector*). In this context the behavior of an output may be likened to a Boolean formula in disjunctive normal form, and learning may be defined as the determination of that formula over all possible input vectors. For the problem of texture discrimination, the inputs are statistical measures of the sample texture, and each output unit is active if a particular texture type is present.

If the DNF expression has n conjunctive subexpressions (joined by "or"'s) then it may be simulated in parallel by a machine with n processors, with the speedup associated with such distribution. Since the expression is in disjunctive normal form, if any of the subexpressions are true then the entire expression is true. Each subexpression is the conjunction of a subset of the inputs (or their negations), and is true only if all of those inputs are in the correct state.

In this sense, each input which is a part of a processor's conjunctive subterm is a *constraint* for that processor. Consider a particular input that corresponds to the binary variable "A". If the processor's conjunctive term contains A, the input must have the value 1 for the processor to be active. If it contains $\overline{A}$, the input must have the value 0. If it contains neither (which we represent as "-"), then its value does not matter to the processor.

As with most learning algorithms, we will assume that there exists some form of *supervision* that gives the network the correct classification (as *target texture* or *non-target texture*) for any given input texture. This supervision may not always be accurate, of course. Sometimes it may report that the target texture is present when it is not, and vice-versa. A learning system with applications in the real world must be able to cope with such errors.

3.1. Representing the statistics

Since the learning algorithm is designed for binary input, the continuous-valued (from 0 to 8) texture statistics were converted by assigning each value 5 binary inputs (giving a total of 130 binary inputs) whose activation depended on whether its value

S fell within a certain range:

input 0 active if $S \leq 4$ input 1 active if $1 \leq S \leq 5$

input 2 active if $2 \leq S \leq 6$ input 3 active if $3 \leq S \leq 7$

input 4 active if $4 \leq S$

This representation has the advantage of being able to represent many *ranges* of acceptable values for a statistic, depending on the number of active constraints and how their ranges overlap. For example, a statistic with value 2.4 would be represented as 11100. From this, we could derive a very restrictive set of constraints (if all 5 were constraints, then the value of the statistic would have to lie between 2 and 3 in order for the processor to be activated), less restrictive sets of constraints (for instance, 11--- would confine the statistic to lie between 1 and 4), or no constraints (----- would mean that the statistic is unimportant to the texture).

4. The learning algorithm

In this section we outline a system called *motion in constraint space* that is designed to learn texture classifications from a set of supervised examples. While it will be described in terms of texture discrimination, a more general and complete description of its capabilities may be found in (Sullins 1988).

Initially, each processor chooses a *seed texture* from the input texture examples receiving positive indication from an output. These input vectors correspond to images containing the texture type of that output (the *target texture* of that output). The states of the input units, which represent the values of the statistics of the seed texture, are the *potential constraints* for that processor. When all these constraints are enforced, the processor is restricted to detecting only those textures that have identical statistics to the seed texture. Focusing on a particular seed texture at each processor and eliminating input statistics that have little or no effect on it helps the system to properly distribute the task of learning the behavior.

The processor will then *generalize* to detect the subclass of the target texture class which contains the seed texture by removing these constraints. Eliminating constraints widens the range that a statistic may lie in in order for it to be accepted. In many cases we will eliminate *all* constraints for a particular statistic, making it completely irrelevant to the processor. As long as each major subclass is represented by a seed texture at at least one processor, this algorithm will form a network that correctly detects the target texture in most cases.

The core of this system is this addition and subtraction of constraints on the statistics at the processors in order to minimize the difference between the expected and the actual texture classifications -- that is, the *constraint motion*. Simply put, we will want to place restrictions on the values of statistics (by adding constraints) when doing so would prevent non-target textures from being accepted and we will want to remove restrictions (by removing constraints) that would prevent target textures from being accepted. The performance of the network formed by this algorithm will not be perfect, of course, as no natural texture can be represented by a simple DNF expression. However, the processors will tend to choose those constraints that maximize the correctness of the texture classifications.

4.1. Adding constraints

Generally speaking, constraints should be added to processors that accept too many non-target input textures. Because of the DNF structure of the network, a texture incorrectly accepted by any processor is also incorrectly accepted by the entire network. A non-target texture must differ from a processor's seed texture in the value of one or more inputs (otherwise, that seed texture would not have received positive indication). Each of those inputs is a potential constraint that would prevent the texture from being accepted in the future, as they would restrict the values of certain statistics to a point where the input texture would no longer lie within their ranges. Each of these potential constraints receives a *positive vote* for change at that processor.

On the other hand, we do not wish to add a constraint to a processor if the constraint would prevent too many target texture images from being accepted, specifically those target textures *not accepted by any other processor*. If that processor were no longer able to accept such "uniquely accepted" textures, then the entire network would incorrectly reject them as well. For all target textures uniquely accepted by the processor, each input that has a different value from that of the seed texture (and thus would cause the texture to be rejected if it were to become a constraint) receives a *negative vote* against change.

Positive and negative votes are collected for each potential constraint over a large sampling of the input vectors. This insures that supervisor error will have little effect on the system, as the incorrect data will usually be outvoted by the correct data. At the end of that time, potential constraints with significantly more positive than negative votes are added to the processor, narrowing the acceptance range of the processor for those statistics.

4.2. Removing constraints

The removal of constraints is somewhat similar to the addition of them, removing constraints that prevent target texture images from being accepted and not removing those that prevent non-target texture images from being accepted. The main difference has to do with the conjunctive structure of the processors. A texture may have values that lie outside of the accepted ranges of many of its statistics, so it might not meet *many* existing constraints of a processor. We want to remove constraints when doing so would cause more target textures to be accepted, but for most textures removing a single constraint will not make any difference.

We allow all textures that are not accepted by the network to influence the removal of constraints, but using an exponential function to give "near miss" textures more influence than others. If a target texture is not accepted by the network, then each constraint at processor i that prevented it from being accepted at that processor is given a positive vote for removal proportional to σ^{-n}, where n is the number of constraints that the vector failed to meet at processor i. That is, each processor is changed in proportion to how close it already is to accepting the texture; this helps the system to properly distribute the responsibility for accepting textures by assuring that only a few processors are forced to learn each one. Initially σ is 1 (each texture has an equal effect on the voting) and over time it is increased. This forces a processor to focus on a particular subclass of the target texture and helps to stabilize the behavior of the network in the long run.

Negative votes against removing constraints are collected in a similar manner. If a constraint helps to prevent a non-target texture from being accepted, it receives a negative vote against removal proportional to σ^{-n_i}. As in the case of adding constraints, the input is sampled over a certain period of time. If the number of positive votes is greater than the number of negative votes, the constraint is removed, widening the acceptance range of the processor for that statistic.

5. Experimental results

Eight 96×96 texture images (shown in Figure 1) were chosen for the tests. In order to have co-occurrence arrays P_2 of manageable sizes, the number of grey levels was reduced from 256 to 9 using the Isodata algorithm. A particular target texture was chosen for each run represented by a single output which was active when the target texture was present, inactive otherwise. This output was assigned 10 intermediate processors. An example texture image was chosen from the target texture with probability 50%, from one of the other textures otherwise. The image was a randomly chosen 16×16 subpicture of the texture. This gave a total of 6400 possible examples for each texture.

5.1. Learning ability of the system

Several runs of the learning system were made (with no supervisor error) for each of the eight target textures shown in Figure 1. Figure 3 gives the average learning curves for each texture. The X axis represents the total number of examples presented to the network at that point, and the Y axis represents the percentage of the time that the network correctly classified the input as target or non-target texture, over samplings of 1000 input vectors.

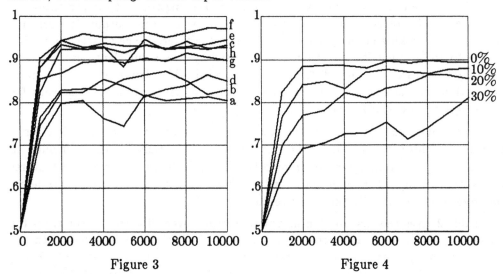

Figure 3   Figure 4

The network reached a certain level of correctness very quickly (2000 - 3000 sample input vectors), and made any further improvements very slowly. The system was able to make generalizations about the textures after seeing a small percentage of the 51,200 possible texture samples. The textures with the least correctness were those most easily confused with the others. The average correctness over all of the textures quickly stabilizes at close to 90%. This is a very good result, especially for sample images of size 16×16. The typical correctness of other feature classification systems is considered to be around 90% for 64×64 images.

5.2. Dealing with supervisor error

Several runs were also made for each texture at various levels of *supervisor error*. Figure 4 gives the learning curves for the average correctness over all textures for error percentages of 0%, 10%, 20%, and 30% in the supervision. In this case, the Y-axis is a measure of the *true* correctness of the responses -- how well they matched the actual desired output value, before any corruption by supervisor error. The learning is very resilient to levels of error less than 30% percent, and even for 30% the correctness seems to approach that of other levels. In fact, the correctness of the system for these high levels was greater than the correctness of the supervision itself, indicating that the network was able to find good features despite the error.

5.3. Importance of distribution

As mentioned above, this system creates networks capable of recognizing textures from a very small sample, 16×16 versus the usual 64×64. This is due in large part to the system's ability to properly *distribute* the detection of the different types of samples about the network. The smallness of the sample size would often cause some of the expected properties of the target texture to be absent for a particular sample. Distribution of the target description allows the creation of different detectors for each of these situations.

We measure the importance of distribution by keeping track of the percentage of time that each intermediate processor is active when the output is correctly activated, and looking at the processor with the highest percentage of activation. If it is close to 100%, then the set of features detected by it is a sufficient description of the target texture and no distribution is needed. The lower it is, the more distribution was necessary. Table 1 gives percentages taken from the test runs for each target texture at 0% supervisor error.

Comparing this with Figure 3 shows that the amount of distribution for a texture was directly related to the detection error -- that is, the difficulty in discriminating the target texture from other textures. In the cases where textures were similar, many different detectors with tighter constraints were needed to detect the target texture and only the target texture.

Table 2 gives this distribution measure, averaged over all target textures, for the different levels of supervisor error. The amount of distribution increases with the error. As the process of texture classification becomes more confused, tighter constraints are needed to discriminate the textures. Since this decreases the number of textures that a processor can detect, more processors are needed to cover them all. This distribution is one of the main reasons for the system's good performance at high levels of supervisor error.

fieldstone	82.19
ice	79.60
marble	84.28
pebbles	78.58
pigskin	88.43
ricepaper	94.89
straw	87.24
tree	82.08

0%	84.66
10%	75.57
20%	64.76
30%	60.95

Table 1   Table 2

5.4. Understanding the networks

Table 3 shows what the constraints of a typical processor might look like after a while. This particular processor was one of those set up to learn the texture "marble". The table gives the constraints for the features described in section 3 (with angles and distances for the second order constraints), formed after 10,000 input vectors. In effect, it characterizes "marble" as having low mean, high energy and low inertia at distance 1 in the direction $\pi/2$, and low correlation at distance 1 in all directions.

statistic	constraints					seed	statistic	constraints					seed
mean	-	1	-	-	-	11000	variance	-	-	-	-	-	10000
$E(1,0)$	-	-	-	-	-	00111	$E(2,0)$	-	-	-	-	-	00111
$E(1,\pi/4)$	-	-	-	-	-	00111	$E(2,\pi/4)$	-	-	-	-	-	00111
$E(1,\pi/2)$	-	0	0	-	-	00011	$E(2,\pi/2)$	-	-	-	-	-	00011
$E(1,3\pi/4)$	-	-	-	-	-	00111	$E(2,3\pi/4)$	-	-	-	-	-	00111
$I(1,0)$	-	-	-	-	-	00011	$I(2,0)$	-	-	-	-	-	00001
$I(1,\pi/4)$	-	-	-	-	-	00111	$I(2,\pi/4)$	-	-	-	-	-	00001
$I(1,\pi/2)$	1	-	-	-	0	11100	$I(2,\pi/2)$	-	-	-	-	-	11100
$I(1,3\pi/4)$	-	-	-	-	-	00111	$I(2,3\pi/4)$	-	-	-	-	-	00001
$C(1,0)$	-	-	-	0	-	11100	$C(2,0)$	-	-	-	-	-	10000
$C(1,\pi/4)$	-	-	-	0	-	11100	$C(2,\pi/4)$	-	-	-	-	-	10000
$C(1,\pi/2)$	-	-	-	0	-	11100	$C(2,\pi/2)$	-	-	-	-	-	11000
$C(1,3\pi/4)$	-	-	-	0	-	11000	$C(2,3\pi/4)$	-	-	-	-	-	10000

Table 3

This table demonstrates another advantage of the constraint representation. Besides producing networks capable of correct texture discrimination, it can also show us things about the textures themselves. The statistical measures used here (and elsewhere) to classify textures usually do not correspond to our intuitive measures (such as "roughness"), so it is less than obvious how to simply describe them in those terms. The final states of the constraints in processors such as Table 3 can give us those descriptions.

6. Conclusions

We have presented and tested a distributed system that learns texture discrimination in terms of simple first and second order statistics. The networks formed by this learning system were capable of performing this task quite well, at a level comparable to that of more complex measures of texture. This high level of correctness was maintained despite small sample size and high supervisor error.

In addition, the creation of distributed networks of different, specialized detectors (versus a single, general detector) was shown to be important for the texture discrimination problem. This was especially true for textures that were very similar, or under conditions of small sample size or high supervisor error, situations that often occur in the real world.

6.1. Acknowledgements

This work was supported by the Defense Advanced Research Projects Agency and the U. S. Center for Night Vision and Electro-Optics under Contract DAABo7-86-KF073. The author wishes to thank John Aloimonos and Azriel Rosenfeld for their advice and constructive criticism.

Bibliography

Aloimonos, J. (1988). Shape from texture. *Biological Cybernetics, 58*, 345-360.

Fu, K. S. (1974) *Syntactic Methods in Pattern Recognition* New York: Academic Press.

Fukunaga, K. (1972) *Introduction to Statistical Pattern Recognition.* New York: Academic Press.

Hall, E. L., Kruger, R. P., Dwyer, S. J., Hall, D. L., McLaren, R. W., and Lodwick, G. S. (1971). A survey of preprocessing and feature extraction techniques for radiographic images. *IEEE Transactions on Computers, 20.*

Haralik, R. M., Shanmugam, R., and Dinstein, I. (1973). Textural features for image classification. *IEEE Transactions on Systems, Man and Cybernetics, 3,* 610-621

Kruger, R. P., Thompson, W. B., and Twiner, A. F. (1974). Computer diagnosis of pneumoconiosis. *IEEE Transactions on Systems, Man, and Cybernetics, 45,* 40-49.

Laws, K. I. (1980). Textured image segmentation. Ph.D. dissertation, Department of Engineering, University of Sothern California.

Sullins, J. R. (1988). Distributed learning: motion in constraint space. University of Maryland Technical Report CAR-412.

Tou, J. T. and Gonzalez, R. C. (1974) *Pattern Recognition Principles.* Reading, MA: Addison-Wesley.

Fast Shape from Shading

Richard Szeliski

Digital Equipment Corporation, Cambridge Research Lab

One Kendall Square, Bldg. 700, Cambridge, MA 02139

Abstract

Extracting surface orientation and surface depth from a shaded image is one of the classic problems in computer vision. Many previous algorithms either violate integrability, i.e., the surface normals do not correspond to a feasible surface, or use regularization, which biases the solution away from the true answer. A recent iterative algorithm proposed by Horn overcomes both of these limitations but converges slowly. This paper uses a new algorithm, hierarchical basis conjugate gradient descent, to provide a faster solution to the same problem. This approach is similar to the multigrid techniques which have previously been used to speed the convergence, but it does not require heuristic approximations to the true irradiance equation. The paper compares the accuracy and the convergence rates of the new techniques to previous algorithms.

1 Introduction

The variation in brightness due to changes in surface orientation across a surface, or its *shading*, is an important visual cue for interpreting three-dimensional scenes. Recovering the surface orientation and surface shape (height) from shaded images—the *shape from shading* problem—has been one of the classic research problems in computer vision [3, 5]. Shape from shading is a useful technique for determining the shape of smooth man-made objects of known reflectivity. It is most often used in conjunction with other computer vision algorithms such as shape from stereo or shape from contour, since it provides information that is complementary to these other techniques.

In the usual formulation, the surface *reflectance map* $R(p, q)$ is known, so that the shaded image provides a single constraint on the surface gradient (p, q) at each point in the image. Additional constraints, such as requiring that the gradient vary smoothly and/or that the gradient field be integrable, are then used to find a unique solution for the surface.

Most shape from shading algorithms formulate the task as a large non-linear optimization problem, and then use iterative schemes to compute the surface orientation and height estimates. These algorithms may require many iterations to converge towards the optimum solution. In this paper, we develop a new iterative algorithm, hierarchical basis conjugate gradient descent, that has a much higher convergence rate. Our approach is similar to the multigrid techniques which have previously been studied in this context. Compared to multigrid relaxation, our technique is easier to implement, is more generally applicable, and does not suffer from the same limitations. Before we describe our new approach, however, we first review the previous work on which it is based.

In the field of computer vision, the shape from shading problem was first studied by Horn [3]. This early work was based on the solution of the *image irradiance equation*. Later algorithms have used the reflectance map to formulate the problem and numerical relaxation to find its solution [6]. In most algorithms, the gradient field (p, q) is first computed, and the height field z is then computed from the gradient. A more detailed review of these algorithms can be found in [4].

One of the problems with the iterative approaches commonly used for shape from shading is that they converge very slowly towards the optimal solution. Multigrid techniques, which create a hierarchy of problems at different resolution levels, have been used successfully to speed this convergence [10]. As pointed out by Ron and Peleg [7], however, the non-linearity in the reflectance map $R(p, q)$ means that the brightness obtained by blurring the high resolution image is not the same a the brightness computed from a blurred version of the high resolution surface. Ron and Peleg propose a non-linear method for deriving a low resolution brightness image which produces a better approximation to the image corresponding to a low resolution surface. The approach described in this paper does not require this approximation, since it never explicitly builds a multiresolution pyramid of brightness images.

Recently, Horn devised a new algorithm which simultaneously computes both the height field z and gradient field (p, q) from shaded images [4]. His formulation combines three constraints used in previous algorithms—the brightness constraint, the gradient smoothness constraint, and the integrability constraint—into a single functional which is then minimized. Horn also uses a local linear expansion of the reflectance function to stabilize the algorithm and speed its convergence.

The algorithm which we develop in this paper is based on Horn's continuous problem formulation. However, instead of using the calculus of variations to derive the Euler equations for the solution [4], we discretize the functional using finite differences and then minimize the new energy function directly using conjugate gradient descent [1]. This results in a faster algorithm, as we will demonstrate using numerical experiments.

To obtain even greater speedups, we use hierarchical basis representations of the height and gradient fields in conjunction with conjugate gradient descent [8]. This approach is similar to multigrid relaxation, but does not require a hierarchy of problems at different resolutions. We thus avoid the need to compute lower resolution brightness images (as in [7]). Hierarchical basis functions are as simple to implement as pyramid-based smoothing, and do not require the additional storage of a pyramidal representation. We can thus obtain the speedups associated with multiresolution techniques while minimizing the original (fine resolution) energy function (see Section 5).

2 Shape from shading: continuous formulation

The ability to recover surface shape from shaded images depends on the systematic variation of image brightness with the surface orientation. This variation can be captured in the *reflectance map* $R(p, q)$, where $p = \partial z/\partial x$ and $q = \partial z/\partial y$ are the surface gradient components of the height field $z(x, y)$. In shape from shading, we must solve the inverse problem, i.e., given a shaded image, find the surface which generated it. To do this, we place three different constraints on the solution: brightness, integrability, and smoothness.

The *brightness constraint* requires that the intensity image $E(x, y)$ agree with the local reflectance $R(p(x, y), q(x, y))$. We measure the total squared brightness error over the image

$$C_1(p, q) = \frac{1}{2} \int \int \left(E(x, y) - R(p, q) \right)^2 dx \, dy, \tag{1}$$

and use this as part of a functional which is minimized to find the solution.

To ensure that the gradient field (p, q) computed by the shape from shading algorithm corresponds to a valid surface, we must satisfy (as best we can) the defining equations for the surface gradient. This can be turned into a weak *integrability constraint*

$$C_2(z, p, q) = \frac{1}{2} \int \int (z_x - p)^2 + (z_y - q)^2 \, dx \, dy, \tag{2}$$

where z_x and z_y are the partial derivatives of the height field z.

To help stabilize the iterative shape from shading algorithm and to ensure that it has a unique minimum, we can add a weak *smoothness constraint*

$$C_3(p, q) = \frac{1}{2} \int \int p_x^2 + p_y^2 + q_x^2 + q_y^2 \, dx \, dy, \tag{3}$$

where p_x, p_y, q_x, and q_y are the partial derivatives of the surface gradient. This measure works best with smooth surfaces, and favors constant (p, q) fields, which define planar surfaces.

The three constraints presented above have been used in various combinations in previous shape from shading algorithms. Horn's new coupled height and gradient scheme [4] combines all three into a single functional

$$C(z, p, q) = C_1(p, q) + \mu C_2(z, p, q) + \lambda C_3(p, q). \tag{4}$$

Minimizing this functional (or solving the associated Euler equations) produces both a gradient field (p, q) which is nearly integrable, and a surface $z(x, y)$ which best fits this gradient. Putting the integrability constraint directly into the functional allows this constraint to influence the solution for the gradients, rather than relying solely on the smoothness constraint to disambiguate the possible solutions. In fact, the smoothness constraint can be made negligible by setting λ to a small value, thereby avoiding the bias introduced by regularization [4].

3 Discrete formulation

To convert the analytic cost function $C(z, p, q)$ into a discrete form, we use a set of regular rectangular (fine-grained) grids and finite difference approximations for the derivatives. This leads to simple energy and updating equations. It also leads naturally to massively parallel algorithms that can be implemented on mesh-connected SIMD architectures.

In this paper, we have chosen to use a discretization which corresponds to that used by Horn [4] so that we can compare our results to his. In this discretization, the p and q discrete fields lie on a *cell-centered* dual lattice coincident with the brightness field E [9]. Each p and q value is therefore a linear combination of the four discrete z values at the corners of the cell. With this discretization, we can define the discrete energy components

$$C_1 = \frac{1}{2}\sum (E - R_0)^2 \tag{5}$$

$$C_2 = \frac{1}{2}\sum (\overline{z_x} - p)^2 + (\overline{z_y} - q)^2 \tag{6}$$

$$C_3 = \frac{1}{2}\sum p_x^2 + p_y^2 + q_x^2 + q_y^2 \tag{7}$$

where $R_0 = R(p,q)$, $\overline{z_x}$ and $\overline{z_y}$ are the z gradient values at the center of the cell, and p_x, p_y, q_x, and q_y are simple first differences [9]. The global energy is then

$$C(\mathbf{x}) = C_1 + \mu\, C_2 + \lambda\, C_3 \tag{8}$$

where $\mathbf{x}$ is the *state vector* formed by concatenating the $\mathbf{z}$, $\mathbf{p}$, and $\mathbf{q}$ fields.

4 Energy minimization

Once we have defined our discrete energy function, we must devise an efficient algorithm for finding its minimum. The traditional approach [6] is to set the gradient of the energy function to zero

$$\nabla_{\mathbf{x}} C = \mathbf{Ax} - \mathbf{b} = 0 \quad \text{or} \quad \mathbf{Ax} = \mathbf{b}. \tag{9}$$

In the above equation, the terms linear in $\mathbf{x}$ have been grouped into $\mathbf{Ax}$, while the constant and non-linear terms are put into $\mathbf{b}$. Because of the large size of the resulting system of equations, iterative algorithms (relaxation) are used to find their solution [12].

One of the simplest forms of parallel relaxation is the *weighted Jacobi* method,

$$\mathbf{x}^{(n+1)} = \mathbf{x}^{(n)} + \omega \mathbf{D}^{-1} \mathbf{r}^{(n)}, \tag{10}$$

where

$$\mathbf{r} = \mathbf{b} - \mathbf{Ax}$$

is the current *residual* (negative gradient), and the matrix $\mathbf{D}$ is the diagonal matrix obtained from $\mathbf{A}$. The choice of ω determines how fast the above algorithm converges [2, 12]. A closely related relaxation method is *Gauss-Seidel*, where only a subset of the nodes is updated at each step (this helps convergence) [2, 12].

4.1 Local linear approximation of the reflectance map

To compute the element of $\mathbf{D}$ for the shape from shading problem, we must evaluate the second partial derivatives of the cost function $\partial^2 C/\partial z^2$, $\partial^2 C/\partial p^2$, and $\partial^2 C/\partial q^2$. If we ignore terms involving partial derivatives of R, these values for $\mathbf{D}$ lead to updating equations very similar to those given in [4, p. 26]. As Horn has remarked, this method does not converge very well.

A better estimate for $\mathbf{D}$ can be obtained by using a local linear approximation of the reflectance map [4], i.e., evaluating R_p and R_q at each iteration, but ignoring R_{pp} and R_{qq}. If we also include the partial derivative $\partial^2 C/\partial p \partial q$ from the Hessian matrix $\mathbf{A}$, we obtain a new block-diagonal matrix $\mathbf{B}$. Using this block-diagonal matrix in our updating step

$$\mathbf{x}^{(n+1)} = \mathbf{x}^{(n)} + \mathbf{B}^{-1} \mathbf{r}^{(n)}$$

results in a set of updating rules similar to Horn's [4, p. 30]. Intuitively, $\mathbf{B}^{-1}$ rotates and scales the gradient so that most of the brightness error is reduced to zero. As we will see in our experiments (Section 6), using this block diagonal approximation for $\mathbf{A}^{-1}$ can speed up the relaxation.

4.2 Conjugate gradient descent

Conjugate gradient descent [1] is similar to Jacobi relaxation, except that we take a step along the direction vector $\mathbf{d}$, which need not be coincident with the residual $\mathbf{r}$. The updating rule is

$$\mathbf{x}^{(n+1)} = \mathbf{x}^{(n)} + \alpha^{(n)} \mathbf{d}^{(n)}, \tag{11}$$

where the step size α is chosen to minimize the new energy. Successive directions are chosen so that they are *conjugate* with respect to $\mathbf{A}$, i.e., $(\mathbf{d}^{(n)})^T \mathbf{A} \mathbf{d}^{(n-1)} = 0$. This is achieved by setting

$$\mathbf{d}^{(n)} = \mathbf{r}^{(n)} - \beta^{(n)} \mathbf{d}^{(n-1)} \tag{12}$$

for a suitable choice of β. For a complete description of this algorithm, see [1, 8, 9].

Conjugate gradient normally converges much faster than Jacobi relaxation. To further speed up the convergence, we can use *preconditioning* [1], which involves a change of variables

$$\mathbf{y} = \mathbf{S}\mathbf{x}. \tag{13}$$

The conjugate gradient algorithm is modified to use the transformed residual $\tilde{\mathbf{r}} = \mathbf{S}\mathbf{S}^T\mathbf{r}$ to compute the new direction [1]. If $\mathbf{S}$ is properly chosen, the new algorithm converges much faster because of it smaller condition number [8]. Two choices for $\mathbf{S}$ that we have investigated are $\mathbf{S} = \mathbf{D}^{-1/2}$, which resembles Jacobi relaxation, and $\mathbf{S} = \mathbf{B}^{-1/2}$, which converges faster than both Horn's method and standard conjugate gradient (see Section 6).

5 Hierarchical basis functions

While conjugate gradient descent is much faster than regular relaxation, it is still slow compared to multiresolution techniques such as multigrid. This is because the solution to the problem requires long-range (low frequency) error to be reduced through strictly local interactions. This situation arises mainly from the local nature of the representation used for our height and gradient fields. Consider the alternative of using a pyramidal representation for these fields. Changing the value of nodes at higher levels in this pyramid will affect larger regions of the solution, so, intuitively, the algorithm should convergence faster.

More concretely, we define a *hierarchical basis* representation [13] of a field—such as z, p, or q—by first constructing a pyramid in the usual fashion (subsampling every other point in both the x and y dimensions). However, since we do not wish to increase the number of degrees of freedom in our new representation, we use only some of the nodes at each level [8]. To convert from the hierarchical basis representation to the usual fine-level (*nodal*) representation, we start at the coarsest (smallest) level of the pyramid and interpolate the values at this level (thus doubling the resolution). These interpolated values are then added to the hierarchical representation values at the next lower level, and the process is repeated until the nodal representation is obtained [8, 9].

We can write the above process algebraically as

$$\mathbf{x} = \mathbf{S}\mathbf{y} = \mathbf{S}_1\mathbf{S}_2 \dots \mathbf{S}_{L-1}\mathbf{y}. \tag{14}$$

Using a hierarchical basis representation for the height and gradient fields can thus be viewed as special kind of preconditioner (c.f. (13)). As we discussed in the previous section, using a preconditioner for conjugate gradient descent simply requires the ability to evaluate the transform $\mathbf{S}$ (defined above) and its adjoint $\mathbf{S}^T$, which can be defined just as simply [8, 9]. We have also tried combining hierarchical basis preconditioning with Horn's local linear approximation (block preconditioning) [9].

6 Experimental results

For the experiments described in this section, we used shaded images generated from two different surfaces: a 64×64 elevation map cropped from a digital terrain model of the Rockies in Colorado, and a 64×64 map of a spherical bump on a flat plane. The contour maps for the two surfaces are shown in Figure 1. These two surfaces were chosen because they represent the two typical applications of shape from shading, i.e., cartography/photoclinometry and the extraction of shape from piecewise smooth man-made objects.

The synthetically generated shaded images corresponding to these surfaces are shown in Figures 2 and 3. The left image in each pair is input to the shape from shading algorithm. The right image is used to monitor the progress of the relaxation algorithm, since looking at the brightness image corresponding to the solution alone may not be very informative [4].

To evaluate the performance of our algorithm, we must choose some error measure. Horn [4] discusses a number of possible quality measures and methods for displaying the algorithm's progress. In this research, we chose to study four different measures:

- The cost (or energy) $C^{(n)}$ associated with the current estimate.
- The magnitude of the residual $|\mathbf{r}^{(n)}|$.
- The magnitude of the gradient error $|\mathbf{p}^{(n)} - \mathbf{p}^*| + |\mathbf{q}^{(n)} - \mathbf{q}^*|$.
- The magnitude of the height error $|\mathbf{z}^{(n)} - \mathbf{z}^*|$.

Due to space limitations, we only present the results using the cost measure in this paper. A more complete set of experiments is described in [9].

Figure 4 shows how these four error measures decrease over time for five different relaxation algorithms: (1) Horn's method (Gauss-Seidel with block preconditioning, $\omega = 1$); (2) conjugate gradient (CG); (3) conjugate gradient with block preconditioning (PCG); (4) hierarchical basis conjugate gradient, $L = 2$ (HBCG); and (5) hierarchical basis conjugate gradient with block preconditioning, $L = 2$ (HBCG). The regularization parameters for these runs were $\lambda = \mu = 1.0$. The hierarchical basis preconditioners (2 levels) were applied to each of the $\mathbf{z}$, $\mathbf{p}$, and $\mathbf{q}$ fields separately. From these plots, we see how conjugate gradient descent has a much faster convergence rate than Gauss-Seidel (Horn's method). Using the block preconditioner increases the convergence rate, as does preconditioning with hierarchical basis functions.

How does the number of smoothing levels in the hierarchical basis pyramid (L) affect the rate of convergence? Figure 5 shows the performance of block preconditioned hierarchical basis conjugate gradient for $L = 1 \dots 4$ (HBCG with $L = 1$ is the same as plain

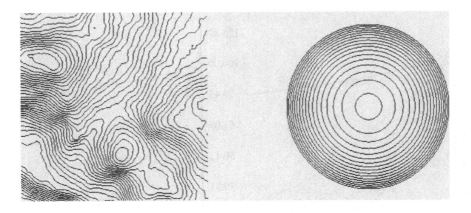

Figure 1: Contour maps of terrain model and bump

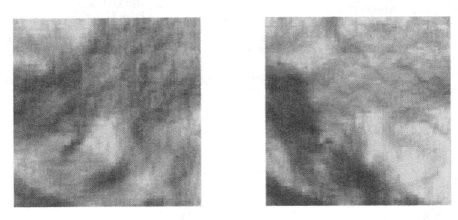

Figure 2: Shaded images of terrain model (illumination from NW and NE)

Figure 3: Shaded images of bump (illumination from NW and NE)

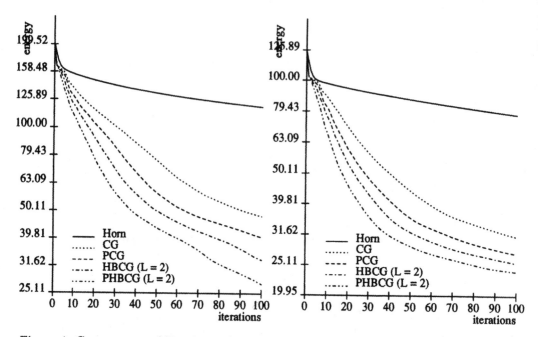

Figure 4: Convergence of Horn's method vs. conjugate gradient: (a) terrain map image, (b) bump image.

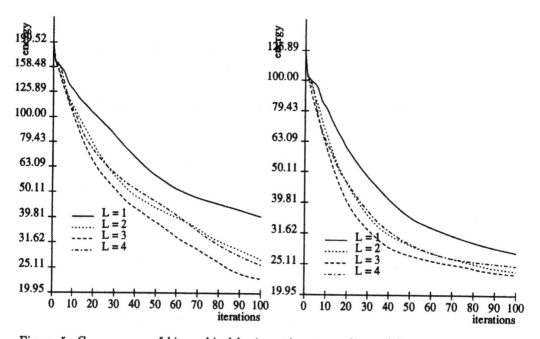

Figure 5: Convergence of hierarchical basis conjugate gradient: (a) terrain map image, (b) bump image.

conjugate gradient). From these plots, it looks like a three-level pyramid ($L = 3$) performs the best. These results are similar to the ones we observed for surface interpolation [8], where the optimum number of levels was found to depend on the smoothness of the solution and the density of the sample points. We would thus expect in our case that the optimum number of levels would vary with the regularization parameter λ.

A variety of additional experiments have been performed on our new shape from shading algorithm, and are reported in [9]. When we decrease the regularization parameter λ, hierarchical basis conjugate gradient descent still performs the best, but block preconditioning does not seem to help. The algorithms also gets trapped in local minima (or at least steep local ravines) when the smoothness constraint is weakened. We have also tired slowly decreasing λ. The best performing algorithm is still hierarchical basis conjugate gradient ($L = 2$), but plain conjugate gradient performs almost as well.

7 Discussion

Our experimental results show that using hierarchical basis conjugate gradient descent dramatically improves the performance of shape from shading algorithms. Our general experience with developing the shape from shading algorithm has been that it is a remarkably difficult problem to solve numerically. This is due to two main factors: the long-range interactions which arise from the local smoothness and integrability constraints (which make the system very stiff), and the non-linearity of the brightness constraint.

Multigrid algorithms are very good at speeding up the long-range emergent interactions (or, in their terms, at smoothing the low-frequency error [2]). However, for the naive implementation of multigrid to work, we must satisfy two conditions: the solution must be smooth, and we must have an easy way to obtain a coarse set of equations. In natural scenes, we usually have multiple objects, so the first condition is violated because the solution is only *piecewise* smooth or continuous. In shape from shading, the second condition is complicated by the non-linear nature of the reflectance map, which necessitates non-linear methods for deriving the coarse equations [7].

Hierarchical basis conjugate gradient overcomes both of these problems. Because the direction of descent is being conjugated, the algorithm can recover from systematic errors induced by the hierarchical basis. Since the energy function being minimized corresponds to the fine level discretization (and the hierarchical basis is used only to suggest a new direction for descent), we do not need to derive coarse equations or even explicitly build a pyramidal representation. Previous experience also suggests that this algorithm is less sensitive to the choice of interpolators than are multigrid algorithms [8].

Much work remains to be done in refining and extending this algorithm. An automatic method for adjusting the regularization parameter λ using scale-space continuation [11] needs to be developed. We must also study the effect of varying the strength of the integrability constraint (μ) and the number of smoothing levels (L), and determine how to adjust them automatically. The question of which discretization for the problem yields the quickest and most accurate solutions must be systematically explored. The noise sensitivity of this algorithm should also be measured by adding noise to the input image and observing the increase in solution error, as well as the power spectrum of this error. Finally, our shape from shading algorithm should be integrated with other vision modules such as shape from contour and shape from stereo.

8 Conclusions

In this paper, we have developed a new algorithm for computing shape from shading and analyzed its performance. To derive this algorithm, we start with the formulation introduced by Horn [4] which combines a brightness constraint, an integrability constraint, and a smoothness constraint into a single functional. We discretize this functional using finite differences to obtain an energy equation which is then minimized.

The algorithm which we develop in this paper uses conjugate gradient descent together with a hierarchical basis representation [8]. This requires smoothing the residual using a pyramid before choosing a new descent direction. Our algorithm thus acquires multi-scale characteristics similar to multigrid relaxation without requiring the construction of a hierarchy of problems. This results in a simpler implementation, and avoids the need to non-linearly blur the surface reflectance [7]. In general, we expect our algorithm to have a much wider range of applicability than multigrid techniques since it can be applied to other non-linear problems in computer vision such as 3-D energy-based models.

Our experimental results indicate that the new algorithm converges much faster than single-level relaxation. Much work remains to be done in exploring the effects of the algorithm parameters and how to automatically adjust them. We are confident that this research will lead to even faster and more robust shape from shading algorithms.

References

[1] O. Axelsson and V. A. Barker. *Finite Element Solution of Boundary Value Problems: Theory and Computation.* Academic Press, Inc., Orlando, 1984.

[2] W. L. Briggs. *A Multigrid Tutorial.* Society for Industrial and Applied Mathematics, Philadelphia, 1987.

[3] B. K. P Horn. Obtaining shape from shading information. In P. H. Winston, editor, *The Psychology of Computer Vision*, pp. 115–155. McGraw-Hill, New York, 1975.

[4] B. K. P. Horn. Height and gradient from shading. A. I. Memo 1105, Artificial Intelligence Laboratory, Massachusetts Institute of Technology, May 1989.

[5] B. K. P. Horn and M. J. Brooks. *Shape from Shading.* MIT Press, Cambridge, Massachusetts, 1989.

[6] K. Ikeuchi and B. K. P. Horn. Numerical shape from shading and occluding boundaries. *Artificial Intelligence*, 17:141–184, 1981.

[7] G. Ron and S. Peleg. Multiresolution shape from shading. In *Conf. Comp. Vision Patt. Recog. (CVPR'89)*, pp. 350–355, San Diego, June 1989.

[8] R. Szeliski. Fast surface interpolation using hierarchical basis functions. In *Conf. Comp. Vision Patt. Recog. (CVPR'89)*, pp. 222–228, San Diego, June 1989.

[9] R. Szeliski. Fast shape from shading. Technical report, Digital Equipment Corporation, Cambridge Research Lab, January 1990.

[10] D. Terzopoulos. Image analysis using multigrid relaxation methods. *IEEE Trans. Patt. Anal. Mach. Intell.*, PAMI-8(2):129–139, March 1986.

[11] A. Witkin, D. Terzopoulos, and M. Kass. Signal matching through scale space. *International Journal of Computer Vision*, 1:133–144, 1987.

[12] D. M. Young. *Iterative Solution of Large Linear Systems.* Academic Press, New York, 1971.

[13] H. Yserentant. On the multi-level splitting of finite element spaces. *Numerische Mathematik*, 49:379–412, 1986.

INVERSE PERSPECTIVE OF A TRIANGLE:
NEW EXACT AND APPROXIMATE SOLUTIONS

Daniel DeMenthon and Larry S. Davis

Computer Vision Laboratory
Center for Automation Research
University of Maryland, College Park, MD 20742 USA

1 Introduction

One of the techniques in model-based pose estimation of 3D objects consists of locating "interest points" on models of the objects, detecting these points in the image, and matching subsets of these image points against subsets of the interest points of the models. Valid matches determine a similar object pose, which can be found by clustering techniques.

Solving for the position and orientation of an object knowing the images of n points at known locations on the object is called the n-point perspective problem (see [5] for a review and a four points solution). The three-point problem, also called the triangle pose problem [9], has been solved in various ways. A review of the major direct solutions for three points under exact perspective is provided in [4]. Another direct solution is described in this paper. Computation speed is important when many or all possible combinations of triples of image feature points and triples of model interest points are considered. Direct methods require quite a few floating point operations. Some researchers have proposed faster methods based on scaled orthographic projection [8,6,10]. We introduce an alternative approximation which we call orthoperspective, a *local* scaled orthographic projection using a plane normal to one of the rays, which causes smaller errors for off-center images. The angular terms of the pose of a given triangle depend then only on two parameters of the image triangle, and these can be precomputed in a two dimensional lookup table resulting in a very fast pose estimation algorithm.

2 Preliminaries

Figure 1 shows a triangle $M_1 M_0 M_2$ of known dimensions. Its sides $M_0 M_1$ and $M_0 M_2$ have known lengths D_1 and D_2 and the angle $M_1 M_0 M_2$ at vertex M_0, which we call α, is also known. This triangle projects on the image plane in $m_1 m_0 m_2$. The projection on the left drawing of the figure is an exact perspective projection. The projection on the right is an approximation that we call *orthoperspective* (Section 4). With both projections, the position of the triangle in space is completely determined by the location of its image, by the range R_0 of the vertex M_0 along the line of sight Om_0, and by the angles θ_1 and θ_2 that the sides $M_0 M_1$ and $M_0 M_2$ make with the line of sight Om_0. Once the three unknowns R_0, θ_1 and θ_2 are determined, the triangle vertices are given by

$$\overrightarrow{OM_0} = R_0 \overrightarrow{e_0}, \quad \overrightarrow{OM_1} = [R_0 + D_1(\cos\theta_1 - \tfrac{\sin\theta_1}{\tan\gamma_1})]\overrightarrow{e_0} + D_1\tfrac{\sin\theta_1}{\sin\gamma_1}\overrightarrow{e_1}, \quad \overrightarrow{OM_2} = [R_0 + D_2(\cos\theta_2 - \tfrac{\sin\theta_2}{\tan\gamma_2})]\overrightarrow{e_0} + D_2\tfrac{\sin\theta_2}{\sin\gamma_2}\overrightarrow{e_2}$$

where $\overrightarrow{e_0}$, $\overrightarrow{e_1}$ and $\overrightarrow{e_2}$ are the known unit vectors along the lines of sight Om_0, Om_1 and Om_2, and γ_1 and γ_2 are the angles between $\overrightarrow{e_0}$ and $\overrightarrow{e_1}$ and $\overrightarrow{e_0}$ and $\overrightarrow{e_2}$ respectively.

3 Triangle Pose Solutions for Exact Perspective

In the exact perspective projection (Figure 1, left), the law of sines for triangle $OM_0 M_1$ yields

$$\frac{\sin(\theta_1 - \gamma_1)}{R_0/D_1} = \sin\gamma_1 \tag{1}$$

A similar equations is written for θ_2. Dividing the two equations eliminates R_0

$$\frac{\sin(\theta_1 - \gamma_1)}{\sin(\theta_2 - \gamma_2)} = K, \quad \text{with } K = \frac{\sin\gamma_1/D_1}{\sin\gamma_2/D_2} \tag{2}$$

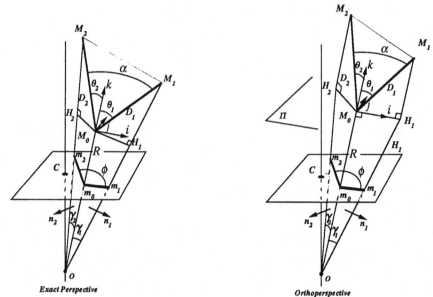

Figure 1: Exact perspective and orthoperspective for a triangle

K is a combination of known parameters. We call K the *foreshortening ratio*, because it is a ratio of foreshortening measures for the two sides of the triangle.

A second relation between θ_1 and θ_2 is obtained by expressing the cosine of α, the angle at vertex M_0 of the world triangle, as a dot product [7]:

$$\cos \alpha = \sin \theta_1 \sin \theta_2 \cos \phi + \cos \theta_1 \cos \theta_2 \qquad (3)$$

where ϕ is the angle between the plane containing O, m_0 and m_1 and the plane containing O, m_0 and m_2. This equation also applies to the orthoperspective approximation (Figure 1, right).

Eliminating θ_2 yields a fourth degree equation in $\sin^2(\theta_1 - \gamma_1)$. The coefficients of the equation and details are given in [3]. A total of two triangle poses compatible with the original equations is generally found. Representative pose surfaces are presented graphically in Figure 2. The top row of Figure 2 shows the three diagrams of one set of solutions (R, θ_1, θ_2), the bottom row the second set. However, this separation into two sets is artificial. The step discontinuity of the surfaces of one set matches the step discontinuity of the surfaces of the other set so that the resulting surface does not possess any discontinuity. For example, the general shape of the surface giving θ_1 is the double valued surface shown in Figure 3; the surface for θ_2 is similar. The surface for R_0 also combines two layers which cross each other, but are very close together.

A startling consequence is that if the image of a triangle deforms smoothly (corresponding to a smooth 3-D motion of the 3D triangle) so that the final image is identical to the initial image, then the corresponding world triangle has not necessarily returned to its original position. It would require a second cycle of the image deformation sequence to bring back the world triangle to its initial position (Figure 3, right). This occurs when the trajectory corresponding to the triangle poses on the solution surface describes a closed cycle around the point $(K = 1, \phi = \alpha)$.

4 Orthoperspective

Figure 1 (right) describes the orthoperspective approximation. A plane Π through point M_0 perpendicular to the line of sight Om_0 is considered. Points M_1 and M_2 project onto this plane at H_1 H_2 using an orthogonal projection; the image points m_0, m_1 and m_2 are the perspective images of M_0, H_1 and H_2.

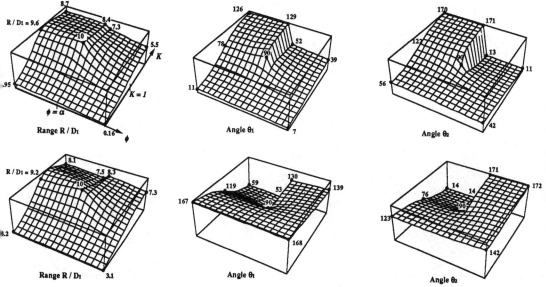

Figure 2: Solutions for triangle pose using exact perspective for $\tan \gamma_1 = 0.1$, $D_1/D_2 = 0.5$ and $\alpha = \pi/4$. Range of K from 1/4 to 4. Range of $\tan(\phi/2)$ from one quarter to four times $\tan(\alpha/2)$.

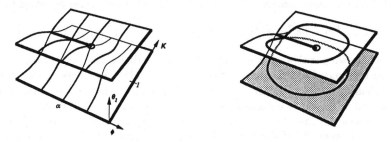

Figure 3: Solution surface for angle θ_1 or θ_2. Two cycles may be necessary in the image for the world triangle to return smoothly to its original position

Notice that if we rotate the camera around the center of projection O to bring the optical axis to the line of sight Om_0 (a *standard rotation*[7]), the image plane becomes parallel to the plane Π on which we performed the orthogonal projection. After this camera rotation, orthoperspective is simply a scaled orthographic projection. The camera rotation removes the dependence of the construction on the offset of the world object from the optical axis. For this reason orthoperspective is a better approximation to perspective than classical scaled orthographic projection for image elements which are not centered in the image plane [1,3].

In the right triangle OM_0H_1 of Figure 1 (right)

$$\frac{\sin \theta_1}{R_0/D_1} = \tan \gamma_1 \tag{4}$$

Writing a similar equation for θ_2 and dividing the two equations eliminates R_0, yielding

$$\frac{\sin \theta_1}{\sin \theta_2} = K, \text{ with } K = \frac{\tan \gamma_1/D_1}{\tan \gamma_2/D_2} \tag{5}$$

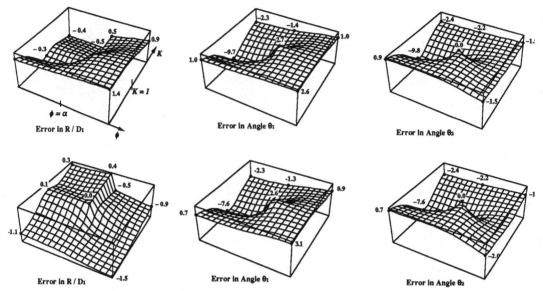

Figure 4: Errors for triangle pose using orthoperspective

Eliminating θ_2 between Equation 5 and Equation 3 yields a second degree equation in $\sin^2 \theta_1$

$$\sin^2 \phi \sin^4 \theta_1 - \left(K^2 - 2K \cos \alpha \cos \phi + 1\right) \sin^2 \theta_1 + K^2 \sin^2 \alpha = 0 \qquad (6)$$

Equation 6 always has two positive solutions, but only the smaller solution has magnitude less than 1 and can thus be equated to the square of the sine of an angle. The single solution for $\sin \theta_1$ results in two complementary solutions for θ_1 corresponding to mirror image directions of M_0M_1. The corresponding value for $\sin \theta_2$ obtained from Equation 5 also yields two complementary solutions for θ_2. Equation 3 has to be used to discard two of the four combinations of θ_1 and θ_2. Finally, the distance R_0 of the vertex M_0 from the center of projection is computed using Equation 4. This yields a single solution for R_0. Thus the two solution triangles share a common vertex M_0, whereas in exact perspective the two solution triangles have slightly distinct M_0 vertices. Therefore a single-valued surface is obtained for R_0/D_1. Otherwise the diagrams for the approximation are almost identical to those obtained for exact perspective (Figure 2) and are omitted for lack of space. They can be found in [3].

5   Comparison of Triangle Pose obtained by Exact and Approximate Perspective

We numerically compare the results provided by the exact perspective and the approximate perspectives for the parameter values used in Figure 2. The results are plotted in Figure 4. The numbers given on the 3D plots of Figure 4 are nondimensional errors in R_0/D_1 and angular errors in degrees. The following characteristics of the errors can be observed:

1. The largest range errors occur for small values of K, i.e. when the side M_0M_1 is much more foreshortened than the other side (M_0M_1 and M_0M_2 do not play symmetric roles in these diagrams because the size of the image of M_0M_1 is fixed since γ_1 and K are fixed).

2. The largest angular errors occur along the line $(K = 1, \phi < \alpha)$ and reach almost 10 degrees in these valleys of the surfaces. At the edges of the diagrams the angular errors are only around two or three degrees. Other error diagrams are provided in [3].

6 Discussion

One advantage of using approximate perspective is that small lookup tables can be constructed. The solutions for the angles θ_1 and θ_2 depend only on two parameters, K and ϕ. For each triangle of features of an object a two dimensional table can be generated, in which the possible values for θ_1 and θ_2 are stored for a range of the observable parameters K and ϕ. From an image of the triangle, the parameters K and ϕ are calculated and the angles are read from the table. Then the range R_0 is obtained using Equation 4. On a 16K Connection Machine without floating point processors, this technique programmed in StarLisp provides a pose estimate of a single polyhedra with 40 triangles with a smooth background in around one second [2]. This involves matching all pairs of image and model triangles, and clustering the resulting set of pose estimates.

The error diagrams shown in Figure 4 are useful for increasing the accuracy of pose estimation by lookup tables. The table cells for which the approximate pose is very different from the exact pose (for example for K close to one and $\phi < \alpha$, as seen in Figure 4) can be flagged, and the pairs of model and image triangles corresponding to these table cells can be disregarded.

Acknowledgements

The support of the Defense Advanced Research Projects Agency and the U.S. Army Engineer Topographic Laboratories under Contract DACA76-88-C-0008 (DARPA Order No. 6350) is gratefully acknowledged.

References

[1] J. Aloimonos and M. Swain, "Paraperspective Projection: Between Orthography and Perspective", Center for Automation Research CAR-TR-320, University of Maryland, College Park, May 1987.

[2] L.S. Davis, D. DeMenthon, T. Bestul, D. Harwood, H.V. Srinivasan, S. Ziavras, "RAMBO—Vision and Planning on the Connection Machine", Proc. 1989 DARPA Image Understanding Workshop, 631-639.

[3] D. DeMenthon, L.S. Davis, "New Exact and Approximate Solutions of the Three-Point Perspective Problem", Center for Automation Research CAR-TR-471, University of Maryland, College Park, November 1989.

[4] R.M. Haralick, "The Three Point Perspective Pose Estimation Problem", Internal Note, Dept. of Electrical Engineering, FT-10, University of Washington.

[5] R. Horaud, B. Conio, O. Leboulleux, B. Lacolle, "An Analytical Solution for the Perspective Four-Point Problem", to appear in *Computer Vision, Graphics, and Image Processing*, Academic Press, 1989.

[6] D. Huttenlocher and S. Ullman, "Recognizing Solid Objects by Alignment", Proc. 1988 DARPA Image Understanding Workshop, 1114-1122.

[7] K. Kanatani, "Constraints on Length and Angle", *Computer Vision, Graphics and Image Processing*, 41, 1988, 28-42.

[8] Y. Lamdan, J.T. Schwartz, and H.J. Wolfson, "On Recognition of 3-D Objects from 2-D Images", Proc. 1988 IEEE Int. Conf. on Robotics and Automation, 1407-1413.

[9] S. Linnainmaa, D. Harwood, and L.S. Davis, "Pose Determination of a Three-Dimensional Object using Triangle Pairs", *IEEE Trans. on Pattern Analysis and Machine Intelligence*, 10, 1988, 634-647.

[10] D.W. Thompson and J.L. Mundy, "Model-Directed Object Recognition on the Connection Machine", Proc. DARPA Image Understanding Workshop, 1987, 98-106.

Finding Geometric and Relational Structures in an Image

Radu Horaud, Françoise Veillon
LIFIA–IMAG, 46, avenue Félix Viallet
38031 Grenoble, FRANCE
and
Thomas Skordas
ITMI, Filiale de CAP–SESA, ZIRST Chemin des Prés
38240 Meylan, FRANCE

Abstract   *We present a method for extracting geometric and relational structures from raw intensity data. On one hand, low-level image processing extracts* isolated *features. On the other hand, image interpretation uses sophisticated object descriptions in representation frameworks such as semantic networks. We suggest an intermediate-level description between low- and high-level vision. This description is produced by grouping image features into more and more abstract structures. First, we motivate our choice with respect to what should be represented and we stress the limitations inherent with the use of sensory data. Second, we describe our current implementation and illustrate it with various examples.*

1 Introduction and motivation

One of the fundamentals goals of computer vision is to interpret an image in terms of scene objects and relationships between these objects. One major difficulty associated with this task is that the image to be analysed and the objects to be recognized are embedded in two different representation frameworks.

On one hand, an image is a collection of pixels, each pixel being the 2D projection of a scene event. Hence, the task of the low-level processing is to make explicit the local image geometric and relational properties that are likely to represent local scene geometric structure.

On the other hand, an object is a collection of 3D primitives and relationships between these primitives. This representation must be transformed such that *visually salient* object features are made explicit. Then, the task of high-level vision is to match object descriptions against image descriptions.

In this paper we propose to investigate the image description the best suited for high-level tasks. We analyse the desired properties of this description from two viewpoints:

1. What *should* be represented in an image for successful scene interpretation?

2. What *could* be properly detected in an image?

Therefore our work is at the crossroads of low-level and high-level vision. We call this image description an "intermediate-level description," or InterLevel, because it embeds both feature detection and feature grouping. Feature detection is essentially a bottom-up process while feature grouping makes some assumptions about the geometry of the scene.

The description that we envision is a hierarchy of linear segments, relationhips between these segments, and more abstract geometric and relational structures. The latter are groups (collections) of segments satisfying some sets of properties. The InterLevel description currently produced by our system is best illustrated on Figures 1 and 2. Figure 1 (left) is the grey level image of a relatively simple object on a textured background. Figure 1 (right) shows the straight line segments detected in the previous image using a classical paradigm: edge detection, edge linking, and piecewise polygonal approximation. The locations shown on Figure 2 (left) with a small circle are "terminations at a common point" or junctions. Segments and junctions are mapped in a graph. The image may be further searched for groups of segments satisfying various constraints. Figure 2 (right) shows some parallelograms and some curved contours (convex chains of connected, smoothly turning lines).

The remainder of this paper is organized as follows. Section 2 contains an overview of previous research efforts in the domain of feature grouping and structural description. Section 3 contains a description of the properties of the intermediate-level representation. Section 4 describes the current implementation. Finally we discuss the usefulness of our description with respect to ongoing research in computer vision.

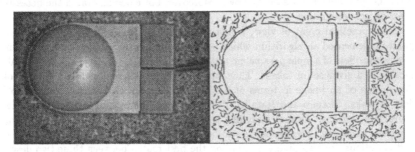

Figure 1: A grey level-image and the detected straight lines

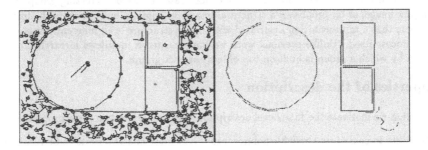

Figure 2: The relational graph extracted from the above image and some simple image structures

2 Background

The interest for feature grouping stems from Gestalt psychologists' figure/ground demonstrations, [13]: Certain image elements are organized to produce an emergent figure. Zucker, [13] distinguishes two types of grouping. The first corresponds to the inference of one-dimensional contours. The second involves orientation information that is densely distributed accross areas rather than curves. In this paper we are concerned with the first type of grouping.

Our work is also closely related to Marr's statement that an image contains two types of information [8]: Changes in intensity and image local geometry. Then, grouping is a process that makes both these pieces of information explicit.

Witkin and Tenenbaum [12] defined perceptual organization as a process able to detect *primitive structure*: Those relationships we are able to perceive even when we can't interpret them in terms of familiar objects. Our work may well be viewed as an attempt to build such primitive structures.

Connell and Brady [2] built a semantic network description of the image of a single object. The description contains both contour and region properties and relationships embedded in the "smoothed local symmetry" representation proposed by Brady and Asada [1]. The description thus derived is used for learning visual object representations. With respect to the work of Brady and Asada, and Connell and Brady we advocate a description incorporating, among others, smoothed local symmetries. Unlike their semantic network, our graph contains lower-level information, i.e., primitive structure. Nevertheless we believe that the semantic network advocated by Connell and Brady can be inferred from our description.

Lowe [6] argued that perceptual groupings are useful to the extent that they are unlikely to have arisen by accident of viewpoint or position, and therefore are likely to reflect meaningful structure in the scene. Such groupings include collinearity, curvilinearity, terminations at a common point, parallel curves, etc. We will argue that the properties embedded in our description include (explicitly or implicitly) the groupings advocated by Lowe. Moreover, Lowe demonstrated the necessity of such groupings for object recognition.

From a more practical point of view, Dolan and Weiss [3] attacked the problem of perceptual grouping and implemented an algorithm which performed linking, grouping, and replacement in an iterative procedure. Sets of simple parametric tests were used to determine which structural relations were applicable to a given set of tokens. Then, a set of tokens was replaced by a simple token.

The description of an image in terms of a structural description has been proved to be useful for a variety of tasks. Shapiro and Haralick [10] introduced the idea of inexact matching between two structural descriptions. Inexact matching is an important concept. Consider for example the problem of stereo matching. The two images to be matched are different and hence, their associated structural descriptions are also different. Finding the best match between these two descriptions is an inexact matching problem. Skordas [11], and Horaud and Skordas [5] showed how to cast the stereo matching problem into a double subgraph isomorphism problem and how to solve it.

In conclusion previous research efforts have concentrated either on detecting groups of features in a complex image, or on proposing a structural image description, but not on both. In this paper we will argue that a representation associated with image structure is an inherent component of the grouping process itself. Unlike previous work we implemented a multilevel hierarchical structural description by which a group is built on top of less abstract groups.

3 Properties of the description

In this section we motivate the InterLevel description with respect to two main properties:

- its ability to constitute a reliable pointer into the memory of 3-D objects to be recognized, and

- its robustness with respect to the image formation process and the low-level segmentation process.

An image description in terms of its 2-D features is rarely an end in itself. Therefore, the vocabulary chosen for describing an image must contain those features and relationships that are likely to be associated with 3-D shapes. Moreover, among the properties of the latter, one must choose those which are invariant under projection.

Consider a 3-D object. In general, the boundary of such an object is a C^3 piecewise-smoothed surface. Unfortunately these surfaces do not appear directly in an image. Instead one has to deal with contours. There are two types of contours arising from surfaces: discontinutity and extremal contours. If two C^3 surfaces intersect such that at each point along the intersection the surfaces have distinct tangent planes, it can be shown that the intersection of the two surfaces is a C^3 edge. A discontinuity contour is the projection of such an edge. An extremal contour occurs whenever a smooth surface turns away from the viewer.

A point of intersection of two or more edges form an object vertex. The projection of edges and vertices give rise to image junctions which constitute important elements for describing an image. Interestingly, the range of possible configurations of contours and junctions is quite restricted. Malik [7] suggested an analysis based on the projection of edge and vertex neighbourhoods to predict the possible image configurations. Figure 3 shows a catalog of image junctions, where each junction represents a combination of *linear* and *curved* contours. Then, the task of our system is to detect these junctions and to classify them accordingly.

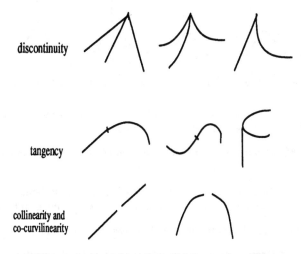

Figure 3: A catalog of possible image junctions

Another important property encountered in many natural and man-made objects is local symmetry. The relationship between locally symmetric 3-D shape and image symmetry has been stressed by Marr [8], Horaud and Brady [4], and Ponce, Chelberg, and Mann [9]. It has been shown that under certain constraints object symmetry projects in the image to give rise to contour symmetry. Therefore, the detection and representation of image local symmetry is another important task of our system. Figure 4 shows a catalog of possible pairwise symmetric image contours.

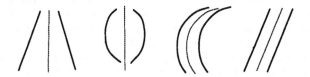

Figure 4: A catalog of possible image symmetric contours

Unfortunately, the accurate detection of contours and their relationships (junctions and symmetries) are affected by a certain number of undesirable and unavoidable phenomena.

First, the image formation process is a complex one and surface discontinuities are not the only scene features giving rise to contours. Color, textures, shadows, and specularities give rise to equally important[1] contours. We believe that the process of grouping, i.e., finding junctions and symmetries, partially throughs away uninteresting contours (contours that are not tight into a sought configuration). Hence, groups of contours are potentially more reliable than isolated ones. The "filtering" properties of grouping are stressed on the example shown on Figures 1 and 2.

Second, another unavoidable phenomenon is occlusion. For example, a circular edge at the intersection of two object surfaces may appear in an image decomposed in several pieces with relatively important gaps in between the pieces. This is the reason for which collinearity and co-curvilinearity are explicitly detected and added to our junction catalog, e.g., Figure 3.

---

[1]In terms of their image local structure.

Third, low-level image segmentation (feature detection) is known to be a non robust process. Figure 5 shows perfect junctions and symmetries and the configurations altered by segmentation.

Figure 5: A "perfect" image junction and junctions altered by noise

To conclude this section, we have tried to build a collection of image features and relationships that are well suited for describing an image. In particular:

- Linear and curved contours, junctions and local symmetries are good candidates for describing 3-D shapes,

- Groups of image features are intrinsically less ambiguous than isolated ones. Therefore there will be a dramatic reduction of the complexity associated with image interpretation,

- The process of grouping eliminates undesired isolated features, and

- Noise and imperfect segmentation augments the difficulty of detecting groups of features. Hence, robust techniques are needed for feature grouping.

4 Implementation

We concentrate now on the process of detecting geometric and relational structures. This process takes as input the list of straight-line segments (shortly, segments) produced by the feature detection process and outputs a data structure which is the relational graph associated with the InterLevel representation. This graph is built on top of the initial list of segments and makes explicit relationships between these segments (junctions, collinearity, symmetry) as well as groups of segments satisfying some geometric constraints (curves, parallelograms, etc.). This graph is best illustrated in Figure 6. The graph is divided into two parts: The bottom level of this graph is constituted of a set of nodes where a node could be either a segment or an endpoint (as it will be explained later in this section, an endpoint is in fact a small region). Each segment has two endpoints associated with it. Two or several segments sharing a common endpoint form a junction. Junctions are represented implicitly in the bottom level of the graph or explicitly, in the top level. More generally, the top level graph contains such structures (i.e., combinations of nodes and arcs) as:

- junctions or termination at a common endpoint;

- curves which are sequences of connected segments;

- ribbons which are pairs of symmetric segments with the region in between them and

- other objects that can be built on top of these basic ones: pairs of symmetric curves, junctions combining curves and segments, parallelograms, sets of equally distant paraller lines or curves, chains of collinear lines, chains of co-curvilinear curves, groups of ribbons sharing a common symmetry axis, etc.

The organization of this graph is motivated by practical rather than theoretical considerations. Indeed, robust techniques are available for line detection. This is not the case for curve detection. A line has a simple analytical representation while a curve has not: A curve may be described by equations of various degrees. An alternative for describing a curve may well be "a smooth piece

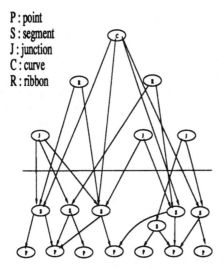

P : point
S : segment
J : junction
C : curve
R : ribbon

Figure 6: The InterLevel relational graph which describes image structure

of contour with no first or second order discontinuity." This is the description that we'll use in the graph. If specific knowledge about the exact shape of an object is available, one may take the Inter-Level curve description and fit it with the desired model. In the following paragraphs we describe in detail junction, symmetry, curve, parallelogram detection, and symmetry grouping.

4.1 Junction detection

An image junction is a set of lines terminating at a common point. In practice however, the lines rarely terminate exactly at the same point. Instead they terminate at a "small common region," i.e., Figure 7. Hence the difficulty of detecting junctions is twofold:

- the complexity of considering all subsets of lines as potential junctions, and

- the lack of a simple mathematical definition of a set of lines terminating at a common region.

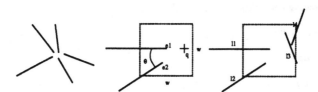

Figure 7: A junction is characterized by a set of lines terminating at a common region. The parameters of angle detection. A local configuration of segments.

In order to reduce the complexity mentioned above we build a simple data structure which consists of associating with each image pixel the information of whether there is a segment located at that pixel or not. If there is a segment, the data structure indexes it in the segment list. Then, for any image neighborhood one can quickly determine the list of segments with their endpoints terminating at this neighborhood.

It is easy now to determine the complexity of junction detection. Let n be the number of image lines and a be the average number of endpoints in the neighborhood of a line. The complexity is bounded by $2 \times a \times n$ since each segment has two endpoints. Next we describe a two-stage method for detecting junctions: angle detection and angle grouping. Then we consider the special case of collinear lines.

Angle detection. An angle is a two-line junction. Consider an endpoint e_1 associated with a segment l_1 and a neighborhood of size $w \times w$ pixels centered at this endpoint. Let l_2 be another segment such that it has an endpoint e_2 lying in the neighborhood. Let q be the point of intersection of l_1 and l_2. Let d_1 be the distance from e_1 to q and d_2 be the distance from e_2 to q. Let θ be the value of the angle between the two segments. Segments l_1 and l_2 form an angle if the following conditions are satisfied, Figure 7:

1. $\epsilon_1 \leq \theta \leq \pi - \epsilon_1$

2. $d_1 \leq w/2$

3. $d_2 \leq w/2$

4. *q falls outside the segments l_1 and l_2*

5. *Test that there is no other line crossing $e_1 q$ and $e_2 q$*

Condition 1 guarantees that the lines are not collinear or do not form a very sharp angle (which is unlikely to represent a scene property). Conditions 2 and 3 guarantee that the point of intersection of the two considered segments lies within the "small region." For example, when applied to the configuration of Figure 7, only the pair l_1 and l_2 will be selected as an angle.

Angle grouping. Once all the angles have been detected, junction detection is quite straightforward. The principle of grouping is that two angles that share a common line and whose tips are not too faraway are fused into a unique junction. First a junction is initialised with an angle. Second the list of remaining angles is examined and angles are added, one by one, whenever they satisfy the above condition.

Detecting collinearity. Colinear lines are treated separately because one cannot always find the intersection of two collinear or almost collinear lines, i.e., Figure 8. Sometimes this intersection is simply not significant. Let again θ be the value of the angle between two segments, l_1 and l_2 and let e_1 and e_2 be their closest endpoints. Let d_1 be the distance from e_1 to l_2 and d_2 be the distance from e_2 to l_1. Let l be the amount of overlap of the two lines, i.e., Figure 8. The two segments are collinear if the following conditions are satisfied:

1. $\theta \leq \epsilon_1$ or $\pi - \epsilon_1 \leq \theta \leq \pi$

2. $l \leq max\_overlap$

3. $\max(d_1, d_2) \leq w/2$

Figure 8: Pairs of "almost" collinear lines to be detected and the parameters of detecting these pairs.

4.2 Curve detection

Once the junctions are properly detected and mapped into the graph representation described above, one can build other image structures such as curves. A curve may be defined as a sequence of connected segments (the extremity of a segment and the origin of the next one in the sequence belong to the same junction). Let θ_i be the value of the angle between two consecutive segments in such a sequence. The sequence is a curve if the following conditions are satisfied:

1. $\epsilon_2 \leq \mid \theta_i \mid \leq \pi/2 - \epsilon_2$

2. $\mid \theta_i - \theta_{i+1} \mid \leq \epsilon_3$

3. $\theta_i \theta_{i+1} > 0$

4. *The curve is maximal*

Conditions 1 and 2 guarantee that the sequence turns "smoothly" and condition 3 gurarantees that there is no convexity change along the sequence. A maximal curve is a curve that cannot be extended to include another segment. The technique for detecting sequences with the above properties is based on a classical recursive graph traversal algorithm. This algorithm starts with a two-segment sequence and tries to extend it by exhaustively exploring the graph.

Let j be the number of junctions in the image and d be the average number of segments of each junction. The number of two-segment sequences is $j \times d \times (d - 1)$. Hence the complexity of finding a curve with L segments is: $j \times d \times (d - 1) \times (L - 2) \times (d - 1)$. In practice, d is bounded by 5 and the most likely value for d is 2. Therefore $d \times (d-1) \times (d-1)$ is bounded by a constant value.

4.3 Symmetry detection

Following the "smoothed local symmetry" definition in [1], the symmetry axis of two lines is the bisector of their angle. For two segments to be symmetric, they must have some amount of overlap. Referring to Figure 9, where l_1 is the length of one segment, l_2 is the length of the other one, and l is the length of their overlap, we must have:

1. $\mid l_2 - l_1 \mid \leq \epsilon_4$

2. $\mid (l_2 + l_1)/2 - l \mid \leq \epsilon_5$

Two segments together with the region in between them are called a "ribbon". One may further combine ribbons and build more complex symmetries.

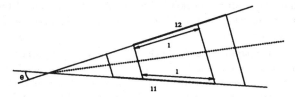

Figure 9: The definition of the local symmetry associated with two straight-line segments

4.4 Finding parallelograms

The task of finding all pairs of symmetric segments in the image is a costly one, i.e., proportional to the squared number of image segments. The detection of parallelograms is an example of a simple use of symmetry.

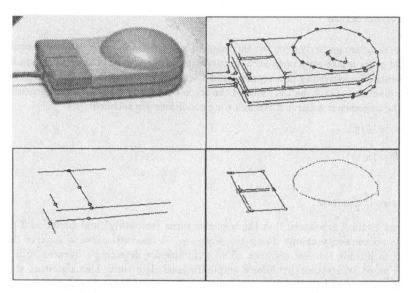

Figure 10: Another image of the mouse and the obtained structures.

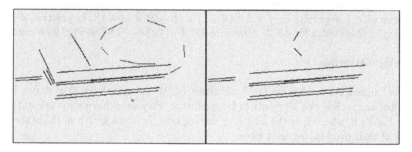

Figure 11: The pairs of symmetric lines (left) and a group of line-pairs with parallel symmetry axes (right).

We define a parallelogram as a sequence of four segments in the graph such that the first one is symmetric and parallel with the third one and the second one is symmetric and parallel with the fourth one. Notice that "symmetric and parallel" is just a particular case of the symmetry defined above. In our implementation we allow the side of such a parallelogram to be a meta-segment, that is, a sequence of collinear segments.

The technique for finding parallelograms is as for curve detection, a variation of a recursive graph traversal algorithm. For each junction and for each pair of segments (or meta-segments) within each junction we investigate all the segments (or meta-segments) that are connected to the initial pair and we select those four segments (or meta-segments) which could form a parallelogram.

To compute the complexity of this parallelogram search, let again j be the number of junctions and d be the average number of segments forming a junction. The number of segment pairs is (different than the number of two-segment sequences): $j \times d/2 \times (d-1)$. The cost of searching parallelograms is: $j \times d/2 \times (d-1) \times 2 \times (d-1) = j \times d \times d \times (d-1)$. Since d is bounded by a small constant value, the cost is a linear function of the number of image junctions.

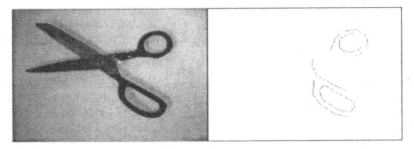

Figure 12: The image of a pair of cisors and the detected curves.

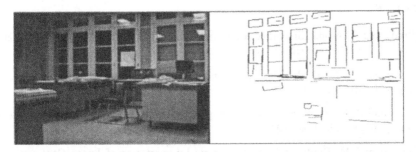

Figure 13: The INRIA office image and the rectangles found by our system. Notice that the system have detected many rectangles that do not have a semantic interpretation.

4.5 Examples

We have run our system over a large database of images. The mouse of Figure 1 is shown from a different viewpoint on Figure 10 (top-left). There are also shown on this Figure the segment-and-junction graph (top-right), the sets of collinear segments, or meta-segments (bottom-left), and curves and parallelograms finally detected (bottom-right). Figure 11 shows all the pairs of symmetric lines (left) and a set of line-pairs with parallel symmetry axes (right).

Two more interesting examples are provided by the image of a pair of cisors, e.g., Figure 12 with the detected curves, and by the INRIA office image and the rectangles detected here, e.g., Figure 13. It is important to notice that the same processing (the same edge detector and line finder) with the same set of parameters have been used over the whole image set. It takes approximatively 10 seconds to entirely process a 256 by 256 image on a Sun 4/60: 8 seconds for edge detection and line finding and 2 seconds for grouping.

5 Conclusions

In this paper we presented a method for extracting geometric and relational structures under the form of a graph representation from raw intensity data. This graph is shown on Figure 6 and some image structures are shown on Figures 2, 10, 12, and 13. This graph may well be viewed as an intermediate representation between low- and high-level vision. The information embedded in the graph is useful for a variety of tasks.

Object recognition is often mapped into a graph matching problem: The sensory data is compared with object models. Currently we are investigating techniques for performing sub-graph isomorphism.

Another domain of computer vision for which our graph is a useful representation is visual learning. Consider for example the mouse object shown on both Figure 1 and Figure 10, and the graphs associated with these images. Is it possible, by comparing these two graphs, to derive a description of the "mouse"?

Finally, we investigate techniques for combining the representation advocated here with camera motion in order to infer a 3-D version of the graph. This may be useful for modelling 3-D objects or for building scene descriptions in general.

Acknowledgements. This research has been sponsored by the "Ministère de la Recherche et de la Technologie" and by the "Centre National de la Recherche Scientifique" through the ORASIS project as part of the PRC Communications Homme/Machine and by CEC through ESPRIT-BRA 3274 (The FIRST project). The authors would like to thank Emmanuel Arbogast and Roger Mohr for their invaluable comments.

References

[1] M. Brady and H. Asada. Smoothed local symmetries and their implementation. *International Journal of Robotics Research*, 3(3):36–61, 1984.

[2] J. H. Connell and M. Brady. Generating and generalizing models of visual objects. *Artificial Intelligence*, 31:159–183, 1987.

[3] J. Dolan and R. Weiss. Perceptual Grouping of Curved Lines. In *Proc. Image Understanding Workshop*, pages 1135–1145, 1989.

[4] R. Horaud and M. Brady. On the Geometric Interpretation of Image Contours. *Artificial Intelligence*, 37(1–3):333–353, December 1988.

[5] R. Horaud and T. Skordas. Stereo Correspondence Through Feature Grouping and Maximal Cliques. *IEEE Transactions on Pattern Analysis and Machine Intelligence*, PAMI-11(11):1168–1180, November 1989.

[6] D. Lowe. *Perceptual Organization and Visual Recognition*. Kluwer Academic Publisher, 1985.

[7] J. Malik. Interpreting Line Drawings of Curved Objects. *International Journal of Computer Vision*, 1(1):73–103, 1987.

[8] D. Marr. Representing and Computing Visual Information. In Patrick Henry Winston and Richard Henry Brown, editors, *Artificial Intelligence: An MIT Perspective*, pages 17–82, MIT Press, 1979.

[9] J. Ponce, D. Chelberg, and W. B. Mann. Invariant Properties of Straight Homogeneous Generalized Cylinders and Their Contours. *IEEE Transactions on Pattern Analysis and Machine Intelligence*, PAMI-11(9):951–966, September 1989.

[10] L. G. Shapiro and R. M. Haralick. Structural descriptions and inexact matching. *IEEE Transactions on Pattern Analysis and Machine Intelligence*, PAMI-3(5):504–519, September 1981.

[11] Th. Skordas. *Mise en correspondance et reconstruction stéréo utilisant une description structurelle des images*. PhD thesis, Institut National Polytechnique de Grenoble, October 1988.

[12] A. P. Witkin and J. M. Tenenbaum. On perceptual organisation. In Alex B. Pentland, editor, *From Pixels to Predicates*, chapter 7, pages 149–169, Ablex Publishing Corporation, Norwood, New Jersey, 1986.

[13] S.W. Zucker. The Diversity of Feature Grouping. In Michael Arbib and Allen Hanson, editors, *Vision, Brain, and Cooperative Computation*, pages 231–262, MIT Press, 1988.

SHAPE DESCRIPTION

Recovery of Volumetric Object Descriptions From Laser Rangefinder Images

F.P. Ferrie   J. Lagarde   P. Whaite

Abstract

This paper describes a representation and computational model for deriving three dimensional, articulated volumetric descriptions of objects from laser rangefinder data. What differentiates this work from other approaches is that it is purely bottom-up, relying on general assumptions cast in terms of differential geometry.

1   Introduction

The ability of a robot to correctly perceive its environment is essential to tasks involving interaction and navigation. Descriptions computed by the perceptual system must reflect the characteristics of the world; that objects take up space, are often composed of many parts, and can be articulated in a number of different ways. This paper is about computing such descriptions from the bottom up. That is, beginning with estimates of surface points obtained with a laser rangefinder, we will describe how an articulated volumetric description of an object can be obtained through a succession of intermediate representations and computational steps. Some of the building blocks that we shall use are well-known, but it is the way in which these are tied together and the computational aspects of the problem that are the principal contributions of this paper.

Our approach follows a traditional bottom-up transition from surfaces to parts to objects in a hierarchical fashion [11, 12, 19]. Darboux frames, which describe the orientation, principal curvatures and directions at a point on a surface [7, 22, 23] are used as a local representation for a surface at each sample point. Initial estimates of these measures are readily computable using a number of different approaches [3, 8, 12]. But the technical difficulty is to refine the initial estimates, which are often corrupted by noise and quantization error, into robust measures. We have adopted a computational strategy based on the minimization of a residual form that measures the total deviation from an implicit model of surface curvature [22, 23, 9, 10]. This approach allows us to reconstruct a surface from sensor data and, more importantly, to make the features that are needed for the parts decomposition explicit (e.g. negative local minima of curvature, orientation and jump discontinuities) [11, 12, 15].

Contours formed by these features can serve to partition a surface into regions corresponding to different parts. However this grouping problem is often difficult, especially when features are sparse or when adjacent contours are in close proximity. An elegant solution can be obtained by exploiting the directional properties of the frame[1] associated with each feature point. Following the scheme devised by Zucker et al. [26], the direction vectors of each frame are used to generate a potential field which acts on a covering of unit length snakes (energy minimizing splines) [16, 24, 25]. When this system reaches a steady-state, contours are obtained that smoothly interpolate the data,

---

[1]For brevity we will use frame to refer to Darboux frame or augmented Darboux frame, depending on the context.

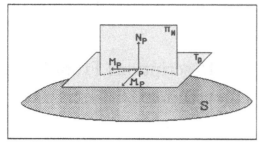
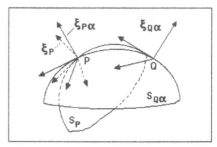

Figure 1: Local surface representation – the augmented Darboux Frame.

Figure 2: Local extrapolation using a parabolic quadric approximation.

and correspond to the requisite part boundaries. What is important about this scheme is that the parameters of the minimization are determined primarily by the data [26].

Each surface partition is taken to be the visible part of a single object. Part geometry is determined by a process of inference in which a suitable volumetric primitive is fit to each surface patch obtained from the parts decomposition [12, 10]. Our approach is in contrast to the minimal length encoding strategy used by Pentland [19]. He begins with a with a fixed parametric model (a superquadric) and uses a minimization that seeks a best fit to data using a minimal set of primitives. We view Pentland's approach as a top down strategy and ours bottom up. The advantages offered by the latter scheme are a reduction in computational complexity, and a representation that is not tied to any specific model. Depending on what needs to be made explicit, any volumetric model can be used to characterize a part without changing the interpretation of the object.

The paper is structured along the computational steps that define our procedure for building volumetric object models from sensor data. Section 2 describes the local representation for surfaces and the minimization procedure used to compute reliable estimates of its parameters. The task of identifying the feature points that make up the part boundaries and the interpolation procedure for aggregating these into contours using the potential field method are described in Section 3. To obtain comparative results with Pentland's approach, we used superquadric primitives to represent part geometry. The technique used to fit the primitives is similar to others [2, 19, 1, 6, 14], and is described in Section 4. Finally, the performance or our scheme on real data is shown in Section 5. Articulated models of two objects are derived from range maps acquired with a laser rangefinder.

2   Local Representation of a Surface

The local surface representation at a point P, the augmented Darboux frame $\mathcal{D}(P)$ [7, 22, 23] serves three purposes: (1) It facilitates the task of surface reconstruction, (2) makes explicit the features necessary for identifying putative boundary points, and (3) provides the direction frames used to smoothly interpolate contours on the surface. It is explained as follows. Let the local neighbourhood of a point P on a surface S be represented by a parabolic quadric of the form $w = au^2 + cv^2$, with origin at P and the w axis aligned with the surface normal at P, Np, as shown in Figure 1. The orientation of this local frame is such that the u and v coordinate axes align with two special directions on S at P. These are the directions for which the normal curvature at P (a directional property) takes on maximum and minimum values, κ_{MP} and $\kappa_{\mathcal{M}P}$, and are referred to as the principal directions M_P and $\mathcal{M}_P$ respectively [7]. The scalar quantities κ_{MP} and $\kappa_{\mathcal{M}P}$ are similarly referred to as the principal curvatures at P. Following the convention of [22, 23], we refer to $\mathcal{D}(P) = (p, M_P, \mathcal{M}_P, N_P, \kappa_{MP}, \kappa_{\mathcal{M}P})$ collectively as the augmented Darboux Frame at P. The problem is to estimate $\mathcal{D}(P)$ from laser rangefinder images of the form $z = f(x, y)$. Local least-squares estimation is sometimes sufficient to determine κ_{MP}, $\kappa_{\mathcal{M}P}$, and N_P, but rarely M_P and $\mathcal{M}_P$ [3, 12]. However, the latter directional properties are essential to the inference of discontinuities, occluding contours, and part boundaries

[26]. Our approach is to use local methods to obtain a first estimate of $\mathcal{D}(P)$, and then apply a second stage of minimization to obtain a stable reconstruction of S. The former problem is not addressed in this paper, but typical approaches are described in [3, 8, 12].

2.1 Iterative Refinement of the Darboux Frame

Local consistency of curvature, subject to orthogonality constraints on $\mathcal{D}(P)$, is the basis of our minimization algorithm [18, 23]. The method was first introduced by Sander and Zucker in the context of C.T. image reconstruction [22, 23]. Aside from the application to range data, there are a number of important technical details which differentiate our work from that of Sander and Zucker. However, the motivation is similar and can be explained with the aid of Figure 2. Because $\mathcal{D}(P)$ is a dual form of a parabolic quadric, one can extrapolate outward from a point Q to its neighbour P to get an idea of what the surface at P looks like according to its neighbour Q. By performing this operation for each neighbour of P, one obtains a set of frames, $\xi_{P\alpha}$, each providing an estimate of P from its associated neighbour [22, 23]. This provides a mechanism for setting up a minimization which enforces the local model over the surface, somewhat analogous to the constant curvature assumption in [18]. What the algorithm does in practice is to iteratively update each $\mathcal{D}(P)$ with the least-squares estimate computed from $\xi_{P\alpha}$.

Among the considerations in formulating the algorithm are, (i) the particular form of extrapolation along the surface (related to parallel transport [7, 22, 23]) to obtain $\xi_{P\alpha}$ and (ii) the form of minimization functional applied in updating $\mathcal{D}(P)$. For the results presented in this paper, we followed [22, 23] and used a parabolic quadric to estimate $\xi_{P\alpha}$ from the surrounding neighbourhood[2]. But this does not enforce the constant curvature constraint according to [17]. A better choice is a toroidal patch that has constant curvature along its principal directions. However, with densely sampled range data, the particular form of parallel transport does not appear to be critical.

The functional itself is set up to minimize the variation in $\mathcal{D}(P)$ subject to constraints on M_P, $\mathcal{M}_P$, and N_P. These are,

$$(N_P \cdot N_P) = 1 \quad (M_P \cdot M_P) = 1 \quad (M_P \cdot N_P) = 0. \tag{1}$$

As formulated in [22, 23], the minimization consists of two terms corresponding to (1) the surface normal N_P and principal curvatures κ_M and $\kappa_{\mathcal{M}}$, and (2) the principal direction M_P[3]. To simplify the analysis, each is minimized independently. The first term, E_1, follows directly from [22, 23]:

$$E_1 = \sum_{\alpha=1}^{n} \|N_P - N_{P\alpha}\|^2 + (\kappa_M - \kappa_{MP\alpha})^2 + (\kappa_{\mathcal{M}} - \kappa_{\mathcal{M}P\alpha})^2 + \lambda((N_P \cdot N_P) - 1) \tag{2}$$

where $\xi_P = (P, \kappa_{MP}, \kappa_{\mathcal{M}P}, M_P, \mathcal{M}_P, N_P)$ and $\xi_{P\alpha} = (P_\alpha, \kappa_{MP\alpha}, \kappa_{\mathcal{M}P\alpha}, M_{P\alpha}, \mathcal{M}_{P\alpha}, N_{P\alpha})$. Using standard methods, one obtains the following updating functionals for N_P, κ_{MP}, and $\kappa_{\mathcal{M}P}$:

$$N_P^{(i+1)} = \frac{(\sum_{\alpha=1}^{n} N_{xP\alpha}^{(i)}, \sum_{\alpha=1}^{n} N_{yP\alpha}^{(i)}, \sum_{\alpha=1}^{n} N_{zP\alpha}^{(i)})}{\sqrt{(\sum_\alpha N_{xP\alpha}^{(i)})^2 + (\sum_\alpha N_{yP\alpha}^{(i)})^2 + (\sum_\alpha N_{zP\alpha}^{(i)})^2}} \quad \kappa_M^{(i+1)} = \sum_{\alpha=1}^{n} \frac{\kappa_{MP\alpha}^{(i)}}{n} \quad \kappa_{\mathcal{M}}^{(i+1)} = \sum_{\alpha=1}^{n} \frac{\kappa_{\mathcal{M}P\alpha}^{(i)}}{n} \tag{3}$$

where the superscript i refers to the current iteration step.

Because M_P and $\mathcal{M}_P$ are *directions*, there is a 180° ambiguity in orientation. For this reason the formulation of minimization term E_2 needs to be re-cast from that described in [22, 23]. We avoid the ambiguity by minimizing the difference of directions in the tangent plane at point P as follows. Express M in tangent plane coordinates as

$$M_P = \bar{b}_1 \cos\theta + \bar{b}_2 \sin\theta, \ (0, 2\pi) \text{ s.t. } \bar{b}_1, \bar{b}_2 \in T_P, \ \|\bar{b}_1\| = \|\bar{b}_2\| = 1, \text{ and } (\bar{b}_1 \cdot \bar{b}_2) = 0 \tag{4}$$

---

[2]The technical details of how this is accomplished are described in [22].

[3]Since $\mathcal{M}_P$ is orthogonal to both M_P and N_P, it need not be considered.

Then

$$E_2 = \min_\theta \sum_{\alpha=1}^n \left[1 - (M_P(\theta) \cdot M_{P\alpha})^2 \right] \qquad (5)$$

$M_P^{(i+1)}$ is found by substituting the value of θ that minimizes (5), back into (4). Again, using standard methods, one obtains the following updating functional for θ:

$$\theta^{(i+1)} = \tan^{-1} \left[\frac{(A_{22} - A_{11}) + \sqrt{(A_{11} - A_{22})^2 + 4A_{12}^2}}{2A_{12}} \right], \quad A_{ij} = \sum_{\alpha=1}^n (M_{P\alpha} \cdot \bar{b}_i)(M_{P\alpha} \cdot \bar{b}_j). \qquad (6)$$

Note that this also determines the solution for $M_P^{(i+1)}$ for the reason cited earlier.

Control over iteration is maintained by tracking the convergence of the derivative of a composite measure R_S, which is the sum of local difference measures computed over the surface,

$$R_S^{(i)} = \sum_j R_j(\xi_P^{(i)}, \xi_{P\alpha}^{(i)}) = \sum_j E_{j1}^{(i)} + E_{j2}^{(i)}, \quad P_j \in S. \qquad (7)$$

The algorithm is allowed to interate until the difference $|R_S^{(i)} - R_S^{(i-1)}|$ falls below a specified threshold. A discussion of the convergence properties is beyond the scope of this paper and is addressed in [17, 22, 23]. However, we have confirmed empirically over a large number of experiments that the algorithm produces stable results quite rapidly, generally within 5 iterations.

2.2 Identifying Feature Trace Points

The point of the minimization strategy is to obtain a description of the surface S that is *stable* with respect to further interpretation [4]. This allows for a more direct interpretation of features and specifically avoids having to deal with the problem at the level of feature interpretation, e.g. [5]. The determination of features used to partition the surface is a case in point.

Hoffman & Richards [15] argue that a natural basis for surface decomposition is the principle of transversality regularity. Simply stated, the interpenetration of two arbitrarily shaped surfaces (i.e. corresponding to different parts) results in a contour of concave discontinuity of their tangent planes. In the context of smooth surfaces this translates into the partitioning of S into parts at loci of negative minima of each principal curvature along its associated family of lines of curvature [15]. We will refer to such loci as *critical points*. Thus, it is important to have stable estimates of the principal curvatures and directions at each point on S. The following procedure is used to determine the loci of critical points on smooth surfaces.

Let $\kappa_M(x,y)$ and $\kappa_{\mathcal{M}}(x,y)$ represent stable estimates of the principal curvatures of S sampled on the discrete grid (x,y), with corresponding principal directions $M(x,y)$ and $\mathcal{M}(x,y)$. The directional derivatives in these directions are $\kappa'_M(x,y)|_M$ and $\kappa'_{\mathcal{M}}(x,y)|_{\mathcal{M}}$ respectively. Then P is deemed to be a critical point iff

$$\kappa'_M(x,y)|_M = 0 \quad \text{AND} \quad \kappa_M(x,y) < 0 \qquad \text{OR} \qquad \kappa'_{\mathcal{M}}(x,y)|_{\mathcal{M}} = 0 \quad \text{AND} \quad \kappa_{\mathcal{M}}(x,y) < 0. \qquad (8)$$

As presently implemented, the curvature consistency algorithm does not have an explicit representation for orientation discontinuities, but does make use of such information (i.e. an externally computed discontinuity map) in the updating procedure. For example, a local edge operator can be used to provide an estimate of surface discontinuities [13]. While this does not solve the problem of correctly localizing all discontinuities on a surface, it can be used to significantly reduce edge smoothing in the reconstruction procedure. As far as the identification of critical points due to concave discontinuities is concerned, these will be smoothed into negative local minima and can be identified as outlined above. But jump discontinuities, caused either by self-occlusions of the object or occlusions by other objects in the scene, are also necessary for the partitioning task.

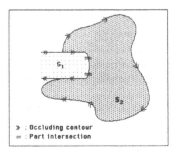

Figure 3: Partitioning contours on a surface

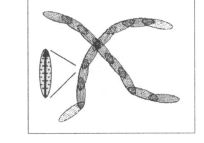

Figure 4: Tangents and the potential field

Such points can be identified from ξ_P by considering the angle between the surface normal N_P and the view vector V. Furthermore, if S is assumed to be smooth and in orthographic projection to the viewer, then the Z component of N_P will roll off to zero along occluding contours. It is this latter property that we use in identifying the trace points of the occluding contour. In fact, because of the stability of ξ_P as a result of reconstruction, even a thresholding of Z component values can suffice in localizing these points.

3   Piecing Together Partitioning Contours

Critical points, orientation discontinuities, and jump discontinuities are not themselves a solution to the parts decomposition problem, but they can provide the materials necessary to infer the partitioning contours. We will refer to such points collectively as the set of *trace* points $\{P_n^t\}$, where n is an index on the set. The second computational task is, given $\{P_n^t\}$ and the associated set of frames $\{\xi_{P_n}^t\}$, find the set of integral curves, $v_k(s) = (x_k(s), y_k(s), z_k(s))$, $k \in (1, \text{no. contours })$ that partition the surface S (Figure 3).

The approach used to solve this problem involves the use of energy-minimizing spline fitting, [16, 24, 25], using the strategy devised by Zucker et al. [26] for finding a global covering of plane curves through a 2-D tangent field. This strategy is best understood by analogy. Consider what happens when iron filings are distributed on a piece of paper with a magnet placed beneath. With a bit of shaking the filings eventually align with the magnetic field and smoothly interpolate the lines of flux. In the tangent field model, the iron filings become unit length snakes that align to a potential field generated by the set of trace points and their tangent directions. That is, for each trace point, one associates a local potential with an asymmetric Gaussian profile as shown in Figure 4. The sum of these local potentials defines the field.

In principle, one can extend the tangent field model to three dimensions by associating a 3D asymmetric Gaussian envelope with each frame of $\{\xi_{P_n}\}$ and including a torsion component in the minimization functional. However, for surfaces acquired from single viewpoint and without large foreshortening, it is often sufficient to assume that $z_k(s) \approx 0$. That is, one can use the planar model by projecting each frame ξ_P onto the view plane[4], where the principal direction M is used to generate a planar potential field as in the tangent field model. The resulting plane curves are used to segment a depth map of the surface into regions corresponding to parts.

The formulation of the minimization problem is as follows. Let $v(s,t) = (x(s,t), y(s,t))$, $0 \leq s \leq 1$ represent a deformable curve with kinetic energy functional $T(v)$ defined as

$$T(v) = \frac{1}{2} \int_0^1 \mu |v_t|^2 ds, \tag{9}$$

---

[4]The XY plane, assuming orthographic projection.

where μ is the (constant) mass density, and the potential energy functional $U(v)$ defined as

$$U(v) = \frac{1}{2}\int_0^1 \Big(\omega_1(s)|v_s|^2 + \omega_2(s)|v_{ss}|^2 + I(v) + S(v)\Big)ds, \tag{10}$$

where $\omega_1(s)x_s$ controls the tension of the curve, $\omega_2(s)x_{ss}$ controls the rigidity of the curve, $I(v)$ is the potential field coming from $\{\xi_{P_n}^t\}$, and $S(v)$ is a force between neighbouring curves that operates when they are in close proximity.

The space curves that we seek are described by those functions $x(s,t)$ and $y(s,t)$ for which

$$\int_{t_0}^{t_1} T(v) - U(v) \quad dt \tag{11}$$

is a minimum. Zucker et al. [26] describe a solution for the 2D case obtained from the calculus of variations. This method was used in the experiments described in Section 5. Surface regions corresponding to parts are obtained from the set of contours $\{v_k(s)\}$ using a conventional region labeling algorithm.

This model of surface partitioning will not suffice for a single view without an additional assumption. That is, the silhouette contour is assumed to close a part boundary for those cases where the endpoints of the boundary terminate on the contour. In other words, parts can be cut out of surface by a segment drawn between two points on the silhouette contour. This will become apparent in the experimental results presented in Section 5.

4 Fitting Part Models

The set of contours $\{v_k(s)\}$ partition the surface S into a set of regions $\bigcup_l S_l$. For each $l = (1, \text{no. parts})$, we now seek to infer a corresponding volumetric element V_m that best characterizes the 3-D shape of the part. Different subscripts are used to signify the fact that one or more surface patches S_l can map to a single volumetric element V_m, e.g. where a surface is occluded or where multiple viewpoints are involved. For the purposes of this paper, however, it is assumed that $l = m$. Given a set of volumetric primitives Γ, the final computational task consists of (1) finding $V_m \in \Gamma$ that best characterizes a particular S_l, and (2) determining the parameters of V_m by minizing

$$|V_m(x,y,z) - S_l(x,y,z)|. \tag{12}$$

There are a number of different approaches to solving this problem. They range from simple geometric approximations [12] to more complex fits using superquadric models [1, 6, 14, 19]. Pentland demonstrated how superquadrics could be used to advantage in representing a wide variety of shapes. His success with this representation has motivated others, including ourselves, to investigate the use of superquadrics for object and part models. But the task is not at all straightforward. The minimization represented by (12) does not necessarily possess a single global minimum, i.e. the problem is underdetermined, and additional constraints are needed to select a suitable minimization. For example, it is not possible to determine the depth of a only one face, but if we add the additional constraint that the volume be minimized, then the only possible volumetric solution is a thin plate.

Our approach is somewhat of a hybrid method. A simple fit to an ellipsoidal model is used to provide the starting point for a subsequent iterative fit to a superquadric

$$F = \left(\left|\frac{x}{a_x}\right|^{\frac{2}{\epsilon_2}} + \left|\frac{y}{a_y}\right|^{\frac{2}{\epsilon_2}}\right)^{\frac{\epsilon_2}{\epsilon_1}} + \left|\frac{z}{a_z}\right|^{\frac{2}{\epsilon_1}} = 1. \tag{13}$$

For a superquadric in an arbitrary position 11 parameters have to be found: three translation parameters; three rotation parameters; two shape parameters (ϵ_1 and ϵ_2); and three extent parameters (a_x, a_y, and a_z). The goal is to attempt to restrict the search and solution space by starting the minimization in the correct neighbourhood.

4.1 Initial Fit to an Ellipsoid Model

In our experience the iterative fitting procedure will converge to an acceptable solution provided it has a good initial estimate of the rotation and translation parameters of the volumetric primitive V_m. Only a rough estimate of the extent parameters is required. The shape parameters are not critical and can be initialized so that the superquadric starts as an ellipsoid ($\epsilon_1 = 1$, $\epsilon_2 = 1$).

We initialize the translation parameters to the centroid of the points in the surface patch S_l. Like Ferrie [11] the initial rotation parameters are found by aligning the axes of the ellipsoid along the principal moments of inertia of S_l about the centroid. The extent parameters are set to the maximum projected length of a vector from the centroid to a point on S_l for each ellipsoid axis.

No attempt has been made to compensate for the fact that the patch corresponds to only a partial view of the surface. Others, e.g. [11], have imposed symmetry constraints to improve the initial estimate, but in practice such methods have little impact on the final result.

4.2 Iterative Fit to a Superquadric Model

We have used the Levenberg-Marquardt gradient descent procedure [20] to minimize the error of fit between a superellipsoid surface V_m and a surface patch of range data S_l. The method requires calculation of the Jacobian of an error of fit function with respect to the adjustable parameters for each point of range data. Unlike others, [1, 6], we have not used Poisson distributed "jitter" to avoid local minima during the fitting process. Generally the solutions reached have been acceptable but in a small number of cases the procedure would benefit from the technique.

We have tried the true euclidean measure of fit error as suggested by Gross and Boult [14] but found that while it tended to fit the range data well, there was a tendency for the volume of the superquadric to become very large, especially when the surface patch was flat. We have found that the minimum volume measure motivated by Bajcsy and Solina produces more intuitive results [1]. Gross and Boult [14] defined a modified error of fit based on this measure as follows

$$R = \sqrt{a_x a_y a_z}\left(F^{\epsilon_1}(\bar{x}_w, \bar{a}) - 1\right), \tag{14}$$

where $F(\bar{x}_w, \bar{a})$ is the translated and rotated version of the inside-outside function shown in (13) and is 1 when the data lies exactly on the surface. The factor $a_x a_y a_z$ is a measure of the superquadric volume, so, given any set of superquadrics models that fit data equally well as measured by $F(\bar{x}_w, \bar{a})$, the smaller members of that set will have the lower overall error and will be preferred by the fitting procedure. Raising the inside-outside function to the power ϵ_1 shapes the error of fit to be more quadratic and more suited to rapid convergence under the assumptions of the Levenberg-Marquardt method.

5 Experiments

The results of two experiments are now presented that show how our strategy works for computing articulated volumetric descriptions of objects from laser rangefinder data. Range images used in these experiments are part of a standardized database available through the National Research Council of Canada [21]; each has been averaged down to a 256×256 by 12-bit resolution. The majority of the computation performed was done on a Symbolics 3650 Lisp machine. Rendering of the object models was done using the SuperSketch modeling system.

5.1 The Doll

Figure 5a shows an image rendered from a depth map of a toy doll which is lying face down. The first stage of processing uses the curvature consistency algorithm outlined earlier to reconstruct

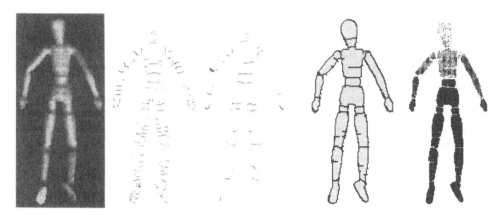

Figure 5: (a) Range image of the doll (b) Initial trace points (c) Trace points after 5 iterations

Figure 6: (a) Snake contours at steady state. (b) Regions computed from the snake contours

the surface, computing a stable intermediate description in the process. This is best illustrated by comparing Figures 5b and 5c, which show the trace points computed from the initial frame estimates, and after 5 iterations of the algorithm respectively. Other elements of ξ_{P_α} are similarly stabilized. In examining Figure 5c, notice how the trace points serve to delineate the part boundaries.

The frames $\{\xi_{P_n}\}$ associated with the set of trace points $\{P_n^i\}$ are used to generate a potential field for the second stage of processing. Accurate determination of the frame directions is important because this field is generated from a sum of local potentials, each oriented in its respective M_P. Snakes are then deposited at each trace point and are allowed to evolve in the potential field according to (11) until the system reaches a steady state. Figure 6a shows the resulting contours obtained after running the algorithm. Because of the dense covering of trace points at the part boundaries, the result is not much different from that shown in Figure 5c. The interpolating behaviour of the snakes does become important, however, when trace points become sparser. A region map, Figure 6b, is derived from the snake contours by clustering the interior regions. Using this procedure, the algorithm was able to correctly locate the parts of the doll with one exception. The right elbow joint was not located, largely because the arm is straight. For this reason the forearm and upper arm are grouped together as a single part.

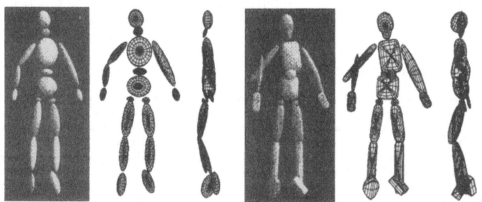

Figure 7: (a) Shaded image of the initial ellipsoid fit (b) Wire frame showing parts (front) (c) (side)

Figure 8: (a) Shaded image of the final superquadric fit (b) Wire frame showing parts (front) (c) (side)

The final stage of processing consists of fitting superquadric primitives to each of the regions located above. This proceeds in two stages, beginning with an initial approximation using ellipsoids to define initial starting points. Figure 7a shows a rendering of these initial fits as a shaded image; front and side profiles of the same model are rendered as wire frames in Figures 7b and 7c respectively. Superquadrics are then fit to each region using the gradient descent algorithm described earlier, with initial conditions and volume constraints provided by the initial ellipsoid fits. Figures 8a–8c show the results obtained by this procedure from the same viewpoints as shown earlier in Figures 7a–7c. From a qualitative viewpoint, the results are very pleasing and capture the essential structure of the doll.

5.2 A Toy Horse

The second example is of a toy horse (actually a unicorn) shown in Figure 9a. What is particularly interesting about this example is the complex nature of its surfaces. A comparison between the initial (Figure 9b) and final (Figure 9c) principal direction fields, i.e. $\mathcal{M}_P$, shows how the curvature consistency algorithm is able to correctly recover directional properties of the surface. Using the same procedure as above for the statue, the volumetric model shown in Figure 10a is obtained. A wire frame corresponding to this model showing each part is shown in Figure 10b, and a depth map generated from the volumetric model is shown in Figure 10c.

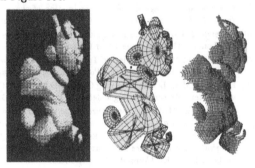

Figure 9: (a) Range image of the toy horse (b) Principal direction field computed from initial estimates (c) Principal direction field after 5 iterations

Figure 10: (a) Superquadric model rendered as a shaded image (b) Wire frame of the model showing the parts (c) Depth map computed from the model

6   Discussion and Conclusions

Although the approach we have just presented is advertized as being a method for computing volumetric models from laser rangefinder data, it is intended to provide a more general framework that includes other kinds of sensors. The augmented Darboux frame used in reconstruction can be estimated from either orientation or depth. It has already been applied to shape-from-shading [9]. In fact, sensor fusion is possible using this kind of minimization with a minimal change in the formulation. At present, no attempt is made to use cues provided by the silhouette contours in the parts decomposition. It is assumed that the observer is sufficiently active so as to choose a vantage point where part intersections are visible. A more comprehensive strategy must include analysis of contour as well as representation at multiple scales. Finally, the methodology for fitting part models needs to be extended. One of strengths of Pentland's [19] method is that model fitting takes the structure of neighbouring parts into account. By running the fitting procedure in parallel, constraints from neighbouring parts can be used to improve the local fit. These topics are currently under investigation in our laboratory.

References

[1] R. Bajcsy and F. Solina. Three dimensional object recognition revisited. In *Proceedings, 1ST International Conference on Computer Vision*, London,U.K., June 1987. Computer Society of the IEEE, IEEE Computer Society Press.

[2] A. H. Barr. Superquadrics and angle preserving transformations. *IEEE Computer Graphics and Applications*, 1(1):11–23, Jan. 1981.

[3] P. Besl and R. Jain. Segmentation through symbolic surface description. In *Proceedings IEEE Conf. Computer Vision and Pattern Recognition*, pages 77–85, Miami Beach, Florida, June 1986.

[4] A. Blake and A. Zisserman. *Visual Reconstruction*. MIT Press, Cambridge, Massachusetts, 1987.

[5] P. Boulanger. Label relaxation technique applied to the topographic primal sketch. In *Proceedings, Vision Interface 1988*, pages 158–162, Edmonton, Canada, June 1988.

[6] T. Boult and A. Gross. Recovery of superquadrics from depth information. In *Proceedings of the AAAI workshop on spatial reasoning and multisensor integration*, pages 128–137. American Association for Artificial Intelligence, 1987.

[7] M. do Carmo. *Differential Geometry of Curves and Surfaces*. Prentice-Hall, Inc., Englewood Cliffs,New Jersey, 1976.

[8] T. Fan, G. Medioni, and R. Nevatia. Description of surfaces from range data using curvature properties. In *Proceedings IEEE Conf. Computer Vision and Pattern Recognition*, pages 86–91, Miami Beach, Florida, June 1986.

[9] F. Ferrie and J. Lagarde. Robust estimation of shape from shading. In *Proceedings 1989 Topical Meeting on Image Und. and Machine Vision*, Cape Cod, Massachusetts, June 1989.

[10] F. Ferrie, J. Lagarde, and P. Whaite. Towards sensor-derived models of objects. In *Proceedings. Vision Interface '89*, London, Ontario, June 19-23 1989.

[11] F. Ferrie and M. Levine. Piecing Together the 3-D Shape of Moving Objects: An Overview. In *Proceedings IEEE Conf. on Computer Vision and Pattern Recognition*, pages 574–584, San Francisco, CA., June 1985.

[12] F. Ferrie and M. Levine. Deriving Coarse 3D Models of Objects. In *IEEE Comp. Soc. Conf. on Computer Vision and Pattern Recognition*, pages 345–353, University of Michigan, Ann Arbor, Michigan, June 1988.

[13] G. Godin and M. Levine. Structured edge map of curved objects in a range image. In *Proceedings IEEE Comp. Soc. Conf. on Computer Vision and Pattern Recognition*, San Diego, California, June 4-8 1989.

[14] A. D. Gross and T. E. Boult. Error of fit measures for recovering parametric solids. In *Proceedings, 2ND International Conference on Computer Vision*, pages 690–694, Tampa,Florida,UK, Dec. 1988. Computer Society of the IEEE, IEEE Computer Society Press.

[15] D. Hoffman and W. Richards. Parts of recognition. *Cognition*, 18:65–96, 1984.

[16] M. Kass and D. Terzopoulos. SNAKES: active contour models. *International Journal of Computer Vision*, 1:321–332, 1988.

[17] J. Lagarde. Constraints and their satisfaction in the recovery of local surface structure. Master's thesis, Dept. of E.E., McGill Univ., 1989. in preparation.

[18] P. Parent and S. Zucker. Curvature consistency and curve detection. *J. Opt. Soc. Amer., Ser. A*, 2(13), 1985.

[19] A. Pentland. Recognition by parts. In *Proceedings, 1ST International Conference on Computer Vision*, London,UK, June 1987. Computer Society of the IEEE, IEEE Computer Society Press.

[20] W. Press, B. Flannery, S. Teukolsky, and W. Vetterling. *Numerical Recipes in C - The Art of Scientific Computing*. Cambridge University Press, Cambridge, 1988.

[21] M. Rioux and L. Cournoyer. The nrcc three-dimensional image data files. National Research Council of Canada, CNRC No. 29077, June 1988.

[22] P. Sander. *Inferring Surface Trace and Differential Structure from 3-D Images*. PhD thesis, Dept. Elect. Eng., McGill University, Montréal, Québec,Canada, 1988.

[23] P. Sander and S. Zucker. Inferring surface trace and differential structure from 3-d images. *IEEE Trans. PAMI*, 1990. To appear.

[24] D. Terzopoulos. On matching deformable models to images. *Topical Meeting on Machine Vision, Technical Digest Series*, 12:160–163, 1987.

[25] D. Terzopoulos. On matching deformable models to images: Direct and iterative solutions. *Topical Meeting on Machine Vision, Technical Digest Series*, 12:164–167, 1987.

[26] S. Zucker, C. David, A. Dobbins, and L. Iverson. The Organization of Curve Detection: Coarse Tangent Fields and Fine Spline Coverings. In *Proceedings, 2ND International Conference on Computer Vision*, Tampa,Florida,USA, Dec. 1988. Computer Society of the IEEE, IEEE Computer Society Press.

EXTRACTION OF DEFORMABLE PART MODELS

Alex Pentland

Vision and Modeling Group, The Media Lab, Massachusetts Institute of Technology
Room E15-387, 20 Ames St., Cambridge MA 02139[1]

Many important objects consist of approximately rigid parts connected by hinges and other sorts of joints, so that the obvious way to describe these objects is in terms of the shapes of the component parts. Furthermore, if we are interested in the *behavior* of these parts, or if they are not completely rigid, then we must also account for their non-rigid shape and dynamics using a technique such as the finite element method.

Use of a 3-D dynamic model based on the finite element method was first suggested by Terzopoulos, Witkin, and Kass [1]. This approach to modeling is also known as the "thin plate" model. I will begin, therefore, by reviewing the finite element method.

THE FINITE ELEMENT METHOD

The finite element method (FEM) is the standard technique for simulating the dynamic behavior of an object. In the FEM energy equations (or functionals) are derived in terms of nodal point unknowns and the resulting set of simultaneous equation iterated to solve for displacements as a function of impinging forces:

$$\mathbf{M\ddot{u}} + \mathbf{D\dot{u}} + \mathbf{Ku} = \mathbf{f} \tag{1}$$

where $\mathbf{u}$ is a $3n$ x 1 vector of the (x, y, z) displacements of the n nodal points relative to the objects' center of mass, $\mathbf{M}$, $\mathbf{D}$ and $\mathbf{K}$ are $3n$ by $3n$ matrices describing the mass, damping, and material stiffness between each point within the body, and $\mathbf{f}$ is a $3n$ x 1 vector describing the (x, y, z) components of the forces acting on the nodes. This equation can be interpreted as assigning a certain mass to each nodal point and a certain material stiffness between nodal points, with damping being accounted for by dashpots attached between the nodal points. Normally the damping matrix $\mathbf{D} = s_1\mathbf{M} + s_2\mathbf{K}$ for some scalars s_1, s_2.

One of the major drawbacks of the finite element method is its large computational expense, due to the large size of the $\mathbf{M}$, $\mathbf{D}$, and $\mathbf{K}$ matrices. For instance, an object whose geometry is defined by 100 points produces 300 x 300 matrices, corresponding to the 300 unknown coordinates of the 100 nodal points, (x_i, y_i, z_i). Furthermore, for 3-D models the computation scales as $O(n^3)$ as the number of points n defining the object geometry increases.

---

[1]This research was made possible in part by National Science Foundation, Grant No. IRI-8719920.

Another related drawback of the finite element method, at least as applied to vision, is the large numbers of unknowns that must be solved for. Because the number of unknowns is typically much larger than the number of measurements available from sensor data, external information such as axis direction and shape, and heuristics such as symmetry and smoothness must be used to achieve useful extraction of shape.

A final drawback of the finite element approach is the non-unique and unstable nature of the descriptions produced. Because the number of degrees of freedom in the model is at least as large as the number of sensor measurements available, the final position of a particular nodal point is strongly constrained in only in the direction perpendicular to the object's surface. Thus it is not in general possible to compare the shape of two finite element models directly; instead, one must sample the surface of one model, synthesize 3-D points, and then measure the distance between those points and the surface of the second model. Further, unless special care is taken in the sampling step, the comparison is not transitive or unique with respect to rotation and translation.

MODAL DYNAMICS

A better method of describing non-rigid object behavior — at least for vision — is by use of *modal dynamics*, that is, by describing an object's behavior in terms of its natural *strain* or *vibration* modes. The modal method is equivalent to the finite element or thin-plate method in expressiveness and accuracy, but has the additional virtue that it separates non-rigid object behavior into independent modes of deformation, each of which may be separately analyzed and (often) solved in closed form. This in turn can lead to a much more efficient and stable computational scheme.

An object's strain modes may be found by simultaneously diagonalizing $\mathbf{M}$, $\mathbf{D}$, and $\mathbf{K}$. Because these matrices are normally positive definite symmetric, and $\mathbf{D}$ is a linear function of $\mathbf{M}$ and $\mathbf{K}$, Equation 1 can be transformed into $3n$ independent differential equations by use of the *whitening transform*, which is the solution to the following eigenvalue problem:

$$\lambda\phi = \mathbf{M}^{-1}\mathbf{K}\phi \tag{2}$$

where λ and ϕ are the eigenvalues and eigenvectors of $\mathbf{M}^{-1}\mathbf{K}$.

Using the transformation $\mathbf{u} = \phi\bar{\mathbf{u}}$ we can then re-write Equation 1 as follows:

$$\phi^T\mathbf{M}\phi\ddot{\bar{\mathbf{u}}} + \phi^T\mathbf{D}\phi\dot{\bar{\mathbf{u}}} + \phi^T\mathbf{K}\phi\bar{\mathbf{u}} = \phi^T\mathbf{f} \quad . \tag{3}$$

In this equation $\phi^T\mathbf{M}\phi$, $\phi^T\mathbf{D}\phi$, and $\phi^T\mathbf{K}\phi$ are diagonal matrices, so that if we let $\bar{\mathbf{M}} = \phi^T\mathbf{M}\phi$, $\bar{\mathbf{D}} = \phi^T\mathbf{D}\phi$, $\bar{\mathbf{K}} = \phi^T\mathbf{K}\phi$, and $\bar{\mathbf{f}} = \phi^T\mathbf{f}$ then we can write Equation 3 as $3n$ independent equations:

$$\bar{M}_i\ddot{\bar{u}}_i + \bar{D}_i\dot{\bar{u}}_i + \bar{K}_i\bar{u}_i = \bar{f}_i \quad , \tag{4}$$

where $\bar{M}_i$ is the ith diagonal element of $\bar{\mathbf{M}}$, and so forth.

What Equation 4 describes is the time course of one of the object's *strain* or *vibration* modes. The constant $\bar{M}_i$ is the generalized mass of mode i, that is, the inertia of the ith vibration mode. Similarly, $\bar{D}_i$, and $\bar{K}_i$ describe the damping and spring stiffness associated with mode i, and $\bar{f}_i$ is the amount of force coupled with this vibration mode. The ith row of ϕ describes the *deformation* the object experiences as a consequence of the force $\bar{f}_i$, and the eigenvalue λ_i is proportional to the natural resonance frequency of that vibration mode.

To obtain an accurate simulation of the dynamics of an object one simply uses linear superposition of these modes to determine how the object responds to a given force. Because Equation 4 can be solved in closed form, we have the result that for objects composed of linearly-deforming materials *the non-rigid behavior of the object in response to a simple force can be solved in closed form for any time t*. In complex environments, however, numerical solution is preferred.

Number Of Modes Required. The modal representation decouples the degrees of freedom within the non-rigid dynamical system of Equation 1, but it does not by itself reduce the total number of degrees of freedom. However modes associated with high resonance frequencies (large eigenvalues) normally have little effect on object shape. This is because (on average) the displacement amplitude for each mode is *inversely* proportional to the *square* of the mode's resonance frequency, and because damping is proportional to a mode's frequency. The combination of these effects is that high-frequency modes generally have very little amplitude. Experimentally, I have found that most commonplace multi-body interactions can be adequately modeled by use of only[2] rigid-body, linear, and quadratic strain modes [2].

Although discarding high-frequency modes has little effect on accuracy, it can have a profound effect on computational efficiency. Not only does it result in having to solve fewer equations in fewer unknowns, but (because of Nyquist considerations) we can also employ a much larger time step. For a problem involving 100 nodal points, for instance, the modal method (using 30 modes) will be roughly *two orders of magnitude* more efficient than the standard finite element approach, and yet will typically have roughly the same accuracy.

Another equally important benefit of using only low-order modes to describe object deformations is that they change very slowly as a function of object shape. Consequently the same modes ϕ may be used for a *range* of different — but similar — undeformed shapes without incurring substantial error. This allows the modes to be precomputed, avoiding the expense of solving for ϕ at run time.

Advantages of the Modal Representation. The modal representation has several advantages over a representation based on the standard finite element method. First, of course, it is much more efficient — typically one or two orders of magnitude more efficient — and scales as $O(n)$ rather than the $O(n^3)$ scaling of the standard finite element method.

Perhaps more importantly, however, it provides a natural hierarchy of scale, so that we can smoothly vary the level of detail by adding in or discarding high-frequency modes. That is, the modal representation provides a natural multi-scale representation for three-dimensional object shape in much the same manner that the Fourier transform provides a multi-scale representation for images. By matching the level of detail (the number of modes) to the number of sensor measurements available the shape recovery problem can be kept overconstrained without resorting to heuristics or external knowledge.

Further, by keeping the number of modes less than the number of sensor measure-

---

[2]Note, however, higher-order modes are required to accurately model the objects whose dimensions differ by more than an order of magnitude.

ments, we can calculate a shape description that is *unique* with respect to sampling and viewpoint (assuming, of course, that a sufficient distribution of surface measurements is available). In the finite element/thin-plate approach a recovered description is not unique because changes in sampling or viewpoint cause the nodal points to move about on the object's surface; this is a direct consequence of having more degrees of freedom than surface measurements. Polynomial and spline representations also suffer from these same problems. In the modal representation, however, the high-frequency modes that allow nodal points to move relative to one another have been discarded, and as a consequence the representation is insensitive to sampling and viewpoint.

Because of this uniqueness property, the modal representation is well-suited for object recognition and other spatial database tasks. To compare two objects described using a modal representation one simply compares the vector of mode values $\bar{u}$; if the dot product of the $\bar{u}$ for each object is small, then the objects are similar (excepting some degenerate conditions).

USE OF VOLUMETRIC MODELING PRIMITIVES

A disadvantage of any vertex or knot based representation is the expense of calculating the distance between 3-D points (e.g., sensor data) and the modeled surface; this calculation has a computational complexity of $O(Nn)$ where N is the number of data points and n is the number of points defining the object's geometry. Consequently this distance calculation is often a large fraction of the total computational cost.

One method of reducing the cost of computing the data error term is to use a representation that has an inside-outside distance function $f(x, y, z) = d$. For such implicit function representations the distance d between a point (x, y, z) and the surface can be found by simply substituting the point into the distance function $f(x, y, z)$. The computational complexity of this computation is $O(N)$, a significant improvement.

The modal representation of shape can be combined with analytic shape primitives by first describing each mode by an appropriate polynomial function, and then using global deformation techniques to warp the shape primitive into the appropriate overall form. The polynomial deformation mappings that correspond to each of the modes are determined by a linear regression of a polynomial with m terms in appropriate powers of x, y, and z, against the n triples of x, y and z that compose ϕ_i, a 3n x 1 vector containing the elements of the ith row of ϕ:

$$\alpha = (\beta^T \beta)^{-1} \beta^T \phi_i \quad , \tag{5}$$

where α is an m x 1 matrix of the coefficients of the desired deformation polynomial, β is an 3n x m matrix whose first column contains the elements of $u = (x_1, y_1, z_1, x_2, y_2, z_2, ...)$, and whose remaining columns consist of the modified versions of u where the x, y, and/or z components have been raised to the various powers. See reference [2] for more details.

By linearly superimposing the various deformation mappings one can obtain an accurate accounting of the object's non-rigid deformation. In the Thingworld modeling system [2] the set of polynomial deformations is combined into a 3 by 3 matrix of polynomials that is refered to as the *modal deformation matrix*. Because low-order modes change slowly as a function of object shape, the matrix can be used for a *range* of similar shapes, and thus may be precomputed.

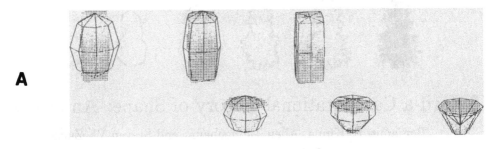

A

B

Figure 1: (a) Time-lapse sequence of a sphere being deformed to fit a vertical box, and (b) a sphere being deformed to fit a hollow vase.

FITTING 3-D MODELS

Given a segmentation into parts (such as produced by the algorithm described in reference [3]), the next step is to fit a deformable model to each part using available data. For simple objects this can be accomplished directly from the object's axes. This is because the planes of symmetry are the singularities (zero points) of the various low-order modal deformations, so that along axes and planes of symmetry the various modes are effectively decoupled. Consequently, the unknown polynomial coefficients in the modal deformation matrix can be solved for by measuring axis length, direction, and bending. From these coefficients the modal amplitudes $\bar{u}$ can be solved for directly. Such model recovery from symmetry analysis may be useful for understanding human vision, or for constructing efficient "first-pass" machine vision systems.

More accuracy can be obtained by fitting the model to range data. This can be accomplished by assigning a gravity-like potential field to each data point, so that the deformable part's surface is attracted to the data points by the resulting forces. Figure 1 shows two examples of this fitting process. In Figure 1(a) the range data is of a vertical box, and in Figure 1(b) the range data is of a hollow vase. In both cases the original spherical shape is progressively deformed by the attractive forces exerted by the range measurements, until the surface exactly fits the data. Typical fitting times on a Sun 4 using this formulation is a few seconds per part, depending upon the number of data points. These examples also illustrate that a wide variety of shapes can be generated by applying first and second order deformations to a basic spherical shape.

REFERENCES

[1] Terzopolis, D., Witkin, A., and Kass, M., (1987) Symmetry-Seeking Models for 3-D Object Reconstruction, *Proc. First International Conf. on Computer Vision*, pp. 269-276, London, England.

[2] Pentland, A., and Williams, J. (1989), Good Vibrations: Modal Dynamics for Graphics and Animation, *Computer Graphics ,.* Vol. 23, No. 3, pp. 215-222.

[3] Pentland, A., (1989) Part Segmentation for Object Recognition, *Neural Computation*, Vol. 1, No. 1, pp. 82-91.

Toward a Computational Theory of Shape: An Overview

Benjamin B. Kimia, Allen Tannenbaum and Steven W. Zucker

McGill Research Center for Intelligent Machines

Abstract

Although the shape of objects is a key to their recognition, viable theories for describing shape have been elusive. We propose a theory that unifies the competing elements of shape—parts and protrusions—and we develop a framework for computing them reliably. The framework emerges from introducing conservation laws to computational vision, and has application in areas ranging from robotics to the psychology and physiology of form. [1] [2] [3]

Introduction: How should the shape of objects be described to enable recognition? This is one of the key problems in perception, and two views have emerged. One view holds that composite objects are formed when distinct components interpenetrate each other [9], as when two lumps of clay are put together. We refer to this as the *part* view because it suggests that shapes are broken into "parts" at the junctions between lumps. The other (*protrusion*) view holds that existing parts should be deformed, as when clay is drawn out (or pushed in) from a lump [17]. While each of these views has some intuitive appeal, taken in the pure form neither seems completely right nor completely wrong. For example, a key missing ingredient is that of "necks", or the nature of the join between parts. Rather, this part vs. protrusion distinction has emerged as one of the frustrating dilemmas around shape; others are discussed in figure 1.

We propose an approach to representing shape, based on a reaction-diffusion equation, which resolves these dilemmas. Observe that, for two-dimensional curves, slightly deformed shapes are visually similar. We therefore study the evolution of shapes under general deformations, and show that they decompose into two types, a deformation that is constant (along the normal) and corresponds to a non-linear, hyperbolic (wave) type of process; and a deformation which varies with the curvature and corresponds to a quasi-linear diffusive one. The two types of processes interact, analagously to the way forces in physics interact at interfaces, and related questions involving conservation laws and entropy arise. Together the two processes give rise to shocks, the singularities of shape, which then provide a hierarchical decomposition of a shape into our proposed shape elements, parts and protrusions. Intuitively, necks then emerge as intersecting protrusions connecting coupled parts. Examples show that our proposed scheme is reliably computable. Moreover, the requirements of the algorithm are compatible with a physiologically-plausible model of curve detection [21] and with psychophysical evidence [1].

Shape from an Evolutionary Sequence: Since slightly deformed shapes are visually similar, we begin by studying the evolution of a shape under various deformations. Our immediate goal is to demonstrate that deformations which depend on the local geometry of the objects can be regarded

[1] AT is with Department of EE at University of Minnesota. BBK and SWZ are with Research Center for Intelligent Machines, McGill University, 3480 University Street, Montréal, Québec, Canada H3A 2A7.

[2] This article is an overview intended to convey a general perpective on our research on shape representation. More developed presentations as well as full references may be found in "B.B. Kimia, A. Tannenbaum, and S.W. Zucker, McRCIM tech report CIM-89-13" and "B.B. Kimia, A. Tannenbaum, and S.W. Zucker, On the Evolution of Curves via a Function of Curvature, I: The Classical Case, J. Math. Anal. and Appl., Submitted."

[3] We wish to thank Allan Dobbins, Lee Iverson, Frederic Leymarie, and John Tsotsos for valuable discussions and programming support. This research was supported by grants from NSERC, MRC, FCAR and AFOSR.

Figure 1: a) The *part* vs. *protrusion* dilemma: Some objects are naturally described as the result of composition of parts [9], e.g. the overlapping discs (left), while others are more naturally described as deformations [17], or protrusions, on a basic component (right). These two views are taken as competitive, but, intuitively, each has a certain appeal. Our theory provides a framework in which they both participate, eliminating the need to arbitrarily trade one off against the other.   b) The *boundary* versus *region* dilemma: There are two complementary ways to approach a figure, either as a collection of boundary points or as a collection of interior points, and representations of shape have been based on each of these approaches; for example, boundary representations have been based on the chain code [5] and Fourier descriptors [20], while interior representations have been based on skeletons and medial-axis transformations [2]. Although the two representations are equivalent, in that one may be derived from the other, they each make different information explicit. This leads to trade-offs in stability and efficiency of computations. For example, while the structure of a "neck" at points A and B is explicit in a region-based representation, it is implicit in a boundary-based representation. The computation of a neck is local in a region-based representation, but global (thereby unstable in presence of occlusion and noise) in a boundary-based one. Our scheme makes both kinds of information explicit simultaneously, thereby enjoying much greater stability properties.

as the linear sum of two basis deformations: constant motion and motion proportional to curvature. This will then lead us naturally into the study of a PDE and finally to its application to shape.

Consider the most general deformation of a curve $\mathcal{C}$, namely a deformation of some arbitrary amount along the tangent $\vec{T}$ and some other arbitrary amount along the normal $\vec{N}$ (Fig. 2), $\frac{\partial \mathcal{C}}{\partial t} = a(u,t)\vec{T} + b(u,t)\vec{N}$, where u denotes position along the curve and t is the evolutionary step (time). Without loss of generality, this deformation can be written as a deformation along the normal by some other magnitude [6]. In addition, for a theory of shape the deformation should be restricted to a local function of the geometry of the curve, and should be time invariant. Now, since the local geometry of the curve is completely determined by its curvature function [4], a time-invariant, local deformation is equivalent to a deformation along the normal as a function of curvature

$$\frac{\partial \mathcal{C}}{\partial t} = \beta(\kappa)\vec{N}. \qquad (2)$$

Qualitatively, the behaviour of this deformation is governed by the first two terms in the Taylor expansion of $\beta(\kappa) \approx \beta_0 + \beta_1\kappa$. The first term describes constant motion outwards or inwards along the normal (fig 2ii), and the second term describes a motion along the normal that is proportional to the curvature (fig 2iii). Observe that, for the curvature term, highly curved segments will move faster than slightly curved ones [12].

Conservation Laws We now show that a deformation composed of constant motion and curvature motion satisfies a *viscous conservation law*. In particular, constant motion along the normal satisfies a hyperbolic conservation law for the slope of the boundary $u_t + \beta_0[H(u)]_x = 0$ where u is the slope in an extrinsic cartesian coordinate system (with horizontal axis x), $H(u) = -\sqrt{1 + u^2}$ is the slope-flux [10], and β_0 is the extent of constant motion. When curvature motion is introduced, "viscosity" is added to the system $u_t + \beta_0[H(u)]_x = \beta_1\left[\frac{u_x}{1+u^2}\right]_x$, where β_1 is the extent of curvature motion.

This viscous conservation law is a parabolic equation ($\beta_1 \neq 0$), and contains two terms [19]. The β_0 term, which is hyperbolic and corresponds to the constant motion, is the *wave* part. The β_1 term, which is parabolic and corresponds to the curvature motion, is the *diffusion* part. The diffusion term is quasi-linear, and tends to "dampen" and smooth u, while the wave term is *non-linear* and tends to produce large solutions, steep gradients, and discontinuities. Alternatively,

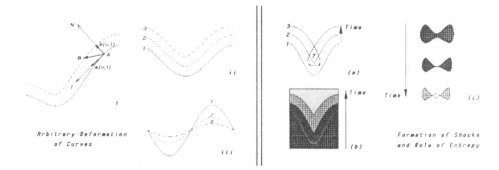

Figure 2: (left) An arbitrary deformation of a curve is captured by two basis deformations: a constant motion (reaction) and a motion proportional to curvature (diffusion). Figures (ii) and (iii) illustrate the constant and curvature motions, respectively. Note that for constant motion all points move with the same speed, so that concave segments become more curved while convex ones become less so. However, under the curvature term, highly curved segments will move faster than slightly curved ones, so that all curved segments become smoother.

Figure 3: (right)The formation of shocks and the role of entropy. Nonlinear processes can transform initially smooth functions to functions with singularities. (a) shows a curve with a negative curvature extremum which, when evolved by constant motion along the normal, leads to a singularity. This evolution can be based entirely on boundary information until the singularity arises. However, at this point the entropy condition is required to further control evolution, so that the curve does not cross over itself and the swallowtail configuration can be properly handled (b). The entropy condition is region-based, and controls how interior information interacts with the boundary. It plays another key role in controlling topological evolution, by globally managing the splitting of a single boundary into two closed boundaries (c).

curvature satisfies the intrinsic evolution equation $\kappa_t + \beta_1\kappa_{ss} + \beta_0\kappa^2 + \beta_1\kappa^3 = 0$, where s is the arclength parameter along the curve and κ is curvature. This equation is a *reaction-diffusion* equation, a common model of chemical and biological phenomena. Observe that for $\beta_1 = 0$, the only effect is that of *reaction*. However, when $\beta_1 \neq 0$ *diffusion* is introduced to the system.

Entropy and Shocks: In order to solve these equtions one must address the question of the space of solutions to these equations. In order to deal with formed singularities the space of measurable and bounded functions (*generalized functions*) is employed [14]. To restrict solutions to physically significant ones and further to constrain them to satisfy conservation across singularities, notion of *entropy* and *jump conditions* were developed [18, 15, 16]. Singularities which satisfy both conditions are called *shocks*. Generalized functions whose only discontinuities are shocks enjoy existence and uniqueness properties as solutions to conservation laws [14, 19, 18, 3]. To relate the concepts of entropy, jump condition, and shocks to the problem of shape representation see [13, 10]. For example, the role of entropy, is one of handling discontinuities by explicitly introducing region-based information into the boundary-based approach whereas shocks support the decomposition of shapes into parts and protrusions, figure 3.

The Reaction-Diffusion Space: Thus far, we have modelled the deformation of a curve as a viscous conservation law and a reaction-diffusion equation. We now view reaction and diffusion as two complementary forces acting on shape, where the relative strength of the hyperbolic wave process to the parabolic diffusive one determines the nature of deformation. Together with the time of evolution, they give rise to a two-dimensional space, the *reaction-diffusion space*, spanned by the ratio β_1/β_0 and time t.

A pure *diffusion* process (no wave $\beta_0 = 0$) is a *quasi-linear* heat process. It is formally equivalent to the coordinates of the parametrized equation of the curve satisfying the heat equation [7]. Thus evolution in time under pure diffusion is tantamount to filtering the coordinates by a Gaussian kernel. Its role therefore is one of *smoothing* the boundary of the shape. In fact, the boundary converges to a circle [8]. Since the heat equation spreads information globally with infinite speed, diffusion is a *global* process. Finally, since diffusion operates soley on the boundary curvature information it is a *boundary process*.

On the other hand, a pure *reaction* process (no diffusion $\beta_1 = 0$) is a *non-linear* hyperbolic

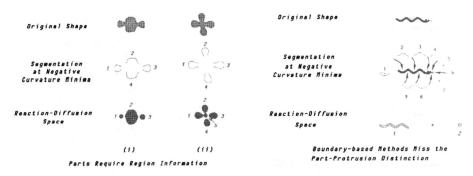

Figure 4: a) Indentification of parts requires region information. Although the objects in this figure clearly have different parts, each is segmented into four pieces based only on the boundary information and segmented at negative curvature minima. The reaction-diffusion space captures the natural difference between them, and also illustrates the non-linear nature of shape descriptions by the transition from "two petals stuck onto a blob" to "four petals composed around a common center". Such families of shapes can also serve as stimuli for psychological and physiological experimentation, and our theory makes both quantitative and qualitative predictions about when transitions like that between i) and ii will occur.   b) Pure boundary-based methods miss the part-protrusion distinction. The segmentation of the snake at negative curvature minimma leads to inappropriate parts. The reaction-diffusion space, however, correctly distinguishes the body of of the snake as one deformed part.

process. It can be shown to create *singularities* from the negative minima in curvature [10]. It is a *local* process in that only a limited portion of the shape affects the evolution of any single point. Finally, it requires no boundary information and is a *region process*.

The pure reaction and pure diffusion processes are extremes along one axis of the reaction-diffusion space; intermediate combinations of reaction and diffusion are compromises on their various features. Traversing along the other axis in the reaction-diffusion space, namely time, the process has the effect of simplifying the shape: diffusion spreads information instantaneously and globally along the boundary, while the reaction process removes information non-linearly and locally through the region. This provides the basic structure for a scale-space for shape, as we show in the next section.

In summary, then, reaction and diffusion contrast and complement each other on issues of linearity vs. nonlinearity, smoothness vs. singularity, global vs. local, and boundary vs. region.

Parts and Protrusions: The reaction-diffusion space's significance is in the segmentation of a shape into pieces. More precisely, a shape is hierarchically analysed into a composition of parts and protrusions, our proposed computational elements of shape. Examples show how certain parts protrude into one another, thus naturally giving rise to necks, the neglected aspect of shape.

But these notions are formal ones within our framework, and differ somewhat from standard usage. To clarify, recall that some traditional approaches to shape representation argue for a decomposition into "parts" or components by segmentation at the negative minima of curvature of the boundary [9, 1]. This appears reasonable because, when two distinct objects interpenetrate, the intersections are almost always transversal, projecting to negative minima in curvature. However, figure 4 shows that boundary curvature is not sufficient to determine "parts"; region information must also be taken into account. Moreover, although the statement "parts are bounded by negative curvature minima" is true, the converse does not necessarily hold. In fact, deformations of objects can give rise to negative curvature extrema, as is illustrated in figure 4. Observe that, when strictly applied the decomposition at negative curvature minima leads to counterintuitive results.

Therefore, in addition to the negative curvature minima criterion, a further condition is needed to recover parts. The intuition must be captured that parts are bounded by *pairs* of negative curvature minima that are close in the distance through the region (and not necessarily along the boundary) forming a "neck". Such a partitioning of objects along necks makes sense because it is easiest, physically, to break objects at their narrowest regions, namely the necks. Furthermore, for objects with moving parts, the joints are often narrower than the components, and joints map onto necks. For further support of this argument see [10].

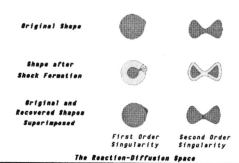

Figure 5: The Reaction-Diffusion space. The deformed disc forms a first order singularity, while the peanut-shaped object splits into two parts by forming a second order singularity. Thus, the order of the singularity differentiates between parts and protrusions. Observe the neck that develops between the two parts; this is what distinguishes the peanut shape from overlapping circles of figure 1.

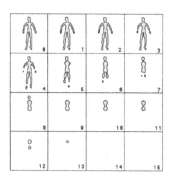

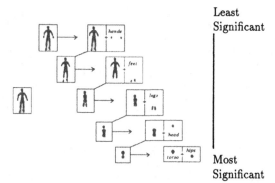

Least Significant

Most Significant

Figure 5: a) The evolution of shocks leads to parts and protrusions. This figure shows the development of an image of a doll (National Research Council of Canada Laser Range Image Library CNRC9077 Cat No 422; 128X128). The contour shown in box N corresponds to some time step. Observe that the "feet" partition from the "legs" (via second-order shocks) between frames 3 and 4, and the "hands" from the "arms" between frames 2 and 3. Following these second-order shocks, first-order shocks develop as the "arms" are "absorbed" into the chest. Running this process in the other direction would illustrate how the arms "protrude" from the chest.   b) The Hierarchical decomposition of a doll into parts. Selected frames were organized into a hierarchy according to the principle that significance of part is directly proportional to its survival duration.

How does one then interpret the unpaired curvature extrema? Consider a circular ring of flexible material which is deformed as if someone had attempted to push a finger through it, figure 5 (left). This deformation creates a single curvature extremum. It is plausible, then, to associate the unpaired curvature extrema with deformations.

The reaction-diffusion space provides the framework for making these arguments precise. The key is in the formation of shocks which are the singularities in the slope of the outline. Note that, for the deformed circle (Fig 5) a single shock of the first order develops. For the peanut, figure 5 (right), however, a topological split occurs, and second order shocks are formed. *Therefore, the distinction between parts and protrusions is in the order of formed singularities in the reaction-diffusion space: first-order singularities signal protrusions while second-order singularities signal parts.* To completely recover either part or protrusion, the evolved shape is run backwards through the reaction-diffusion space so it may be compared with the original shape, Fig 5. For a rigorous treatment see [10].

Discussion: In summary, we have presented a framework for a theory of shape based on the geometry of curves and their interiors. This framework resolves some of the classical dilemmas of shape perception, and results from viewing shape as a tension between reaction and diffusion in the context of a conservation law. It defines a hierarchy of parts and protrusions as singularity types (shocks) in a reaction-diffusion space, and elucidates a mechanism for decomposing shapes into them reliably and consistently. Furthermore, a notion of scale naturally arises within this mechanism [11]. Finally, a whole family of qualitative predictions are opened up by the reaction-diffusion approach to shape, e.g. regarding the similarity between shapes expressed as a metric over the reaction-diffusion space [10].

References

[1] I. Biederman. Recognition by components. *Psych. Review*, 94:115–147, 1987.

[2] H. Blum. Biological shape and visual science. *J. Theor. Biol.*, 38:205–287, 1973.

[3] M. Crandall and P. Lions. Viscosity solutions of Hamilton-Jacobi equations. *Trans. Amer. Math. Soc.*, 277:1–42, 1983.

[4] M. P. do Carmo. *Differential Geometry of Curves and Surfaces*. Prentice-Hall, New Jersey, 1976.

[5] H. Freeman. Computer processing of line drawing images. *Computer Surveys*, 6(1):57–98, March 1974.

[6] M. Gage. On an area-preserving evolution equation for plane curves. *Contemp. Math.*, 51:51–62, 1986.

[7] M. Gage and R. S. Hamilton. The heat equation shrinking convex plane curves. *J. Differential Geometry*, 23:69–96, 1986.

[8] M. A. Grayson. The heat equation shrinks embedded plane curves to round points. *J. Differential Geometry*, 26:285–314, 1987.

[9] D. D. Hoffman and W. A. Richards. Parts of recognition. *Cognition*, 18:65–96, 1985.

[10] B. B. Kimia. Conservation laws and a theory of shape. Ph.D. dissertation, McGill Centre for Intelligent Machines, McGill University, Montreal, Canada, 1989.

[11] B. B. Kimia, A. Tannebaum, and S. W. Zucker. An entropy-based scale space for shape. In Preparation, 1990.

[12] B. B. Kimia, A. Tannebaum, and S. W. Zucker. On the evolution of curves via a function of curvature, I: The classical case. *JMAA*, Submitted, 1990.

[13] B. B. Kimia, A. Tannebaum, and S. W. Zucker. Parts and protrusions: The computational elements of shape. In Preparation, 1990.

[14] P. Lax. *Hyperbolic Systems of Conservation Laws and the Mathematical Theory of Shock Waves*. SIAM Regional Conference series in Applied Mathematics, Philadelphia, 1973.

[15] P. D. Lax. Hyperbolic systems of conservation laws II. *Comm. Pure Appl. Math.*, 10:537–566, 1957.

[16] P. D. Lax. *Shock Waves and Entropy*, pages 603–634. Academic Press, New York, 1971.

[17] M. Leyton. A process grammar for shape. *Artificial Intelligence*, 34:213–247, 1988.

[18] O. Oleinik. Discontinuous solutions of nonlinear differential equations. *Amer. Math. Soc. Transl. Ser. 2*, 26:95–172, 1957.

[19] J. Smoller. *Shock Waves and Reaction-Diffusion Equations*. Springer-Verlag, New York, 1983.

[20] C. T. Zahn and R. Z. Roskies. Fourier descriptors for plane closed curves. *IEEE Transactions on Computers*, C-21:269–281, 1972.

[21] S. W. Zucker, A. Dobbins, and L. Iverson. Two stages of curve detection suggest two styles of visual computation. *Neural Computation*, 1:68–81, 1989.

STABILIZED SOLUTION FOR 3-D MODEL PARAMETERS

David G. Lowe*

Computer Science Dept., Univ. of British Columbia
Vancouver, B.C., Canada V6T 1W5

One important component of model-based vision is the ability to solve for the values of all viewpoint and model parameters that will best fit a model to some matching image features. This is important because it allows some tentative initial matches to constrain the locations of other features, and thereby generate new matches that can be used to verify or reject the initial interpretation. The reliability of this process can be greatly improved by taking account of all available quantitative information to constrain the unknown parameters during the matching process. In addition, parameter determination is necessary for identifying object sub-categories, for interpreting images of articulated or flexible objects, and for robotic interaction with the objects.

Our solution for unknown viewpoint and model parameters is based on Newton's method of linearization and iteration to perform the non-linear minimization. This is augmented by a stabilization method that incorporates a prior model of the range of uncertainty in each parameter and estimates of the standard deviation of each image measurement. This allows useful approximate solutions to be obtained for problems that would otherwise be underdetermined or ill-conditioned. In addition, the Levenberg-Marquardt method is used to always force convergence of the solution to a local minimum. These techniques have all been implemented and tested as part of a system for model-based motion tracking, and have been found to be reliable and efficient.

Previous approaches

Attempts to solve for viewpoint and model parameters date back to the work of Roberts [14], but his solution methods were specialized to certain classes of objects such as rectangular blocks. In 1980, the author [8] presented a general technique for solving for viewpoint and model parameters using Newton's method for nonlinear least-squares minimization. Since that time the method has been used successfully in a number of applications, and it also provides the starting point for the work presented in this paper. The application of the method to robust model-based recognition has been described by Lowe [9, 10], McIvor [12], and Worrall, Baker & Sullivan [16]. Verghese & Dyer [15] have used this method for model-based motion tracking. Ishii et al. [5] describe the application of this work to the problem of tracking the orientation and location of a robot hand from a single view of LED targets mounted on the wrist. Their paper provides a detailed analysis that shows good accuracy and stability. Recently, Liu et al. [7] and Kumar [6] have examined alternative iterative approaches to solving for the viewpoint parameters by separating the solution for rotations from those for translations. However, Kumar shows that this approach leads to worse parameter estimates in the presence of noisy data, so he adopts a similar simultaneous minimization as is used in the work above.

Much work has been published on characterizing the minimum amount of data needed to solve for the six viewpoint parameters (assuming a rigid object) and on solving for each of

* Contact the author (lowe@cs.ubc.ca) for a full-length version of this paper. The author is supported as a Scholar of the Canadian Institute for Advanced Research.

the multiple solutions that can occur when only this minimum data is available. Fischler and Bolles [2] show that up to four solutions will be present for the problem of matching 3 model points to 3 image points, and they give a procedure for identifying each of these solutions. A solution for the corresponding 4-point problem, which can also have multiple solutions under some circumstances, is given by Horaud *et al.* [3]. Huttenlocher and Ullman [4] show that the 3-point problem has a simple solution for orthographic projection, which is a sufficiently close approximation to perspective projection for some applications. In the most valuable technique for many practical applications, Dhome *et al.* [1] give a method for determining all solutions to the problem of matching 3 model lines to 3 image lines. This could be particularly useful for generating starting positions for the iterative techniques used in this paper when there are multiple solutions.

While this work on determining all possible exact solutions will no doubt be important for some vision applications, it is probably not the best approach for practical parameter determination in general model-based vision. One problem with these methods is that they do not address the issue of ill-conditioning. Even if a problem has only one analytic solution, it will often be sufficiently ill-conditioned in practice to have a substantial number and range of solutions. Secondly, all these methods deal with specific properties of the six viewpoint parameters, and there is little likelihood that they can be extended to deal with an arbitrary number of internal model parameters. In addition, these methods fail to address the problem of what to do when the solution is underconstrained. The stabilization methods described in this paper allow an approximate solution to be obtained even when a problem is underconstrained, as will often be the case when models contain many parameters. Possibly the most convincing reason for believing that it is not necessary to determine all possible solutions is the fact that human vision apparently also fails to do so. The well-known Necker cube illusion illustrates that human vision easily falls into a local minimum in the determination of viewpoint parameters, and seems unable to consider multiple solutions at one time.

Stabilizing the solution

As long as there are significantly more constraints on the solution than unknowns, Newton's method will usually converge to a stable solution from a wide range of starting positions. However, in both recognition and motion tracking problems, it is often desirable to begin with only a few of the most reliable matches available and to use these to narrow the range of viewpoints for later matches. Even when there are more matches than free parameters, it is often the case that some of the matches are parallel or have other relationships which lead to an ill-conditioned solution. These problems are further exacerbated by having models with many internal parameters.

All of these problems can be solved by introducing prior constraints on the desired solution that are used in the absence of further data. In many situations, the default solution will simply be to solve for zero corrections to the current parameter estimates. However, for certain motion tracking problems, it is possible to predict specific final parameter estimates by extrapolating from velocity and acceleration measurements, which in turn imply non-zero preferences for parameter values in later iterations of non-linear convergence. The general form of this process for motion tracking would be equivalent to the use of the extended Kalman filter [17], but the predictive component does not play a role in recognition applications.

Any of these prior constraints on the solution can be incorporated by simply adding rows to the linear system constructed on each iteration of Newton's method. Let $\mathbf{J}$ be the Jacobian matrix of partial derivatives with respect to each model or viewpoint parameter, and $\mathbf{e}$ be the vector of error measurements from the current solution to corresponding image features (see the full-length version of this paper for efficient methods for calculating these). Then Newton's method solves the following matrix equation on each iteration for the vector of parameter corrections, $\mathbf{x}$:

$$\mathbf{J}\mathbf{x} = \mathbf{e}$$

The solution can be stabilized by adding rows to this equation specifying prior desired parameter values, $\mathbf{d}$, in the absence of constraints from the image. If we assume that the errors, $\mathbf{e}$, have unit standard deviation, then the prior estimates of the parameter values should be weighted by a diagonal matrix $\mathbf{W}$ in which each weight is inversely proportional to the standard deviation, σ_i, for parameter i:

$$W_{ii} = \frac{1}{\sigma_i}$$

Incorporating these new constraints, we wish to minimize the following stabilized system:

$$\begin{bmatrix} \mathbf{J} \\ \mathbf{W} \end{bmatrix} \mathbf{x} = \begin{bmatrix} \mathbf{e} \\ \mathbf{Wd} \end{bmatrix}$$

We will minimize this system by solving the corresponding normal equations (see full-length paper for discussion of the numerical stability of the normal equations):

$$\begin{bmatrix} \mathbf{J}^T & \mathbf{W}^T \end{bmatrix} \begin{bmatrix} \mathbf{J} \\ \mathbf{W} \end{bmatrix} \mathbf{x} = \begin{bmatrix} \mathbf{J}^T & \mathbf{W}^T \end{bmatrix} \begin{bmatrix} \mathbf{e} \\ \mathbf{Wd} \end{bmatrix}$$

which multiplies out to

$$\left(\mathbf{J}^T \mathbf{J} + \mathbf{W}^T \mathbf{W} \right) \mathbf{x} = \mathbf{J}^T \mathbf{e} + \mathbf{W}^T \mathbf{Wd}$$

Since $\mathbf{W}$ is a diagonal matrix, $\mathbf{W}^T \mathbf{W}$ is also diagonal but with each element on the diagonal squared. This means that the computational cost of the stabilization is trivial, as we can first form $\mathbf{J}^T \mathbf{J}$ and then simply add small constants to the diagonal that are the inverse of the square of the standard deviation of each parameter. If $\mathbf{d}$ is non-zero, then we add the same constants multiplied by $\mathbf{d}$ to the right hand side. If there are fewer rows in the original system than parameters, we can simply add enough zero rows to form a square system and add the constants to the diagonals to stabilize it.

Forcing convergence

Even after incorporating this stabilization based on a prior model, it is possible that the system will fail to converge to a minimum due to the fact that this is a linear approximation of a non-linear system. We can force convergence by adding a scalar parameter λ that can be used to increase the weight of stabilization whenever divergence occurs. The new form of this system is

$$\begin{bmatrix} \mathbf{J} \\ \lambda \mathbf{W} \end{bmatrix} \mathbf{x} = \begin{bmatrix} \mathbf{e} \\ \lambda \mathbf{Wd} \end{bmatrix}$$

This system minimizes

$$\|\mathbf{Jx} - \mathbf{e}\|^2 + \lambda^2 \|\mathbf{W}(\mathbf{x} - \mathbf{d})\|^2$$

This as an example of regularization using a Tikhonov stabilizing functional, as has been applied to many areas of low-level vision (Poggio et al. [13]). In this case, the parameter λ controls the trade-off between approximating the new data, $\|\mathbf{Jx} - \mathbf{e}\|^2$, and minimizing the distance of the solution from its original starting position, $\mathbf{d}$, prior to non-linear iteration, $\lambda^2 \|\mathbf{W}(\mathbf{x} - \mathbf{d})\|^2$.

The use of this parameter λ to force iterative convergence for a non-linear system was first studied by Levenberg and later reduced to a specific numerical procedure by Marquardt [11]. Marquardt did not assume any prior knowledge of the weighting matrix $\mathbf{W}$, but instead estimated each of its elements from the euclidean norm of the corresponding column of $\mathbf{J}^T \mathbf{J}$. In our case, the availablity of $\mathbf{W}$ allows the algorithm to perform much better when a column of $\mathbf{J}^T \mathbf{J}$ is near zero. When the solution fails to improve on any iteration, increasing the value of λ (by factors of 10, as suggested by Marquardt) will essentially freeze the parameters having the lowest standard deviations and therefore solve first for those with higher standard deviations.

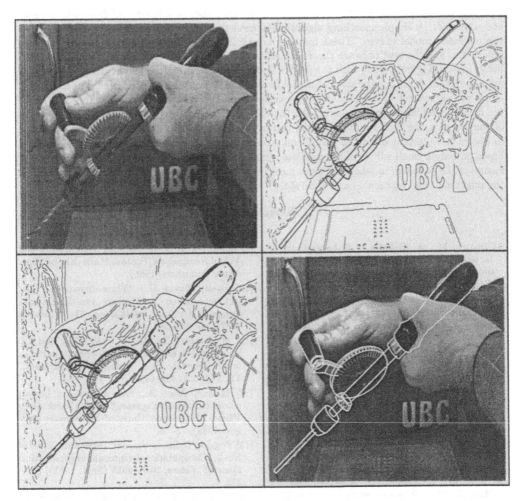

Figures 1–4: Parameter solving during a motion tracking sequence.

Results of implementation

One initial application of these methods has been to the problem of motion tracking. A Datacube image processor is used to implement Marr-Hildreth edge detection in real time on 512 by 485 pixel images. The image containing these edge points is transferred to a Sun 3/260, where the edges are linked into lists on the basis of local connectivity. A fairly simple matching technique is used to identify the image edges that are closest to the current projected contours of a 3-D model. The few best initial matches are used to perform one iteration of the viewpoint solution, then further matches are generated from the new viewpoint estimate. Up to 5 iterations of this procedure are performed. For simple models with straight edges, all of these steps can be performed in less than 1 second, resulting a system that can perform robust but rather slow real-time motion tracking. Full details of the components of this system other than parameter solving will be published in a separate paper.

Figures 1–4 show the operation of the system for one frame of motion tracking. However, due to the complexity of the model, this version requires about 6 seconds of processing per frame and does not operate in real time. Figure 1 shows an image of a hand drill from which edges are extracted. A simple matching algorithm is used to identify image edges that are close to the projected model curves. These matches are ranked according to their length and average separation, and the best ones are chosen for minimization. The selected matches are shown with heavy lines in Figure 2 along with the perpendicular errors between model and image curves that are minimized. After one iteration of model fitting, the new model position is shown in Figure 3 along with a new set of image matches generated from this position. Note that the rotation of the handle is a free parameter along with the viewpoint parameters. After this second iteration of convergence, the final results of model fitting are shown superimposed on the original image in Figure 4. Note that due to occlusion and errors in low-level edge detection, this final result is based on only a small subset of the predicted image edges. However, due to the overconstrained nature of the problem, in which far more measurements are available than unknown parameters, the final result can be reliable and accurate.

References

[1] Dhome, M., M. Richetin, J.T. Lapresté, and G. Rives, "Determination of the attitude of 3-D objects from a single perspective view," *IEEE PAMI*, **11**, 12 (1989), 1265–78.

[2] Fischler, Martin A. and Robert C. Bolles, "Random sample consensus: A paradigm for model fitting with applications to image analysis and automated cartography," *Communications of the ACM*, **24**, 6 (1981), 381-395.

[3] Horaud, R., B. Conio, O. Leboulleux, and B. Lacolle, "An analytic solution for the perspective 4-point problem," *Proc. Conf. Computer Vision and Pattern Recognition*, San Diego (June 1989), 500–507.

[4] Huttenlocher, Daniel P., and Shimon Ullman, "Object recognition using alignment," *Proc. First Int. Conf. on Computer Vision*, London, England (June 1987), 102–111.

[5] Ishii, M., S. Sakane, M. Kakikura and Y. Mikami, "A 3-D sensor system for teaching robot paths and environments," *The International Journal of Robotics Research*, **6**, 2 (1987), pp. 45–59.

[6] Kumar, Rakesh, "Determination of camera location and orientation," *Proc. DARPA Image Understanding Workshop*, Palo Alto, Calif. (1989), 870–879.

[7] Liu, Y., T.S. Huang and O.D. Faugeras, "Determination of camera location from 2-D to 3-D line and point correspondences," *IEEE PAMI*, **12**, 1 (1990), 28–37.

[8] Lowe, David G., "Solving for the parameters of object models from image descriptions," *Proc. ARPA Image Understanding Workshop* (College Park, MD, April 1980), 121–127.

[9] Lowe, David G., *Perceptual Organization and Visual Recognition* (Boston, Mass: Kluwer Academic Publishers, 1985).

[10] Lowe, David G., "Three-dimensional object recognition from single two-dimensional images," *Artificial Intelligence*, **31**, 3 (March 1987), pp. 355-395.

[11] Marquardt, Donald W., "An algorithm for least-squares estimation of nonlinear parameters," *Journal. Soc. Indust. Applied Math.*, **11**, 2 (1963), 431–441.

[12] McIvor, Alan M., "An analysis of Lowe's model-based vision system," *Proc. Fourth Alvey Vision Conference*, Univ. of Manchester (August 1988), 73–78.

[13] Poggio, Tomaso, Vincent Torre and Christof Koch, "Computational vision and regularization theory," *Nature*, **317**, 6035 (Sept. 1985), 314–319.

[14] Roberts, L.G., "Machine perception of three-dimensional solids," in *Optical and Electro-optical Information Processing*, eds. J. Tippet et al. (Cambridge, Mass.: MIT Press, 1965), 159-197.

[15] Verghese, G. and C.R. Dyer, "Real-time model-based tracking of three-dimensional objects," *Univ. of Wisconsin, Computer Sciences TR 806* (Nov. 1988).

[16] Worrall, A.D., K.D. Baker and G.D. Sullivan, "Model based perspective inversion," *Image and Vision Computing*, **7**, 1 (1989), 17–23.

[17] Wu, J.J., R.E. Rink, T.M. Caelli, and V.G. Gourishankar, "Recovery of the 3-D location and motion of a rigid object through camera image," *Inter. Journal of Computer Vision*, **3** (1988), 373–394.

a way that it maximizes an energy function or measure of fit. The apparent dynamic interaction between the template and the image causes the template to be mapped on to the desired object. We have found that the concept of deformable templates can be applied to create a dynamic and robust algorithm for the localization of medical structures from digitized Computer Tomography images.

We have utilized an extension of the theoretical template method in an application to localize automatically the vertebral trabecular bone and then to measure the attenuation coefficients of this structure as well as the Cann-Genant calibration samples contained within CT images. The measurements are used to determine the trabecular bone density and fat percentage. Bone mineral content and fat percentage are used clinically to diagnose and monitor patients suffering from bone loss and abnormal marrow fat content, as occur in osteoporosis and Gaucher disease. Two factors which are crucial to the successful long term maintenance of these patients are accuracy and consistency in the measurement techniques. Presently, time-consuming and tedious manual methods are used to obtain the trabecular bone and calibration samples' attenuations. In order to reduce radiologists' time commitments and to enhance patient maintenance, we have developed a PACS based image analysis program which automatically performs accurate and reproducible measurements upon dual energy CT images.

In section 2 we describe deformable template models in general. In section 3 we show how to construct the deformable template model for this specific example and describe its dynamics. Section 4 describes the details of the implementation and section 5 gives the results on the database of 552 images (69 patients with 8 images per patient).

This work is described in more detail in Lipson et al (1989) with many illustrations of the deformable template in action.

2 Deformable Templates

Deformable templates (DTs) have some similarities with elastic deformable models (Burr 1981a, 1981b, Durbin and Willshaw 1987, Durbin, Szeliski and Yuille 1988), and to snakes (Kass, Witkin and Terzopolous 1987, Terzopolous, Witkin and Kass 1987) (see also applications to medical imaging, principally to construct 3D objects, by Ayache et al. 1989, Chen et al. 1989, Vandermeulen et al. 1989). There are, however, three important differences: (i) DTs embody *a priori* knowledge about the feature being detected, (ii) the structural forces on the DTs are global rather than local (preventing many local minima), and (iii) DTs can be easily implemented (since they involve only a small number of parameters) and give a compact description of the feature.

Grenander and his collaborators (Chow et al. 1989, Knoerr 1989) have represented the boundaries of two-dimensional natural objects in terms of a Markov model. This is somewhat similar to the snake approach, except they use *a priori* knowledge about the features, found by statistical analysis of the shapes of the objects, rather than assuming the structures are thin plates and membranes. Staib and Duncan (1989) also recommend this approach.

3 The Ellipsoidal Template and the Energy Function

Axial cross sectional CT images of the spine yield approximately elliptical vertebral contours. Thus, we have utilized an isomorphic ellipsoidal template as a feature detector. This ellipsoidal template is mapped on to the vertebra and is constrained within the boundaries of the cortical bone. The region anterior to the spinal cord within the elliptical template accurately describes the vertebral trabecular bone.

An ellipse can be represented by five parameters. These parameters include the major and minor axes (a and b), the x and y coordinates of the center point ($\vec{c} = (c_x, c_y)$), and the orientation angle (θ). The equation for an ellipse with these parameters is

Deformable Templates for Feature Extraction from Medical Images

P.Lipson (A.I. Lab, M.I.T.), A.L.Yuille (D.A.S. Harvard), D.O'Keeffe (M.G.H.)
J.Cavanaugh (M.G.H.), J.Taaffe (M.G.H.), D.Rosenthal (M.G.H.)

Abstract

We propose a method for detecting and describing features in medical images using deformable templates, for the purpose of diagnostic analysis of these features. The feature of interest can described by a parameterized template. An energy function is defined which links edges in the image intensity to corresponding properties of the template. The template then interacts dynamically with the image content, by evaluating the energy function and accordingly altering its parameter values. A gradient maximization technique is used to optimize the placement and shape of the deformable template to fit the desired anatomical feature. The final parameter values can be used as descriptors for the feature. Measurements of intensity values within a region of the template can be used as inputs to a medical diagnostic system. We have developed a Picture Archive and Communication System (PACS) based image analysis program which employs the technique of deformable templates to localize features in dual energy CT images. Measurements can then be automatically made which can be used for maintenance of patients suffering from bone loss and abnormal marrow fat content. This system has been successfully tested on 552 (69 × 8) images and is currently in use at Massachusetts General Hospital, Boston, MA. Statistical comparisons between the system and previously used manual techniques show that their performances are practically equivalent and that the system has several advantages over the human operator, for example, consistency, accuracy and cost.

1 Introduction

Feature extraction of structures from medical images is a significant and complex problem. Automated localization of anatomical parts may be useful to clinicians by relieving labor intensive processes and increasing the accuracy, consistency, and reproducibility of image interpretations. However, such automated localization processes are often hindered by the irregular attributes of the medical structures and the imperfect scanning procedures.

It is desirable to have representations for the anatomical parts and a recognition process that are able to adapt to irregularities in both the object structure and in the image content (i.e. noise). Thus, both the feature representation and the localization procedure must be sufficiently general in order to accept variations of the same object and sufficiently specific in order to be able to differentiate the desired object from all others. One idea, developed by Yuille, Cohen, and Hallinan (1989), is to represent the desired object by a deformable template, an isomorphic parametric model that embodies the ideal characteristics of, and the *a priori* knowledge about, the object. The localization process involves altering the parameters of the template in such

$$\{\frac{(\vec{x} - \vec{c}) \cdot \vec{n}}{a}\}^2 + \{\frac{(\vec{x} - \vec{c}) \cdot \vec{n}^*}{b}\}^2 = 1, \quad where \quad \vec{n} = (\cos\theta, \sin\theta) \quad \vec{n}^* = (-\sin\theta, \cos\theta) \qquad (1)$$

The value of the energy function that is maximized during the feature detection is a function of the template's five parameters.

Let $I(x, y)$ be the intensity at the point (x, y) in an (filtered, edge extracted) image. The *Total Energy* TE_p with respect to the perimeter of the ellipse is

$$TE_p(\vec{c}, \theta) = \int_0^{2\pi} I(x, y)\{a^2 \sin^2 \psi + b^2 \cos^2 \psi\}^{1/2} d\psi \qquad (2)$$

where $x = c_x + a\cos\psi\cos\theta - b\sin\psi\sin\theta$ and $y = c_y + a\cos\psi\sin\theta + b\sin\psi\cos\theta$.

Each parameter is updated according to its contribution to the change in total energy function with respect to the change in time. For the five parameters the update rules are

$$\frac{da}{dt} = KE_a\frac{\partial TE_p}{\partial a}, \quad \frac{db}{dt} = KE_b\frac{\partial TE_p}{\partial b}, \quad \frac{dc_x}{dt} = KE_{c_x}\frac{\partial TE_p}{\partial c_x}, \quad \frac{dc_y}{dt} = KE_{c_y}\frac{\partial TE_p}{\partial c_y}, \quad \frac{d\theta}{dt} = KE_\theta\frac{\partial TE_p}{\partial \theta}.$$
$$(3)$$

Lipson et al (1989)A gives explicit forms for the derivatives of TE_p with respect to the parameters a, b, c_x, c_y, θ.

The values for the constants KE are dependent upon the size and steepness of the gaussian field. Typical initial values of the constants, for an image processed with a large steep gaussian field (for long range interactions), are $(KE_a, KE_b, KE_{c_x}, KE_{c_y}, KE_\theta) = (1/7290, 1/7290, 1/7290, 1/7290, 1/36000000)$. Typical initial values of the constants for an image processed with a small shallow gaussian field (for precise localization) are $(KE_a, KE_b, KE_{c_x}, KE_{c_y}, KE_\theta) = (1/20250, 1/20250, 1/20250, 1/20250, 1/26000000)$.

4 The Algorithm

The automatic analysis program must localize and measure five areas of interest in each image. The first area is the vertebral trabecular bone. The other four areas are contained in the Cann-Genant calibration phantom. The phantom is easily recognized by psearching the image for the two high attenuation steel rods placed at either end of the phantom as markers. The vertebral trabecular bone is localized using the method of deformable templates.

The algorithm to localize the vertebral trabecular bone has several stages. First, the vertebra must be detected. A preprocessing stage estimates the size and location of the vertebra by analyzing the image edge data. These estimates are used to determine initial values for the parameters of the ellipsoidal template. The ellipsoidal template is then applied to the binary edge data, extracted using a sobel operator, which has been convolved with a Gaussian filter, with large standard deviation, to give an edge field. This field is largest at the original edge data and acts to attract the template towards these edges. The template, within the edge field, is deformed in such a way as to maximize the total energy or goodness of fit over the perimeter of the ellipse. Deformation takes place over 40 iterations. The mapping of the template to the vertebral contour is further refined, for better localization, by deforming the ellipse on the same binary image which has been convolved with a smaller shallower Gaussian field. This second deformation takes place over 15 iterations. Once the template is fitted to the vertebral contour, it is automatically scaled down to avoid overlap with any edges of the high intensity cortical outline. A measure of the trabecular bone is invalid if made in a region containing both trabecular and cortical bone. The final stage transfers the template from the processed image to the original CT image. The average attenuation coefficient of the region anterior to the spinal cord within the template is measured. This region, which represents the vertebral trabecular bone, is marked in black on the original image for a later visual quality control inspection.

5 The Results

Application of the technique to 69 patients (8 images per patient) led to correct localization in all but two cases (see later this section). It yielded attenuation coefficient measurements (in hounsfield units) differing by less than 1 percent from manually derived values.
$:(ENERGY(kVP), SLICE, CORRELATION) = (80, A, 0.970), (140, A, 0.968), (80, B, 0.988),$

$(140, B, 0.988), (80, C, 0.984), (140, C, 0.986), (80, D, 0.986), (140, D, 0.990).$

The calculated values for individual bone density in mg/c.c. show a correlation of .989 for single energy bone mass (80kVp), .990 for dual energy bone mass(140kVp), and a .980 for the 2 line linear fit between the 80kVp and 140kVp. Values for the vertebral fat percentage show a correlation of 0.868. Concentration based on fit to 80kVp data is 0.989. Concentration based on fit to 140kVp data is 0.990. Concentration based on linear fit between 80kVp and 140kVp is 0.980. Fat percentage is 0.868.

Plots of the automatic vs. manually derived values, in addition to analysis of variance and parameter estimation (slope and intercept of plots, mean of the difference between the two sets of values), are given in Lipson et al (1989).

There were two clear sets of cases where the algorithm was unable to correctly localize the vertebral trabecular bone. The first case occurs when the patient shows signs of a calcified aorta. The calcified aorta appears as a high intensity closed ring anterior to the vertebra. The second case occurs when the part of the spine, connected to a vertebra from a higher level, appears as a disconnected, large, high intensity cluster. In both cases, the ellipsoidal template is attracted to the anomalous features. Clearly, the current template is an insufficient model for feature detection in these instances.

The model could be extended to contain smaller ellipses (not greater than two cm on either axis) anterior and posterior to the original template. These two new additions would be constrained to have the same orientation as the original ellipse. Additional parameters include the distance from each of the two new templates to the original template.

The automated analysis program is currently being used by the Department of Radiology at the Massachusetts General Hospital, Boston, MA.. It was incorporated into a Picture Archive and Communications System (PACS), which has been patented with a trade name uRSTAR (O'Keeffe et al. 1989).

6 Conclusion

In conclusion, deformable templates and energy functions can be easily adapted to the irregularities both of the imaging process and the geometric variabilities of anatomical features in medical images. Deformable templates have proved, in a sample of over 552 images, to be a sufficient method for localization of the vertebral trabecular bone in order to measure its average attenuation coefficient. For the most part, it seems that the template representation and dynamic interaction process can be easily adapted to other medical structures.

The detailed relevance of this system to medical imaging and diagnostics is reported elsewhere (Lipson et al. 1990). This describes how statistical comparisons between the system and previously used manual techniques show that their performances are practically equivalent. Additionally, the advantages of the automated technqiue over manual techniques, including consistency, accuracy, and cost, are described.

Acknowledgements

A.L.Y. would like to thank the Brown, Harvard and M.I.T. Center for Intelligent Control Systems for an United States Army Research Office grant number DAAL03-86-C-0171.

References

Ayache, N, Boissonnat, J.D., Brunet, E., Cohen, L., Chieze, J.P., Geiger, B., Monga, O., Rocchisani, J.M. and Sander, P. "Building Highly Structured Volume Representations in 3D Medical Images". *Comp. Ass. Rad., Proc. Int. Symp.*. Berlin .pp 765-772. 1989.

Burr, D.J. "A Dynamic Model for Image Registration". *C.G.I.P.* 15, pp 102-112. 1981.

Burr, D.J. "Elastic Matching of Line Drawings". *IEEE Trans. P.A.M.I.* PAMI-3, No. 6, pp 708-713. 1981.

Chen, S-Y, Lin, W-C, Liang, C-C and Chen, C-T. "Improvement on Dynamic Elastic Interpolation Technique for Reconstructing 3-D Objects from Serial Cross Sections". *Comp. Ass. Rad.,, Proc. Int. Symp.* Berlin. pp 702-706. 1989.

Chow, Y., Grenander, U. and Keenan, D.M. "Hands: A Pattern Theoretic Study of Biological Shape". Research Mongraph. Brown University, Providence, R.I. 1989.

Durbin, R and Willshaw, D.J. "An Analogue Approach to the Travelling Salesman Problem using an Elastic Net Method". *Nature.* 386. pp 689-691. 1987.

Durbin, R., Szeliski, R. and Yuille, A.L. "The Elastic Net and the Travelling Salesman Problem".*Neural Computation.* In press. 1989.

Kass, M., Witkin, A. and Terzopoulos, D. "Snakes: Active Contour Models". *Proc. First Int. Conf. Comp. Vis.* London. June 1987.

Knoerr, A. "Global Models of Natural Boundaries: Theory and Applications". *Pattern Analysis Tech. Report No. 148.* Brown University, Providence, R.I. 1989.

Lipson, P., Yuille, A.L., O'Keeffe, D., Cavanaugh, J., Taaffe, J. and Rosenthal, D. Harvard Robotics Laboratory Technical Report. No. 89-14. 1989.

Lipson, P., Yuille, A.L., O'Keeffe, D., Cavanaugh, J., Taaffe, J. and Rosenthal, D. "Automated Bone Density Calculation using a PACS Workstation Based Image Processing Technique of Deformable Templates". In preparation. 1990.

O'Keeffe, D., Taaffe, J. and Rosenthal, D. "The Application of uRSTAR, a Personal Computer-Based PACS to Quantitative CT : Early Experience", Department of Radiology, Massachusetts General Hospital, Boston, MA., 1989.

Pentland, A. "Recognition by Parts". *Proc. First Int. Conf. Comp. Vis.* London. June 1987.

Staib, L.H. and Duncan, J.S. "Parametrically Deformable Contour Models". *Proc. Comp. Vis. 89 and Patt. Recog.* San Diego. 1989.

Terzopoulos, D., Witkin, A., and Kass, M. "Symmetry-seeking Models for 3D Object Recognition". *Proc. First Int. Conf. Comp. Vis.* London. June 1987.

Vandermeuelen, D., Suetens, P., Gybels, J., Marchal, G. and Oosterlinck, A. "Delineation of Neuroanatomical Objects using Deformable Models". *Comp. Ass. Rad.: Proc. Int. Sym.* Berlin. pp 645-650. 1989.

Yuille, A.L., Cohen, D.S, and Hallinan, P.W. " Feature Extraction from Faces using Deformable Templates". *Proc. Comp. Vis. 89 Patt. Rec..* San Diego. 1989.

Charting Surface Structure

Peter T. Sander
INRIA — Rocquencourt
78153 Le Chesnay Cedex, France
and
INRIA — Sophia-Antipolis
06565 Valbonne Cedex, France

Steven W. Zucker*
McRCIM
McGill University
3480 University St.
Montréal QC, H3A 2A7, Canada

Abstract

Computing surface curvature would seem to be a simple application of differential geometry, but problems arise due to noise and the quantized nature of digital images. We present a method for determining principal curvatures and directions of surfaces estimated from three-dimensional images. We use smoothness constraints to *connect* different surface points and by then comparing information over local neighbourhoods we iteratively update the information at each point to ensure that this information is consistent over the estimated surface.

1 Introduction

Differential geometry supplies the tools necessary for the computation of surface tangent and curvature. However, applying these tools to the determination of the structure of surfaces in digital three-dimensional images is problematic since surfaces are only implicit in image intensities (the "edge detection" problem), and computations are adversely affected by image quantization and noise. In this paper, we are concerned with estimating the following from 3-D images: the surface trace (that is, the image points lying on a surface); the surface normal (or tangent plane) at each surface trace point; and the principal curvatures and directions of the surface.

A simple paradigm for computing surface structure consists of the following steps: (i) extract the surface trace points from the image; (ii) fit some surface(s) to the points; (iii) compute tangent and curvature for the fit surface(s); and (iv) take this information as belonging to the underlying surface. In practice, this straightforward approach generally leads to unsatisfactory results due to well-known problems with the so-called "edge detectors" for computing (i), and to instabilities in the surface fitting process (ii). The general conclusion has been that curvature computations are somewhat delicate.

Support for this research was partially provided by NSERC Grant A4470 and MRC Grant MA 6125.
*Fellow, Canadian Institute for Advanced Research

We show that this need not be the case, that curvature computation can indeed be made reliable. We largely follow the above scheme, but consider that it yields only an initial estimate of local surface characteristics which must be further refined. In particular, we consider that local surfaces fit in (ii) correspond to the differential geometric notion of a *chart*, and that the surface itself consists of overlapping charts. The fundamental idea then is to look for such a surface within a class of admissible surfaces which minimizes and appropriate functional ensuring that local curvature information is consistent over neighbourhoods. The implementation of these ideas of local curvature consistency is by iterative minimization with similarities to relaxation labeling procedures [1]. At each step of the iterative refinement process, information at each trace point is updated by making it more consistent with information over local neighbourhoods. This leads to more consistent information over the whole surface.

In this paper, we present experiments in applying these methods to clinical 3-D magnetic resonance (MR) imagery. Work related to ours is found primarily in the domain of laser rangefinder image understanding, e.g., [2,3,4], with the difference that their surface trace points, albeit noisy, are available *a priori* (see Ferrie *et al* [5] for an application of methods based on ours to rangefinder images).

2 Local surface parametrization

Our working definition of a surface S (technically a *2-dimensional differentiable manifold* [6]) is a set of points such that for every $r \in S$, there exist open sets $U \subset \mathbb{R}^2$, and $V \subset \mathbb{R}^3$ with $r \in V$, and a diffeomorphism $\phi : U \to V \cap S$ (recall that a diffeomorphism is a smooth bijective map with a smooth inverse). The pair (ϕ, U) is a *parametrization of S at r*, $\phi(U) = \mathcal{O}$ is a *coordinate neighbourhood*, and ϕ^{-1} is a *chart* which assigns coordinates at r. Note that when r is in two different coordinate neighbourhoods $r \in \mathcal{O}_1 \cap \mathcal{O}_2$ (we also say that r is in overlapping charts or parametrizations), then the *change of coordinate* mapping taking (part of) $\mathbb{R}^2$ into $\mathbb{R}^2$

$$\phi_2^{-1} \circ \phi_1 : \phi_1^{-1}(\mathcal{O}_1 \cap \mathcal{O}_2) \to \phi_2^{-1}(\mathcal{O}_1 \cap \mathcal{O}_2)$$

is smooth with a smooth inverse. Our refinement methods of §3 are based in an essential way on the change of coordinate mappings at overlapping parametrizations.

We depart slightly from the above notation by explicitly including a set of right-handed orthonormal axes in the notion of local chart, writing (ϕ, U, ξ) where $\xi = (\vec{P}, \vec{Q}, \vec{N})$, $\vec{P}, \vec{Q}, \vec{N} \in \mathbb{R}^3$, and $U \subset \mathrm{span}(\vec{P}, \vec{Q})$. The vector $\vec{N}$ is always the normal of the surface at r, and $(\vec{P}, \vec{Q})$ spans the tangent plane so that ξ is a *Darboux trihedron* or *frame* [6] of S bound to r. Note that, as defined, the tangent plane basis $(\vec{P}, \vec{Q})$ is arbitrary up to orthonormality constraints — when $\vec{P}$, $\vec{Q}$ are the principal directions of the surface at r, then ξ is the *principal direction frame*, and we term such a chart a *principal chart*.

In this paper, we can only present a brief sketch of the process of instantiating the local charts from the input image (the reader is referred to [7] for details). Following the scheme outlined in the introduction, step (i) estimates surface trace points as the thresholded output of a 3-D gradient operator [8]. The operator also returns an estimate of the surface normal which allows us to set up a frame representing the local tangent plane coordinate system. There is a wide choice for the mapping ϕ of step (ii), and we take it to be a *parabolic (or non-central) quadric* surface

$$\phi(p,q) = (p,q,n(p,q)); \qquad n(p,q) = \frac{1}{2}\left(ap^2 + 2bpq + cq^2\right), \qquad (1)$$

where p, q, n are in the local coordinates $(\vec{P}, \vec{Q}, \vec{N})$. This is the simplest smooth surface from which curvature can be computed. We instantiate a chart at each trace point r by fitting a parabolic quadric via least-squares error minimization to the positions and normals of neighbouring estimated trace points. Note that for the quadric surface fit at r, $\phi(0,0) = r = (0,0,0)$. In step (iii), we compute principal curvatures and directions of the chart from Eq. (1). We use the principal directions as a new basis for the tangent plane at r so that the frame ξ becomes $(\vec{M}, \vec{m}, \vec{N})$ where $\vec{M}, \vec{m}$ are the principal directions corresponding to the maximum and minimum principal curvatures κ, λ respectively. In this principal chart, the mapping ϕ assumes a particularly significant form with parameters $a = \kappa, b = 0, c = \lambda$.

As we show in §4, the results obtained by steps (i)-(iii) do not warrant step (iv) directly — they should only be treated as rough estimates of the structure of the underlying surface. In the following section, we develop a method for refining such estimates.

3 Refining estimated local information

Our goal is to determine a surface of the form introduced above which minimizes a residual functional based on local tangent and curvature compatibilities. In practice, our approach to refining estimated surface structure is similar to relaxation labeling [1] in that the information at each candidate surface point is updated based on information from neighbouring surface points. It differs in that we don't use explicit quantized labels but keep everything in the continuous domain.

We will develop a rule $(\phi_r, U_r, \xi_r)^i \mapsto (\phi_r, U_r, \xi_r)^{i+1}$ for iteratively updating the principal chart at each estimated surface point $r \in S^i$ so that the outcome of iteration i is to effectively update the whole estimated surface $S^i \mapsto S^{i+1}$. Three issues must be specified:

1. which neighbouring points s should contribute to the support of the principal chart at r, i.e., which points determine its *contextual neighbourhood* $\mathcal{N}_r$;

2. how the information at point $s \in \mathcal{N}_r$ supports the principal chart at r;

3. given 1 and 2, how to update the principal chart at r.

For consistent notation, let r refer to the point being refined and s be a point of its supporting contextual neighbourhood.

3.1 Contextual neighbourhood

Recall that the principal chart (ϕ_s, U_s, ξ_s) at $s \in S$ determines the quadric surface patch $\mathcal{O}_s = \phi_s(u,v) = \left(u, v, \frac{1}{2}(\kappa_s u^2 + \lambda_s v^2)\right)$, $(u,v) \in U_s \subset \operatorname{span}(\vec{M}_s, \vec{m}_s)$. We take the contextual neighbourhood of $r \in S$ to be the set

$$\mathcal{N}_r = \{s \in S : \|r - s\| \le d \text{ and } r \in \mathcal{O}_s\},$$

for some neighbourhood radius d ($\|\bullet\|$ is the Euclidean norm).

3.2 Local support (compatibility)

Given that s is in the contextual neighbourhood of r, we wish to determine how compatible the (principal) chart at r is with the (principal) chart at s. This cannot be computed directly since the charts are bound to different points. (Note the crucial fact that, e.g., the inner product $\langle \vec{M}_r, \vec{M}_s \rangle$ is meaningless since $\vec{M}_r, \vec{M}_s$ are bound to different surface points, r and s respectively — computing the inner product would assume that the surface was flat. See [7] for a discussion of the underlying notion of parallel transport.) Thus, we set up a mapping "transporting" the chart at s to the point r, which we denote $(\phi_s, U_s, \xi_s) \mapsto (\phi_{(s,r)}, U_{(s,r)}, \xi_{(s,r)})$, as explained below.

Support is developed with reference to the following commutative diagram. γ, δ are the principal curvature and direction computations respectively in a chart, e.g., from Eq. (1, and R is the rotation taking the principal directions from tangent plane coordinates into the image space coordinate system.

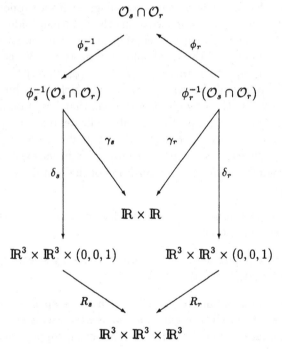

By the assumption that $s \in \mathcal{N}_r$ we have $r \in \mathcal{O}_s \cap \mathcal{O}_r$, and if we consider that $r \in \mathcal{O}_r$ then coordinates assigned by $\phi_r^{-1}(r)$. In this chart, the principal curvatures and directions at r are computed along the right-hand side of the above diagram as $\gamma_r \circ \phi_r^{-1}(r)$ and $R_r \circ \delta_r \circ \phi_r^{-1}(r)$ respectively. We have "direct" access to the principal curvatures and directions here since these are updated from the previous iteration, and are available in r's principal chart itself as the coefficients κ_r, λ_r of ϕ_r, and as ξ_r respectively.

On the other hand, $r \in \mathcal{O}_s$ and hence its coordinates in the chart at s are given by $\phi_s^{-1}(r)$, with the corresponding principal curvatures and directions computed along the left-hand side of the diagram as $\gamma_s \circ \phi_s^{-1}(r)$ and $R_s \circ \delta_s \circ \phi_s^{-1}(r)$ respectively. These are less direct to determine than along the right-hand side since $r \neq \phi_s(0,0)$; details may be found in [7]. We consider this computation as the determination of the mapping

$(\phi_s, U_s, \xi_s) \mapsto (\phi_{(s,r)}, U_{(s,r)}, \xi_{(s,r)})$ since it says what r's structure ought to be as seen from s.

Now we have the principal curvatures and directions at r in its own principal chart along the right-hand side of the diagram, and also as a point of s's chart as computed along the left-hand side of the diagram. These principal curvature measures are directly comparable as scalars — indicated in the diagram by mappings γ into the same $\mathbb{R} \times \mathbb{R}$ space. The principal directions computed by δ as frames $\xi_r = (\vec{M}_r, \vec{m}_r, \vec{N}_r)$ at r as a point of $\mathcal{O}_r$, and $\xi_{(s,r)} = (\vec{M}_{(s,r)}, \vec{m}_{(s,r)}, \vec{N}_{(s,r)})$ at r as a point of $\mathcal{O}_s$ consist of vector components now bound to r and, once transformed by R into the common image space coordinate system, permit the computation of inner products $\langle \vec{M}_r, \vec{M}_{(s,r)} \rangle, \langle \vec{N}_r, \vec{N}_{(s,r)} \rangle$.

3.3 Updating local information

The mapping $(\phi_s, U_s, \xi_s) \mapsto (\phi_{(s,r)}, U_{(s,r)}, \xi_{(s,r)})$ transports the principal curvatures and directions along the left-hand side of the above diagram from supporting point s to the point r being refined. Curvatures computed along the right-hand side say what the current estimated data is at r. The contextual neighbourhood of r consisting of k points, say, thus contributes k estimates of the local information at r. We now use this support $\left\{ (\phi_{(s,r)}, U_{(s,r)}, \xi_{(s,r)})^i, s \in \mathcal{N}_r \right\}$ to map $(\phi_r, U_r, \xi_r)^i \mapsto (\phi_r, U_r, \xi_r)^{i+1}$, producing a better local estimate of the principal direction chart at r which contributes to updating the surface $\mathcal{S}^i \mapsto \mathcal{S}^{i+1}$. Briefly, the new chart is the one that best fits, in the natural least-squares error sense, what the supporting points indicate that the principal chart at r "ought" to be. Some care must be taken with the fit due to the possibility of umbilic points (where $\kappa = \lambda$), since no unique principal direction frame exists there. However, the normal direction remains valid for the computation of the best fit $\vec{N}$. Thus we decompose the fit into two steps:

1. all k surface normals from $\mathcal{N}_r$ are used to determine $\vec{N}$, the surface normal component of the new frame; then

2. the $\hat{k}$ non-umbilic points determine the new frame by selecting principal direction $\vec{M}$ ($\vec{m}$ follows by orthonormality of $\vec{N}$, $\vec{M}$).

Principal curvatures can be computed at all k points in step 1, since only the principal directions are singular at umbilics. Figure 2 shows the increasing support of a representative r from the MR image in Fig. 1 with its neighbouring support points over the course of several iterations.

We will make the common assumption that the noise in the data is roughly zero mean Gaussian i.i.d. (independent and identically distributed), so that a linear least-squares estimator can be used (this assumption is supported in [7]).

Let us write the k charts from the $s_\alpha \in \mathcal{N}_r, \alpha = 1, \ldots, k$ as $(\phi_{(\alpha,r)}, U_{(\alpha,r)}, \xi_{(\alpha,r)})$, with $(N_{(\alpha,r)x}, N_{(\alpha,r)y}, N_{(\alpha,r)z}) = \vec{N}_{(\alpha,r)} \in \xi_{(\alpha,r)}$ being the components of the surface normal, etc. Step 1 above determines their best fit unit normal $\vec{N}$ as

$$\vec{N} = (N_x, N_y, N_z) = \left(\frac{\sum_{\alpha=1}^k N_{(\alpha,r)x}}{\sqrt{d}}, \frac{\sum_{\alpha=1}^k N_{(\alpha,r)y}}{\sqrt{d}}, \frac{\sum_{\alpha=1}^k N_{(\alpha,r)z}}{\sqrt{d}} \right)$$

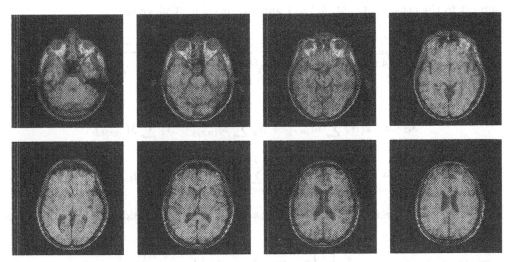

Figure 1: Eight images from a sequence of 41 slices from an MR image of the head. The images are 256×256 pixels in the xy plane, with a resolution of 0.2 cm. in the z direction. (Courtesy Dr. T. Peters, Montreal Neurological Institute.)

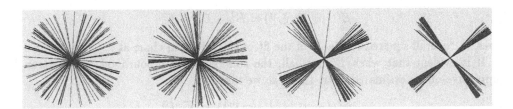

Figure 2: Increasing support over the contextual neighbourhood for the two principal directions at a typical point of the MR image. Initial estimate, one, three, and five iterations.

where

$$d = \left(\sum_{\alpha=1}^{k} N_{(\alpha,r)x} \right)^2 + \left(\sum_{\alpha=1}^{k} N_{(\alpha,r)y} \right)^2 + \left(\sum_{\alpha=1}^{k} N_{(\alpha,r)z} \right)^2,$$

by minimizing

$$E_N^2 = \sum_{\alpha=1}^{k} \left\| \vec{N} - \vec{N}_{(\alpha,r)} \right\|^2$$

subject to the constraint $\|\vec{N}\| = 1$ (by the method of Lagrange multipliers, see [7] for details). Step (2) subsequently determines the principal direction corresponding to the maximal principal curvature, subject to unit length and orthogonality with the surface normal. We set this up as the problem of determining the $\vec{M} = (M_x, M_y, M_z)$ which minimizes

$$E_M^2 = \sum_{\alpha=1}^{\hat{k}} \left\| \vec{M} - \vec{M}_{(\alpha,r)} \right\|^2$$

subject to the constraints $\langle \vec{M}, \vec{M} \rangle = 1$ and $\langle \vec{M}, \vec{N} \rangle = 0$. This gives

$$\vec{M} = \left(\frac{\lambda_G N_x - 2\sum_{\alpha=1}^{k} M_{(\alpha,r)x}}{\sqrt{4d - \lambda_G^2}}, \frac{\lambda_G N_y - 2\sum_{\alpha=1}^{k} M_{(\alpha,r)y}}{\sqrt{4d - \lambda_G^2}}, \frac{\lambda_G N_z - 2\sum_{\alpha=1}^{k} M_{(\alpha,r)z}}{\sqrt{4d - \lambda_G^2}} \right)$$

where

$$\lambda_G = 2(N_x \sum_{\alpha=1}^{k} M_{(\alpha,r)x} + N_y \sum_{\alpha=1}^{k} M_{(\alpha,r)y} + N_z \sum_{\alpha=1}^{k} M_{(\alpha,r)z}),$$

$$d = \left(\sum_{\alpha=1}^{k} M_{(\alpha,r)x} \right)^2 + \left(\sum_{\alpha=1}^{k} M_{(\alpha,r)y} \right)^2 + \left(\sum_{\alpha=1}^{k} M_{(\alpha,r)z} \right)^2.$$

A measure of the quality of the new chart is given by the residual errors E_N^2, E_M^2 of the frame fit, and

$$E_\kappa^2 = \sum_{\alpha=1}^{k} \left[\frac{\left(\kappa - \kappa_{(\alpha,r)}\right)^2}{\max(|\kappa - \kappa_{(\alpha,r)}|)} + \frac{\left(\lambda - \lambda_{(\alpha,r)}\right)^2}{\max(|\lambda - \lambda_{(\alpha,r)}|)} \right]$$

of the best fit principal curvatures κ, λ, so that

$$E^2((\phi_r, U_r, \xi_r)^i) = E_N^2 + E_M^2 + E_\kappa^2$$

gives an "overall squared residual" of the fit of the principal chart at r.

It is evident that when E^2 is small, the contextual neighbourhood of r is strongly supportive of the estimated chart there, so we take

$$\Phi(\{(\phi_r, U_r, \xi_r)^i, r \in S^i\}) = \sum_{r \in S^i} E^2$$

as the functional to be minimized by updating the surface $S^i \mapsto S^{i+1}$.

4 Experiments

The Gaussian curvature at a point of a surface is the product $K = \kappa\lambda$ of the principal curvatures, and its sign can be used to segment the surface into regions of elliptic $(K > 0)$ and hyperbolic $(K < 0)$ points. Figures 3, 4 show the iterative improvement of the Gaussian curvature segmentation in the face and lateral venticle regions of the MR image (lateral venticles are internal structures in the brain seen as the dark grey areas near the centre of the images in the lower panel of Fig. 1).

Figures 5, 6 show the progressive improvement in the principal direction estimates over the course of several iterations until (approximate) convergence of the algorithm.

5 Discussion

In this paper, we have shown that surface trace, tangent and curvature can indeed be computed from digital images. However, that most commonly used tool of differential

 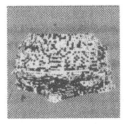  

Figure 3: Coarse segmentation by Gaussian curvature classes of the face region of the MR image. Bright points are elliptic and dark are hyperbolic. Shaded tangent plane orientations, initial segmentation, one, and seven refinement iterations.

Figure 4: Curvature labeling of the top surface of the lateral ventricles. Shaded tangent plane orientations, initial labeling, two, and five refinement iterations.

geometry, differentiation, is insufficient for the task, and other notions must be exploited as well. Specifically, the problem is how to develop a *connection* between different points, how to relate estimated information at different estimated surface points.

We have presented an approach to this problem by considering the surface as a collection of local surface patches with smoothness constraints between overlapping patches imposing curvature consistency over neighbourhoods. Our methods are implemented as an iterative functional minimization process seeking a surface whose information at each point is maximally consistent with structure over a local neighbourhood.

We presented experiments showing that this approach works well in computing surface structure from clinical three-dimensional magnetic resonance imagery.

Figure 5: One of the principal direction field in the nose region of the MR image. Shaded tangent plane orientations, initial estimates, iterations one, and five. Apparent multiple principal directions are due to projection.

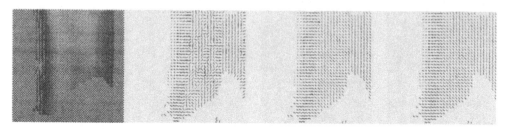

Figure 6: Principal direction field on a region of the surface of the lateral ventricles. Shaded tangent plane orientations, initial field, two, and three iterations.

References

[1] Robert A. Hummel and Steven W. Zucker. On the foundations of relaxation labeling processes. *IEEE Transactions on Pattern Analysis and Machine Intelligence*, PAMI-5:267–287, 1983.

[2] Paul Besl and Ramesh Jain. Intrinsic and extrinsic surface characteristics. In *IEEE Proceedings on Computer Vision and Pattern Recognition*, pages 226–233, June 1985.

[3] Michael Brady, Jean Ponce, Alan Yuille, and Haruo Asada. Describing surfaces. In Hideo Hanafusa and Hirochika Inoue, editors, *Proceedings of the Second International Symposium on Robotics Research*, pages 5–16, MIT Press, Cambridge, Mass., 1985.

[4] O.D. Faugeras and M. Hebert. The representation, recognition, and locating of 3-D objects. *International Journal of Robotics Research*, 5(3):27–52, Fall 1986.

[5] Frank P. Ferrie, Jean Lagarde, and Peter Whaite. Recovery of volumetric object descriptions from laser rangefinder images. These proceedings.

[6] Manfredo P. do Carmo. *Differential Geometry of Curves and Surfaces*. Prentice-Hall, Englewood Cliffs, 1976.

[7] Peter T. Sander and Steven W. Zucker. Inferring surface trace and differential structure from 3-D images. *IEEE Transactions on Pattern Analysis and Machine Intelligence*. To appear.

[8] S.W. Zucker and R.M. Hummel. A three-dimensional edge operator. *IEEE Transactions on Pattern Analysis and Machine Intelligence*, PAMI-3(3):324–331, May 1981.

Projectively Invariant Representations Using Implicit Algebraic Curves

David Forsyth, Joseph L. Mundy, Andrew Zisserman, and Christopher M. Brown[1]

Robotics Research Group
Department of Engineering Science
Oxford University
England

We demonstrate that it is possible to compute polynomial representations of image curves which are unaffected by the projective frame in which the representation is computed. This means that:

> The curve chosen to represent a projected set of points is the projection of the curve chosen to represent the original set.

We achieve this by using algebraic invariants of the polynomial in the fitting process. We demonstrate that our procedure works for plane conic curves. We show that for higher order plane curves, or for aggregates of plane conics, algebraic invariants can yield powerful representations of shape that are unaffected by projection, and hence make good cues for model based vision. Tests on synthetic and real data have yielded excellent results.

1   Introduction

It is common in machine vision to wish to represent a set of image points, $S = (x_i, y_i)$, $i = 1 \cdots M$, by a polynomial curve, thus compressing its information content. This representation makes it possible to define a set of projectively invariant shape descriptors.

Representation and approximation are different goals. A curve that is a poor approximation of a set of points may serve as an excellent representation, if the procedure for choosing the curve is stable and is independent of the frame in which the curve is chosen. For machine vision, a good representation for a set of points has the frame independence property:

> Given an observation of the set in a transformed frame, the representation computed for this set is exactly the original representation transformed according to the change of frame.

A representation with this independence property need not be a good *approximation* to the data. This property is essential to guarantee that descriptive features of the curve are unaffected by image transformation.

Stability is a second important property of a representation. A stable representation has the property:

> A small change in the data will result in a small change in the representation.

---
[1]DAF acknowledges the support of Magdalen College, Oxford. AZ acknowledges the support of the Science and Engineering Research Council. JLM acknowledges the support of the General Electric Coolidge Fellowship. CMB acknowledges the support of the DARPA U.S. Army Engineering Topographic Laboratories Grant DACA76-85-C-0001 and the Air Force Systems Command (RADC, Griffiss AFB, NY) and Air Force OSR Contract F30602-85-C-0008, which supports the Northeast Artificial Intelligence Consortium.

This definition is meaningful for point sets and polynomial curves, because it is possible to specify what a small change is.

In this paper, we show that algebraic invariants can be used to achieve projectively invariant curve fitting, and that invariants can be useful in matching and identification tasks. This work considers the case where the set, S, can be represented by a planar curve, C of known order. Furthermore, we assume that the "segmentation problem" has been solved, so that it is known which sets of points require distinct representations. The curve C is represented implicitly as a polynomial. This representation has the advantage that the invariant theory for such curves is well established, and that a projectively invariant error metric exists. This measure of error is called algebraic distance, and is described below.

The discussion in this paper is strongly focused on conics, because they are familiar curves which have been widely addressed in the vision literature. The mathematical development holds, however, for implicit polynomial curves and surfaces of any degree.

1.1 Background

Space does not permit a full review of the literature in this area. The interested reader is referred to [4]. A number of papers have contributed significantly to the line of thought presented here [1,2,11,12,13].

We write an implicit polynomial curve as $Q(x, y, \mathbf{p}) = 0$, where $\mathbf{p}$ is the vector of coefficients of the polynomial. In particular, for a conic we have $\mathbf{p} = [A\,B\,C\,D\,E\,F]$ and $Q(x, y, \mathbf{p}) = Ax^2 + Bxy + Cy^2 + Dx + Ey + F$. The *algebraic distance* of a point (x_i, y_i) from an implicit polynomial curve $Q(x, y, \mathbf{p})$ is $Q^2(x_i, y_i, \mathbf{p})$. The algebraic distance of a set of points from a curve is the mean of their individual algebraic distances. Algebraic distance is often used as a measure of the deviation of a point (x_i, y_i) from a polynomial curve because this error metric can lead to a linear system of equations for the curve fitting problem, where euclidean distance leads to complex, non-linear equations. Furthermore, when data can be viewed in different projections, euclidean distance is no longer a useful error measure, because a pair of points that are widely separated in one projection may be arbitrarily close together in a second. However, algebraic distance is not unique, since $kQ(x_i, y_i, \mathbf{p}) = 0$ defines the same polynomial curve. In order to make the value of $Q^2(x_i, y_i, \mathbf{p})$ unique, it is necessary to define a normalization function, $N(\mathbf{p})$ which is held constant during the fitting process. This constraint can be introduced by the use of Lagrange multipliers [2]. Note that if $N(\mathbf{p})$ is not a homogenous, positive (or negative) definite function (i.e. one that is always positive, and zero only if the argument is zero) of $\mathbf{p}$, then it may not be possible to fit certain curves. In particular, if we fit a curve requiring $N(\mathbf{p}) = 1$, if $N(\mathbf{p})$ is not positive (or negative) definite, then there will be some $\mathbf{p}$ such that $N(\mathbf{p}) = 0$, and this curve is clearly unattainable by our fitting process. This effect can cause poor fits.

In this paper, we show that normalizing by an algebraic invariant (defined below) of the group of changes of frame, means that the fitted curve has the frame independence property described above. Furthermore, we demonstrate invariant curve fitting leading to curve descriptors that are unaffected by the frame in which the curve is viewed, and demonstrate frame invariant curve descriptors.

2 Algebraic invariants

An invariant of a transformation is defined as follows:

Definition An invariant, $I(\mathbf{p})$, of a function $f(\mathbf{x}, \mathbf{p})$ subject to a group, $\mathcal{G}$, of transformations acting on the coordinates $\mathbf{x}$, is transformed according to $I(\mathbf{p}') = I(\mathbf{p})h(g)$. Here $g \in \mathcal{G}$ and $h(g)$ is a function only of the parameters of the transformation and does not depend on the coordinates, $\mathbf{x}$, or on the parameters, $\mathbf{p}$. $I(\mathbf{p})$ is a function only of the parameters, $\mathbf{p}$.

For linear coordinate transformations, $\mathbf{x} = \mathbf{T}\mathbf{x}'$, the form of the invariance relation becomes $I(\mathbf{p}') = I(\mathbf{p})|\mathbf{T}|^w$ where $\mathbf{T}$ is the transformation matrix and $|\mathbf{T}|$ indicates the determinant of $\mathbf{T}$. In this case, it can be shown that $\mathbf{p}' = \mathcal{T}_g \mathbf{p}$, where $\mathcal{T}_g$ is a transformation matrix for the polynomial coefficients corresponding to the group element g. The exponent, w, is called the weight of the invariant. An invariant with $w = 0$ is called a *scalar* invariant. The function $h(g)$ is represented by the determinant of the transformation matrix. The simplest example of an algebraic invariant is the discriminant of a quadratic equation of one variable, which is an invariant of weight 2, under translation and scaling. Algebraic invariants formed a major research topic of 19th century mathematics: an introduction can be found in, for example, [6]. A modern perspective of some of this work is given by [8], or by [14].

A conic polynomial can be expressed as:

$$Q(\mathbf{x}, \mathbf{p}) = \mathbf{x}^T \mathbf{P} \mathbf{x}, \text{ where } \mathbf{P} = \begin{bmatrix} A & \frac{B}{2} & \frac{D}{2} \\ \frac{B}{2} & C & \frac{E}{2} \\ \frac{D}{2} & \frac{E}{2} & F \end{bmatrix} \tag{1}$$

$\mathbf{P}$ is the coefficient matrix, and $\mathbf{x} = [x_1, x_2, x_3]^T$. Note that the standard expression for the conic in euclidean coordinates is obtained by performing the indicated matrix operations and then setting $x_3 = 1$. In a different frame, where $\mathbf{x} = \mathbf{T}\mathbf{x}'$, this conic will be represented by a different matrix, $\mathbf{P}'$. However, the value of the conic at some point on the plane is fixed, whatever the coordinates of this point, so $Q(\mathbf{x}, \mathbf{p}) = \mathbf{x}^T \mathbf{P} \mathbf{x} = Q(\mathbf{x}', \mathbf{p}') = \mathbf{x}'^T \mathbf{P}' \mathbf{x}' = \mathbf{x}'^T \mathbf{T}^T \mathbf{P} \mathbf{T} \mathbf{x}'$, and this is true for all $\mathbf{x}$, so that $\mathbf{P}' = \mathbf{T}^T \mathbf{P} \mathbf{T}$.

In effect, $\mathbf{T}^T$ and $\mathbf{T}$ "strip" the effect of the change in coordinate system, so that for a given point on the projective plane, although the point has different coordinates in the different coordinate system, the value of the polynomial is the same. For transformations corresponding to general matrices in homogenous coordinates (i.e. projective transformations combined with a choice of scale), it is easily seen that $|\mathbf{P}|$ is an invariant of weight 2. This is in fact the only invariant of a single conic under projective transformations.

3 Normalization by invariants

Bookstein [2] considers using algebraic distance as an error measure, and shows that, in this case, if one normalises a curve by a *scalar* invariant of the transformation group that will act on it, the curve chosen to fit a set of points will be unaffected by the frame in which the curve is chosen. In fact a stronger theorem is possible: it is sufficient to use an invariant *of any weight*.

Theorem 1 Let $I(\mathbf{p})$ be an invariant of the polynomial form $Q(\mathbf{x}, \mathbf{p})$ under a group of linear transformations $\mathcal{G}$. Assume I is homogenous of degree n, with weight w. Let $<\mathbf{p}>$ be the parameter vector determined by minimizing $\sum_i Q^2(\mathbf{x}_i, \mathbf{p})$ over a set of points, $\mathbf{x}_i$, subject to the constraint, $N(\mathbf{p}) = I(\mathbf{p}) = constant$. If the point set is transformed under $\mathcal{G}$, i.e., $\mathbf{x} = \mathbf{T}_g \mathbf{x}'$, let $\mathcal{T}_g$ be the corresponding transformation matrix *for the coefficients* $\mathbf{p}$. The coefficients of the polynomial fitted to the point set in the new frame are given by $<\mathbf{p}'>$. Assume that n is odd or that w is even (or both). Under these conditions, we have:

$$<\mathbf{p}> = k_g.\mathcal{T}_g <\mathbf{p}'>$$

where k_g is a scalar depending on $g \in \mathcal{G}$.

The proof of this theorem is given in appendix A. The curve chosen by the approximation process is then effectively decoupled from the frame in which we make observations, and has the desired frame independence property. Note that this property is true of the *curve*, $\{\mathbf{x} \mid Q(\mathbf{x}, \mathbf{p}) = 0\}$, and not of the *polynomial* $Q(\mathbf{x}, \mathbf{p})$. It is not difficult to convince oneself that this theorem must be true for the case where the normalization is a scalar invariant, where the algebraic distance of a curve from a set of observations is unaffected by an admissible transformation. Intuitively, it is true for the case of an invariant of any weight because, although the algebraic distance varies for an admissible transformation, the *curve* with minimum algebraic distance remains the same.

4 Fitting conics using algebraic distance

The algebraic distance from a set of points to a curve is:

$$\overline{Q^2(\mathbf{x}_i, \mathbf{p})} = \frac{1}{M} \sum_{i=1}^{i=M} \left(\mathbf{x}_i^T \mathbf{P} \mathbf{x}_i \right)^2 = \mathbf{p}^T \mathbf{S} \mathbf{p}$$

with $\mathbf{P}$ the conic coefficient matrix, $\mathbf{p}$ the vector of conic coefficients, M the number of data points. The sum is over all the data points. $\mathbf{S}$ is referred to as the scatter matrix for the conic, and its form can be explicitly reconstructed from the above expression. Note that the scatter matrix is positive, and unless an exact fit is possible, positive definite.

The error is minimized subject to the normalizing constraint, $N(\mathbf{p}) = constant$, where $N(\mathbf{p})$ is the determinant of the scatter matrix. It can be shown that for this particular problem, the value of this constant is not important [4]. We chose to use $N(\mathbf{p}) = 1$ throughout this work.

It is possible to show that a global minimum exists by demonstrating that this problem is equivalent to maximising a cubic form confined to the five dimensional sphere, which is compact [4]. Extensive experiments with conventional iterative methods, which halt at local extrema, suggest that the function may not always be convex. These methods perform poorly as a result, because this technique requires a global extremum.

We use a method that evaluates the function on increasingly fine subdivisions of the five-sphere, and retains the largest value found. This method avoids combinatorial explosion by pruning the search space, using the fact that there is an upper bound, say B, on the norm of the gradient of the objective function. In particular, if the most recent estimate of the global maximum exceeds the value at the centre of a cell whose radius is r by Br, then the cell need not be subdivided further because the cell cannot contain the global maximum. At each subdivision, because the domain of the function under consideration has been pruned, the upper bound on the size of the gradient is recomputed. This process leads to an algorithm that in practice finds very good global extrema reasonably fast, and in practice appears to avoid the worst effects of combinatorial explosion. A detailed discussion of this method appears in [5].

5 Results

The invariant fitting procedure can be checked for frame independence by comparing fits achieved in different frames. This procedure involves constructing a data set, fitting to that set in one frame, projecting the set, fitting in a second frame, and comparing the fit in the second frame with the projection of the fit in the first frame. Any discrepancy is due to a dependence of the fitting procedure on the frame in which the fit is achieved.

Data sets lying on conics and scattered data sets were tested in this way. We do not show the results for data sets drawn from a conic, because any fitting procedure based on algebraic distance that does not have an unreasonable normalisation will be frame independent on such data sets. A result for a scattered data set is shown in figure 1. In particular, the fitting process was wholly unaffected by the frame in which the curve was fitted.

Stability can be investigated by adding gaussian noise to the data sets *in each frame*. Thus, in the example shown, the data set is projected into the appropriate frame, and noise is added after projection. This means that the data sets are no longer exactly within a projection of one another. One then fits a curve to the noisy data in each frame, and checks that the curves are within projection of one another. Using this procedure gives a good indication of performance in the presence of quantization and imaging noise. From figure 2, our technique appears to degrade gracefully in the presence of limited amounts of gaussian noise. These tests suggest that the fitting technique is frame independent and reasonably stable. A measure of the stability can be obtained by computing the joint scalar invariants of a pair of conics, which are described below. These joint invariants are nearly constant for pairs of conics fitted to different views of the same pair of curves (see table 1). These results come from image data, where noise effects mean that the data sets are no longer within projection of one another.

5.1 Joint scalar invariants of conics

A pair of conics admits two joint scalar invariants, described in detail in [4] or in [16]. For two conics, Q_a and Q_b, in addition to $I_{a3} = |\mathbf{P}_a|$ and $I_{b3} = |\mathbf{P}_b|$, there are the joint invariants of weight 2,

$$I_{ab1} = \sum_i \sum_j P_a{}^{ij} P_{bij}, \quad I_{ab2} = \sum_i \sum_j P_b{}^{ij} P_{aij}$$

where $\mathbf{P}_a$ and $\mathbf{P}_b$ are the coeficient matrices of Q_a and Q_b respectively. $P_a{}^{ij}$ and $P_b{}^{ij}$ are the cofactors of the corresponding matrix elements. Two independent scalar invariants for a configuration of two conics can then be formed by,

$$I_{ab3} = \frac{I_{ab1}}{|\mathbf{P}_a|}, \quad I_{ab4} = \frac{I_{ab2}}{|\mathbf{P}_b|}$$

Conics	First joint invariant	Second joint invariant
Conics a and b from figure 3a	3.419	3.546
Conics a and b from figure 3b	3.418	3.543
Conics a and b from figure 3c	3.414	3.538
Conics a and b from figure 3d	3.407	3.528

Table 1: The joint scalar invariants computed for conics a and b for the four different images of the tape from different positions and angles, shown in figure 3. Note that the joint invariants for the coplanar conics a, b from the four images are effectively constant.

Conics	First joint invariant	Second joint invariant
Conics a and c from figure 3c	-7.547	-6.359
Conics a and c from figure 3d	-9.144	-6.778

Table 2: The joint scalar invariants computed for the non-coplanar conics a and c in figure 3c and figure 3d. Because these curves are not coplanar, the joint scalar invariants change as the viewpoint changes.

For two sets of coplanar data points, the joint scalar invariants should be constant whatever the camera viewpoint. Figure 3 shows fitted curves superimposed on image data. Table 1 shows the joint invariants for these curves, observed in different images from different angles. The joint invariants for this pair of coplanar curves, observed from different viewpoints in different images, agree very closely. However, if one data set is not coplanar with a second, then the joint scalar invariants computed should vary with the camera position, because they are invariant only under planar projection. This yields an elegant test for coplanarity for sets of points, demonstrated in table 2.

Figure 4 shows an instance of a model found in a complex scene by fitting conics to all the curves in that scene, and then marking those curves in the scene that have the correct pair of joint scalar invariants. These results argue that the invariant fitting technique is successful and stable in computing projectively invariant representations for pairs of curves. The test data shown in the figures, and other test sets, can be obtained from the first author.

6 Discussion and conclusions

This work has raised a number of issues that remain open. Our work leads us to believe that stability of fit and frame independence are the most important properties of a representation to be used for recognition in vision applications. Although the invariant representation we used is frame independent, we do not yet understand the manner in which uncertainties resulting from, for example, quantization error, reflect in the invariants. Furthermore, because the fitting process chooses a curve that best represents the entire dataset, it is sensitive to occlusion. It is not clear at this stage how to build a projectively invariant representation of a curve that is insensitive to occlusion.

Our results indicate that the combination of algebraic distance and invariant normalization leads to representations of data sets that are unaffected by projection. Theorem 1 shows that this will be true for curves of a higher order than conics. Projectively invariant representations have major applications. Three examples appear below.

In recovering three dimensional object contours by integrating multiple image views, it is important that curve representations are transformationally consistent so that object boundaries seen from different viewpoints can be matched.

We have demonstrated that algebraic invariants may be used to describe curves by projectively invariant signatures (e.g. the joint scalar invariants for pairs of conic curves). Weiss [16] first proposed using these invariants for vision, but computing such invariant signatures successfully requires an invariant fitting procedure for conics such as the one we have developed. The invariance of the ratio of areas of concentric squares and concentric triangles, used by Nielson [10] for robot navigational landmarks follows from the conic pair invariants by considering concentric circles. Our results do not require the invariants to be defined using the point cross ratio, which Nielson used in his analysis.

One of the most important issues in model-based object recognition is efficient indexing of image features onto corresponding three dimensional model features. The bulk of current research is focussed on polyhedral models and image features comprised of groups of image edges and vertices, [7,9,15]. As figure 4 indicates, algebraic invariants present a potentially efficient and reliable index between two dimensional image curves and corresponding models.

Although we have demonstrated our techniques with conics, higher order plane curves have richer sets of invariants. For example, a single cubic has two invariants, from which a single scalar invariant can be computed. Thus, a projectively invariant shape description can be computed for a single cubic curve, while for conics two curves are necessary. The techniques we have shown here are relevant in their present form only for coplanar curves. Geometrical considerations suggest that, for example, rigidly coupled pairs of plane curves, subjected to euclidean actions and then projected, admit scalar invariants, meaning that there is ample scope for these ideas to be extended.

Acknowledgements

Our thanks to Michael Brady and the Robotics Research Group, Oxford, for providing a pleasant environment in which to think and work. Our thanks to Andrew Blake for a number of stimulating discussions. Thanks to GE for the phone calls that made this paper possible.

References

[1] Agin, G.J. "Fitting ellipses and general second order curves," Carnegie Mellon University CMU-RI-TR-81-5, undated.

[2] Bookstein, F. "Fitting conic sections to scattered data," *CVGIP*, 9, 56-91, 1979.

[3] Canny, J.F. "Finding edges and lines in images," TR 720, MIT AI Lab, 1983.

[4] Forsyth, D.A., Mundy, J.L., Zisserman, A.P. and Brown, C.M. "Projectively invariant data representation by implicit algebraic curves: a detailed discussion," Internal report 1814/90, Department of Engineering Science, Oxford University, 1990.

[5] Forsyth, D.A. "Finding the global extremum of a function with bounded derivative over a compact domain," Internal report 1815/90, Department of Engineering Science, Oxford University, 1990.

[6] Grace, J.H. and Young, A. *The algebra of invariants*, Cambridge University Press, Cambridge, 1903.

[7] Grimson, W.E.L. and T. Lozano-Pérez, "Localizing overlapping parts by searching the interpretation tree," IEEE T-PAMI, **PAMI-9**, 4, 469-481, 1987.

[8] Kung, J.P.S. and Rota, G.-C. "The invariant theory of binary forms," *Bull. Amer. Math Soc.*, **10**, 27-85, 1984

[9] Lowe, D. *Perceptual Organisation and Visual Recognition*, Kluwer, Boston, 1985.

[10] Nielson, L. "Automated guidance of vehicles using vision and projectively invariant marking," *Automatica*, **24**, 2, 135-148.

[11] Porrill, J. "Fitting ellipses and predicting confidence envelopes using a bias corrected Kalman filter," Proc. 5'th Alvey vision conference, 1989.

[12] Pratt, V. "Direct least-squares fitting of algebraic surfaces," *ACM SIGGRAPH*, **21**, 145-151, 1987.

[13] Sampson, P.W. "Fitting conic sections to "very scattered" data: an iterative refinement of the Bookstein algorithm," *CVGIP*, **18**, 97-108, 1982.

[14] Springer, T.A. *Invariant theory*, Springer-Verlag lecture notes in Mathematics, **585**, 1977.

[15] Thompson, D.W. and J.L. Mundy, "3D model matching from an unconstrained viewpoint," *Proc. IEEE Conf. on Robotics and Automation*, 1987.

[16] Weiss, I. "Projective invariants of shapes," *Proc. DARPA IU workshop*, 1125-1134, 1989.

Appendix: The invariant fitting theorem

Theorem 1 Let $I(\mathbf{p})$ be an invariant of the polynomial form $Q(\mathbf{x}, \mathbf{p})$ under a group of linear transformations $\mathcal{G}$. Assume I is homogenous of degree n, with weight w. Let $<\mathbf{p}>$ be the parameter vector determined by minimizing $\sum_i Q^2(\mathbf{x}_i, \mathbf{p})$ over a set of points, $\mathbf{x}_i$, subject to the constraint, $N(\mathbf{p}) = I(\mathbf{p}) = constant$. If the point set is transformed under $\mathcal{G}$, i.e., $\mathbf{x} = \mathbf{T}_g \mathbf{x}'$, let $\mathcal{T}_g$ be the corresponding transformation matrix *for the coefficients* $\mathbf{p}$. The coefficients of the polynomial fitted to the point set in the new frame are given by $<\mathbf{p}'>$. Assume that n is odd or that w is even (or both). Under these conditions, we have:

$$<\mathbf{p}> = k_g . \mathcal{T}_g <\mathbf{p}'>$$

where k_g is a scalar depending on $g \in \mathcal{G}$.

Proof: With $\mathbf{p}^T = [A\,B\,C\,D\,E\,F]$, we have $\overline{Q^2(\mathbf{x}_i, \mathbf{p})} = \mathbf{p}^T \mathbf{S}\mathbf{p}$, where $\mathbf{S}$ is the positive definite scatter matrix defined above. $<\mathbf{p}>$ is determined by minimizing the mean algebraic distance subject to the constraint of constant normalization. The number of polynomial coefficients depends on the degree r of Q, and will be denoted by $M(r)$. For example, for a conic $M(2) = 6$ and for a cubic, $M(3) = 10$. The appropriate Lagrangian is $\mathcal{L} = \mathbf{p}^T \mathbf{S}\mathbf{p} + \lambda N(\mathbf{p})$

At any local minimum of mean squared algebraic distance $\nabla_{\mathbf{p}}\mathcal{L} = 0$. This gives $M(r)$ equations in $M(r) + 1$ unknowns. The final equation is given by $N(\mathbf{p}) = constant$. Suppose that the point set is transformed by some element, $g \in \mathcal{G}$ so that $\mathbf{x} = \mathbf{T}_g \mathbf{x}'$. The polynomial coefficients are transformed linearly by a $M(r) \times M(r)$ transformation matrix, $\mathcal{T}_g$, such that $\mathbf{p}' = \mathcal{T}_g \mathbf{p}$, and the scatter matrix transforms to $\mathbf{S}'$.

The form of $\mathbf{S}'$ can be determined by observing that for any polynomial with coefficients $\mathbf{p}$ in some frame, the algebraic distance at a given point on the plane is not affected by the particular frame in which the point is expressed. Thus

$$\mathbf{p}'^T \mathbf{S}' \mathbf{p}' = \mathbf{p}^T \mathcal{T}_g^T \mathbf{S}' \mathcal{T}_g \mathbf{p} = \mathbf{p}^T \mathbf{S} \mathbf{p}$$

This is true for all $\mathbf{p}$, so that $\mathbf{S} = \mathcal{T}_g^T \mathbf{S}' \mathcal{T}_g$.

The normalization is an invariant under $\mathcal{G}$ so $N(\mathbf{p}') = |\mathcal{T}_g|^w N(\mathbf{p})$, where w is the weight of the invariant. Applying these transformation rules to the Lagrangian,

$$\mathcal{L} = \mathbf{p}^T \mathbf{S} \mathbf{p} + \lambda N(\mathbf{p}) = \mathbf{p}^T \mathcal{T}_g^T \mathbf{S}' \mathcal{T}_g \mathbf{p} + \lambda \frac{1}{|\mathcal{T}(g)|}^w N(\mathcal{T}_g \mathbf{p}) = \mathbf{p}' \mathbf{S} \mathbf{p}' + \lambda' N(\mathbf{p}') = \mathcal{L}'$$

Hence the Lagrangian for $\mathbf{p}'$ is the same as the Lagrangian for $\mathcal{T}_g \mathbf{p}$. It follows that $<\mathbf{p}'>$ can differ from $\mathcal{T}_g <\mathbf{p}>$ by at most a scalar factor, $k(g)$. The final equation, $N(<\mathbf{p}>) = constant$, can be satisfied by scaling $<\mathbf{p}>$. This will fail only when N is homogenous of even degree, and the signs of $N(<\mathbf{p}>)$ and $N(\mathcal{T}_g <\mathbf{p}>)$ differ. Thus, the theorem is proven.

List of Figures

1. The diagonal crosses show a scattered data set. The crosses show that set in a different projection. Curve a was fitted to the diagonal crosses. Curve b was fitted to the crosses, and curve c is the appropriate projection of curve a. The two curves are indistinguishable at the resolution of laser-printer output. This demonstrates the frame independence of the fitting process.

2. The crosses show the projected data set of figure 1, and line up with the crosses in figure 1. Curve a was fitted to the crosses. Isotropic gaussian noise of $\sigma = 3$ was added to this set *in this frame*, to give the points plotted as diagonal crosses. The data set is thus no longer a projection of the original set. Curve a is the curve fitted to the diagonal crosses, curve b is the projection of curve a of figure 1. These curves now differ slightly, as a result of the noise. The small difference indicates a degree of stability for fitting circles and ellipses, and suggests that the frame independence property is stable.

3. Four images of a computer tape, with two fitted conics in overlay. The data for the conics in these images was obtained by acquiring the image edges using a local implementation of Canny's [3] edge finder, linking edges, and then choosing corresponding curves by hand. In these images, the conics have been drawn three pixels thick to make them visible. These conics were used to obtain the joint scalar invariants shown in tables 1 and 2.

4. The joint scalar invariants of a pair of conics can be used to find instances of models in scenes, when the objects involved have plane conic curves which lie on their surfaces. Here we show an instance of a computer tape found in a cluttered scene by fitting conics to all of the curves, and marking those pairs of conics with the correct joint scalar invariants. The data for the conics in this image was obtained by acquiring the image edges using a local implementation of Canny's [3] edge finder and then linking these edges.

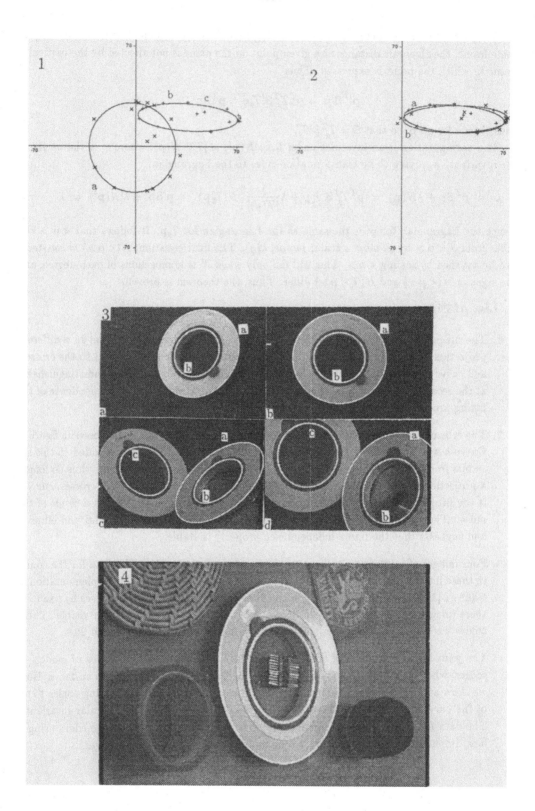

Shape from Contour Using Symmetries

Shiu-Yin Kelvin Yuen *
Cognitive Studies Programme,
University of Sussex,
Brighton BN1 9QH
UK
Janet : kelviny@cogs.susx.ac.uk

Abstract

This paper shows that symmetry is essential to shape from contour and indicates problems with existing measures, based on energy and information.

The paper is divided into two parts : The first part establishes the importance of shape from contour using symmetry from a computational theoretic viewpoint. The second part proposes algorithmic solutions to the problem of symmetry finding.

We have omitted all proofs and some parts from this paper due to lack of space. They may be found in [1]. It also contains some more references on shape from contour, reflectional and rotational symmetry. We encourage you to read it.

1 Introduction

1.1 Shape from contour

Shape from contour is the interpretation of a single line drawing as the projection of a three dimensional entity. Following Levitt [22], we model a line drawing or a binary digital image L as a two dimensional point set S. A pixel belongs to S if it is part of L. Then for a line drawing with n pixels,

$$S = \{(c_i, d_i)/i = 1, ..., n\} \tag{1}$$

where (c_i, d_i) is the (known) x-y position of the $i^{th} pixel$. The *shape from contour problem* is to solve for a three dimensional point set S' such that

$$S' = \{(c_i + u_i\, t_x, d_i + u_i\, t_y, u_i\, t_z)/i = 1, ..., n\} \tag{2}$$

*partially supported by the Croucher Foundation

where u_i is the unknown parameter and (t_x, t_y, t_z) is the known projection vector (assuming parallel projection), such that some measure of simplicity is extremized.

There have been two streams of research on shape from contour : "top down" and "bottom up".

1.2 Top down vision

Roberts [3], in his seminal work, considered shape from contour as discovering a set of model instances whose projection gives rise to the input line drawing. This research program was extended significantly by Pentland [4, 5]. It consists of three steps. (1) It finds a parameterized model that spans all geometrical objects in the world. (2) It uses a generate-and-test strategy to predict all possible appearances of the objects. (3) It compares the given appearance and finds the best matching model parameters. There are two main difficulties : Firstly, the parameterized models require too many parameters. For example, a superquadric with (restricted) bending and tapering requires 14 parameters ([5], p. 614). Not withstanding this, the superquadric has difficulties representing some simple shapes, eg. the hexagonal prism. More general parameterized models are needed. The hyperquadrics [6], which includes the superquadrics as a special case, needs many more parameters, however. Secondly, complex nonconvex solids should be represented by more than one parameterized model. This gives rise to a "part-whole" problem : Local best model match needs not be global best. (See Mackworth's conterexamples to Roberts' program in [7]). The top down approach does not appear computationally feasible if we have no knowledge about the visual environment.

1.3 Bottom up vision

Marr [8] and others proposed a radically different approach. These researchers envisage that shape from contour may be accomplished without the use of models. Insteads, it is achieved through some primarily data driven process involving only information processing and physical principles. Shape from contour is cast as a quest for finding a *shape metric*. Ideally, the metric should allow us to tell which of any two shapes are more complex. Shape from contour is then the problem of finding the backprojection that extremizes the shape metric. Under this research paradigm, many shape measures have been reported in the past decade. Almost all of them (1) use some form of energy or information as their basis and (2) assume that the given drawing has been "labelled" [1]. One of the main aims of this paper is to show that (1) is misguided. Moreover, it is well known that it is extremely difficult to label a line drawing in general. So far, "labelling" has been successful only in the simple domains (polyhedra and simple curved solids) with stringent assumptions ("perfect" line drawing, etc). This leads one to suspect whether labelling is a neccessary step. As will be shown in the rest of this paper, we may do away with labelling altogether if we use the more naive but intuitive concept of symmetry.

---

[1]We are not aware of any shape from contour work (including Kanade's) that does not depend on a labelled drawing. Stevens [9] described a method which uses lines of curvature to infer shape. The problem is that the lines are not given. Tsuji and Xu [10] reported a method to construct a net of similiar lines from a drawing. However, their method is heuristic and depends on first labelling the image into "connect" and "occlude" edges. The recent work on symmetries by Ulupinar and Nevatia [11] also implicitly assumed a labelled drawing (for example, see their shared boundary constraint). The works of Witkin [12] and Kanatani [13], however, have the potential of being labelling independent

1.4 Middle out vision

We believe that shape from contour is neither top down nor bottom up. In natural images as well as in line drawings, there are some *invariants*. These invariants have three dimensional implications. Perception of shape from contour proceeds by identifying these invariants. The perception is the result of selecting a backprojection which best *preserves* the invariants. One of the aims of this paper is to show that the (parallel) projection of reflectional and rotational symmetries form invariants that are robust and may be detected in a highly parallel fashion. Moreover, the amount of effort required to detect these invariants (2 and 3 degrees of freedom respectively) are very small compared with that required to detect parameterized models. These invariants are very useful at inferring the three dimensional shape as well as predicting the hidden and occluded parts. In this paper, we shall present the theoretical results in support of our contention.

This theory echos nicely with the direct perception idea of Gibson [14]. The idea of using invariants has been suggested by Ballard, who proposed that "a major function of the perceptual system is to compute collections of invariants at different levels of abstraction" (pg. 89, [15]) and applied it to the recognition of polyhedra.

It is important to realize that this paradigm is not bottom up vision in another guise. The theory does not limit ourselves to general invariants from images. Invariants may also arise from specific models (eg. the relative spatial dispositions of two projected model lines [16]). In fact, recent works on model-based vision has begun to explore these invariants [15-20]. This paper would however concentrate on image invariants, which are more generally applicable. Theorists have long recognized that both bottom up and top down vision are part of a whole theory. Middle out vision attempts to unify both paradigms through the common notion of invariants.

Part I

Symmetries

2 Symmetries and shape from contour

2.1 A symmetry measure

Symmetry is a prolific phenomenon in the world [21]. It may be defined in terms of three linear transformations in n-dimensional Euclidean space E^n : reflection, rotation and translation. Formally, a subset S of E^n is *symmetric* with respect to a linear transformation T if $T(S) = S$. We shall concentrate on reflectional and rotational symmetries. A reflectional symmetry has a reflectional plane, for which the left half space is a mirror image of the right half. A rotational symmetry has an axis of rotation A and an angle of rotation θ. A rotation of θ about A will give an identical figure. Trivially, a rotation of $0°$ will produce an identical figure. We shall count this as one rotational symmetry. Let k be the sum of the number of reflectional symmetries and rotational symmetries. Then because a figure has at least one trivial rotational symmetry, $k > 0$. A sphere is reflectionally symmetric about all reflectional planes passing through the center and rotationally symmetric about all axes passing through the center. Hence for a sphere, $k = \infty$.

This measure was originally proposed by Levitt [22].

Human beings perceive a three dimensional object as the putting together of "perceptual parts" [4]. Suppose S' is decomposed into m "parts" $S'_1, ..., S'_m$ ($m \geq 1$). Let $k(S'_i)$ be the sum of the reflectional and rotational invariance in S'_i. We wish to have as few parts as possible. However, the more points S'_i includes, the lower is $k(S'_i)$. A compromise is needed between the number of parts and the k of each part. Levitt [22] achieved this by minimizing

$$E = \frac{m}{\sum_{i=1}^m k(S'_i)} \, (m \geq 1) \tag{3}$$

where m is the number of parts. This measure is used by Levitt to decompose an 2-dimensional point set into a few salient subsets. Some good decomposition results have been obtained by Levitt.

In this paper, we shall use (3) to solve shape from contour problems.

2.2 Interpreting skewed symmetry as real symmetry

A "skewed symmetry" is a planar point pattern such that iff (x, y) exists, $(-x, y)$ exists. The x axis is called the "skewed transverse axis" t_k; whilst the y axis is called the "skewed symmetric axis" s_k. If t_k and s_k are not orthogonal, then we have a skewed symmetry. If they are orthogonal, then the skewed symmetry degenerates to a real (reflectional) symmetry, the reflectional axis of which is s_k.

Stevens [9] 's psychological experiments indicated that human beings will interpret a skewed symmetry as a real symmetry. Although it is possible to find counterexamples (for example, we would not interpret a door as a square), this is almost always true. Hence a good measure should likewise interpret skewed symmetry as real symmetry.

Brady and Yuille [24] showed that maximizing their compactness measure will interpret skewed symmetry correctly as real symmetry. Liang and Todhunter [25] proved that minimizing Barrow et. al. 's equality of angle measure [2] will also interpret skewed symmetry correctly.

These works [24, 25] assume that the skewed symmetry is a parallel projection of a *planar* pattern. That is, the underlying transformation is affine. In the following, we shall also assume that the transformation is affine.

We identify a skewed symmetric subset of the point set S of line drawing L as a part (later, we report how such a part may be identified directly from the line drawing). Hence what we need is the general equation (3) applied to a part S'_i, which is

$$E_i = 1/k(S'_i) \tag{4}$$

This means that the three dimensional interpretation of a part S_i maximizes the sum of the number of reflectional and rotational symmetries of S'_i.

Proposition 1 *Minimizing (4) will interpret a skewed symmetric figure as a real symmetric figure if the projection is an affine transformation.*

<u>Outline of the Proof</u>

It can be proved that the real symmetric interpretation will not have fewer rotational (reflectional) symmetries than any other interpretation.

3 Analysis of shape from contour methods

omitted.

4 Shape from contour measures revisited

In this section, the disadvantages of existing shape measures will be discussed. The symmetry measure has none of these disadvantages, but on the contrary has many delightful advantages. Many of them are unique.

The disadvantages of existing measures are :

d1) The shape measures do not suggest a process to extremize them. Though they are quite simple, it is not apparent how to extremize them. Frequently, brute force quantization is resorted to [24]. A good measure, on the contrary, should suggest a process for computing it.

d2) Measures based on angles are very sensitive to small changes in boundary shape.

d3) The extremization of these measures usually gives two global extrema only. (There are two because of Neckar reversal). Other values of the measures do not have physical meaning (see the section 3.2).

d4) Different dimensions require different measures. For example, the three dimensional version of the compactness measure is

$$(volume)^2/(surface\ area)^3 \qquad (5)$$

Some other measures have no obvious higher dimensional generalization at all.

d5) The measures are not applicable to point sets composed of isolated points.

The symmetry measure has none of the above disadvantages. On the other hand, it has the following advantages :

The following are also shared by most existing shape measures :

a1) An n-dimensional sphere has $k = \infty$. The measure considers an n-dimensional sphere as the simplest shape.

a2) It is scale invariant.

The following are unique :

a3) It is applicable to point sets as well as line drawings.

a4) It is "dimension invariant". It stays the same for all dimensions. In n dimensions, a reflectional plane is an (n-1)-dimensional hyperplane and an axis of rotation is an (n-2)-dimensional hyperplane. For example, in two dimensions, they are the line of reflection and the center of rotation respectively. In three dimensions, they are the plane of reflection and the axis of rotation.

a5) It is a *process theoretic* measure. It suggests a two level process theory : Level 1 : Look for symmetries in the image. Level 2 : Find the optimal combination of the symmetries found. Symmetry finding will be discussed in Part 2.

a6) Symmetries are global and robust. They are insensitive to noise as well as occlusion.

a7) It does not rate irregular and random figures. This is consistent with the limits of human ability.

a8) Multiple percepts are allowed. Alternative percepts are those with fewer symmetries. They may still turn out to be favourite, particularly if motivated by a priori preference or beliefs. Allowing multiple percepts is very important. For example, it has been shown [26] that human beings may imagine different hidden part completions given the same line drawing. We also believe that a good measure should not only give the ideal best output, but exhibit graceful degradation [27].

Part II

A computational theory of shape from contour using symmetries

5   Introduction

There are two radically different approaches to solving (3) : Feedback and Generate-and-test.

The main idea behind the first approach is to cast the problem as a self organization problem. The spirit of feedback is behind [28, 29].

However, this approach has three serious difficulties : (1) the feedback function may not have physical significance; (2) the feedback function usually introduces some extra coefficients to the system; (3) it is very difficult to find a suitable feedback function that converges to a good equilibrium. It is refreshing to remember that to grow a good crystal, one has to start at near equilibrium conditions.

As we shall show, in the line drawing, there are *invariants* of symmetries that are preserved under projection. They are usually readily apparent (though the computational method to find them is not apparent). An example is skewed symmetry, which is the projection of a reflectional symmetry. One approach to find these invariants is by a generate-and-test strategy. Two important advantages of this strategy are high parallelism and reliability [30]. The apparent difficulty is the combinatorial explosion. For example, naively, to generate all possible reflectional and rotational symmetries requires 3 and 5 degrees of freedom (d.f.) respectively. One of the key results of this paper is showing that they may be done with 2 and 3 d.f. ! In the rest of the paper, we shall explore the potential of the generate-and-test approach.

Our theory is in two levels :

Level 1 : find invariants due to symmetries in the given image;

Level 2 : find the backprojection that provides the best combination of the invariants.

In this paper, efficient algorithms are reported to tackle Level 1.

In the rest of the paper, we shall assume parallel projection.

The next section will deal with two interesting properties of symmetries.

6 Two interesting properties

Proposition 2 *A reflectional symmetry will project as a skewed symmetry or a reflectional symmetry. Moreover, if it is a skewed symmetry, then if and only if the backprojected reflectional symmetry is planar, the skewed symmetric axis will be straight.*

This leads to two interesting corollaries :

Corollary 2.1

If a skewed symmetric figure is to be interpreted as the projection of a reflectional symmetry, then the reflectional symmetry must be planar.

Corollary 2.2

If a reflectional (real) symmetric figure is to be interpreted as the projection of another reflectinal symmetric figure, then the figure need not necessarily be planar. An example is Fig 1. (pointed out to us by Prof. Harry Barrow).

Note : We are using skewed symmetry in a more general sense. Kanade's initial definition of skewed symmetry has a straight axis.

Proposition 3 *A rotationally symmetric figure may have neither skewed symmetric nor reflectional symmetries.*

An example is shown in Fig 2.

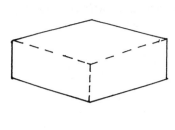

Fig 1   Fig 2

It is well known that a rotational symmetry is a composite of two reflections. However, the last proposition suggests that it is not helpful to look for skewed and reflectional symmetries when finding rotational symmetries.

7 Reflectional symmetry

7.1 Finding skewed symmetry under parallel projection — mapping pair

Levitt [22] proposed an elegant method to detect *reflectional* symmetry in a point set, without requiring a priori that the point set is symmetric. Suppose there are n points. For each of the $_nC_2$ pairs, the midpoint is found and assigned a direction perpendicular to the line joining the pair. To this is added the n points with no direction assigned. A Hough transform is then used to find straight reflectional symmetry axes.

We may extend Levitt's method to detect skewed symmetry in the following way. Suppose P_1, P_2 is a pair of points. Then insteads of storing the angle of the perpendicular, we may store the angle β of line $\overline{P_1 P_2}$ with some reference axis on a third dimension of the Hough space. High counts on the Hough space correspond to a skewed symmetric axis s_k where the skewed transverse axes are at the same angle with respect to s_k.

This method requires a 3-d Hough space. Below, we describe a variant which requires only a 2-d Hough space.

We point out in proposition 2 that skewed symmetry is the projection of reflectional symmetry. Suppose p', q' are two three dimensional points and p' reflects onto q' (and vice versa). Then we shall say that they are *mapping points* or a *mapping pair*.

Suppose r', s' are another mapping pair. Then clearly $\overline{p'q'}//\overline{r's'}$. Now since a parallel projection is an affine transformation, parallelism is preserved. Hence upon projection $\overline{pq}//\overline{rs}$, where p is the projected point of p' and etc. We note that this is an invariant.

7.2 Extending Kanade's skewed symmetry constraint to parallel projection

We now show how to make use of this invariant to find skewed symmetry in the image and the constraint of the skewed symmetry on the backprojected reflectional symmetry.

Suppose p and q are mapping points. Let the known projection vector be (t_x, t_y, t_z) and σ and τ be the slant and tilt (as defined in [24]) of the reflectional plane . Let

$$\begin{bmatrix} a_x \\ a_y \end{bmatrix} //(q - p) \tag{6}$$

and

$$\begin{bmatrix} a'_x \\ a'_y \end{bmatrix} = \begin{bmatrix} cos\tau & -sin\tau \\ sin\tau & cos\tau \end{bmatrix} \begin{bmatrix} a_x \\ a_y \end{bmatrix} \tag{7}$$

then it can be shown [1] that

$$\sigma = \frac{\pi}{2} - tan^{-1}\left(\frac{t_z a'_y}{t_x a'_y - t_y a'_x}\right) \tag{8}$$

(8) may be understood this way. Mapping pairs must be parallel to (a_x, a_y). Hence (a_x, a_y) defines a "mapping direction". Given the mapping direction, we may use (8) to calculate all feasible slant σ and tilt τ of the reflectional plane, assuming a known parallel projection. Once the equation of the reflectional plane is fixed, it is an easy matter to calculate the three dimensional positions of any mapping points.

Recall that proposition 2 states that if the skewed symmetric axis is straight, the reflectional (real) symmetric interpretation must be planar. Hence given a skewed symmetric figure (with straight skewed symmetric axis) in the image plane and a mapping direction, we may apply (8) to find all reflectional planes that will give a symmetric interpretation. We note that 1 d.f. is required for the mapping direction. This is consistent with Friedberg's result [31] that 1 d.f. is needed to detect skewed symmetry. Moreover, the gradient of the reflectional plane has 1 d.f., which is again consistent with the 1 d.f. of Kanade's skewed symmetric heuristic.

However, our formulation is more general because it is applicable to parallel projection, whilst Kanade's formulation assumes orthographic projection. It may also be noted that given τ, the gradient of the reflectional plane is constrained to lie on a line in the gradient space. This may be constrasted with Kanade's heuristic, which specifies that the gradient of the plane *on which the backprojected figure lies on* is on a hyperbola in the gradient space.

Under orthographic projection, the solution is particularly simple, whence it can be shown [1] that the mapping direction is τ.

7.3 Extremum symmetry

We shall define the more general notion of "extremum symmetry" and show that Ulupinar and Nevatias' "parallel symmetry" and "mirror symmetry" (see below) are special cases.

We say that a curve is "mapped" to another curve in the image if a one-one mapping may be defined which maps one curve to another. Of course, the "mapping directions" have to be the same. It is very important to recognize the following :

Lemma 4 *Let c_1, c_2 be image curves and c'_1, c'_2 their backprojection. If a mapping exists which maps c_1 to c_2, a c'_1 and c'_2 may always be found such that c'_1 is reflectionally symmetric to c'_2.*

This lemma indicates that reflecional symmetry may be imposed on almost any two curves. However, this is not so with human beings. Though a mapping may be defined for the two curves in Fig 3, no one would seriously consider that their backprojection is reflectionally symmetric. What extra constraints do we exploit then ?

Ulupinar and Nevatia [11] proposed two special forms of symmetry : "parallel symmetry" and "mirror symmetry". Two analytic curves c_1, c_2 have parallel symmetry if

$$\theta_1(s) = \theta_2(as + b) \tag{9}$$

where $\theta_i(s)$ is the angle the curve made with the positive x axis at curve length s, and a, b are constant. Since a is merely a scaling factor, in the rest of the paper, we shall assume $a = 1$. Then one curve is a translation of the other.

The mirror symmetry is identical to Kanade's skewed symmetry. Recall that Kanade's skewed symmetry has a straight skewed symmetric axis.

Fig 4 shows a reflectional symmetry. The two curves, which are reflective to each other, may be described by

$$(\pm x(t), y(t), z(t)) \tag{10}$$

where

$$c(t) = (0, y(t), z(t)) \tag{11}$$

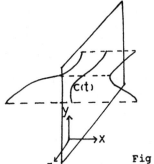

Fig 3   Fig 4

It is not hard to show that

(i) if x(t) = constant, the projected image has parallel symmetry;

(ii) if c(t) is a straight line, the projected image has mirror symmetry

Hence it may be concluded that both parallel and mirror symmetries are special cases of the (parallel) projection of reflectional symmetries.

However, it is incorrect to suppose that parallel and mirror symmetries are the only symmetries that human beings perceive. An example which neither fits parallel nor mirror symmetry is the top face of Fig 5.

We define below the concept of "extremum symmetry", which we believe is the most general form of projected reflectional symmetry that human beings perceive.

Two image curved fragments c_1 and c_2 have "extremum symmetry" if a one-to-one map may be found between points in c_1 and points in c_2 such that

(i) if a point in c_1 is an extremum (curvature maximum, curvature minimum, corner with m incident lines), then the mapping point in c_2 must be an extremum;

(ii) if a point in c_1 is not an extremum, then the mapping point in c_2 is not an extremum.

It is clear that our extremum symmetry subsumes both the parallel and the mirror symmetry. Under this definition, Fig 3 is not a projection of reflectional symmetry whilst Fig 5 is. Once an extremum symmetry is found, the three dimensional position of the backprojected curves may be found using the result of the last subsection.

The concept most closely related to extremum symmetry is Brooks ribbon [32]. A Brooks ribbon is generated by line segments (of varying length) centered on a curved axis. The angle of the line segment is variable. The angle between the line segment and the tangent to the axis is constant. For an extremum symmetry, the angle of the line segment is constant whereas the angle between the line segment and the tangent is variable. If the axis is straight, then both give skewed symmetry with straight axis. But extremum symmetry is in general not Brooks ribbon. Thus we may also say that Brooks ribbon in general does not backproject to reflectional symmetry.

In general, the skewed symmetric axis may be defined as the locus of the midpoints of the mapping pairs of an extremum symmetry. Mirror symmetry restricts the axis to be straight, whilst for extremum symmetry the axis may be an arbitrary connected curve. It is interesting to ask if there are any notions of symmetry intermediate between these two extremes. For example, is "quadric symmetry" — that is, the axis is a quadric curve — possible ? Fig 6 shows an example for which the axis is a circular arc. However, it is not apparant that the two curves have any symmetry relations to a human observer. Indeed, the lower curve has a curvature minimum which is absent in the upper curve, thereby violating the concept of extremum symmetry. We conclude that quadric symmetry is not perceived by human beings.

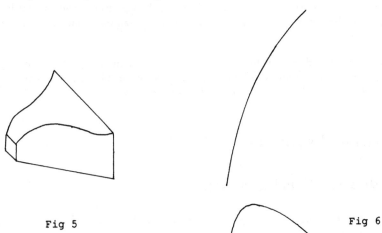

Fig 5   Fig 6

7.4 An algorithm for finding skewed symmetry

It is simple to detect parallel symmetry. Given a mapping, the distance between the two mapping points will remain constant. (One curve is a translation of the other). Hence in this subsection, we shall concentrate on finding an efficient algorithm for detecting skewed symmetry.

Let the mapping direction make an angle α with the positive x axis.

Algorithm 1 (Skewed symmetry finding)

Input : a two dimensional point set.

1. $\alpha \leftarrow 0$;
    ;;; the initial mapping direction is parallel to the x axis

2. clear the midpoint_array;
    rotate the point set by $-\alpha$ about the origin;
    ;;; after the rotation, the mapping direction is parallel to the x axis

3. for all lines L parallel to the x axis do
      if two points p, q lie on the L then
        record the midpoint of p and q on midpoint_array;
      endif;
    endfor;

4. use a Hough transform to find straight lines in midpoint_array
    and store symmetry axes found and their associated point subset;

5. increment α;

6. if $\alpha > \pi$, then exit else goto 2.

Output : a set of symmetry axes and their point subsets.

We have 1 d.f. in α. Finding lines using the Hough transform involves 1 d.f. [1]. Hence the algorithm has two d.f.

The extremum symmetry finding is a special case of the algorithm. The straight line finding is unnecessary. Neighboring midpoints of the mapping pairs of an extremum symmetry may simply be connected together.

It should also be noted that we have not exploited extremum at all in the above algorithm. In actual facts, only α which maps an extremum to another extremum needs be considered. Hence one d.f. may be shrinked. We have not pursued this possibility.

8 Rotational symmetry

8.1 Finding rotational symmetry

Reflectional symmetry does not capture all symmetries in the world. Fig 2, for example, has no global reflectional symmetry. It looks pleasing because it is invariant under rotations of multiples of $\pi/3$. Proposition 3 suggests that rotational symmetry should not be detected by detecting skewed nor reflectional symmetry. This motivates us to study an independent method for detecting rotational symmetry. It should be simple, robust, computationally inexpensive, highly parallel, and should not presuppose a symmetric point set. Last but not least, it should be able to detect rotational symmetry which is not symmetric in the image plane, but is so after a backprojection.

Let us consider the problem of detecting rotational symmetry on the image plane first. Suppose our input is a point set S. Let p, q be two points in S. If p and q are rotationally symmetric, then we may define a rotation $R(o, \theta)$ which rotates p into q. o is center of rotation and θ is the angle of rotation. It is clear that o must lie on the perpendicular of $\overline{pq}$ passing through the midpoint of $\overline{pq}$. Denote this perpendicular as $\overline{pq}'$. Now suppose we have many pairs of points which are rotationally symmetric with the *same* $R(o, \theta)$. Then the perpendiculars must intersect at the origin $o(x_0, y_0)$.

Use the x-y image plane as an accumulator array. If we accumulate all points on $\overline{pq}'$ for every pair p, q on the x-y accumulator array, then there will be a high count at the common origin $o(x_0, y_0)$ (Fig 7).

This transform converts detection of *concentricity* to detection of high counts in the x-y accumulator array. But a rotational symmetry is concentricity and *equality of subtended angles*. That is, the angle subtended by p and q on o, $\angle poq$, must be equal to other angles subtended. Clearly, this common angle is the angle of rotation θ.

One solution is to accumulate in a three dimensional accumulator array (x,y,θ). However, there is a better way to do it.

Scott et. al. [33] reported a brilliant algorithm for detecting "smoothed local symmetries" (SLS). The problem is to find the locus of the SLS axis, points of which are the centers of bitangent circles. Scott used the following analogy. Each point is considered as a small stone dropped on a pond which sets off a one-off ripple. The SLS axis are those points where the wavefronts meet !

We may use the wave propagation analogy to detect rotational symmetry, except that a "wavefront" would be replaced by a "particle" (as dictated by quantum mechanics !). Look at Fig 8. The "particle" is initially at m_0, the midpoint of p and q. The "particle" then splits into two and moves away from m_0 in two directions along $\overline{pq}'$. At "time" $t = \theta$, the "particle" is at m_θ. The position of m_θ may be readily calculated using the right-angled triangle $\Delta m_\theta m_0 p$. If there is a rotational symmetry with angle of rotation θ at o, then many "particles" will meet there and it will have a high count at that instant.

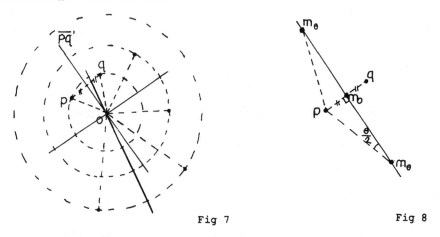

Fig 7   Fig 8

It is clear that this algorithm involves only 1 d.f. (θ) and the space complexity is the x-y plane, which is two dimensional insteads of three dimensional as with the Hough transform.

However, there may exist some rotational symmetries on a slanted and tilted plane. If so, the projection need not have any rotational symmetry. To detect such symmetries, we first backproject the image point set onto a slanted and tilted plane, then perform the symmetry detection above *on that plane*. To specify the slant and tilt entails 2 d.f. Hence our algorithm has a total of 3 d.f. (slant, tilt, θ). Note that the naive way to specify a rotational symmetry axis and its angle of rotation needs 5 d.f. (2 for the x-y projection of the axis, 2 for the x-z projection of the axis, 1 for the angle of rotation). Hence our method represents a significant improvement. It should also be noted that 3 d.f. is well within the computing cabability of today's serial computer. This is even more so with a parallel computer. It is possible that "extremum" may shrink the d.f. further, as with reflectional symmetries. We have not explored this possibility.

So far, our rotational symmetries must lie on planes, which may be slanted and tilted. How about subsets of points which have rotational symmetries but are not necessarily planar ? We note that the normal vector of the slanted and tilted plane is parallel to the rotational axis. Now suppose for such a plane P', there exists two centers of rotation c_1', c_2' with the same angle of rotation, whose projections are c_1, c_2 respectively. Let S_1' and S_2' be the associated three dimensional point sets (which lie on P'). Let S_2'' be a translation of S_2' along the projection vector. Now if $\overline{c_1 c_2}$ on the image plane I is parallel to the gradient vector g of plane P' (Fig 9), a rotational axis exists which S_1' and S_2'' are both rotational symmetrical with.

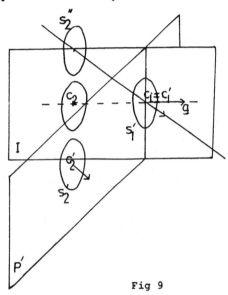

Fig 9

8.2 An algorithm for finding projected rotational symmetry

Algorithm 2 (Projected rotational symmetry finding)

    Input : a two dimensional point set; the projection vector

    for each slant and tilt do

   1.   backproject input point set to slanted and tilted plane passing through origin;

   2.   for each angle of rotation θ do
              for each pair of backprojected point do
                  compute m_θ;
                  project m_θ to image plane and accumulate
                  on the image plane;
              endfor;
              record high points on the image plane, θ, slant and tilt;
            endfor;

    endfor;

Output : a set of projected rotational symmetries.

In the algorithm, m_θ is projected back to the image plane. This allows us to use the image plane as the accumulator array over and over without worrying about re-quantization.

9 A computational solution for the combination of symmetries found

omitted.

Conclusions

We propose a middle out paradigm for shape from contour problem, which is composed of two stages : Level 1 : find invariants in the given images. Level 2 : find the backprojection that provides the best combination of the invariants. We argue that bottom up vision, which uses low level primitives, does not capture the essential properties of the world. On the other hand, top down vision, which uses full models, are computationally impossible. We believe that some intermediate approach like our middle out paradigm captures the best of both paradigms, yet has none of their attendant disadvantages. It may be regarded as a describe-and-match approach.

We have provided theoretical as well as experimental evidence that suggests that energy/information content based approach to shape from contour is a problematic research program. On the other hand, we are able to show that symmetry has a lot of nice and unique advantages, including the interpretation of skewed symmetry as reflectional symmetry.

Next, we show that symmetry may be implemented with highly parallel, robust and computationally inexpensive algorithms (compared with models and the iterative minimization needed for existing shape measures). We report two such algorithms which also substantially extend the state of the art in skewed symmetry finding and rotational symmetry detection. We have also extended Kanade's skewed symmetry heuristic to parallel projection. We propose a general concept for reflectional symmetry known as "extremum symmetry", which we believe is the most general form of invariance due to reflectional symmetry that human beings perceive. All of our results in this paper are applicable to parallel projection.

This paper is also a step towards a theory of shape [34].

Acknowledgement

First and foremost I wish to thank for the supervision of Dr. David Hogg and for his very careful reading of a draft of this paper. I am indebted to the following people for helpful discussions : Prof. Harry Barrow, Dr. David Hogg, Dr. Hlavac Vaclav, Dr David Young. My research at the Dept. of Electronic Engg, Hong Kong Polytechnic laid the seeds of the present work. Prof. Aaron Sloman's lectures on "Philosophical Foundations on Computing and AI" was helpful in shaping some of my thoughts. I thank Alistair, David and Hilary for help with the SUN workstation.

References

1. Yuen, S.Y., Shape from contour using symmetries, CSRP 141, School of Cognitive and Computing Sciences, University of Sussex (1989).

2. Barrow, H.G. and Tenenbaum, J.M., Interpreting line drawings as three dimensional surfaces, Artificial Intelligence **17** (1981) 75-116.

3. Roberts, L.G., Machine perception of three-dimensional solids, Optical and Electro-Optical Information Processing (1965) 159-197.

4. Pentland, A.P., Perceptual organization and the representation of natural form, Artificial Intelligence **28** (1986) 293-331.

5. Pentland, A.P., Recognition by parts, Proceedings 1st International Conference on Computer Vision, London (1987) 612-620.

6. Hanson, A.J., Hyperquadrics : smoothly deformable shapes with convex polyhedral bounds, Comput. Vision Graph. Image Process. **44** (1988) 191-210.

7. Mackworth, A.K., How to see a simple world : an exegesis of some computer programs for scene analysis, Machine Intelligence **8** (1977) 510-537.

8. Marr, D., <u>Vision</u> (Freeman and Company, New York, 1982).

9. Stevens, K.A., The visual interpretations of surface contours, Artificial Intelligence **17** (1981) 47-73.

10. Tsuji, S. and Xu, G., Inferring surfaces from boundaries, Proceedings 1st International Conference on Computer Vision (1987) 716-720.

11. Ulupinar, F. and Nevatia, R., Using symmetries for analysis of shape from contour, Proceedings 2nd International Conference on Computer Vision (1988) 414-426.

12. Witkin, A.P., Recovering suface shape and orientation from texture, Artificial Intelligence **17** (1981) 17-45.

13. Kanatani, K., Detection of surface orientation and motion from texture by a stereological technique, Artificial Intelligence **23** (1984) 213-237.

14. Gibson, E.J., Constrasting emphasis in gestalt theory, information processing, and the ecological approach to perception, in: J. Beck (Ed.), <u>Organization and Representation in Perception</u> (Lawrence Erlbaum Associates, 1982) 159-166.

15. Ballard, D.H., Form perception using transformation networks : polyhedra, in: C. Brown (Ed.), <u>Advances in Computer Vision</u> vol. **2** (Lawrence Erlbaum Associates, 1988) 79-120.

16. Bray, A., Tracking objects using image disparities, Proceedings 5th Alvey Vision Conference (1989) 79-84.

17. Goad, C., Special purpose automatic programming for 3d model-based vision, Proceedings Image Understanding Workshop (1983) 94-104.

18. Weiss, I., Projective invariants of shapes, Proceedings Image Understanding Workshop (1988) 1125-1134.

19. Rao, K. and Nevatia, R., Generalized cone descriptions from sparse 3-D data, Proceedings IEEE Conference on Computer Vision and Pattern Recognition (1986) 256-263.

20. Ponce, J., Chelberg, D. and Mann, W., Invariant properties of the projections of straight homogeneous generalized cylinders, Proceedings 1st International Conference on Computer Vision, London (1987) 631-635.

21. Rosen, J., Symmetry Discovered (Cambridge Univ. Press, 1975).

22. Levitt, T.S., Domain independent object description and decomposition, Proceedings AAAI-84 (1984) 207-211.

23. Kanade, T., Recovery of the three-dimensional shape of an object from a single view, Artificial Intelligence 17 (1981) 409-460.

24. Brady, M. and Yuille, A., An extremum principle for shape from contour, IEEE Trans. Pattern Anal. Mach. 6 (1984) 288-301.

25. Liang, P. and Todhunter, J.S., Three dimensional shape reconstruction from image by minimum energy principle, Proceedings 2nd Conference on Artificial Intelligence Applications (1985) 100-105.

26. Yuen, S.Y., Computer perception of hidden lines, MPhil thesis, Dept. of Electronic Engg., Hong Kong Polytechnic (1988).

27. Marr, D., Early processing of visual information, Phil. Trans. R. Soc. Lond. 275(942) (1976) 483-534.

28. Attneave, F., Prägnanz and soap bubble systems : a theoretical exploration, in: J. Beck (Ed.), Organization and Representation in Perception (Lawrence Erlbaum Associates, 1982) 11-29.

29. Terzopoulos, D., Witkin, A. and Kass, M., Constraints on deformable models : recovering 3D shape and nonrigid motion, Artificial Intelligence 36 (1988) 91-123.

30. Hogg, D.C., Interpreting images of a known moving object, DPhil thesis, Univ. of Sussex (1984).

31. Friedberg, S.A., Finding axes of skewed symmetry, Comput. Vision Graph. Image Process. 34 (1986) 138-155.

32. Ponce, J., Ribbons, symmetries and skewed symmetries, Proceedings Image Understanding Workshop (1988) 1074-1079.

33. Scott, G.L., Turner, S.C. and Zisserman, A., Using a mixed wave/diffusion process to elicit the symmetry set, Proceedings of the 4th Alvey Vision Conference (1988) 221-228.

34. Mumford, D., The problem of robust shape descriptors, Proceedings 1st International Conference on Computer Vision, London (1987) 602-606.

Using Occluding Contours for 3D Object Modeling*

Régis Vaillant

INRIA

Sophia-Antipolis 2004 Route des Lucioles

06565 Valbonne Cedex France

Abstract

This paper presents an algorithm for detecting the occluding contours generated by a surface and for reconstructing depth along them. It also describes an algorithm for computing the two main curvatures of the surface in the neighborhood of the occluding contours. We have used these algorithms on synthetic and real data.

1 Introduction

One of the aims of computer vision is to extract concise surface descriptions from several images of a scene. The descriptions can be used for the purpose of object recognition and for geometric reasoning (such as obstacle avoidance).

Stereovision is often used for recovering the structure of the 3D world. Standard techniques can determine the depth of edges on a surface. These techniques fail with extremal boundaries as these change according to the viewpoint.

Nonetheless, in some cases, they are the only source of 3D information (imagine a white sphere on a black background), if we are not willing to exploit shape from shading techniques. In all cases, they are a rich source of 3D information as will be shown here. In this paper, we propose a new method for detecting extremal boundaries. We also propose an algorithm for reconstructing exactly the curves observed by each camera and computing the principle curvatures of the object surface in their vicinity.

In the first part, we briefly describe the main characteristics of the experimental setup used and we present the theoretical framework of our algorithms. In the second part, we present a method for detecting and reconstructing the extremal boundaries. The third part is devoted to the study of the computation of the two first fundamental forms of the surface in the neighbourhood of the extremal boundary. In the last part, we present results on real and synthetic data and a discuss their accuracy.

---

*This work was supported in part by Esprit project P2502,Voila

2 Background

We assume that after calibration our cameras can be accurately modeled as pinholes. We suppose that we are looking at a smooth object, i.e., whose surface is at least C^2. For a given position of the camera, we can draw the optical rays tangent to the surface of the object. These rays cut on the retinal plane a curve, the *occluding contour*, and touch the object along a smooth curve on its surface, the *rim*.

Several questions can be asked at this point. First what kind of information can be obtained from one occluding contour and second what kind of information can be obtained from several, possibly many, occluding contours obtained from a number of different viewpoints. The first question has been dealt with by Koenderink [5]. In his paper, he proves that concavities and convexities of the visual contour allow to draw implications about the local shape of the surface looked at: convexity of the contour corresponds to a convex patch of the surface while a concavity correspond to a saddle-shaped patch. These conclusions fall from a nice theorem which has also been derived later by Brady [1].

The second question has been addressed by a number of authors, among which Giblin [3], who worked theoretically and Basri and Ullman [6] who worked on the positioning of objects from their occluding contours. Giblin proposes to consider the surface of the object as the envelope of its tangent planes. There are two problems with this: how to compute the envelope of a family of planes and how to handle inflection points. Giblin and Weiss propose to solve the problem by assuming a planar motion of the camera. In a second approach, they derive information about the surface from singular points of the occluding contours.

We also consider the surface of the object as the envelope of its tangent planes but make no assumption about the camera motion or about the projection on the retina plane being orthographic. In fact we deal with the full perspective projection case.

2.0.1 Definitions and notations:

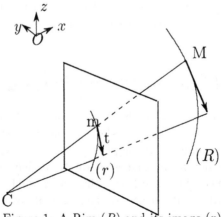

Figure 1: A Rim (R) and its image (r)

As shown in figure 1, we consider a fixed coordinate system $(Oxyz)$; the optical center is at C. The camera looks at the rim (R) which produces the occluding contour (r). A point m on (r) is the image of a point M on (R) at which the optical ray determined by Cm is tangent to the object surface. The tangent plane to the surface at M is defined by the optical ray and the tangent t to the occluding contour at m. Let n be the unit length normal vector to this plane, defined by its Euler angles θ and ϕ and $p(\theta, \phi)$ the distance from the origin to the tangent plane. The equation of this plane can be written as:

$$n(\theta, \phi)^T X - p(\theta, \phi) = 0 \qquad (1)$$

where X is the vector $(x, y, z)^T$ and $n = (\cos(\theta)\cos(\phi), \sin(\theta)\cos(\phi), \sin(\phi))$.

2.0.2 The envelope theorem

Now consider the mapping

$$(\theta, \phi) \to p(\theta, \phi)$$

which associates to every direction the distance from the origin to the plane tangent to the surface whose normal is in the direction (θ, ϕ). In fact, we know that this mapping is locally one to one for elliptic and hyperbolic points [4] but not for parabolic points.

The envelope of the two parameters family of planes defined by equation (1) is obtained by eliminating θ and ϕ between equation (1) and

$$\frac{\partial n(\theta, \phi)}{\partial \theta}^T X - \frac{\partial p(\theta, \phi)}{\partial \theta} = 0 \qquad (2)$$

$$\frac{\partial n(\theta, \phi)}{\partial \phi}^T X - \frac{\partial p(\theta, \phi)}{\partial \phi} = 0 \qquad (3)$$

The physical interpretation of this is that the point M where the plane of equation (1) is tangent to the surface is obtained as the intersection with the planes defined by equations (2), and (3).

Mathematically, there are no difficulties; it is from the practical standpoint that they arise. Indeed, in practice we measure pieces of the surface $(\mathcal{P})$ from which we have to estimate first and second order derivatives which in turn yield properties of the object surface.

3 Detection of the occluding contours

In the previous part, we have assumed that we can detect the extremal boundaries. In fact this is not an easy problem. We will show in this part that sophisticated models are needed. This investigation is interesting as it provides us with some ideas about the numerical stability of the algorithms that we want to implement. We use an algorithm which is a simplified version of the general algorithm which will be detailed in the following part. This simplified algorithm, allows us test the feasability of this kind of computations.

3.1 Edge classification

The most interesting property of occluding edges is that they do not correspond to a physical marking on the surface. This means that they do not correspond to a discontinuity of the normal of the surface or in the reflectance properties of the surface. In spite of this, they are detected, as the other edges by the edge detection process.

We suppose that we have matched segments among different images. We want to verify if they belong to an extremal boundary. One way to proceed is to assume that they belong to one and to write the corresponding equations. We make the hypothesis that the observed surface is part of a cylinder. This provides us, with a number of equations that can be used to compute the parameters of the hypothetical cylinder: its axis and its radius. Fortunately this computation can be divided into two independent parts:

- the direction of the axis.

- the position of the axis and the radius.

▷ **Computation of the direction of the axis of the cylinder**

We know the optical plane corresponding to an image line. The axis of the cylinder is solution of a linear equation which is a function of the normal to the optical plane. The problem turns to be equivalent to finding the smallest eigen-vector for a symmetric matrix.

▷ **Computation of the position of the axis and the radius of the cylinder**

These computations are very simple if we perform them in the right coordinate system. A good one is $(Ouvw)$, where w is the direction of the axis of the cylinder, u and v define an arbitrary frame in the plane $\mathcal{P}$ which is perpendicular to w. The projection of the cylinder is a circle $\mathcal{C}$ and the parameters of this cylinder can be obtained by solving linear equations.

We have a set of equations which can be used to compute the parameters of a cylinder such that the observed line segments are the image of its rim, as seen from each camera. We need a criterion to check whether our hypothesis is correct i.e. do we observe the rim of something which is locally cylindrical or a normal edge. We can first notice that the model we used is still correct if we suppose that the radius of the cylinder is zero. A cylinder of zero radius is physically equivalent to a normal edge. The occluding edges and the normal edges can be classified by performing a test on the value of the radius. There is still a problem: we have to fix a threshold for taking a decision.

We want to estimate the uncertainty on the measure of the radius of the cylinder. We can consider that we have constructed a function f such that

$$(c_1, c_2, r) = f(u_1, v_1, \cdots, u_n, v_n)$$

where (u_i, v_i) are the coordinates of the extremities of the image-segments. We suppose that these values are corrupted by a Gaussian noise of variance σ_{u_i} and σ_{v_i}. In this case, we can express the uncertainty on r by the formula:

$$\sigma = \sqrt{\sum_{u_i} \frac{\partial f(u_i, v_i, \ldots)}{\partial u_i}^2 \sigma_{u_i}^2 + \sum_{v_i} \frac{\partial f(u_i, v_i, \ldots)}{\partial v_i}^2 \sigma_{v_i}^2}$$

The expression of f is

$$f = (A^T A)^{-1} A^T B$$

where A is a matrix $n \times 3$ and B is a matrix of $n \times 1$. They are constructed with linear equations. These matrices depend on (u_i, v_i). We can infer a good criterion for checking if an edge is an occluding edge or not. Since a normal edge is characterized by a zero radius, the criterion is based on the probability for zero to be in the interval of confidence: $r - 2\sigma < 0 < r$.

3.2 Results of the implemented system

We have tested the algorithm on synthetic and real data.

3.2.1 Synthetic data

The test on synthetic data aims at testing the software and the verification of the noise model that we have used.

The principle of this test is to take a description of a system of real cameras and to simulate the observation of a cylinder. In fact, we only compute the image-segment of the extremal boundary of the cylinder. We add some noise to the endpoints of this image-segment. We use a Gaussian noise with a variance of one pixel.

The next table shows the value of the following parameters for a set of five cylinders: the radius of the cylinder, the measured radius, the value of the uncertainty σ and the criterion $\mathcal{C}$.

Real Radius	Estimated Radius	σ	$\mathcal{C}$
0	1.9	24.5	< 0
50	59.2	23.2	0.21
100	98.3	22.1	0.55
130	114.5	21.3	0.62
160	175.2	20.0	0.77

The baseline is approximately 250 millimeters wide and the distance from the optical center of the cameras to the objects is about 800 millimeters.

Nonetheless, we have to keep in mind that the uncertainty on the calibration of the cameras and the determination of the motion have not modelized.

3.2.2 Real data

For the test on real data, we have used small toys. The results are presented in figure (2). On the first image, we show a tea box, a mug and a cylinder with on the left side a tape. We then show two images:

- On the first one, the width of the line depends on the value of the criterion $\mathcal{C}$.

- On the second one, the width of the line depends on the value of the curvature radius.

The dotted lines correspond to segments which have not been matched between at least three images.

We can consider that we have detected an extremal boundary if the radius is non-null with a criterion which is sufficient. The results are satisfactory for the longer segments. There are occasional mistakes for the smaller segments.

One can note for example, that the segment-image of the cylinder are characterized by an important radius and a great value for the criterion $\mathcal{C}$.

The vertical segments are in general more precise because they have a better orientation with respect to the epipolar lines.

 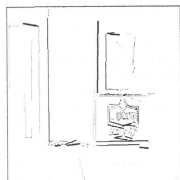

Scene 1, mode 1   Scene 1, mode 2

Figure 2: Scene 1

3.3 Conclusion

The main output of this first part of our work is that it is possible to detect the extremal boundaries. We have to use a model of uncertainty. It is clear that this process can provide false results if the observed occluding contour does not correspond to a cylinder.

The following algorithm handles this case by dealing locally with the occluding contour results are corrected by the algorithm which deals locally with the occluding contour. Supplementary details about the cylinder case can be found in [8].

4 Estimating the object curvature along an occluding edge

In this part, we suppose that we have detected an extremal boundary and we want to compute some properties of the surface in the neighbourhood of the rim. We are interested by the differential properties of order up to two of the surface. Fundamental theorems of Differential Geometry [2] assert that these properties are sufficient to characterize the surface.

- The zero order differential property is the simple estimation of the position of the point. It means that we have to compute the exact position of the contact point between the surface and the optical ray for each of the cameras.

- The first order differential property is the estimation of the tangent plane to the surface. It is the easiest to obtain as we are observing an extremal boundary. In this case the tangent plane is the optical plane.

- The second order properties are the more difficult to obtain as they require the evaluation of second order derivatives. Such computation can be sensitive to noise.

4.1 Estimation of the position of the points

We first notice that the rim (R) of a surface is a curve, and thus the image of the rim (r) must be a curve. It is always true in a generic position. So, we can suppose that we have detected a curve (r_i) in each image. For each of these curves, it is possible to compute the tangent vectors at each of their points. The key idea is to neglect the apparent curvature and to use only the radial curvature. This can be realized by estimating the osculating circle of the radial curve. We have explained in [7] how this estimation can be carried out.

4.2 Computation of the second differential properties of the object shape

In this part, we show that it is possible to compute the two main curvatures of the surface near the points M_i.

4.2.1 Some useful equations

We have previously established that if we observe an occluding boundary and that if we suppose that (θ, ϕ) is an admissible parametrization of the surface in the neighborhood of M_i, located on the rim, we can obtain $X = X(\theta, \phi)$ from equations (1 - 3).

The solution of these equations gives us the analytical expression of $X(\theta, \phi)$. The two fundamental quadratic forms of a surface, which is represented by an admissible parametrization, can be derived from this expression.

At the end of all these computations we obtain that the evaluation of the first and second fundamental quadratic forms requires an estimation of the value of θ, ϕ, $p(\theta, \phi)$, $\frac{\partial p(\theta, \phi)}{\partial \theta}$, $\frac{\partial p(\theta, \phi)}{\partial \phi}$, $\frac{\partial^2 p(\theta, \phi)}{\partial \theta^2}$, $\frac{\partial^2 p(\theta, \phi)}{\partial \phi^2}$, $\frac{\partial^2 p(\theta, \phi)}{\partial \theta \partial \phi}$.

This is very interesting since these values can be estimated with sufficiently good accuracy for the points belonging to an extremal boundary. In [7], we detail the estimation of these values.

4.2.2 A particular case: zero Gaussian curvature

The previous algorithm is exact under the assumption that (θ, ϕ) is an acceptable parametrization of the observed surface in the neighborhood of M_i. For a generic position, this assumption fails if and only if one of the two main curvatures is equal to zero or in other words if the Gaussian curvature is equal to zero. But this is precisely the case where our cylinder model yields directly the answer: the computed radius of the cylinder gives us the first main curvature and the second is zero. The detection of this situation is performed directly by testing the curvatures of the image curves.

5 Experimental Results

We have tested the algorithm mostly on synthetic data. In fact we should say "almost" synthetic since, even though we have been using synthetic models (and one real ball), their rims have been projected on real 512 by 512 images and quantization noise is therefore present in the data. The reason why we have used synthetic images at this stage is that computing the curvatures requires calculating the second order derivatives and the process of differentiation is well known to be noise sensitive. Hence we have decided to isolate the possible sources of error by testing the algorithm on synthetic images first and to delay the experiments using real objects after we gain a better understanding of the numerical stability of the algorithm.

We have conducted a first set of experiments on synthetic images corresponding to a torus and a one-sheet hyperboloid. In order to analyze the results, we have focused on two things, the reconstructed points and the estimate of the curvatures.

Figures (3 - 5) represent the reconstruction obtained for a synthetic torus, a one-sheet hyperboloid and the image of a real volley-ball ballon. The figures show the points which are reconstructed from each camera, and the position of the centers of the circles found when applying the cylinder method.

We notice that we have reconstructed three different chains. On some parts, they are a bit noisy. These parts correspond to points where the epipolar plane (C_i, m_i, C_j) is tangent to the rim (r_j). In this case, the cylinder model fails to reconstruct the rim (R_i). This is not a problem of the method, but a general problem: there are two images which provide the same information. This situation is detected by the algorithm through the test on the variance on the radius of the cylinder.

The other figures represent the computed radii of curvature for the points which belong to the reconstructed rim. These curves are a little more difficult to interpret. Two of them

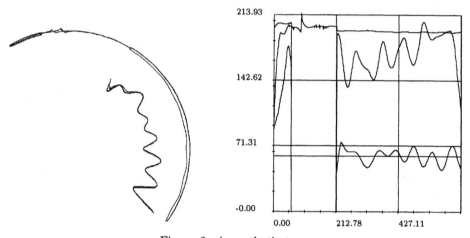

Figure 3: A synthetic torus

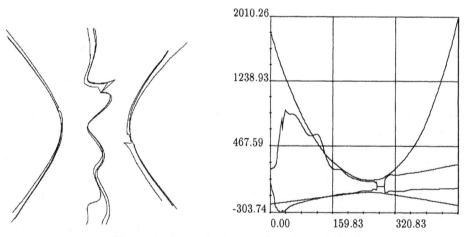

Figure 4: A synthetic one sheet hyperboloid

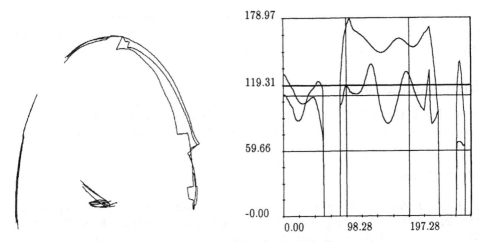

Figure 5: A real volley-ball ballon

correspond to the values of each of the two main curvatures along the extremal boundary. The two other curves correspond to the "theoretical" radii of curvature. Several parts can be distinguished:

- The two curvatures are equal to zero. For these points, the algorithm fails because of the epipolar plane problem.

- One of the two curvatures is equal to zero. This means that we have detect that the Gaussian curvature was equal to zero. This detection is obtained by testing the curvature of the image curve. In this case, the other curve in the figure shows the radius of the cylinder.

- The two curvatures are different from zero. They can be of the same sign (a positive Gaussian curvature) as in the case of the volley-ball or the synthetic sphere. The corresponding points on the surface are elliptic. They can have different signs (a negative Gaussian curvature) as in the case of the synthetic one-sheet hyperboloid. The corresponding points are hyperbolic.

6 Conclusion

In this article, we have shown that occluding edges were a robust source of 3D information. Points on the rim can be accurately reconstructed and good estimates for the second order differential properties of the surface in the vicinity of the rim can be reliably computed. More work needs to be done to test our algorithms further on a larger variety of shapes, study degenerate cases, and include this kind of processing in the framework of an active exploration of an object shape. We are actually testing our algorithm on a large number of images representing several different shapes of occluding edges. We also want use the algorithm with more than three views. The supplementary views will be obtained by moving the object with a known motion. We think that this will improve accuracy greatly.

Acknowledgements

We would like to thank Pr Olivier D. Faugeras and Michel Schmitt for their many helpful conversations and suggestions and Luc Robert for drawing the figures of this article.

References

[1] M. Brady, J. Ponce, A. Yulle, and H. Asada. Describing Surfaces. In Cambrdige MA USA MIT Press, editor, *Proceedings 2nd International Symposium Robotics Research*, 1986.

[2] M. P. Do Carmo. *Differential Geometry of Curves and Surfaces*. Prentice-Hall, Inc., Englewood Cliffs, New Jersey, 1976.

[3] Peter Giblin and R. Weiss. Reconstruction of Surfaces from Profiles. In *First International Conference on Computer Vision*, december 1986.

[4] D. Hilbert and S. Cohen-Vossen. *Geometry and the Imagination*. Berlin : Springer, 1932.

[5] Jan Koenderink. What Does the Occluding Contour Tell Us About Solid Shape? *Perception*, 13:321–330, 1984.

[6] Basri Ronen and Ullman Shimon. The Alignement of Objects with Smooth Surfaces. In *Second International Conference on Computer Vision*, december 1988.

[7] Régis Vaillant and Olivier Faugeras. Using Extremal Boundaries for Recrovering Shape Properties of Objects. In *IEEE Workshop on Interpretation of 3D Scenes*, pages 26–32, november 1989.

[8] Régis Vaillant and Olivier Faugeras. Using Occluding Contours for 3D Object Modelling. In *International Advanced Robotics Programme*, october 1989.

Robust Estimation of Surface Curvature from Deformation of Apparent Contours

Andrew BLAKE   Roberto CIPOLLA

Department of Engineering Science

University of Oxford

Oxford OX1 3PJ, U.K.

Abstract

Surface curvature along extremal boundaries is potentially useful information for navigation, grasping and object identification tasks. Previous theories have shown that qualitative
information about curvature can be obtained from a static view. Furthermore it is known
that, for orthographic projection, under planar viewer-motion, quantitative curvature information is available from spatio-temporal derivatives of flow. This theory is extended here to
arbitrary curvilinear viewer-motion and perspective projection.

We show that curvatures can actually be computed this way in practice, but that they
are highly sensitive to errors in viewer-motion estimates. Intuitively, relative or *differential*
measurements of curvature might be far more robust. Rather than measuring the absolute
deformation of an apparent contour, *differential* quantities depend on the rate at which
surface features are swept over an extremal boundary as the viewer moves. It is shown that,
theoretically, such *differential* quantities are indeed far less sensitive to uncertainty in viewer-
motion. *Ratios* of differential measurements are less sensitive still. In practice sensitivity is
reduced by about two orders of magnitude. We believe this represents a significant step in
the development of practical techniques for robust, qualitative 3D vision.

1   Introduction

The deformation of an apparent contour (the silhouette of a smooth surface or the image of the
extremal boundary) under viewer-motion is a potentially rich source of geometric information
for navigation, motion-planning and object-recognition. Barrow and Tenenbaum [*Barrow78*]
pointed out that surface orientation along an extremal boundary can be computed directly
from image data. Koenderink [*Koenderink84*] related the curvature of an apparent contour
to the intrinsic curvature of the surface (Gaussian curvature); the sign of Gaussian curvature
is equal to the sign of the curvature of the contour. Convexities, concavities and inflections
of an apparent contour indicate, respectively, convex, hyperbolic and parabolic surface points.
Giblin and Weiss [*Giblin87*] have extended this by adding viewer motions to obtain quantitative
estimates of Gaussian and mean curvature. A surface can be reconstructed from the envelope
of all its tangent planes, which in turn are computed directly from the family of silhouettes of
the surface, obtained under planar motion of the viewer. By assuming that the viewer follows
a *great circle* of viewer directions around the object they restricted the problem of analysing
the envelope of tangent planes (a 2-parameter family) to the less general one of computing the
envelope of a family of lines in a plane. Their algorithm was tested on noise-free, synthetic data
(on the assumption that extremal boundaries had been distinguished from other image contours)
demonstrating the reconstruction of a planar curve under orthographic projection.

This paper extends these theories further, to the general case of arbitrary non-planar camera motion under perspective projection. The Gaussian curvature of a surface, at a point on its silhouette, can be computed given some known local motion of the viewer. Curvature is computed from spatio-temporal derivatives (up to second order) of image-measurable quantities. The theory can, of course, be applied to detect extremal boundaries and distinguish them from surface markings or discontinuities. Experiments show that, with adequate viewer-motion calibration, itself computed from visual data [Tsai87], it is possible to obtain curvature measurements of useful accuracy.

A consequence of the theory, representing an important step towards qualitative vision, concerns the robustness of *relative* or *differential* measurements of curvature at two nearby points. Intuitively it is relatively difficult to judge, moving around a smooth, featureless object, whether its silhouette is extremal or not — whether the Gaussian curvature along the contour is bounded or not. This judgment is much easier to make for objects with feature-rich surfaces. Under small viewer-motions, features are "sucked" over the extremal boundary, at a rate which depends on surface curvature. Our theory reflects this intuition exactly. It is shown that relative measurements of curvature across two adjacent points are entirely immune to uncertainties in the viewer's rotational velocity. This is somewhat related to earlier results showing that relative measurements of this kind are important for depth measurement from optic flow [LHiggins80, Weinshall89] and for curvature measurements from stereoscopically viewed highlights [Blake88]. Furthermore, they are relatively immune to uncertainties in translational motion in that, unlike single-point measurements, they are independent of the viewer's acceleration. Only dependence on velocity remains. Experiments show that this theoretical prediction is borne out in practice. *Differential* or relative curvature measurements prove to be more than an order of magnitude less sensitive than single-point measurements to errors in viewer-motion calibration. There is some theoretical evidence that *ratios* of differential curvature measurements are less sensitive. In our experiments absolute measurements of curvature were so sensitive that they became unreliable for viewer motion errors of 0.5mm in position and 1mrad in orientation. For ratios of *differential* measurements of curvature the corresponding figures were about 50mm and 70mrad respectively.

2 Theoretical framework

2.1 Surface Geometry

Consider a point P on the extremal boundary of a smooth, curved surface in R^3 and parameterised locally by a vector valued function $\mathbf{r}(s, t)$. The parametric representation can be considered as covering the surface with 2 families of curves: $\mathbf{r}(s, t_0)$, and $\mathbf{r}(s_0, t)$ where s_0, t_0 are fixed for a given curve in the family. A one-parameter family of views is indexed by the time parameter t and s, t are defined so that the s-parameter curve, $\mathbf{r}(s, t_0)$, is an extremal boundary for a particular view t_0. A t-parameter curve $\mathbf{r}(s_0, t)$ can be thought of as the 3D locus of points grazed by a light-ray from the viewer, under viewer-motion. Such a locus is not uniquely defined.

Local surface geometry can be specified in terms of the basis $\{\mathbf{r}_s, \mathbf{r}_t\}$ for the tangent plane ($\mathbf{r}_s$ and $\mathbf{r}_t$ denote $\partial \mathbf{r}/\partial s$ and $\partial \mathbf{r}/\partial t$ respectively) and the surface normal - a unit vector $\mathbf{n}$.

2.2 Imaging model

The imaging model is a spherical pin-hole camera of unit radius. The image of the world point, P, with position vector $\mathbf{r}(s, t)$ is a unit vector $\mathbf{T}(s, t)$ defined by

$$\mathbf{r}(s, t) = \mathbf{v}(t) + \lambda \mathbf{T}(s, t), \tag{1}$$

where λ is the distance along the ray to the point P (figure 1).

For a given vantage position t_0 the apparent contour is a continuous family of rays $\mathbf{T}(s, t_0)$ emanating from the camera's optical centre which touch the surface so that $\mathbf{T}.\mathbf{n} = 0$. The moving observer at position $\mathbf{v}(t)$ sees a 2 parameter family of apparent contours $\mathbf{T}(s, t)$.

2.3 Properties of the extremal boundary and its projection

In [Blake89] we derive for perspective projection the following well-known properties of the extremal boundary and its projection [Barrow78, Koenderink82, Brady85, Giblin87].

1. The orientation of the surface normal, $\mathbf{n}$, can be recovered by measuring the direction of the ray $\mathbf{T}$ of a point on an extremal boundary and the tangent to the apparent (image) contour, $\mathbf{T}_s$.

$$\mathbf{n} = \frac{\mathbf{T} \wedge \mathbf{T}_s}{|\mathbf{T} \wedge \mathbf{T}_s|}. \tag{2}$$

2. The ray direction, $\mathbf{T}$, and the tangent to the extremal boundary, $\mathbf{r}_s$, are in *conjugate* directions with respect to the second fundamental form

$$\mathbf{T}.\mathbf{n}_s = 0 \tag{3}$$

The ray direction and the extremal boundary will only be perpendicular if the ray $\mathbf{T}$ is along a *principal* direction.

3. The curvature of the apparent contour, κ^p, (more precisely the geodesic curvature of the curve, $\mathbf{T}(s, t_0)$ which can be computed from spatial derivatives in image measurements), can be written in terms of the *normal curvature* of the extremal boundary, κ^s:

$$\kappa^p = \frac{\mathbf{T}_{ss}.\mathbf{n}}{|\mathbf{T}_s|^2} \tag{4}$$

$$= \lambda \frac{\kappa^s}{1 - (\mathbf{T}.\frac{\mathbf{r}_s}{|\mathbf{r}_s|})^2} \tag{5}$$

Equations (4) and (5) shows that an apparent contour is smooth except for a special viewing geometry when the ray direction runs along the extremal boundary. At such points the apparent contour may have a cusp. For opaque surfaces only one branch of the cusp is visible, however, corresponding to a contour-ending [Koenderink82, Koenderink84].

2.4 Choice of Parameterisation

There is no unique spatio-temporal parameterisation of the surface. The mapping between extremal boundaries at successive instants is undetermined. The problem of choosing a parameterisation is an "aperture problem" for contours on the spherical perspective image ($\mathbf{T}(s,t)$) or on the Gauss sphere ($\mathbf{n}(s,t)$), or between space curves on the surface $\mathbf{r}(s,t)$). A natural parameterisation is the "epipolar parameterisation" defined by

$$\mathbf{r}_t \wedge \mathbf{T} = 0. \tag{6}$$

For this parameterisation the tangent-plane basis vectors $\mathbf{r}_s$ and $\mathbf{r}_t$ are in *conjugate* directions (from (3)).

Differentiating (1) with respect to time and enforcing (6) leads to the "matching" condition[1]

$$\mathbf{T}_t = \frac{(\mathbf{v}_t \wedge \mathbf{T}) \wedge \mathbf{T}}{\lambda}. \tag{7}$$

Points on different contours are "matched" by moving along great-circles on the image sphere with poles defined by the direction of the viewer's instantaneous translational velocity $\mathbf{v}_t$. This induces a natural *correspondence* on the surface between extremal boundaries from different viewpoints. If the motion is linear corresponding points on the image sphere will lie on an epipolar great-circle. This is equivalent to Epipolar Plane matching in stereo. For a general motion, however, the epipolar structure rotates continuously as the direction of $\mathbf{v}_t$ changes and the space curve, $\mathbf{r}(s_0, t)$, generated by the movement of a contact point will be non-planar.

The parameterisation will be degenerate when $\{\mathbf{r}_s, \mathbf{r}_t\}$ fails to form a basis for the tangent plane. This occurs if the contour is not an extremal boundary but a 3D rigid space curve (when $\mathbf{r}_t = 0$) or at a cusp/contour-ending in the projection (when $\mathbf{r}_s \wedge \mathbf{r}_t = 0$, see earlier)[Blake89]. The parameterisation degrades gracefully and hence these conditions pose no special problems.

2.5 Information available from the deformation of the apparent contour

We show in [Blake89] that local surface geometry can be recovered from spatio-temporal derivatives (up to 2nd order) of image measurable quantities and known viewer motion. We summarise the most important results below.

1. **Recovery of depth**

   Depth (distance along the ray, λ) can be computed from the rate of deformation ($\mathbf{T}_t$) of the apparent contour under known viewer motion (translational velocity $\mathbf{v}_t$)[Bolles87]. The depth is given by

   $$\lambda = -\frac{\mathbf{v}_t.\mathbf{n}}{\mathbf{T}_t.\mathbf{n}} \tag{8}$$

---

[1]If we choose the reference frame to be the instantaneous camera co-ordinate system we can express $\mathbf{T}$ and $\mathbf{T}_t$ in terms of an spherical image position vector, $\mathbf{Q}$, and image velocities (optic flow) $\mathbf{Q}_t$. Namely $\mathbf{T} = \mathbf{Q}$ and $\mathbf{T}_t = \mathbf{Q}_t + \omega \wedge \mathbf{Q}$. Equation (7) reduces to the equation of motion and structure from motion from optic flow [Maybank85].

This formula is an infinitesimal analogue of triangulation with stereo cameras. The numerator is analogous to baseline and the denominator to disparity. The result also holds for a rigid space curve or an occluding boundary. It is independent of choice of parameterisation [Blake89].

2. Local surface curvature

The normal curvature at P in the direction of the ray $\mathbf{T}$, κ^t (which is the same as the normal curvature of the space curve, $\mathbf{r}(s_0, t)$) can be computed from the rate of deformation ($\mathbf{T}_t$) of the apparent contour under viewer motion, and its temporal derivative. This requires knowledge of viewer motion (translational and rotational velocity and acceleration)

$$\kappa^t = \frac{(\mathbf{T}_t.\mathbf{n})^3}{(\mathbf{T}_{tt}.\mathbf{n})(\mathbf{v}_t.\mathbf{n}) + 2(\mathbf{T}.\mathbf{v}_t)(\mathbf{T}_t.\mathbf{n})^2 - (\mathbf{v}_{tt}.\mathbf{n})(\mathbf{T}_t.\mathbf{n})} \qquad (9)$$

The sign and magnitude of Gaussian curvature can then be computed from the product of the normal curvature κ^t, and the curvature of the apparent contour, κ^p,(measured by (4)) scaled by inverse-depth [Koenderink84]

$$\kappa_{gauss} = \frac{\kappa^p \kappa^t}{\lambda} \qquad (10)$$

3 Experimental Results: Determining curvatures from absolute measurements

Figure 2 shows 3 views from a sequence of a scene taken from a camera mounted on a moving robot-arm whose motion has been accurately calibrated from visual data [Tsai87]. Using a numerical method for estimating surface curvatures from 3 discrete views (see [Blake89]) we can estimate the radius of curvature of the *normal section* R (where $\kappa^t = 1/R$) for a point on an extremal boundary of a cup, B. The method is repeated for a point which is not on an extremal boundary but is a surface marking, A. This is a degenerate case of the parameterisation. A surface marking can be considered as the limiting case of a point with infinite curvature and hence ideally will have zero "radius of curvature". If the measurements are error-ridden and the motion is not known accurately, however, surface markings will appear as extremal boundaries on surfaces with high curvature.

	measured (mm)	actual (mm)	error (mm)
surface marking A	1.95	0.0	1.95
extremal boundary B	45.7	44.4	1.3

Table 1. Radius of curvature (of *normal section* defined by the ray direction) estimated from 3 distinct views of a point on a surface marking and a point on an extremal boundary.

The radius of the cup was measured using calipers as 44.4 ± 2mm. The estimated curvatures agree with the actual curvatures. However, the results are very sensitive to perturbations in the motion parameters (figure 3a and 3b).

4 Differential measurement of curvature

We have seen that although it is perfectly feasible to compute curvature from the observed deformation of an apparent contour, the result is highly sensitive to motion calibration errors. This may be acceptable for a moving camera mounted on a precision robot-arm or when a grid is in view so that accurate visual calibration can be performed. In such cases it is feasible to determine motion to the accuracy of around 1 part in 1000 that is required. However, when only crude estimates of motion are available another strategy is called for. It is sometimes possible in such a case to use the crude estimate to bootstrap a more precise visual egomotion computation [*Harris87*]. However this requires an adequate number of identifiable corner features, which may not be available in an unstructured environment. Moreover, if the estimate is too crude the egomotion computation may fail; it is notoriously ill-conditioned [*Tsai84*].

The alternative approach is to seek qualitative measurements of geometry that are much less sensitive to error in the motion estimate. In this section we show that relative or *differential* measurements of curvature have just this property. Differences of measurements at two points are insensitive to errors in rotation and in translational acceleration. Typically, the two features might be one point on an extremal boundary and one fixed surface point. The surface point has infinite curvature and therefore acts simply as a stable reference point for the measurement of curvature at the extremal boundary. Intuitively the reason for the insensitivity of differential curvature is that global additive errors in motion measurement are cancelled out.

Consider two visual features whose projections on the image sphere are $\mathbf{T}(s_i, t)$, $i = 1, 2$ which we will abbreviate to $\mathbf{T}^i$, $i = 1, 2$. Think of them as two points on extremal boundaries, which trace out curves with (normal) curvatures κ^{t1} and κ^{t2} as the viewer moves. The first temporal derivatives of $\mathbf{T}^i$ are dependent only on image position, viewer velocity and rotation and depth:

$$\mathbf{T}_t^i = \frac{(\mathbf{v}_t \wedge \mathbf{T}^i) \wedge \mathbf{T}^i}{\lambda} \tag{11}$$

Second order temporal derivatives are,

$$\mathbf{T}_{tt}^i \cdot \mathbf{n} = \frac{1}{\lambda} \left[-\frac{(\mathbf{T}_t^i \cdot \mathbf{n})^2}{\kappa^{ti}} + 2(\mathbf{T}^i \cdot \mathbf{v}_t)(\mathbf{T}_t^i \cdot \mathbf{n}) - \mathbf{v}_{tt} \cdot \mathbf{n} \right] \quad i = 1, 2. \tag{12}$$

Let us define two relative quantities. The *differential curvature* $\Delta \kappa^t$ of the feature pair is defined by

$$\frac{1}{\Delta \kappa^t} = \frac{1}{\kappa^{t1}} - \frac{1}{\kappa^{t2}}. \tag{13}$$

Note that it is not an infinitesimal quantity but a difference of inverse curvature. The *relative view vector* is defined to be

$$\delta(t) = \mathbf{T}(s_2, t) - \mathbf{T}(s_1, t) \tag{14}$$

Consider the two features to be instantaneously spatially coincident, that is, initially, $\mathbf{T}(s_1, t) = \mathbf{T}(s_2, t)$. Moreover assume they lie at a common depth λ, and hence, instantaneously, $\mathbf{T}_t^1 = \mathbf{T}_t^2$. In practice, of course, the feature pair will only coincide exactly if one of the points is a surface marking which is instantaneously on the extremal boundary. Now, taking the difference of equation (12) for $i = 1, 2$ leads to a relation between these two *differential* quantities:

$$\delta_{tt}.\mathbf{n} = \frac{(\mathbf{v}_t.\mathbf{n})^2}{\lambda^3} \frac{1}{\Delta\kappa^i}. \tag{15}$$

From this equation we can obtain *differential curvature* $\Delta\kappa^t$ as a function of depth λ, viewer velocity $\mathbf{v}_t$, and the 2nd temporal derivative of δ. Absolute measurement of curvature (12) depended also on the viewer's translational acceleration $\mathbf{v}_{tt}$. Uncertainty from practical measurements (based on finite differences, for example) of the lower derivative will, of course, be much reduced. Hence the relative measurement should be much more robust. Moreover, because δ is a relative measurement on the projection sphere, unlike the image vectors $\mathbf{T}^i$ which occur in the absolute measurement of curvature, it is unaffected by errors in viewer rotation.

In the case that $\mathbf{T}^1$ is known to be a fixed surface reference point, with $1/\kappa^{t1} = 0$, then $\Delta\kappa^t = \kappa^{t2}$ so that the differential curvature $\Delta\kappa^t$ constitutes an estimate, now much more robust, of the normal curvature κ^{t2} at the extremal boundary point $\mathbf{T}^2$. Of course this can now be used in equation(10) to obtain a robust estimate of gaussian curvature.

Our experiments confirm this. Figures 4a and 4b show that the sensitivity of the differential curvature to error in position and rotation computed between points A and B (2 nearby points at similar depths) is reduced by an order of magnitude. This is a striking decrease in sensitivity even though the features do not coincide exactly as the theory required.

Further robustness can be obtained by considering the ratio of differential curvatures. Ratios of two-point differential curvature measurements are, in theory, completely insensitive to viewer motion [Blake89]. This is because in a *ratio* of $\Delta\kappa^t$ measurements for two *different* point-pairs, terms depending on absolute depth λ and velocity $\mathbf{v}_t$ are cancelled out in equation (15). This result corresponds to the following intuitive idea. The rate at which surface features rush towards or away from an extremal boundary is proportional to the (normal) curvature there. The constant of proportionality is some function of viewer-motion and depth; it can be eliminated by considering only ratios of curvatures. Results (figures 5a and 5b) show another striking decrease in sensitivity – of another order of magnitude.

5 Conclusion

We conclude that in theory and practice given just one surface reference point, highly robust relative curvature measurements can be made at points on apparent contours. Moreover, the technique can achieve by motion analysis something which has so far eluded photometric analysis: namely discrimination between fixed surface features and points on extremal boundaries.

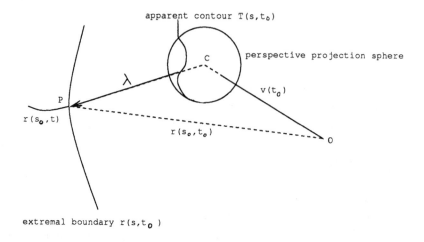

Figure 1. Surface and Viewing Geometry.

P lies on a smooth surface which is parameterised by $\mathbf{r}(s,t)$. For a given vantage point, $\mathbf{v}(t_0)$, the family of rays emanating from the viewer's optical centre (C) that touch the surface defines an s-parameter curve $\mathbf{r}(s,t_0)$ – the extremal boundary from vantage point t_0. The spherical perspective projection of this extremal boundary – the apparent contour, $\mathbf{T}(s,t_0)$ – determines the direction of rays which grazes the surface. The distance along each ray, CP, is λ. A moving observer at position $\mathbf{v}(t)$ sees a 2 parameter family of extremal boundaries $\mathbf{r}(s,t)$ whose spherical perspective projections are represented by a 2 parameter family of apparent contours $\mathbf{T}(s,t)$. t-parameter curves ($\mathbf{r}(s_0,t)$ and $\mathbf{T}(s_0,t)$) are not uniquely defined.

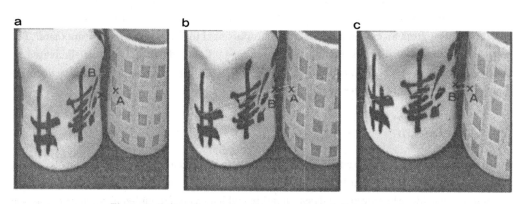

Figure 2. Estimating surface curvatures from 3 discrete views

Points are selected on image contours in view 1 (a), indicated by crosses A and B for points on a surface marking and extremal boundary respectively. For *Epipolar parameterisation* of the surface corresponding features lie on epipolar lines in views 2 and 3 (figures 2b and 2c). Measurement of 3 view vectors lying in an epipolar plane can be used to estimate surface curvatures.

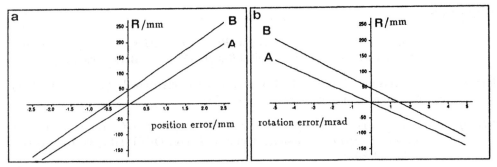

Figure 3. Sensitivity of curvature estimated from absolute measurements to errors in motion.
The radius of curvature (mm) for both a point on a surface marking (A) and a point on an extremal boundary (B) is plotted against error in the estimate of position (a) and orientation (b) of the camera for view 2. The estimation is very sensitive and a perturbation of 1mm in position produces an error of 190% in the estimated radius of curvature for the point on the extremal boundary. A perturbation of 1mrad in rotation about an axis defined by the epipolar plane produces an error of 70%.

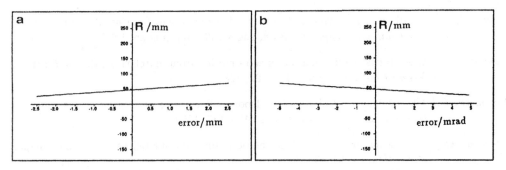

Figure 4. Sensitivity of *differential curvature*
The difference in radii of curvature between a point on the extremal boundary and the nearby surface marking is plotted against error in the position (a) and orientation (b) of the camera for view 2. The sensitivity is reduced by an order of magnitude to 17% per mm error and 8% per mrad error respectively.

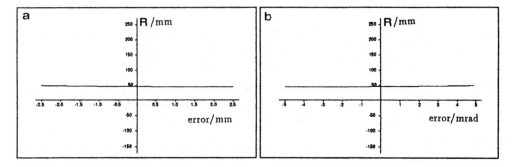

Figure 5. Sensitivity of ratio of *differential curvatures*
The ratio of *differential curvatures* measurements made between 2 points on an extremal boundary and the same nearby surface marking is plotted against error in the position (a) and orientation (b) of the camera for view 2. The sensitivity is further reduced by an order of magnitude to 1.5% error for a 1mm error in position and 1.1% error for 1mrad error in rotation. The vertical axes are scaled by the actual curvature for comparision with figures 3 and 4.

References

[Barrow78]   H.G. Barrow and J.M. Tenenbaum. *Recovering Intrinsic Scene Characteristics from Images*. A.I Center Technical Report 157, SRI International, 1978.

[Blake88]   A. Blake and G. Brelstaff. Geometry from specularities. In *Proc. 2nd Int. Conf. on Computer Vision*, pages 394–403, 1988.

[Blake89]   A. Blake and R. Cipolla. *Robust Estimation of Surface Curvature from Deformation of Apparent Contours*. Technical Report OUEL 1787/89, University of Oxford, 1989.

[Bolles87]   R.C. Bolles, H.H. Baker, and D.H. Marimont. Epipolar-plane image analysis: an approach to determining structure. *International Journal of Computer Vision*, vol.1:7–55, 1987.

[Brady85]   M. Brady, J. Ponce, A. Yuille, and H Asada. Describing surfaces. *Computer Graphics Image Processing*, 32:1–28, 1985.

[Giblin87]   P. Giblin and R. Weiss. Reconstruction of surfaces from profiles. In *Proc. 1st Int. Conf. on Computer Vision*, pages 136–144, London, 1987.

[Harris87]   C.G. Harris. Determination of ego - motion from matched points. In *3rd Alvey Vision Conference*, pages 189–192, 1987.

[Koenderink82] J.J. Koenderink and A.J. Van Doorn. The shape of smooth objects and the way contours end. *Perception*, 11:129–137, 1982.

[Koenderink84] J.J. Koenderink. What does the occluding contour tell us about solid shape? *Perception*, 13:321–330, 1984.

[LHiggins80]   H.C. Longuet-Higgins and K. Pradzny. The interpretation of a moving retinal image. *Proc.R.Soc.Lond.*, B208:385–397, 1980.

[Maybank85]   S.J. Maybank. The angular velocity associated with the optical flow field arising from motion through a rigid environment. *Proc. Royal Society, London*, A401:317–326, 1985.

[Tsai84]   R.Y. Tsai and T.S. Huang. Uniqueness and estimation of three-dimensional motion parameters of a rigid objects with curved surfaces. *IEEE Trans. on Pattern Analysis and Machine Intelligence*, 6(1):13–26, 1984.

[Tsai87]   R.Y. Tsai. A versatile camera calibration technique for high-accuracy 3d machine vision metrology using off-the-shelf tv cameras and lenses. *IEEE Journal of Robotics and Automation*, RA-3(4):323–344, 1987.

[Weinshall89]   D. Weinshall. *Direct computation of 3D shape and motion invariants*. AI Memo 1131, MIT, 1989.

Acknowledgments

The authors acknowledge discussions with Professor Mike Brady, Dr Andrew Zisserman, Dr Peter Giblin, Dr David Murray and Dr David Forsyth. Roberto Cipolla acknowledges the support of the IBM UK Scientific Centre.

SPATIAL LOCALIZATION OF MODELLED OBJECTS OF REVOLUTION IN MONOCULAR PERSPECTIVE VISION

M.DHOME, J.T.LAPRESTE, G.RIVES & M.RICHETIN

Electronics Laboratory, UA 830 of the CNRS
Blaise Pascal University of Clermont-Ferrand
63177 AUBIERE CEDEX (FRANCE)

tel: 73.40.72.28 ; fax : 73.40.72.62
email : EVALEC @ FRMOP11.BITNET

1 INTRODUCTION

In monocular vision, recognition and spatial localization are often difficult problems because the appearance of the observed objects varies according to the point of view. The various aspects of an object can be radically different as shown in Figure 1.

For polyhedral objects, i.e. the block-world, many works have been done on the labelling of the visible edges and on the interpretation of the contours in terms of shape, of relative positions of plane facets, deduced from this labelling at the intersection of the edges [SHI-87]. Moreover it has been proved that localization of modelled polyhedra can be obtained from one single image [DHO-89].

For curved objects, the contours which can be detected in an image and which result from different physical phenomena in the image formation process, are curves whose geometrical characteristics depend on the shape of the object. They are the projections of space curves whose position on the surface of the object is related with the point of view.

For objects of revolution, much information can be derived from a single perspective view, especially their localization in the viewer coordinate system.

Figure 1 presents differents aspects of the image of a vase. The shape of the contours which could be extracted would be very variable from one image to an other. But if we look at the external contours of such an object, i.e. the contours which separate the vase from the background, it is worth noting that they are of two different kinds :
- either they come from the perspective projection of angular edges on the object surface which are obviously circles,
- or they are the projections of limbs i.e. of space curves on the object surface at each point of them the normal of the surface is orthogonal to the viewing direction of this point.

On figure 1, the recurrence of some shape or special points is noticeable in the external contour : ellipses, zero-curvature points, angular points. These three geometrical features are very useful to solve the localization problem as it will be shown in the next sections.

For modelled curved objects, especially generalized cylinders, J. Ponce [PON-89b] has recently proposed a very general algebraic approach for recognition and localization. He made successful experiments on objects of revolution. But we believe that it will be always valuable to exploit geometrical properties of the contours of an object since we will see that the resolution of the localization problem is quite simple in this way. Moreover this geometrical approach gives a deep understanding of the perspective projection of curves and surfaces.

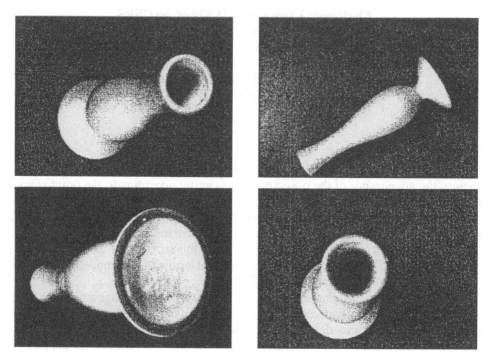

Figure 1. Various aspects of the image of a vase.

2 INTERPRETATION OF ELLIPTICAL CONTOURS

2.1 The cyclic plane problem

Let us consider an object of revolution having a circle as an angular edge, for which the support plane, the center and the radius are known in an object coordinate system. Let us suppose also that the perspective projection of this circle is viewed in the image. From this projection, which is an ellipse in the general case, it is possible to recover the spatial attitude of the object.

If the ellipse has been estimated by the following equation,

$$a_0 x^2 + b_0 xy + c_0 y^2 + d_0 x + e_0 y + 1 = 0$$

the equation of the cone defined by the optical center of the camera and by the viewing directions of all the points of the ellipse is :

$$a_0 x^2 + b_0 xy + c_0 y^2 + \frac{d_0}{f} xz + \frac{e_0}{f} yz + \frac{1}{f^2} z^2 = 0 \tag{1}$$

with f being the focal length of the camera.

The location problem is then a cyclic plane problem, a cyclic plane being a plane whose intersection with a cone is a circle. Moreover this circle must have a given radius here.

2.2 Determining the cyclic planes

In order to simplify this determination, it is useful to choose a viewer coordinate system such that the equation of the cone is reduced to :

$$\frac{x_1^2}{a_1^2} + \frac{y_1^2}{b_1^2} - \frac{z_1^2}{f^2} = 0 \quad \text{with} \quad \frac{1}{a_1^2} \geq \frac{1}{b_1^2} \tag{2}$$

This new coordinate system is obtained after a rotation R_D of the original one which makes diagonal the following matrix :

$$\begin{pmatrix} a_0 & \frac{b_0}{2} & \frac{d_0}{2f} \\ \frac{b_0}{2} & c_0 & \frac{e_0}{2f} \\ \frac{d_0}{2f} & \frac{e_0}{2f} & \frac{1}{f^2} \end{pmatrix}$$

In this coordinate system, the great axis of the ellipse can always be horizontal (X axis). To solve the cyclic plane problem, then it is sufficient to find the rotation R_α of angle α around the horizontal axis for which the intersection of the cone by a vertical plane is a circle.

After the application of rotation R_α, the equation of the cone becomes :

$$\frac{x_1^2}{a_1^2} + \left(\frac{\cos^2\alpha}{b_1^2} - \frac{\sin^2\alpha}{f^2} \right) y_1^2 + \left(\frac{\sin^2\alpha}{b_1^2} - \frac{\cos^2\alpha}{f^2} \right) z_1^2 + 2 \left(\frac{\sin\alpha\cos\alpha}{b_1^2} + \frac{\sin\alpha\cos\alpha}{f^2} \right) y_1 z_1 = 0 \tag{3}$$

The intersection of the cone by a vertical plane (constant z_1) is a circle if and only if the coefficients of x_1^2 and y_1^2 are equal. This gives equation (4) :

$$\frac{1}{a_1^2} = \frac{\cos^2\alpha}{b_1^2} - \frac{\sin^2\alpha}{f^2} \tag{4}$$

which implies the following determinations for $\cos\alpha$ and $\sin\alpha$:

$$\cos^2\alpha = \frac{b_1^2(a_1^2 + f^2)}{a_1^2(b_1^2 + f^2)} \qquad \sin^2\alpha = \frac{f^2(a_1^2 - b_1^2)}{a_1^2(b_1^2 + f^2)}$$

and a multiple determination for angle α :

$$\alpha_1 \quad , \quad \alpha_2 = \pi - \alpha_1 \quad , \quad \alpha_3 = \pi + \alpha_1 \quad , \quad \alpha_4 = -\alpha_1.$$

A quadratic cone has then two families of cyclic planes. For any plane of each family there are two positions of the object since it can placed in each of the two half spaces delimited by this plane. Then there are four families of positions of the object.

To solve the localization problem completely, the center of the circle has to be determined. Thus four spatial attitudes for the object will be got.

2.3 Finding the center of the circle

From equations (3) and (4), the cone equation is :

$$x_1^2 + y_1^2 + a_1^2 \left(\frac{1}{b_1^2} - \frac{1}{a_1^2} - \frac{1}{f^2} \right) z_1^2 \pm 2 \frac{\sqrt{a_1^2 + f^2} \sqrt{a_1^2 - b_1^2}}{b_1 f} y_1 z_1 = 0 \tag{5}$$

The symbol $\pm$ corresponds with the sign of the product $\cos \cdot \sin$ for the considered angle α_i.

As the space position of a circle of radius ρ must compatible with the cone equation, this circle must be in a vertical plane and its center C must be in the plane $(y_1, 0, z_1)$. The coordinates of C are then of the form $(0, y_{1c}, z_{1c})$ which corresponds with the following cone equation :

$$x_1^2 + y_1^2 - \frac{\rho^2 + y_{1c}^2}{z_{1c}^2} z_1^2 + 2\frac{y_{1c}}{z_{1c}} y_1 z_1 = 0 \tag{6}$$

The coordinates of C are derived from equations (5) and (6).

The sign $\pm$ has the same meaning as in equation (5).

$$\begin{cases} x_{1c} = 0 \\ y_{1c} = \pm\rho\dfrac{\sqrt{a_1^2 + f^2}\sqrt{a_1^2 - b_1^2}}{a_1^2} \\ z_{1c} = \dfrac{\rho b_1 f}{a_1^2} \end{cases}$$

The position of the circle center C in the original viewer coordinate system is obtained by applying the rotation $R_{\alpha_i}^{-1}$ and the rotation R_D^{-1}.

2.4 Experiment

The preceding method allows to calculate the four spatial attitudes of a given radius circle defined by the position of its center and by the normal to its support plane, which is compatible with an elliptical contour. Knowing the position of a circular edge on the surface of an object of revolution, it is then straightforward to determine the rigid transform to apply to the object model to bring it in position such that the projection of the circular edge will give the observed elliptical contour.

Figure 2 (left) shows the detected contours of a brightness image of a vase. One of the detected ellipses appears in figure 2 (right). Figure 3 presents the four attitudes of the object model which have been determined from the interpretation of this ellipse.

It is quite visible that three of them are not valid. The quality of the covering of the external contours of the projected model image by the contours of figure 2 (left) can be evaluated [RIC-89]. A score between 0 and 1 is calculated (see figure 3) and the selected solution has the highest score. Both the brightness image and the projected model image for the selected solution are presented in figure 4.

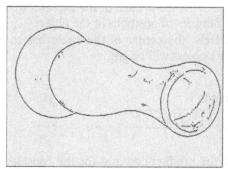 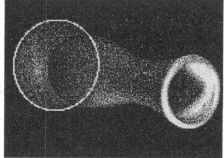

Figure 2. Detected Contours and selected detected ellipse.

A localization error occurs in the area of the top of the vase. It is due an inaccurate determination of the values of the ellipse parameters. But an iterative adjustement procedure can be used to refine the attitude. It involves matching of straight segments (straight contours after polygonal segmentation of the contour image and projections of the external edges of the object polyhedral model), a bucketing technique to fasten this matching, and the Lowe's algorithm [LOW-85] to calculate a new attitude after the matching. At the end of this procedure, a better attitude is obtained (figure 4).

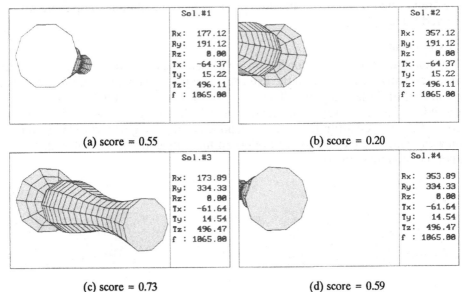

(a) score = 0.55   (b) score = 0.20

(c) score = 0.73   (d) score = 0.59

Figure 3. The four compatible attitudes and the corresponding scores.

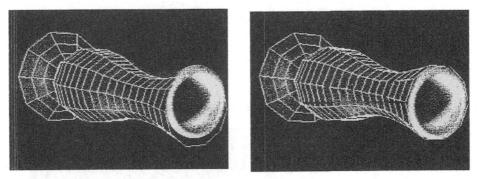

Figure 4. Selected and final model attitudes.

3 INTERPRETATION OF ZERO-CURVATURE CONTOUR POINTS

3.1 Principle

In this section, it is recalled how the zero-curvature of the contours of an object of revolution can be used to locate it.

This method relies on the following theorem [KOE-84, PON-89a] :

Theorem : *The points of the limbs of an object of revolution which correspond with zero-curvature points of the scaling function of the object, are viewed under perspective projection as zero-curvature points of the contours corresponding to the projection of the limbs.*

Thus if the scaling function of an object of revolution has at least one zero-curvature point and if the point of view is such that the limbs go through at least one section corresponding to a zero-curvature point of the scaling function, then the external contours of the object contains at least one pair of zero-curvature points.

The two points of a pair and the tangents to the contours at these points define a triangle. It has been shown [RIC-89] that this triangle is the perspective projection of a cone of revolution defined by the corresponding section and by the tangents to the surface of the object at each point of this section. It has also been demonstrated that there are in general two space attitudes of the cone of revolution compatible with the observed triangle.

3.2 Experiment

This approach can be used for the image of figure 2.a. Figure 6 shows the selected pair of zero-curvature points on the external contours of the vase, and the tangents to the contours at these two points. The two possible attitudes of the cone of revolution are visible on figure 5. A superposition score is calculated in a same way as in the former section. The selected attitude which corresponds with the highest score leads to figure 7 on which one can see the good fit between the projection of the polyhedral model and of the brightness image.

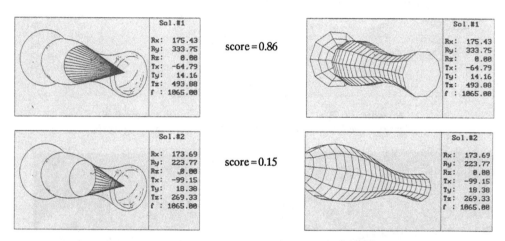

Figure 5. The two space attitudes of the cone of revolution and of the object model.

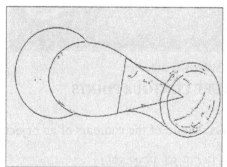

Figure 6. The triangle defined by a pair of zero-curvature contour points and the tangents to the contour at these points.

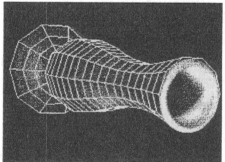

Figure 7. The superposition of the projection of the model in its calculated attitude and of the brightness image.

The result obtained with this approach is better than the one derived from the interpretation of an ellipse. This could be surprising as it is known that it is quite difficult to locate zero-curvature points accurately. But the triangle which is interpreted is made of two segments which are the tangents to the contours at these points. Then their orientation is very precise and the triangle is not so badly defined.

4 INTERPRETATION OF LIMBS FROM PAIRED POINTS

4.1 Overview

The method presented here is a generalization of the preceding one. Using a procedure described in the next sub-section, two points of the projection of the limbs which belong to the projection of a circular section of the object are selected. As before, these two points and the tangents to the contours at these points define a triangle. But here the position of this section in the object coordinate system is supposed not to be known. So it is necessary to scan all the possible sections of the object and to calculate all the possible attitudes compatible with this triangle. In order to select the best one, a superposition score is calculated.

4.2 Choice of a judicious viewer coordinate system

At first, it must be noted the plane going through the optical center of the camera and the axis of revolution of the object is a plane of symmetry for the object. Thus the limbs which are situated apart this plane are symmetrical curves on the surface of the object with respect to this plane. The perspective projection of these curves are symmetrical if and only if the optical axis intersects the axis of revolution of the object.

If this is the case then the problem is much simpler since two symmetrical contour points in the image belong to the same section of the object. Fortunately it is always possible to be in this situation thanks to a transform which allows to calculate a virtual image in which the external contours are symmetrical.

For that it is necessary to know two points belonging to the same section. A pair of zero-curvature points could have been used. Here angular points have been retained. They are double points of the projection of the limbs and they can be easily obtained since the curvature of a contour reaches a maximum absolute value in a neighbourhood of these points.

Having chosen a pair of such points (see figure 8), a rotation R_0 is applied to the viewer coordinate system in order to make symmetrical the two points with respect to the vertical axis of the image (see figure 9). Transforming the projection of the limbs by this rotation leads to symmetrical external contours with respect to the vertical axis.

4.3 Compatible attitudes

Let us consider an object of revolution whose axis coincides with the Z axis of a reference system (O, X, Y, Z). The coordinates of a point P of the surface of this object are $(X, Y, Z) = (r\cos\theta, r\sin\theta, Z)$ with $r = r(Z)$ being the scaling function or the generating curve of the object, and θ being the azimuth angle in the plane (O, X, Y). The normal $\vec{N}$ to the surface at point P is $\frac{\partial \overrightarrow{OP}}{\partial \theta} \wedge \frac{\partial \overrightarrow{OP}}{\partial z}$ and its components are $(n_x, n_y, n_z) = (\cos\theta, \sin\theta, -r')$ with r' being the derivative of $r(Z)$ with respect to Z.

Let us suppose also that O is the optical center of the camera, that the Z axis is the optical axis, and that the X and Y axes are respectively the horizontal and the vertical axes of the camera.

The projection of the limbs after the rotation R_0 being symmetrical (figure 9), to find the space attitude of the object it is sufficient to apply to the object model a rotation R_α of angle α around the X axis, and a translation T_{ovw} of components $(0, v, w)$ to bring it in a position compatible with the observed contours.

After these two transforms, P and $\overline{N}$ become P' and $\overline{N'}$ respectively. They are defined by the following equations :

$$\begin{pmatrix} X' \\ Y' \\ Z' \end{pmatrix} = \begin{pmatrix} 1 & 0 & 0 \\ 0 & \cos\alpha & -\sin\alpha \\ 0 & \sin\alpha & \cos\alpha \end{pmatrix}\begin{pmatrix} X \\ Y \\ Z \end{pmatrix} + \begin{pmatrix} 0 \\ v \\ w \end{pmatrix} = \begin{pmatrix} r.\cos\theta \\ r.\cos\alpha.\sin\theta - Z.\sin\alpha + v \\ r.\sin\alpha.\sin\theta + Z.\cos\alpha + w \end{pmatrix}$$

$$\begin{pmatrix} n'_x \\ n'_y \\ n'_z \end{pmatrix} = \begin{pmatrix} 1 & 0 & 0 \\ 0 & \cos\alpha & -\sin\alpha \\ 0 & \sin\alpha & \cos\alpha \end{pmatrix}\begin{pmatrix} n_x \\ n_y \\ n_z \end{pmatrix} = \begin{pmatrix} \cos\theta \\ \cos\alpha.\sin\theta + \sin\alpha.r' \\ \sin\alpha.\sin\theta - \cos\alpha.r' \end{pmatrix}$$

Let us choose a point p_0 of the limbs projection (figure 11). Its coordinates in the viewer coordinate system are (x_0, y_0, f) with f being the focal length of the camera. Let $\overrightarrow{t_0}$ be the tangent to the contour at point p_0. Its components are $(a_0, b_0, 0)$. Let us suppose that p_0 is the perspective projection of a point P'. Then equations (7), (8) and (9) hold.

$$\overrightarrow{N'}.\overrightarrow{Op_0} = 0 \quad (7) \qquad \overrightarrow{N'}.\overrightarrow{t_0} = 0 \quad (8) \qquad \overrightarrow{Op_0} = \lambda\overrightarrow{OP'} \quad (9)$$

Combining equations (7) and (8) leads to the system of equations (10) where $c_0 = (a_0 y_0 - b_0 x_0)/f$:

$$\begin{cases} \sin\theta &= (a_0 r'\cos\alpha - c_0 r'\sin\alpha)/(a_0\sin\alpha + c_0\cos\alpha) \\ \cos\theta &= -b_0 r'/(a_0\sin\alpha + c_0\cos\alpha) \end{cases} \quad (10)$$

But as $\cos^2\theta + \sin^2\theta = 1$, if $t = \mathrm{tg}\,\alpha$, equation (11) is easily obtained from sytem (10).

$$(r'^2(b_0^2 + c_0^2) - a_0^2)t^2 - 2a_0 c_0(r'^2 + 1)t + r'^2(a_0^2 + b_0^2) - c_0^2 = 0 \quad (11)$$

Let us consider a particular section of the object model. For it $r(Z)$ and $r'(Z)$ are known. Then equation (11) gives the possible values for α. There are four determination for α because if α_1 and α_2 are solutions of (11) then $\alpha_1 + \pi$ and $\alpha_2 + \pi$ are solutions too.

Equation (9) gives the following system (12) :

$$\begin{cases} \cos\theta &= x_0(Z + w\cos\alpha - v\sin\alpha)/r(f\cos\alpha - y_0\sin\alpha) \\ \sin\theta &= (y_0 w - fv + \cos\alpha y_0 Z + \sin\alpha f Z)/r(f\cos\alpha - y_0\sin\alpha) \end{cases} \quad (12)$$

If solutions α_i are set in equations (10) and (12), the following system is obtained which is linear in u and w.

$$\begin{cases} (x_0\cos\alpha_i)w - (x_0\sin\alpha_i)v &= r\cos\theta(f\cos\alpha_i - y_0\sin\alpha_i) - x_0 Z \\ y_0 w - fv &= r\sin\theta(f\cos\alpha_i - y_0\sin\alpha_i) - \cos\alpha_i y_0 Z - \sin\alpha_i f Z \end{cases}$$

Then rotation R_α and translation T_{ovw} have been determined. It can be proved that they are at most two real possible solutions because w must be positive. They are such that the projection of two points belonging to the same section (fixed Z) of the object model is compatible with the triangle in the image defined by p_0, $\overrightarrow{t_0}$ and their symmetrical with respect to the vertical axis. It must be noted that the same result as in section 3 has been obtained but with a different approach.

At this stage in order to locate the object it is necessary to find the good section.

4.4 Evaluation of an attitude

For that let us choose some control points $(p_1, p_2, \ldots p_i, \ldots p_n)$ along the projection of the limbs (figure 9). For each of them, parameters x_i, y_i, a_i, b_i and c_i are known. If any particular compatible attitude A_j is considered, its parameters α_j, v_j and w_j are known too. From systems (10) and (12) the following system, linear in Z and $r(Z)r'(Z)$, can be derived :

$$
\begin{cases}
(a_i\sin\alpha_j + c_i\cos\alpha_j)(y_i\cos\alpha_j + f\sin\alpha_j)Z & + & (c_i\sin\alpha_j - a_i\cos\alpha_j)(f\cos\alpha_j - y_i\sin\alpha_j)rr' \\
& = & (a_i\sin\alpha_j + c_i\cos\alpha_j)(fv_j - y_iw_j) \\
(a_i\sin\alpha_j + c_i\cos\alpha_j)x_iZ & + & (f\cos\alpha_j - y_i\sin\alpha_j)b_irr' \\
& = & (a_i\sin\alpha_j + c_i\cos\alpha_j)(v_j\sin\alpha_j - w_j\cos\alpha_j)x_i
\end{cases}
$$

A value Z_{ij} is obtained which corresponds with the section of the object such that a limb point of this section will be projected in p_i if the considered attitude A_j is the good one.

The limbs equation for attitude A_j is :

$$\overline{OP'}.\overline{N'} = r(Z) - r'(Z)Z(v_j\sin\alpha_j - w_j\sin\alpha_j) + \sin\theta(v_j\cos\alpha_j + w_j\sin\alpha_j) = 0 \quad (13)$$

Feeding equation (13) with Z_{ij} the value of $\sin\theta_{ij}$ is obtained, and then the value of θ_{ij} since the sign of $\cos\theta_{ij}$ must be the same as the sign of the abscissa x_i of point p_i.

Using $\theta_{ij}, Z_{ij}, \alpha_j, v_j$ and w_j, the coordinates of a point P'_{ij} can be calculated. This point must be projected in p_i if attitude A_j is correct. But if it is not the case the projection of P'_{ij} will give point p_{ij} and the quadratic distance $|\overline{p_ip_{ij}}|^2$ between p_i and p_{ij} is used to evaluate the quality of the attitude A_j.

It is preferable to consider more than one point p_i to do that. So the criterion which is used is $C(Z) = \underset{j=1}{\overset{m}{\text{Min}}} \left(\sum_{i=1}^{n} |p_ip_{ij}|^2 \right)$ where m is the number of real possible attitudes for a given Z. Thus m is equal to 2 or 1. Let us recall that this criterion is a function of Z, Z being the Z coordinate of the section, in the object reference system, for which it is supposed that p_0 is the projection of one of its points. So all the values of Z have to be scanned. The selected attitude of the object model is the one for which $C(Z)$ is minimal.

4.5 Experiment

This method has been also experimented in case of the image of figure 2 (left). Figure 8 gives the two selected angular points. After the application of R_0, symmetrical contours are obtained as shown in figure 9. On this figure, the four control points used in this experiment are also visible.

The height of the vase is nearly 250mm. We chose to sample the Z axis each millimeter. The value of the criterion $C(Z)$ is drawn in figure 11. Obviously this curve has two relative minima for $Z = 158$mm ($C(Z) = 21$) and $Z = 60$mm ($C(Z) = 1341$). The corresponding attitudes for the object model in the virtual image are presented in figure 10.

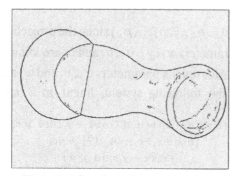

Figure 8. Selected angular contour points.

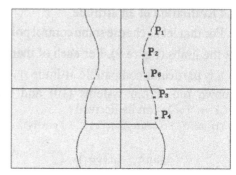

Figure 9. External contours in the virtual image.

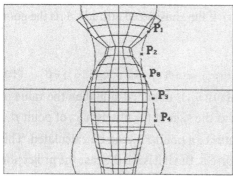

(a) criterion = 1341

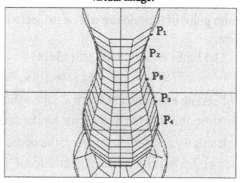

(c) criterion = 21

Figure 10. Vase attitudes for the control point p_0 and for $Z = 60$ et 158.

After the best solution is selected, it is easy to compute the rigid transform to apply to the object model which corresponds with the interpretation of a pair of angular contour points. The superposition of the projected object model and of the brightness image is presented in figure 12. The quality of the mutual covering of the two images is quite good. The only noticeable differences appear to be due to the polyhedral approximation.

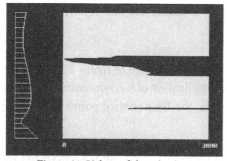

Figure 11. Values of the criterion.

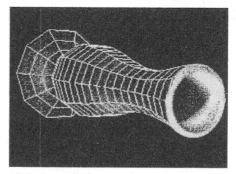

Figure 12. Projected model on the image plane.

5 CONCLUSION

It has been demonstrated here how some geometrical features extracted from the brightness image of an object of revolution and coming from the perspective projection of lines or points situated on the surface of the object, can be used to find the spatial attitude of the object in the viewer coordinate system.

The various methods presented in this paper which involve different kinds of geometrical features are complementary since they could not be all present for some aspect of the object. For example, if the angle between the viewing direction and the axis of revolution is too small then elliptical contours will surely be seen meanwhile the projection of zero-curvature points of the scaling function will probably be hidden.

Before the design of an automatic system for localization of objects of revolution, the problem of the matching of similar contour points (zero-curvature or angular ones in the present case) has to be solved efficiently. So it is necessary to built a robust procedure to find the projection of the axis of revolution of these objects.

At last, it must be noted that for the three methods the localization problem is much simpler if at first the original viewer coordinate system is changed for a more judicious one. In fact with this new reference system, the direction of view always goes through the axis of revolution of the objects. It is worth noting that human vision proceeds in this way since the cone angle for accurate vision is rather small and consequently since the look is approximatively centered on the objects to be recognized or localized.

6 BIBLIOGRAPHY

[DHO-89]   DHOME M., RICHETIN M, LAPRESTE J.T. & RIVES G., *Determination of the Attitude of 3D Objects from a Single Perspective View*, IEEE Trans. on PAMI, vol 11, n°12, pp1265-1278, December 1989.

[KOE-84]   KOENDERINK J.J., *What Does The Occluding Contours Tell Us About Solid Shape?*, Perception, 13, pp321-331, 1984.

[LOW-85]   LOWE D.G. *Perceptual Organization and Visual Recognition*, Kluwer, Boston, 1985.

[PON-89a] PONCE J., CHELLBERG D. & MANN W., *Invariant Properties of Straight Homogeneous Generalized Cylinders and Their Contours*, IEEE Trans. on PAMI, vol 11, n°9, pp951-966, September 1989.

[PON-89b] PONCE J. & KRIEGMAN D.J., *On Recognizing and Positioning Curved 3D Objects from Image Contours*, Proc. of IEEE Workshop on Interpretation of 3D Scenes, pp61-67, Austin, Tx, November 1989.

[RIC-89]   RICHETIN M., DHOME M. & LAPRESTE J.T., *Inverse Perspective Transform from Zero-Curvature Curve Points. Application to the Localization of Some Generalized Cylinders*, Proc. of IEEE CVPR '89, pp 517-522, San Diego, Ca, June 1989.

[SHI-87]   SHIRAI Y, *Three Dimensional Computer Vision*, Springer-Verlag, Berlin, 1987.

RECOGNITION - MATCHING

On the Verification of Hypothesized Matches in Model-Based Recognition[1]

W. Eric L. Grimson
MIT Artificial Intelligence Laboratory
545 Technology Square, Cambridge, Mass. 02139

Daniel P. Huttenlocher
Department of Computer Science
Cornell University, Ithaca, NY 14853

Abstract

Model-based recognition methods often use *ad hoc* techniques to decide if a match of data to a model is correct. Generally an empirically determined threshold is placed on the fraction of model features that must be matched. We instead rigorously derive conditions under which to accept a match. We obtain an expression relating the probability of a match occurring at random to the fraction of model features accounted for by the match, as a function of the number of model features, the number of image features, and a bound on the degree of sensor noise.

Our analysis implies that a proper matching threshold must vary with the number of model and data features, and thus should be set as a function of a particular matching problem rather than using a predetermined value. We analyze some existing recognition systems and find that our method predicts thresholds similar those determined empirically, supporting the technique's validity.

1. Introduction

A central problem in machine vision is recognizing partially occluded objects from noisy data. Recognition systems generally search for a match between elements of an object model and instances of those elements in the data, thereby recovering a transformation that maps part of the model onto part of the image. Approaches to model-based recognition (see [3, 2] for reviews) include clustering in parameter space (e.g. [17, 18]), searching a tree of corresponding model and image features (e.g., [9, 13, 5, 16, 1, 6, 15]), and directly searching for possible model-to-image transformations (e.g.,[8, 14]). These approaches all must decide if an object is present or absent on the basis of geometric evidence acquired from the sensory input. Here, we analyze this decision process and develop a formal means for deciding when a match should be accepted as correct.

Most recognition systems use *ad hoc* methods to determine what constitutes an acceptable match of a model to an image. For example, many systems order the possible interpretations of the data in terms of some measure of completeness, (e.g. the percentage of the model accounted for), and accept the best interpretations under this measure. If instances of the object model are present in the scene, this approach generally will find them. If no instance of the object is present, the interpretations that best account for the data are in fact incorrect. In this case, one must either accept false interpretations or have some means of deciding if the object is present. To reduce the computational complexity of recognition, the measure of completeness is often used to terminate the search once an interpretation is found that exceeds some empirically determined threshold. Once again, it is necessary to determine if an interpretation is good enough to accept as correct.

[1]Research funded in part by ONR URI grant N00014-86-K-0685, in part by NSF Grant IRI-8900267, and in part by DARPA under Army contract DACA76-85-C-0010 and ONR contract N00014-85-K-0124.

Current methods for deciding if a match is correct are based on empirically determined thresholds. In this paper we instead rigorously analyze what constitutes a good match of a model to an image. Specifically, we address the following question:

> Suppose that we are given a model with m features, a set of s data features, and bounds ϵ_p and ϵ_a on the positional and orientational error in the data. Further, suppose that some recognition method has found a match accounting for a fraction f ($f \in [0,1]$) of the m model features. What is the relation between f and the likelihood δ that such a match can occur at random?

We use this relation to set a threshold on the minimum fraction of model features that must be matched, f_0, so that the likelihood of such a match occurring at random is small (e.g., $\delta < .001$). Note that there may not be a value of f_0 for all choices of δ (e.g., as δ gets very small, or as m, s, ϵ_p, or ϵ_a get very large there may be no fraction of model features that limits the probability of a random match to δ).

There are three basic steps to our analysis. First, given a particular feature type, the type of transformation from model to image, and a bound on the sensor error, we characterize the set of transformations consistent with a single pairing of a model and image feature. This set of transformations defines a volume V in the *transformation space*, $\mathcal{P}$ (a d-dimensional space with one dimension for each of the d parameters of the transformation). We then use an occupancy model [7] to determine the probability, $Pr\{v \geq l\}$ that the number of volumes intersecting at a common point in transformation space is at least l. This provides an estimate of how often a match of l features will occur at random. Finally, the probability that l volumes will intersect at random is used to set a threshold on the minimum fraction of model features, f_0, that must be matched in order for an interpretation to have a small probability of occuring at random.

2. The Space of Transformations

A rigid object's pose is characterized by a transformation from model to sensor coordinates. We focus on the case of a similarity transformation (i.e., a translation, rotation, and scaling). The set of possible poses can be viewed as a *transformation space* having one dimension for each parameter of the transformation from model to sensor coordinates. A point in this transformation space defines a pose of an object, which in turn defines a possible solution to the recognition problem. For example, with a 2D image and world, the transformation space is 4D (translation in x and y, planar rotation, and scaling).

A match of a model feature and an image feature (e.g., an edge or vertex) defines a range of possible transformations, i.e., a volume in the transformation space. The size and shape of this volume depends on the type and accuracy of the feature. In this section we present an analytic expression for the size of this volume. The development is similar to that in [11], however here we consider a continuous space as opposed to one that is uniformly tessellated. The discussion is limited to the case of 2D problems where the transformation is an isometry (translation and rotation without scaling), and the features are linear edge fragments or points. A similar analysis holds for 3D problems and for problems involving change of scale, and is described in [12].

Consider the problem of recognizing a two-dimensional polygonal model from noisy, occluded data. We let $\mathbf{M}_J$ be the vector from the origin to the midpoint of the Jth model edge, $\hat{\mathbf{T}}_J$ be the unit tangent of the edge, and L_J be the length of the edge, all measured in the model coordinate system, $\mathcal{M}$. We let $\mathbf{m}_j, \hat{\mathbf{t}}_j, \ell_j$ denote similar parameters for the

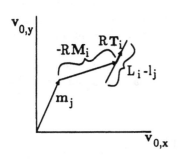 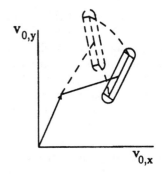

Figure 1: The range of feasible translations. Left: fixed θ without error, the line in direction RT_i denotes set of feasible translations. Right: allowing error, the region enclosed in solid lines shows slice $S(\theta, j, J)$ for a particular θ. Dashed region shows helical path of slice as θ varies.

---

j^{th} data edge, measured in the sensor coordinate system, $\mathcal{I}$. (Note that we use upper case for model parameters and lower case for data parameters.)

The transformation from model to sensor coordinates may be represented by

$$\mathbf{v}_s = R_\theta \mathbf{V}_M + \mathbf{V}_0$$

where $\mathbf{V}_M$ is a vector in model coordinates, R_θ is a rotation matrix of angle θ, $\mathbf{V}_0$ is a translation offset, and $\mathbf{v}_s$ is the corresponding vector in sensor coordinates.

What transformations will map a model edge to a data edge? If $\ell_j > L_J$, the two edges cannot match. Thus, suppose that $\ell_j \leq L_J$. Then the rotation matrix R_{θ_m} is defined by the angle θ_m between $\hat{\mathbf{T}}_J$ and $\hat{\mathbf{t}}_j$. Many translations will cause the edges to overlap, because $\ell_j \leq L_J$. If one endpoint of the data edge coincides with a transformed model edge endpoint, the translation is the difference between them:

$$\mathbf{V}_0 = \mathbf{m}_j - R_{\theta_m} \mathbf{M}_J \pm \frac{L_J - \ell_j}{2} R_{\theta_m} \hat{\mathbf{T}}_J$$

where the $\pm$ indicates the two cases. Because any intermediate position is also acceptable, the set of translations consistent with matching model edge J to data edge j is

$$\left\{ \mathbf{m}_j - R_{\theta_m} \mathbf{M}_J + \gamma R_{\theta_m} \hat{\mathbf{T}}_J \quad \middle| \quad \gamma \in \left[-\frac{L_J - \ell_j}{2}, \frac{L_J - \ell_j}{2} \right] \right\}. \tag{1}$$

Hence, matching model edge J to data edge j gives a set of points in transformation space, with a single value for the rotation and a set of values for the translation, corresponding to a line of length $L_J - \ell_j$, with orientation $R_{\theta_m} \hat{\mathbf{T}}_J$ in the x–y plane (Figure 1 Left).

This ignores the issue of noise in the measurements. In practice, we may only know the position of the data edge's endpoints to within some ball of radius ϵ_p, and the orientation to within an angular error of ϵ_a. For 2D lines, these error ranges are related. Given endpoint variations of ϵ_p, the maximum angular variation occurs when the correct line is tangent to circles of radius ϵ_p about the two endpoints, and provided $\ell > 2\epsilon_p$, is given by

$$\epsilon_a = \tan^{-1} \left(\frac{2\epsilon_p}{\sqrt{\ell^2 - 4\epsilon_p^2}} \right).$$

Inclusion of error effects on position measurements imply that the line of feasible translations, for a given rotation, (equation (1)), must be expanded to include any points in the parameter space within ϵ_p of that line. Further, this expansion into a region must be repeated for each value of θ in $[\theta_m - \epsilon_a, \theta_m + \epsilon_a]$. Note that this carves out a skewed volume in transformation space [4], because the region's center and orientation are functions of θ, as illustrated in Figure 1 Right.

Thus given $\mathbf{M}_J, \hat{\mathbf{T}}_J, L_J, \mathbf{m}_j, \hat{\mathbf{t}}_j$, and ℓ_j, if $\ell_j - 2\epsilon_p > L_J$ then there are no consistent transformations, otherwise the set of feasible transformations is denoted by the volume

$$\mathcal{V}(j, J) = \bigcup_{\theta \in [\theta_m - \epsilon_a, \theta_m + \epsilon_a]} \mathcal{S}(\theta, j, J)$$

where an individual set of translations is denoted by:

$$\mathcal{S}(\theta, j, J) = \left\{ (\theta, \mathbf{V}_0) \,\middle|\, \exists \gamma, |\gamma| \leq \frac{L_J - \ell_j}{2}, \|\mathbf{m}_j - R_\theta \mathbf{M}_J + \gamma R_\theta \hat{\mathbf{T}}_J - \mathbf{V}_0\| \leq \epsilon_p \right\}.$$

Since each slice $\mathcal{S}(\theta, j, J)$ consists of two hemicircles and a rectangle, it is easy to show that the volume of the region $\mathcal{V}(j, J)$ is given by

$$c_{jJ} = 2\epsilon_a \left[2\epsilon_p(L_J - \ell_j) + \pi \epsilon_p^2 \right].$$

The term in braces is the area of one slice. Integration over a range of angles yields the $2\epsilon_a$ term. For simplicity, we let the data edge have a length $\ell_j = (1 - \alpha_{jJ})L_J$ where α_{jJ} denotes the amount of occlusion of the edge, so that the volume is

$$c_{jJ} = 2\epsilon_a \left[2\epsilon_p \alpha_{jJ} L_J + \pi \epsilon_p^2 \right]. \tag{2}$$

If we are dealing with point features, rather than extended edges, the above result can be specialized. Here $L_j \to 0$ so that equation (2) becomes $c_{jJ} = 2\epsilon_a \pi \epsilon_p^2$. What is ϵ_a in the case of a point feature? For a vertex, its orientation is the direction of the bisector of the two edges defining the vertex, and hence ϵ_a is a bound on the error in measuring that orientation. For a curvature extremum or inflection, the local tangent of the curve defines the orientation, and ϵ_a is again defined by a bound on measuring this orientation. For truly isolated points, $\epsilon_a = \pi$. In any event, our analysis provides estimates for c_{jJ} both for edge features and for vertices.

For the case of a rigid 2D isometric transformation, we have characterized in (2) the volume of transformation space, c_{jJ}, consistent with a single data-model pairing (j, J). The expression is a function of the noise in the data measurements, ϵ_p and ϵ_a, and in the case of edges is further a function of the amount of occlusion, α_{jJ}, and the length of the model edge, L_J. In [12] we consider adding scaling to the transformation as well as the case of 3D transformations. We now turn to the question of how these volumes interact.

3. The Probability of a Conspiracy

The intersection, if any, of two volumes in transformation space defines the set of transformations consistent with both pairings. Thus a correct match of a model to an image will lie in the intersection of several volumes. In this section we consider the likelihood that l volumes in transformation space will intersect at random. Such an event corresponds to an arrangement of image features that happens to be consistent, within error, with l of the model features, but which does not actually correspond to an instance of the object.

The likelihood that l transformation space volumes will intersect at random is a function of their number and size. The number depends on the number of model and image features. The size depends on the noise, the feature type, and for edge features, the amount of occlusion. To be confident that a match with l model features is correct, we would like l to be large enough that a random matching of that size is very unlikely.

To characterize the likelihood that several volumes will intersect at random we use a statistical occupancy model. In the discrete case, if r events are uniformly randomly distributed across n buckets, an occupancy model can be used to estimate the probability that a given bucket will contain k events. The events in our case are points in the volumes in transformation space, and the buckets are points in the transformation space itself. These quantities are continuous, and thus we consider the limiting case as $n, r \rightarrow \infty$.

The volume of transformation space defined by each incorrect model and image feature pairing is independent of the correct match. We assume that the image features are also independent of one another, so we can model the volumes in transformation space as independent random events. The distribution of these volumes depends on the image features, which are unknown, so we assume the uniform distribution as an approximation.

While the volumes in transformation space can reasonably be viewed as independent random events, we are modeling the probability of events occurring at points in these volumes. As the number of volumes, R, gets large (compared with the ratio of the total size of the transformation space to the size of each volume, V/c) the overall distribution of points in the space also is random. For the cases of interest here $Rc \gg V$, so the assumption of independent random pointwise events is a reasonable approximation.

Given a uniform random distribution of r events into n cells, a number of different statistical occupancy models can be used to characterize the likelihood, p_k, that a given cell will contain exactly k events. We use the Bose-Einstein statistic, where it is assumed that each assignment of counts to cells occurs with equal probability [7]. Under this model, for large r and n, where $\frac{r}{n} \rightarrow \lambda$, the limiting case is the geometric distribution,

$$p_k \approx \frac{\lambda^k}{(1+\lambda)^{k+1}} \approx \frac{1}{1+\lambda} \left(\frac{1}{1+\frac{1}{\lambda}} \right)^k. \tag{3}$$

We are interested in establishing conservative bounds on the likelihood that a large number of volumes will intersect at random, thus we use the Bose-Einstein statistic because it provides a higher estimate of this likelihood than do other models.

The parameter λ of the occupancy model is the ratio of the occupied volumes of the transformation space to the total size of the transformation space. From equation (2) we know that each pair of model and image features defines a volume of size c_{jJ} in transformation space. There are ms such volumes for m model features and s image features, so the occupied volume of the transformation space is given by the sum of c_{jJ} for $j = 1, ..., s$ and $J = 1, ..., m$.

The total size of the transformation space is just the product of the ranges for the space's dimensions. Each rotational dimension ranges over $[0, 2\pi]$, and each translational dimension ranges over $[0, D]$, where D is the linear extent of the image. Thus for a 2D isometry (translation and rotation) we get

$$\lambda = \frac{\sum_{j=1}^{s} \sum_{J=1}^{m} c_{jJ}}{2\pi D^2} = ms\bar{c},$$

where $\bar{c}$ is the average normalized volume size. For 2D edges, from equation (2) we obtain

$$\bar{c} = \frac{2\epsilon_a \left[2\epsilon_p \bar{\alpha} \bar{L} + \pi \epsilon_p^2 \right]}{2\pi D^2} = \frac{\epsilon_a}{\pi} \left[2\bar{\alpha} \frac{\epsilon_p}{D} \frac{\bar{L}}{D} + \pi \left(\frac{\epsilon_p}{D} \right)^2 \right] \tag{4}$$

where $\bar{L}$ is the average edge length and $\bar{\alpha}$ is the average amount of occlusion of the edges (the average value of α_{jJ}). As expected $\bar{c}$ increases as the noise ϵ_a, ϵ_p increases, and as the average amount of occlusion of the edges $\overline{\alpha}$ increases. In the case of 2D points (with associated orientations), the average normalized volume size is $\overline{c} = \epsilon_a \epsilon_p^2 / D^2$. Note that we can restrict $\epsilon_a \leq \pi$ and $\epsilon_p \leq \frac{D}{2}$. In the extreme, this can lead to $\overline{c} > 1$, which does not make physical sense. We should really take the minimum of the above expressions and unity, but in practice $\overline{c} \ll 1$ and hence we ignore this special case.

A particular recognition task thus defines a value for λ, based on the type of transformation from model to image, the type of features, the number of model features, m, and data features, s, and a bound on the positional and angular error, ϵ_p and ϵ_a. Given a value for λ, the probability that l or more of the volumes intersect at random is given by

$$Pr\{v \geq l\} = 1 - \sum_{k=0}^{l-1} p_k. \tag{5}$$

This corresponds to an arrangement of data features occurring at random such that l pairs of model and data features are consistent with one another (within the error bounds). From $Pr\{v \geq l\}$ we can determine the fraction of model features, f_0, such that the probability of $m f_0$ features being matched at random is less than some predefined level, δ. This value is just the smallest f such that $Pr\{v \geq mf\} \leq \delta$, i.e.,

$$f_0 = \min\{f \mid Pr\{v \geq mf\} \leq \delta\}. \tag{6}$$

4. Deriving Formal Thresholds

We have used an occupancy model to determine an expression for the probability that l or more volumes in transformation space will intersect at random, as a function of the number of features, the type of features, and bounds on the sensor error. The expression was then used to set a threshold, f_0, on the fraction of model features that must be matched in order to limit the probability of a random matching to some level. In this section we derive a closed-form expression for f_0.

The probability that there will be l or more events occurring at random at a point in transformation space is given by (5). Thus to distinguish a correct interpretation from a random one we set the threshold, f_0, such that the probability of $l = m f_0$ events coinciding at random is less than δ. Substituting $m f_0$ for l and equation (3) for p_k in (5), we obtain

$$Pr\{v \geq mf\} = 1 - \sum_{k=0}^{m f_0 - 1} \frac{\lambda^k}{(1+\lambda)^{k+1}} \leq \delta.$$

Using the geometric series relationship, we can isolate f_0 by appropriate algebra:

$$f_0 \geq \frac{\log\left(\frac{1}{\delta}\right)}{m \log\left(1 + \frac{1}{ms\overline{c}}\right)}. \tag{7}$$

Thus to obtain a value for the fraction of model features that must be matched in order to limit the probability of a random conspiracy to δ, we simply need to compute $\overline{c}$ for the particular parameters of our recognition task, and then use (7) to compute f_0. The value of $\overline{c}$ depends on the particular type of feature being matched and the bounds on the sensor error. In the case of 2D edge fragments, we derived $\overline{c}$ in (4).

Note that equation (7) exhibits the expected behavior. If the noise in the data increases, then $\overline{c}$ increases, and so does the bound on f_0. Similarly, as the amount of occlusion increases, then so does $\overline{c}$ and thus the bound on f_0. As either m or s increases so does the bound on f_0, and as δ decreases f_0 increases. Also note that for large values of ms, one gets the approximation $f_0 \geq s\overline{c} \log 1/\delta$. Thus, in the limit, the bound on the

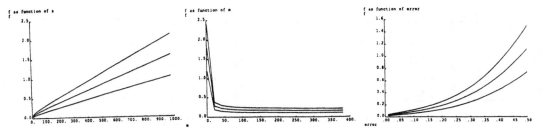

Figure 2: Bounds on threshold. Left: f_0 as a function of s. Middle: f_0 as a function of m. Right: f_0 as a function of error. Percentage of error along horizontal axis p defines sensing errors of $\epsilon_a = p\pi$ and $\epsilon_p = p\overline{L}$. The three plots in each case are for $\delta = .0001, .001, .01$ from top to bottom respectively.

fraction of the model is linear in the number of sensory features, linear in the average size of the volumes in transformation space, and varies logarithmically with the inverse probability of a false match.

The expression for f_0 in (7) can yield values that are greater than 1.0, which makes no sense as a *fraction* of the model features. When f_0 is greater than 1.0 it means that for the given number and type of features, and the given bounds on sensor error, it is not possible to limit the probability of a false match to the chosen δ (even if all the model features are matched to some sensor feature).

There are several possible choices for δ. One could simply set δ to be some small number, e.g., $\delta = .001$, so that a false positive is likely to arise no more than one time in a thousand. One could also set δ as a function of the scene complexity, e.g., some multiple of the inverse of the total number of data model pairings, (i.e., $\frac{\beta}{ms}$.) A third possibility is to set δ so that the likelihood of a false positive, integrated over the entire transformation space, is small (e.g., < 1). The idea is to determine the appropriate value of δ such that one expects no random matches to occur. If we let ν be a measure of the system's sensitivity in distinguishing transformations, then we could choose $\delta = \nu/2\pi D^2$. For example, we could set ν to be a function of the noise in the data measurements, given by the product in uncertainties: $(2\epsilon_a)(\pi\epsilon_p^2)$. In this case, we get

$$f_0 \geq \frac{\log\left(\frac{D^2}{\epsilon_a \epsilon_p^2}\right)}{m \log\left(1 + \frac{1}{m s \bar{c}}\right)}. \tag{8}$$

We graph examples of f_0 in Figure 2. Figure (2 Left) displays values of f_0 as a function of s, with $m = 32$, $c = .0002215$ (these numbers are taken from the RAF system analyzed in section 5). Each graph is for a different value of δ. Note that as s gets large, the graphs become linear, as expected. Figure (2 Middle) displays f_0 as a function of m for different values of δ. Here, $s = 100$, $c = .0002215$. Note that as expected, when m becomes large, f_0 becomes a constant independent of m. Figure (2 Right) displays f_0 as a function of the sensor error, for different values of δ. Here, $s = 100$, $m = 32$. The percentage of error along the horizontal axis, p, is used to define sensing errors of $\epsilon_a = p\pi$ and $\epsilon_p = p\overline{L}$. As expected, the threshold on f_0 increases with increasing error.

We can modify our preceding analysis to handle weighted matching methods as well. One common scheme is to use the size of each data feature as a weight. In the case of 2D edges, for example, a data-model pairing (j, J) would carry a weight of ℓ_j (the length of the data edge), so that transformations consistent with pairings of long data edges to

Occlusion	f, eqn (8)	f, $(\delta = .001)$	f, $(\delta = .0001)$	f, $(\bar{l} = \bar{L})$	f, $(\bar{l} = .75\bar{L})$	f, $(\bar{l} = .5\bar{L})$
0.0	0.225	0.173	0.230	0.119	0.091	0.062
0.2	0.263	0.202	0.270	0.153	0.116	0.079
0.4	0.301	0.231	0.308	0.188	0.142	0.097
0.6	0.337	0.259	0.346	0.222	0.168	0.114
0.8	0.374	0.287	0.383	0.257	0.194	0.131
1.0	0.409	0.315	0.420			

Table 1: Predicted bounds on termination threshold, as a function of amount of occlusion. First three cases are unweighted, second three use edge length as a weight.

model edges would be more highly valued than those involving short data edges. Working through similar algebra [12], where $\bar{l}$ is the average length of the data edges, leads to:

$$f_0 \geq \frac{\log\left(\frac{1}{\delta}\right)}{m\bar{L}\log\left(1 + \frac{1}{ms\bar{l}\bar{c}}\right)}. \tag{9}$$

5. Some Real World Examples

To demonstrate the utility of our method, we analyze some working recognition systems that utilize a threshold on the fraction of model features needed for a match. The analysis predicts thresholds close to those determined experimentally, suggesting that the technique can be profitably used to analytically determine thresholds for model-based matching. Because our analysis shows that the proper threshold *varies* with the number of model and data features, it is important to be able to set the threshold as a function of a particular matching problem rather than setting it once based on experimentation.

We first consider the interpretation tree method [13, 5, 16] for recognizing sets of 2D parts. In this approach, a tree of possible matching model and image features is constructed. Each level of the tree corresponds to an image feature. At every node of the tree there is a branch corresponding to each of the model features, plus a special branch that accounts for model features that do not match the image. A path from the root to a leaf node maps each image feature onto some model feature or the special "no-match" symbol. The tree is searched by maintaining pairwise consistency among the nodes along a path. Consistency is checked using distance and angle relations between the model and image features specified by the nodes. If a given node is inconsistent with any node along the path to the root then the subtree below that point is pruned from further consideration.

A consistent path from the root to a leaf that accounts for more than some fraction of the model features is accepted as a correct match. This threshold is chosen experimentally. In our analysis of thresholds for the interpretation tree method, we use the parameters for the examples presented in [13]. These values are substituted into equation (2), and then a threshold f_0 is computed using equations (7) and (8). In the experiments reported in [13], the following parameters hold: $m = 32$, $s = 100$, $\bar{L} = 23.959$, $\epsilon_p = 10$, $\epsilon_a = \frac{\pi}{10}$.

We have computed $\bar{c}$ as a function of the amount of occlusion $\bar{\alpha}$, and then determined the corresponding threshold f_0 on the fraction of model features. Note that an occlusion of 1 represents the limiting case in which only a point on the line is visible. The results are given in Table 1. The first column of the table shows the values of f_0 computed using equation (8), where $\delta = \epsilon_a \epsilon_p^2 / D^2$. For comparison, the second and third columns of the table are computed using equation (7), with the probability of a random match, δ, set to .001 and .0001, respectively.

Occlusion	Object-1		Object-2	
	$f, (\epsilon_a = \frac{\pi}{10})$	$f, (\epsilon_a = \frac{\pi}{15})$	$f, (\epsilon_a = \frac{\pi}{10})$	$f, (\epsilon_a = \frac{\pi}{15})$
0.0	0.224	0.185	0.206	0.168
0.2	0.261	0.212	0.243	0.195
0.4	0.297	0.238	0.280	0.221
0.6	0.333	0.264	0.316	0.247
0.8	0.368	0.289	0.353	0.273

Table 2: Predicted bounds on termination threshold, as a function of the amount of occlusion, for the HYPER system.

As expected, the bound on f increases as the amount of occlusion increases. Note that for occlusions ranging from none to all (0 to 1), the bound on f only varies over a range of 0.225 to 0.409. Empirically, running the RAF system on a variety of images of this type [13] using thresholds of $f = .4$ resulted in no observed false positives, while using thresholds of $f = .25$ would often result in a few false positives. Since the occlusion was roughly .5, this observation fits nicely with the predictions of Table 1, i.e., a threshold of .4 should yield no errors, while a threshold of .25 cannot guarantee such success.

If we use the lengths of the data features to weight the individual feature matchings then using equation (9) in place of (8) leads to the predictions shown in the second part of Table 1. Again, this agrees with empirical experience for the RAF system, in which weighted matching using thresholds of $f = .25$ almost always led to no false positives, while using thresholds of $f = .10$ would often result in a few false positives.

Second, we consider the HYPER system [1]. HYPER also uses geometric constraints to find matches of data to models. An initial match between a long data edge and a corresponding model edge is used to estimate the transformation from model to data coordinates. This estimate is then used to predict a range of possible positions for unmatched model features, and the image is searched over this range for potential matches. Each potential match is evaluated using position and orientation constraints, and the best match within error bounds is added to the current interpretation. The additional model-data match is used to refine the transformation estimate, and the process is iterated.

Although not all of the parameters needed for our analysis are given in the paper, we can estimate many of them from the illustrations in the article. Given several estimates for the measurement error, a range of values for the threshold f are listed in Table 2. Object-1 and Object-2 refer to the object labels used in [1]. In these examples, we use errors of $\epsilon_a = \pi/10$ and $\pi/15$ radians, and $\epsilon_p = 3$ pixels.

In HYPER, a threshold of .25 is used to discard false positives, and no false positives are observed during a series of experiments with the system. For the two objects listed in Table 2, HYPER found interpretations of the data accounting for a fraction of .55 of the model for Object-1 and accounting for a fraction of .40 of the model for Object-2. Both these observations are in agreement with the thresholds predicted in Table 2, for different estimates of the data error.

Thus for two different recognition systems (RAF and HYPER), using both weighted and unweighted matching schemes, we see that the technique developed here yields matching thresholds similar to those determined experimentally by the systems' designers.

6. Conclusion

In order to determine what constitutes an acceptable match of a model to an image, most recognition systems use an empirically determined threshold on the fraction of model features that must be matched. We have developed a technique for analytically determining the fraction of model features f_0 that must be matched in order to limit the probability of a random conspiracy of the data to some level δ. This fraction f_0 is a function of the feature type, the number of model features m, the number of sensor features s, and bounds on the translation error ϵ_p and the angular error ϵ_a of the features.

Our analysis shows that the proper threshold varies with the number of model and data features. A threshold that is appropriate for relatively few data features is not appropriate when there are many data features. Thus it is important to set the threshold as a function of a particular matching problem, rather than setting a single threshold based on experimentation. Our technique provides a straightforward means of computing a matching threshold for the values of m and s found in a given recognition situation.

References

[1] Ayache, N. & O.D. Faugeras, 1986, "HYPER: A new approach for the recognition and positioning of two-dimensional objects," *IEEE Trans. PAMI* **8**(1), pp. 44–54.

[2] Besl, P.J. & R.C. Jain, 1985, "Three-dimensional object recognition," *ACM Computing Surveys* **17**(1), pp. 75–154.

[3] Chin, R.T. & C.R. Dyer, 1986, "Model-based recognition in robot vision," *ACM Computing Surveys* **18**(1), pp. 67–108.

[4] Clemens, D.T., 1986, The recognition of two-dimensional modeled objects-in images, M. Sc. Thesis, Massachusetts Institute of Technology, Electrical Engineering and Computer Science.

[5] Ettinger, G.J., 1988, "Large Hierarchical Object Recognition Using Libraries of Parameterized Model Sub-parts," *IEEE Conf. on Comp. Vis. & Pattern Recog.*, pp. 32–41.

[6] Faugeras, O.D. & M. Hebert, 1986, "The representation, recognition and locating of 3-D objects," *Int. J. Robotics Research* **5**(3), pp. 27–52.

[7] Feller, W., 1968, *An Introduction to Probability Theory and Its Applications*, New York, Wiley.

[8] Fischler, M.A. & R.C. Bolles, 1981, "Random sample consensus: A paradigm for model fitting with applications to image analysis and automated cartography," *Commun. ACM* **24**, pp. 381–395.

[9] Grimson, W.E.L., 1989, "On the Recognition of Parameterized 2D Objects," *International Journal of Computer Vision* **2**(4), pp. 353–372.

[10] Grimson, W.E.L., 1989, "The Combinatorics of Object Recognition in Cluttered Environments Using Constrained Search," *Artificial Intelligence*, to appear.

[11] Grimson, W.E.L. & D.P. Huttenlocher, 1990, "On the Sensitivity of the Hough Transform for Object Recognition," *IEEE Trans. PAMI*, to appear.

[12] Grimson, W.E.L. & D.P. Huttenlocher, 1989, On the Verification of Hypothesized Matches in Model-Based Recognition, MIT AI Lab Memo 1110.

[13] Grimson, W.E.L. & T. Lozano-Pérez, 1987, "Localizing overlapping parts by searching the interpretation tree," *IEEE Trans. PAMI* **9**(4), pp. 469–482.

[14] Huttenlocher, D.P. & S. Ullman, 1990, "Recognizing solid objects by alignment with an image," *Intl. J. of Computer Vision*, to appear.

[15] Ikeuchi, K., 1987, "Generating an interpretation tree from a CAD model for 3d-object recognition in bin-picking tasks," *International Journal of Computer Vision* **1**(2), pp. 145–165.

[16] Murray, D.W. & D.B. Cook, 1988, "Using the orientation of fragmentary 3D edge segments for polyhedral object recognition," *Intern. Journ. Computer Vision* **2**(2), pp. 153–169.

[17] Stockman, G., 1987, "Object recognition and localization via pose clustering," *Comp. Vision, Graphics, Image Proc.* **40**, pp. 361–387.

[18] Thompson, D.W. & J.L. Mundy, 1987, "Three-dimensional model matching from an unconstrained viewpoint," *Proc. Intern. Conf. Robotics & Automation*, Raleigh, NC, pp. 208–220.

Object Recognition Using Local Geometric Constraints: A Robust Alternative To Tree-Search

Alistair J Bray

University of Sussex

Janet: `alib@cogs.susx.ac.uk`

Abstract

A new algorithm is presented for recognising 3D polyhedral objects in a 2D segmented image using local geometric constraints between 2D line segments. Results demonstrate the success of the algorithm at coping with poorly segmented images that would cause substantial problems for many current algorithms. The algorithm adapts to use with either 3D line data or 2D polygonal objects; either case increases its efficiency. The conventional approach of searching an *interpretation* tree and pruning it using local constraints is discarded; the new approach accumulates the information available from the local constraints and forms match hypotheses subject to two global constraints that are enforced using the *competitive paradigm*. All stages of processing consist of many extremely simple and intrinsically parallel operations. This parallelism means that the algorithm is potentially very fast, and contributes to its robustness. It also means that the computation can be guaranteed to complete after a known time.

Many algorithms have been developed in recent years that rely upon dynamically growing and pruning an interpretation tree that describes the correspondence between object features and image features. The features may be 3D [1, 2, 3] or 2D [4, 5, 6]; likewise, objects may be 3D polyhedra or 2D polygons. For example, Goad's algorithm matches 2D line segments in the image to 3D segments describing the object; the algorithm is an efficient one given a low degree of noise in the image and a good segmentation, and has been well-tested in a commercial environment. This work examines such algorithms based upon the tree-search formulation with respect to *robustness* and *speed* under noisy conditions, and concludes that their performance does not *degrade gracefully*. A new formulation is presented that exploits the same information in the image, but is inherently parallel and transforms the problem from one of tree-search to signal detection.

The search-tree paradigm

In this paradigm possible matches define a *search tree* which can be pruned using local constraints. A solution-state is defined by a leaf-node of the search tree. Heuristics can provide an efficient way of searching this tree to find solutions quickly and so providing a sophisticated *sequential* traversal of the search tree. For example, Goad describes describes various ways his algorithm can be unwound and how optimal search paths can be pre-computed for models off-line; these are all extensions and modifications to what is essentially a *depth-first sequential search* [4, 5]. There are weaknesses with this formulation in terms of the sensitivity to image noise, the speed of the search, and the unpredictability of the search times involved:

Sensitivity to Noise

Tree-search is sensitive to error in the segmentation of the image because *one* incorrect evaluation of a local constraint is sufficient to prevent a correct solution being found. Local constraints will be described later (See Section 3) that *can* cope with the type of error in the data commonly associated with occlusion, poor lighting or bad edge-detection. However, such measures are ineffective when coping with the problem of image features being wholly obscured eg. if predicted lines are not present the algorithm becomes inefficient. Solutions to this problem have been suggested such as the *Null Face Hypothesis* [4, 3]. However, these solutions lead to a large expansion of the search-tree and are therefore undesirable. Such algorithms are demonstrating *ungraceful degradation*; a single missing line causes substantial problems for the search, and multiple missing lines makes the search space hopelessly large.

Speed of Computation

Tree-search algorithms are essentially sequential and well-suited to conventional Von Neumann machines. However, given the best of optimisation techniques we can only expect a sequential algorithm to achieve certain levels of speed. On the other hand, if an algorithm is inherently parallel the potential speed of a parallel implementation is extremely high; this sort of parallelism can be exploited by many modern architectures today and it seems certain that it is the way that hardware is progressing. It is certain that a tree-search can always be performed using a parallel algorithm in such a way as to exploit parallel facilities. However, this post-hoc exploitation of potential parallelism is usually harder to control and less flexible than when dealing with an *intrinsically* parallel algorithm; time-gains are not always as great as anticipated.

Predictability

Despite the effect of pruning, the size of the search tree involved, given noisy data, can be quite substantial. The time taken for a correct solution to be found can vary enormously depending upon where in the tree it lies and how the search is guided. Heuristics can be exploited for guiding the search eg. ways of pre-computing optimal search strategies, unwinding the search tree. These can be very effective. However, they are only heuristics and, by their nature, will fail to prove effective some of the time. They are also often model-specific, and the pre-computation of the optimal search path for a new model is itself time-consuming. There is therefore an element of *uncertainty* in whether the system will arrive at a correct solution within a given time. This can cause problems in a real-time system that requires the best solution available after a given degree of processing. It is also true that many heuristics require parameters to be set and how to determine the optimal setting for these parameters is often opaque; trial and error is often ineffective and always time consuming. *How many failures to find a predicted line should be tolerated before pruning a branch of the tree? How many matches are sufficient for a correct match? How many matches must be found before search is terminated?* The answer to these questions is not obvious and it would be far preferable if the questions could simply be side-stepped.

Aims of New Formulation

It is apparent that a serial search algorithm can be optimised, but still encounters critical problems in the above areas. It might be hoped that a new parallel formulation can be derived that exploits to the full the advantages of a parallel approach. If this can be achieved then the algorithm might possess the following properties:

1. **Graceful Degradation:** The ability to compute the correspondence set and an estimation of object orientation with a degree of accuracy comparable to the accuracy of the data.

2. **Predictability:** The ability to compute the correspondence set and an estimation of object orientation within a particular fixed time.

3. **Speed:** The ability to compute the correspondence set and an estimation of object orientation in very short periods of time.

4. **Heuristic Independent:** The ability to compute the correspondence set and an estimation of object orientation in a way that does not rely heavily on *ad hoc* parameter determination.

1 New Formulation

The new formulation consists of four stages of processing:

1. **Accumulation:** In the accumulation stage all possible binary relations between image segments are compared with all possible binary relations between model segments over all possible viewpoints. If the two relations are consistent then support is accumulated for the two assignments of model to image that the relations represent.

2. **Cooperation:** Once accumulation is complete there is a cooperative stage in which support in the accumulators is distributed between locally connected regions of the viewsphere.

3. **Competition:** In the competitive stage support is distributed *within* the accumulators for each viewpatch so as to enforce uniqueness constraints ie. only one model line can match one image line and vice versa. The result is an accumulator with a small number of high peaks that corresponds to an interpretation for that viewpoint.

4. **Signal Extraction:** The final stage of signal extraction finds the interpretation for each accumulator and finds the largest maximal clique within the interpretation that is wholly consistent in terms of the local constraints. These cliques represent the best interpretation for that viewpoint. Those of sufficient size are ordered according to the signal/noise ratio in the accumulator and then passed to the model test stage in the usual way.

1.1 Data Structures

Consider an image $[I]$ consisting of $\mathcal{I}$ 2D line segments and a model $[M]$ consisting of $\mathcal{M}$ 3D line segments. Let the Viewsphere be quantised into a set of $\mathcal{V}$ viewpatches – $[V]$. For each viewpatch a subset of $[M]$ will be visible. Consider a single viewpatch – V_v where $1 \leq v \leq \mathcal{V}$; for V_v a binary constraint array can be constructed that has dimension $\mathcal{M} \times \mathcal{M}$ – call this MA_v. An element of MA_v, $MA_{v_{(m,m\prime)}}$ where $1 \leq m, m\prime \leq \mathcal{M}$, describes the local relationship between model line M_m and model line $M_{m\prime}$ over viewpatch V_v. If either M_m or $M_{m\prime}$ is obscured at V_v then $MA_{v_{(m,m\prime)}}$ will be empty; otherwise it will contain the bounds upon the binary constraints between M_m and $M_{m\prime}$ over V_v. Of course all diagonal elements $MA_{v_{(m,m\prime)}}$ where $1 \leq m = m\prime \leq \mathcal{M}$ will also be empty as they represent the relation between a model segment and itself. Hence MA_v will be a sparse array due to visibility constraints; also, if the local constraints being used are symmetric then only a triangular half of the array need be stored ie. $MA_{v_{(m,m\prime)}}$ where $1 \leq m < m\prime \leq m$.

In a like manner a single binary constraint array can be constructed for the image segments – call this IA. This contains the actual values for the binary constraints between any pair of image segments. If the constraints being measured rely upon the segments being *directed*, then this will be of dimension $\mathcal{I} \times 2 \times \mathcal{I} \times 2$; otherwise it will have dimension of simply $\mathcal{I} \times \mathcal{I}$. This is basically decomposing an undirected image segment into two possible directed image segments of opposite directions – this effectively doubles the size of the image. Since the two constraints to be used both use directed image segments we will assume a dimension of $\mathcal{I} \times 2 \times \mathcal{I} \times 2$ from here on. This array will not be sparse except that the diagonal elements $IA_{(i,d,i\prime,d\prime)}$, where $1 \leq i = i\prime \leq \mathcal{I}$ and $d, d\prime \in [1,2]$ will be empty since they represent the relationship between an image segment and itself.

For the same viewpoint V_v it is also possible to construct a *Correspondence* array. Let us call it CA_v. This array will act as an *accumulator*. It will store information concerning possible correspondences between model and image segments for V_v. It will therefore have dimensions of $\mathcal{M} \times \mathcal{I}$. Since it acts as an accumulator each element is an accumulator cell such that a high count in $CA_{v_{(m,i)}}$ where $1 \leq m \leq \mathcal{M}$ and $1 \leq i \leq \mathcal{I}$, implies strong support for model segment m corresponding to image segment i in Viewpatch V_v.

1.2 Updating The Accumulators

Given the above representation of data-structures it is easy to see how the $\mathcal{V}$ accumulator arrays can be updated. Given a particular viewpatch V_v, each non-empty element of IA – $IA_{(i,d,i\prime,d\prime)}$ – can be checked against each non-empty element of MA_v – $MA_{v_{(m,m\prime)}}$ – to give a boolean output that will depend upon whether the values of $IA_{(i,d,i\prime,d\prime)}$ fall inside the bounds defined by $MA_{v_{(m,m\prime)}}$. If $IA_{(i,d,i\prime,d\prime)}$ is consistent with $MA_{v_{(m,m\prime)}}$ then the two cells $CA_{v_{(m,i)}}$ and $CA_{v_{(m\prime,i\prime)}}$ in the accumulator array CA_v can be incremented. If it is inconsistent then no update occurs. Therefore for V_v the number of possible updates to CA_v is equal to the twice the product of the number of elements in MA_v and the number of elements in IA. In practice it will be much less than this since updates only occur for consistent matches.

1.3 Competition and Co-operation

Let us assume that the accumulation stage is complete on the $\mathcal{V}$ accumulator arrays. The problem is how to extract the correspondence set and an estimation of viewpoint from these arrays. This is a signal detection problem. Each accumulator array consists of a mixture of signal and noise. At viewpoints far away from the correct one the signal is more or less non-existent; as the viewpoint tends towards the correct one the signal becomes much stronger and it is expected that it should be detectable from the background noise. In this sort of problem it is the ratio of signal to noise that is significant. The signal is defined by correct correspondences satisfying the local constraints, and the noise is defined by incorrect correspondences satisfying the constraints by chance. For instance, error in sensor data will lead to a reduction in *signal*; irrelevant image segments on the other hand lead to and increase in *noise*; missing image lines will again lead to signal reduction.

Competition

However, there is further information that can be applied to the accumulators that helps to amplify the signal and reduce the noise. Two assumptions can be made which allow the competitive paradigm to be applied within each accumulator array. The assumptions are:

- **No image segment can correspond to more than one model segment**

- **No model segment can correspond to more than one image segment**

The first of these assumptions is certainly one that we would wish to incorporate in a system, since it is only under very exceptional circumstances that an image line corresponds to more than a single model line. The second is not so certain since a model line can produce more than a single image line; this may occur in cases of occlusion or poor edge-detection. As will be seen, this formulation provides an elegant way of incorporating the two different constraints in terms of competition in two orthogonal directions.

The accumulator has both a model and an image dimension. It is simple to implement a form of competition independently in each of these dimensions. Consider the raster in the accumulator defined by $CA_{v_{(m,K)}}$ where K is a constant $(1 \leq K \leq \mathcal{I})$ and $1 \leq m \leq \mathcal{M}$. This raster represents the support for the various model lines matching image line K at viewpoint v. Since only one model line is allowed to match that image line competition can be introduced between the elements of this raster. That is to say, one of these cells will *win* at the expense on the other $\mathcal{M} - 1$ lines. If competition was in the model dimension alone this competition would be simple to implement. Obviously the element in the raster with the highest count would win, and maybe if this count was insufficient then there would be no winner. However, there is also similar competition in the image dimension ie. those rasters defined by $CA_{v_{(K,i)}}$ where K is a constant $(1 \leq K \leq \mathcal{M})$

and $1 \leq i \leq \mathcal{I}$. Therefore a way on implementing this competition in both dimensions simultaneously must be adopted that provides a solution consistent with both the above assumptions.

Co-operation

There is one more assumption that can be incorporated into this formulation. So far we have quantised the viewsphere into $\mathcal{V}$ viewpatches. We have then considered each of these independently. In doing this we are ignoring the topology of the viewsphere. As stated above, we expect the degree to which the signal will dominate over the noise in a particular viewpoints accumulator to be correlated with how close that viewpoint is to the correct viewpoint. It is therefore to be expected that traces of the signal will not only show up in the correct viewpatch but also in neighbouring viewpatches. It is expected that the signal will trail off as the viewpoint moves away from the correct one. It is therefore possible to introduce a degree of topographical cooperation between like cells of the accumulator for neighbouring viewpatches. In this way the signal is being picked up over a larger portion of the viewsphere.

1.4 Global Consistency

One of the main problems with the approach as formulated is that of *global inconsistency*. There can be two types of global inconsistency:

- **Within Local Constraints:** An interpretation is globally inconsistent *within* local constraints when one or more of the correspondences is inconsistent with one or more different correspondences in terms of the local constraints used.

- **Beyond Local Constraints:** An interpretation is globally inconsistent *beyond* local constraints when all the local constraints between correspondences are satisfied, and yet there is still no single 3D position of the object that will map each model feature onto each image feature.

In the tree search formulation, the interpretation is guaranteed to be globally consistent within the local constraints, and consistency beyond local constraints is guaranteed by some form of model test. In the formulation proposed here local consistency within constraints is *not* guaranteed; inconsistent correspondences may result from incorrect peaks resulting from noise. In such interpretations it may be that many of the correspondences are correct but a minority are not (eg. along rasters corresponding to missing model lines). This minority is easily sufficient to force the model test to fail since these matches may be radically wrong in terms of the geometry of 3 space [1]. One solution to this problem is to have some form of filtering in the model test that removes matches that are preventing convergence. It is to be expected that this would be computationally expensive. A better solution is to ensure consistency within local constraints by finding *maximal cliques*.

---

[1] A model test performing a least squares solution can cope with error as long as incorrect matches are "close" to correct. Interpretations that are consistent within constraints usually meet this criterion of closeness. However, those that are inconsistent in terms of local constraints often fail to meet this criterion and subsequent *convergence* is impossible

The result of the competitive layer is an interpretation that is expected to be largely consistent within constraints but not totally so. This can be represented as a graph in which each node is a correspondence and a connection represents consistency. The problem of ensuring global consistency within constraints is reduced to that of finding maximal cliques in the graph. The study of this problem is well-developed, and given that the interpretation in expected not to be that large (it is bounded by $max(\mathcal{I}, \mathcal{M})$), there are very efficient algorithms for finding these cliques. Bolles & Cain [7, 8] discuss this problem in the domain of model-based vision, and state that they have found Johnston's algorithm [9] to be quite satisfactory. It is to be expected that the largest maximal clique will represent the correct interpretation, although smaller cliques could be considered worth investigating if the model test still fails.

1.5 Diagrammatic Representation

The above formulation is represented in Figure 1. The lowest layer represents the input or stimulus – this is an $\mathcal{I} \times \mathcal{I}$ array of binary relations generated from the image segments. The second layer represents the constraints generated from the model – an $\mathcal{M} \times \mathcal{M}$ array for each of the $\mathcal{V}$ viewpoints. There is a mapping (or *connection*) between each element in Layer 1 and each element in Layer 2. The third layer represents the accumulator space. It consists of an $\mathcal{M} \times \mathcal{I}$ array for each of the $\mathcal{V}$ viewpoints. For each connection between the first and second layers there is a corresponding pair of connections between the second layer and two elements in the third (given binary constraints). These connections will "fire" depending upon the consistency of model with image. The fourth layer has the same dimensions as the third and represents the co-operative process. Each element in the fourth layer is connected to the set of similar elements for neighbouring viewpatches in the third layer. The final layer again has the same dimensions as the previous and contains the results of the competitive stage. From these $\mathcal{V}$ arrays of dimension $\mathcal{M} \times \mathcal{I}$ the best globally consistent signal is extracted.

2 Algorithm

The first stage is to perform pre-computation of the model bounds array for each viewpoint on those model lines visible at that viewpoint [2]. This can be computed off-line using the chosen set of constraints. Obviously, the model constraints need only be computed once and from then on can be loaded up before run-time. Likewise it is possible to compute the image constraint array that describes the relationship between any two image lines in terms of the local constraints chosen.

2.1 Accumulation

The iterative procedure for doing this is outlined in Figure 2. As can be seen it is wholly parallel in nature – a simple outline of the procedure described in Section 1.2. Figure 3 shows a typical set of accumulator arrays after the accumulation process.

---

[2]It is assumed that the visibility of model lines over the viewpatches has already been pre-computed

2.2 Co-operation

The co-operative activity between neighbouring patches on the viewsphere is simulated by simply making the counts in an accumulator array for a viewpatch a function of both themselves and their corresponding counts over neighbouring viewpatches ie. a weighted average over neighbouring viewpatches. An example result is shown in Figure 4 of a typical set of accumulator arrays after this cooperative stage has been executed.

2.3 Competition

The competitive activity along the model and image dimensions of the accumulator arrays is simulated by re-adjusting counts within a raster line in favour of the winning cell. That is to say, given a particular raster line the counts are adjusted so that the winning count takes a small amount from each of its losing competitors. This adjustment is performed iteratively, adjustments alternatively along the model rasters and then along each of the image rasters. This procedure iterates until the adjustments being made are insignificantly small. The usual result is a new accumulator array with a few very large peaks, and these peaks tend to be the only peaks along any raster line (ie. along model or image raster lines). An example result is demonstrated in Figure 5 of a typical accumulator after the competitive stage. It is easy to see that it is very simple using this technique to relax the assumptions that the competition is representing; for example, allowing competition only in the image dimension removes the assumption that a model line can match only a single image segment. It is also easy to see how the assumptions can be incorporated to different degrees by the amount of *relative* competition in the two dimensions. Both the rules governing the re-distribution of the counts along rasters, and the integration of the iteration between the two dimensions can easily be adjusted to favour either of the two assumptions.

2.4 Signal Extraction: Maximal Cliques

The final stage of the algorithm consists of extracting the signal from the accumulators; that is to say, deciding which viewpatch is displaying the strongest consistent pattern of activation and what that pattern means in terms of the correspondence set. This is achieved in a sequence of three stages:

In the first stage of **Basic Extraction** the accumulator after competition is examined; all cells that are peaks in *both* their corresponding model and image dimensions are taken to represent correct correspondences for the given viewpoint ie. they are considered "on". The correspondence is obviously denoted by the indices of the accumulator cell. This gives an interpretation that meets both the assumptions that a model line matches a single image line and vice versa. However, this interpretation can still be globally inconsistent in terms of the local constraints used.

In the second stage **Maximal Cliques** are extracted. The interpretation arising from basic extraction can be considered to be a graph in which a node represents a correspondence and connectivity is defined by consistency in terms of local constraints. The maximal clique algorithm will list all combinations of correspondences that are mutually consistent. The largest maximal clique is the largest set of combinations. Applying this algorithm is therefore enforcing global consistency within constraints in that each correspondence must be locally consistent with every other. The algorithm used for maximal clique extraction is that of Johnston [9], and as used by Bolles & Cain [8] in their *Local Feature Focus* method for locating 2D objects [3]. The algorithm is highly recursive and has been found to run very quickly on the short interpretations commonly found. The result is that the smallest number of nodes in the original interpretation are turned "off" to make the interpretation globally consistent within local constraints.

The final stage is that of **Ordering Viewpoints**. The signal strength for a viewpoint is measured as a function of the mean value and the standard deviation of the cells in the interpretation that remain "on" after Stage 2 (ie. are part of the largest maximal clique). The viewpatches are ordered in terms of the number of "on" cells (if less than 3 cells are "on" for a viewpoint it is ignored since at least 3 correspondences are necessary to invert the perspective transform); those viewpoints with the same number of "on" cells are ordered in terms of signal strength. The result is an ordered set of viewpoints and their corresponding interpretations that can be passed to a further model test stage to check for global consistency *beyond* local constraints and to extract the 6 parameters that determine object position. Due to the global consistency already enforced this set of viewpoints usually represents a small proportion of the whole viewsphere.

3   Local Constraints

In the implementation that produced the results in the next section the scale factor is unknown; the object may therefore appear at any size in the image. There are two approaches to coping with this problem in the formulation proposed: either accumulation operates over a set of likely scales as well as the set of viewpoints, or else the local constraints used are *scale independent*. The latter approach was adopted. The two geometric constraints between segments used were:

- **The Angle Constraint** states that for two image lines to match two model lines at a particular viewpatch the angle between the image lines must be within the range defined the model lines over that viewpatch (within a certain error bounds)

- **The Direction Constraint** states that for two image lines to match two model lines at a particular viewpatch the bounds on the angle defined by the arbitrary vector connecting the first image segment to the second, relative to the first segment, must be within the similar bounds defined by the model lines for that viewpatch.

These two constraints are illustrated in Figure 6. It can be seen that both these constraints are scale independent, since the constraints are defined by angles that are wholly independent of segment size. It can also be seen that both constraints are insensitive to error in the length of the image segments. The Angle constraint is wholly independent of segment length. The Direction constraint is independent as long as the error makes the segment too small rather than too large; that is to say, the range of angles defined by any subsegments of the true image segments will be a subset of the range of angles defined by the true image segments (or model segments). The result is that another dimension

---
[3]The actual algorithm is listed in the Appendix of [8] and described in greater detail in [7]

of scale does not need to be added to the accumulator, and the constraints used will also be robust to errors in segmentation caused by occlusion, poor edge-detection, or poor lighting.

4 Results

The results were obtained from an implementation of the above algorithm. It has been run on noisy simulated data to illustrate its potential. The implementation is actually an iterative serial one, the iteration reflecting its parallel nature, and is written in the high-level language POP11, part of the Sussex University POPLOG system.

Two models were used. The first consists of 6 3D segments connected as an irregular tetrahedron; the second consists of 10 randomly generated 3D segments that are unconnected. No account is taken in the model concerning the visibility of the segments; all lines are considered visible. In this formulation invisible lines should provide low counts in the accumulator and therefore be ignored. Visibility can be incorporated very effectively into this formulation when the model constraints are compiled; constraints are only computed for $MA_{v_{(m,m\prime)}}$ if both model lines M_m and $M_{m\prime}$ are visible at V_v. When accumulation occurs, connections involving elements of $MA_{v_{(m,m\prime)}}$ which are empty are simply ignored. This way the two rasters in CA_v corresponding to m and $m\prime$ are guaranteed to be totally empty and will be ignored in further computation.

In this implementation a very weak form of the above was incorporated. It was considered that if two model lines were projected into the image and found to intersect at a particular viewpoint then one of them was actually invisible at this viewpoint. Therefore the constraint box corresponding to these lines at this viewpoint was left empty and ignored in subsequent accumulation [4].

Noisy Images were generated from the model. The model was projected at random orientations to produce a basic image and then noise would be added. Three types of noise have been added to varying degrees:

- **Segment Reduction:** As in poor feature extraction or occlusion [5].

- **Segment Addition:** As in non-object features.

- **Segment Removal:** As in missing object features.

The two Scale Independent and Noise Robust constraints discussed in the previous section – the Angle and the Direction constraint – were used. These were both implemented as Direction Dependent and Symmetric.

Figures

The 5 figures 7 to 11 illustrate some results obtained. In the top half of the figure the accumulator arrays are displayed for the final viewpoints that provide an interpretation

---

[4]Intersection is easy to detect since when two segments intersect the computed bounds on the Direction Constraint will be greater than $180°$

[5]Segmentation reduction was set throughout these examples at 40% ie. each segment was reduced to 60% of its true length.

and that can be verified by a model test. A maximum of 25 viewpoints are displayed; at the course resolution this corresponds to all viewpatches except one; at the fine resolution it corresponds to about $11\frac{1}{2}\%$ of the viewsphere. For display purposes the values in *each* accumulator have been normalised to take advantage of the full range of grey-values available; therefore absolute intensities are lost, but relative intensities *within* accumulators are preserved.

In the examples using the coarsely quantised viewsphere the cooperative stage was left out since it was considered that the signal would not be strongly present in neighbouring viewpatches as they are so far away. The left-hand side displays the accumulator cells for the final viewpatches in ascending order *before* competition. The right-hand side displays the same accumulators for the same viewpatches *after* competition. The horizontal represents the *model* dimension, and the vertical represents the *image* dimension. As can be seen, no peak in the competitive layer shares a raster line with another peak, so enforcing the two assumptions described above. The cells displayed in the competitive layer are not necessarily all used in the model test since some are filtered out by the maximal clique test.

The lower half of the Figure displays the results of the model test using an interpretation extracted from one of the viewpatches. Firstly the perfect image is shown – this consists of the projection of the model at the appropriate viewpoint (ignoring visibility), along with the display of the extra noise segments added to the image. Secondly, the image is degraded by reducing the length of all segments (both those that are part of the object and those that are not) – this is the actual image that is given to the matching algorithm. In the third portion the model is displayed from the viewpoint selected after matching, but having undergone a 2D rotation about this viewpoint that makes a rough approximation at the missing degree of freedom. The final projection shows the position of the model after Lowe's iterative model test [10, 11, 12] has been applied to the interpretation to match the model as accurately as possible to the corresponding line segments.

- **Figure 7:** This first figure displays results for the *ideal* situation using the model of random lines and the fine quantisation. All model lines are present but no extraneous image lines have been added. The model segments have been degraded by up to 40%.

- **Figure 8:** The small viewsphere has been used (only twenty six) viewpatches). Only five of the six model lines are present along with nine extra noisy image lines. The correct interpretation has been found in only a single viewpatch. This is not perfect since the solution should be found in at least two patches - that corresponding to the correct viewpoint, and that on the opposite side of the viewsphere corresponding to the Necker reversal [6]. The interpretation found for this viewpoint (view 17) is correct and seen to be very close to the actual viewpoint. The model test positions the model in perfect position.

- **Figure 9:** This Figure uses the small viewsphere and the random model. Only

---

[6]Since near parallel projection has been assumed throughout there is very minimal difference between projection at one viewpoint and that on the opposite side of the viewsphere. Since visibility constraints have been ignored the system cannot discriminate between the two possibilities.

nine of the ten model lines are present and nine extra image lines have been added. Two interpretations are found, the first of which is correct, despite the fact that the signal is masked considerably (as can be seen from the accumulators displayed). The model test is successful and the model is correctly located.

- **Figure 10:** The fine quantisation of the viewsphere has been used. Only four of the six model lines are present, and five extra image lines have been added. Three solutions are found and the model test succeeds on the first of these.

- **Figure 11:** This demonstrates this simple implementation of the algorithm being pushed to its limits. The random model has been used and the fine quantisation of the viewsphere. Five model lines have been removed and five extra image lines have been added. It can be seen from the co-operative layer that the signal/noise ratio is very weak. Seven interpretations are found and the third one is correct and passes the model test; the model test had failed on the two previous attempts using the "better" interpretations.

5 Discussion

5.1 Signal/Noise Ratio

The main factor determining the success of the algorithm described above is the signal/noise ratio; increasing its strength has a very positive effect upon the efficiency of the algorithm. This can be done only by altering the constraints checked in the accumulation stage. In the tree-search formulation the computation of the bounds on the local constraints is determined by the quantisation of the viewsphere and the degree of noise anticipated (eg. [4, 5]). In this formulation similar bounds may be computed but not all of them *need necessarily* be used. Given a correct match of two model lines to two image lines at a particular viewpoint, it is not essential that the corresponding accumulations $CA_{v(m,i)}$ and $CA_{v(m',i')}$ need either be considered or, if considered, that they should pass the constraint check. What is essential is the signal/noise ratio.

For instance, a local constraint bound corresponding to $MA_{v(m,m')}$ may be so large that it not only lets in the signal but also a large degree of noise; in this case it might be considered that it is not worth being considered in the accumulation stage. In other words the constraint bounds may be *tightened* so that both more of the signal and more of the noise is excluded from the accumulators; it is the ratio of this exclusion that is significant. This method of exclusion could be used as a method of introducing a measure of *saliency* in that it is not accumulating upon constraints that could easily be met by chance, but only on those that are likely to be met by correct correspondences.

Of course, another method of changing the signal/noise ratio is to alter the constraints considered. The advantage of the constraints used in this implementation (Direction and Angle) is that they are Scale Independent, and so they can detect the object at any distance or scale without having to have a Scale dimension in the accumulator. If the scale/distance of the object was already known in advance then stronger constraints could be used that would substantially decrease the amount of noise in the accumulator. The signal/noise detection problem would therefore be simplified.

5.2 Advantages

The approach outlined above possesses three advantages over the original search-tree formulation of the matching problem:

1. **Parallelism:** The main advantage offered by this this approach is its inherent parallelism. The original tree-formulation is naturally serial, although such a search can be implemented to some degree in parallel. This parallelism means:

   - **Graceful Degradation:** The algorithm has robust qualities, meaning that no particular parts of the image are of especial significance. This is in distinct contrast to the tree-search formulation that has problems coping with either occlusion or degraded data.

   - **Predictable Speed:** Given a degree of serialism, it is possible to predict quite accurately how long the process will take to complete (Given *complete* serialism this may be a very long time).

   - **Potential Speed:** There is no reason why, given appropriate hardware facilities, such an algorithm could not be made to run as fast as required.

2. **Scale Independence:** The system of constraints used allows the algorithm to cope satisfactorily with the scale problem without extending the accumulation space into another dimension. This is at the expense of a weaker signal/noise ratio. If scale is already determined, the best of both worlds can be had.

3. **Reduced "Ad Hoc" Parameter Settings:** The tree-search formulation is heavily reliant on "ad hoc" parameter settings that control the search strategies and how the tree can grow. This formulation is less dependent upon such settings.

5.3 Coarse-to-Fine Strategy

It is very simple to see how this new formulation could be redesigned to employ a coarse-to-fine resolution strategy. The results above show that a resolution of 26 viewpatches is sufficient in some cases to yield a correct solution; even if it does not give a perfect interpretation it often points to the correct viewpatch. The correct viewpatch could be quantised to a finer resolution and then the accumulation repeated anew over this limited portion of the viewsphere. Extraction of the interpretation need only take place at the end. For example, given the two resolutions of $\mathcal{V}_1$ and $\mathcal{V}_2$, where $\mathcal{V}_1 < \mathcal{V}_2$, the ratio of the accumulations necessary for a coarse-to-fine strategy to the accumulations necessary for the finer resolution would be:

$$(\mathcal{V}_1 + \frac{\mathcal{V}_2}{\mathcal{V}_1}) : \mathcal{V}_2$$

In the case of the two resolutions 26 and 218 this represents savings of about 85%; this must be offset against time for dynamic configuration and potential error [7].

---

[7] The recursive quantisation of the viewsphere necessary for this sort of strategy comes naturally from the description of the viewsphere in terms of the icosahedron

Acknowledgements

Thanks to Dave Hogg, Vaclav Hlavac and Kelvin Yuen for their help with this work. Thanks also to Peter North for a critical insight.

References

[1] **Grimson W E L and Lozano-Perez T**. Model-based recognition and localisation from sparse range or tactile data. *International Journal of Robotics Research*, 3(3):3–35, 1984.

[2] **Grimson W E L**. The combinatorics of object recognition in cluttered environments using constrained search. *Second International Conference on Computer Vision,IEEE*, pages 218–227, 1988.

[3] **Grimson W E L**. On the recognition of curved objects. *IEEE Pattern Analysis and Machine Intelligence PAMI*, 11(6):632–643, 1989.

[4] **Goad C**. Special purpose automatic programming for 3d model-based vision. *Proc. Image Understanding Workshop,Virginia,USA*, pages 94–104, 1983.

[5] **Goad C**. Fast 3d model-based vision. In Pentland A P, editor, *From Pixels to Predicates*, pages 371–391. 1984.

[6] **Bray A J**. Tracking objects using image disparities. *Proceedings of the 6th Alvey Vision Conference, Reading*, 1989.

[7] **Bolles R C**. Robust feature matching through maximal cliques. *Proc. SPIE Technical Symposium on Imaging and Assembly*, April, 1979.

[8] **Bolles R C and Cain R A**. Recognizing and locating partially visible objects, the local-feature-focus method. *International Journal of Robotics Research*, 1(3):57–82, 1982.

[9] **Johnston H C**. Cliques of a graph - large or small. *Draft, Queen's University of Belfast*, 1975.

[10] **Lowe D G**. *Perceptual Organisation and Visual Recognition*. PhD thesis, Stanford University, Dept. of Computer Science, 1984.

[11] **Lowe D G**. *Perceptual Organisation and Visual Recognition*. Boston: Kluwer, 1985.

[12] **Lowe D G**. Three-dimensional object recognition from two-dimensional images. *Artificial Intelligence*, 31(3), 1987.

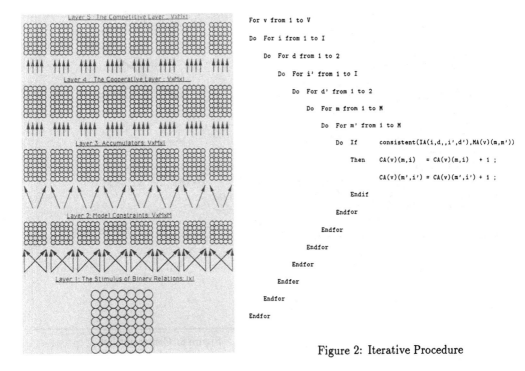

Figure 1: Diagrammatic Formulation

```
For v from 1 to V
Do  For i from 1 to I
    Do  For d from 1 to 2
        Do  For i' from 1 to I
            Do  For d' from 1 to 2
                Do  For m from 1 to M
                    Do  For m' from 1 to M
                        Do  If    consistent(IA(i,d,,i',d'),MA(v)(m,m'))
                            Then  CA(v)(m,i)  = CA(v)(m,i)   + 1 ;
                                  CA(v)(m',i') = CA(v)(m',i') + 1 ;
                            Endif
                        Endfor
                    Endfor
                Endfor
            Endfor
        Endfor
    Endfor
Endfor
```

Figure 2: Iterative Procedure

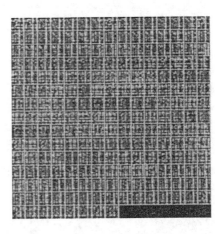

Figure 3: Typical Accumulator   Figure 4: Accumulator after Cooperation

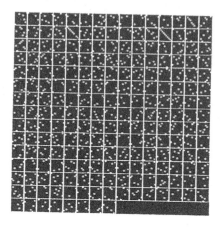

Figure 5: Accumulator after Competition

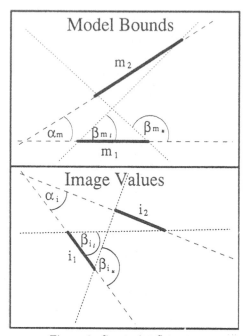

Figure 6: Constraint Set

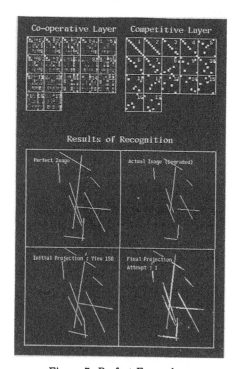

Figure 7: Perfect Example

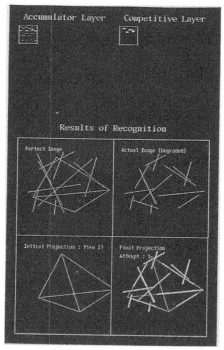

Figure 8: Small Viewsphere 1

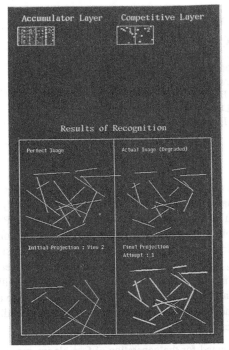

Figure 9: Small Viewsphere 2

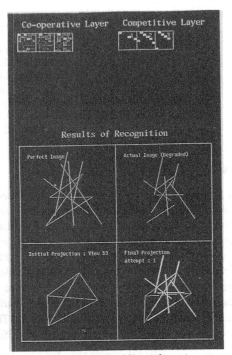

Figure 10: Large Viewsphere 1

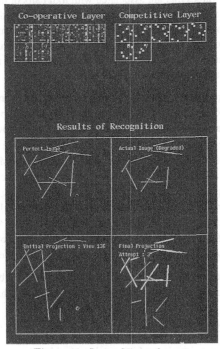

Figure 11: Large Viewsphere 2

SIMD GEOMETRIC MATCHING

Derrick Holder
City of London Polytechnic, Department of C.M.S.M.S.,
100 Minories, London EC3N 1JY.
and
Hilary Buxton
Queen Mary Westfield College, Department of Computer Science,
Mile End Road, London E1 4NS.

INTRODUCTION

It is generally agreed that the primary task for any true machine vision system, when confronted with geometric data (i.e. location, orientation and hence shape) from low level vision sources, is to identify what is where within the field of view. One approach, that has been adopted by Grimson and Lozano-Pèrez[1], Faugeras, Ayache and Faverjon[2], Murray and Cook[3] and others, is to consider objects in the form of separate, possibly non-convex, polyhedra, for which there are accurate geometric models. First they generate feasible interpretations by means of simple, generally pairwise, geometric comparisons between object models and sensor data. Then they test the interpretations, in detail, for compatibility with the surface equations of a particular object model, bearing in mind the fact that an object may have up to six degrees of freedom relative to the robot's sensors. The method is thus based on the hypothesis, prediction and verification paradigm that is widely used in Artificial Intelligence.

Numerous sequential algorithms have been implemented for the generation of feasible interpretations. Measurements involving location vectors and surface normals at m data points, considered in pairs, are compared with corresponding values that are associated with nxn pairs of object model faces. It is found that, when simple geometric constraints that are independent of the coordinate frame of reference are applied to sparse data, the possible asignments of data points to object model faces can generally be represented by just a few, frequently only one, feasible interpretation. This is done entirely without resort to a detailed solution of the surface equations. Even so, the algorithms are not generally fast enough, in sequential form, to offer a practical solution to the problem.

A parallel algorithm for the generation of feasible interpretations has been implemented by Flynn and Harris[4] on the Connection Machine at the Massachusetts Institute of Technology. This algorithm exploits the parallelism in the problem at the expense of processor numbers which grow exponentially with problem size. However a similar degree of parallelism has been achieved by the present authors[5,6] with a processor set that is only quadratic in the problem size. Using a distributed array SIMD processor, the AMT DAP 510, problems are handled that would previously have far outstripped the capacity of the Connection Machine. The algorithm can equally well be applied to measurements relating to the edges of a polyhedron. Instead of using a small number of discrete measurements, edge matching generally involves the processing of a substantial volume of grey level data, and the production of a $2^1/2$D sketch. Nevertheless, this form of input is efficiently provided by the ISOR system[7,8,9] developed at GEC Hirst Research Centre, and currently being implemented on the AMT DAP at Queen Mary Westfield College.

The purpose of this paper is to present an overview of the algorithms for both face matching and edge matching interpretations of visual data, together with some of the results that have been achieved to date.

THE GENERATION OF FEASIBLE INTERPRETATIONS

The generation of feasible interpretations in the face matching problem proceeds as follows:-

(i)   For each pair of data points, trial assignments to the faces of a particular object model are recorded in an interpretation tree, with each node representing a given assignment, and with alternative paths representing the sequences of assignments embodied in different interpretations of the data set.

(ii)   A geometric match is said to be achieved when the values of certain primitives, such as the distance between two points or the angle between two surface normals, associated with a given pair of data points, are compatible with the ranges of values associated with the object model faces to which they have been assigned. The interpretation tree is pruned, i.e. the path representing a given interpretation is terminated, wherever there is a failure to achieve a geometric match.

(iii)   Finally the interpretation tree is pruned wherever a trial assignment would be inconsistent with assignments already made at higher levels in the interpretation tree.

By far the most important single step in the quest for parallelism is to note that pairwise geometric matching is totally independent of the preceding partial interpretations and can be implemented as a parallel process, leaving the global consistency of interpretations to be taken into account at a later stage. We note that the sub-trees from the nodes at a particular level in the interpretation tree are all the same until they are pruned for consistency, and will be reproduced many times over. The results of the geometric matching process may therefore be best represented by a network rather than a tree structure, and stored compactly in an array such as is illustrated in Figure 1, where all paths downwards through true values have to be explored.

		face_1 face_2	1 12345	2 12345	3 12345	4 12345	5 12345
datum_1	datum_2						
1	2		T·T··	·····	·····	··T··	···T·
1	3		·····	···T·	·····	·····	·T···
1	4		·····	·····	·····	·····	··T··
2	3		···T·	·····	··T··	·T···	·····
2	4		·····	·T···	·····	·T···	·····
3	4		·····	·T···	·····	·····	·····

Figure 1 *The Matching Array*

In fact pairs of data points may be considered in any order, and we note that there is only one feasible assignment of data points 1 and 4, in the hypothetical example above, namely to object model faces 5 and 1, respectively, so this is obviously a good place to start. We can generally avoid a proliferation in the number of alternatives to be considered at a given level in the interpretation, by sorting the data pairs into ascending order of geometric match, and a simple tag sort procedure using standard functions in DAP FORTRAN may be used for this purpose.

Although at any stage the check for consistency is dependant on the preceding partial interpretation, it can be performed as a parallel process within a recursive procedure, and the subsequent assignments of data points to object model faces can be made conditional on the outcome. The conditional processing within the loop is of a sequential nature, but this seems inevitable if the demands on processing elements are to be kept within reasonable bounds. Nevertheless, highly effective pruning of alternative interpretations is thus achieved, because the matrix of consistent matches at a given level in the interpretation is generally very sparse, and the selection of true values is efficiently implemented in DAP FORTRAN.

In the case of edge matching, sensory data expressed in terms of position vectors and edge direction vectors are assigned to particular edges of an object model, but the object model database and the method of generating feasible interpretations are essentially the same.

VALIDATION

Having generated an interpretation in which sensory data have been provisionally assigned to particular faces of a given polyhedral object model, on the basis of simple geometric constraints, there is no guarantee that the object model description will be entirely consistent with the data.

The validation process involves the following three steps:-

(i)   establishing the location and orientation of the object model that is most compatible with the data;
(ii)   confirming that every data point then lies sufficiently close to and within the perimeter of the object model face to which it has been assigned;
(iii)   confirming that every data point is visible in the given interpretation.

Now, a rigid body rotation and translation may be expressed in terms of a 3x3 orthogonal rotation matrix R, and a translation vector r_0, and we may determine R and r_0 in such a way that first the object model surface normals after rotation, and then the perpendicular distances from the origin after translation, match the data as closely as possible. The orthogonality condition $R^T R = I$ imposes 6 non-linear constraints on the elements of R, and a further three equations are obtained when the method of constrained least squares is applied to the residual differences between normal directions. The solution for r_0 is obtained more easily, with the method of least squares applied to the residual differences in perpendicular distance from the origin.

It has been demonstrated[10] that the solution of the equations for R may be expressed in terms of singular value decomposition, with the best result selected from 4 possible rotations. However, the Newton-Raphson process readily lends itself to a parallel implementation, with a good first approximation obtained from the relationship that applies when the data exactly fit the object model The process converges to sufficient accuracy after just one or two iterations. Faugeras, Ayache and Faverjon, work rather more compactly with quaternions to determine R and r_0, achieving what appears to be an equivalent result, but presumably their algorithm is implemented in sequential form.

Having established the appropriate location and orientation of the given object model, we may easily determine whether the locations of the data points are consistent with the object model face equations, but it remains to be verified that every data point lies within the perimeter of the face to which it has been assigned, and that it is not hidden from view by another part of the object model. For a given data point to be visible from the position of the sensor, it must lie in a face that is not directed away from the sensor, and its projection on the viewing plane must not fall within the perimeter of another face that is nearer to the sensor. We consider the intersections with the edges of a polygon when a line is drawn from a given data point to some external point. There will be an odd number of intersections if the first point is inside the polygon, and an even number of points if it is outside. We note that the equations for intersections take a particularly simple form when the external point is located at an infinite distance along the positive x-axis.

The first task in an SIMD implementation of the validation process is to map the object model against the data and to set the unused rows of the mapped object model and data matrices to zero. The initial rotation matrix, the solution of the Newton Raphson equations and the translation vector for best fit are then computed using standard DAP FORTRAN Library subroutines, and Standard operations are used to maximise parallelism in setting up the equations for the Newton-Raphson process. The rotation matrix and the translation vector are replicated before their application to the object model.

Before proceeding with the validation of individual data points, the Cartesian coordinates of the vertices associated with given faces, originally stored sequentially in rows, are moved into the columns of separate DAP matrices, and the coordinates of the data points are replicated in columns using a simple but effective binary algorithm. As a consequence, when a given row of vertex coordinates is replicated and related to a matrix of data coordinates, the process simultaneously relates every data point to every object model face, and mxm parallelism is thus achieved. The organisation of the information within DAP matrices at this stage is illustrated in Figure 2.

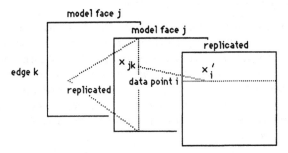

Figure 2. *The Organisation of Information within DAP Matrices*

Transforming into viewing coordinates and initialising a DAP logical matrix, we set up a logical matrix **inside**, and proceed to investigate intersections of the line joining each data point to the given external point with the edges of every face, successive edges of each face being considered in turn face. We switch an element of **inside** between TRUE and FALSE whenever an intersection occurs, and it is thus rapidly established which data points fall inside which faces when these are all projected onto the viewing plane. The perpendicular distances from data points to the object model faces, and the backface condition , are determined from straightforward parallel calculations, and a standard function collates results within a given row. In this way, the process efficiently determines which data points lie sufficiently near to the face to which they have been assigned, and which if any are not visible from the position of the sensor.

Essentially the same algorithms apply to the edge matching problem, with perpendicular vectors from the origin to observed edge segments used in computing r_0, but it then has to be established that every data point is sufficiently close to the edge segment to which it has been assigned.

TEST RESULTS

The method works well with synthetic data related to simple object models, and a representation of a three-pin electric plug, similar to that used by Murray and Cook[3], has been adopted with a view to further performance tests. The plug is viewed from three different positions, with data points at the centre of each visible face. The first view, looking towards the face of the plug, has 14 visible faces and 91 pairwise comparisons are involved in the generation of feasible interpretations. The second view, looking towards the back, has 12 visible faces requiring 66 comparisons, whereas the third view, looking directly down on the pins, has only 4 faces that are clearly visible involving only 6 comparisons. The three views are illustrated in Figure 3.

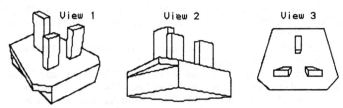

Figure 3 *The Three Views of the Electric Plug*

There are more that 10^{20} possible interpretations to be considered with regard to View 1, and the method of Flynn and Harris would require either a separate processor for each one or totally unacceptable segmentation of the problem. In the meantime, the present method allows the problem to be accomodated easily on a 32x32 DAP, and the process converges quickly to a single interpretation for View 1 and View 2, with run times of 65ms and 37ms, respectively. However, not too surprisingly, the process fails to distinguish between the ends of the two short pins in View 3.

Three additional back face points were included with the View 1 test data for the purpose of testing the validation process, and we note that the point at the centre of the visible side of the plug was in fact obscured by the flange. Four back face data points were included with the data for View 2, two points in faces of the earth pin were obscured by the neutral pin, and the one in the underside of the flange on the far side of the plug was obscured by the rest of the plug. For View 3, three data points were in back faces but none were obscured otherwise.The location and orientation of the plug were determined, in each case, within about 26.5 milliseconds, and the back faces and obscured data points were identified by the validation process in a further 12 milliseconds.

Further tests were then made with simulated errors in the spatial coordinates of the data points, and the surface normal directions. It was found that, whereas coordinate errors of about 0.05 inches might simply result in the rejection of the offending data points, with the electric plug being about 1.5 inches across and viewed from a distance of about 5 inches, errors of the order of 0.25 inches resulted in substantial errors in r_0, leading to the rejection of several valid points. The orientation of the plug, and the run times for validation, were not affected by errors in spatial coordinates. On the other hand, errors ranging from 0.1 to 0.2 in the direction cosines of the surface normals led to errors in both R and r_0, with the subsequent rejection of several valid points. Again, there was no change in run times, because the errors were not sufficient to provoke further iterations of the Newton Raphson process, in computing R.

CONCLUDING REMARKS

Interpretation and validation, with obscurred data points and simulated errors, is achieved in about 90 milliseconds, for the given exemplar. Work is continuing with regard to the interpretation of real, as opposed to synthetic, edge matching data, and the subsequent validation of interpretations, to meet the demands of interfacing with the parallel version of the ISOR system.

REFERENCES

1. Grimson W.E.L. and Lozano-Pèrez T. "Model-Based Recognition and Localisation from Sparse Range or Tactile Data". *International Journal of Robotics Research*, Vol 3 (1984) pp 3 -35.
2. Faugeras O.D., Ayache N. and Faverjon B. "A Geometric Matcher for Recognising and Positioning 3-D Rigid Objects". *Proc.. A.I. Applications Conference*, IEEE (1984) pp 218 - 224.
3. Murray D.W. and Cook D.B. "Using the Orientation of Fragmentary 3D Edge Segments for Polyhedral Object Recognition". *International Journal Computer Vision*, Vol 2 (1988) pp 147 - 163.
4. Flynn A.M. and Harris J.G. "Recognition Algorithms for the Connection Machine". *Proc. Ninth International Joint Conference on A. I.*, Morgan and Kaufman (1985) pp 57 - 60.
5. Holder D.R. and Buxton H. "Polyhedral Object Recognition with Sparse Data in SIMD Processing Mode". *Proc. Fourth Alvey Vision Conference*, (September 1988) pp 103 - 110, also published in *Image and Vision Computing* (February 1989).
6. Holder D.R. and Buxton H. "Polyhedral Object Recognition with Sparse Data - Validation of Interpretations". *Proc. Fifth Alvey Vision Conference* (September 1989).pp 19 - 24
7. Castelow D.A., Murray D.W., Scott G.L. and Buxton B.F. "Matching Canny Edgels to Compute the Principal Components of Optic Flow". *Proc. Third Alvey Vision Conference* (September, 1987) pp 193-200.
8. Murray D.W., Castelow D.A., and Buxton B.F. "From an Image Sequence to a Recognised Polyhedral Model". *Proc. Third Alvey Vision Conference* (September, 1987) pp 201 - 210.
9. Buxton H. and Wysocki J. "The SIMD Parallel Image Sequence Object Recognition (SPISOR) System" (In draft).
10. Buxton B.F. *Personal communication* (February, 1989).

A 3D Interpretation System Based on Consistent Labeling of a Set of Propositions. Application to the Interpretation of Straight Line Correspondences[†]

Robert Laganière   Amar Mitiche

INRS Télécommunications

3, Place du Commerce, Ile des Soeurs, Qc, Canada H3E 1H6

Abstract: We propose a 3D interpretation system where knowledge is represented by a set of propositions, and where interpretation and truth maintenance are based on a consistent labeling of this set of propositions. The basic concepts are illustrated on the problem of 3D interpretation of image straight line correspondences.

1. Introduction

A number of 'truth maintenance' systems have been proposed [1]-[3] that keep track of the dependencies between propositions (statements resulting from observations) by associating to each proposition a *justification* which is the set of propositions that have allowed its derivation, or an *origin* which is the minimal set of assumptions that must hold for the proposition to be valid. Because of this structure, the interpretation of the resulting data base is often quite complex. Special care must also be taken to avoid contradictions. To facilitate the process of inference, some systems use *context*, which is a subset of *beliefs* under which the analysis is currently made. In this case, the problem is to determine how and when a context switching should be made.

KNOBIS is not a justification-based or an assumption-based system. It does not use complex dependency pointers between propositions. Rather, recording of dependencies is directly incorporated in a dedicated database. Futhermore, KNOBIS clearly separates the process of inference from the one of interpretation, and it generalizes the notion of uncertainty by considering both *data uncertainty* and *rule uncertainty*.

2. General structure of KNOBIS

KNOBIS contains three knowledge bases: a *rule base*, a *data base*, and a *constraint base* (Fig. 1). These are supervised by two distinct schemes, one to manage inferencing, the other to interpret the current data base.

[†] This work was supported in part by the Natural Sciences and Engineering Research Council of Canada under grant NSERC-A4234

In the present system, propositions are represented by labeled nodes (Fig. 2). Each label of a node contains conjunctions of primitive propositions. A node is not a representation of individual propositions but, rather, the set of all nodes constitute a non-disjoint segmentation of the data base in which each element contains jointly derived propositions. The assignment of a particular label to a node validates the associated propositions. *Null* labels are allowed to indicate that no valid proposition is associated with a node.

Rules can be *certain* or *uncertain*. Uncertain rules have a consequent which may not be a logical consequence of its antecedent. In addition, rules can have several distinct consequents, allowing various alternate decisions. The parsing of these rules can be done using standard resolution methods that are applied on the data base formed by the union of all propositions present in the different labels. The only restriction is that, in the satisfaction of an antecedent, each node can contribute to only one label. Therefore, the application of one rule results in the creation of a node i having L_i labels $\Psi_1^i,...,\Psi_{L_i}^i$. These propositions come from the different labels of existing nodes in the data base. Consider now the set Γ_i which is the union of all propositions included in the different labels used in the derivation of node i:

$$\Gamma_i = \Psi_{\ell_1}^{i_1} \cup ... \cup \Psi_{\ell_K}^{i_K}$$

which means that propositions included in the label ℓ_k of node i_k has contributed to the derivation of node i. This set, called the *support set*, includes all the propositions that satisfied the condition expressed by the antecedent of the rule and, eventually, other propositions that are also included in the selected labels (recall that a given label may include several propositions). This support set is only used during constraint recording; once done it does not have to be memorized anymore.

Constraints make explicit the possible interrelationship between each proposition (label). Two types of constraints are considered. The *dependency constraint* expresses the fact that the validity of a given proposition depends upon the validity of the propositions used in its derivation. The *compatibility constraint* is used to determine if two labels are compatible or not. Two labels are incompatible if they include contradictory propositions.

An *interpretation* can be obtained at any moment by resolving the corresponding constraint satisfaction problem. This can be realized by assigning a unique label to each node of the data base. The labeling thus found must be consistant according to the constraints recorded in the constraint base. Possible interpretations will therefore be all the consistent assignments thus found. The theory associated with an interpretation is the union of all propositions included in the selected labels.

More details on the structure and semantics of KNOBIS can be found in [4][5].

3. Interpretation of image line correspondences

Various mathematical formulations of the problem of interpreting image straight line correspondences have been proposed [6] that are often unstable and sensitive to image measurement errors [7]. This is in part due to the fact that these formulations are mainly concerned with general cases, treating special cases with marginal interest. However, special cases abound in man-made environments; their occurrence, if ascertained, simplifies drastically the task of 3D interpretation.

We proceed to show how KNOBIS can effectivly interpret image straight line correspondences in such a context [8]. By hypothesizing plausible special configurations in the scene, it will be possible to suggest a number of interpretations. Ideally, we end up with a single interpretation. The role of KNOBIS is to control the process of inference and to propose possible interpretations under the various sources of uncertainty (data, rules). A proposition designates, here, either a relation existing between a number of lines or a particular numerical assignment for a given attribute of a line. We use orientation as the attribute. In this case, the negation of a proposition would be the assignment of a different orientation (within some tolerance) to a given line. Rules that will be used for our particular example are:

Hypothesizing parallel lines rule: if two lines are nearly parallel in at least one image, then these lines are hypothesized to be parallel in the scene.

Hypothesizing orthogonal lines rule: if three lines meet at one point in both images, then these three lines are hypothesized to be orthogonal.

Orthogonal lines rule: if three lines are orthogonal, then their orientation can be computed by the corresponding computational unit [7].

Parallel lines rule: if two lines are parallel, then their orientation can be computed by the corresponding computational unit [7].

Propagation rule: if the orientation of two non-parallel lines is known over two views, then the orientation of all the other lines can be found by propagation [7].

Resolving special configurations such as parallel lines and orthogonal lines, and spreading computation from one configuration to another are simple operations [8].

Figure 3a and 3b show two views of a wedge. With these images as input, the following relations are hypothesized:

```
(parallel  0  5)    (parallel  1  4)       (parallel  3  5)
(parallel  2  4)    (orthogonal  0  1  3)  (orthogonal  1  2  5)
```

To each of these hypotheses corresponds a node; the hypothesis itself is one label of this node and a null label is another. This null label will be assigned if the hypothesis is rejected. For each activated hypothesis a corresponding computational unit can be applied [8]. The creation of a node causes the updating of the constraint base.

Once inferencing is completed, the resulting constraint satisfaction problem is solved. A total of eleven consistent labelings are thus found for our example. In the absence of any other information, each of these interpretations is acceptable. However, if another view is available (Fig. 3c), the application of the same process on the second and third views can disambiguate the problem. The acceptable interpretation becomes the only one that assigns the same attributes (orientations) to lines of the second view, i.e.

```
Interpretation #0

(parallel  1  4)    (parallel  3  5)    (orthogonal  0  1  3)

              image 1                 image 2                 image 3
0:    ( 0.500,  0.000, -0.866)  ( 0.117, -0.321, -0.940)  ( 0.754, -0.133, -0.643)
1:    (-0.150,  0.985, -0.087)  ( 0.019,  0.947, -0.321)  ( 0.004,  0.980, -0.198)
2:    (-0.215, -0.743,  0.634)  (-0.091, -0.504,  0.859)  (-0.498, -0.652,  0.571)
3:    ( 0.853,  0.174,  0.492)  ( 0.993,  0.019,  0.117)  ( 0.656,  0.147,  0.740)
4:    (-0.150,  0.985, -0.087)  ( 0.019,  0.947, -0.321)  ( 0.004,  0.980, -0.198)
5:    (-0.853, -0.174, -0.492)  (-0.993, -0.019, -0.117)  (-0.656, -0.147, -0.740)
```

These are, indeed, the actual orientations.

Summary: We have presented KNOBIS, an 'intelligent system' that has the capability of reasoning under uncertainty. In this system, propositions are represented by labeled nodes. The label of a node is the set of the propositions that can be derived from the satisfaction of the antecedent of a given rule. Rules, which can be certain or uncertain, have a special format which allows the use of several consequents associated with a given antecedent. Each step of inference creates a new node in the data base and each time a node is created the constraint base is updated. Constraints are used to record compatibilities and dependencies between these nodes. The problem of finding an interpretation is then reduced to the one of finding a consistent labeling of the resulting network. KNOBIS has been applied to the line interpretation problem.

References

[1]   J. Doyle, "A Truth Maintenance System", *Artificial Intelligence*, vol. 12, pp. 231-272, 1979.

[2]   J. de Kleer, "An Assumption-based TMS", *Artificial Intelligence*, vol. 28, pp. 127-162, 1986.

[3]   J.P. Martins, S.C. Shapiro, "A Model for Belief Revision", *Artificial Intelligence*, vol. 35, pp. 25-79, 1988.

[4] R. Laganière, A. Mitiche, "A Knowledge-based Intelligent System for Real World Interpretation", *Proc. of Intelligent Autonomous Systems Conference*, Amsterdam, The Netherlands, pp. 772-782, december 1989.

[5] R. Laganière, "Système intelligent pour l'interprétation de séquences d'images", *Master thesis*, INRS-Télécommunications, 1989.

[6] A. Mitiche, O. Faugeras, J. Aggarwal, 'Counting Straight Lines,' *Computer Vision Graphics and Image Processing*, no 47, pp. 353-360, 1989.

[7] A. Mitiche, G. Habelrih, 'Interpretation of Straight Line Correspondences Using Angular Relations,' *Pattern Recognition*, 22, No.3, pp. 299-308, 1989.

[8] A. Mitiche, R. Laganière, "Interpreting 3-D Lines", *Traditional and Non-traditional Robotic Sensors*, Springer Verlag 1990, (proc. of the *NATO Advanced Research Workshop on Traditional and Non-traditional Robotic Sensors*, Maratea, Italy, august 1989).

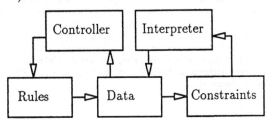

Fig. 1 Structure of KNOBIS.

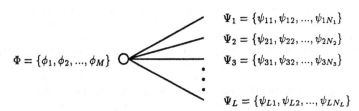

$$\Psi_1 = \{\psi_{11}, \psi_{12}, ..., \psi_{1N_1}\}$$
$$\Psi_2 = \{\psi_{21}, \psi_{22}, ..., \psi_{2N_2}\}$$
$$\Psi_3 = \{\psi_{31}, \psi_{32}, ..., \psi_{3N_3}\}$$
$$\Psi_L = \{\psi_{L1}, \psi_{L2}, ..., \psi_{LN_L}\}$$

$$\Phi = \{\phi_1, \phi_2, ..., \phi_M\}$$

Fig. 2 A node and its associated propositions.

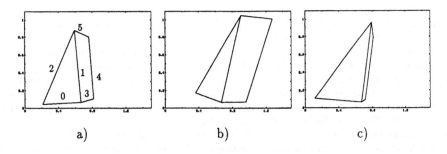

a)   b)   c)

Fig. 3 Images of a wedge.

Model-Based Object Recognition by Geometric Hashing

Haim J. Wolfson

Computer Science Dept., School of Math., Tel Aviv University, wolfson@math.tau.ac.il
and
Robotics Lab, Courant Inst. of Math., New York University, wolfson@acf8.nyu.edu

Abstract

The *Geometric Hashing* paradigm for model-based recognition of objects in cluttered scenes is discussed. This paradigm enables a unified approach to rigid object recognition under different viewing transformation assumptions both for 2-D and 3-D objects obtained by different sensors, e.g. vision, range, tactile. It is based on an intensive off-line model preprocessing (learning) stage, where model information is indexed into a hash-table using minimal, transformation invariant features. This enables the on-line recognition algorithm to be particularly efficient. The algorithm is straightforwardly parallelizable. Initial experimentation of the technique has led to successful recognition of both 2-D and 3-D objects in cluttered scenes from an arbitrary viewpoint. We, also, compare the *Geometric Hashing* with the *Hough Transform* and the *alignment* techniques. Extensions of the basic paradigm which reduce its worst case recognition complexity are discussed.

1 Introduction

Object recognition is a major task in computer vision. The ability to recognize objects in cluttered scenes is essential to the functionality of a flexible robotic system. In most practical industrial applications we may assume that the objects to be recognized are known in advance. This model-based approach proved to be promising, and the best currently known object recognition systems are model-based (for comprehensive surveys see [4, 6]). Ideally, such systems should be able to deal with substantial occlusion, and recognize rigid, articulated and even deformable objects under different viewing transformations. The recognition algorithms should be efficient enough to enable on-line recognition of objects belonging to large libraries.

In order to discuss object recognition algorithms one has to give some mathematical description of an object. Let us assume that some *interest features* can be extracted from the images, so that both the model-objects and the observed (cluttered) scene can be represented by sets of these *interest features*. In the most general (and least informative) case the *interest features* will be just points or lines. In such a situation, the recognition of a partially occluded object in a scene amounts to the discovery of a match between a subset of the scene *interest features* and a subset of the *interest features* of some model object. The discovered match should be consistent with some (viewing) transformation. Given m *interest features* on a model and n *interest features* in the scene, there are $O(n^m)$ ways to match the model features to the scene features. Since such an exponential complexity is unacceptable for object recognition algorithms, various approaches were suggested to prune the space of possible matches. Some of them employ efficient tree search techniques, where the pruning is based on geometric constraints ([8]). However, for recognition of partially occluded objects, the search still remains exponential in the number of scene features (see [9]).

To overcome this exponential complexity, one may observe, that a transformation of a **rigid** object is usually defined by the transformation of a small number of the object's features (points). This geometric observation is at the core of the, so called, Hough Transform or *pose clustering* ([2, 27, 22]), the *alignment* ([13, 14]), and the *Geometric Hashing* ([19, 18, 20]) techniques.

In this paper we address the recognition of objects in cluttered scenes using the *Geometric Hashing* technique. In its present form, the *Geometric Hashing* paradigm enables a unified approach to rigid object recognition under different viewing transformation assumptions both for 2-D and 3-D objects, obtained by different sensors, e.g. vision, range, tactile ([20]). This technique is especially suitable for dealing with

partially occluded objects in cluttered scenes due to its representation scheme which encodes geometric constraints between local features. The *Geometric Hashing* paradigm is based on an intensive off-line model preprocessing (learning) stage, where model information is indexed into a hash-table using minimal, transformation invariant features. This enables the on-line recognition algorithm to be particularly efficient. The recognition time depends directly on the complexity of the scene to be recognized, and increases only sub-linearly with the number of model-objects in the data base. It is also straightforwardly parallelizable, and can be quite easily implemented on a fast special purpose hardware. Initial experimentation of the technique has led to successful recognition of both 2-D and 3-D objects in cluttered scenes from an arbitrary viewpoint ([19, 18, 20]).

In this paper we first survey the basic ideas of the *Geometric Hashing* technique. Then, we compare it with two other efficient model based object recognition techniques, the, Hough Transform paradigm ([2, 27, 22]), and the *alignment* technique ([13, 14]). We discuss the advantages and deficiences of these different techniques compared to each other. Finally, we present some extensions of the basic technique, which can further reduce its time complexity.

2 The *Geometric Hashing* Paradigm

Recently, the, so called, *Geometric Hashing* technique for model based object recognition was introduced by Lamdan, Schwartz and Wolfson. Efficient algorithms were developed for recognition of flat rigid objects assuming the affine approximation of the perspective transformation ([19, 18]), and the technique was also extended to the recognition of arbitrary rigid 3-D objects from single 2-D images ([20]).

In a model based object recognition system one has to address two major interrelated problems, namely, *object representation* and *matching*. The *representation* should be rich enough to allow reliable distinction between the different objects in the data-base, yet terse to enable efficient *matching*. A major factor in a reliable representation scheme is its ability to deal with partial occlusion.

The *Geometric Hashing* paradigm presents a unified approach to the *representation* and *matching* problems, which applies to object recognition under various geometric transformations both in 2-D and 3-D. The objects are represented as sets of geometric features, such as points or lines, and their geometric relations are encoded using minimal sets of such features under the allowed object transformations. This is achieved by standard methods of *Analytic Geometry* invoking *coordinate frames* based on a minimal number of features, and representing other features by their coordinates in the appropriate frame.

In the sequel we present the *Geometric Hashing* method mainly for point matching. This is done for two reasons. First, the 'point matching' case allows a succinct mathematical representation of the basic method, without getting overburdened with extraneous details. Second, since a single point is a 'least informative feature', we present the method in the most difficult case. Clearly, better results can be achieved by using more stable and more informative features such as lines and various groupings (see Section 4.1).

2.1 Recognition under the Affine Transformation

To illustrate the method we first discuss flat object recognition under the affine transformation (rotation, translation, scale, and shear). It is well known that the perspective projection is well approximated by a parallel projection with a scale factor (see [16, 12, 23, 28]). This approximation is especially suitable for flat bodies, which are relatively far from the camera. Hence, we may assume, that two different images of the same flat object are in an affine 2-D correspondence, namely, there is a **non singular** 2×2 matrix **A** and a 2-D (translation) vector **b**, such that each point **x** in the first image is translated to the corresponding point $\mathbf{Ax} + \mathbf{b}$ in the second image.

Let us assume that the model objects and the scene are described by sets of *interest points*, which are invariant under the affine transformation. The choice of the interest operator is not essential to our method. Corners, points of sharp concavities and convexities, or points of inflection are suitable candidates (see [19]). From now on we rephrase the model-based recognition problem to the point-set matching task, where one is given a set of known (model) point-sets and an observed (scene) point-set. The recognition task can be stated as the following subset isometry problem :

Is there a transformed (rotated, translated, scaled, and sheared) subset of some model point-set which matches a subset of the scene point-set ?

As was mentioned before we have to address two major interrelated problems: *representation* and efficient *matching*.

2.1.1 Representation of geometric constraints

Our goal is to represent a set of planar points belonging to a rigid body by few intrinsic parameters. This representation should efficiently encode the geometric constraints of a rigid body, be affine invariant, and enable handling of occlusion.

Assume that we have an arbitrary set of m points belonging to a rigid body. An affine transformation of a planar rigid body is uniquely defined by the transformation of three ordered non-collinear points ([16, 29]). Moreover, one can pick any triplet of non-collinear points in the set and represent all the other points using this triplet. Specifically, let e_{00}, e_{10}, e_{01} be an ordered triplet of such points. Then, any point v in the plane can be represented in this **affine basis**, namely, there is a pair of scalars (α, β), such that $v = \alpha(e_{10} - e_{00}) + \beta(e_{01} - e_{00}) + e_{00}$.

The above coordinates are invariant under an affine transformation. Accordingly, we will represent the m points of our set by their coordinates in the affine basis triplet (e_{00}, e_{10}, e_{01}). Naturally, the coordinates of the basis points are $(0,0)$, $(1,0)$ and $(0,1)$, respectively.

This representation allows comparison of occluded objects, since the point coordinates of an occluded object in the scene will have a partial overlap with the coordinates of the stored model , if both are represented in a coordinate frame which is based on the same triplet of points. This dependence of the representation on a specific basis triplet may, however, preclude recognition when at least one of the basis points is occluded. Hence we represent the object points by their coordinates in **all** possible affine basis triplets. More specifically, given a model object, the following preprocessing is applied to each model object:

a) Extract the objects *interest points*. (Assume that m *interest points* were found.)

b) For each ordered non-collinear triplet of model points (affine basis) compute the coordinates of all other $m - 3$ model points in the affine coordinate frame defined by the basis triplet. Use each such coordinate (after a proper quantization) as an address (index) to a hash-table, and record in the table the pair *(model, basis)*, namely, the model and the affine basis at which the coordinate was obtained.

The complexity of this preprocessing step is of order m^4 per model. Each of the m points is represented in m^3 different bases. The major advantage of this somewhat redundant representation is its ability to allow efficient recognition of objects in an occluded scene.

Note, that the preprocessing step is done without any knowledge of the scene to be recognized. Hence, it can be executed off-line, so that its execution time does not add to the actual recognition time. New models added to the data-base can be processed independently without recomputing the hash-table.

The hash table preparation stage may be viewed as a *learning* stage of the algorithm. In this stage relevant information of the models is memorized in its various representations. The triplet of points, which serves as an (affine) basis, for a given representation may be viewed as a (geometric) *focus of attention*. The hash table serves as a memory of the model-objects under different foci of attention.

2.1.2 Matching

The matching stage of the algorithm uses the hash table, prepared in the representation (learning) stage. Given a scene of *interest points* one chooses an affine basis (focus of attention) in the scene, and tries to match the coordinates of the scene points to those memorized in the hash table. Specifically, given an image of a scene with partially occluded objects one does the following :

a) Extract its *interest points*. (Assume we have n such points.)

b) Choose an arbitrary ordered triplet of non-collinear points in the scene and compute the coordinates of the scene points referring to this triplet as an affine basis. (If appropriate, one might, of course try some 'intelligent' choice of the affine basis, rather then chosing it at random.)

c) For each such coordinate check the appropriate entry in the hash-table, and for every pair *(model, affine basis)*, which appears there, tally a vote for the model and the affine basis as corresponding to the pair which was chosen in the scene. (The accumulator of the votes will have $\sum_{i=1}^{N} m_i^3$ entries, where N is the number of models and m_i is the number of *interest points* on the i'th model.)

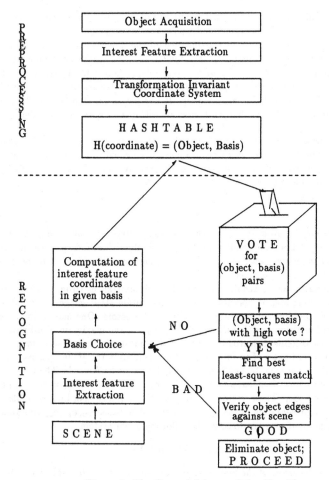

Figure 1: The General Scheme of the Algorithm

d) If a certain pair *(model, affine basis)* scores a large number of votes, conjecture that this pair corresponds to the one chosen in the scene. The affine transformation between the coordinate frames based on these triplets is assumed to be the transformation between the appropriate model and the scene.

e) Consider all the model-image point pairs which voted for the affine transformation of Step (d), and find the affine transformation giving the best least-squares match between all these corresponding point pairs (see [19]). Since the computation of this transformation is based on more than 3 point pairs, it will, hopefully, be more accurate.

f) Transform all the edges of the model according to the affine transformation of Step (e) and **verify** them versus the scene edges. If the verification fails, go back to Step (b) and begin the procedure for a different image affine basis. This final verification is done not only for the *interest points* but for all the relevant model edges. (See Fig. 1 for a general scheme of our procedure.)

It is important to mention that in general we do not expect the voting scheme to give only one candidate solution (see [21]). Its goal is to reduce significantly the number of candidates for the verification step (f), which might be quite tedious and time consuming (see [10]).

Since the voting is done simultaneously for all models and all possible bases on a model, **for the algorithm to be successful it is enough to pick three points in the scene, belonging to some model.** In such a case the model with the appropriate affine basis gets a high score in the voting procedure.

The voting process, per basis-triplet, is linear in the number of points in the scene. Hence, the overall recognition time is dependent on the 'density' of model points in the scene.

The presented method assumes no *a-priori* classification of the model and scene points to reduce the number of candidates for matching basis-pairs. In this difficult case its worst case complexity is $O(n^4)$. If some classification or perceptual grouping of features is available, it can be incorporated into this method by concentrating only on some special basis-triplets ([18]). We further discuss this issue in Section 4.

The method is parallel in a straightforward manner. It has few serial steps as can be seen from the diagram of Fig. 1. A prototype of the affine matching algorithm has been implemented on the highly parallel Connection Machine by Bourdon and Medioni ([5]).

Because of the duality between lines and points in the affine plane, all the algorithms, that were developed for point matching, apply directly to line matching. Namely, given a line $ax + by = 1$, one may view it as a point (a, b) in the dual space. Since line extraction is, usually, more reliable than point extraction, one may benefit from this dual approach.

2.2 Error Analysis

So far, we have discussed our algorithm in its 'noiseless' version. However, since real scene measurements are noisy, this noise will affect both the implementation of the algorithm and its performance. We give here a short description of the issues raised by noise analysis, and their handling. A more extensive noise analysis, including simulation results, is given in ([21]).

We assume that the models can be acquired under 'ideal' circumstances (from a CAD model, for example), hence the preprocessing step is noiseless. In the recognition step, image coordinates of *interest points* are measured. These coordinates are represented by $2 - D$ vectors. One may define a norm on this $2 - D$ vector space. We will usually use either the Euclidean (L_2), or the maximum coordinate (L_∞) norm. Assume that image point measurements introduce an error of at most ϵ in the given norm.

The computation of the coordinates of an *interest point* $\mathbf{v} = (v_1, v_2)$ in the affine basis $\mathbf{e}_{00}, \mathbf{e}_{10}, \mathbf{e}_{01}$ can be formulated as a solution of the linear system of 2 equations in 2 unknowns $\mathbf{Ax} = \mathbf{b}$. The two columns of the matrix $\mathbf{A}$ are the difference vectors of the basis *interest points* $\mathbf{e}_x = \mathbf{e}_{10} - \mathbf{e}_{00}$ and $\mathbf{e}_y = \mathbf{e}_{01} - \mathbf{e}_{00}$ respectively, and the free vector is $\mathbf{b} = \mathbf{v} - \mathbf{e}_{00}$. These vectors are represented in image coordinates, while the solution vector $\mathbf{x} = (\alpha, \beta)$ gives the representation of the point $\mathbf{v}$ in the affine basis $\mathbf{e}_{00}, \mathbf{e}_{10}, \mathbf{e}_{01}$ coordinates .

Taking the errors into account, one has to solve the following linear system :

$$(\mathbf{A} + \delta\mathbf{A})(\mathbf{x} + \delta\mathbf{x}) = \mathbf{b} + \delta\mathbf{b}$$

where $\delta\mathbf{A}$, $\delta\mathbf{x}$, and $\delta\mathbf{b}$ are the errors of the matrix $\mathbf{A}$ and the vectors $\mathbf{x}$ and $\mathbf{b}$ respectively. By the nature of our point measurements, we may assume that the absolute values of entries of the matrix $\delta\mathbf{A}$ and the vector $\delta\mathbf{b}$ are less than some given measurement error ϵ. The stability and accuracy of solutions to systems of linear equations is a well investigated problem of Numerical Analysis. An extensive treatment of such 'approximate' linear systems is given, for example, in [17]. A good estimate of the maximal error in the k'th coordinate of $\mathbf{x}$ ([17], p. 45) is $Delta x_k \leq (1 + x)c_k\epsilon$, $(k = 1, 2)$, where $x = |\mathbf{x}_1| + |\mathbf{x}_2|$, is the sum of the absolute values of the unknowns, and $c_k = |\mathbf{c}_{k1}| + |\mathbf{c}_{k2}|$, is the sum of the absolute values of the k'th row in the inverse matrix $\mathbf{C} = \mathbf{A}^{-1}$. This inequality gives an estimate of the maximal error which can be introduced by the image measurement noise into the coordinates of the hash-table address $\mathbf{x}$. Hence, the voting procedure reflects this noise. Namely, for an address $\mathbf{x} = (\mathbf{x}_1, \mathbf{x}_2)$, all the bins with addresses $(\mathbf{x}_1 \pm \Delta\mathbf{x}_1, \mathbf{x}_2 \pm \Delta\mathbf{x}_2)$ participate in the voting. This ensures that votes for a correct model basis are not missed due to noise. In practice, tighter bounds, usually, apply ([17]). Note, that the amount of error ϵ is defined by the known imaging process, hence the worst case $\delta\mathbf{A}$ and $\delta\mathbf{b}$ can be computed in advance. Since for a given basis triplet and an additional image point both $\mathbf{C}$ and x can be computed, we can evaluate the relative merit of voting for a given coordinate, and eliminate those (unstable) coordinates which are going to introduce excessive noise. It should be noted that if a certain basis-triplet belonging to some model did not get enough votes, we still have a chance to recover this model from another basis-triplet.

Since, the appropriate voting bins for each address can be evaluated in advance, we do not expect a correct basis triplet to achieve less votes than the corresponding number of unoccluded model points (except the eliminated 'unstable' votes). There still remains the possibility of a 'random' basis-triplet achieving a large number of votes. Such a 'wrong' candidate will be discovered by two verification procedures that are incorporated in the algorithm (see (e) and (f) of Section 2.1.2). Although 'wrong' candidates will be discovered in the verification steps and discarded, we would still like to show that the probability of a 'random' configuration to get a high vote is small. Such a discussion and simulation results are presented in [21].

2.3 General Framework

The algorithm that was illustrated in the previous section is suitable not only for the affine 2-D case. It represents a unified approach which applies also to many other useful transformations encountered in object recognition problems. The only difference from one application to another is the number of features (points) that have to be taken as a basis for the coordinate frame. This, of course, affects the complexity of the algorithm in the different cases.

The following list gives a number of examples in which this general paradigm applies. Almost in all the cases we will discuss point matching. Use of other features, such as lines, can be understood by analogy. We discuss first recognition of 2-D objects from 2-D data.

1) *Translation in 2-D* - the technique applies with a *one point* basis. This point may be viewed as the origin of the coordinate frame. Hence, the worst case complexity of the recognition step is $O(n^2)$.

2) *Translation and rotation in 2-D* - a two point basis can be used, however, one point with a direction (obtained, say, from an edge segment) has enough information for a unique definition of a basis.

3) *Translation, rotation and scale in 2-D* - the key observation here is that since the similarity transformation is orthogonal, two points (or, a line segment) are enough to form a basis which spans the 2-D plane. (The first point is assigned coordinates $(0,0)$ and the second $(1,0)$. The third basis point $(0,1)$ is uniquely defined by these two points.) Hence, the *Geometric Hashing* technique can be applied with a two point basis, resulting in a recognition stage of complexity $O(n^3)$.

4) *Affine transformation in 2-D* - was discussed in Section 2.1. See [19, 18] for some recognition results.

5) *Projective transformation in 2-D* - a 4 point basis is needed to recover a projective transformation between two planes (see, for example, [7]).

In case 3-D data (such as range or stereo) of the objects is available the recognition of 3-D objects from 3-D images has to be considered. Development of techniques for this case is especially important in an industrial environment, where 3-D data can be readily obtained and used.

6) *Translation in 3-D* - exactly as the 2-D case. One point basis will suffice.

7) *Translation and rotation in 3-D* - this is the *interesting* case representing a rigid motion. A two non-collinear line basis is enough.

8) *Translation, rotation and scale in 3-D* - a two non-collinear and non-planar line basis (or a point and a line) gives a minimal set.

In the previous discussion we considered cases where both the data of the object and the data of the scene have been given in the same dimension, either 2-D or 3-D. However, in the *recognition of 3-D objects from single 2-D images* we have the additional problem of the reduced dimension in the image space compared with the model space. A number of methods have been developed to tackle this problem by the *Geometric Hashing* technique (see [20]). One of this methods, based on the recovery of the correct viewing angle and the appropriate similarity transformation, has been implemented and successfully tested (see [20]).

Specifically, in the model learning stage the viewing sphere is tesselated into (few hundred) discrete viewing angles. For a fixed angle, the viewing transformation is well approximated by the orthographic projection. Hence, two different images of the same object taken from the same viewing angle are in a 2-D similarity correspondence (case (3)). The usual model representation scheme is applied here, except for the additional information of the viewing angle, which is also memorized in the hash-table. Namely, given a two point basis, the coordinates of all other *interest points* are computed and the information *(model, basis, viewing angle)* is stored in the hash-table at the addresses defined by the appropriate coordinates. In the recognition stage, the usual procedure for the 2-D similarity case is applied, although the votes this time are accumulated for the triplets *(model, basis, viewing angle)* (see [20] for details). Since the number of possible viewing angles is fixed, the method retains the recognition complexity of $O(n^3)$ (2-D similarity).

3 Comparison with Other Methods

In this section we compare the *geometric hashing* with the *alignment* technique and with the *generalized Hough Transform*. Since all these techniques share some common ingredients it is important to understand their differences and relative advantages.

3.1 Comparison with the Hough Transform

Since the *Geometric Hashing* method involves a voting procedure it immediately invokes association with the *generalized Hough Transform* (see [2] or [3] pp. 128-131) or, so called, *pose clustering* techniques ([27, 22]). We consider in detail the differences between the *Geometric Hashing* and *pose clustering* techniques.

3.1.1 The Pose Clustering (Hough) Method

In the *pose clustering (Hough Transform)* approach, recognition of an object in a scene is achieved by finding a transformation between a model-object and the scene, which maps a large enough number of the model *interest features* into scene *interest features*. The transformation is discovered by voting for its parameters, which are consistent with hypothetical pairings of subsets of object and image *interest features*.

Let us consider the example of object recognition under the affine (rotation, translation, scaling, and shearing) transformation using the Hough paradigm. Assume as in Section 2.1 that both the models and the scene are represented by *interest points*. An affine transformation can be represented by six independent parameters. Hence, the *Hough* technique regards an affine transformation as a point in the 6-D space, each parameter describing a coordinate of this 6-D point. Although the parameters may achieve a continuum of values, each coordinate (parameter) axis is quantized into a finite number of values, thus creating 6-D bins (volumes) of admissible transformations.

Since an affine transformation is uniquely defined by the correspondence of triplets of points, each correspondence of a triplet of model points with a triplet of scene points provides a *candidate* affine transformation. Hence, in the *Hough* technique each triplet of model points is compared with each triplet of scene points, an affine transformation is computed for each such comparison, and votes are cast for the appropriate bins (according to some noise model) in the 6-D transformation table. For a model of m points and a scene of n points, $m^3 n^3$ transformations are computed. If both m and n are of order n, the voting complexity is $O(n^6)$ (assuming constant time for each table access).

A well known problem of the *Hough* technique is the large size of the transformation table (6-D in the affine case). Hence, the usual practice (see, for example, [22]) is to recover the transformation parameters sequentially by first projecting the votes of the transformation table onto a small number (usually, one) of parameters, recovering this parameter, and proceeding in a similar manner for the remaining parameters. This, in turn, introduces the problem of false peaks in the collapsed parameter space.

3.1.2 Differences between the Hough technique and Geometric Hashing

At first glance there are noticeable similarities between the *Geometric Hashing* technique as described in Section 2.1 and the *pose clustering (Hough)* technique as described in the previous subsection. Both use large tables, and both use voting procedures to recover the correct transformation. In this subsection we will state the major differences between both techniques, resulting in a different complexity, and different approaches in their statistical evaluation.

One of the main differences is the independent processing of the model and scene information by *Geometric Hashing*. The model hash-table is prepared without any knowledge of the scene, due to the transformation invariant shape representation (see subsection 2.1.1). This allows the recognition to be performed on all the models and all the different encodings of models (by different bases) simultaneously, while in the *pose clustering* technique one has to check sequentially each model against the scene, and each model basis against the considered image basis, making the recognition time linearly dependent on the number of models in the data base, and cubically dependent on the number of model points. Given N models, order of m points on a model, and order of n points in the scene, the recognition time of the *Hough* technique is $O(N \times m^3 \times n^3)$, while the worst case time complexity of *Geometric Hashing* is $O(n^4)$[1].

---

[1] Assuming constant time for each hash-table bin processing. This is a reasonable assumption, as can be seen from the remark on *weighted* voting at the end of this section.

Another major difference is in the voting procedure. In the *pose clustering* one is voting for certain *a-priori* unknown transformation parameters, hence a large table (6-D for affine transformation) is kept. In *Geometric Hashing* one is voting only for *a-priori* known model object representations (by bases). Thus, for example, an m-point model has only m^3 admissible representations (and thus possible transformations) under the affine transformation. It should be noted that in *Geometric Hashing* the equivalent of the transformation parameter table of the Hough technique is not the hash-table, but the discrete accumulator of the votes for different bases. The size of this accumulator is relatively small and well defined in advance (see (c) of section 2.1.2). Also, the voting in this accumulator is done **only for one image basis** at a time. This voting procedure compares one given image basis versus all possible model bases, looking for peaks in the compatibility score. On the other hand, in the Hough procedure one compares **all** image bases against **all** model bases, each of such pairings voting for one possible transformation. This introduces more background noise and makes it more difficult to discover the peaks.

There is no equivalent for the hash table of the *Geometric Hashing* in the *pose clustering* technique. However, one should note, a number of properties. First, the dimensionality of this table is a function of the dimensionality of the image. It is, usually, 2-dimensional for intensity images, and 3-dimensional for range data. Second, this table is prepared in advance for a given database. Hence, its bin size can be controlled as a function of the specific database. In particular, one may introduce a *weighted* voting scheme (see [15, 11]). In such a scheme bins which are not informative, namely, bins, representing coordinates with a big number of *(model, basis)* candidates, will achieve a small weight, or, even, zero weight to preclude their time consuming processing, while votes of 'small' bins will have bigger weights.

3.2 Comparison with Alignment

Recently considerable work was done using the, so called, *alignment* method ([13, 14]). As the two other methods mentioned, this method utilizes minimal sets of features which suffice to establish a unique transformation between a model and its alleged instance in a scene.

3.2.1 The Alignment

In the *alignment* technique a transformation is hypothesized based on the correspondence of minimal sets of features, and then the *candidate* correspondence is verified in regard to the other features.

Let us, again, consider the case of affine 2-D matching ([13]), and assume that both the model and the scene are represented by *interest points*. As in the *pose clustering* technique one picks a triplet of points is the scene and a triplet on the model, and computes the affine transformation between these two triplets. Given the *candidate* affine transformation, all the other model points are transformed and matched with the image points. This match should be done according to an error model analogous to the one given in Section 2.2 (see [25]). If the number of matching *interest points* is above a certain threshold, additional edge verification is invoked.

The time complexity of alignment for the affine case is $O(N \times n^3 \times m^4)$, where N, m, n are as in the previous subsection.

3.2.2 Differences between the Alignment and the Geometric Hashing

The *alignment* technique and the *geometric hashing* are based on the same geometric principles. If they employ the same error model and *candidate* transformation acceptance thresholds, both techniques should eventually accept (or reject) the same *candidate* transformations. The basic difference between the methods is in their algorithmic approach. While in the *alignment* method an exhaustive enumeration is applied over all the possible pairings of minimal sets of model and image features, in the *geometric hashing* the recognition stage is significantly sped up by using the previously prepared hash-table which encodes the relevant information about the model objects. Another major advantage of the *geometric hashing* algorithm is its ability to process all the model objects simultaneously. By picking a 'correct' basis in the scene the voting procedure of *geometric hashing* discovers both the model it belongs to, and the appropriate transformation between this model and the scene, while in the *alignment* method one has to process the 'candidate' models and triplet bases sequentially. For example, in the affine matching case, which was mentioned before, the worst case complexity of recognition is of order $N \times n^7$ for the *alignment* method, and $O(n^4)$ for *geometric hashing*. Here we assume that the number of model points, and scene points are of order n and the number of models in the data-base is N.

A frequently expressed concern about hash-table voting is the possibility of having a large number of *(model, basis)* candidates for a certain index (coordinate). Note, that if the same error model is used, the *alignment* procedure will have to deal with exactly the same number of candidate transformations.

On the other hand, when memory is a concern, it is better to use the *alignment* procedure which has almost no storage requirements, except the image and model edge data. The efficiency of the *geometric hashing* is achieved by memorizing the model information into a hash-table using appropriate representations. It gives this method the ability to determine for a given scene minimal feature set (basis) a corresponding feature set on one of the models, by considering only the other scene features which 'vote' for the correct interpretation. This 'voting' procedure requires existence of few additional model features in the scene image except the basis. In the extreme case, when such additional features do not exist, the algorithm will first try interpretations which scored high but are uncorrect. Since these interpretations will be rejected by the verification step, in its fast version the algorithm will fail to recognize the model object. In such a case one may decide to backtrack and check candidate solutions with less votes. Eventually, the correct solution will be found after an exhaustive search resulting in the same worst case complexity as the *alignment* method.

To conclude, the *geometric hashing* is considerably more efficient than the *alignment*, when the scene contains 'enough' model features for efficient recognition by voting ('enough' usually means about 6-10). It is also efficient for multiple model processing. In case the number of model features is exceptionally small (for example, only one basis appears in the scene), both methods will have the same worst case complexity.

The above analysis holds for the techniques in their 'pure' form. Obviously, various 'intelligent' basis choice methods, can speed up both the *alignment* and the *geometric hashing*. For example, in the HYPER alignment method ([1]) the bases for the similarity transformation are defined by 'privileged' physical segments in the image. A version of *Geometric Hashing* (see [26]) exploits, so called, super-segments. The basic observation is that all these improvements can be equally well incorporated in both techniques.

4 Reduction of Complexity

In the previous sections we have represented the *Geometric Hashing* technique in the most general case. We have used the technique for 'minimal information' features, such as points or lines. Obviously, if more informative features are available, the complexity of the algorithms presented can be further improved. In this section we examine some of the complexity reductions for the 2-D affine matching case. Obvious analogues exist for other cases as well.

4.1 Informative Features

The discussion of the previous sections assumed existence of *interest points* in the image. One needs three such features to define uniquely an affine 2-D transformation. However, more informative features may be available. For example, if one can reliably extract a segment, its endpoints supply two basis points. If one can extract a triangle, the basis is determined completely. One possibility of reliably extracting triangles, based on concavities is discussed in the next subsection.

4.1.1 Concavities

In the case of an affine transformation each concavity supplies us with a stable feature from which an affine invariant basis can be recovered. A concavity is, usually, bounded by a single segment of the convex hull which we call (following [24]) the *concavity entrance*. It is a simple geometric observation that the concavity entrance segment is **invariant** under affine transformations (see [18]). Moreover, if the concavity is unoccluded, it uniquely defines an affine basis triplet, since the point of the concavity, which is on the most distant tangent line, parallel to the concavity entrance, is also affinely invariant (if this point is not unique one may choose the leftmost such point). The computation of a concavity based affine triplet is stable and computationally efficient, since it is based on convex hull extraction.

Now, if one chooses affine basis triplets defined by concavities, the worst case computational burden of recognition becomes linearly dependent on the number of concavities in the scene, instead of having a cubic dependence on the number of *interest points*. This computational burden can be even further reduced by introducing affine invariant *shape signatures* of concavities, and comparing only those concavities having similar signatures (see [18]).

Even if a concavity is partially occluded, we can still extract relevant information from it. For example, if only the *entrance* segment is visible, its endpoints serve as two points of the affine basis (they suffice, for example, for a similarity invariant basis). Hence, one may choose affine triplets based on a *concavity entrance* segment plus one additional *interest point*, thus still significantly reducing the worst case recognition time complexity, compared to the general case of three *interest point* affine bases.

4.2 Shape Signatures

In the previous section we mentioned that one may distinguish different concavities by using affine invariant *shape signatures* so that only affine bases derived from similar concavities have to be matched. This *shape signature* technique applies also to other cases, where the examined feature has more than minimal information. To illustrate this point, let us consider the *geometric hashing* procedure for the case of planar rigid motion (rotation and translation) using two point bases.

In many industrial applications one is faced with the restricted problem of object recognition under translation and rotation only (case (2) of section 2.3). This might be the case in recognition of flat objects moving on a conveyor belt under a stationary camera. One may apply the *geometric hashing* technique for this case, using two point bases. However, in this case two points have additional information which remains invariant under rigid motion. This information is the distance between the points. Thus, different 2-point bases can be initially distinguished by the distance between the basis points. This is the *shape signature* of such a basis. Hence, one may introduce an additional coordinate into the hash-table address (index), which specifies the *shape signature* (or, one may call it, geometric color) of the basis for which the coordinate was computed. This allows to reduce the voting procedure to relevant bases only.

Analogous *shape signature* coordinates can be introduced also in other cases when the basis supplies more than the minimal information needed. Of special interest is the case of 3-D object recognition from range data. There, a 3 point basis can be used. However, since the possible transformation is rotation and translation only, each 3 point basis has a unique signature, which is defined by the measures of the basis triangle. Only identical triangles have to be matched. This case has an exciting application in *Molecular Biology* for the problem of structural comparison between protein molecules ([?]).

4.3 Projection to a Subspace

When the number of *interest points* on the models is large, one may exploit invariant properties of subspaces, assuming that enough *interest points* are located on such subspaces. For example, in the 2-D affine case, one may use affine invariant properties of lines such as the one given in the following

Lemma (see p.73 in [16]): Two straight lines which correspond in an affine transformation are 'similar', i.e. corresponding segments on the two lines have the same length ratio.

Moreover, the same statement holds for sets of parallel lines. Hence, if we have a set of points, which are located on parallel lines in a model, and another set of points on parallel lines in the scene, we can efficiently check the conjecture, that some of these points correspond.

In this case the information stored in the hash-table for each coordinate will be *(model, line, basis-pair)*, and the voting will be done accordingly. Note that a pair of points define an affine invariant line basis, hence the complexity of recognition is reduced by a factor of n. The price for this reduction, is the restriction to points on parallel lines, which may be practical, when a large number of *interest points* is involved. This is exactly the case, when complexity reduction is most desirable.

5 Conclusions and Future Research

We have surveyed the *Geometric Hashing* paradigm, and compared it with the *Hough Transform* and *Alignment* techniques. Certain extensions of the *Geometric Hashing* allowing further reduction of recognition complexity have been discussed. Future research should include additional extensions of the technique. In particular, recognition of articulated objects composed of rigid parts with internal degrees of freedom is under current investigation.

References

[1] N. Ayache and O. D. Faugeras. HYPER: A New Approach for the Recognition and Positionning of Two-Dimensional Objects. *IEEE TPAMI*, 8(1):44–54, 1986.

[2] D. H. Ballard. Generalizing the Hough Transform to Detect Arbitrary Shapes. *Pattern Recognition*, 13(2):111–122, 1981.

[3] D. H. Ballard and B. C. M. *Computer Vision*. Prentice-Hall, 1982.

[4] P. J. Besl and R. C. Jain. Three-Dimensional Object Recognition. *ACM Computing Surveys*, 17(1):75–154, 1985.

[5] O. Bourdon and G. Medioni. Object Recognition using Geometric Hashing on the Connection Machine. Technical report, Inst. for Robotics and Intell. Systems, USC, 1989.

[6] R. T. Chin and C. R. Dyer. Model-Based Recognition in Robot Vision. *ACM Computing Surveys*, 18(1):67–108, 1986.

[7] B. N. Delone and D. A. Raikov. *Analytic Geometry*, volume 2. Moscow, 1949.

[8] W. E. Grimson and T. Lozano-Pérez. Localizing overlapping parts by searching the interpretation tree. *IEEE TPAMI*, 9(4):469–482, 1987.

[9] W. E. L. Grimson. The Combinatorics of Object Recognition in Cluttered Environments using Constrained Search. In *Proc. of ICCV*, pages 218–227, Tampa, Florida, Dec. 1988.

[10] A. Heller and J. Stenstrom. Verification of Recognition and Alignment Hypothesis by Means of Edge Verification Statistics. In *Proc. of the DARPA IU Workshop*, pages 957–966, Palo Alto, Ca., 1989.

[11] J. Hong and H. J. Wolfson. An Improved Model-Based Matching Method Using Footprints. In *Proc. of ICPR*, pages 72–78, Rome, Italy, Nov. 1988.

[12] B. K. P. Horn. *Robot Vision*. MIT Press, 1986.

[13] D. P. Huttenlocher and S. Ullman. Object Recognition using Alignment. In *Proc. of ICCV*, pages 102–111, London, 1987.

[14] D. P. Huttenlocher and S. Ullman. Recognizing Solid Objects by Alignment. In *Proc. of the DARPA IU Workshop*, pages 1114–1122, Cambridge, Massachusetts, Apr. 1988.

[15] E. Kishon and H. Wolfson. 3-D Curve Matching. In *Proc. of AAAI Workshop on Spatial Reasoning and Multisensor Fusion*, pages 250–261, St. Charles, Illinois, 1987.

[16] F. Klein. *Elementary Mathematics from an Advanced Standpoint ; Geometry*. Macmillan, New York, 1925 (Third edition).

[17] I. B. Kuperman. *Approximate Linear Algebraic Equations*. Van Nostrand, 1971.

[18] Y. Lamdan, J. T. Schwartz, and H. J. Wolfson. Object Recognition by Affine Invariant Matching. In *Proc. of CVPR Conf.*, pages 335–344, Ann Arbor, Michigan, June 1988.

[19] Y. Lamdan, J. T. Schwartz, and H. J. Wolfson. On Recognition of 3-D Objects from 2-D Images. In *Proc. of IEEE Int. Conf. on Robotics and Automation*, pages 1407–1413, Philadelphia, Pa., Apr. 1988.

[20] Y. Lamdan and H. J. Wolfson. Geometric Hashing: A General and Efficient Model-Based Recognition Scheme. In *Proc. of ICCV*, pages 238–249, Tampa, Florida, Dec. 1988.

[21] Y. Lamdan and H. J. Wolfson. On the Error Analysis of 'Geometric Hashing'. Technical report, Robotics Lab, Courant Inst. of Math., NYU, 1989.

[22] S. Linnainmaa, D. Harwood, and L. Davis. Pose Determination of a Three-Dimensional Object Using Triangle Pairs. *IEEE TPAMI*, 10(5):634–647, 1988.

[23] Y. Ohta, K. Maenobu, and T. Sakai. Obtaining Surface Orientation from Texels under Perspective Projection. In *Proc. of IJCAI*, pages 746–751, Vancouver, B.C., Canada, Aug. 1981.

[24] J. Schwartz and M. Sharir. Some Remarks on Robot Vision. In *Trans. of 3'rd Army Conf. on Applied Math. and Computing*, pages 1–36, Atlanta, Ga., May 1985.

[25] D. Shoham and S. Ullman. Aligning a Model to an Image using Minimal Information. In *Proc. of ICCV*, pages 259–263, Tampa, Florida, Dec. 1988.

[26] F. Stein and G. Medioni. Graycode Representation and Indexing: Fast Two Dimensional Object Recognition. Technical report, Inst. for Robotics and Intell. Systems, USC, 1989.

[27] G. Stockman. Object Recognition and Localization via Pose Clustering. *J. of Computer Vision, Graphics, and Image Processing*, 40(3):361–387, 1987.

[28] D. Thompson and J. Mundy. Three-Dimensional Model Matching from an Unconstrained Viewpoints. In *Proc. of IEEE Int. Conf. on Robotics and Automation*, pages 208–220, Raleigh, N. Carolina, 1987.

[29] H. J. Wolfson and R. Nussinov. Efficient Detection of Motifs in Biological Macromolecules by Computer Vision Techniques. Technical report, Tel Aviv University, 1990. *in preparation*.

[30] I. Yaglom and V. Ashkinuze. *Ideas and Methods of Affine Projective Geometry*. Moscow, 1962.

AN ANALYSIS OF KNOWLEDGE REPRESENTATION SCHEMES FOR HIGH LEVEL VISION

Gregory M. Provan

Department of Computer Science   University of British Columbia
Vancouver, BC   Canada V6T 1W5

Abstract This paper analyses the criteria necessary for a knowledge representation (KR) language for implementing high level vision (HLV) recognition systems. We show the importance of introducing a specific KR language for specification, and possibly for implementation of HLV systems. In particular, we examine the adequacy, tractability and suitability of implementing a HLV system using logic, the KR language most commonly used in areas of Artificial Intelligence isomorphic to HLV. In addition, we use this analysis of classical logic to identify the criteria necessary for any HLV KR language. Logic is seen to be at least as good a language for specification of HLV systems as any other KR language. However, using evidence obtained from an object recognition system implemented using propositional logic, evidence which is supported by theoretical analyses, we argue that classical logic is an inadequate KR language for implementing HLV systems. It cannot identify preferred interpretations, and is computationally intractable, even for simple propositional languages.

1   INTRODUCTION

Although progress is being made in high level vision (HLV) on many different fronts, it is becoming apparent that some foundations must be established to provide coherence for the different research areas. Currently, it is difficult to compare different systems, even if they attempt to solve the same problem. Such systems use a large number of widely divergent knowledge representation schemes, which include constraints (algebraic, geometric, symbolic), graphs, logics, rules and neural nets; they also require different inputs (e.g. edges, orientations, etc.) and produce different outputs. What is needed is: (1) a metric for judging the efficiency of the different techniques; and (2) a means of testing and guaranteeing the completeness and correctness of the systems (e.g. to determine if *all correct* interpretations are identified).

A logical framework for depiction and image interpretation has recently been introduced "as a foundation for the specification, design and implementation of vision" systems [10]. This framework ensures correctness with respect to task and algorithm levels, and, in conjunction with the criteria proposed by Mackworth [4], is a means of system specification at least as good as any other KR language. This paper analyses the criteria necessary for an implementation language, criteria which have not been analysed as closely in vision as in artificial intelligence (AI). In addition to the criteria of [4], we assume polynomial-time algorithms to be a necessary criterion, since scene understanding must almost always done quickly.

In examining the criteria necessary for a HLV KR language, we study logic in particular, for several reasons: (1) logic is the most important AI KR language; (2) the isomorphism between diagnostic reasoning (a sub-field within AI) and HLV ([10], [9]) suggests that the tools (specifically logic) used for diagnostic reasoning can be used for HLV as well; and (3) it is important to know whether logic is suitable not just for specification but also for implementation of HLV systems.

2 LOGIC-BASED OBJECT RECOGNITION SYSTEM

2.1 Objectives

We define high level vision loosely as the process of ascribing an interpretation to a set of image primitives. We assume that these image primitives could include edges, surfaces, textures, etc. We simplify the image primitives in order to focus on the reasoning necessary to interpret images composed of the image primitives.

The object identification task on which we focus is "proving" the existence of instances of a class of models, and involves: (1) processing a set of image primitives by filters; (2) creating hypotheses about the existence of "seed" subparts which are used to initiate the object identification process; (3) generating logical clauses based on the spatial relationships of the image primitives; and (4) identifying (partially) complete, consistent interpretation(s) of the image by using a theorem prover to obtain the logical interpretation(s) of the set of clauses together with a set of axioms for the object-models. The filtering of image primitives is domain dependent, but the processing of logical clauses is domain *independent*.

We study whether logic satisfies the descriptive and procedural criteria proposed in [4]. We note two assumptions underlying the choice of logic: (1) a complete axiomatisation of the image and scene domains and of the scene-image transformations is necessary; and (2) absolute correctness (e.g. all quantities are precisely specified, and no noise is present).

We study an articulated model, which is one method of generalising the rigid models currently used in many vision systems. To avoid confusion with logical models, we use the term *r-model* to denote recognition models, which are composed of a specification of the geometrical properties of an object or object class (e.g. a cup class may contain coffee mugs with handles, styrofoam cups without handle, or cups for drinking water), and a specification of the mapping between scene and image.

The model class studied is an articulated puppet, which is broken into multiple sub-parts,[1] each with rotatory, translational and scaling degrees of freedom with respect to the part to which it is joined. Each puppet subpart is a rectangle in both scene and image domains. Puppet models are defined hierarchically: a puppet consists of four major subcomponents, head, neck, trunk and limb. The limb subcomponent is the subdivided into legs and arms, each of which have further subcomponents, etc. This hierarchical structure enables definition of figures of varying complexity. For example, a hand can be subdivided into a palm and five fingers. It is hoped that this analysis will provide a basis with which to extend the models currently used in most model-based systems, and analyse the computational costs of identifying completely articulated objects.

2.2 Logical Formulation

The problem the Visual Constraint Recognition System (VICTORS) solves is as follows: given a set of n 2D randomly overlapping rectangles and a relational and geometric description of a figure, find the best figures if any exist. We define the figure using a set of constraints over the overlap patterns of $k \leq n$ rectangles. VICTORS is a simple implementation of an object recognition task which we argue is the basis for almost all model-based HLV systems. This high-level description is broken down into two distinct components: preprocessing and logical encoding.

Preprocessing: Preprocessing filters out unlikely image primitives, and even though it could be formulated logically, is *not* considered logical manipulation. A logical encoding would only be for formalisation purposes (e.g. proving correctness), and not for implementation purposes. We argue that logically modeling the complete set of image primitives of a typical image is computationally expensive and unnecessary, primarily because the uncertainty introduced by noise and the large number of primitives possible will overwhelm any theorem prover. Preprocessing consists of passing

---

[1]For example, a 7-part puppet consists of head, neck, trunk and 4 limbs. More detailed figures of 15 or more parts are defined similarly. In general, a library of classes of figures can easily be defined.

the input data (image primitives) through a set of filters before encoding the logical clauses. A *filter* is a test of the geometric properties of a rectangle. A *constraint* is a set of filters. Each constraint places restrictions on acceptable assignments of puppet parts to rectangles based on the overlap patterns of the rectangles. For example, one of the filters for a trunk is that there are at least 5 smaller rectangles overlapping it (which could be a neck and four limbs).

Logical Axiomatisation: The r-model axioms show the underlying logical formulation of the object models typically used in visual interpretation tasks. The formalisation of the axioms for the image domain, scene domain and image-scene domain is based on the axiomatisation framework of Reiter and Mackworth [10].

The *image domain axioms* formalise the spatial relationships among the image primitives, of which there is one kind, a rectangle. The *scene domain axioms* formalise the puppet figure. There are axiom sets for both complete and incomplete figures, reflecting the basic and extended modes modes of operation possible in VICTORS, respectively. In the *basic mode*, logical descriptions of the part hierarchy and of joints are made for complete puppets.[2] Joints define how the puppet parts are connected. The *extended mode* formalisation consists of a similar set of axioms which are extended to allow incomplete or occluded figures, etc. The *Image-Scene Domain axioms* formalise the relationships between image elements and scene elements, and can account for occlusion. For example, an edge in the image maps to a specific corner of a block, or the edge of a ball in the scene produces a curved line in the image.

For a particular image, hypotheses are made about the interpretation of image primitives using the logical axiomatisation, and a consistent subset of this initial hypothesis set is derived[3] using an ATMS [2] as the theorem prover. First, *seed hypotheses* are made: seeds are specific puppet part-hypotheses which are chosen because they are good hypotheses for "growing" puppet figures and for constraining the search space. (Only image primitives which pass the constraints are encoded logically.) For example, a trunk is a seed because it is tightly constrained by having 4 attached limbs; in contrast, a foot is loosely constrained with one attached limb. Based on the seeds, secondary clauses defining limbs attached to previously-hypothesised puppet parts are created, thus "growing" the puppets.

An assumption, a special propositional literal, is assigned to each clause to denote the source of the clause. If assumptions are denoted by subscripted A's, a seed clause can be $A_1 \Rightarrow R_2 :$ $scene - element - x$, meaning that rectangle R_2 is hypothesised to be $scene - element - x$ under assumption A_1; a secondary clause can be $A_2 \wedge R_4 : scene - element - x \Rightarrow R_5 : scene - element - y$, meaning that rectangle R_5 is hypothesised to be $scene - element - y$ under assumption A_2, by virtue of its overlap with rectangle R_4, currently hypothesised to be $scene - element - x$. This set of clauses is passed to the theorem prover. All logical models for this clause set are derived, such that each logical model corresponds to a puppet hypothesis.

We refer the reader to [6] and [9] for descriptions of system capabilities, such as identifying complete puppets, puppets with ambiguous interpretations, missing pieces, occluded pieces, puppets amid clutter, etc. Other primary features include robustness given noise, explanation generation, studying many different alternative interpretations by simple database changes.

VICTORS has been extended with an uncertainty calculus, Dempster Shafer (DS) theory, to test how a preference ordering affects system performance. DS theory is used because it has an underlying logical semantics [7], and hence can be implemented in a propositional logic system in a straightforward manner.

The ATMS is extended by assigning [0,1] weights to assumptions. Assigning weights to assumptions entails replacing the symbolic token associated with each assumption by a [0,1] measure which indicates the degree of acceptability of the particular constraint. Computing such measures does not require significantly more processing than is necessary with the traditional ATMS. This is because the rectangle data that exists already is used to define criteria for "quality" of part acceptability. Thus, instead of testing a constraint that the overlap of rectangle C, identified as *trunk*, with rect-

---

[2]We note that the articulation of the puppet adds significant complexity to the identification task.
[3]This is called the abductive approach in diagnostic reasoning.

angle D either qualifies D to be a *thigh* or not, a weight or probability with which the constraint could be true is calculated.

For the purposes of this paper, it is sufficient to note that the assignment and updating of the measures assigned to puppet parts and puppet figure hypotheses is done in a coherent manner, according to the semantics of DS theory. The measures used introduce a partial ordering on the puppet figure interpretations, the effect of which is described in the following section.

3 SYSTEM PERFORMANCE

In practice, the performance of VICTORS degraded significantly with (1) increased complexity within a given model class, and (2) density of rectangles (i.e. the degree to which the rectangles overlapped each other) in the input image. Although results are good for simple inputs and simple puppet models, the system is quite slow for images with a large number of partial interpretations and for complicated models. This is because an enormous search space could be generated when exploring all possible visual interpretations, given that there is no ordering to prune unlikely (partial) interpretations. Analysis of the complexity of the modules of VICTORS shows the TMS (i.e. the means of determining logical interpretations) to be the most computationally expensive module, as the other modules have at worst linear or polynomial complexity.

The use of DS Belief functions has enabled interpretation ranking, such that the best interpretation can be found. However, the computational costs of deriving DS measures so undermine its advantages that the use of approximation DS measures is necessary. In fact, the problem of computing exact DS measures is intractable [7]. The use of heuristic approximation algorithms for DS uncertainty measures, as done in VICTORS, trades off intractability for theoretical accuracy.

Results to date indicate that even simple weight assignments prove useful in generating an ordering of partial interpretations equivalent to the theoretically accurate ordering. However, for more complicated input data these approximation techniques are too inaccurate. Indeed, we anticipate that real, sensor-derived data will require sophisticated weight manipulation. Even so, there are domains in which efficiently-computed approximate weight assignments can provide the partial ordering necessary for directing search and improving the efficiency of the ATMS.

The poor system performance is corroborated by theoretical results about the complexity of the problems underlying VICTORS. The problem underlying VICTORS can be formalised as computing all logical models of the axiom and assumption sets. It is well known that a set of axioms may have an exponential number of logical models; but even when there are few models the only known method of solving this problem is exponential in the number of assumptions or axioms, on average [8].

In addition, the hope that simple, propositional preference logics might be more efficient than the corresponding classical logics has proved to be false: even propositional Horn default logics are NP-hard [11][4] and preference logics are less tractable than comparable classical logics [8]. Moreover, preference logics are not well understood in terms of their semantics and the definitions of model minimality [8], and is it not possible to define a consistent, universal preference logic [3].

This theoretical analysis supports the practical experience of VICTORS, namely that the two main drawbacks of classical logic for HLV are intractability and the lack of a preference ordering. In fact, it has been argued that logic cannot be the basis for any form of human reasoning because of its intractability [1]. The restrictions (placed on preference logics) to ensure tractability are so severe that too much expressiveness is lost. What is needed is a meta-logic to manipulate the preference orderings (e.g. updating and combining preferences). In essence, it may be argued that what is needed is an evidential calculus.

---

[4]In contrast, determining satisfiability for propositional Horn (classical) logic is $O(n)$.

4 CONCLUSIONS

In this paper we have identified many of the criteria necessary for a KR language for HLV. We have argued that specification is crucial, a task for which logic is at least as good as any other language. For implementation, we have identified, in addition to the procedural and descriptive adequacy criteria of [4], the need for polynomial-time algorithms and a preference ordering. We showed that logic fails the last two criteria, and that preference logics do not posses polynomial-time algorithms for all but the most trivial restrictions. These negative results indicate that significant effort must be directed to approximation, control and problem decomposition issues. We argue that the KR scheme is determined by efficiency criteria and control issues, and not vice versa.

A big question is whether any KR language can satisfy all these criteria. Probability theory, to date, is the only other possibility. It has a well-defined semantics, defines a preference ordering and has linear-time algorithms for a large class of problems,[5] but probability theory has many unique, and different problems [9].

The other alternative is not to implement a HLV system with a single KR language. Subject to a well-defined system specification, efficient, task-dependent techniques which transform specified inputs to specified outputs can be developed. It is possible that this, in fact, may be the most promising future research, and also the approach most similar to biological vision systems.

This analysis also addresses the speculation regarding the integration of high level vision and lower level vision. Since most high level KR schemes are purely symbolic, as opposed to intermediate and low level vision KR schemes being purely numeric, the symbolic/numeric interface has been seen to be problematic. We argue that high level visual reasoning must have an important numeric component, hence avoiding a problematic numeric/symbolic integration of lower level and high level vision.

Acknowledgements: Thanks to Mike Brady, David Lowe and Alan Mackworth for useful comments. The author completed this research with the support of a scholarship from the Rhodes Trust, Oxford, and of the University of British Columbia Center for Integrated Computer Systems Research, BC Advanced Systems Institute and NSERC grant A9281 to A.K. Mackworth.

References

[1] C. Cherniak. *Minimal Rationality*. MIT Press, 1986.

[2] J. de Kleer. An Assumption-based TMS. *Artificial Intelligence*, 28:127–162, 1986.

[3] J. Doyle and M.P. Wellman. Impediments to Universal Preference-Based Default Theories. In *Proc. of the Conf. on Principles of Knowledge Representation and Reasoning*, pages 94–102, 1989.

[4] A. Mackworth. Adequacy Criteria for Visual Knowledge Representation. In Z. Pylyshyn, editor, *Computational Processes in Human Vision: An Interdisciplicary Perspective*, pages 464–476. Ablex Publishers, Norwood, NJ, 1988.

[5] J. Pearl. Fusion, Propagation, and Structuring in Belief Networks. *Artificial Intelligence*, 29:241–288, 1986.

[6] G. Provan. The Visual Constraint Recognition System (VICTORS): Exploring the Role of Reasoning in High Level Vision. In *Proc. IEEE Workshop on Computer Vision*, pages 170–175, 1987.

[7] G. Provan. A Logic-based Analysis of Dempster Shafer Theory. *International Journal of Approximate Reasoning*, to appear, 1990.

[8] G. Provan. An Analysis of Model Minimisation Methods of Computing AI Theories. Technical Report to appear, University of British Columbia, Department of Computer Science, 1990.

[9] G. Provan. Model-based Object Recognition using an Extended ATMS. Technical Report to appear, University of British Columbia, Department of Computer Science, 1990.

[10] R. Reiter and A.K. Mackworth. A Logical Framework for Depiction and Image Interpretation. *Artificial Intelligence*, 41:125–155, 1990.

[11] B. Selman and H. Kautz. The Complexity of Model-Preference Default Theories. In *Proc. Conf. Canadian Soc. Computational Studies of Intelligence*, pages 102–109, 1988.

---

[5]The problems must be tree-structured [5], a class for which preference logics are NP-hard [11].

Experiments on the use of the ATMS
to label features for object recognition

R.M. Bodington[†] **G.D. Sullivan**[††] **K.D. Baker**[††]

1. Abstract

Experiments are reported on the use of an Assumption-based Truth Maintenance System (ATMS) [6] to establish a match between a 3-d model and a single 2-d image. We show that the ATMS improves the efficiency of the search for maximal combinations of consistently labelled features. A memory cost is incurred, associated with the recording system of the ATMS; this can be reduced by simple heuristics. Empirical evidence is presented quantifying the costs and benefits of the method.

2. The consistent labelling of image features

Features extracted from images are usually ambiguous and uncertain. Object recognition depends on the discovery of extended combinations of image features which are mutually compatible with a known object. Grimson and Lozano-Perez [9,10] have shown that simple binary constraints between pairs of model features may be sufficient to reject most mislabellings. If the number of model features is small, it is feasible to store all binary constraints explicitly, in tables compiled in advance. Furthermore, it is only necessary to compute the measures once for a given set of image features, and these too can be stored. As the search proceeds, the consistency of a pair of labels can be tested simply by comparing the measured datum against values in the constraint table. The space of all possible combinations of labels may then be searched by a depth-first expansion of the interpretation tree.

Grimson and Lozano-Perez have shown that if the scene contains a single isolated object, very little of the interpretation tree need be explored. The strong constraints invalidate large sub-trees, and the search rapidly collapses to the single solution. However, a depth-first search with back-tracking is inherently very redundant. At each expansion of the interpretation tree, checks must be made that the newly labelled datum is pair-wise consistent with all existing labels. Identical checks are therefore duplicated throughout the tree. For example, two data features separated from each other by n levels in the tree, will be cross-checked $m^{(n-1)}$ times (where m is the number of model features - i.e. the "fan-out" of the tree). Such duplication of effort is only acceptable if the constraint checks are extremely cheap to evaluate. This is the case in [10] where the constraints compare pre-compiled geometrical measurements of 3-d model features with 3-d sensory data. The duplication is not acceptable in systems where the sensory data under-constrains the model, for example when matching 2-d data against 3-d models, as is considered in this paper.

† *British Aerospace, Sowerby Research Centre, PO Box 5, Filton, Bristol, BS12 7QW. UK*
†† *Dept. of Computer Science, University of Reading, Reading, RG6 2AX. UK.*

Two types of constraint are available between 2-d data and 3-d models:

(1) Qualitative constraints, derived from 2-d geometrical relationships between image features, such as coincidence, adjacency, enclosure, similarity. Changes of viewpoint cause large changes between image features, so these constraints are often weak, and erroneous matches may not be detected.

(2) Quantitative constraints, using 3-d knowledge, requiring all data to be geometrically compatible with a single perspective view of the object (see e.g. Lowe, [12]). This requires the viewpoint to be solved iteratively, followed by the evaluation of combinations of features. It provides strong, metrical constraints, but they are very expensive to apply.

Both types of constraint are inherently view-dependent, i.e. the exact relationship between the image features is strongly dependent on the pose of the object in front of the camera. Therefore 2-d to 3-d constraints cannot be recorded in advance in simple numerical look-up tables, and must be evaluated dynamically as the model becomes instantiated. The computational cost of applying such constraints is often large, and a highly redundant depth-first search with backtracking is impractical. Instead, a record of all partial results must be kept, so that unnecessary repetition of the work can be avoided. Truth Maintenance Systems have been proposed for this purpose in other areas of Artificial Intelligence. We demonstrate below how the ATMS can be used within a model-based vision system for recognising vehicles within unconstrained single images. The main features of the implementation have been reported elsewhere [1,2]. In this paper we report an experimental investigation into the costs and benefits of the ATMS for the consistent labelling of image cues. A more detailed report of this material is also available [3].

3. Outline of CARRS

CARRS (CAR Recognition System) is an experimental system for finding and locating vehicles in images such as Figure 1(a). CARRS adopts an "hypothesis-and-test" strategy which (in brief) consists of the following stages.

Stage 1: Data-driven determination of feature groups.

S1.1 Feature Extraction. Connected edges (Figure 1(b)) from a single scale Canny operator [5] are segmented into straight lines at curvature maxima, to form polygonal approximations.

S1.2 Cue Identification. Polygons (and fragments of polygons) are extracted to form cues for the labelling process. Cues are application-specific features which facilitate heuristic methods. At present these consist of U-shaped triples, S-shaped triples, and closed quadrilaterals. Each type of cue is associated with known components of the model which may give rise to it, e.g. U-shapes and quadrilaterals may arise at any of the windows, S-shapes may occur at the near- or off-side pillars of the windscreen. Figure 1(c) (central box) shows the cues found in Figure 1(b).

S1.3 Identification of Areas of Interest. Each cue is considered in turn as a seed feature (SF), and is associated with a subset of all cues likely to be due to a single car. The set is based on proximity to the SF in the image, conditioned by rules dependent on the type of SF, and its size and orientation in the image. The SFs are then ordered according to the cardinality of the proximity sets (Figure 1(c)). Note the overlap between proximity sets.

Stage 2: Search for maximal consistent labellings

S2.1 Application of 2-d Constraints. Taking each SF in rank order, a search is made for maximal subsets of its proximity set, which may be labelled as model features, such that they are consistent with 2-d constraints. The constraints used are heuristic and domain-dependent; they are chosen to be fairly independent of viewpoint, and are not strongly specific to particular models of vehicles. They express requirements such as:

• The windows on each side of the car must be aligned and closely adjacent.

• A near-side window and an off-side window cannot be visible simultaneously.

These 2-d constraints use metrical concepts such as distance and orientation, which are scaled by measurements between junctions within cues.

S2.2 Hypothesis instantiation. Any maximal set of a SF which contains a sufficient number of cues is passed immediately to Stage 3 for verification. All other maximal subsets are stacked, pending the examination of the remaining SFs. When all SFs have been so treated any surviving maximal set is then passed to Stage 3.

Stage 3: Model-based verification of hypotheses.

S3.1 Viewpoint inversion. Labelled maximal sets allow the pose of a known object to be determined, using a 3-d model. This is carried out in three stages, each of which either invalidates the hypothesis, or progressively refines the viewpoint estimate.

(1) Labelled features identify a patch on the viewsphere from which all the features are visible - the "viewpatch" - which is represented as a quad-tree [14].

(2) The "roll-consistency constraint" [16] is applied, to reject parts of the viewpatch in which the angles (in the image) of labelled lines are inconsistent.

(3) The perspective transformation is inverted by iterating from the current best estimate of view [15].

If an hypothesis is rejected at any of these stages then the solution set is invalidated, and is removed from further consideration. The previously satisfied set which led to this invalid set is then reconsidered as a maximal set.

S3.2 Iconic verification. The vehicle hypothesis is evaluated by using the view estimate to project the entire model back into the image. An hypothesis-driven check is carried out for the existence of the projected model features in the image [4].

S3.3 Success Pruning. The acceptance of an hypothesis "consumes" the SF and cues in its maximal set. All constraints involving a consumed cue are declared invalid, and the cue is removed from all existing sets. This prunes the search carried out in Stage 2.

Stage 2 and Stage 3 are repeated as necessary, until all maximal partial labellings have been confirmed or invalidated.

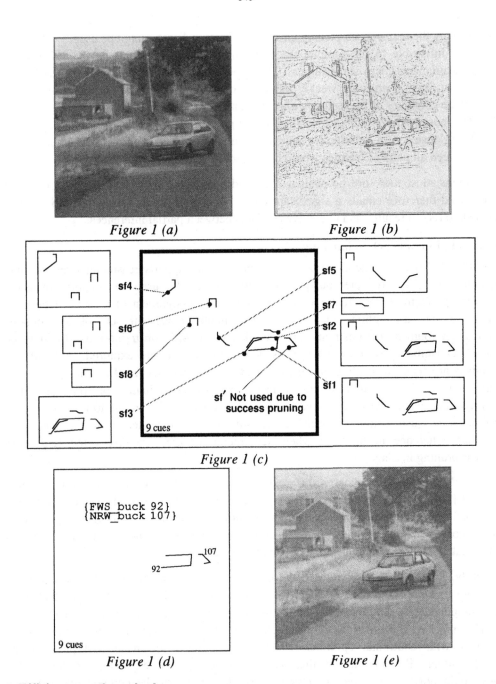

Figure 1 (a)

Figure 1 (b)

Figure 1 (c)

{FWS_buck 92}
{NRW_buck 107}

9 cues

Figure 1 (d)

Figure 1 (e)

4. Efficient search methods

All three stages of the system impose significant computational burdens. This study concerns the search strategy used in Stage 2, and we take as fixed the cue sets delivered by Stage 1, and the verification algorithms used in Stage 3. The current problem is how to make efficient use of the weak 2-d constraints. There are two main reasons why the methods adopted by Grimson & Lozano-Perez cannot be used.

(1) It is infeasible to pre-compile the constraints, and to pre-compute the pairwise measurements between features. The measurements between features carried out in Stage 2 are only weakly independent of view and must be parameterised by the geometry of the feature within the image.

(2) The constraints which are applied are highly specific to the particular labels being tested. An unselective initial calculation of all possible measurements between all pairs of features would be absurdly expensive.

Constraints must therefore be evaluated as and when required. Any evaluation which is repeated therefore entails a significant cost. A form of Truth Maintenance System is needed, to record results, and to resolve conflicts between partial inferences.

5. An ATMS approach to labelling

The ATMS [6] is a general purpose mechanism that supports reasoning over multiple hypotheses. We have previously shown that it can be used to support consistent labelling problems [1,2]. In brief, we represent each assignment of a model label to a feature as an assumption, and search for mutually consistent sets of assumptions (called environments). Constraints are recorded as ATMS constraint-nodes (akin to de Kleer's consumers), which are tested as new environments are explored. Invalid sets of labels are reduced to minimum inconsistent subsets, and recorded as "no-goods". All partial results of the search are stored, to avoid repeated evaluation of constraints; this is important where SFs have overlapping proximity sets. In addition, the ATMS maintains a "justification network" to record the interdependency of the data; this enforces coherence in the data when new inconsistencies are discovered (e.g. after success pruning in Stage 3).

The ATMS offers a radically different approach to that of the interpretation tree method [10] and can be used to solve a wider class of problems. However, there is a significant computational cost to pay: the recording mechanism consumes a great deal of memory, and the up-dating of the justification network is time-consuming. It is not immediately clear when the benefits outweigh the costs.

Recently, Provan [13] has criticised the use of an ATMS on the grounds that it fails to scale up to complex problems effectively. He considered the visual task of identifying the components of a humanoid puppet thrown onto a table-top. The simulated "sense-data" comprised the positions of rectangles, representing different parts of the puppet, together with additional spurious rectangles introduced to create multiple possible interpretations. Provan showed that as environments were expanded in the search, they could not be disproved until very late, when most of the components had been labelled. This resulted in large no-goods which are ineffective at invalidating other environments. The number of consistent partial labellings therefore increased exponentially, and created an unacceptable storage cost. Provan argued that the cost greatly outweighed the slight benefits, and that the ATMS approach is unsuited even to quite small labelling problems in vision.

Provan's example is only partly relevant to CARRS. It is true that strong constraints on the sensory data cannot be applied at early stages of the search. However, unlike Provan's example, CARRS can invoke strong (viewpoint-dependent) constraints at the later stages of processing (Stage 3 above). The use of "success pruning" (S3.3) then allows much of the ATMS memory load to be avoided, since it consumes features, and reduces the number of environments needing to be considered. Success pruning has an effect similar to the use by Grimson and Lozano-Perez [10] of a cut-off, when some fractional match has been found. It reduces the amount of the search tree explored and the number of environments which need to be maintained.

6. Experimental results

In this section the costs and savings of using the ATMS are considered, by using CARRS to analyse five representative images, IM1-IM5. Typical behaviour is illustrated using IM1.

6.1 Analysis of constraint savings

Where a significant portion of the search space is explored repeatedly, the encoding of constraints and recording of no-goods in the ATMS will result in a reduction in the number of constraints evaluated. The savings can be measured by recording how many constraints were necessary to check the consistency of an environment the first time it is explored. Each time an environment is re-explored, that number of constraint checks will have been avoided by the use of an ATMS. A count of the total savings is then increased by that amount.

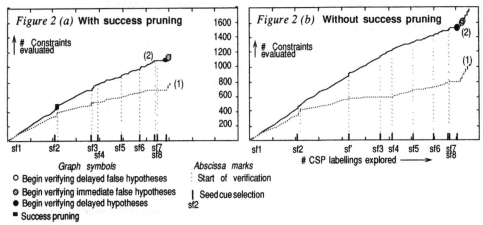

Figure 2: The affect of success pruning on the size of the search space

These savings are illustrated in Figure 2(a). The abscissa of the graph shows the number of labellings explored, i.e. the amount of the search space explored. The point where a new SF is selected is shown on the graph, as well as the points where hypotheses testing takes place. Graph (1) shows the number of constraints evaluated during the search and graph (2) shows the number that would have been evaluated if no recording mechanisms were used. The distance between the two graphs is an indication of the savings in constraint evaluation gained by using the ATMS.

The results show that significant amounts of the search space are re-explored and that constraints are re-evaluated repeatedly. This may occur during the search within a proximity set, as happens during the search from SF1. However, significantly more re-exploration occurs when proximity sets overlap as is the case with SF2 and SF3 which have a large intersection with SF1, the first SF explored (see Figure 1(c)).

Images	IM1	IM2	IM3	IM4	IM5
Constraints					
(a.1) # evaluated:	799	467	1 007	1 344	1 126
Labellings explored					
(b.1) # explored:	2 366	3 198	3 382	5 399	5 030
(b.2) # re-explored:	1 372	2 004	1 996	3 280	3 012
(b.3) Percentage re-explored:	58%	63%	59%	61%	60%
Reduction in constraints evaluated					
(c.1) Total reduction:	374	833	655	1 460	1 356
(c.2) Percentage reduction:	32%	64%	39%	52%	55%
$(c.2) = 100 \times (c.1) / ((c.1) + (a.1))$					

Table 1 Summary of Constraint Savings

The savings in constraint evaluation gained by using the ATMS are summarised for the 5 test images in Table 1. Entry c.2 is the percentage of constraint evaluations that have been avoided by use of the ATMS encoding. The results show that using the ATMS does offer significant savings by avoiding constraint repetition. On aggregate over the 5 test images approximately 50% of the constraint evaluations are saved.

6.2 Analysis of results of success pruning

The effects of success pruning can be illustrated in CARRS by disabling the cue consumption. The result is shown graphically for image IM1 in Figure 2(b), where it is seen that a considerably greater number of constraints are evaluated.

Images	IM1	IM2	IM3	IM4	IM5
(a) # of consumed cues:	2	-	2	2	3
(b) # of labellings taken OUT:	66	-	55	63	106
(c) Immediate verification reduction:	3	-	0	2	2
(d) Delayed verification reduction:	1	-	8	7	12

Table 2 The effects of success pruning for each test image

The effects of success pruning for the 5 test images is summarised in Table 2. Entry (a) shows the reduction in the number of seed features being considered. Entry (b) shows the number of consumed image features that were previously labelled as model features. Entries (c) and (d) show the number of maximal labellings pending verification that were invalidated as a consequence of success pruning.

6.3 Memory requirements

The memory required by the ATMS increases with the search and limits the size of application that can be considered. This limit can be estimated by measuring the size of the ATMS data structures, constraint nodes and justifications, environments and no-goods as each image is analysed. These are summarised in Table 3.

Images	IM1	IM2	IM3	IM4	IM5
Nodes,					
(a.1) # of constraint nodes:	789	927	1 093	1 219	1 155
(a.2) # of model feature nodes:	366	454	518	698	521
(a.3) # of feature labelling assumptions:	7	0	7	7	10
Justifications,					
(b.1) # stored:	2 772	2 768	3 774	5 323	4 405
Environments,					
(c.1) Final # stored:	931	3 778	2 221	2 071	2 041
(c.2) Largest # stored:	1 195	4 375	5 216	8 606	3 500
Nogoods,					
(d.1) Final # stored:	317	410	431	680	504
(d.2) Largest # stored:	317	410	431	680	504

Table 3 Storage requirement of the ATMS

Of these data structures, those that increase the most are the consistent environments and the minimal no-goods. Their growth for image IM1 is shown in the graphs in Figure 3 which shows the effect of success pruning clearly (occurring at the verification of SF1) in reducing the number of consistent environments and minimal no-goods stored. These results indicate that in CARRS the storage overhead associated with the use of an ATMS does not grow excessively.

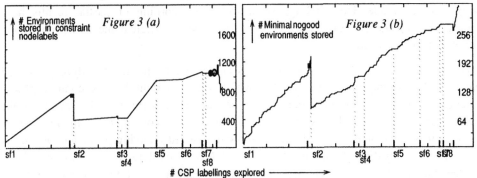

Figure 3: Storage of Environments for test image IM1

6.4 CPU costs of using the ATMS

A further disadvantage of the ATMS is the computational overhead incurred in maintaining the consistency of the justification network and environments. This can be estimated by recording the CPU time spent on ATMS operations. However, such measurements are strongly influenced by minor implementation details, such as the choice of data representation and the need for garbage collection. It is shown elsewhere [3] that in CARRS up to 73% of the time spent evaluating constraints concerns updating the ATMS to reflect newly discovered inconsistencies.

7. Conclusions

Our demonstrations based on the CARRS program have shown that an ATMS can be used effectively to determine all maximal consistent sets in a constrained labelling task. This approach overcomes a crucial defect inherent in the backtracking algorithm of Grimson and Lozano-Perez, and is therefore able to cope with the less well constrained problems of matching 2-d image features to 3-d models. In our experiments we have used relatively simple sets of features. Even here it has been shown that the recording mechanism of the ATMS allows approximately 50% of the constraint evaluations to be avoided. In more complex examples, having greater potential overlap between feature sets, the savings would be correspondingly greater.

We have also shown that the storage burden needed to maintain the ATMS is not excessive - this contradicts Provan's [13] findings for vision tasks such as that addressed by CARRS. One important factor in limiting the number of environments is the use of "success pruning", based on the ability to verify hypotheses by 3-d model-based methods. This has a dramatic effect both on the memory requirements and the up-dating costs of the ATMS. The most time-consuming component of CARRS is the maintenance of the consistency of environments. Meaningful CPU-time estimates are difficult, but our experiments suggest that the ATMS update time may completely cancel out the savings provided by the avoidance of constraint re-evaluation.

Our experiments on the consistent labelling problem using the general-purpose mechanism provided by the ATMS have allowed us to identify where significant savings can be made. Our findings show that the recording of partial results has a strong impact on the cost of the search, provided that an efficient means of maintaining a coherent record is used.

However, we note that the ATMS represents an "over-kill" for the labelling problem in CARRS. The properties of the ATMS which are important in this context may be implemented in a far more efficient way, especially if, as here, only binary constraints between labels are used. The success or failure of each constraint evaluation may be stored, as it is calculated, in a 2-d matrix allowing hashed indexing by means of the cue-labels. The breadth-first search used by the ATMS to explore the environment lattice can then be imitated by a systematic breadth-first expansion of the interpretation tree. Repeated evaluation of constraints can then be avoided by first checking each constraint against the evolving results matrix. We are currently investigating this strategy. It should be noted that this method will only work for binary constraints. To use higher order constraints a more general mechanism is needed - such as de Kleer's ATMS.

Acknowledgments

CARRS was developed under Alvey grant MMI-007, while all three authors were at the University of Reading. We are grateful to British Aerospace for continued support, which allowed this investigation to be completed.

8. References

[1]   R.Bodington and P.Elleby "Justification and Assumption-based Truth Maintenance Systems: When and How to use them in Constraint Satisfaction," in *Reason Maintenance Systems and their Applications*, Ed. B.Smith and G.Kelleher, Ellis Horwood, 1988

[2]   R.Bodington, G.D.Sullivan and K.D.Baker, "The Consistent Labelling of Image Features using an ATMS", *Image Vision & Computing*, vol. 7, no. 1, February 1989.

[3]   R.Bodington, G.D.Sullivan and K.D.Baker, "Experiments on the use of the ATMS to label features for object recognition" *University of Reading, unpublished report* December 1989

[4]   K.Brisdon, G.D.Sullivan and K.D.Baker, "Feature Aggregation in Iconic Model Evaluation," *Proceedings of the Alvey Vision Conference, AVC88*, Sept. 1988.

[5]   J.Canny, "Finding edges and lines in images.", Ph.D. AI-laboratory, MIT, Cambridge, MA, 1983

[6]   J.De Kleer "An Assumption-Based TMS", *Artificial Intelligence*, vol. 28, no. 2, March 1986.

[7]   J.De Kleer "Extending the ATMS", *Artificial Intelligence*, vol 28, no. 2, March 1986.

[8]   C.Goad "Special purpose automatic programming for 3-d model-based vision", *Proc. ARPA Image Understanding Workshop, 1983.*

[9]   W.Grimson, and T.Lozano-Pérez, "Model-Based Recognition and Localization from Sparse Range or Tactile Data." A.I. Memo 738, MIT, Cambridge, MA, August 1983.

[10]   W.Grimson and T.Lozano-Pérez, "Localizing overlapping parts by searching the interpretation tree," *IEEE Trans. on Pattern Analysis and Machine Intelligence*, vol. PAMI-9, no. 4, July 1987.

[11]   W.Grimson, "The combinatorics of object recognition in cluttered environments using constrained search," *2nd International Conference on Computer Vision, ICCV88*, Dec. 1988.

[12]   D.G.Lowe, "Three-dimensional Object Recognition from Single Two-dimensional Images.," *Artificial Intelligence*, no. 31, 1987.

[13]   G.Provan, "Efficiency Analysis of Multiple-Context TMSs in Scene Representation." *Proceedings 6th National Conference on Artificial Intelligence, AAAI-87*, 1987.

[14]   A.Rydz, G.D.Sullivan and K.D.Baker, "Model-based Vision Planar Representation of the Viewsphere,"*Proceedings of the Alvey Vision Conference, AVC88*, Sept. 1988.

[15]   A.Worrall, G.D.Sullivan and K.D.Baker, "Model-based Perspective Inversion," *Proceedings of the Alvey Vision Conference, AVC88*, Sept. 1988.

[16]   A.Worrall, G.D.Sullivan and K.D.Baker, "The Roll Angle Consistency Constraint" *Proceedings of the Alvey Vision Conference, AVC89*, Sept. 1989

The Combinatorics of Heuristic Search Termination for Object Recognition in Cluttered Environments

W. Eric L. Grimson [1]

MIT Artificial Intelligence Laboratory

545 Technology Square, Cambridge, Mass. 02139

Abstract

Earlier work on using constrained search to locate objects in cluttered scenes showed that the expected search is quadratic in the number of features, if all the data comes from one object, but is exponential if spurious data is included. Consequently, many methods terminate search once a "good" interpretation is found. Here, we show that correct termination procedures can reduce the exponential search to quartic. This analysis agrees with empirical data for cluttered object recognition. These results imply that one must select subsets of the data likely to have come from one object, before finding a correspondence between data and model features.

Constrained tree search [e.g. 6,7,10], which identifies data/model feature pairings consistent with a rigid coordinate transformation, is a common approach to object recognition and localization in noisy cluttered environments. Formal analysis of these methods [2] shows that if all of the data are known to have come from one object, the expected amount of search is quadratic, while if spurious data is allowed, the expected search is exponential.

Hence, a hard part of recognition is isolating, from the spurious data, a subset likely to belong to one object. While grouping methods (e.g. generalized Hough transform, or [9,8]) can reduce the search space size [4], they cannot, in general, select sets of data features all from one object, without also encurring a high false positive rate [4].

An alternative is to terminate the search [e.g. 1,6,7,9] once a measure of an interpretation's "goodness" (fraction of the object accounted for) exceeds some threshold. The threshold can be set based on scene clutter and model size [5], so that no false positive solutions are expected. Here, we show how termination reduces the expected search.

1. The constrained search model

Constrained search finds pairings of geometric data and model features, consistent with a rigid model transformation. To find feature matches, we search an interpretation tree depth-first. Nodes at the first level of the tree match the first data feature to each model feature, or to the null character indicating the data feature is not part of the object. Each node then branches to $m+1$ other nodes, where the next data feature is matched to each model feature or the null character, and so on, so that a level n node and its ancestors define a matching of the first n data features. We search the tree depth first, testing each node's consistency with unary and binary constraints [6,7] based on properties like length, relative orientation and relative separation of features. Any constraint involving the null character is always consistent. If any other constraint is false, we backtrack. If we reach a leaf, we verify the data/model pairings by solving for a rigid transformation and testing that it maps all the model features into their matched data features. If so, we save the interpretation, backtrack and continue, until all interpretations are found.

[1]Research funded in part by ONR URI grant N00014-86-K-0685, in part by NSF Grant IRI-8900267, and in part by DARPA under Army contract DACA76-85-C-0010 and ONR contract N00014-85-K-0124.

2. Previous results

Empirically [6,7], this method is very efficient when all the data features are known to come from one object. With spurious data, however, the method slows down considerably. If sets of data/model pairings consistent with similar model transformations are isolated before the search, efficiency improves. If termination is added, the method improves even more. Some of these empirical observations are supported by formal analysis [2]:

- If all s data features lie on one object with m equal size features, the noise is small, and the data is uniformly distributed, then the expected search is bounded by

$$m^2 \leq N_s \leq m^2 + ams.$$

- If only c of the s sensory features lie on an object and the other conditions above hold, then the expected search is bounded above and below by expressions of order

$$O(N_s^*) = m[1 + \gamma]^s + ms2^c + \delta m^6 + m^2 s^2[1 + \mu]^c \quad \text{and} \quad o(N_s^*) = m2^c + ms.$$

Here, a, γ, δ, μ are constants that depend on the object and the sensor noise, $\gamma, \mu < 1$. Hence, constrained search is polynomial (quadratic) when all of the data is known to come from a single object, but is exponential when spurious data is included. Here, we consider the effects of heuristic search termination in reducing the exponential cost.

3. Setting up the formal termination model.

The probability that matching the ith data and the Ith model feature is consistent is

$$q_{i,I} = \begin{cases} 1 & \text{if } i \mapsto I \text{ is correct, or if } I \text{ is the null character,} \\ p_1 & \text{otherwise.} \end{cases}$$

The probability that the matches $i \mapsto I, j \mapsto J$ are consistent is

$$q_{i,j;I,J} = \begin{cases} 1 & \text{if } i \mapsto I, j \mapsto J \text{ is correct, or if either } I \text{ or } J \text{ are the null character,} \\ p_2 & \text{otherwise.} \end{cases}$$

Given a partial interpretation at a search tree node, the probability of consistency is [2]:

$$\prod_i q_{i,I} \prod_{i \neq j} q_{i,j;I,J}.$$

Given these definitions, one can derive an explicit expression for the expected number of nodes in the tree [3]. For the case of terminating the search once the number of actually matched data features in a valid interpretation exceeds a predetermined threshold, some messy manipulations (which in the interest of space are deferred to [3]) give:

Proposition 1: Given a uniform distribution of correct data features among the spurious, with density $\delta = c/s$, m model features, s data features and a termination threshold t, and since $p_2 = (\kappa/m)^2$ [2], the expected amount of search is bounded by

$$N \leq t\frac{1}{\delta} + \frac{mp_1}{\delta}\left[t^{j_0+1}\mu^{j_0-1}\left(1 + (t-1)\frac{\kappa^2}{m^2}\right) + \gamma^{i_0}f(t-1)\left(\frac{1-p_2^t}{1-p_2} + \beta\frac{1-p_2^{(3-\delta)t}}{1-p_2^{(3-\delta)}}\right)\right.$$

$$-\gamma^{i_0}\left[\left(\frac{1}{\delta}-2\right)(t-1)+f\right]\left(\left[\frac{p_2(1-p_2^t)}{(1-p_2)^2}-\frac{tp_2^t}{1-p_2}\right]+\beta\left[\frac{p_2^{(3-\delta)}(1-p_2^{(3-\delta)t})}{(1-p_2^{3-\delta})^2}-\frac{tp_2^{(3-\delta)t}}{1-p_2^{3-\delta}}\right]\right)$$

$$N \geq \frac{t}{\delta}+mp_1\left[a\frac{1-p_2^{2t}}{1-p_2^2}-b\frac{p_2^2(1-p_2^{2t})}{(1-p_2^2)^2}+\frac{btp_2^{2t}}{1-p_2^2}\right.$$

$$+\alpha\left(a\frac{1-p_2^{(3-\delta)t}}{1-p_2^{(3-\delta)}}-b\frac{p_2^{(3-\delta)}(1-p_2^{(3-\delta)t})}{(1-p_2^{(3-\delta)})^2}+\frac{btp_2^{(3-\delta)t}}{1-p_2^{(3-\delta)}}\right)\right].$$

where

$$\mu=\frac{\kappa^2}{m}, \qquad \gamma=(s-3)\mu, \qquad i_0=\lfloor(s-3)\mu-1\rfloor,$$

$$j_0=\lfloor\kappa^2-1\rfloor, \qquad \beta=mp_1^{1-\delta}p_2^{\frac{2-\delta^2+\delta}{2}}, \qquad f=s-t-\frac{1}{2}\left(\frac{1}{\delta}+1\right)$$

$$a=(s-t+2)\left(\frac{1}{\delta}-\frac{1}{2}\right)-\frac{1}{2\delta}\left(\frac{1}{\delta}-1\right), \quad b=\left(\frac{1}{\delta}-1\right)\left(\frac{1}{\delta}-\frac{1}{2}\right), \quad \alpha=mp_1^{1-\delta}p_2^{\frac{3\delta-\delta^2+2}{2}}. \blacksquare$$

Corollary 1.1: The expected search is of order:

$$o(N)=ms\frac{s}{c} \quad \text{and} \quad O(N)=mts\frac{s}{c}\left(1+\frac{\kappa^2}{m}\right)^2\left(\kappa^2\frac{s}{m}\right)^{\lfloor\frac{s}{m}\kappa^2-1\rfloor}.$$

Proof: For both bounds, we identify, then simplify, the dominant terms:

$$o(N)=m\left(s-t+2-\frac{s}{2c}\right)\frac{s}{c}, \quad O(N)=mt\left(s-t-\frac{c+s}{2c}\right)\frac{s}{c}\left(\frac{m+\kappa^2}{m}\right)^2\left(\frac{\kappa^2s}{m}\right)^{\lfloor\frac{s}{m}\kappa^2-1\rfloor}. \blacksquare$$

$$\tag{1}$$

Corollary 1.2: If $s\kappa^2<2m$ then termination has an expected search of order

$$O(N)=amts\frac{s}{c} \quad \text{and} \quad o(N)=ms\frac{s}{c}. \blacksquare$$

4. Implications of the results

By Cor. 1.1, terminated search need not be polynomial, although it is reduced from normal constrained search. Cor. 1.2 implies that if the scene clutter is small enough relative to the model size, we do get a polynomial algorithm. When the scene clutter increases, however, we need to select (e.g. [9,8,11]) subsets of data features of size $s<\frac{2m}{\kappa^2}$ while still having at least t features from the object in the set.

This extends earlier results on the role of selection in efficient object recognition. For pure constrained search [2], knowing that all the data features are from a given object reduces the expected search to polynomial, but general constrained search is exponential [2]. This suggests that selection must perfectly isolate relevant data subsets, since if even one spurious point is included, either an exponential search results, or the entire feature subset is rejected. With termination, however, selection can allow an amount of spurious data bounded by the conditions of Cor. 1.2 and still have an efficient search method.

The constant κ depends on properties of the object model and the sensor [2]. Since κ increases with increasing data noise, the expected search also increases, and the amount of

	th = .3 m	.4 m	.5 m	.6 m	Full Search
Predicted lower bound	1,234	1,152	1,069	987	5.4×10^6
Actual nodes, average case, in theory	1,776	1,635	1,498	1,364	
Predicted upper bound	7,017	8,675	9,992	10,969	3.2×10^8
Median, using features	2,689	2,993	2,605	2,143	
Mean, using features	6,223	6,610	9,536	15,340	10^7
Deviation, using features	9,440	9,345	30,278	47,872	
Median, using perimeter	6,627	8,834	8,977	9,479	
Mean, using perimeter	19,437	16,307	23,297	38,362	10^7
Deviation, using perimeter	50,199	34,215	50,062	104,662	

Table 1:

spurious data tolerable decreases. Typically $\kappa \approx .2\frac{P}{D}$ where P is the total object perimeter (for 2D objects) and D is the image dimension. Given this, the conditions for a polynomial search are $s \leq 50m \left(D/P\right)^2$ so that considerable spurious data is still tolerable.

4.1 Comparing search results
We can extend the earlier analysis [2] as follows (proof in [3]):

Proposition 2: If the data from a correct interpretation are uniformly distributed among the spurious data, then normal constrained search is bounded by

$$m\frac{s}{c}2^c \leq N_{occ} \leq m\frac{s}{c}2^c + \frac{m}{\epsilon}\left[1 + \epsilon\right]^s \left[1 + \frac{p_2}{1+\epsilon}\right]^{c-1} + \frac{m^3 s}{\kappa^2}\frac{s}{c}\left[1 + p_2\right]^c . \blacksquare \qquad (2)$$

From Cor. 1.2, for small scene clutter, the expected search reduces to order

$$ms\frac{s}{c} \leq N_{term} \leq mts\frac{s}{c}.$$

Comparing with Prop. 2, heuristic search termination significantly reduces the search.

4.2 Consistency with real data
We also compare this analysis with real data. Features from a cluttered image were placed in 100 random orderings, and the RAF [6,7] system was for thresholds of $.3m, .4m, .5m$ and $.6m$, where m is the number of model features, and thresholds of $.3P, .4P, .5P$ and $.6P$, where P is the model perimeter, with appropriate measures of an interpretation. In this example, $m = 20, s = 35, c = 17$. Table 1 lists the predicted bounds (eqn (1)) and actual number of nodes, statistics for each case, and the predicted and observed number of nodes with no termination (eqn (2)).

Note that the derived bounds on the search correctly contain the actual search. Also, the median number of nodes searched, using number of features matched as a termination procedure, lies within the predicted bounds and is in close agreement with the actual theoretical number. The mean search is higher, as expected, since the analysis assumed a uniform distribution of correct data features among the spurious. The increase in search when more spurious data are among the first features examined is much larger than the decrease in search when more of the correct features are among the first few features.

We also applied RAF to 10 real images, with threshold $.3m$. Figure 1 plots (top to bottom) the predicted upper bound, observed median, predicted actual search, and predicted

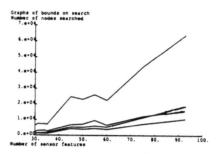

Figure 1:

lower bound, based on 100 trials, all as a function of the number of data features. While other factors can influence both the actual and predicted search, these graphs demonstrate that the predicted number is always between the bounds and is close to the lower bound, and that the observed number of nodes closely follows the prediction.

5. Conclusion

Heuristic termination of constrained search dramatically reduces the expected search in object recognition in cluttered noisy data. To obtain polynomial time algorithms, the ratio of scene clutter to object size must be small enough, and this implies that for significantly cluttered scenes, a selection method is needed to select out data subsets that are likely to include a subset arising from an instance of a known object. Moreover, such methods lead to low order polynomial performance, and to fast practical methods.

References

[1] Ayache, N. & O.D. Faugeras, 1986, "HYPER: A new approach for the recognition and positioning of two-dimensional objects," *IEEE Trans. PAMI*, **8**(1) , pp. 44–54.

[2] Grimson, W.E.L., 1989, "The combinatorics of object recognition in cluttered environments using constrained search," *Artificial Intelligence*, to appear.

[3] Grimson, W.E.L., 1989b, "The combinatoric of heuristic search termination for object recognition in cluttered environments," MIT AI Lab Memo 1111.

[4] Grimson, W.E.L. & D.P. Huttenlocher, 1988, "On the sensitivity of the Hough transform for object recognition," *Second Intl. Conf. on Computer Vision*, Tarpon Springs, FL., pp. 700–706.

[5] Grimson, W.E.L. & D.P. Huttenlocher, 1989, "On Choosing Thresholds for Terminating Search in Object Recognition," 1110, M.I.T. Artificial Intelligence Laboratory.

[6] Grimson, W.E.L. & T. Lozano-Pérez, 1984, "Model-based recognition and localization from sparse range or tactile data," *Int. Journ. Robotics Res.*, **3**(3) , pp. 3–35.

[7] Grimson, W.E.L. & T. Lozano-Pérez, 1987, "Localizing overlapping parts by searching the interpretation tree," *IEEE Trans. PAMI*, **9**(4) , pp. 469–482.

[8] Jacobs, D.W., 1988, "The Use of Grouping in Visual Object Recognition," 1023, M.I.T. Artificial Intelligence Laboratory.

[9] Lowe, D.G., 1985, *Perceptual Organization and Visual Recognition*, Boston, Kluwer Academic Publishers.

[10] Murray, D.W. & D.B. Cook, 1988, "Using the orientation of fragmentary 3D edge segments for polyhedral object recognition," *Intern. Journ. Computer Vision*, **2**(2) , pp. 153–169.

[11] Sha'ashua, A. & S. Ullman, 1988, "Structural saliency," *Second Intl. Conf. on Computer Vision*, Tarpon Springs, FL., pp. 321–327.

POSTERS

Combinatorial Characterization of Perspective Projections from Polyhedral Object Scenes

Jens Damgaard Andersen

Department of Computer Science

University of Copenhagen

DK-2100 Copenhagen Ø, Denmark

Abstract

Model-based computer vision systems which recognize objects in single gray-scale images require the matching of stored object models and the image data resulting from the perspective projection. If objects may be located arbitrarily in relation to each other occlusions can occur, thereby creating different line configurations in the projected image with different viewing directions. For a collection of polyhedral objects containing n vertices, there are of the order $O(n^3)$ different general views and $O(n^2)$ degenerate views and the algoritmic complexity for constructing a view list is $O(n^3)$, where n is the sum of polyhedral bounding faces part of the assembly convex hull and planes arising from visual interaction of polyhedral parts inside the convex hull.

1 Introduction

Recognition of 3-D objects from single perspective views in model based vision requires matching of 2-D line pictures of object scenes with stored 3-D model descriptions. By perspective transformation information is lost and automatic recognition is made difficult because objects may occlude one another in the projected image. Some object recognition methods match image with model by generating all possible line configurations arising from projection and checking for possible correspondence.

The aim of this communication is a combinatorial analysis of topological properties of perspectively projected scenes containing assemblies of polyhedral objects. Combinatorial properties of multi-object vertex-edge enumeration are investigated, using methods from computational geometry.

2 Method

A labelled planar graph corresponding to a line drawing of the object assembly obtained by perspective (central) projection is called an *image structure graph* [2] of the given viewpoint. For each junction in the line drawing there is a corresponding vertex in the graph, and for each line segment there is an edge between the vertices that corresponds to the endpoints of the line segment. Each vertex in the graph is labelled by the names of the edges of the object whose projections meet at the vertex, and each edge is labelled by the labels of its endpoint vertices. The *aspect* of the viewpoint is the topological structure of the image structure graph of that viewpoint. Thus, two different viewpoints have the same aspect if and only if the corresponding image structure graphs are isomorphic.

The method employed to assess the number of different aspects for a given polyhedral assembly is enumeration of configurations based on cell complexes induced by planes. The planes are 1) defined by the bounding faces of convex hull of the polyhedral assembly and 2) vertex–edge combinations from interior faces, i.e. not being part of the convex hull. The division of the viewing space by ruled quadrics resulting from interaction of three or more non-adjacent lines which are pairwise skew in $\mathcal{R}^3$ is not considered here, because they are rare compared to ordinary vertex-edge combinations.

3   Arrangements

A finite set H of planes in $\mathcal{R}^3$ induces a cell complex called the *arrangement* $\mathcal{A}(H)$ of H [1]. $\mathcal{A}(H)$ consists of four types of components: vertices (0-components), edges (1-components), faces (2-components), and cells (3-components). An arrangement $\mathcal{A}(H)$ of planes in $\mathcal{R}^3$ is called *simple* if 3 any three planes of H intersects in a point but four or more never do.

Upper bounds on the number of components in an arrangement is given in Edelsbrunner [1]. If $|H| = n$ then $\mathcal{A}(H)$ consists of at most $\binom{n}{3}$ vertices, $3\binom{n}{3} + \binom{n}{2}$ edges, $3\binom{n}{3} + 2\binom{n}{2} + n$ faces and $\binom{n}{3} + \binom{n}{2} + n + 1$ cells. These bounds are attained only if $\mathcal{A}(H)$ is simple. Each component can be identified by a position vector or *component word* $w(f) = (d_1, d_2, \cdots, d_n)$, where n is the number of planes in the arrangement. The r-th element of the position vector d_r indicates location above (+), on (0) or below (-) the r-th plane of the arrangement. Figure 1 shows a simple arrangement of four planes marked with position vectors of three cells, one face, one edge and one vertex.

4   Cell Carving Algorithm

The cell carving algorithm permits enumeration of aspects for non-convex polyhedra and polyhedral assemblies. Figure 2 illustrates the algorithm for a twodimensional example. The algorithm consists of the following steps:

1. Convex hull formation. Construct the convex hull of the vertex point set $V = \{v_1, \ldots, v_n\}$. During the procedure mark vertices as external points (e) if all surrounding faces are part of the convex hull, or as border points (b), if some surrounding faces are part of the hull, others not. In this case it is necessary to bridge a gap

to form the convex hull, the bridging polygon is called an *entrance*. Store entrances in a list and mark them when included.

2. Vertex points which are neither external nor border points are marked interior points (*i*).

3. Form the arrangement of the point set consisting of interior and border points. While forming the arrangement check if planes intersect at least one entrance, and if they do not, they are not included in the arrangement. The planes included are called the *interaction planes*.

4. For each entrance in turn: Proceeding by levels in the arrangement ([1], p. 47) check a visibility predicate for each new cell. A cell in level k is not visible if there are no visible neighbours in the k-level or (k-1)-level.

5. Finally the space outside the convex hull is divided into aspect regions by the arrangement consisting of planes defined by faces being part of the convex hull and interaction faces.

5 Complexity

Assume that the polyhedral assembly is initially provided as a list of edges for each object. Each list contains pairs of points defining an edge along with coordinate information for points. Points are sorted according to their first coordinate (degenerate cases are handled as in [1]) and the convex hull of the point set corresponding to the object assembly vertices can then be constructed in $O(n_v log n_v)$ time and $O(n_v)$ storage, where n_v is the total number of vertex points. After the procedure interior and border points are known. Each subset of 3 points from the set of interior and border points defines a plane. Assuming n_i interior and n_b border points, and n_f bounding faces being part of the convex hull, there are $n = n_f + \begin{pmatrix} n_i + n_b \\ 3 \end{pmatrix}$ planes defining the aspect arrangement. Constructing an arrangement from these n planes in $\mathcal{R}^3$ can be done optimally in time $O(n^3)$ ([1], theorem 7.6, page 135).

From the upper bounds on cells and faces (in section 3) it follows that the viewing space is partitioned in $O(n^3)$ aspects regions and since there are $O(n^2)$ dividing planes the order of degenerate views is $O(n^2)$.

References

[1] Edelsbrunner, H. *Algorithms in Combinatorial Geometry*. EATCS Monographs on Theoretical Computer Science, Volume 10. Berlin: Springer-Verlag 1987.

[2] Gigus, Z., Canny J., and Seidel, R. Efficiently Computing and Representing Aspect Graphs of Polyhedral Objects. *Second Int. Conf. on Computer Vision, Tarpon Springs, Florida.* 30-39 (1988).

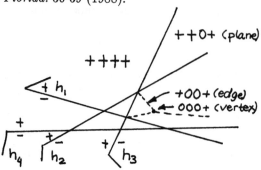

Fig. 1. Simple arrangement consisting of four planes with three cells, one face, one edge, and one vertex labelled with component words.

Interaction lines: lines arising from interaction in the interior:

v_{11}, v_{12}	side	1
v_{11}, v_{13}	discarded	-
v_{11}, v_{21}	entrance	2
v_{11}, v_{22}	discarded	-
v_{11}, v_{23}	arrangement line	3
v_{12}, v_{13}	side	4
v_{12}, v_{21}	discarded	-
v_{12}, v_{22}	arrangement line	5
v_{12}, v_{23}	arrangement line	6
v_{13}, v_{21}	discarded	-
v_{13}, v_{22}	discarded	-
v_{13}, v_{23}	entrance	7
v_{21}, v_{22}	discarded	-
v_{21}, v_{23}	discarded	-
v_{22}, v_{23}	side	8

Convex hull: lines formed by bounding sides and entrances:

v_{11}, v_{14}	side	9
v_{14}, v_{13}	side	10
v_{13}, v_{23}	entrance	7
v_{23}, v_{24}	side	11
v_{24}, v_{21}	side	12
v_{21}, v_{11}	entrance	2

Fig. 2. Illustrating convex hull, external points (e), border points (b), and interior points (i).

Estimation of Curvature in 3D Images Using Tensor Field Filtering

H. Bårman   G.H. Granlund   H. Knutsson

Linköping University
Department of Electrical Engineering
Computer Vision Laboratory
S-581 83 Linköping Sweden

Abstract

This paper describes an algorithm for estimation of directionality in 2D and 3D vector fields and how that feature relates to the curvature of curves in 2D images and surfaces in 3D images.

One of the main properties of the method is that no thresholding is required. It consists of two steps. First the grey level image/volume is filtered with a number of filters to obtain a tensor description of the local orientation. Secondly the tensor image/volume is filtered with a number of filters to achieve the local direction description.

1   Introduction

Earlier papers (e.g [2]) have presented an methodology where a vector data representation is used for curvature estimation in 2D images. Here this algorithm is modified to work on data represented as tensors. This modification enables the generalization from 2D to 3D.

The curvature concept is well–known from vector analysis and differential geometry [7]. We will denote the curvature of a 2D curve with κ and the tangent of the curve with $\vec{t}$. Surfaces have a direction of most curvature as well as a direction of least curvature, and these directions are, apart from being perpendicular to the normal vector of the tangent plane of the surface, also perpendicular to each other. We will denote the two directions (*the principal directions*) as $\vec{k_1}$ and $\vec{k_2}$, while the amounts of curvature (*the principal curvatures*) will be denoted κ_1 and κ_2.

There exists a variety of different curvature estimation and description algorithms, e.g. [1, 3, 5, 6]. The new algorithm presented here differs from standard curvature algorithms in two very important aspects. First no thresholding is required. Secondly the detection is done hierarchically in two steps, where erroneous local orientation information is suppressed (as opposed to 'eliminated' which is the case in other two–step algorithms with thresholding) before the

actual curvature estimation takes place.

2   Orientation Estimation

The first step is to achieve a local orientation estimate. The algorithm utilizes the observation that a neighbourhood with one dominant orientation has the energy in the Fourier domain concentrated around a line through the origin orientated at the orientation (or gradient direction) $(x, y)^T$ ($(x, y, z)^T$ for 3D images). A number of *quadrature filters* are applied on the grey level image/volume, where each filter is concentrated in a specific partition of the Fourier domain. The dominant local orientation is achieved by

$$f_1(\xi_x, \xi_y) = \sum_{k=1}^{K} q_k(\xi_x, \xi_y)(\mathbf{T}_k - \alpha\mathbf{I}) \qquad (1)$$

where f_1 denotes the obtained tensor image, ξ_x and ξ_y are spatial coordinates (add ξ_z for 3D), K is the number of filters, q_k denotes the magnitude of the filter response and $\mathbf{T}_k$ is the direction of the filter in the representation domain. This results in a tensor representation which for a dominant orientation $(x, y)^T$ (or $(x, y, z)^T$) equals

$$\mathbf{T}_{2D} = \begin{pmatrix} x^2 & xy \\ xy & y^2 \end{pmatrix} \quad \mathbf{T}_{3D} = \begin{pmatrix} x^2 & xy & xz \\ xy & y^2 & yz \\ xz & yz & z^2 \end{pmatrix} \qquad (2)$$

$\mathbf{T}_k$ in Eq. (1) is computed by expressing the Fourier domain direction of the filter (the direction where the filter is concentrated) as a cartesian vector and put it into Eq. (2). The filters are evenly spaced over a half of the Fourier space. [4] describes the algorithm in detail. It should be noted that this representation is continuous and contains a certainty measure.

3   Curvature Estimation in 2D

The curvature property of curves in the original image is transferred into the orientation image

$f_1(\xi_x, \xi_y)$. It turns out that estimation of curvature direction can be made in the same manner as the estimation of local orientation, i.e. by a summation of the magnitude responses of a number of filters, where each filter is concentrated in a partition of the Fourier domain.

It is in [2] shown that a 2D–neighbourhood with one dominant curvature will have a local Fourier spectra where the centre of gravity is located in the direction of $\vec{t}$ provided that the pixel values are complex and equals

$$(x^2 - y^2) + i2xy. \tag{3}$$

This implies that the following formula (see Fig. 1) can be used

$$f_2(\xi_x, \xi_y) = \sum_{k=1}^{K} q_k(\xi_x, \xi_y)\mathbf{n}_k \tag{4}$$

The vector achieved from Eq. (4) will coincide with the direction of the tangent of the curve in the grey level image. The magnitude of the vector relates to the curvature κ of the curve. The magnitude q_k of the tensor field filtering in Eq. (4) is computed as

$$\begin{aligned} q &= [(h_e * x^2 - h_e * y^2 + 2h_o * xy)^2 + \\ &\quad (h_e * xy - h_o * x^2 + h_o * y^2)^2]^{\frac{1}{2}} \end{aligned} \tag{5}$$

where h_e and h_o are the even and odd parts of the quasi–quadrature filter. Observe that the entire Fourier domain is covered by filters as opposed to the case of orientation estimation, where it is enough to cover half of the Fourier domain. A quasi–quadrature filter is defined in the Fourier domain as

$$H(\mathbf{u}) = \mathbf{H}_\rho(\mathbf{u}) \cos^{2A} \frac{\phi}{2} \tag{6}$$

where $\mathbf{u}$ is the frequency coordinate vector, u is the length of the vector and

$$\phi = \arccos(\frac{\mathbf{n}_k \cdot \mathbf{u}}{u}) \tag{7}$$

and $\mathbf{n}_k$ is a unit vector determining the main direction of the filter. H_ρ describes the frequency characteristics. The parameter A specifies how concentrated the filter is with respect to its main direction.

Note that the simple formulation of Eq. (4) results in the following features.

- The magnitudes in the neighbourhood are taken into account so that only relevant parts (pixels on the curve) have effect on the computation.

- the gradient of $\varphi(x, y)$ is estimated for those pixels.

- The magnitude of the estimate $f_2(\xi_x, \xi_y)$ contains information about the certainty of the input data, i.e. the quality of the orientation estimates in the neighbourhood, as well as information about the fit to the curvature model.

- The curvature magnitude κ is implicitly reflected through the magnitude of $f_2(\xi_x, \xi_y)$ and the frequency characteristics of the filters used.

The algorithm can be modified to take into account that the tangent and gradient of a curve are perpendicular, i.e.

$$\arg(f_1'(\xi_x, \xi_y)) = 2\arg(f_2(\xi_x, \xi_y)) + \pi \tag{8}$$

Neighbourhoods not fulfilling Eq. (8) are not of curve/curvature type and the direction of the vector field corresponds to another type of event, e.g. line ends.

4 The algorithm in 3D

The interpretation of the orientation tensor as a complex number (Eq. (3) can in the 3D–case be done in three different ways. (Substitute x or y in Eq. 3 with z.) Applying Eq. (4) on the three interpretations results in three different 3D–vectors. It can be shown (proof omitted) that the vectors will point in the principal direction (or in the opposite direction) of most curvature provided that there is one dominant curvature direction in the neighbourhood.

Experiments have shown correct estimation of curvature direction and a reasonable total (the three estimates combined) magnitude invariance of the curvature direction. The scheme has been able to keep track of the weaker 'least curvature direction'. Even the surface of a sphere, with two equal strength curvatures, is handled correctly.

5 The Inverse

The inverse (the transformation from the curvature description to the principal directions $\vec{k_1}$ and $\vec{k_2}$) is obtained by summing the outer products of the three 3D–vectors (denoted $\mathbf{b}_x, \mathbf{b}_y$ and $\mathbf{b}_z$) and computing the eigenvalues of the obtained matrix.

$$\sum_{k=x}^{z} \mathbf{b}_k \mathbf{b}_k^T \tag{9}$$

The eigenvector of the largest eigenvalue determines, apart from the sign, $\vec{k_1}$. The eigenvector of the second largest eigenvalue determines, also apart from the sign, $\vec{k_2}$. The third eigenvalue should for a well defined surface be close to zero and the local normal

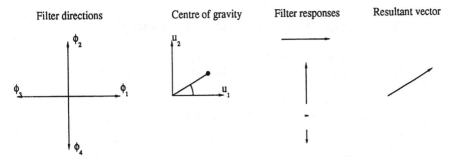

Figure 1: A stylized example of Eq. (4).

vector should be perpendicular to the two curvature directions. The sign of $\vec{k_1}$ and $\vec{k_2}$ is obtained by checking the directions of the three 3D-vectors.

6 Conclusion

A new algorithm for 3D curvature description has been presented. It is a hierarchical non-thresholding method, where the curvature is estimated on a gradient-equivalent image derived from the grey level volume (or time sequence). Both steps are performed without thresholding and with convolution as base operation. The algorithm output consists of a continuous representation constituted by three different 3D vectors. This representation can be translated into the two principal curvature directions.

Acknowledgment

This work was supported by the National Swedish Board for Technical Development.

References

[1] H. Asada and M. Brady. The curvature primal sketch. *IEEE Trans. Pattern Anal. Machine Intell.*, PAMI-8(1):2–14, January 1986.

[2] H. Bårman, G. H. Granlund, and H. Knutsson. A new approach to curvature estimation and description. In *3rd International Conference on Image Processing and its applications*, pages 54–58, Warwick, Great Britain, July 1989. IEE. ISBN 0 85296382 3 ISSN 0537–9989.

[3] E.D. Dickmanns and A. Zapp. A curvature-based scheme for improving road vehicle guidance by computer vision. In William J. Wolfe and Nelson Marquina, editors, *Mobile Robots*, pages 161–168. SPIE, Bellingham, 1987. vol. 727.

[4] Hans Knutsson. Representing local structure using tensors. In *The 6th Scandinavian Conference on Image Analysis*, pages 244–251, Oulu, Finland, June 1989. Report LiTH-ISY-I-1019, Linköping University, Sweden.

[5] Jan J. Koenderink. An internal representation for solid shape based on the topological properties of the apparent contour. In Whitman Richards and Shimon Ullman, editors, *Image Understanding 1985–86*, chapter 9, pages 257–285. Ablex Publishing Corporation, 1987.

[6] P. Parent and S. W. Zucker. Trace inference, curvature consistency, and curve detection. *Pattern Analysis and Machine Intelligence*, PAMI–11(8), August 1989.

[7] Michael Spivak. *A Comprehensive Introduction to Differential Geometry*, volume 2. Publish or Perish, Inc., 2nd edition, 1979.

TRANSPARENT-MOTION ANALYSIS

James R. Bergen   Peter J. Burt
Rajesh Hingorani   Shmuel Peleg*

David Sarnoff Research Center
Princeton, NJ 08543-5300, U.S.A.

Abstract

A fundamental assumption made in formulating optical flow algorithms is that motion at any point in an image can be represented as a single pattern component undergoing a simple translation: even complex motion will 'look like' uniform displacement when viewed through a sufficiently small window. This assumption fails for a number of situations that commonly occur in real world images. For example, transparent surfaces moving past one another yield multiple motion components at a point.

We propose an alternative formulation of the local motion assumption in which there may be two distinct patterns undergoing different motions within a given local analysis region. We then present an algorithm for the analysis of transparent motion.

1   Models for Local Motion

Motion estimation is based, ultimately, on an assumed model relating motion to observed image intensities. The traditional model used in optical flow computation postulates a single pattern moving uniformly within any local analysis region [7, 9, 4]. We introduce a new model that postulates two such components.

Let $I(x, y, t)$ be the observed gray scale image at time t. Let R be the analysis region in which we wish to estimate motion. The traditional model used in optical flow analysis [8, 1, 6] assumes that within the region R, $I(x, y, t)$ may be represented as a pattern $P(x, y)$ moving with velocity $\mathbf{p} = (p_x, p_y)$.

$$I(x, y, 0) = P(x, y) \qquad \text{and} \qquad I(x, y, t) = P(x - tp_x, y - tp_y) = P^{t\mathbf{p}} \tag{1}$$

where $P^{t\mathbf{p}}$ denotes the pattern P transformed by the motion $t\mathbf{p}$.

We introduce an alternative model for local motion. Within the analysis region the image is assumed to be a combination of two distinct image patterns, P and Q, having independent motions of $\mathbf{p}$ and $\mathbf{q}$:

$$I(x, y, 0) = P(x, y) \oplus Q(x, y) \qquad \text{and} \qquad I(x, y, t) = P^{t\mathbf{p}} \oplus Q^{t\mathbf{q}}. \tag{2}$$

Here the $\oplus$ symbol represents an operator such as addition or multiplications to combine the two patterns.

2   Estimating Multiple Motion

We now consider the analysis of motion described by the multiple component model. Alternative approaches have been proposed that simultaneously estimate multiple component motion without segmentation. Examples include the use of Hough transform techniques and cross correlation [4, 5]. However, these are computationally complex, and do not provide precise results.

---

*Permanent address: Dept. of Computer Science, The Hebrew University of Jerusalem, 91904 Jerusalem, Israel.

A key observation for the present approach is that if one of the motion components and the combination rule $\oplus$ are known, it is possible to compute the other motion using a single component motion algorithm without making any assumptions about the nature of the patterns P_i. In what follows we will assume that the combination operation is addition.

Suppose, for the moment, that motion $\mathbf{p}$ is known, so that only motion $\mathbf{q}$ must be determined. The pattern component P moving at velocity $\mathbf{p}$ can be removed from the image sequence by shifting each image frame by $\mathbf{p}$ and subtracting it from the following frame. The resulting sequence will contain only patterns moving with velocity $\mathbf{q}$.

Let D_1 and D_2 be the first two frames of this difference sequence. From Equation (2):

$$
\begin{aligned}
D_1 &\equiv I(x,y,2) - I^{\mathbf{p}}(x,y,1) = (P^{2\mathbf{p}} + Q^{2\mathbf{q}}) - (P^{2\mathbf{p}} + Q^{\mathbf{q}+\mathbf{p}}) \\
&= Q^{2\mathbf{q}} - Q^{\mathbf{q}+\mathbf{p}} = (Q^{\mathbf{q}} - Q^{\mathbf{p}})^{\mathbf{q}}
\end{aligned} \tag{3}
$$

$$
\begin{aligned}
D_2 &\equiv I(x,y,3) - I^{\mathbf{p}}(x,y,2) = (P^{3\mathbf{p}} + Q^{3\mathbf{q}}) - (P^{3\mathbf{p}} + Q^{2\mathbf{q}+\mathbf{p}}) \\
&= Q^{3\mathbf{q}} - Q^{2\mathbf{q}+\mathbf{p}} = (Q^{\mathbf{q}} - Q^{\mathbf{p}})^{2\mathbf{q}}
\end{aligned}
$$

The sequence $\{D_n\}$ now consists of a new, fixed, pattern $Q^{\mathbf{q}} - Q^{\mathbf{p}}$ moving with a single motion $\mathbf{q}$, that is: $D_n = (Q^{\mathbf{q}} - Q^{\mathbf{p}})^{n\mathbf{q}}$. Thus the motion $\mathbf{q}$ can be computed from the two difference images D_1 and D_2 using a single motion estimation technique.

In an analogous fashion the motion $\mathbf{p}$ can be recovered when $\mathbf{q}$ is known. The observed images $I(x,y,t)$ are shifted by $\mathbf{q}$, and a new difference sequence is formed:

$$
D_n = I(x,y,n+1) - I^{\mathbf{q}}(x,y,n).
$$

This sequence is the pattern $P^{\mathbf{p}} - P^{\mathbf{q}}$ moving with velocity $\mathbf{p}$: $D_n = (P^{\mathbf{p}} - P^{\mathbf{q}})^{n\mathbf{p}}$, so $\mathbf{p}$ too can now be recovered using a single motion estimation.

In practice, of course, neither motions $\mathbf{p}$ or $\mathbf{q}$ are known a priori. Still it is possible to recover both motions precisely if we start with even a crude estimate of either. Multiple component motion analysis can therefore be formulated as a two (or n) component iterative refinement procedure. Let $\mathbf{p}_n$ and $\mathbf{q}_n$ be the estimates of motion after the nth cycle. Estimates alternate between $\mathbf{p}$ and $\mathbf{q}$, so if $\mathbf{p}$ is obtained on even numbered cycles, $\mathbf{q}$ is obtained on odd cycles. Steps of the procedure are:

1. Set an initial estimate for the motion $\mathbf{p}_0$ of pattern P.

2. Form the difference images D_1 and D_2 as in Equation (3) using the latest estimate of $\mathbf{p}_n$.

3. Apply a single motion estimator to D_1 and D_2 to obtain an estimate of $\mathbf{q}_{n+1}$.

4. Form new difference images D_1 and D_2 using the estimate $\mathbf{q}_{n+1}$.

5. Apply the single motion estimator to the new sequence D_1 and D_2 to obtain an update $\mathbf{p}_{n+2}$.

6. Repeat starting at Step 2.

In the cases that we have tried convergence of this process is fast: with artificially generated image sequences, the correct transformations are recovered to within roughly 1% after three to five cycles regardless of the initial guess of $\mathbf{p}_0$.

3   Examples of Multiple Motion Analysis

We have tested the transparent motion algorithm with several examples. In all examples, the analysis region R is taken to be the entire image.

  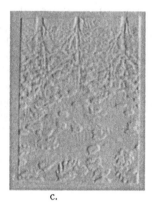

a.   b.   c.

Figure 1: Transparent motion. A viewer face is reflected in a picture, with images taken from a moving camera.

a) One frame from the sequence.
b) Subtracting two consecutive frames after compensating for $\mathbf{p}$, the motion of the picture.
c) Subtracting two consecutive frames after compensating for $\mathbf{q}$, the motion of the reflected face.

Example 1: Transparent Motion

An example involving additive transparency is shown in Figure 1. In this case a sequence was captured with a moving video camera showing a face reflected in the glass covering a print of Escher's "Three Worlds". A single frame from this sequence is shown in Figure 1a. As the camera moved, the image reflected in the glass and the image in the print moved differently. These two motions were computed from this sequence and used to produce the compensated difference images shown in Figure 1b and Figure 1c. In Figure 1b the reflected image (barely visible in Figure 1a) is revealed showing that the other component was accurately registered. In Figure 1c, the reflected image has been nulled.

Example 2: Multiple Aperture Effect

An example involving both transparency and multiple aperture effects is shown in Figure 2. The image sequence in this case consists of the sum of two squares moving diagonally in opposite directions. In this case, the actual motions were $(2.0, 2.0)$ and $(-2.0, -2.0)$. The estimated velocities after 2 iterations were accurate to machine precision. For comparison, an optical flow computation was also made on this sequence using a previously described 'warp motion' technique [2]. The resulting flow field is shown in the lower right. Note that only at the corners of the square do the estimated velocities correspond to the actual motions. Attempts to resolve this complex flow field into correct estimates of the motions would be complicated by the presence of the two differently moving objects.

4   Concluding Remarks

A method has been presented for detecting multiple components of motion within an image region. This technique is based on a two component model of local image motion, which is a generalization of the single component model implicit in standard optical flow computation. The technique does not require segmentation to obtain precise motion estimates. Instead, it relies on an iterative process in which each estimate of one component of the motion is used to improve the accuracy of the other. This allows the motions to be estimated accurately without explicitly knowing their corresponding

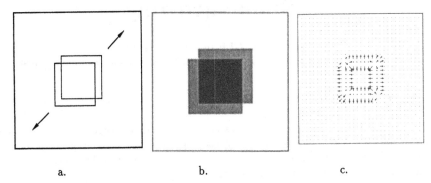

a.   b.   c.

Figure 2: Multiple motions with aperture effects.

a) Input configuration.

b) One frame from sequence. The multiple motion analysis technique accurately extracts both motions.

c) Optical flow field computed from two frames of the sequence. Note aperture effects.

pattern components. This work has been extended to additional cases of multiple motion as near occluding edges [3].

References

[1] P. Anandan. A unified perspective on computational techniques for the measurement of visual motion. In *International Conference on Computer Vision*, pages 219–230, London, May 1987.

[2] J.R. Bergen and E.H. Adelson. Hierarchical, computationally efficient motion estimation algorithm. *J. Opt. Soc. Am. A.*, 4:35, 1987.

[3] J.R. Bergen, P.J. Burt, R. Hingorani, and S. Peleg. Multiple component motion: Motion estimation. Technical report, David Sarnoff Research Center, January 1990.

[4] C.L. Fennema and W.B. Thompson. Velocity determination in scenes containing several moving objects. *Computer Graphics and Image Processing*, 9:301–315, 1979.

[5] B. Girod and D. Kuo. Direct estimation of displacement histograms. In *Image Understanding and Machine Vision*, pages 73–76, Cape Cod, June 1989. Optical Society Of America.

[6] D.J. Heeger. Optical flow using spatiotemporal filters. *International Journal of Computer Vision*, 1:279–302, 1988.

[7] B.K.P. Horn. *Robot Vision*. MIT Press, 1986.

[8] B.K.P. Horn and E.J. Weldon. Direct methods for recovering motion. *International Journal of Computer Vision*, 2(1):51–76, June 1988.

[9] J.O. Limb and J.A. Murphy. Estimating the velocity of moving images in television signals. *Computer Graphics and Image Processing*, 4(4):311–327, December 1975.

Snake growing

Marie-Odile BERGER (e-mail: berger@loria.crin.fr)

CRIN / INRIA Lorraine Campus scientifique, B.P. 239 , 54506 Vandœuvre Cedex , France

1   Introduction

Standard edge detectors usually run in two steps: (i) Detection of edge points; (ii) linking of those points to make a coherent edge feature. Optimal edge detectors were designed ([Can 86]) but this was mainly done on the basis of signal processing aspects and such detectors are often criticized for their failure to detect the most salient edges.

New algorithms based on Active Contour Models have been introduced in [Kas 88]. They provide a global view of edge detection. An active contour model, called a snake, is an elastic curve C which is moving under the influence of the potential energy created by the image gradient. The minima of the snake functional energy are the edges. Using the regularization theory [Pog 85], they are searched for in the restricted class of controlled continuity splines[Ter 86].

The snake functional energy is the sum of two terms: $E = w_1 E_{int} + w_2 E_{ext}$. The Internal energy E_{int} describes features of the curve $C = v(t), E_{int} = \int_C (\alpha|v'|^2 + \beta|v''|^2)$. (Parameters α and β influence the elasticity or stiffness of the curve). The External energy E_{ext} depends on the feature which is searched for in the image (dark lines, white lines, edges, termination of line segments). In edge detection, we use $E_{ext} = \int_C (-|\nabla I(v(t))|)dt$ where I is the image intensity.

The numerical minimization of E is however problematic [Ami 88] because of numerical instability and of numerous parameters in the functional E.

2   Snake growing

Most encountered problems and particularly the instability, are due to the application of the method in bad conditions; for instance, when several parts of the curve have very different behaviors or when several contours (minima of E) lie in the vicinity of the initialization. The snake is influenced by those edges and reaches an apparent equilibrium position. The algorithm is then stopped without reaching a minimum.

2.1   Bootstrapping

Nevertheless, some parts of the curve can be considered as an edge when the whole curve is not. According to an assessment criterion, the snake is then cut into edge subcurves which will be used as bootstrapping edges. So, we develop a method based on the active contour principle using local strategies: a sequence S_i of snakes is built incrementally by successive lengthenings from a small part of an edge, which is either interactively given or inferred by cutting the snake as mentioned above. The sequence S_i is built so as to be always in the neighborhood of a unique contour.

2.2   The basic idea

Let us suppose that we have a curve S_0, even small, which is a contour or is very close to a contour. Then, lengthening the extremities of the curve in the tangent direction gives rise to a curve S_1 which is also close to the contour. Using S_1 as an initialization, the method will converge quickly towards Cl_1 as S_1 is near a local minimum of E. Cl_1 quality is then estimated and the method is iterated.

The Snake growing algorithm can be summarized as:

- *start from a curve S_0 close to a contour,*
- *run the snake algorithm with S_0 as initialization, which yields C_0.*
- *While lengthening is possible, do:(build a sequence C_i of contours whose lengths increase)*
    - *lengthen C_i in the tangent direction to have the initialization curve S_i.*
    - *running the traditional algorithm which converges towards Cl_{i+1}*
    - *assessment of the curve Cl_{i+1} which yields C_{i+1} and the result quality.*

The lengthening of the snake can be interactively selected. It can also be chosen in an adaptive manner in accordance with the snake quality at the previous steps: If the snake quality is good, lengthening can be increased, whereas it is decreased when the estimation quality is bad.

The main advantage of snake growing is that we are at each step in conditions such that the algorithm converges in a satisfying and fast manner because we built an initialization very close to a unique contour. The interest of such a method is obvious when rectilinear contours or contours which present a weak curvature variation have to be detected. This method is also well suited for contour tracking.

2.3 Results and multisnake strategy

Results of snake growing are shown in (Fig.1) on noised and badly contrasted angiographic images. We have chosen to follow an artery using the growing method. Some significant results are exhibited at different steps of the growth.

Automatic Speech Recognition can take advantage of the active contour models. (Fig.2a) shows a cepstrally smoothed spectrogram of the phoneme sequence "soleil". One of the major challenges in acoustic-phonetic decoding is to track the vocal tract resonance frequencies which appear as black lines (maxima of energy) on the spectrogram. (Fig.2) illustrates a **multiple snake growing**. A rectilinear initialization and the resulting curve C after 5 iterations can be seen (Fig.2a). The gradient along this curve is produced in (Fig.2b). The assessment procedure leads to cut the curve C into tree subcurves which are used as initialization of the growth process. The three formant tracks have been correctly caught by this "multisnake" approach as (Fig.2c) shows.

2.4 Discussion

A simple but efficient improvement consists in lengthening the snake in the discrete curvature direction [Lux 85]. This method is particularly efficient when curves whose curvature varies regularly as in smoothed shapes like fingerprint images have to be detected.

Finally, what are the drawbacks of the method ? Our method seems to be more expensive than the traditional algorithm because we have to inverse a matrix at each lengthening (whose size is the snake length), whereas we have a unique matrix to inverse in the traditional method. However, as the sequence S_i is built in such a manner that S_i is close to a contour, the algorithm is more stable and converges more quickly. Nevertheless, it seems difficult to compare the theoretical value of the costs of the two algorithms as they depend on image and on initialization.

3 Conclusion

In this article, we have shown that some local strategies (snake cutting, snake growing, multiple snake) can be used in order to enhance the algorithm efficiency, even in noisy surrounding. This kind of method is particularly powerful if we have at disposal knowledge on the searched contours which permit to guide the local strategies. Nevertheless, some problems still remain unsolved in active contour models. The influence of the miscellaneous parameters on the snake behavior and on the accuracy must be clarified. It would be also suitable to be able to merge snake and modeling in order to detect edges whose features are given.

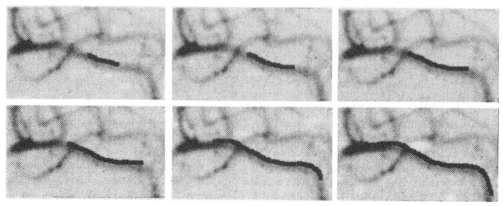

Figure 1: Growing snake on angiographic images.

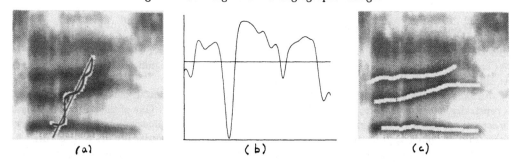

(a)   (b)   (c)

Figure 2: Formant tracking using multiple lengthenings and the gradient curve along the snake.

Acknowledgment: I would like to thank Roger Mohr for his fruitful suggestions.

References

[Ami 88] A. Amini, S. Tehrani, and T. Weymouth. Using Dynamic Programming for minimizing the energy of active contours in the presence of hard constraints. In *Second International conference on computer vision*, pages 95–99, Tampa Florida (USA), 1988.

[Can 86] J. Canny. A Computational Approch to Edge Detection. *IEEE Transactions on Pattern Analysis and Machine Intelligence*, 679–698, november 1986.

[Kas 88] M. Kass, A. Witkin, and D. Terzopoulos. Snakes: Active Contour Models. *International Journal of Computer Vision*, 321–331, september 1988.

[Lux 85] A. Lux. *Algorithmique et controle en vision par ordinateur*. Thèse d'Etat, Institut National Polytechnique de Grenoble, 1985.

[Pog 85] T. Poggio, V. Torre, and C. Koch. Computational Vision and Regularization Theory. *Nature*, 314–319, 1985.

[Ter 86] D. Terzopoulos. Regularization of Inverse Visual Problems Involving Discontinuities. *IEEE Transactions on PAMI*, 8:413–424, 1986.

Local Cross-Modality Image Alignment Using Unsupervised Learning

Öjvind Bernander and Christof Koch

Computation and Neural Systems Program, 216-76
California Institute of Technology, Pasadena, Ca 91125

Abstract

We propose a method for automatically aligning images with local distortions from different sensors, using real images instead of calibration objects. The algorithm has three components. First, we extract intensity discontinuities, because this is a feature that is likely to show up across modalities. Second, we use a correlation scheme that averages over time rather than space, for high precision. Third, we propose an architecture and a learning scheme that learn the correlation surfaces over time and implement the image coordinate transform.

Introduction and problem definition

Fusion, or integration, of information from different sensors is believed to facilitate object recognition. The sensors may be of different modalities, e.g. video, infra-red or laser range cameras. Before fusion can occur, however, the images must be properly aligned. Thus the problem is defined: *given two cameras at two positions with overlapping fields-of-view, find the coordinate transform that will align the overlapping portions.* We represent the transform using a *shift field*, a vector field sharing some formal properties with the optical flow field. Note that the misalignment generally will vary across the image, due to rotation, zoom, and local distortions. The simplest approach is to use special calibration objects, e.g. *hot corners*, to produce calibration images, and then interpolate between these points. This approach presents problems in remote-control or autonomous situations (e.g. for a Mars rover or for biological visual systems) when calibration objects are not likely to be at hand. This is the motivation for developing an algorithm that achieves image alignment using natural images.

A correlation scheme for image alignment

One problem with using different modalities is that there is expected to be little correlation between intensity values, since reflectance, temperature, and distance correlate only to a very small degree. We therefore need to extract features that are likely to show up in any modality. Edges, or intensity discontinuities, have this property, since they often arise at object boundaries. This was recognized by Barniv and Casasent [1], who studied *full-image* correlations between images that were captured with the same camera, but with different filters. Correlating edge maps yielded a clear peak, the position of which gives any shift between the two images.

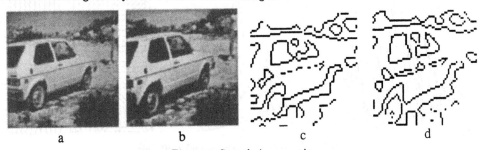

a   b   c   d

Figure 1: Sample image pairs.
(a) and (b) depict the same scene, but are captured using two video cameras of different makes. (c) and (d) show the thresholded zero-crossings of (a), after filtering with the Laplacian of a Gaussian.

In the general case, however, the two sensors do not have identical fields-of-view and resolutions. While a full-image correlation will work for pure translations, it will not work for rotations and zooms (different resolutions). To remedy this we average in time rather than space. Let us define the *atomic correlation function* $\mathbf{a}_{i,j} = m_{i,j}^{(1)} * m_{i,j}^{(2)}$, where $m_{i,j}^{(1)}$ is the pixel value at (i,j) in image 1 and $m_{i,j}^{(2)}$ is a patch of pixels in image 2. $\mathbf{a}_{i,j}$ defines a correlation surface that is local in both space and time. Averaging $\mathbf{a}_{i,j}$ over the whole image, we get a full-image correlation. Averaging $\mathbf{a}_{i,j}$ over a patch gives patch correlation. We average $\mathbf{a}_{i,j}$ in time over a set of image pairs, sometimes in combination with a small amount of spatial averaging over 2x2 or 3x3 squares, in order to reduce the number of image pairs we need. Hence we get one correlation surface at every pixel position in image 1. We use a 2-D parabolic fit to find the position of the peak of the correlation surface, using sub-pixel accuracy. This position vector gives the local misalignment and the set of all position vectors defines a *shift field*, which gives us the desired coordinate transform.

To generate a database of 180 video image pairs, we directed two cameras of different makes towards a screen upon which vacation slides were projected. The digitized images were filtered with the Laplacian of a Gaussian, and edges were marked at the zero-crossings. A sample image pair is shown in figure 1. Figure 2(a-b) shows sample correlation surfaces and 2(c) the shift field displayed as a needle diagram. One camera was zoomed-in compared to the other, which clearly shows in figures 1 and 2(c). The average error in peak position was very low. For auto-alignment, i.e. when images 1 and 2 were identical and the true misalignment was known, the average error in peak position was less than 0.1 pixels. Figure 3 shows how the average error in peak position varies with the number of images used and with the amount of spatial averaging. For cross-alignment, i.e. when images 1 and 2 were different and the true misalignment was not known, we estimated the standard deviation to $\sigma = 0.2$ pixels. To assess the robustness to noise we generated artificial Mondrian images and added various amounts of salt-and-pepper noise. We found that even at very low signal to noise ratios (SNR = 1.0) the average position error was less than 0.2 pixels and the width of the correlation peak increased by 50%.

An architecture for learning and implementing the coordinate transform

One way to implement the shift would be to use the setup depicted in figure 4. A patch $\mathbf{m}^{(2)}$ of neurons in image 2 project via a set of weights $\mathbf{w}$ (referred to as a receptive field) to a neuron with output $y = \mathbf{m}^{(2)} \cdot \mathbf{w}$. If only one component of $\mathbf{w}$ is non-zero, the corresponding component of $\mathbf{m}^{(2)}$ will be shifted into y. We can approximate the ideal receptive field $\mathbf{w}$ with the correlation surface. A similar approach was suggested by [4]. With $\mathbf{a}$ defined as above, i.e. $\mathbf{a}_{i,j} = m_{i,j}^{(1)} * m_{i,j}^{(2)}$, the learning rule $\dot{\mathbf{w}} = \alpha\mathbf{a} + \beta y\mathbf{w}$ (α and β positive constants) will converge to $\mathbf{w}(t = \infty) \propto \mathbf{a}_{average}$ [3],

a   b   c

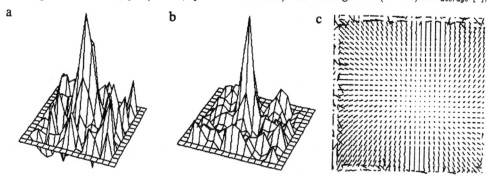

Figure 2: Correlation surfaces and shift field.

Using a set of 180 edge image pairs, an example of which is given in figure 1 (c-d), the time-averaged correlation was calculated at every pixel. (a) A typical correlation surface at a sample pixel position. (b) The average correlation surface of four neighbors. The peak is somewhat more pronounced, and its position better defined. (c) Shift field. Parabolic fits were used to find the peak positions which are represented as needles in the diagram. The zoom effect is obvious. For clarity only every fourth needle is shown.

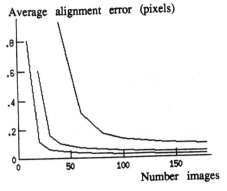

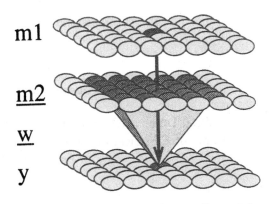

Figure 3: Accuracy of alignment.
The alignment error decreases rapidly as more images are used. The curves represent different amounts of spatial averaging. From top to bottom, no averaging, 2x2 averaging, 3x3 averaging.

Figure 4: Learning Architecture. See text for details.

i.e. the time-averaged correlation surface. We used the same image set as in the previous section to test this learning algorithm, presenting the images in random order. Starting from random weights or a peak in the wrong position, we get convergence after 300–500 iterations.

If this algorithm runs continuously on an autonomous vision system, it would rapidly adapt to changes in camera positions or other distortions. Such dynamic realignment has been shown to occur in the barn owl optic tectum [2]. A VLSI implementation of the algorithm would be useful for integrated early vision modules now under development.

Further discussion

Our problem definition translates to finding correspondences between image pairs. This is related to the problem of binocular stereo and image motion, and algorithms developed in these areas could conceivably be used. However, the problem to be solved *is not* analogous, and our algorithm has two advantages. First, stereoscopic effect are a major source of error, since a close object would be interpreted as a local distortion in the imaging equipment. Averaging over several image pairs is a necessity and reduces stereoscopic errors at the cost of a wider peak. Second, time-averaging over many images yields very high precision and allows for a straightforward implementation of the learning algorithm as described in the previous section.

We would like to thank the Hughes Aircraft AI Center for partial support for this research.

References

[1] Barniv, Y. and Casasent, D. (1981) *Multisensor image registration: experimental verification*, SPIE 292: Processing of images and data from optical sensors, 160-171.

[2] Knudsen, E.I. (1983) *Early auditory experience aligns the auditory map of space in the optic tectum of the barn owl* , Science 222, 939-942.

[3] Kohonen, T. (1984) *Self-organization and associative memory* , Springer-Verlag, Berlin.

[4] Pearson, J.C., Sullivan, W.E., Gelfand, J.J., Peterson, R.M. (1987) *A computational map approach to sensory fusion* , AOG/AAAIC Proc. Joint conf. on merging tomorrow's technologies with defense readiness requirements

A Heterogeneous Vision Architecture

A.W.G. Duller, R.H. Storer, A.R. Thomson, M.R. Pout, E.L. Dagless
Dept. of Electrical and Electronic Engineering, University of Bristol
Bristol BS8 1TR, UK

Real-time computer vision needs huge computing power and diverse programming approaches. As the processing progresses away from the pixel level, we expect to deal with fewer, more complex objects. No one architecture or language seems able to meet these demanding requirements, but rather an integration of different architectures and software techniques - a heterogeneous vision architecture.

GLiTCH

GLiTCH (**G**oes **L**ike **T**he **C**lappers, **H**opefully) [1] [2] is a full custom VLSI associative processor array [3] designed for computer vision by the authors in collaboration with the SERC IC Design and Test Centre at UMIST.

The GLiTCH chip has 64 processing elements (PEs), each with 68 bits of content addressable memory (CAM). Any number of GLiTCH chips can be linked to form a 1-D array of PEs with communication hardware which can simulate a 4, 6 or 8 connected mesh. The array has a common microcode sequencer, program and data memories all under the control of a single transputer. A video shift register runs down the edge of the array, having one stage in each PE. By shifting concurrently with PE operation, it enables an image region to be loaded into the array while a region loaded previously is being processed.

Research already completed [4] shows GLiTCH to be suitable for a number of typical vision tasks including Laplacian of Gaussians filtering, histogram equalisation, connected component labelling, Hough and fast Fourier transforms and image resampling.

A Proposed Heterogeneous Vision Architecture

We envisage an architecture where feedback from high level processes directs the nature of the lower level processes and the regions of the image on which they work.

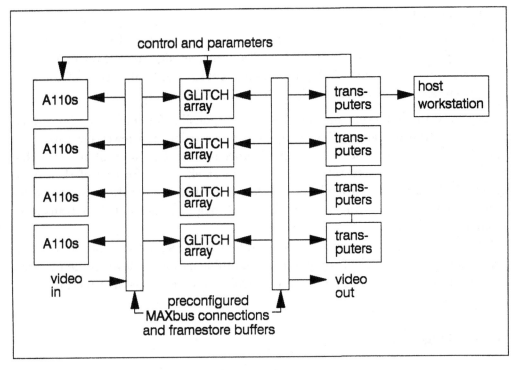

A Heterogeneous Vision Architecture

This type of flexibility becomes increasingly important as improved technology provides us with higher resolution images; the high resolution is of benefit in areas of the image containing information we require, but a hindrance in areas which are receiving, unneccesarily, the same amount of processing.

In order to provide an environment where processing power can be applied to selected regions of interest, an architecture is proposed (see figure) containing compatible modules of three different types. The three types of module can be connected in a variety of configurations, usually with intermediate framestores as buffers, allowing modules to work on different sized regions or load data at different rates. Each module contains at least one controlling transputer which communicates parameters with other modules.

It is important that the modules are combined in a way which does not limit their functionality by imposing too rigid or inappropriate a structure; a point-to-point video bus allows us complete flexibility of interconnection between modules.

- A110 modules: These comprise a number of INMOS (TM) A110 chips fed by MAXbus (TM) video interfaces. They perform fast, low-level processing on a stream of video data.

- GLiTCH modules: These contain an array of GLiTCH chips, described above, with MAXbus interfaces. The controlling transputer coordinates the transfer of video data over MAXbus and communicates non-video results with other modules over transputer links.

- Transputer modules: These assimilate and analyse data from the modules above, processing higher-level information such as edge tokens, region descriptions etc. The results of this processing provide the final output of the subsystem and feedback to the GLiTCH and A110 modules.

Typically, the top level transputers will be responsible for controlling (via transputer links) which regions of an image are processed in the lower level modules and the parameters to be used in those processes. In this way all the available parallelism can be concentrated where it is most effective. Several of these subsystems will operate in parallel, each performing different operations on the entire image, each communicating with its fellows through the links of their transputer modules.

The aim of building the system described here is to discover the optimum configurations for a wide range of computer vision problems and to develop the powerful software tools needed to program and control a heterogeneous architecture.

Acknowledgements

We wish to acknowledge the financial support of the Science and Engineering Research Council, BP International Limited, the MOD at RSRE and the EEC Esprit programme for this work.

INMOS is a trademark of the INMOS group of companies.

MAXbus is a trademark of Datacube.

References

[1] Duller A.W.G., Storer R., Thomson A.R & Dagless E.L.
   An Associative Processor Array for Image Processing
   Image and Vision Computing Vol. 7 No. 2 (Butterworth Press) May 1989.

[2] Duller A.W.G., Storer R., Thomson A.R., Dagless E.L., Pout M.R., Marriott A.P., & Goldfinch J.
   Design of an Associative Processor Array
   Proceedings IEE Vol. 136 Part E, No. 5 pp 374-382, September 1989.

[3] Foster C.C.
   Content Addressable Parallel Processors
   Van Nostrand Reinhold, New York 1976.

[4] Duller A.W.G., Storer R., Thomson A.R., Pout M.R., & Dagless E.L.
   Image Processing Applications Using An Associative Processor Array
   Proceedings of the 5th Alvey Vision Conference, pp289-292 September 1989.

SPATIAL CONTEXT IN AN IMAGE ANALYSIS SYSTEM

Philippe Garnesson, Gérard Giraudon
INRIA Sophia Antipolis, 2004 route des Lucioles
06565 Valbonne cedex, France

Abstract

"Scene Analysis", especially for real data, is a complex problem. There are two main explanations which interest us in this paper.

The first one is that sometimes an object identification needs informations about his spatial context ([GAR76,OHTA89]). We define the spatial context of an object as topological relations beetween this object and the other objects of the world.

The second explanation is that the detection of an object implies to solve simultaneously two problems, the localization and the identification. This is really difficult because sometimes for the same object class, we must consider variations, for instance shape variation or colour variation. To solve these problems, we use generic models of objects [FUA87,GAR89] which can be expensive with computing time if they explore the whole scene.

In this paper, we increase the formalization of spatial context and we show how it allows to focus the search objects in a limited aera of the scene.

Introduction

This paper is talking about a part of the development of MESSIE, a Multi Expert System for Scene Interpretation. In the same way as [RIS89,ANT89,LAA89], the specialists co-operate in a Blackboard architecture [HAY83]. In MESSIE, a specialist is a system whose knowledge (declarative statements or procedures) allows to localize and identify objects of a certain class. The specialists co-operation is based on the spatial context utilization. Spatial context is used to send request to the specialists. To have more details refer to the paper [GAR89,MON89].

We illustrate this research with results in the domain of the aerial imagery interpretation (image with a resolution of 80cm/pixels).

Spatial context

The concept of spatial context is a knowledge which can be expressed with **Localization Heuristics**. A localization heuristic is a rule which expresses a spatial relation between objects under constraints. For instance, *"There are often cars on the main roads."* Different concepts are used in a localization heuristic:

1. **The goal of the Heuristic.**
   Two different goals are distinguish.
   - On the one hand, an object instance is used **to infer** a request to search another one. For instance, if we have found a road, we can try to find cars on it.
   - On the other hand, the goal is **to validate** an hypothesis object using his context. For instance we try to find a shadow to validate a building hypothesis. It may be more than one localization heuristic to validate an hypothesis or, at least, increase his confidence.

2. **The localization operator.**
   The localization operator allows to use a spatial relation (for instance "near" or "inside") to compute an aera in the scene. This aera determines the search aera of a specialist.
   For instance, "cars are **on** roads" implies that the car specialist will try to find cars on the aera roads.
   The calculation of the search aera is done using basic operators (or a combination of basic operators).
   In the aerial imagery domain, we have defined four groups of operators. These operators have parameters which permits to adapt an operator to different heuritics.
   - The proximity operator: it allows to define a search aera around an object.
   - The inside operator: the search aera is defined **inside** the aera of an object.
   - The projection operator: the search aera is calculated using a projection depending on an orientation and a distance.
   - The sub-window operator: it computes the smallest rectangular envelope around an object. It is usually used by the low level process.

3. **The constraints.**
   There are constraints which say if the heuristic can be used. For instance "searching cars only on **main** roads". There are other constrainsts which specify conditions on the objects to search or how to search them. For instance it allows to specify parameters of the segmentation process according to current hypothesis.

4. **The probability of success of an localization heuristic.**
   This knowledge allows to choose an heuristic. We can associate to this choice a cost of time execution. A learning mechanism can calculate these values. Today these ideas are research subjects that we have not implemented.

Example of utilization of the localization heuristic "shadow near building"

| Figure 1 | Figure 2 | Figure 3 | Figure 4 | Figure 5 |

The building specialist has found an hypothesis (fig.1), the supervisor tries to confirm it with the heuristic "shadow near building". Using the localization operator *near*, it projects the building (fig.2) (according to the direction of sunlight and a distance depending of a maximal height) and computes a sub-window (fig.3). Then the supervisor sends a request to find a shadow in the sub-window and with constrainsts depending of the shape building. The shadow specialist responds to the request, it uses a low level process of segmentation in the sub-window and finds one shadow which corresponds to the shape constraints (fig.4). The supervisor increases the confidence of the building and put the shadow in the context slot of the building (fig.5).

Example of utilization of the localization heuristic "cars on road"

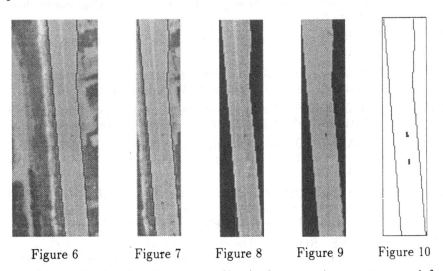

| Figure 6 | Figure 7 | Figure 8 | Figure 9 | Figure 10 |

The road specialist has found an object (fig.6), the supervisor attempts to infer cars with the heuristic "cars on road". Using the localization operator *on*, it computes a sub-window (fig.7) and then computes the road aera (fig.8). Then the supervisor sends a request to find cars in the sub-window. The car specialist responds to the request. It uses an algorithm of diffusion [FAI89] (fig.9) and finds two cars on this road (fig.10). In fact it finds two anomalies on the road because the low resolution of image.

Conclusion

In this work, we have attempted to bring to light the different kinds of knowledge of the concept of spatial context:

- The two ways to use this concept: inference or validation. They are the base mechanisms of the control strategy, especially well adapted to a Multi Specialist System. We must quote that inference from spatial context is sometimes the only mean to localize an object. We must also quote that validation from spatial context is an interesting way to solve the conflict problem which can arrive during scene analysis.

- The operator: it is a declarative knowledge which allows to define and limite the search space. Actually the system is based on two dimensional operator, but noting in the system architecture and in the control strategies precludes the inclusion of three dimensional information.
- The constraints: according to the current results of interpretation, they allow to specify parameters used by the specialists and the low level processes.

References

[ANT89]   J.L ANTOINE, O. CORBY, M. PUGGELI: "ERASME, a multi expert system for pavement rehabilitation" International Conference on Applications of Advanced Technologies in Transportation Engineering, San Diego, February 89.

[FAI89]   J. FAIRFIELD: "An Algorithm of diffusion" Rapport de Recherche INRIA à paraître.

[FUA87]   P. FUA, A.J. HANSON : " Using geometric models for intelligent Shape extraction", Image Understanding Workshop, February 87.

[GAR76]   T.D. GARVEY : "Perceptual strategies for purposive vision". Tech. Note 117. Menlo Park, Calif.: SRI Artificial Intelligence Center.

[GAR89]   P. GARNESSON, G. GIRAUDON, P. MONTESINOS : "MESSIE: application à l'interprétation en imagerie aerienne" Rapport de Recherche INRIA n 1012, Mars 89.

[GIR89]   P. GARNESSON, P. MONTESINOS, G. GIRAUDON, "MESSIE: Un système multi spécialiste en vision" 7ème congrès Reconnaissance des formes et Intelligence Artificielle, AFCET Novembre 89.

[HAY83]   B. HAYES-ROTH : " The Blackboard Architecture : A General Framework for Problem Solving ?", Stanford University, Report HPP-83-30, May 83

[LAA89]   H. LAASRI, B. MAITRE : "Coopération dans un univers multi-agents basée sur le modèle du blackboard : études et réalisations". These, Université de Nancy I, 89.

[MON89]   P. MONTESINOS, G. GIRAUDON, "A Rule Interpreter for Handling Image Primitives in Scene Analysis", Rapport de Recherche INRIA, A paraître.

[OHTA89]   Y. OHTA :"A region oriented image-analysis system by computer" Ph.D. thesis, Kyoto University Departement of Information Science.

[RIS89]   B.A. DRAPER, R.T. COLLINS, A.R. HANSON, E.M.RISEMAN :"The Schema System". International Journal of Computer Vision 89.

OBJECT DETECTION AND IDENTIFICATION BY HIERARCHICAL SEGMENTATION

Patrick HANUSSE and Philippe GUILLATAUD
Centre de Recherche Paul Pascal, CNRS
Château Brivazac, F-33600 PESSAC

One of the traditional problems of vision is to understand the devices of the tremendous compression of information that is performed so rapidly and apparently so easily by human vision. To extract meaning from numbers, hierarchical analysis is a possible answer. It has been used in many contexts as a way to deal with complex situations and systems [1,2]. Its most important feature resides in its factorization property, which recursively splits difficult tasks into simpler ones. To do so, one should be able to design an operational procedure to construct some meaningful hierarchical structure from a complex observation.

To avoid ambiguity and undesirable references to broad, fuzzy or extensively used terminology, and to stress its quite universal status, we define as "dendronic analysis" a general method that is able to produce a significance coding tree structure from a set of measurements of any kind, an a "dendrone" or "dendronic structure", the tree structure that is produced.

We will deal here with segmentation as an object self-detecting method, i.e. that extracts objects without requiring any a priori knowledge of the objects, an idea that we have already explored [3].

Definition of dendronic analysis

We shall consider thresholding as a process in which segmented regions are less important than the way they change with threshold value. We are more interested in the qualitative changes, which are structural changes, rather than in quantitative ones.

Considering the picture intensity profile as a seascape and the threshold value as sea level, we initially start with the sea covering everything. We let the sea level go down, and we record the position and tide level of appearing islands. From one tide level to the other, three events may occur: (i) a new island appears, (ii) an existing island grows in area, (iii) two nearby separated islands merge into a larger one. From a strictly structural point of view, the last event is the most important. It defines a branching process, a qualitative change. The structure of the collected information is one of a tree, the nodes of which may be labelled by various types of information. In this process a large amount of less pertinent information is left out. This information is of morphological nature. The dendronic analysis aims at separating topology from morphology.

Properties of dendronic analysis

Discretization effects: There are two effects due to the discrete nature of a digital picture, in space and amplitude, but also to physical cutoffs related to noise amplitude and correlation. They can be controlled by two parameters of the analysis, namely, the Minimum Island Size (MIS), and the Sea Level Decrement (SLD). The MIS defines the smallest island area that is detectable and can be added to the dendrone. The SLD defines the intensity resolution, and controls the local contrast that is required to form an island.

Filtering properties: These very same parameters can also have a role at picture comprehension level as they may be used to specify a possible a priori knowledge of minimum object size and contrast. Consequently, they act as filtering parameters in the dendronic analysis itself. Filtering could act on the dendrone itself, to reduce its complexity and increase its significance. In this case we would speak of dendrone driven pattern recognition.

The efficiency of pattern recognition controlled by the dendronic structure is increased by its higher significance content as compared to the original picture or region and contour segmentation.

Adaptability and Invariance properties: The topology of a dendrone is not dependent on local intensity levels, so that two objects with the same structure located in different lightening or contrast conditions will appear with the same, simply shifted, dendronic contribution. The method is thus naturally adaptative.

It is also easy to show that the same structural topology will be invariant under translation, rotation, and scaling. This is again a consequence, and benefit, of the separation between the structural content of a picture and any local contextual dependence on lightening, position, orientation and magnification.

Robustness: Its adaptative character is already providing robustness to long range defects. The dendronic contribution of noise is unambiguously detectable. It can be eliminated by adjusting the two already mentioned parameters, MIS and SLD.

Finally, another reason for robustness is due to the structural stability of the topology of a picture. The dendronic structure emphasizes the mutual topological relationship between regions, rather than the regions themselves. This is not much perturbated by random noise or context variations.

Significance extraction: Filtering processes acting on a dendrone increase its significance. But, the most important aspect of dendronic analysis resides certainly in self-building significance. By this, we mean that, besides being adaptative and robust, this method reveals the internal segmentation, or the hierarchical levels of description, that are present in the picture, without having to formulate explicitly a model of the scene and of the objects it contains. It is easy to setup a numerical discrimination criterion on the dendrone itself that would implement rules such as "*noise branches at all levels, objects don't*". This is what we call significance self-detection.

First example: Disks (Fig. 1)

This noisy artificial picture (256x256) represents two disks on a non uniform background (fig.1a). Figure 1b displays a diagonal cross-section showing backgound slope and disk amplitude which is lower than noise level. This noise is spatially correlated over 5 pixels. This is a very difficult situation for thresholding or any region or contour segmentation. Figure 1d presents a representation of the dendrone (MIS=100,SLD=2) with node horizontal base size proportional to region area. Dendrone structure inspection, along with several quantitative parameters collected during dendronic analysis, allows us to separate unambiguously, at two different levels, the two objects present in the picture (see thick lines in fig 1d). This is what we call self-detection. Figure 1c gives the two extracted disks at their own self-determined levels (not a plain picture thresholding). This example shows the robustness and adaptativity of the method.

Second example: Bolts (Fig. 2)

A group of 9 bolts is set on a more or less uniform background (Fig 2a). Lightening conditions are not very good. There are shadows between the objects, which would make it difficult to find a good segmentation threshold. A global view of the corresponding dendrone (half of it) is given in fig. 2b. The representation uses the base area has horizontal scale. Five bolts are very easily detectable. The varying local base threshold reveals again adaptativity. It is noticeable that there is a structure within the bolts, and that this structure can be found in each bolt. Fig. 2c presents two magnified objects dendrones, here using plain structure representation (horizontal scale as no physical meaning). The bolt head groups (H1,H3) and thread pairs (T1,T3) can be identified. This structure is invariant under translation and rotation. It would also be invariant under scaling. Getting rid of these possible sources of contextual dependences obviously greatly simplifies the identification step.

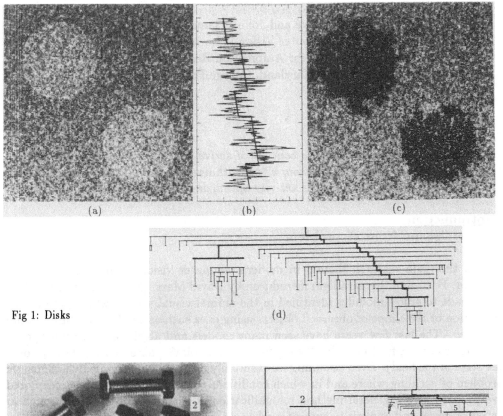

Fig 1: Disks

Fig 2: Bolts

References

[1] R. DUDA and P. HART, *"Pattern classification and scene analysis"*, Wiley (1973).

[2] H. SAMET and R.E. WEBBER, *"Hierarchical data structures. A review"*. Proceeding of the International Electronic Image Week, Nice, April 1986, CESTA Paris, (1987) and references therein.

[3] P. HANUSSE, *"Analyse morphologique de trajectoires en dynamique chimique"* J.Chim.Phys. 84,1315-1327, (1987).

Shading into Texture and Texture into Shading: an Active Approach

Jean-Yves Hervé and John (Yiannis) Aloimonos
Computer Vision Laboratory
Center for Automation Research
University of Maryland, College Park, Md 20742

Abstract
We present a theory, based on the concept of active vision, for the fusion of information provided by different "shape from x" algorithms (here shading and texture), and a method for choosing the observer's activity optimizing the recovery of the shape.

1 Introduction
1.1 The "shape from x" problem

One of the main research themes in modern computer vision is the "shape from x" problem. Following the paradigm introduced by David Marr ([Marr82]), attempts have been made to study modules identified in the animal visual system which perform the extraction of the shape of observed objects using cues such as shading, texture, motion or stereo. The last few years have seen many models and algorithms being proposed, still the problem is far from solved (see [Horn86] or [Aloim88] for extensive description of the state of the field). Even assuming we eventually dispose of satisfying algorithms, the problem of deciding where and in which conditions to apply them remains unexplored. We propose here a general method for unification of shading and texture through an active observer and apply it to current models of these modules. The use of the resulting algorithm is shown to be appropriate regardless of the intensity distribution in the image.

1.2 The active observer

The concept of Active Perception was recently introduced in [Bajcs86] and further analyzed in [Aloim87]. An active observer, when engaged in an activity (self motion, tracking, focusing,...), modifies the constraints underlying a given phenomenon (and the equations describing them) and thus creates new information that will help in eliminating ambiguities, make the solution easier to get and often more reliable, that is, more robust.

2 Unification of the two modules

The observer considered here is a monocular optical system (camera) we will represent by a classical pinhole model (optical center O, optical axis OZ, focal length f). A point M of the object projects in the image plane as m. Slightly abusing the notation, we will identify the coordinates vector $\mathbf{M}$ (resp. $\mathbf{m}$) with the position vector $\overrightarrow{OM}$ (resp $\overrightarrow{Om}$).

The following relations hold between M and m :

$$\mathbf{m} = \frac{f}{Z}\mathbf{M} \quad \text{and} \quad \frac{\partial \mathbf{m}}{\partial \mathbf{M}} = \frac{1}{Z}\mathbf{D} \quad \text{where} \quad \mathbf{D} = \begin{pmatrix} f & 0 & -x \\ 0 & f & -y \\ 0 & 0 & 0 \end{pmatrix}. \quad (1)$$

The observer is moving with a known rigid motion composed of a translation $\mathbf{T}$ and a rotation ω (of associate skew-symmetric matrix $\mathbf{\Omega}$). A point M from the surface is seen with the apparent velocity $\dot{\mathbf{M}} = -\mathbf{\Omega}\mathbf{M} - \mathbf{T}$. The motion of its image (*optical flow*) is:

$$\dot{\mathbf{m}} = \frac{d\mathbf{m}}{dt} = \frac{\partial \mathbf{m}}{\partial \mathbf{M}}\frac{d\mathbf{M}}{dt} = \frac{1}{Z}\mathbf{D}(-\mathbf{\Omega}\,\mathbf{M} - \mathbf{T}). \quad (2)$$

Since we want to reconstruct the shape of the surface (i.e. recover $Z(x,y)$), an apparently simple idea would be to compute the optical flow $\dot{\mathbf{m}}$ and report it in (2) to get Z for each (x, y) in the image.

The traditional optical flow constraint used to relate image intensity data to the unknown flow is

$$\frac{dI}{dt} = \nabla I \cdot \dot{\mathbf{m}} + \frac{\partial I}{\partial t} = 0. \quad (3)$$

This equation merely states that corresponding points have the same intensity. Even if we consider this model acceptable (for example, it doesn't take specularity into account), the problem of calculating the flow from this equation is known to be ill-posed. In order to compute a regularized solution we would have to add unnatural constraints (smoothness of the flow). Still, this method requires pointwise calculations of differential operators while the data (intensity) are locally inaccurate . On the other hand, the reliability of averages computed on image windows is quite satisfying. This is why we introduce linear features ([Amari86]) in our theory.

A *linear feature* is a 3-tuple $LF_l = (\mu_l, f_l, \Sigma_l)$ where Σ_l is an image window, the *measuring function* μ_l is a nice (differentiable) function defined over Σ_l, f_l is defined as an integral over Σ_l

$$f_l = \iint_{\Sigma_l} \mu_l\, I\, ds = \iint_{\Sigma_l} \mu_l(x,y)\, I(x,y)\, dx dy.$$

It can be proven that, in the case considered here,

$$\dot{f}_l = \frac{d}{dt}\left(\iint_{\Sigma_l} \mu_l\, I\, ds\right) = \iint_{\Sigma_l} \frac{d}{dt}(\mu_l\, I)\, ds = \iint_{\Sigma_l} I\nabla\mu_l \cdot \dot{\mathbf{m}}\, ds.$$

If we now replace $\dot{\mathbf{m}}$ by the expresssion given in (2), we obtain:

$$\iint_{\Sigma_l} \frac{1}{Z}I\nabla\mu_l \cdot (\mathbf{D}\,\mathbf{T})\, ds = -\dot{f}_l - \frac{1}{f}\iint_{\Sigma_l} I\nabla\mu_l \cdot (\mathbf{D}\,\mathbf{\Omega}\,\mathbf{m})\, ds. \quad (3)$$

We model $1/Z$ as a linear combination of simple differentiable functions (polynomials, Gabor functions,...): $1/Z = \sum_{k=0}^{m} \alpha_k\, \varphi_k(x,y)$.

Reporting in (3), we obtain the final equation:

$$\sum_{k=0}^{m} \left[\iint_{\Sigma_l} \left(I\,\varphi_k\,\nabla\mu_l \cdot \mathbf{D\,T} \right) ds \right] \alpha_k = -\dot{f}_l - \frac{1}{f} \iint_{\Sigma_l} \left(I\,\nabla\mu_l \cdot \mathbf{D\,\Omega\,m} \right) ds.$$

For each linear feature we obtain a linear equation in function of the α_k's. The coefficients of these equations are determined by summations over whole image windows. In particular, no pointwise derivatives are computed, no a priori assumption is made on the variation of the intensity in the image: shading and texture are allowed to be mixed.

3 Search of the best activity

We have seen that an active observer can create new information and will be able to solve problems which were ill-posed for a passive observer. But all activities are not equally good. The "best" possible activity is one which maximizes the stability of the solution (with respect to perturbations of the images or of the apparatus). Since the final equation (obtained by least squares minimization) is linear, a good criterion for measuring the stability of the result is the condition number of the equation's matrix: the determination of its maximum gives the best move for the observer.

4 Conclusion

We have presented here a theory for the extraction of the shape of observed objects. Our approach combines the modules of shape from shading and shape from texture by fusing the information relevant to each of them (without segmentation). By using linear features, we avoid the pointwise computation of unreliable operators. Finally, we showed how to determine a motion that optimizes the stability of the equation, and therefore, the reliability of the solution.

References

[Aloim89] J. Aloimonos, "Unifying Shading and Texture through an Active Observer," in *proceedings Royal Society*, London, ser. B., in press.

[Aloim88] J. Aloimonos and A. Basu, "Combining Information in Low-Level Vision," in *proceedings DARPA Image Understanding Workshop*, Cambridge, Massachusetts, 1988, pp 862–906.

[Aloim87] J. Aloimonos, I. Weiss and A. Bandyopadhay, "Active Vision," in *proceedings DARPA Image Understanding Workshop*, Los Angeles, 1987, pp 552–573.

[Amari86] S. Amari, Personnal Communication, 1986.

[Bacjs86] R. Bacjsy, "Passive Perception vs. Active Perception," in *proceedings IEEE Workshop on Computer Vision*, Ann Arbor, 1986.

[Horn86] B.K.P. Horn, *Robot Vision*, M.I.T. Press, Cambridge, Massachussets, 1986.

[Marr82] D. Marr, *Vision*, Freeman, San Francisco, 1982.

3-D Curve Matching Using Splines

Eyal Kishon and *Trevor Hastie*

AT&T Bell Laboratories Murray Hill, NJ 07974

Haim Wolfson

Robotics Research Laboratory New York University, and

Computer Science Department Tel Aviv University Tel Aviv 69 978, Israel

Abstract

A machine vision algorithm to find the longest common subcurve of two 3-D curves is presented. The curves are represented by splines fitted through sequences of sample points extracted from dense range data. The approximated 3-D curves are transformed into 1-D numerical strings of rotation and translation invariant *shape signatures*, based on a multi-resolution representation of the curvature and torsion values of the space curves. The *shape signature* strings are matched using an efficient hashing technique that finds longest matching substrings. The results of the string matching stage are later verified by a robust, least-squares, 3-D curve matching technique, which also recovers the Euclidean transformation between the curves being matched. This algorithm is of average complexity $O(n)$ where n is the number of the sample points on the two curves. The algorithm has applications in assembly and object recognition tasks. Results of assembly experiments are included.

1 Introduction

Curves in 3-space carry a large amount of information about the scenes they appear in, and can successfully characterize objects drawn from large sets of candidates. The curves can either be 'painted curves' (i.e. curves that correspond to changes of reflectivity without any rapid local depth changes), curves of intersection between two surfaces (either convex or concave), curves of occlusion (i.e. object boundaries), or curves of maximum curvature ('ridges'). The data required to characterize objects by this method reduces to a small group of curves extracted from the object, and is is extremely compact as compared to the full 2-D surface specification. Matching algorithms based on 3-D curves are simpler and more efficient than surface matching algorithms, because the points that represent a 3-D curve are naturally ordered as a sequence.

Schwartz and Sharir ([2]) developed a matching algorithm which finds the position and orientation of an observed curve best matching (in the least squares sense) a previously stored model curve. This algorithm was proven to be robust for practical applications, and was extended to support cases in which several subcurves had to be simultaneously matched against the same curve (see [1] for extended bibliography).

The algorithm by Schwartz and Sharir requires the observed curve to be a proper sub-segment of the stored model curve. Such prior segmentation is not always available in composite scenes of overlapping objects. This problem is particularly evident for assembly tasks, where two objects match along a common boundary without any obvious start point and end point for the common subcurve (see figure 1). Thus we were motivated to develop a general curve matching algorithm that solves the following problem:

Given two curves, find the longest matching subcurve which appears in both curves.

We reduce the 3-D curve matching task into a 1-D string matching problem, find the proper matching substrings, transform the problem back to the 3-D domain, and match the appropriate subcurves. Specifically, the 3-D curves are transformed into sequences of *local, rotationally*, and *translationally invariant* shape signatures. We then apply a hashing technique to find long matching substrings. Finally, we apply the least squares algorithm to select the best candidate. As a by-product we obtain the Euclidean transformation aligning both curves along their longest matching subpart.

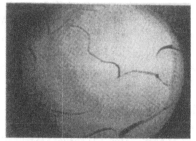

Figure 1: (left) A plastic ball (right) The 'broken' ball assembled from the separate pieces

2 Shape Signatures

In this section we describe the shape signatures which are used in our matching algorithm. A signature should uniquely characterize a relatively short segment of the curve. Hence we require signatures to be:

i) local (calculated at each point),

ii) translationally and rotationally invariant,

iii) stable, so that small changes in the curve induce small effects on the signature.

It is well known from Differential Geometry that smooth space curves can be uniquely reconstructed within a rigid motion (i.e rotation and translation) using three geometric invariants: arc length s, curvature $\kappa(s)$ and torsion $\tau(s)$ as a function of s.

Since curvature and torsion are essentially second and third order derivatives respectively, it is essential to smooth the data a small amount before computing the signatures. Regression splines are a convenient class of smoothers for this task, since they permit one to easily compute derivatives. We chose a quintic spline representation (order 6) since we want the torsion to be continuous, which in turn will make the signature based on the torsion more stable.

3 The Matching Algorithm

This section describes the *shape signature* matching algorithm applied in our experiments.

All the curves in the data-base are preprocessed as follows. The algorithm accepts as its input the curvature-torsion signature strings computed at different levels of resolution. For each signature we record the *curve number* and the *sample point number* at which this signature was generated. The data is stored in a hash-table.

In the matching stage, the observed curve is sampled and signatures (at different levels of details) are computed at the sampling points. For each signature we check the appropriate entry in the corresponding hash-table, and for every (*model curve, sample point*) pair, appearing in the hash table, we add a vote for this model curve and the relative shift between the model curve and the observed curve.

At the end of this process we determine which (*model curve, shift*) pairs received the most votes, and determine the approximate start and end points of the corresponding signature substrings in the observed and model curve.

Given the starting point and endpoint of a signature subsequence, we identify the actual subcurve to which they correspond, and apply the robust least-squares matching algorithm to the corresponding subcurves.

For a detailed description of the matching algorithm see [1]. The algorithm is of average complexity of $O(n)$, and improves the result of Schwartz and Sharir which is of complexity $O(n \log n)$.

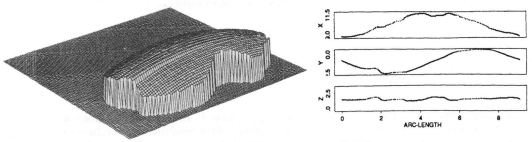

Figure 2: (left) Range image (right) boundary curve X,Y,Z

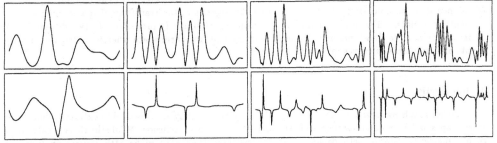

Figure 3: (top) Curvature (bottom) torsion computed at four different resolutions

4 Experimental Results

Figure 2 shows a perspective view of the range image of one of the pieces, and the boundary curve extracted from that piece. Figure 3 shows the curvature and torsion derived from the curve for the 4 different levels of smoothing. Figure 4 show the results of matching individual pieces of the ball and finding the best subcurve match.

References

[1] E. Kishon and H.J. Wolfson. 3-D Curve Matching. In *Proceedings of the AAAI workshop on Spatial Reasoning and Multi-Sensor Fusion*, pages 250–261, St. Charles, Ill., 1987.

[2] J. T. Schwartz and M. Sharir. Identification ot Partially Obscured Objects in Two or Three Dimensions by Matching of Noisy Characteristic Curves. *The International Journal of Robotics Research*, 6(2):29–44, 1987.

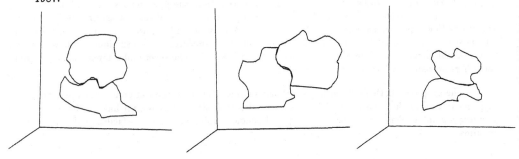

Figure 4: The results of matching four different pieces

The Dynamic Generalized Hough transform

V. F. Leavers

King's College London, Department of Physics
Strand, London WC2R 2LS. VFL@UK.AC.KCL.PH.IPG

A new algorithm for the Generalized Hough transform is presented. The information available in the distribution of image points is used to optimize the computation of the transform. The calculated parameters are those associated with a single image point and all other image points in combinations of the minimum number of points necessary to define an instance of the shape under detection. The method requires only one dimensional accumulation of evidence. Using the algorithm, the transform of sparse images is more efficiently calculated. Dense images may be segmented and similarly processed. In two dimensions, the method provides a feedback mechanism between image and transform space whereby contiguity of feature points and endpoints of curves may be determined.

1 . Introduction

The Hough transform[1], [2] is a powerful tool in shape analysis. It is used to extract global features from shapes and gives good results even in the presence of noise or occlusion. While the theoretical potential of parametric transform methods has been demonstrated they have made little impact on large scale industrial applications because of supposed excessive storage requirements and computational complexity[3] The development of fast, efficient implementations of parametric transformation methods of shape detection has accordingly received much attention in the recent literature[4], [3], [5], [6], [7], [8]. An up-to-date and comprehensive review of the use of the Hough transform is given by Illingworth and Kittler[9].

The Hough transform works by grouping low level feature points (edge image data) into object specific intermediate features (e.g. line segments). This is accomplished by using the low level feature points to generate information concerning all possible groupings of points within the image. The corresponding transform plane is the accumulation of that evidence. The technique is computationally intensive because evidence is generated of all possible groupings of points in the image.

Previous suggested approaches may be divided into two categories. The first seeks to reduce the computational load by using information from the image to reduce the generation of evidence in the transform plane[2], [7], [8], [10]. The second class of methods involves absolute or iterative reductions in resolution of either the transform or the image space[5], [6], [4], [11].

Evaluation of the information generated in the transform space may present difficulties. Problems associated with the detection of maxima in the transform space may be partially solved by the use of matched filtering techniques to detect those maxima[12], [13]. However, a major shortcoming of the technique remains in that all information about feature points contributing to a maxima in the transform space is lost in the transformation process. It is therefore not possible to determine either contiguity of feature points nor end points of curves. Gerig[14] attempts to solve these problems, in the case of circle detection, using a technique which maps information from the parameter space back to the image space. In this way each image point has associated with it a most probable parametrisation. A second transformation is performed where, for each image point, only the cell in parameter space associated with the most probable parametrisation of that image point is incremented. The technique works well in that it is a reliable strategy for interpreting the accumulator space. It is however still computationally complex and offers no reduction in memory allocation.

The proposed method[15] seeks both to cut significantly the computational burden involved in the implementation of the transform, to provide an efficient feedback mechanism linking the accumulated boundary point evidence and the contributing boundary point data and to facilitate the detection of maxima.

2 . The Dynamic Generalised Hough transform

An expression for the Generalized Hough Transform, GHT, may be written in the form suggested by Deans[16]

$$f(\xi,p) = \iint_D F(x,y)\delta\left(p - C(x,y;\xi)\right)dx\,dy \tag{1}$$

where $F(x,y)$ is an arbitrary generalized function[17] defined on the xy plane D and the argument of the delta function defines some family of curves in the xy plane parametrized by the scalar p and the components $\xi_1,\xi_2,\ldots\xi_n$ of the vector ξ.

If, $F(x,y)$, represents a binary image the integral of equation 1 will have a value of 1 when the argument of the delta function evaluates to zero. In computational terms this occurs at all points that are solutions to the discrete equation:

$$p_j = C(x_i,y_i;\xi_j) \tag{2}$$

otherwise it will be zero. Equation 2 is used to calculate the standard GHT. The i, j subscripts refer to ordered pairs in the image and the transform space respectively. For every point, (x_i,y_i), of the image, i is fixed and the values p_j are calculated using stepwise increments of the components of ξ_j. Each point, (p_j,ξ_j), in the transform space will refer to a possible curve in image space which passes through the point (x_i,y_i). The SHT therefore provides a great redundancy of information concerning the image. This is because each image point is treated independently.

The present technique proposes that image points are tested for the most probable, as opposed to all possible, membership of a shape. (In two dimensions shape may refer to a curve and in three dimensions a surface). If, when the image is scanned for candidate feature points, a list of those feature points is maintained, then possible membership of curves/surfaces may be tested. Where n parameters are associated with the shape under detection then a minimum of n points are required to test the membership of a shape of any given point and any other $(n-1)$ image points. The equation of a shape, $p-C(x,y;\xi) = 0$, passing through n non-colinear points may be written in the following way:

$$\begin{pmatrix} c_1 & c_2 & \cdots & c_n & 1 \\ c_1^1 & c_2^1 & \cdots & c_n^1 & 1 \\ \vdots & \vdots & \ddots & \vdots & \vdots \\ c_1^n & c_2^n & \cdots & c_n^n & 1 \end{pmatrix} = 0 \tag{3}$$

and used to calculate the parameters associated with the shape. For an image containing m points, to test each image point in this way would require $C_m^n = \frac{m!}{(m-n)!n!}$ computation cycles Thus, the computational effort is a function of the density of feature points in the image and increases sharply with the increase of m. If the number of feature points is large enough, the number of computations required far exceeds those required when using a standard GHT algorithm. This problem may be resolved in the following way.

Using equation (3), the n parameters associated with the shape under detection are calculated using a single fixed image point and all possible combinations of that image point with sets of $(n-1)$ other image points. The calculated parameters are accumulated in one dimensional histograms. Peaks in the histograms will indicate the parameters associated with the most probable instance of the shape in image space of which the fixed point is a member. This information may be used to detect that shape and to remove the points associated with it from the image. The process is then repeated using the shortened list. In two dimensions, the contiguity of points and endpoints of the detected curve or sections of it may be determined.

Many computer vision tasks require that a particular instance of a shape be recognised and located. Using the constraints inherent in such tasks computational savings may be made. If the calculated parameters are not within the range of possibilities suggested by the shape under detection, this pass of the algorithm may be abandoned. Such a strategy will deal with membership of extraneous shapes accidentally generated by random association of image points. Further computational savings may be made by segmenting the image[18]. The algorithm is inherently parallel thus offering the potential to further decrease computation times.

Particular examples of the algorithm, when used to detect straight lines or circles are given in reference [18]. The algorithm offers a significant improvement to previous suggested implementations of the Hough transform. It is significantly computationally less intensive and is more efficient in memory utilization.

3 . Conclusion

A new algorithm for computing the Hough transform has been presented. It uses information present in the location of the feature points to reduce the generation of evidence in the transform plane. The algorithm gives improved performance compared with the standard Hough transform. The improvement is in computation time and memory allocation. Further advantages of using the algorithm are that peak detection is one dimensional and the end points of curves may be detected. The algorithm is also inherently parallel.

References

[1] Hough P.V.C. *Method and means for Recognising complex patterns.* U.S. Patent No. 3069654, 1962

[2] Ballard D.H. *Generalizing the Hough Transform to detect arbitrary shapes*, Pattern Recognition, Vol 13. No. 2, 111-122, 1981.

[3] Gerig G. and Klein F. *Fast contour identification through efficient Hough transform and simplified interpretation strategies.*Proc. 8th Int. Conf. Pattern Recognition, Vol 1, Paris 1986.

[4] Li H., Lavin M.A. and LeMaster R.J. *Fast Hough Transform*Proc. of 3rd workshop on computer vision: Bellair, Michgan 1985

[5] Li H. *Fast Hough Transform for multidimensional signal processing,*IBM Research report, RC 11562, York Town Heights 1985

[6] Illingworth J. and Kittler J. *The adaptive Hough transform*in press IEEE T-PAMI 1987

[7] Forman A.V *A modified Hough transform for detecting lines in digital imagery*, 151-160, SPIE, Vol 635, Applications of Artificial Intelligence III, 1986.

[8] Jain A.N. and Krig D.B. *A robust Hough transform technique for machine vision*, Proc. Vision 86, Detroit, Michigan, 86.

[9] Illingworth J. and Kittler J. *A survey of the Hough transform*, accepted for publication in IEEE Trans. on Pattern Analysis and Mach. Intell., 1987.

[10] Leavers V.F. and Sandler M. B. *An efficient Radon transform*BPRA 4th International Conf., Cambridge, 1988.

[11] R.S. Wallace *A modified Hough transform for lines*, IEEE conference on computer vision and pattern recognition, San Francisco 1985, 665-667.

[12] Leavers V.F. *Method and Means of Shape Parametrisation*, British Patent App. N0. 8622497. September 1986.

[13] Leavers V.F. and Boyce J.F. *The Radon transform and its application to shape detection in computer vision*, Image and Vision Computing Vol.5, May 1987.

[14] Gerig G. *Linking image-space and accumulator-space. A new approach for object recognition*Proc. of First Int. Conf. on Computer Vision, London, June 1987

[15] Leavers V.F. *Dynamic Generalized Hough Transform*, British Patent Application, April 1989.

[16] Deans S.R. *Hough transform from the Radon transform.*IEEE Trans. Pattern Analysis and Machine Intelligence. Vol. PAMI-3, No., March 1981.

[17] Gel'fand I.M., Graev M.I. and Vilenkin N.Ya. *Generalized functions*Vol 5, Academic Press, New York, 1966.

[18] Leavers V.F., Ben-Tzvi D. and Sandler M.B. *A Dynamic Hough Transform for Lines and Circles*, Alvey Vision Conference, Reading, 1989

The Analysis of time varying image sequences

E. De Micheli[*], S. Uras & V. Torre

[*]Istituto di Cibernetica e Biofisica - C.N.R., Genova, Italy

Dipartimento di Fisica, Genova, Italy

Introduction

The analysis of time-varying image sequences is a classical problem of machine vision [1] [5], which is likely to be rather useful in several fields, such as robotics and passive navigation.

In this paper a general approach to the problem is presented, which is based on the computation of optical flow obtained by a similar procedure to the one proposed by Girosi et al. [2]. By exploiting mathematical properties of the 2D motion field [6] several 3D motion parameters, such as *time-to-collision* and angular velocity, can be recovered with a precision which is dependent on the image texture.

The Computation of Optical Flow

When an opaque object is moving in front of an artificial or a biological eye, it defines (in an appropriate system of reference, possibly solid to the image plane of the eye) a 3D velocity field $\vec{V}(\vec{R}) = (V_x, V_y, V_z)$ where $\vec{R} = (R_x, R_y, R_z)$. Because of the imaging device the 3D velocity field $\vec{V}$ is transformed into a 2D motion field $\vec{v} = (v_x, v_y)$ on the image plane [4] [6]. The available information, however, is not the 2D motion field $\vec{v}$ but the scalar field $E(x, y, t)$ of the image brightness at location (x, y) on the image plane at time t. By optical flow we mean any 2D vector field derived from $E(x, y, t)$ which is close to the 2D motion field $\vec{v} = (v_x, v_y)$. It is therefore evident that many different optical flows exist each of which has different properties and behaviour, according to the computing algorithm and the closeness criteria. It has recently been shown [2] that by assuming $\frac{d}{dt}$ grad $E = 0$ it is possible to obtain an optical flow $\vec{u} = (u_x, u_y)$ computed as:

$$\vec{u} = -H^{-1} \frac{\partial}{\partial t} \text{ grad } E \qquad (1)$$

where $H = \left(\frac{\partial^2 E}{\partial x_i \partial x_j}\right)$ is the Hessian matrix of $E(x, y, t)$ and $\vec{u}$ is related to the 2D velocity field $\vec{v}$, by the equation:

$$\vec{u} = \vec{v} + H^{-1} \left(J_{\vec{v}}^T \cdot \text{grad } E - \text{grad } \frac{dE}{dt} \right) \qquad (2)$$

where $J_{\vec{v}}^T$ is the transpose of the Jacobian matrix $\left(\frac{\partial v_i}{\partial x_j}\right)$ of $\vec{v}$ and $\frac{dE}{dt}$ is the total derivative of the image brightness $E(x, y, t)$. Equation (2), which is just an identity, shows that the optical flow computed from equation (1) usually differs from the true 2D motion field, but also indicates that, when $\frac{dE}{dt}$ and $J_{\vec{v}}^T$ are bounded, the optical flow $\vec{u}$ will approach the true 2D motion field $\vec{v}$ whenever the entries of the matrix H^{-1} are small. Since $\frac{dE}{dt}$ and $J_{\vec{v}}^T$ are likely to be usually bounded, with the exception of those points near object boundaries or motion discontinuities, we can determine whether the computed optical flow $\vec{u}$ is close to $\vec{v}$ by simply evaluating H^{-1}. It is evident that the entries of H^{-1} will be small whenever the two real eigenvalues λ_1 and λ_2 of the symmetric matrix H are large.

An inspection of equation (1), from which the optical flow is derived, shows that numerical stability of the computation of optical flow is guaranteed whenever the inversion of the matrix H is numerically stable. This condition is fulfilled when det H is large and the conditioning number c_H of H is close to 1 [3]. Since the matrix H is symmetric, we have that $c_H = |\lambda_1/\lambda_2|$ where λ_1 and λ_2 are the two real eigenvalues of H with largest and smallest absolute value respectively. Consequently, it is evident that det H large and $c_H \sim 1$ imply both numerical stability in the computation of $\vec{u}$ and similarity between optical flow $\vec{u}$ and 2D motion field $\vec{v}$. As a result when det H is large and $c_H \sim 1$ an optical flow is obtained, which is numerically stable and almost correct (i.e. close to the true 2D motion field).

The Recovery of Motion Parameters

The obvious test of any procedure for motion analysis is the comparison between 3D motion parameters directly measured with those recovered from the analysis of the image sequence. Here we discuss the recovery of 3D motion parameters in two special, but practically relevant cases: pure translation (Fig. 1) and pure rotation (Fig. 2).

It has been shown [6] that in the case of pure translation the 2D motion field has at most one singular point, which is a focus and does not change its location on the image plane with time. Moreover if P_T is the singular point the *time-to-collision* between the image plane and the point projected into P_T is simply $1/\lambda$, where λ is the value of the two coincident eigenvalues of J_v computed at P_T. In the case of a pure axial rotation, that is when the rotation axis is orthogonal to the 3D surface, the angular velocity ω can be obtained from

$$\omega^2 = \det J_v \big|_{P_R} \tag{3}$$

where P_R is the immobile point of the pure rotation [6] which is a center.

By using the sparse optical flow obtained from eq. 1 it is possible to locate the singular point $P = (\bar{x}, \bar{y})$ and to analyse the nature of the singular point by estimating the 6 parameters $\bar{x}, \bar{y}, a, b, c$ and d, such that

$$\begin{aligned} u_x &= a(x - \bar{x}) + b(y - \bar{y}) \\ u_y &= c(x - \bar{x}) + d(y - \bar{y}) \end{aligned}$$

represents the least square approximation of the flow in a suitable neighbourhood of P.

In the case of a pure translation we expect b and c to be negligible and the values of a and d to be very similar, whereas for axial motion a and d close to zero and b and c opposite in sign.

An extensive experimentation on sequences of images of different objects has shown that an accuracy of about 95 % can be obtained in the special case of a highly textured plane parallel to the image plane. This configuration, which is optimal from a theoretical point of view (the 2D motion field becomes linear), also proved experimentally to be the most favourable. For scenes with little texture and strongly departing from a planar structure, the agreement between computed and directly measured 3D motion parameters deteriorates and may become poorer.

We conclude that the proposed technique for the analysis of image sequence is suitable for the vision system of a mobile robot, and for many industrial applications.

References

1 - J.K. Aggarwal, & N. Nandhakumar. On the computation of motion from sequences of images - A review. *Proceedings of the IEEE*, 76, No. 8, 917-935, 1988.

2 - F. Girosi, A. Verri, & V. Torre. Constraints in the Computation of Optical Flow. *Proceedings of the IEEE Workshop on Visual Motion*, Irvine CA, 1989.

3 - C. Lanczos, : Linear Differential Operators. D. Van Nostrand Company, London 1961.

4 - H.C. Longuet-Higgins, & K. Prazdny. The interpretation of a moving retinal image. In *Proc. R. Soc. Lond. B* **208** 385-397, 1980.

5 - S. Ullmann. The Interpretation of Visual Motion. MIT Press, Cambridge, 1979

6 - A. Verri, F. Girosi, and V. Torre. Mathematical properties of the two-dimensional motion field: from Singular Points to Motion Parameters. JOSA A, Vol. 6, No. 5, pp.698-712, 1989.

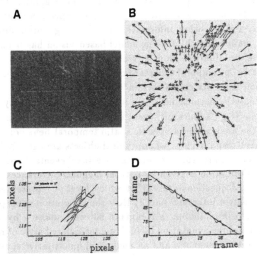

Fig. 1: Pure translation. A: An image of a photograph of two climbers. The image sequence was composed of 44 frames. The camera was slid on the rail of an optical bench by 1 cm between each image acquisition. B: The sparse optical flow relative to the image 15 of the sequence by solving eq. 1. C: The localization of the focus of expansion. The focal length of the camera was 8 mm and the width of a pixel was 0.014 mm so an angular displacement of 1 degree corresponds to a displacement on the image plane of about 10 pixels. D: Comparison between the true *time-to-collision* (straight line) and the computed *time-to-collision* (polygonal line).

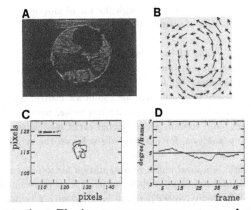

Fig. 2: Pure rotation. The image sequence was composed of 50 frames. The camera was viewing from above and pointing towards a rotating platform, on which different objects were mounted. The platform was rotated by 5 degrees between each frame. B: The smoothed optical flow. C: The localization of the immobile point (i.e. the singular point). D: Comparison between the true (straight line) and the computed (polygonal line) angular velocity.

On the Use of Motion Concepts for Top-Down Control in Traffic Scenes

Michael Mohnhaupt, Bernd Neumann
University of Hamburg, FB Informatik

1. Introduction

The use of models for top-down control is the major strategy to beat the inherent complexity of many visual processes. Fortunately, information to select appropriate models from long-term memory for top-down control is often available for a visual system. For example, it can be provided by previous bottom-up analysis, by the spatio-temporal context, by expectations about a scene, by goals or intentions of the system, and by other information sources, e.g., prior descriptions in natural language and the like. The main research focus in model-based vision has been on the use of static information, for example, the use of object models (see e.g. |Tsotsos 87| for an overview).

In this paper we concentrate on temporal aspects of model-based vision: the use of motion concepts for top-down control. Motion concepts can constrain visual processes in two ways. First, they can provide a spatial focus for analysis, because instances of motion concepts can typically only be found at certain locations in a scene (e.g., a 'turn-off' event can only take place at intersections). Second, motion concepts can focus the analysis on a specific spatio-temporal behavior. We investigate both aspects in the domain of street traffic scenes, where typical objects are cars, pedestrians, trucks, and so on, and typical motion concepts are 'turn-off'-events, 'overtake' events, 'cross'-events and the like. In our examples, top-down information is given by natural language utterances.

Two central and interrelated questions are: 1) At which level of representation should bottom up processes and top-down control interact? and 2) How should motion concepts be represented to support top-down guidance? In |Mohnhaupt + Neumann 90| we propose a hybrid representation of motion concepts. A propositional representation including a logic based style of reasoning is exploited for event recognition and long-term memory, and an analogical quantitative spatio-temporal buffer is used for motion visualization and prediction, for learning object motion and several aspects of spatio temporal reasoning. The buffer facilitates important tasks related to concrete visual scenes. It can be instantiated on demand from long-term memory.

Here, we focus on one aspect of the spatio-temporal buffer: the generation of predictions suited for top-down control of motion analysis. The central idea is to express motion concepts as typicality distributions in the buffer. The buffer is shared between bottom-up and top-down processes. We show that spatio-temporal constraints for motion analysis can be derived, and we sketch how models and bottom-up data can interact to allow for meaningful predictions. Both leads to a significant reduction of complexity: For traffic scenes top-down control through the use of motion concepts can reduce the amount of computation by several orders of magnitude.

2. Motion concepts implied by verbs of locomotion

In street traffic scenes we associate motion concepts with verbs of locomotion like 'drive', 'walk', 'turn-off', etc., as proposed by |Neumann 89| for bottom-up event recognition using propositional event models. These event models are inappropriate for top-down control mainly for two reasons: First, predictions in terms of predicates are unnecessarily imprecise, because typical and atypical instances cannot be distinguished. And second, propositional event models are difficult to adapt to constraints provided by bottom-up analysis, for example obstacles on the street; clearly this should lead to an adapted prediction.

This lead us to consider a spatio-temporal buffer representation for top-down control. It is shared by bottom-up and top-down processes and closely related to perceptual representations. The buffer is fourdimensional (x, y, direction of velocity, speed). It can be filled with a typicality field for motion in a certain subfield of the xy-plane in the scene, for example, a typicality field representing a turn-off model for a particular intersection. The typicality field results from accumulated and processed event instances (see |Mohnhaupt + Neumann 89|, |Mohnhaupt + Neumann 90|). Stationary scene objects like the street shape can also be filled in, from model-based expectations as well as from visual processes. The spatio-temporal buffer is an extension of the purely spatial buffer proposed by [Kosslyn 80]. One can think of it as an internal image-like representation with temporal behavior to simulate events of interest and to derive helpful information.

Given the typicality field of an event and a starting situation resulting from bottom-up analysis, a search space for the likely progression of the event instances can be computed by following all typicality values above a certain threshold. A subsequent motion analysis can focus on this search space which comprises the spatial and spatio-temporal constraints.

3. An example

Consider the task of analyzing a typical street traffic scene as depicted in Figure 1, a synthetic model of a real scene. The number of interesting objects and events, which could in principle be analyzed can be very big. To focus the analysis we assume top-down information given in terms of a natural language question like: *Did a car driving towards Dammtor turn off Schlüterstreet into Bieberstreet in front of the FB Informatik?* Top-down control is now performed in two steps. First, spatial constraints are exploited, and second additional motion constraints are derived.

Location information associated with a motion concept can be derived as follows. We assume the semantic content of the utterance to be represented in a case frame representation, including slots for agent, object, location goal, destination, directional, and the verb. The verb determines an event model (here the event model for 'turn-off'). From information about the applicability of turn-off events it is derived that they can only happen at intersections. The locative of the case frame allows to choose the intersection Schlüterstreet/Bieberstreet as a focus for analysis. In addition, the directional entry of the case frame further constrains the analysis, because a particular turn-off area can be inferred as shown in Figure 1 (dark area).

Our main focus is now on motion information associated with the motion concept to allow for further top-down control. Within the depicted dark area in Figure 1, a certain direction of motion and a certain speed can be expected in case of a turn-off event from Schlüterstreet into Bieberstreet driving towards Dammtor. Hence, the next step is to instantiate the spatio-temporal buffer with the scene geometry and the typicality distribution for turn-off. Given a starting point of the turn-off event in the image sequence, a spatio-temporal search area for subsequent motion analysis can be generated by considering continuations above a certain typicality. The prediction algorithm is local and provides location and velocity information. In our example all the successor cells with typicality values above a certain threshold lead to Figure 2.

In the example in Figure 2 the typicality distribution results from observing several turn-off examples, see Figure 3. xy-traces of observed examples are shown, objects are represented by their center of mass. Note that information about velocity is not visible in Figure 2 and Figure 3 but is part of the model. After recording examples, subsequent local processing leads to generalizations which cover the approximate area represented by the examples (see [Mohnhaupt + Neumann 90] for details of the learning steps and methods to instantiate typicality distributions from long-term memory and from models recorded in a different environment).

To support our considerations a real image sequence was recorded on this intersection. Figure 4 shows one frame of this sequence. The most interesting event for now is the white taxi turning off Schlüterstreet. Other moving objects include cars and pedestrians.

In order to be able to apply the constraining information shown in Figure 2 to the image sequence, the low-level motion representation is based on 3-dimensional Gabor cells [Adelson + Bergen 85]. The implementation is described in [Fleet 88]. The output of spatio-temporal Gabor cells is well suited for top-down control, as by using a Gabor filter bank an image sequence is decomposed into orientation, velocity, and scale specific information. Hence, top-down constraints can be brought to bear by selecting the appropriate subset of cells for an analysis (in the example, only those cells are chosen which are sensitive to motion towards the upper left according to the predictions computed within the buffer). Figure 5 shows the spatio-temporal energy of Gabor cells which are maximally sensitive to an orientation of 45 degrees with a speed of one pixel per frame. The main information about the taxi is within the predicted area, other motions as well as static information are removed.

4. Summary

We showed how to exploit motion concepts associated with verbs of locomotion for top-down control in traffic scenes. Two kinds of constraints could be derived: spatial constraints through knowledge about the applicability of motion concepts, and motion constraints through knowledge about typical motion. We proposed to compute motion constraints using a spatio-temporal buffer as a shared representation for bottom-up and top-down processes. Within the buffer motion concepts are expressed as typicality distributions from which predictions about object motion can be derived. A local prediction algorithm allows for the computation of search areas for low-level motion analysis. A low-level motion representation based on spatio-temporal Gabor cells is well suited for the integration of this kind of top-down information.

We presented an example where this procedure has been implemented. Using top-down guidance, the complexity of computation could be reduced significantly. Instead of analyzing the whole scene at the same level of detail, 1) only a small area could be chosen for an analysis and 2) the analysis could be focussed on specific spatio-temporal behavior within the area of interest.

5. References

[Adelson + Bergen 85]: Spatiotemporal energy models for the perception of motion, Journal of the Optical Society of America, A 2, 1985, pp. 284-299. [Fleet 88]: Implementation of Velocity-Tuned Filters and Image Encoding, Mitteilung, FBI-HH, Universität Hamburg, 1988. [Kosslyn 80]: Image and Mind, Harvard University Press, 1980. [Mohnhaupt + Neumann 89]: Some aspects of learning and reorganisation in an analogical representation, in 'Knowledge representation and organisation in machine learning', K. Morik (Ed.), Lecture Notes in Artificial Intelligence, Springer Verlag 1989, pp. 50-64. [Mohnhaupt + Neumann 90]: Understanding Object Motion: Recognition, Learning and Spatio-Temporal Reasoning, to appear in 'Journal of Robotics and Autonomous Systems', North Holland 1990. [Neumann 89]: Natural Language Description of Time-Varying Scenes, in 'Semantic Structures', David L. Waltz (Ed.), Lawrence Erlbaum, Hillsdale N.Y., 1989, pp. 167-207. [Tsotsos 87]: Image Understanding, in: S. Shapiro (Ed.), The encyclopedia of artificial intelligence, pp. 389-409, John Wiley and Sons, New York.

We thank David Fleet for many interesting comments and discussions.

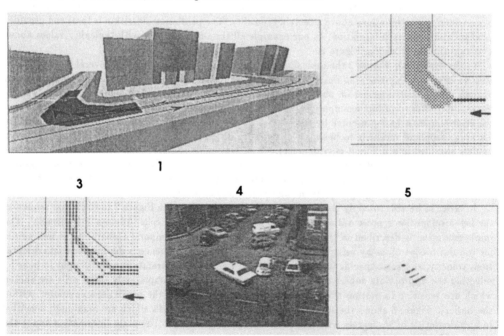

Final Steps Towards Real Time Trinocular Stereovision

Gregory Randall [1] Serge Foret [2] Nicholas Ayache [1]

(1) **INRIA** BP 105, 78153 Le Chesnay Cédex, France

(2) **NOESIS** 5 bis, rue du Petit Robinson, 78350 Jouy-en-Josas, France

1 Introduction

This paper explains the last updates made on the trinocular passive stereovision. This research is done for the ESPRIT Project P940 whose goals are, among others, to integrate this algorithm on a DSP 56000 board in a real time environment. The main idea of these improvements is to reduce the processing time and memory size. However, the basic philosophy keeps being the same as in the past publications:[1,2,3]. The main modifications brought to the algorithm are: (1) a better use of epipolar constraints, and (2) the introduction of a new prediction checking test in the 3rd image (avoiding the neighborhood validation which was memory and time consuming).

The paper only presents shortly the second improvement and results. An exhaustive presentation including the description of the algorithms can be found in a companion report [4].

2 Improvements to the Stereo Matching Algorithm

2.1 Verification

The problem is to compare a predicted 2D segment S_3^* in image 3 with a potential match segment S_3. This is done by the following method:

1. Perform first some simple loose tests to select quickly a subset of potential matches: selection is based on position, orientation and edge contrast comparisons. Buckets are used to cut the algorithm complexity.

2. Perform conservative tests using colinearity and overlapping measures.

2.1.1 Colinearity

Colinearity can be tested using the parameters (a^*, p^*) and (a, p) of the lines supporting S_3^* and S_3 (see appendix). Moreover, it is possible to account for the uncertainty attached to these parameters by introducing the covariance matrix $\Lambda = cov(a, p)$ within a Mahanalobis distance:

$$d^2(S_3^*, S_3) = (\Delta a, \Delta p)\, \Lambda^{-1} \begin{pmatrix} \Delta a \\ \Delta p \end{pmatrix}$$

Matrix Λ is computed from the knowledge of the uncertainty attached to the position of the endpoints of segment S_3. Typically, we assumed independent covariance of 1 pixel [2] for each endpoint coordinate.

The advantage of such an approach, is that a single threshold, chosen from a χ^2 table with two degrees of freedom allows for the homogeneous testing of very disparate segments (in terms of length, orientation and position). This method is similar to the one presented by Skordas et al [5].

2.1.2 Overlapping

First, we compute $\overline{S_3}$, the projection of S_3 onto the line supporting predicted segment S_3^*. Second, we compute the length of the intersection of this projection with S_3^*. Finally, the overlapping ratio is defined by the ratio between this intersection length and the length of S_3^*.

$$Overlapping(S_3^*, S_3) = \frac{\|(\overline{S_3} \cap S_3^*)\|}{\|(S_3^*)\|}$$

If the overlapping ratio is greater than a reasonable threshold, for instance 10%, segment S_3 is considered as a potential match and added to the list of hypotheses.

2.2 Validation

The new above-described prediction scheme yields much better results than the older one. Typically, the number of false matches lies between 0 and 5 % of the total number of matches, the upper bound being attained only on very cluttered scenes. Moreover, it is now possible, due to qualitative improvements, to remove most of the errors by a simpler and cheaper final validation procedure.

It appears that most errors produce multiple matches, but not all multiple matches are errors, because a given segment can be broken differently in two or three images. Our new validation procedure detects multiple matches, and only keeps those corresponding to the correct matching of broken segments. This procedure removes almost all the errors.

3 Experimental results

We have tested this algorithm on a number of industrial and office scenes, and we present in table 1 its performance on 5 typical examples. The programs are written in C and run on a SUN-3 workstation. Note that these results are obtained with fixed point arithmetics (in order to implement the algorithm on dedicated hardware). In the average, memory requirements have been reduced by more than 30 % and computing time has been reduced by a factor three.

For the first scene, Figure 1 presents the 3 camera images and the final results of matching and 3D reconstruction. On the "above view" we can recognize the table, the calibration grid and the edges of a box placed below and left of the table. Figure 1 also shows the reprojection of reconstructed 3D segments on each image plane.

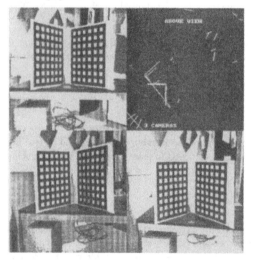

a   b

Figure 1: Scene 1: (a) Triplet of images of an indoor scene and 3d reconstruction results. (b) Reprojection of 3D segments on each image plane

4 Conclusion

We have presented the last developments made on our trinocular stereovision algorithm.

The final version of the presented algorithm is currently being implemented on a multi DSP 56000 board to produce depth maps at 5 Hz rate. This is done within a European *ESPRIT* Project (Project P940 involving ELSAG, GEC, INRIA, MATRA, NOESIS, Univ. of Cambridge, Univ. of Genova) where preprocessing (edge extraction, edge linking, polygonal approximation) is performed by a dedicated hardware at the rate of 24 Hz.

	Nb. Segments			Results	
	image 1	image 2	image 3	Nb. Matches	CPU Sun 3 Time
Scene 1	570	554	630	317	10 2s
Scene 2	495	489	520	298	8.2s
Scene 3	329	361	349	100	4.2s
Scene 4	333	361	282	93	4.2s
Scene 5	333	369	348	107	4.4s
Scene 6	400	342	361	186	6.1s
Scene 7	404	380	365	215	6.6s

Table 1: Performance of the trinocular stereovision algorithm on seven indoor scenes: computing time includes all processing comprising the reading of the image edge segments and the 3D reconstruction of matched triplets. A preliminary simulation on a DSP56000 multiDSP board yielded a computing time of 300 ms on scene 4, before optimization.

A APPENDIX : Mahalanobis Distance and Covariance Matrix

Let us consider a non vertical 2-D segment whose endpoints are $(x_1, y_1), (x_2, y_2)$. Given these two endpoints and their associated uncertainty ,we wish to find the parameters a and p representing the 2-D line as well as their uncertainty. These segment verify the line equation : $ax + y + p = 0$, so, a and p coefficients can be computed from segment coordinates:

$$a = \frac{y_1 - y_2}{x_2 - x_1} \quad p = \frac{x_1 y_2 - x_2 y_1}{x_2 - x_1}$$

The covariance matrix $\Lambda = cov(a, p)$ is given by:

$$\Lambda_{a,p} = \frac{\Delta x^2 + \Delta y^2}{\Delta x^4} \begin{pmatrix} 2 & -(x_1 + x_2) \\ -(x_1 + x_2) & x_1^2 + x_2^2 \end{pmatrix}$$

The determinant of this matrix: $Det = \frac{(\Delta x^2 + \Delta y^2)^2}{\Delta x^6}$ cannot be zero because we assume that the segment is non vertical (resp. non horizontal) and not reduced to a point.

References

[1] N. Ayache and F. Lustman. Fast and reliable passive trinocular stereovision. In *Proc. First International Conference on Computer Vision*, pages 422–427, IEEE, June 1987. London, U.K.

[2] N. Ayache and F. Lustman. Trinocular stereovision for robotics. In *Rapport de Recherche 1086, INRIA, Septembre 1989.*

[3] N. Ayache and C. Hansen. Rectification of images for binocular and trinocular stereovision. In *Proc. International Conference on Pattern Recognition*, October 1988. 9th, Beijing, China.

[4] G. Randall, S. Foret and N.Ayache. Real time implementation of trinocular stereovision. In *INRIA Research Report, to appear, 1990.*

[5] T. Skordas, P. Puget, R Zigmann and N. Ayache. Building 3-D Edge-Lines Tracked in an Image Sequence. In *Proc. Autonomous Intelligent Systems Conference* , December 1989. 2th, Amsterdam, The Netherlands.

B-Spline Contour Representation and Symmetry Detection [1]

Philippe Saint-Marc and Gérard Medioni

Institute for Robotics and Intelligent Systems
University of Southern California
Los Angeles, California 90089-0273
Email: medioni@usc.edu

The detection of edges is only one of the first steps in the understanding of images. Further processing necessarily involves grouping operations between contours. We present a representation of edge contours using approximating B-splines and show that such a representation facilitates the extraction of symmetries between contours. Our representation is rich, compact, stable, and does not critically depend on feature extraction whereas interpolating splines do. We turn our attention to the detection of two types of symmetries, parallel and skewed, which have proven to be of great importance to infer shape from contour, and show that our representation is computationally attractive. As an application, we show how parallel symmetries can be used to infer the 3-D orientation of a torus from its intensity image. Due to lack of space, the reader is referred to [4] for complete mathematical details, survey of previous work, and proper references.

Contour Representation

A very promising idea to represent image contours is to use piecewise polynomials. The advantages are obvious: this representation is rich, compact, analytical and local in the sense that a small change in the original curve does not affect the representation entirely. The approach commonly used consists of first extracting a set of knots from the discrete curve and then to approximate the curve between each pair of knots by polynomials under continuity constraints at the knots. For example, Plass and Stone [2] propose to take the knots as the vertices of a polygonal approximation, then to use dynamic programming in order to select those knots which provide the best approximation by cubics. The main point that we formulate against these methods is that they rely heavily on the always critical segmentation step, which brings up the stability issue. Also, techniques such as dynamic programming can yield a complexity of $O(n^3)$ where n is the number of initial knots [2]. Instead, we propose to use the following B-spline least-squares fitting method which, as we shall see, does not require any knot selection and is relatively insensitive to noise.

A spline can be expressed as a linear combination of B-splines which are themselves piecewise polynomials [1], the coefficients being the vertices of the spline's *guiding polygon*. Thus, a spline can be easily manipulated by modifying its guiding polygon, hence its popularity in CAD/CAM systems. Furthermore, as B-splines are defined locally, modifying the position of a vertex does not affect the spline entirely. In the case of a planar curve, a spline $Q(u) = (X(u), Y(u))$ with $m + 1$ vertices can be defined as $Q(u) = \sum_{j=0}^{m} V_j B_j(u) = \sum_{j=0}^{m}(X_j B_j(u), Y_j B_j(u))$, where (X_j, Y_j) are the *vertices* of the guiding polygon and $B_j(u)$ the B-splines.

Let C be an ordered set of $p + 1$ points $P_i = (x_i, y_i)$, what is the spline which best approximates C? An approach proposed in [1] consists of minimizing the distance $R = \sum_{i=0}^{p} \|Q(u_i) - P_i\|^2 = \sum_{i=0}^{p}(\sum_{j=0}^{m} X_j B_j(u_i) - x_i)^2 + (\sum_{j=0}^{m} Y_j B_j(u_i) - y_i)^2$, where u_i is some parameter value associated with the ith data point. Minimizing R is equivalent to setting all partial derivatives $\partial R/\partial X_l$ and $\partial R/\partial Y_l$ to 0, for $0 \le l \le m$, which yields two linear systems of equations. These are easily solved for all X_j and Y_j respectively using standard linear algebra, yielding the guiding polygon of the spline which best approximates the original curve. In the case of open curves, we have the option to force end-points to be interpolated. In this case, the first and last vertices are simply set to lie at the end-points so that the linear systems are reduced to $m - 1$ equations of $m - 1$ unknowns. In the case of closed curves, the linear systems are over constrained since some vertices are required to be identical. This method has proven to be relatively insensitive to noise [4]. The choice of m (the number of vertices)

---

[1]This research was supported by the Defense Advanced Research Projects Agency under contract F33615-87-C-1436 monitored by the Wright-Patterson Air Force Base.

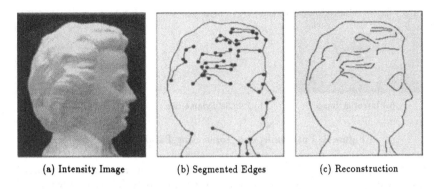

(a) Intensity Image (b) Segmented Edges (c) Reconstruction

Figure 1: Eine Kleine Nachtmusik...

determines how close to the original data the approximation is, which is measured by R (see above). The automatic selection of the number of vertices is not trivial. Our approach is to first set a fitting tolerance r_0, then find the value of m which yields the normalized distance $r = R/(p+1)$ closer to r_0 using a binary search approach.

The input for our system is an edge map produced by an edge detector such as Canny's. Three stages are sequentially considered: linking, corner detection, and spline approximation. Linking of the edgels is done using a simple and fast algorithm which looks for 8-connected components [4]. No gap-bridging or other task is performed since it is our belief that point-wise surgery is too myopic, and that if grouping is needed, it has to be performed at a higher level. Corner detection is performed by detecting tangent discontinuities in connected components after application of the *adaptive smoothing* operator [4]. The final step consists of approximating each elementary curve by a spline. When a closed curve with no corners is considered, a global least-squares approximation is performed. In the case of an open curve or a closed curve with corners, each curve segment between pairs of corners is approximated with the constraint that the end-points be interpolated. This insures the reconstructed curves to be continuous at corner locations. Figure 1 shows the results obtained on a real example. The 167×222 intensity image of a Mozart bust is displayed in figure 1(a) and the contours obtained after edge detection and linking in figure 1(b), with detected corners overlaid. Finally, a quadratic B-spline approximation of each curve segment between corners is done using a fitting tolerance of 0.5 which leads to the reconstruction displayed in figure 1(c).

It is interesting to point out that the method is very tolerant of segmentation errors since, if a corner is missed, more vertices will be used, hence the reconstruction will still be satisfying [4]. In the following section, we show how this piecewise polynomial representation of image contours can be used to detect symmetries in the image plane.

Symmetry Detection

The detection of symmetries is an essential step when inferring shapes from contours. In [5], Ulupinar and Nevatia claim that there exist two kinds of symmetry which give significant information about the surface shape for a variety of 3-D objects: *parallel* and *skewed* symmetry. Most methods for detecting symmetries in edge maps use local properties in order to identify symmetric edge points. In this case, it is necessary to test every possible pair of edge points against the property which leads to an $O(n^2)$ algorithm where n is the number of points [3]. Instead, we propose to use the B-spline representation in order to identify symmetric edge segments. On one hand, the complexity is reduced since n now represents the number of edge segments. On the other hand, the computation is more global, hence less sensitive to noise. Let us go into more details in the case of parallel symmetries knowing that a similar approach is used to detect skewed symmetries [4].

Let $c_1(u)$ and $c_2(v)$ be two parametric planar curves, and $\theta_1(u)$ and $\theta_2(v)$ their respective tangent orientations. These curves are said [5] to be parallel symmetric if there exists a continuous monotonic function f such that $\theta_2(f(u)) = \theta_1(u)$. It is easy to show that when considering two conics, then $f(u)$

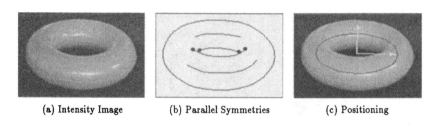

(a) Intensity Image   (b) Parallel Symmetries   (c) Positioning

Figure 2: Positioning of a Torus using Parallel Symmetries

is unique and is simply the ratio of two linear functions of u, with the result that two conics are always parallel symmetric [4]. Now supposing that $c_1(u)$ and $c_2(v)$ are only defined on the interval $[0,1]$, we will say that these two segments are parallel symmetric on $[u_0, u_1] \subseteq [0,1]$ iff $[f(u_0), f(u_1)] \subseteq [0,1]$. The detection of parallel symmetries between quadratic B-splines then follows: A quadratic B-spline can be expressed as a collection of connected conic segments $\mathcal{S} = \{c_i(u)\}$, for $i = 0, \cdots, m$, each defined on the interval $[0,1]$. Given another quadratic B-spline $\mathcal{S}' = \{c'_j(v)\}$, for $j = 0, \cdots, n$, each conic segment of $\mathcal{S}$ is compared against each conic segment of $\mathcal{S}'$ to eventually detect an elementary parallel symmetry between them. Given the simplicity of f and because of the usually small number of conic segments involved, the method is computationally very attractive. Grouping elementary symmetries can then be done by using simple connectivity criteria between segment pairs [4]. As an example, figure 2(b) shows the parallel symmetries detected in the edge map of the intensity image of figure 2(a).

The torus is an interesting example on which to demonstrate the application of parallel symmetry. Assuming that the object is far enough from the camera, and ignoring its actual size, it is reasonable to model the imaging process by an orthographic projection. The torus is a smooth solid of revolution, and the contours generated in its image correspond only to *limbs* or occluding contours, which are unfortunately viewer dependent. Even though it is possible, although complicated, to recover the position and orientation of a torus from its limbs, we propose instead to use the property of the torus that the axes of parallel symmetry in its image are the projection of its circular spine (3-D skeletal axis). This property allows us to recover the 3-D orientation quite simply: we fit an ellipse to the detected parallel symmetry axis, the orientation of the plane on which the torus is lying is given by the eccentricity of the ellipse and the angle of the major axis with the horizontal. Figure 2(c) shows the position of the torus recovered from the detected parallel symmetries of figure 2(b).

Conclusion

We have presented an approach to representing contours using approximating B-splines. It has attractive properties for use in Computer Vision: the representation is rich, compact, stable, local and segmented. We have shown how this representation can be used to extract parallel symmetries from edge maps. Similar ideas are used to extract skewed symmetries [4]. We are currently working on the selection of elementary symmetries, their grouping, and interpretation in order to generate higher level primitives. We also intend to apply these tools to the detection of local symmetries.

References

[1] R. Bartels, J. Beatty, and B. Barsky. *An Introduction to Splines for use in Computer Graphics and Geometric Modeling.* Morgan Kaufmann, Los Altos, CA 94022, 1987.

[2] M. Plass and M. Stone. Curve-Fitting with Piecewise Parametric Cubics. *ACM Transactions on Computer Graphics*, 17(3):229–239, 1983.

[3] J. Ponce. Ribbons, Symmetries, and Skew Symmetries. In *Proceedings of the DARPA Image Understanding Workshop*, pages 1074–1079, Cambridge, Massachusetts, 1988.

[4] P. Saint-Marc and G. Medioni. B-Spline Contour Representation and Symmetry Detection. Technical report, Institute for Robotics and Intelligent Systems, USC, Los Angeles, California 90089-0273, 1990.

[5] F. Ulupinar and R. Nevatia. Using Symmetries for Analysis of Shape from Contour. In *Proceedings of the International Conference on Computer Vision*, pages 414–426, 1988.

Shape and mutual cross-ratios with applications to exterior, interior and relative orientation [*]

Gunnar SPARR

Dept. of Mathematics, Lund Institute of Technology, Box 118, S-22100 Lund, Sweden

Lars NIELSEN

Dept. of Automatic Control, Lund Institute of Technology, Box 118, S-22100 Lund, Sweden

1 Introduction

Essential processes in computer vision consist in drawing quantitative information about a three-dimensional scene by means of two-dimensional perspective images. Here it belongs to everyones subjective experience that depth information is available from comparisons of the "shapes" of corresponding object and image configurations. Unfortunately, however, the concept of *shape* has no canonical mathematical description. In Section 2 a definition, convenient for the above purposes, will be given for discrete point configurations, and its transformation properties, called *mutual cross-ratios*, will be derived. In Section 3 these concepts are applied to various *orientation* problems, as described in e.g. Horn [1], ch. 13, with applications e.g. in robotics, landmark-guided navigation and cartography. Closed form computational schemes for these problems are constructed, invoking only the solution of linear systems of equations and polynomial equations in one variable. In the present article, only the case of *planar objects* will be discussed. No proofs are given. For a more thorough presentation, see Sparr, Nielsen [3].

2 Shape and mutual cross-ratios

While the objects of geometrical study usually are points, we will consider a geometry where the objects are sets of points, treated collectively. More precisely, by an *m-point configuration* in affine space is meant an ordered set of points $\mathcal{X} = \{X^1, \ldots, X^m\}$. Intending to define its shape by means of dual notions, the following lemma is fundamental.

Lemma 1 *Let x^i and $\bar{x}^i$ be the coordinate representations of X^i, $i = 1, \ldots, m$, with respect to two affine coordinate systems, where $\bar{x} = Ax + b$, A non-singular. Then holds, with $\mathcal{N}$ denoting matrix-nullspace,*

$$\mathcal{N} \begin{bmatrix} 1 & 1 & \cdots & 1 \\ x^1 & x^2 & \cdots & x^m \end{bmatrix} = \mathcal{N} \begin{bmatrix} 1 & 1 & \cdots & 1 \\ \bar{x}^1 & \bar{x}^2 & \cdots & \bar{x}^m \end{bmatrix} \tag{1}$$

One consequence of this lemma is that the set (1) is independent of the affine coordinate description of the configuration $\mathcal{X}$, and thus only depends on its intrinsic properties. This makes the following definition meaningful.

Definition. The *shape* of $\mathcal{X} = \{X^1, \ldots, X^m\}$ is the space

$$s(\mathcal{X}) = \mathcal{N} \begin{bmatrix} 1 & 1 & \cdots & 1 \\ X^1 & X^2 & \cdots & X^m \end{bmatrix}$$

[*]The work has been supported by the Swedish National Board for Technical Development (STU).

if non-trivial, where X^i, $i = 1, \ldots, m$, stands for coordinates in an arbitrary affine system. □

Below, non-degenerate 4-point configurations $\mathcal{X} = \{X^1, \ldots, X^4\}$ will play an important role. If e.g. X^1, X^2, X^3 are non-colinear, then $s(\mathcal{X})$ is the homogeneous 4-vector $(\xi_1, \xi_2, \xi_3, -1)$, obtained from the barycentric coordinate representation $X^4 = \xi_1 X^1 + \xi_2 X^2 + \xi_3 X^3$.

Lemma 2 *Let* $\mathcal{Y} = P(\mathcal{X})$, *where* P *is a perspectivity with center* Z, *such that*

$$\overline{ZX^i} = \alpha_i \overline{ZY^i}, \quad i = 1, \ldots, m \qquad (2)$$

Then to every $\xi \in s(\mathcal{X})$ *there exists* $\eta \in s(\mathcal{Y})$ *and a constant, independent of* Z, *such that*

$$\frac{\alpha_i}{\eta_i/\xi_i} = \text{const}, \quad i = 1, \ldots, m$$

This lemma gives a first example of what we call *mutual cross-ratios*. The name is chosen to refer to both the classical cross-ratio, see the remark below, and the fact that it invokes mutual properties of objects and images. The following two theorems show, respectively, how shapes can be used to determine the perspectivety between configurations of known shapes, and how they may be used to define *projective invariants*. (Other such invariants, generalizing the cross-ratio but based on areas, have been developed earlier by the authors, c.f. Nielsen, Sparr [2]. For an exhaustive treatment of projective geometry, see Veblen, Young [4].)

Theorem 1 *Suppose that* $P : \mathcal{X} \longmapsto \mathcal{Y}$, *where* $\mathcal{X}$ *and* $\mathcal{Y}$ *are 4-point configurations, and* P *a perspectivity obeying (2). Let* $\xi \in s(\mathcal{X})$, $\eta \in s(\mathcal{Y})$. *Then* $\alpha = (\alpha_1, \ldots, \alpha_4)$ *is uniquely determined, up to a factor of proportionality, by*

$$\frac{\alpha_1}{\eta_1/\xi_1} = \frac{\alpha_2}{\eta_2/\xi_2} = \frac{\alpha_3}{\eta_3/\xi_3} = \frac{\alpha_4}{\eta_4/\xi_4}$$

Theorem 2 *Let* $P : \{X^1, X^2, X^3, X, \tilde{X}\} \longmapsto \{Y^1, Y^2, Y^3, Y, \tilde{Y},\}$ *be a projectivity between two planar configurations. Suppose that* X^1, X^2, X^3 *and* Y^1, Y^2, Y^3 *are non-colinear. Let* $\xi = s(\{X^1, X^2, X^3, X\})$, $\tilde{\xi} = s(\{X^1, X^2, X^3, \tilde{X}\})$, *and* $\eta = s(\{Y^1, Y^2, Y^3, Y\})$, $\tilde{\eta} = s(\{Y^1, Y^2, Y^3, \tilde{Y}\})$. *Then holds*

$$\frac{\eta_1}{\xi_1} / \frac{\tilde{\eta}_1}{\tilde{\xi}_1} = \frac{\eta_2}{\xi_2} / \frac{\tilde{\eta}_2}{\tilde{\xi}_2} = \frac{\eta_3}{\xi_3} / \frac{\tilde{\eta}_3}{\tilde{\xi}_3} \qquad (3)$$

Remark. The analogue of (3) for linear configurations is the classical cross-ratio

$$\frac{\eta_1}{\xi_1} / \frac{\tilde{\eta}_1}{\tilde{\xi}_1} = \frac{\eta_2}{\xi_2} / \frac{\tilde{\eta}_2}{\tilde{\xi}_2} \iff \frac{\eta_1}{\eta_2} / \frac{\tilde{\eta}_1}{\tilde{\eta}_2} = \frac{\xi_1}{\xi_2} / \frac{\tilde{\xi}_1}{\tilde{\xi}_2} \qquad \square$$

To summarize, the results of this section provide parametrizations of the projective plane, similar to the ones obtained from the cross-ratio on the projective line. In principle, in the planar case algorithms for recognition and for the correspondence problem can be based on comparisons of a 5-point object and its image, by means of Theorem 2. When a correspondence is attained, α is obtained from Theorem 1, and the image of an arbitrary point can be computed. This fact that can be used e.g. to exclude impossible situations in point matching processes.

3 Orientation problems

Shapes and mutual cross-ratios can be used to derive closed formed solutions of orientation problems. For the *exterior/interior* (or *calibration* problem), it can be proved that for two given planar 4-point configurations of known shapes, if there exists a perspectivity between them, then its center is bound to an elliptical space curve. Its equation can be computed explicitly, in terms of the shapes and metrical properties of the two configurations. For details, see Sparr, Nielsen [3]. For the *relative orientation* problem we will be somewhat more detailed, and in fact sketch a

Computational scheme. Let $\mathcal{Y}$ and $\overline{\mathcal{Y}}$ be two images of some unknown planar 4-point configuration $\mathcal{X}$, which has undergone an isometric transformation between two imaging instants, while the camera is held fixed. The problem is to determine this transformation. It is solved once the parameters α and $\overline{\alpha}$ corresponding to the two locations of $\mathcal{X}$ are known. With only 4 image correspondences, the solution is not unique. Having one more correspondence, the results below can be combined with Theorem 2 to eliminate impossible cases.

Let η and $\bar{\eta}$ be the shapes of $\mathcal{Y}$ and $\overline{\mathcal{Y}}$ respectively. If ξ is the (unknown) shape of $\mathcal{X}$, then Theorem 1 says

$$\frac{\alpha_1}{\eta_1/\xi_1} = \ldots = \frac{\alpha_4}{\eta_4/\xi_4}, \quad \frac{\bar{\alpha}_1}{\bar{\eta}_1/\xi_1} = \ldots = \frac{\bar{\alpha}_4}{\bar{\eta}_4/\xi_4}$$

Pairwise division gives

$$\frac{\alpha_1}{\bar{\alpha}_1}\Big/\frac{\eta_1}{\bar{\eta}_1} = \frac{\alpha_2}{\bar{\alpha}_2}\Big/\frac{\eta_2}{\bar{\eta}_2} = \frac{\alpha_3}{\bar{\alpha}_3}\Big/\frac{\eta_3}{\bar{\eta}_3} = \frac{\alpha_4}{\bar{\alpha}_4}\Big/\frac{\eta_4}{\bar{\eta}_4} \tag{4}$$

Let y^k and $\bar{y}^k$ denote the coordinates of Y^k and $\overline{Y}^k$, respectively, with respect to a system with origin in the focus. The metrical camera and image information is contaied in the Gramians $G = (g_{ij})$ with $g_{ij} = y^i \cdot y^j$, $\overline{G} = (\bar{g}_{ij})$ with $\bar{g}_{ij} = \bar{y}^i \cdot \bar{y}^j$. The fact that $\mathcal{X}$ has undergone an isometric transformation between the two imaging instants gives

$$\alpha_i^2 g_{ii} - 2\alpha_i\alpha_j g_{ij} + \alpha_j^2 g_{jj} = \bar{\alpha}_i^2 \bar{g}_{ii} - 2\bar{\alpha}_i\bar{\alpha}_j \bar{g}_{ij} + \bar{\alpha}_j^2 \bar{g}_{jj}$$

for $i, j = 1, 2, 3$, pairwise distinct. Let $\sigma_i = \eta^i/\bar{\eta}^i$. Then, by (4), for some t holds $\bar{\alpha}_i = t\sigma_i\alpha_i$. Hence

$$(g_{ii} - t^2\sigma_i^2\bar{g}_{ii})\alpha_i^2 - 2(g_{ij} - t^2\sigma_i\sigma_j\bar{g}_{ij})\alpha_i\alpha_j + (g_{jj} - t^2\sigma_j^2\bar{g}_{jj})\alpha_j^2 = 0 \tag{5}$$

A necessary and sufficient condition for the existence of solutions to the three equations (5) is the vanishing of their resultant. This gives a polynomial equation in t^2, of degree 9. After insertion of its non-negative solutions, two of the second order equations (5) will give α_1/α_2 and α_2/α_3. In this way, all possible α and $\overline{\alpha}$ are determined, up to proportionality. This settles the problem.

4 Conclusions

An affine concept of shape is introduced, which is independent of the coordinate description. By comparing the shapes of corresponding object and image configurations, it is possible to draw information about the projective mapping that associates them. This is done by means of mutual cross-ratios, which generalize the classical cross-ratio on the line. The geometric machinery is applied to the various orientation problems (exterior, interior and relative). It makes it possible to construct computational schemes, involving only the solutions of systems of linear equations and polynomial equations in one variable.

References

[1] Horn, B.K.P., *Robot Vision*, MIT Press, 1986.

[2] Nielsen, L. and Sparr, G., *Projective Area-Invariants as an Extension of the Cross-Ratio*, in *The 6th Scandinavian Conference on Image Analysis, Oulu*, 969-986, 1989.

[3] Sparr, G. and Nielsen, L., *Shape and mutual cross-ratios with applications to orientation problems*, Technical report, Dept. of Mathematics, Lund, 1990.

[4] Veblen, O. and Young, J.W., *Projective Geometry*, Ginn and Company, Boston, 1910.

Using Neural Networks to Learn Shape Decomposition by Successive Prototypication

Dr Nicholas Walker

Imperial Cancer Research Fund Laboratories, London

Abstract

I describe a neural-network which decomposes a set of inputs into a sequence of generative parameters. It uses a series of coupled parameter finding and removing networks and requires the input to be in a particular temporal format.

Introduction

If a multi-layer neural-network is required to develope its own representations, a good technique is to teach it to reconstruct its input at its output. The hidden layer has to code the inputs in some way and it is this code that can be used as a representation. Hidden layers taught by back-propagation tend to adopt all-or-nothing responses to one or more 'features' in the input. If the particular inputs are examples from a continuous set, as is the case in shape description, we actually want the hidden units to adopt a continuous coding using one or more parametric variables.

Learning to parameterise

There are several techniques which use a combination of competition between hidden units for activation (winner-takes-all) and adoption of similar activity in neighbouring units (where the neighbourhood relation is defined on some topology) and generate exactly these continuous parametric variables [1,2]. I have chosen Saund's technique as it incorporates back-propagation and so can integrate with hidden unit layers which are not being forced to parameterise. He uses an additional error, related to this mixture of competition for activation and spreading of activity between neighbours, which is added to the back-propagation error in the parameterising hidden units.

Figure 1 shows the hidden unit responses to a set of images of the pixel values of circles of varying diameter centred in the image. The first part of the figure shows the result of back-propagation alone, the units respond in an all or nothing fashion to one or more small ranges of circle diameter. Diameter is then coded by bands where a different set of units are switched on or off. It is not possible, just looking at the hidden unit responses, to know how similar two inputs are to each other, except where their coding is identical. The second part of the figure show a parametric coding - the units lie on a 1D array and their activity presents a single parameter which codes the inputs, in this case it corresponds to the diameter. A least-squares fit of a gaussian to the hidden unit responses shows a virtually linear relationship to diameter, except for very small circles where there is probably a quantisation problem.

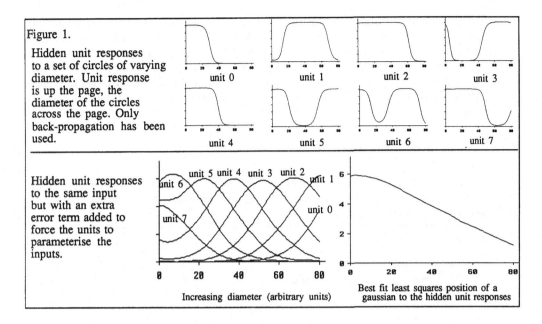

Figure 1.

Hidden unit responses to a set of circles of varying diameter. Unit response is up the page, the diameter of the circles across the page. Only back-propagation has been used.

unit 0   unit 1   unit 2   unit 3

unit 4   unit 5   unit 6   unit 7

Hidden unit responses to the same input but with an extra error term added to force the units to parameterise the inputs.

unit 6   unit 5 unit 4 unit 3 unit 2   unit 1

unit 0

unit 7

Increasing diameter (arbitrary units)

Best fit least squares position of a gaussian to the hidden unit responses

Sequential parameterisation

Dealing with inputs requiring many parameters would only seem to involve extending the topology to the necessary dimensions. However there are problems with this even if we can cope with the computational costs.

• The isotropic nature of the function which spreads activation over the neighbourhood means there is an arbitrary relationship between the axes of the internal parametric space and any particular 'natural parameters' of the input set.

• The distribution of the the internal parametric space depends only on the similarity of the inputs and no account can be taken of relationships such as the fact that two inputs close together in time are likely to require similar internal representations (shape constancy).

An alternative is to perform a sequential parameterisation. The value of a single most explanatory parameter is found and this value is used to 'correct' the input for that parameter, i.e. transform the inputs in a way that corresponds to setting that parameter to a fixed, prototypical, value. For example undoing position means transforming the image so that an object is centred on the origin. Then the next most explanatory parameter is found and so on.

Leyton [3] suggests that shape perception consists of just such a sequential parametric decomposition. The parameters can be considered as the generative processes that have formed the shape.

The assumption that allows learning such a series of parameters is that they have a temporal ordering, with each parameter changing faster than those after it. Over some time interval a parameter can assume that the parameters following it will remain constant. We don't have to worry about the parameters before it in the ordering as they will have been removed by the prototyping networks.

Two networks are required to learn the parameterisation. Consecutive sets of inputs are formed from time slices of a larger input set. The first network, having used Saund's technique to distribute a parameter over the first of these sets then continuously modifies this distribution for the following sets. The parameterising hidden layer responses of this network are used to train a second, 3-layer back-propagation network working from the associated input. Eventually this network will generate the parameter over the full set of inputs.

Another network takes this parameter value to effect the prototypication. The best way to construct a neural-network to perform coordinate axis transformations is not known, but Zipser and Anderson [4] use a 3-layer network and I have done the same. To teach it the transformation which will undo the parameter we have to decide whether it represents a prototypical value or not. If it is the associated input is transferred to a stored training image. The transforming network is taught to produce this stored image as output, with the original input plus the output of the parameter assigning network as input.

A two step parameterisation

This multiple network training scheme was used to prototype lines of different orientation and length. The orientation changed 100 times faster than the length of the lines. The parametric grid that the two stages produce is shown in figure 2.

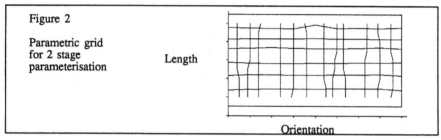

Figure 2

Parametric grid for 2 stage parameterisation

Length

Orientation

Conclusion

The aim of this scheme is to allow a network to generate its own representations of a particular input but to know in advance the properties of the representation it will develope, so it can be used by other networks. The scheme appears to be viable, though only trivial inputs have so far been prototyped.

Its success depends on the data having the right temporal format. An Active Vision System, able to control the position of its sensors, could arrange this. For instance it could arrange that rotations along the axis of the camera are very much less frequent than panning motions of the camera. As the sequential parameters are found it could move its sensors in different ways to correctly format the sensory data.

References

1. Kohonen T. (1984). "Self organisation and Associative memory".
Springer Verlag, Berlin 1984.

2. Saund E. (1989). "Dimensionality Reduction using Connectionist Networks".
IEEE PAMI Vol 11 No 3, pp 304-314.

3. Leyton M. (1985). "Generative Systems of Analysers".
Computer Vision, Graphics and Image Processing 31, pp 201-241.

4. Zipser D. and Anderson R.A. (1988). "A back-Propagation Programmed Network that Simulates Response Properties of a Subset of Posterior Parietal Neurons".
Nature Vol 331, pp 679-684.

Adapting Computer Vision Systems to the Visual Environment: Topographic Mapping[1]

Thomas Zielke, Kai Storjohann, Hanspeter A. Mallot and Werner von Seelen
Institut für Neuroinformatik, Ruhr-Universität
D-4630 Bochum, FRG

1 Introduction

Topographic mappings are neigborhood preserving transformations between two–dimensional data structures. Mappings of this type are a general means of information processing in the vertebrate visual system. In this paper, we present an application of a special topographic mapping, termed the *inverse perspective mapping*, for the processing of stereo and motion. More specifically, we study a class of algorithms for the detection of deviations from an expected "normal" situation. These expectations concern the global space–variance of certain image parameters (e.g., disparity or speed of feature motion) and can thus easily be implemented in the mapping rule. The resulting algorithms are minimal in the sense that no irrelevant information is extracted from the scene. In a technical application, we use topographic mappings for a stereo obstacle detection system. The implementation has been tested on an automatically guided vehicle (AGV) in an industrial environment.

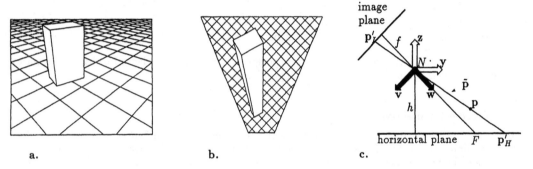

a.   b.   c.

Figure 1: **a.** Perspective view. **b.** Inverse perspective view. **c.** Imaging geometry.

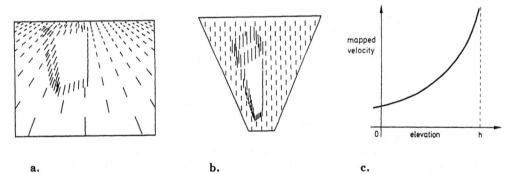

a.   b.   c.

Figure 2: **a.** Flow field of the projected image. **b.** Flow of the mapped image. **c.** Mapped velocity as a function of elevation (cf. Eq. 2).

---

[1]Supported by the German Federal Department of Research and Technology (BMFT)

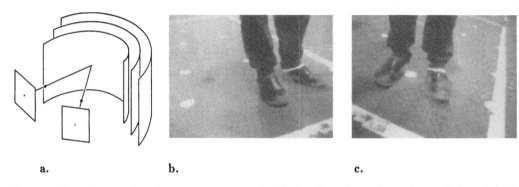

a.   b.   c.

Figure 3: Normal stereo imaging geometry: **a.** Cylindric iso–disparity surfaces. **b., c.** Left and right view of an obstacle.

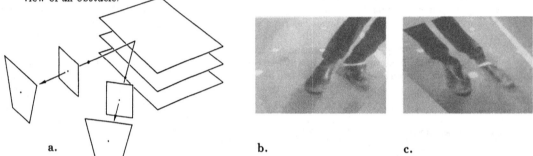

a.   b.   c.

Figure 4: Stereo imaging geometry with inverse perspective mapping. **a.** Planar iso–disparity surfaces. **b., c.** Stereo pair mapped to a common 'cyclopean' plane.

2 Inverse Perspective Mapping

Inverse perspective mappings project the image of a scene onto a plane different from the image plane while retaining the center of projection. An important application of the inverse perspective mapping is to undo the perspective distortion with respect to a certain flat visible surface of which the relative altitude is known (Fig. 1a,b). For camera coordinates $I := \{\mathbf{u}, \mathbf{v}, \mathbf{w}\}$, the inverse perspective mapping $\mathcal{Q} : (u', v') \mapsto (x', y')$ is given by

$$\begin{pmatrix} x' \\ y' \end{pmatrix} := \frac{-h}{u_3 u' + v_3 v' - w_3 f} \cdot \begin{pmatrix} u_1 u' + v_1 v' - w_1 f \\ u_2 u' + v_2 v' - w_2 f \end{pmatrix}, \tag{1}$$

By construction, the composition of perspective mapping $\mathcal{P}_I$ onto the image plane and the inverse perspective mapping $\mathcal{Q}$ from the image plane to the horizontal plane is identical to the projection through the camera nodal point onto the horizontal plane. It is only this projection $\mathcal{Q} \circ \mathcal{P}_I$ that we need to discuss in the applications: $\mathcal{Q} \circ \mathcal{P}_I(\mathbf{p}) = \mathcal{P}_H(\mathbf{p}) = -h/p_3 (p_1, p_2)^\top$.

3 Body–Scaled Obstacle Detection from the Mapped Velocity Field

If a mobile robot is moving in the horizontal plane at a constant translation $\mathbf{m}$, a stationary $3D$ point $\mathbf{r}$ will move with a motion vector $d\mathbf{r}/dt = -\mathbf{m}$ relative to the camera frame. In the image plane, the projected motion vector W_I' at the projection of $\mathbf{r}$ is determined by $W_I'(\mathcal{P}(\mathbf{r})) := d\mathcal{P}_I(\mathbf{r})/dt = -\mathbf{J}_{\mathcal{P}_I}(\mathbf{r}) \cdot \mathbf{m}$, where $\mathbf{J}_{\mathcal{P}_I}$ denotes the Jacobian of the perspective projection (Fig. 2a). In inverse perspective mapping, $\mathbf{J}_{\mathcal{P}_I}$ is replaced by $\mathbf{J}_{\mathcal{P}_H}$. Since egomotion is bound to the plane ($m_3 = 0$), the mapped velocity field W_H' becomes:

$$W_H' = \frac{h}{r_3} \begin{pmatrix} m_1 \\ m_2 \end{pmatrix} \quad ; \quad \|W_H'\| = \left| \frac{h}{h - elev.} \right| \cdot \|\mathbf{m}\|, \tag{2}$$

where $elev. := h + r_3$ is the elevation of the point $\mathbf{r}$ above the ground plane, i.e., its importance as an obstacle. (Note that $r_3 < 0$ in typical cases.) From here, it is easy to detect the obstacle by a local

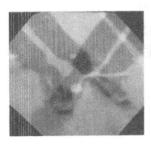  

a.   **b.**   **c.**

Figure 5: **a.** Interlaced representation of cyclopean view. **b.** In the difference image, the obstacle stands out clearly. **c.** An automatically guided vehicle equipped with a visual navigation system

4   Zero Disparity Mapping for Obstacle Detection

For a stereo camera system with the cameras aligned so that the optical axes intersect in some point of fixation, points with constant disparity correspond to horopter surfaces in space (Fig. 3a). The cylindric iso-disparity surfaces can be transformed into parallel planes if we map both the left and the right image to a common 'cyclopean' plane (Fig. 4a). More specifically, it can be shown that for camera nodal points N_R, N_L, the disparity d in the cyclopean image becomes:

$$d \;=\; |elev/(h - elev)| \cdot \|N_R - N_L\| \; ; \;\; elev := h + p_3. \tag{3}$$

Eq. 3 shows that disparity in the mapped image depends only on the z-component of the imaged point. By choosing other mapping functions, different 3D–surfaces can be made 'horopter'-surfaces, as well.

Our general approach to visual obstacle detection is to devise an economical test of whether the image region of interest contains visible objects other than the floor the robot moves on. For automatically guided vehicles (AGVs) that operate on flat floors in an industrial environment, applying the inverse perspective mapping to stereo images offers a powerful alternative to the (monocular) flow field approach. Considering the case when no obstacles are present, two cameras pointing towards the floor will "see" the same picture in the binocular field of vision, exept for a difference in the perspective. Knowing the camera calibration parameters we apply an inverse perspective mapping to both images, projecting the two views into one central cyclopean view.

We denote by Q_L and Q_R the left and right inverse perspective maps. Let $\mathbf{p}'_L$ and $\mathbf{p}'_R$ denote the left and right image of a 3D–point $\mathbf{p}$. If $p_3 = -h$, both images are mapped onto the same point $\mathbf{p}'_H$ in the central view:

$$-1/2(s_1, s_2)^\top + Q_L(\mathbf{p}'_L) = \;\; \mathbf{p}'_H \;\; = 1/2(s_1, s_2)^\top + Q_R(\mathbf{p}'_R) \tag{4}$$

Fig. 3a,b shows a stereo pair taken by the camera system mounted on our AGV, called VISOCAR [2] (Fig. 5c). Fig. 4b,c show the cyclopean images corresponding to the left and right views from Fig. 3b,c. Local mismatches of the two mapped images indicate possible obstacle locations. An interlaced representation of the stereo pair can be used to perform a test of expected correspondences (Fig. 5a). In the technical application, we simply compute the difference of the mapped images, followed by a threshold operation which yields an image segmentation into "free space" (black) and "suspicious regions" (white), cf. Fig. 5b.

References

[1] R. O. Duda and P. E. Hart. John Wiley & Sons, New York, 1973.

[2] H. Frohn & W. von Seelen. *Proc. IEEE Int. Conf. Robotics a. Autom.*, pp 1155 – 1159, 1989.

[3] H. A. Mallot, H. H. Bülthoff, and J. J. Little. Artif. Intell. Lab. Memo. 1067, MIT, 1989.

[4] H. A. Mallot, W. von Seelen, and F. Giannakopoulos. *Neural Networks*, 1990. in press.

[5] K. Storjohann, E. Schulze, W. von Seelen. In: H. Bunke, O. Kübler, Stucki (Eds.) *Mustererkennung 1988*, Proc. 10. DAGM-Symposion, Zürich, Switzerland, 1988.

Stereo Matching Based on a Combination of Simple Features Used for Matching in Temporal Image Sequences

Georg Zimmermann

Fraunhofer-Institut für Informations- und Datenverarbeitung (IITB)

Fraunhoferstraße 1, D-7500 Karlsruhe 1, Tel. (0721)6091-255

Introduction

In image sequence understanding, crucial tasks are the extraction of information about the 3D-content and motions from the depicted scene. Further, the recognition of objects in the scene is necessary. For the first two tasks, an intermediate level of description are displacement vector fields (DVF) which describe in the stereo case the disparity between the objects depicted by the two cameras and in time sequences the displacement due to the motion of the objects in the scene and/or the camera itself. A large amount of work has been done for the estimation of DVF's; for an overview see Nagel /1/.

This paper gives a common framework for the processing of temporal image sequences and stereo pairs. Compared with most of the work done in stereo matching /2/, in this paper no a priori knowledge about the scene and no camera calibration is used. The stereo image pair used has been taken from a stereo image sequence taken under real conditions for car driving in a road scene.

Estimation of displacement vector fields

For the description of images, a multitude of features can be used, such as edges, corners, lines, blobs etc. Here, blobs delivered by the so called monotonicity operator /3/ are used. This operator detects areas of local extrema in the bandpass filtered image and represents them as the center of gravity of binary blobs. There is only one relevant parameter in this method: the center wavelength of the bandpass filter. Due to the bandpass filtering, local extrema in the image are separated at least by the distance of one wavelength, thus giving a valuable hint for the solution of the correspondence problem: If the displacement between two frames is less than half the bandpass wavelength, the phase information of the spatial Fourier components is preserved and therefore unique.

This method has been applied successfuly to a large variety of imagery of indoor and outdoor scenes. In Fig. 1, an example for an DVF is given which was used to detect automatically a starting truck in front of the camera /4/. This field displays only vectors from features, which persisted over 125 frames (5 seconds) and demonstrates besides the longevity, that vectors are found both in natural environment and on man made objects. No vectors are found in the sky region where only fuzzy results could be expected.

Correspondence in Stereo Pairs

In stereo pairs, the displacement between corresponding areas can be of an arbitrary extent. Therefore complex features of a moderate size are needed which are almost unique within the image. Such complex features are constructed here by combining blobs originating from three different wavelengths in a restricted area of the image, thus combining the phase information of three different spatial Fourier components. As these complex features have some similarity to agglomerations of more or less bright stars in the night sky, they are called 'constellations'. The construction of a constellation is done in the following way: **1)** With a large, medium and fine wavelength of the bandpass filter, extract coarse, medium and fine blobs from the image, respectively. **2)** Within a given neighborhood, assign to each coarse blob the medium ones. **3)** Do the same with the medium and fine ones. This results in a list with pointers from the coarse to the medium and from there to the fine blobs. For each blob, the center of gravity and the type of the blob (local maximum or minimum) is entered.

Two constellations are compared by searching through the lists, checking whether the medium and fine blobs are in the same position relative to the coarse one. The measure of similarity used currently simply counts the number of coinciding blobs. A correspondence is assumed if more than three blobs coincide and if the search from one constellation to the other and vice versa leads to the same combination.

Fig. 2 displays the result of a crucial test of the method. With a stereo image pair taken from a car riding with a speed of about 40 km/h through a street in the city of Karlsruhe, correspondences were looked for. No information about the relative position of the cameras was used at all. The search area for corresponding constellations was chosen to be the whole image width horizontally (512 pixels) and +/- 256 pixels vertically, which is nearly half the image area. Despite of these difficulties, 31 correspondences have been found correctly and only one is obviously false. It can be seen, that the false one can be easily eliminated by using epipolar geometry.

Conclusion

The complex image feature 'constellation' has been derived in a natural way from the simple features of the monotonicity-operator. It is demonstrated that the constellations can be used for stereo correspondence analysis. Hardware is available to compute the simple features at TV-rate /5/. Furthermore, it has been shown in /6/ that constellations are good for extracting simple image descriptions automatically.

Acknowledgement: This work was funded in parts by the German MoD.

References

/1/ H.-H. Nagel: Image sequences - ten (octal) years - from phenomenology towards a theoretical foundation. International Journal on Pattern Recognition and Artificial Intelligence; Vol. 2, No. 3 (1988), pp. 459-483.

/2/ A. Rosenfeld: Image Analysis and Computer Vision: 1988. Computer Vision, Graphics and Image Processing, 46, 169-264 (1988)

/3/ R. Kories, G. Zimmermann: A Versatile Method for the Estimation of Displacement Vector Fields from Image Sequences. Proc. of the Workshop on Motion: Representation and Analysis., May 1986, Kiawah Island Resort, Charleston, SC, pp. 101-106.

/4/ R. Kories, N. Rehfeld, G. Zimmermann: Towards Autonomous Convoy Driving: Recognizing the Starting Vehicle in Front. 9th ICPR 1988, Rome, Italy, pp. 531-535.

/5/ D. Paul, W. Haettich, W. Nill, S. Tatari and G. Winkler : VISTA: Visual Interpretation System for Technical Applications - Architecture and Use. IEEE Trans. on Pattern Analysis and Machine Intelligence. Vol.-PAMI- 10, No. 3, May 1988, pp. 399-407.

/6/. G. Zimmermann: Creating and Verifying Automatically a Representation of Three Streets in Karlsruhe City Using Complex Image Features. Proc. Conference on Intelligent Autonomous Systems-2, Dec. 1989, Amsterdam, The Netherlands. pp. 705-714.

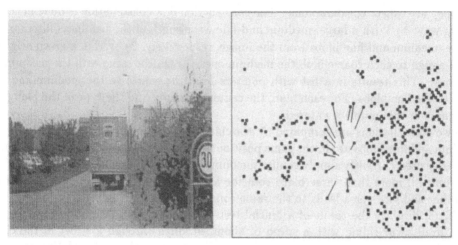

Fig. 1: Displacement vector field from a TV image sequence 5 seconds (125 frames) long. One frame is displayed on the left. On the right hand side, the displacement of only those features is indicated by needles, which could be tracked through the entire sequence. The large vectors in the center are attached to the truck which starts near the camera and becomes smaller in the image while it drives away.

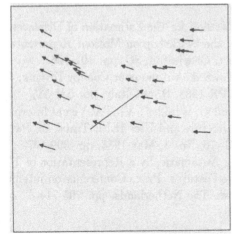

Fig. 2, Upper part: Left and right image of a stereo pair taken by cameras mounted on a moving car.
Lower part: The correspondences found automatically using the complex feature 'constellation'. It can be seen, that the two cameras are not aligned. The only false correspondence indicates the large search space (about half the image area) used in this example.

AUTHOR INDEX

Vol. 379: A. Kreczmar, G. Mirkowska (Eds.), Mathematical Foundations of Computer Science 1989. Proceedings, 1989. VIII, 605 pages. 1989.

Vol. 380: J. Csirik, J. Demetrovics, F. Gécseg (Eds.), Fundamentals of Computation Theory. Proceedings, 1989. XI, 493 pages. 1989.

Vol. 381: J. Dassow, J. Kelemen (Eds.), Machines, Languages, and Complexity. Proceedings, 1988. VI, 244 pages. 1989.

Vol. 382: F. Dehne, J.-R. Sack, N. Santoro (Eds.), Algorithms and Data Structures. WADS '89. Proceedings, 1989. IX, 592 pages. 1989.

Vol. 383: K. Furukawa, H. Tanaka, T. Fujisaki (Eds.), Logic Programming '88. Proceedings, 1988. VII, 251 pages. 1989 (Subseries LNAI).

Vol. 384: G. A. van Zee, J. G. G. van de Vorst (Eds.), Parallel Computing 1988. Proceedings, 1988. V, 135 pages. 1989.

Vol. 385: E. Börger, H. Kleine Büning, M. M. Richter (Eds.), CSL '88. Proceedings, 1988. VI, 399 pages. 1989.

Vol. 386: J.E. Pin (Ed.), Formal Properties of Finite Automata and Applications. Proceedings, 1988. VIII. 260 pages. 1989.

Vol. 387: C. Ghezzi, J. A. McDermid (Eds.), ESEC '89. 2nd European Software Engineering Conference. Proceedings, 1989. VI, 496 pages. 1989.

Vol. 388: G. Cohen, J. Wolfmann (Eds.), Coding Theory and Applications. Proceedings, 1988. IX, 329 pages. 1989.

Vol. 389: D.H. Pitt, D.E. Rydeheard, P. Dybjer, A.M. Pitts, A. Poigné (Eds.), Category Theory and Computer Science. Proceedings, 1989. VI, 365 pages. 1989.

Vol. 390: J.P. Martins, E.M. Morgado (Eds.), EPIA 89. Proceedings, 1989. XII, 400 pages. 1989 (Subseries LNAI).

Vol. 391: J.-D. Boissonnat, J.-P. Laumond (Eds.), Geometry and Robotics. Proceedings, 1988. VI, 413 pages. 1989.

Vol. 392: J.-C. Bermond, M. Raynal (Eds.), Distributed Algorithms. Proceedings, 1989. VI, 315 pages. 1989.

Vol. 393: H. Ehrig, H. Herrlich, H.-J. Kreowski, G. Preuß (Eds.), Categorical Methods in Computer Science. VI, 350 pages. 1989.

Vol. 394: M. Wirsing, J.A. Bergstra (Eds.), Algebraic Methods: Theory, Tools and Applications. VI, 558 pages. 1989.

Vol. 395: M. Schmidt-Schauß, Computational Aspects of an Order-Sorted Logic with Term Declarations. VIII, 171 pages. 1989. (Subseries LNAI).

Vol. 396: T.A. Berson, T. Beth (Eds.), Local Area Network Security. Proceedings, 1989. IX, 152 pages. 1989.

Vol. 397: K.P. Jantke (Ed.), Analogical and Inductive Inference. Proceedings, 1989. IX, 338 pages. 1989. (Subseries LNAI).

Vol. 398: B. Banieqbal, H. Barringer, A. Pnueli (Eds.), Temporal Logic in Specification. Proceedings, 1987. VI, 448 pages. 1989.

Vol. 399: V. Cantoni, R. Creutzburg, S. Levialdi, G. Wolf (Eds.), Recent Issues in Pattern Analysis and Recognition. VII, 400 pages. 1989.

Vol. 400: R. Klein, Concrete and Abstract Voronoi Diagrams. IV, 167 pages. 1989.

Vol. 401: H. Djidjev (Ed.), Optimal Algorithms. Proceedings, 1989. VI, 308 pages. 1989.

Vol. 402: T.P. Bagchi, V.K. Chaudhri, Interactive Relational Database Design. XI, 186 pages. 1989.

Vol. 403: S. Goldwasser (Ed.), Advances in Cryptology – CRYPTO '88. Proceedings, 1988. XI, 591 pages. 1990.

Vol. 404: J. Beer, Concepts, Design, and Performance Analysis of a Parallel Prolog Machine. VI, 128 pages. 1989.

Vol. 405: C.E. Veni Madhavan (Ed.), Foundations of Software Technology and Theoretical Computer Science. Proceedings, 1989. VIII, 339 pages. 1989.

Vol. 407: J. Sifakis (Ed.), Automatic Verification Methods for Finite State Systems. Proceedings, 1989. VII, 382 pages. 1990.

Vol. 408: M. Leeser, G. Brown (Eds.) Hardware Specification, Verification and Synthesis: Mathematical Aspects. Proceedings, 1989. VI, 402 pages. 1990.

Vol. 409: A. Buchmann, O. Günther, T.R. Smith, Y.-F. Wang (Eds.), Design and Implementation of Large Spatial Databases. Proceedings, 1989. IX, 364 pages. 1990.

Vol. 410: F. Pichler, R. Moreno-Diaz (Eds.), Computer Aided Systems Theory – EUROCAST '89. Proceedings, 1989. VII, 427 pages. 1990.

Vol. 411: M. Nagl (Ed.), Graph-Theoretic Concepts in Computer Science. Proceedings, 1989. VII, 374 pages. 1990.

Vol. 412: L.B. Almeida, C.J. Wellekens (Eds.), Neural Networks. Proceedings, 1990. IX, 276 pages. 1990.

Vol. 413: R. Lenz, Group Theoretical Methods in Image Processing. VIII, 139 pages. 1990.

Vol. 414: A.Kreczmar, A. Salwicki, M. Warpechowski, LOGLAN '88 – Report on the Programming Language. X, 133 pages. 1990.

Vol. 415: C. Choffrut, T. Lengauer (Eds.), STACS 90. Proceedings, 1990. VI, 312 pages. 1990.

Vol. 416: F. Bancilhon, C. Thanos, D. Tsichritzis (Eds.), Advances in Database Technology – EDBT '90. Proceedings, 1990. IX, 452 pages. 1990.

Vol. 417: P. Martin-Löf, G. Mints (Eds.), COLOG-88. International Conference on Computer Logic. Proceedings, 1988. VI, 338 pages. 1990.

Vol. 419: K. Weichselberger, S. Pöhlmann, A Methodology for Uncertainty in Knowledge-Based Systems. VIII, 136 pages. 1990. (Subseries LNAI).

Vol. 420: Z. Michalewicz (Ed.), Statistical and Scientific Database Management, V SSDBM. Proceedings, 1990. V, 256 pages. 1990.

Vol. 421: T. Onodera, S. Kawai, A Formal Model of Visualization in Computer Graphics Systems. X, 100 pages. 1990.

Vol. 423: L.E. Deimel (Ed.), Software Engineering Education. Proceedings, 1990. VI, 164 pages. 1990.

Vol. 424: G. Rozenberg (Ed.), Advances in Petri Nets 1989. VI, 524 pages. 1990.

Vol. 426: N. Houbak, SIL – a Simulation Language. VII, 192 pages. 1990.

Vol. 427: O. Faugeras (Ed.), Computer Vision – ECCV 90. Proceedings, 1990. XII, 619 pages. 1990.